Elementary Functions

Andrei Bourchtein • Ludmila Bourchtein

Elementary Functions

Andrei Bourchtein
Institute of Physics and Mathematics
Federal University of Pelotas
Pelotas, Rio Grande do Sul, Brazil

Ludmila Bourchtein
Institute of Physics and Mathematics
Federal University of Pelotas
Pelotas, Rio Grande do Sul, Brazil

ISBN 978-3-031-29074-9 ISBN 978-3-031-29075-6 (eBook)
https://doi.org/10.1007/978-3-031-29075-6

This book is published under the imprint Birkhäuser, www.birkhauser-science.com by the registered company Springer Nature Switzerland AG
The registered company address is: Gewerbestrasse 11, 6330 Cham, Switzerland

To Victoria for amazing elegance and imagination;
To Valentina for invincible courage and perseverance;
To Maxim and Natalia for sagacity and dedication;
To Haim and Maria for everything

Preface

This book covers the topics of Pre-Calculus related to elementary functions at the level corresponding to the introductory university courses of Pre-Calculus and advanced courses of Algebra at the high school.

Differently from other books of Pre-Calculus, this book emphasizes the study of elementary functions with the rigor appropriate for the university courses of mathematics, although the exposition is confined to the pre-limit topics and techniques. This makes the text useful, on the one hand, as an introduction to mathematical reasoning and methods of proofs in mathematical analysis, and on the other hand, as a preparatory course on the properties of elementary functions, studied using pre-limit methods. The performed analytic investigations show how the properties of elementary functions can be revealed in a strict logic way, without unnecessary additional suppositions and mechanical memorization. It prepares the students for the use of mathematical reasoning in subsequent university courses and creates a solid base for the study of the courses of Calculus and Analysis.

The book is divided into five chapters: the first introductory chapter, followed by the three main chapters, and a short epilogue presented in the fifth chapter. The first chapter provides a review of the initial concepts of sets of numbers and their geometric representation on coordinate line and plane. The considered topics include the descriptions of rational and real numbers, the equivalence between the real numbers and the points of the coordinate line, the Cartesian coordinates on the plane and some characteristics of geometric figures, such as distances and symmetries, determined with the use of the coordinates. The second chapter is devoted to the definition and characterization of the main properties of functions, starting with the different modes of definition, proceeding to the properties of symmetry, continuing with monotonicity and extrema and then with concavity and inflection, and finalizing with injective, surjective and invertible functions. Each notion and property is illustrated with corresponding examples of different types of elementary functions. The second chapter is a core part of the text, introducing the basic notions, characteristics and techniques of analysis of elementary functions and showing how these techniques can be used. Further development of the properties of functions and their application to specific classes of functions is found in the next two chapters. The algebraic functions are handled in the third chapter, which specifies the application of the properties considered in the second chapter to the cases of polynomial, rational and irrational functions. In a similar manner, the fourth chapter applies the general characterizations of functions to the cases of transcendental functions: exponential, logarithmic and trigonometric. Each of these two chapters contains both the study of a specific families of functions and investigation of particular functions of a chosen type. Finally, in the fifth chapter, we give some hints about analytic study of functions in Calculus courses and provide simple examples explaining why the Calculus techniques radically simplify an analytic investigation of the properties of functions.

The text contains a large number of examples and exercises, which should make it suitable for both classroom use and self-study. Many standard examples are included in each section to develop basic techniques and test understanding of concepts. Other problems are more intricate and illustrate some finer points of the theory, and/or solution of harder questions and/or nonstandard methods of solutions. Many additional problems are proposed as homework tasks at the end of the introductory chapter and each of the three main chapters. The most tricky and challenging of these problems are marked with an asterisk.

Each chapter is organized in sections and subsections, which are numbered separately within each chapter, bearing the sequential number of section inside a chapter and that of subsection inside a section. For instance, section 3 can be found in each of the five chapters, while subsection 3.2 means the second subsection inside the third section. When referring to a section or subsection inside the current chapter, we do not use the chapter number, otherwise the chapter number is provided. The formulas and figures are numbered sequentially within each chapter regardless of the section where they are found. In this way, formula (2.5) is the reference to the fifth formula in the second chapter, while Figure 3.4 indicates the fourth figure in the third chapter.

The main features of the text are the following:

1. Strict mathematical approach to the study of elementary functions, while keeping the exposition at a level acceptable for the high school students.
2. Gradual development of the material starting from the background topics related to real numbers and functions, which makes the text suitable for both classroom use and self-study.
3. A comprehensive exposition of the traditional topics of study of elementary functions at the Pre-Calculus level.
4. A large number of examples and exercises, solved and proposed, designed both to develop basic techniques and to illustrate solution of harder questions and nonstandard methods of solutions.

The book is aimed at the university freshmen and the high-school students interested in learning strict mathematical reasoning and in preparing a solid base for subsequent study of elementary functions at advanced level of Calculus and Analysis. The required prerequisites correspond to the level of the high school Algebra. All the initial concepts and results related to the elementary functions and required for the development of the material are covered in the initial part of the book and, when necessary, at the starting point of each topic. This makes the text self-sufficient and the reader independent of any other source on this subject. As a consequence, this work can be used both as a textbook for advanced high school/introductory undergraduate courses and as a source for self-study of topics presented in the book.

We hope that for many students interested in mathematics this book will be a helpful, stimulating and encouraging introduction to study of Calculus and Analysis.

Pelotas, Brazil

Andrei Bourchtein
Ludmila Bourchtein

Contents

Chapter 1
Sets of Numbers and Cartesian Coordinates

The subject of Elementary Functions is a preliminary and preparatory material for study of Calculus, which in turn is a simplified version of Real Analysis. One of the main parts and aims of Analysis is a study of general and specific properties of functions starting from their analytic definition (by means of a formula). Following the natural relationship among three disciplines, we can recognize that one of the key problems in the study of Elementary Functions is an investigation of simple functions by elementary analytic methods. Talking about simple functions we have in mind functions of one real variable, whose form is quite comprehensible, frequently known from high school, and referring to elementary analytical methods, we consider those which do not involve finer concepts and results related to the limit and subsequent developing (continuity, derivative, etc.) of a theory of functions.

In Calculus/Analysis, a function is usually determined by its analytic form (through a formula), which is used to discover the relevant properties of a function, including its geometric features reflected in function graph. A geometric visualization of a function is an important tool of evaluation of its behavior, but this visualization is usually obtained as a result of a detailed study of the essential analytic properties of a function.

Having determined the main problem of the course of Elementary Functions, we can see that before to start the analytic study of functions to reveal their geometric properties, we need to have a basic knowledge of some preliminary concepts used in this study. Indeed, among the geometric properties we are looking to know are such as if a graph is climbing up or rolling down, if it has the form of a cup or bell, if there are minimum and maximum points, if it possesses some kind of symmetry, etc. To establish and prove the majority of these properties we need to apply analytic methods of study, working with real numbers, location of points in a coordinate plane, criteria of symmetry, etc. Consequently, we will use a basic knowledge of sets of numbers (especially real numbers and their subsets), Cartesian coordinates on a plane, distances between points, symmetries of planar figures, etc. This chapter is devoted to a concise review of these matters, which are frequently

A. Bourchtein, L. Bourchtein, *Elementary Functions*,
https://doi.org/10.1007/978-3-031-29075-6_1

known at some extent from high school, but require more formal, systematic and precise presentation for the analytic study of functions in the next chapters.

1 General Sets and Operations

In Analysis and Caluclus, and consequently in Elementary Functions, a set is usually considered to be a basic notion, intuitively clear, which is not defined through other simpler concepts. Therefore, we provide below just an intuitive description of a set.

1.1 Description of a Set

Set We consider a *set* to be a primary non-definable concept, which can be described as a *collection of objects* (*elements*) gathered by using some criterion. In particular, this criterion can be a mere belonging to a given set. To distinguish between sets and their elements we will use upper case for the former and lower case for the latter.

Obviously, it is not a formal definition, but just an intuitive description making reference to another undefined concepts (like collection and element), which seem to be intuitively clear, but whose definition can be even more sophisticated than a definition of a set. However, this description is sufficient to study the topics of Elementary Functions (and even to study Calculus and Analysis). In this intuitive description the words like "collection", "family" and "class" can be used as synonyms to a "set".

A set can be *finite* or *infinite*, that is, it can contain a finite or infinite number of elements, respectively.

A specification of a set can be given using a representation of its elements. This representation can be exhaustive (citing all the elements), which is possible only for finite sets, or using a specific characteristic of its elements, which can be made for both finite and infinite sets. In the first case, all the elements are shown in some (arbitrary) order, while in the second case a general form of the elements is presented with specification of their feature, which characterizes belonging to a given set. For example, a set A containing two elements a and b has the following standard notation: $A = \{a, b\}$. A set of all the days of a week can be described as follows: $A = \{a : a = a\ day\ of\ a\ week\}$ (the sign ":" in the description of a set means "such as").

Relation Between a Set and an Element An element a can be in (belong to) A, in which case we write $a \in A$; otherwise we write $a \notin A$.

Examples

1. Finite sets: a set of natural numbers from 1 to 5: $A = \{1, 2, 3, 4, 5\}$; a set B of the letters of the alphabet. For these two sets, $2 \in A$, but $a \notin A$; $a \in B$, but $3 \notin B$.
2. Infinite sets: the set of natural numbers (those used for counting): $\mathbb{N} = \{1, 2, 3, \ldots\}$ (more precisely, this set is defined as the set that contains the first element 1 and each following element is obtained from preceding adding 1); a set of even numbers, that is, $A = \{n \in \mathbb{N} : n = 2k, k \in \mathbb{N}\}$. For the last two sets, $2 \in \mathbb{N}$, but $-2 \notin \mathbb{N}$; $2 \in A$, but $3 \notin A$. (A notation $\mathbb{N}$ is a standard one for the set of natural numbers.)

Special Sets The two special sets are the empty set and the universal set. The *empty set* is the unique set having no elements, its notation is $\varnothing$. The *universal set* is the set containing all the elements of the considered nature, its standard notation is U. Of course, the latter can be different depending on a class of objects under consideration. If we are interested only in the letters of the alphabet, then the universal set is the set containing all these letters; if we consider only the numbers $1, 2, 3, 4, 5$, then the universal set is $U = \{1, 2, 3, 4, 5\}$, while working with natural numbers we have the universal set $\mathbb{N}$.

1.2 Elementary Operations with Sets

Subset If every element of a set A is an element of a set B, we say that A is a *subset* of B (or A is contained in B) and write $A \subset B$ (or equivalently, $A \supset B$). In particular, a set B is its own subset, any set is a subset of the universal set and, by agreement, the empty set is a subset of any set.

If A is a subset of B and B is subset of A, then these two sets are *equal* (*coincide*), in which case we write $A = B$. In other words, two sets are equal if, and only if, they have the same elements. If two sets are not equal we write $A \neq B$. Clearly, it may happen that neither of two sets is contained in another one, even in the case when they share some elements.

Examples
If $A = \{1, 2, 3, 4, 5\}$ and $B = \{2, 3, 4\}$, then $B \subset A$ (B is contained in A), or equivalently, $A \supset B$ (A contains B). If $A = \{a, b, c, d\}$ and B is the set of all the letters of the English alphabet, then $A \subset B$. If A is the set of all the letters of the Cyrillic alphabet and B is the set of all the letters of the English alphabet, then neither of the sets is a subset of another one.

Operations with Sets

1. **Union of sets**. The *union* of two sets A and B (denoted $A \cup B$) is the set C such that $c \in C$ if $c \in A$ or $c \in B$. In other words, the set $C = A \cup B$ contains all the elements of both A and B. (Recall that the elements that belong to both A and B can be taken only once to compose C.)

This definition is easily extended to any number of sets: $C = \bigcup_i A_i$ if for any $x \in C$ there exists a set A_i involved in the union such that $x \in A_i$.

2. **Intersection of sets**. The *intersection* of two sets A and B (denoted $A \cap B$) is the set C such that $c \in C$ if $c \in A$ and $c \in B$. In other words, the set $C = A \cap B$ contains all the elements that belong to both A and B.

   This definition is easily extended to any number of sets: $C = \bigcap_i A_i$ if any $x \in C$ belongs to each of involved sets A_i: $x \in A_i$, $\forall i$.
3. **Difference between sets**. The *difference* of two sets A and B (denoted $A \backslash B$, in this specific order) is the set of all elements $a \in A$ such that $a \notin B$. In other words, $C = A \backslash B$ contains all such the elements of A, that do not belong to B.
4. **Cartesian product**. The *Cartesian product* of two sets A and B (denoted $A \times B$, in this specific order) is the set of all the ordered pairs (a, b) where $a \in A$ and $b \in B$. In other words, we take each element of A and couple it with all elements of B one by one in order to obtain the elements of C.

Examples

1. If $A = \{2, 3, 4, 5\}$ and $B = \{1, 2, 3\}$, then $A \cup B = \{1, 2, 3, 4, 5\}$. If $A = \{a, b, c, d\}$ and $B = \{d, e\}$, then $A \cup B = \{a, b, c, d, e\}$. If $A = \{a, b, c, d\}$ and $B = \{g, h\}$, then $A \cup B = \{a, b, c, d, g, h\}$. If $A = \{2, 3, 4, 5\}$ and $B = \mathbb{N}$, then $A \cup B = \mathbb{N}$.
2. If $A = \{2, 3, 4, 5\}$ and $B = \{1, 2, 3\}$, then $A \cap B = \{2, 3\}$. If $A = \{a, b, c, d\}$ and $B = \{d, e\}$, then $A \cap B = \{d\}$. If $A = \{a, b, c, d\}$ and $B = \{g, h\}$, then $A \cap B = \varnothing$. If $A = \{2, 3, 4, 5\}$ and $B = \mathbb{N}$, then $A \cap B = A$.
3. If $A = \{2, 3, 4, 5\}$ and $B = \{1, 2, 3\}$, then $A \backslash B = \{4, 5\}$ while $B \backslash A = \{1\}$. If $A = \{a, b, c, d\}$ and $B = \{d, e\}$, then $A \backslash B = \{a, b, c\}$ while $B \backslash A = \{e\}$. If $A = \{a, b, c, d\}$ and $B = \{g, h\}$, then $A \backslash B = \{a, b, c, d\}$ while $B \backslash A = \{g, h\}$. If $A = \{2, 3, 4, 5\}$ and $B = \mathbb{N}$, then $A \backslash B = \varnothing$ while $B \backslash A = \{1, 6, 7, \ldots\}$.
4. If $A = \{2, 3, 4\}$ and $B = \{1, 2\}$, then $A \times B = \{(2, 1), (2, 2), (3, 1), (3, 2), (4, 1), (4, 2)\}$ while $B \times A = \{(1, 2), (1, 3), (1, 4), (2, 2), (2, 3), (2, 4)\}$. If $A = \{b\}$ and $B = \{a, b\}$, then $A \times B = \{(b, a), (b, b)\}$ while $B \times A = \{(a, b), (b, b)\}$.

1.3 Elementary Properties of Sets

1. **Commutative law**: $A \cup B = B \cup A$, $A \cap B = B \cap A$. Notice that $A \backslash B \neq B \backslash A$ and $A \times B \neq B \times A$ (the above examples show that the commutative property is not true for the difference and Cartesian product).
2. **Associative law**: $(A \cup B) \cup C = A \cup (B \cup C)$, $(A \cap B) \cap C = A \cap (B \cap C)$.
3. **Distributive law**: $A \cup (B \cap C) = (A \cup B) \cap (A \cup C)$, $A \cap (B \cup C) = (A \cap B) \cup (A \cap C)$.
4. **Relations with the empty set**: $\varnothing \subset A$, $A \cup \varnothing = A$, $A \cap \varnothing = \varnothing$, $A \backslash \varnothing = A$, $A \times \varnothing = \varnothing$. Notice that these properties are true for an arbitrary set A.

5. **Relations with the universal set** U: $A \subset U, A \cup U = U, A \cap U = A, A \backslash U = \varnothing$. Notice that these properties are true for an arbitrary set A.
6. **The number of elements of finite sets**: if A and B are finite sets with m and n elements, respectively, and $A \cap B$ has k elements, then $A \cup B$ has $m + n - k$ elements, $A \backslash B$ has $m - k$ elements, $B \backslash A$ has $n - k$ elements, and $A \times B$ has mn elements (as well as $B \times A$).

Proofs of these properties are quite simple, employing the provided definitions and elementary reasoning. Although this is out of scope of this text, let us provide a few examples of proofs for the interested reader.

The Proof of Commutativity for Union: $A \cup B = B \cup A$ By the definition, any element $x \in A \cup B$ belongs to A or to B (or to both), which is the same as to say that x belongs to B or to A (or to both), which means that $x \in B \cup A$. In the same way, any $y \in B \cup A$ is also the element of $A \cup B$. Thus, $A \cup B$ and $B \cup A$ have the same elements, that is, the two sets coincide. □

The Proof of Associativity for Intersection: $(A \cap B) \cap C = A \cap (B \cap C)$ By the definition, any element $x \in (A \cap B) \cap C$, belongs to both $A \cap B$ and C. Still opening the first inclusion ($x \in A \cap B$), we conclude that x belongs at the same time to the three sets—A, B and C. On the other hand, by the definition, any element $y \in A \cap (B \cap C)$, is contained in A and also in $B \cap C$. Opening the last inclusion ($y \in B \cap C$), we conclude that y belongs simultaneously to the three sets—A, B and C. Therefore, the elements of $(A \cap B) \cap C$ and $A \cap (B \cap C)$ have the same properties, and consequently, these two sets contain the same elements, that is, are equal. □

The Proof of the First Distributive Law: $A \cup (B \cap C) = (A \cup B) \cap (A \cup C)$ Let us start from the right-hand side. Any element x of the set $(A \cup B) \cap (A \cup C)$ belongs to both $A \cup B$ and $A \cup C$. Since these two sets have the identical part, which is the set A, their intersection can be divided into two parts—the first is A and the second is the common part between B and C. Therefore, there are two options for x—it belongs to A or to $B \cap C$, which means that $x \in A \cup (B \cap C)$. Now we start from the left-hand side and consider an arbitrary element y of the set $A \cup (B \cap C)$. By the definition of the union, y either belongs to A or to the common part between B and C. However, we have already seen in the first part of the proof, that this is the property of any element x of $(A \cup B) \cap (A \cup C)$. Consequently, $y \in (A \cup B) \cap (A \cup C)$. Thus, we have shown that any element of $(A \cup B) \cap (A \cup C)$ is also an element of $A \cup (B \cap C)$ and vice-verse. This means that the two sets coincide. □

Examples

1. Show that $A \backslash B = B \backslash A$ if, and only if, $A = B$.
   Solution
   Suppose that the two sets are different. In this case, at least one of them, say A, has at least one element, say a, not contained in another one, B. Then, the difference $A \backslash B$ contains the element a, but $B \backslash A$ does not contain this element,

because a does not belong to B. On the other hand, if $A = B$, then $A\backslash B = B\backslash A = \varnothing$.

2. Let A be a set of the even numbers between 11 and 19, and B be a set of multiples of 3 between the same 11 and 19. Find $A \cup B$, $A \cap B$, $A\backslash B$ and $B\backslash A$.
   Solution
   First, let us specify that $A = \{12, 14, 16, 18\}$ and $B = \{12, 15, 18\}$. Performing the required operations we obtain $A \cup B = \{12, 14, 15, 16, 18\}$, $A \cap B = \{12, 18\}$, $A\backslash B = \{14, 16\}$, $B\backslash A = \{15\}$.

2 Rational Numbers and Their Properties

In this section we review some infinite sets of numbers, which are subsets of the set of real numbers, but have a simpler description.

The Set of Natural Numbers
Let us start with the simplest, but important, infinite set of numbers used in Analysis—the natural numbers ("the counting numbers").

The set of *natural numbers* $\mathbb{N}$ consists of the numbers used for counting, that is, $\mathbb{N} = \{1, 2, 3, \ldots\}$. This intuitive description is far from precise formal definition, but it is sufficient for the purposes of this text.

The Set of Integers
The next simple infinite set, broader than $\mathbb{N}$, is the set of integers, whose description is as follows.

The set of *integers* $\mathbb{Z}$ is composed of the natural numbers, natural negative numbers and zero: $\mathbb{Z} = \{\ldots, -3, -2, -1, 0, 1, 2, 3, \ldots\}$.

The Set of Rational Numbers
Using the sets $\mathbb{N}$ and $\mathbb{Z}$, we can define (in exact way) the next infinite set, still broader than $\mathbb{Z}$—the set of rational numbers.

Definition The set of *rational numbers* $\mathbb{Q}$ consists of *common fractions* (or simply *fractions*) in the form $\frac{p}{q}$, where $p \in \mathbb{Z}$ and $q \in \mathbb{N}$.

Remark If one uses $q \in \mathbb{Z}\backslash\{0\}$ in this definition, the same set $\mathbb{Q}$ is obtained.

Obviously, the relation among the three considered sets is as follows: $\mathbb{N} \subset \mathbb{Z} \subset \mathbb{Q}$.

A natural question that may arise is why we need to "amplify" the set $\mathbb{N}$ and then $\mathbb{Z}$ to arrive at $\mathbb{Q}$? One can recall from elementary school that there are some arithmetic operations which are acting inside the sets $\mathbb{N}$ and $\mathbb{Z}$ (in other words, these sets are closed regarding some operations). For example, $\mathbb{N}$ is closed under addition and multiplication, while $\mathbb{Z}$ is closed under addition, subtraction and multiplication (meaning that these operations result in an element of the same set). However, there are arithmetic operations which take out of the original set. For $\mathbb{N}$ such operations are subtraction and division, and for $\mathbb{Z}$—division.

Let us recall the definition and properties of these operations on the set of rational numbers known in an intuitive form from elementary school. In particular, it will show that the four arithmetic operations are acting inside $\mathbb{Q}$.

Properties of Rational Numbers

1. **Sum of the rationals**. For any two rational numbers $a = \frac{m}{n}$ and $b = \frac{p}{q}$ their sum is the rational number c defined by the formula $c = a + b = \frac{mq+pn}{nq}$.
2. **Product of the rationals**. For any two rational numbers $a = \frac{m}{n}$ and $b = \frac{p}{q}$ their product is the rational number c defined by the formula $c = ab = \frac{mp}{nq}$.
3. **Properties of addition and multiplication**. For any $a, b, c, d \in \mathbb{Q}$ the following properties are satisfied:
    1. commutativity of addition and multiplication: $a + b = b + a$, $ab = ba$.
    2. associativity of addition and multiplication: $(a+b)+c = a+(b+c)$, $(ab)c = a(bc)$.
    3. distributivity: $(a + b)c = ac + bc$.
    4. monotonicity of addition: if $a > b$, then $a + c > b + c$; if $a > b$ and $c > d$, then $a + c > b + d$.
    5. monotonicity of multiplication: if $a > b$ and $c > 0$, then $ac > bc$; if $a > b$ and $c < 0$, then $ac < bc$.
    6. existence of identity (neutral) element of addition and multiplication: $a+0 = 0+a = a$, $\forall a \in \mathbb{Q}$; $a \cdot 1 = 1 \cdot a = a$, $\forall a \in \mathbb{Q}$.
    7. existence of inverse element of addition and multiplication: for any rational number a there exists the rational number $-a$, inverse to a with respect to addition, that is, $a + (-a) = 0$; for any rational number $a \neq 0$ there exist the rational number $a^{-1} = \frac{1}{a}$, inverse to a with respect to multiplication, that is, $a \cdot a^{-1} = 1$.

    The properties 1 and 2 guarantee that the sum and product of two rational numbers $a = \frac{m}{n}$ and $b = \frac{p}{q}$ are again rational numbers determined by the formula $a + b = \frac{mq+pn}{nq}$ and $ab = \frac{mp}{nq}$, respectively. The property 3.7 defines the subtraction and division of two rational numbers $a = \frac{m}{n}$ and $b = \frac{p}{q}$ by the formula $a - b = a + (-b) = \frac{mq-pn}{nq}$ and $a : b = a \cdot \frac{1}{b} = \frac{mq}{np}$, respectively (the latter operation is defined only when $b \neq 0$).

    The following properties of comparison are also known from elementary school at the intuitive level.
4. **Ordering of rationals**. For any two rational numbers a and b exactly one of the following relations is true: $a = b$, $a < b$ or $a > b$. Hence, the set of rational numbers is ordered (frequently called totally ordered).
5. **Properties of ordering** (transitivity of the relations "greater than", "smaller than" and "equal"). If $a > b$ and $b > c$, then $a > c$; if $a < b$ and $b < c$, then $a < c$; if $a = b$ and $b = c$, then $a = c$.

6. **Property of Archimedes**. For any rational number a there exists a natural number n such that $n > a$. Or an equivalent formulation: for any rational number $a > 0$ there exists a natural number n such that $\frac{1}{n} < a$.

Remark A set satisfying the above properties 1-6 is called an *ordered field*.

Examples

1. Show that the expression $\frac{\frac{13}{2}-\frac{7\cdot 2+1}{4}}{(\frac{1}{2})^3+2}$ is a rational number.
   Solution
   First, notice that this expression is composed of arithmetic operations—sum, difference, multiplication, division and power—applied to rational numbers—$\frac{13}{2}$, 7, 2, 1, 4, $\frac{1}{2}$. Since all these operations keep the result in the set of rationals, and all of them are admissible in the given expression (there is no division by 0 or the power 0^0), then this expression respresents a rational number. Another way to show the same result is to simplify the given expression to the form used in the definition of rational numbers: $\frac{\frac{13}{2}-\frac{7\cdot 2+1}{4}}{(\frac{1}{2})^3+2} = \frac{\frac{13}{2}-\frac{15}{4}}{\frac{1}{8}+2} = \frac{\frac{11}{4}}{\frac{17}{8}} = \frac{22}{17}$. Therefore, we arrive to the fraction whose numerator and denominator are natural numbers, which corresponds to the definition of rational numbers.
2. Compare the rational numbers $\frac{5}{2^{10}}$ and $\frac{1}{10^2}$.
   Solution
   A standard way to solve this task is to rewrite both fractions using common denominator and compare their numerators. We have $\frac{5}{2^{10}} = \frac{5\cdot 10^2}{2^{10}\cdot 10^2}$ and $\frac{2^{10}}{2^{10}\cdot 10^2}$. Since $5 \cdot 10^2 = 500 < 1024 = 2^{10}$, the first fraction is smaller then the second. In this example, the fastest solution is to note that $2^{10} = 1024 > 10^3$ and, consequently, $\frac{5}{2^{10}} < \frac{5}{10^3} < \frac{10}{10^3} = \frac{1}{10^2}$.
3. Show that a natural number is even if, and only if, its square is even.
   Solution
   The first implication is evident: if a natural number p is even, then, by definition, $p = 2k$, $k \in \mathbb{N}$, and consequently, $p^2 = 4k^2$ is also an even number. To prove the reverse statement, we argue by contradiction. Consider a natural number p whose square p^2 is even and suppose that p is odd, that is, $p = 2m+1$, $m \in \mathbb{N}$. Therefore, $p^2 = (2m+1)^2 = 4m^2+4m+1 = 2(2m^2+2m)+1 = 2n+1$, where $n = 2m^2 + 2m \in \mathbb{N}$, which means that p^2 is also odd. This contradicts the fact that p^2 is even, and therefore, our supposition that p can be odd is false. Hence, p is even.
   Remark The method used to prove the reverse statement is a powerful technique called the proof by contradiction (reduction to absurd—"reductio ad absurdum").
4. It can be shown that $1 + 2 + \ldots + n = \frac{n(n+1)}{2}$ for any $n \in \mathbb{N}$. In this formula, on the left-hand side we have the sum of natural numbers, which results in a natural number, but on the right-hand side we have a fraction. Explain why this result is not contradictory.

Solution

Let us show that the numerator of the right-hand side is an even number. In fact, if n is even, then $n(n+1)$ is also even. If n is odd, then $n+1$ is even and again $n(n+1)$ is even. Then, for any natural n, the product $n(n+1)$ is always even, and consequently, $\frac{n(n+1)}{2}$ is a natural number.

Remark This formula is known from a middle school and its derivation is quite simple. Denote the sum $S = 1 + 2 + \ldots + n$ and notice that

$$\begin{aligned}
2S = S + S &= (1 + 2 + \ldots + (n-1) + n) + (1 + 2 + \ldots + (n-1) + n) \\
&= (1 + 2 + \ldots + (n-1) + n) + (n + (n-1) + \ldots + 2 + 1) \\
&= (1 + n) + (2 + n - 1) + \ldots + (n - 1 + 2) + (n + 1).
\end{aligned}$$

On the right-hand side the same term $n+1$ is adding up n times, resulting in $2S = n(n+1)$. Solving this relation for S, we obtain the formula given in this example.

3 Real Numbers and Their Properties

3.1 Decimal and Real Numbers

The set of rational numbers is closed regarding the four arithmetic operations, which means that application of any of these four operations to any rational numbers results again in a rational number (except for division by 0, which is not defined). Naturally, the question arises if the set $\mathbb{Q}$ can be considered the universal set for our purposes. The answer is negative, because the study of elementary functions requires measuring the lengths, which involves calculation of roots, and the last operation takes out of $\mathbb{Q}$. The necessity of this operation appears, for example, when one find the length of hypotenuse of a rectangular triangle. According to Pythagoras' theorem, the length h of the hypotenuse of a rectangular triangle with the length of both legs equal to 1 is found by the formula $h^2 = 1^2 + 1^2$, that is, $h = \sqrt{2}$. Let us show that $\sqrt{2}$ is not a rational number.

Proposition $\sqrt{2}$ *is not a rational number.*

Proof Let us suppose, by contradiction, that $\sqrt{2}$ is a rational number and show that this leads to logical contradiction. If $\sqrt{2}$ is rational, it has representation $\sqrt{2} = \frac{p}{q}$, where $p, q \in \mathbb{N}$ (since $\sqrt{2} > 0$, both numbers p and q are naturals). Without loss of generality we may assume that the last fraction is in lowest terms (that is, the numerator and denominator have no common factors other than 1). Then, squaring both sides of the relation $\sqrt{2} = \frac{p}{q}$, we obtain $2 = \frac{p^2}{q^2}$, whence $p^2 = 2q^2$. This means that p^2 is an even number, and consequently, p is even too, that is, $p = 2m, m \in \mathbb{N}$ (see Example 3 of the previous section). Hence, the relation between p and q can be written in the form $4m^2 = 2q^2$, or simplifying, $2m^2 = q^2$. It follows from this that q^2 is an even number and so is q, that is, $q = 2n$, $n \in \mathbb{N}$. Therefore, $\frac{p}{q} = \frac{2m}{2n}$, that is, the original fraction is not in lowest terms, which contradicts our supposition. Thus, the supposition that $\sqrt{2}$ is a rational number is false, which means that $\sqrt{2}$ is not rational. □

Remark The method used to prove this statement is called the proof by contradiction (reduction to absurd—"reductio ad absurdum" in Latin). In general, the algorithm of this method is as follows: to prove some statement, assume first that it is false, and show that this implies something which is known to be false, or leads to a conclusion which contradicts the hypotheses of the statement or the proper supposition; this allows us to conclude that the assumption is wrong, and consequently, the statement is true. In this particular case, a false supposition, contrary to the statement we intend to prove, was that $\sqrt{2}$ is a rational number. After some algebra, this leaded us to contradiction with the proper supposition. The same method was used in the solution of Example 3 of the previous section. There, the false supposition was that p is odd, whence we arrived at the conclusion that p^2 is also odd, which contradicts the hypothesis of Example 3 (that p^2 should be even).

To include all the roots of non-negative numbers (and many other numbers which are not rationals, but still used for length measuring, like π for the length of the unit half-circle) and construct a set of numbers "without holes", let us introduce the concept of *decimal numbers*, which are well known from elementary school and everyday use (as well as rational numbers), although in a less formal and rigorous form then used in this text. Let us start with the definition.

Definition. Decimal Numbers A *decimal number* a or simply a *decimal* is a chain of *decimal digits* represented in the form $a = (\pm)\, a_0.a_1a_2 \ldots a_n \ldots$, where $a_0 \in \mathbb{N} \cup \{0\}$ and the digits $a_i \in \mathbb{N} \cup \{0\}$, $0 \le a_i \le 9$, $i = 1, 2, \ldots, n, \ldots$. The part a_0 (before the decimal separator ".") is called the *integer part* of a and the part after separator is called the *fractional part* of a.

The reference to *decimal system* is due to the choice of 10 as the base for position representation. For the integer part this means that $a_0 = b_n \ldots b_2b_1b_0$, $b_i \in \mathbb{N} \cup \{0\}$, $0 \le b_i \le 9$, $i = 0, 1, 2, \ldots, n$, where b_0 is the number of ones, b_1 is the number of tens, b_2 is the number of hundreds, etc., or expressing by formula: $a_0 = b_0 + b_1 \cdot 10^1 + b_2 \cdot 10^2 + \ldots + b_n \cdot 10^n$. For the fractional part the use of the decimal system means that a_1 shows the number of tenths, a_2—the hundredths, etc., having the following representation: $0.a_1a_2 \ldots a_n \ldots = a_1 \cdot 10^{-1} + a_2 \cdot 10^{-2} + \ldots + a_n \cdot 10^{-n} + \ldots$. The fractional part is more important for us, because the integer part is just a natural number or zero.

A decimal number is *finite* if $a_i = 0$ for all $i \ge m$, where m is a natural number (in this case all zero a_i starting from $i = m$ can be dropped from decimal representation). Otherwise, a decimal number is *infinite*. A decimal is called *periodic or repeating* if $a_{k+i} = a_{k+i+p}$ for some $k \in \mathbb{N} \cup \{0\}$, $p \in \mathbb{N}$, and all $i \in \mathbb{N}$, that is, there exists a finite sequence of digits (of the length p), which is repeated infinitely starting from $(k+1)$-th digit. A finite decimal can be considered a special case of a periodic decimal (with all zero digits starting from a certain position). For periodic decimals the following notation can be used: $a = (\pm)\, a_0.a_1a_2 \dots a_k(a_{k+1}a_{k+2} \dots a_{k+p})$, where the digits $a_1a_2 \dots a_k$ appear before the period, $a_{k+1}a_{k+2} \dots a_{k+p}$ are the digits of the period and the parentheses indicate repetition of the digits inside them. Finally, an infinite non-periodic decimal is called *infinite aperiodic* or simply *aperiodic*.

Examples

1. Finite decimal: $1.34 = 1 + \frac{3}{10} + \frac{4}{100}$.
2. Finite decimal: $-64.84524 = -64 - \frac{8}{10} - \frac{4}{100} - \frac{5}{10^3} - \frac{2}{10^4} - \frac{4}{10^5}$.
3. Infinite periodic decimal: $0.333\ldots = 0.(3) = 0 + \frac{3}{10} + \frac{3}{100} + \frac{3}{10^3} + \ldots + \frac{3}{10^n} + \ldots$.
4. Infinite periodic decimal: $0.1272727\ldots = 0.1(27) = 0 + \frac{1}{10} + \frac{2}{100} + \frac{7}{10^3} + \ldots + \frac{2}{10^{2n}} + \frac{7}{10^{2n+1}} + \ldots$.
5. Infinite aperiodic decimal: $0.1010010001\ldots = 0 + \frac{1}{10} + \frac{1}{10^3} + \frac{1}{10^6} + \frac{1}{10^{10}} + \ldots$.

Definition. Real Numbers The set $\mathbb{R}$ of *real numbers* is the set of all the decimal numbers.

Remark The use of the decimal numbers is an advantageous mode to define the real numbers, because it is intuitively clear and directly connected to the numbers and operations among them which are used in the school and in the day-to-day practice. There are other, more formal, approaches to introduction of the set of real numbers, one of them is the axiomatic approach, when the properties of this set are postulated in the form of axioms. The system of decimal numbers can be considered as one of possible specification, more used in practice, of the axiomatic set of real numbers. These more theoretic issues of the study of the real numbers are out of scope of this text, but it is important to note that all these approaches are equivalent in the sense that whatever formal definition of the set of real numbers is used it can be transformed into the set of decimal numbers possessing the properties considered in the next subsection.

Theorem (Relation Between $\mathbb{Q}$ and $\mathbb{R}$) *Let us show that the set $\mathbb{R}$ is an expansion of $\mathbb{Q}$. More specifically, let us show that $\mathbb{Q}$ is equal to the set of all the finite and infinite periodic decimal numbers, which we temporary denote by $\tilde{\mathbb{Q}}$.*

Proof First, let us take any element of $\mathbb{Q}$ and show that it can be represented through a finite or infinite periodic decimal number. Without loss of generality we may consider only a positive fraction $\frac{p}{q}$, since a generalization to negative numbers is straightforward. Let us perform standard long division of p by q following the rules well known from elementary school. If $p \geq q$ (the fraction is improper), then we start by separating the integer part of the given number. The remaining part is a proper fraction $\frac{m}{q}$ with $m < q$. In the process of long division of the number m by q, it may appear the following remainders: $0, 1, 2, \ldots, q-1$. If at some step of the division the obtained remainder is 0, then the division is terminated and the resulting decimal number is finite. If the remainder is always different from 0, then the division continues infinitely. However, since there are only $q-1$ different non-zero remainders, in at most q steps of the long division we will find the same remainder as appeared before, and consequently, starting from this moment the decimal digits will repeat those obtained previously. Hence, if the remainder never vanishes, we obtain a periodic decimal number with the period less than or equal to $q - 1$ digits.

Now let us prove the reverse: any element $\tilde{\mathbb{Q}}$ can be expressed in the form $\frac{p}{q}$, where $p \in \mathbb{Z}$ and $q \in \mathbb{N}$. Again, it is sufficient to consider only positive numbers. For an arbitrary $a > 0$ from the set $\tilde{\mathbb{Q}}$ there are two options: a is a finite decimal or infinite periodic decimal. Let us consider first the simpler option: a is a finite decimal. Then

$$a = a_0.a_1a_2\ldots a_n = a_0 + \frac{a_1}{10} + \frac{a_2}{10^2} + \ldots + \frac{a_n}{10^n}$$
$$= a_0 + \frac{a_1 \cdot 10^{n-1} + a_2 \cdot 10^{n-2} + \ldots + a_n}{10^n} = a_0 + \frac{a_1a_2\ldots a_n}{10^n}.$$

The last fraction on the right-hand side has the form $\frac{p}{q}$, where $p = a_1a_2\ldots a_n \in \mathbb{N}$ and $q = 10^n \in \mathbb{N}$, that is, this is a rational number. The sum of a natural number with a rational one is again a rational number. Therefore, a finite decimal is a rational number. Take now a periodic decimal: $a = a_0.a_1a_2\ldots a_nb_1b_2\ldots b_kb_1b_2\ldots b_k\ldots = a_0.a_1a_2\ldots a_n\,(b_1b_2\ldots b_k)$ where the parentheses indicate the period containing k digits. We split this number in the two parts—without a period and the proper period: $a = a_0.a_1\ldots a_n + 0.0\ldots 0\,(b_1b_2\ldots b_k)$. It was already shown before that the first part is a rational number. It remains to demonstrate that the second (periodic) part is also a rational number. To do this, denote $s = 0.0\ldots 0\,(b_1b_2\ldots b_k)$ and $t = s \cdot 10^n = 0.\,(b_1b_2\ldots b_k)$. Notice that $t \cdot 10^k = b_1b_2\ldots b_k.\,(b_1b_2\ldots b_k)$, that is, $t \cdot 10^k - t = b_1b_2\ldots b_k$ is a natural number. Then $t = \frac{b_1b_2\ldots b_k}{10^k-1}$ is also rational as the ratio of two natural numbers. Therefore, $s = \frac{t}{10^n}$ is rational, and consequently, a is a rational number. In this way, we show that any element of $\tilde{\mathbb{Q}}$ also belongs to $\mathbb{Q}$. Together with the first part of the proof, this confirms that the sets $\mathbb{Q}$ and $\tilde{\mathbb{Q}}$ contain the same elements, that is, these sets coincide. □

It follows from this result that the set $\mathbb{Q}$ of the rational numbers can be equivalently defined as the set of all finite and infinite periodic decimals. The remaining part of the set of real numbers contains aperiodic decimals called *irrational numbers*. Thus, the set of irrational numbers $\mathbb{I}$ is the set of all aperiodic decimal numbers.

Remark 1 The introduced notation $\mathbb{I}$ of the irrational numbers is not standard one, but for the sake of brevity we will prefer this symbol to more common notation $\mathbb{R}\backslash\mathbb{Q}$.

Remark 2 The following relations between sets follow from the equality between $\mathbb{Q}$ and $\tilde{\mathbb{Q}}$: $\mathbb{N} \subset \mathbb{Z} \subset \mathbb{Q} \subset \mathbb{R}$, $\mathbb{I} \subset \mathbb{R}$, $\mathbb{Q} \cup \mathbb{I} = \mathbb{R}$, $\mathbb{Q} \cap \mathbb{I} = \varnothing$.

Examples

1. Transform the finite and infinite periodic decimals of preceding Examples in common fractions.
   Solution
   For finite decimals 1.34 and -64.84524 the transformation is trivial: $1.34 = \frac{134}{100}$ and $-64.84524 = -\frac{6484524}{10^5}$.
   For aperiodic infinite decimal $a = 0.(3)$, we apply a general algorithm of transformation. First, we multiply a by 10 to bring the first period for the integer part: $10a = 3.(3)$. Then, subtracting a from $10a$, we have: $10a - a = 9a = 3$. Consequently, $a = \frac{1}{3}$.
   For aperiodic infinite decimal $a = 0.1(27)$ we apply the same general procedure. First, we separate the periodic part from the remaining part: $a = 0.1 + 0,0(27) = 0.1 + b$ and denote $t = 10b = 0.(27)$. To transform t, we multiply it by 100 to move the first period to the integer part: $100t = 27.(27)$. Subtracting t from $100t$, we obtain $100t - t = 99t = 27$, or isolating t and simplifying, $t = \frac{27}{99} = \frac{3}{11}$. Therefore, $b = \frac{t}{10} = \frac{3}{110}$ and adding the finite part of the original number, we have $a = 0.1 + b = \frac{1}{10} + \frac{3}{110} = \frac{11+3}{110} = \frac{7}{55}$.
2. Transform the following common fractions into decimal numbers: $\frac{125}{11}$ and $-\frac{13}{7}$.
   Solution
   Using the long division for $\frac{125}{11}$ we have the following result:

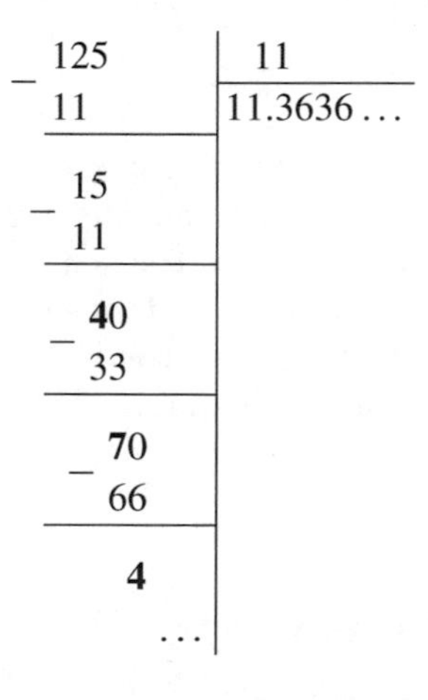

Notice that during division of the proper fraction (4 divided by 11) it may occur at most 10 different (non-zero) remainders. However, in this specific case, the repetition happened already on the third step, because the third remainder $r_3 = 4$ coincides with the first one (all the remainders are marked in bold). Therefore, r_4 coincides with $r_2 = 7$, etc. For this reason, starting from the third step of the division, all the results will repeat those found on the first two steps of the division of the proper fraction. Hence, the final result is $\frac{125}{11} = 11.(36)$.

The long division of 13 by 7 gives the following result:

$$
\begin{array}{r|l}
13 & 7 \\
-\ 7 & 1.857142\ldots \\
\hline
\mathbf{60} & \\
-\ 56 & \\
\hline
\mathbf{40} & \\
-\ 35 & \\
\hline
\mathbf{50} & \\
-\ 49 & \\
\hline
\mathbf{10} & \\
-\ 7 & \\
\hline
\mathbf{30} & \\
-\ 28 & \\
\hline
\mathbf{20} & \\
-\ 14 & \\
\hline
\mathbf{6} & \\
\ldots &
\end{array}
$$

Notice that under the division of the proper fraction (6 divided by 7) it may occur at most 6 different (non-zero) remainders. This maximum chain of different remainders happened to appear for this fraction: $r_1 = 6$, $r_2 = 4$, $r_3 = 5$, $r_4 = 1$, $r_5 = 3$, $r_6 = 2$. The next remainder should coincide with one of the found before. In this example, it equals to the first remainder: $r_1 = r_7 = 6$, and starting from the seventh step all the calculations will repeat. Hence, $\frac{13}{7} = 1.(857142)$ and the final result is $-\frac{13}{7} = -1.(857142)$.

3.2 *Properties of the Real Numbers*

List of Properties: Arithmetic, Ordering, Density and Completeness

Based on the decimal representation (or another equivalent form) it can be shown that the real numbers satisfy the same *arithmetic properties* and *ordering properties* that were listed before for the rational numbers (properties 1–6 of Sect. 2). This means, in particular, that the set $\mathbb{R}$ is an *ordered field*. Additionally, the real numbers have the following properties:

1. **Density** of $\mathbb{Q}$ in $\mathbb{R}$. For any two real numbers a and b such that $a < b$ there exists a rational number r such that $a < r < b$.

   The corollary of this property tells about the density of $\mathbb{I}$ in $\mathbb{R}$: for any two real numbers a and b such that $a < b$ there exists an irrational number t such that $a < t < b$.
2. **Completeness** of $\mathbb{R}$. In very vague terms, the set of real numbers is such that it has no "holes" or "gaps" between numbers (differently from the set of rationals). In terms of the decimal representation, completeness is equivalent to the fact that any infinite string of decimal digits is actually a decimal representation for some real number. A geometric illustration of this property is the result about equivalence between $\mathbb{R}$ and the coordinate line (real number line) which will be shown later.

Remark Just for the curious readers we mention that an exact formulation of the completeness can be done in terms of limits or suprema/infima: the first form says that any Cauchy sequence is convergent, and the second that any non-empty bounded above set has supremum. One can see that a mere formulation of the last property in an exact form requires additional mathematical knowledge and is out of scope of this text. The interested reader is referred to the books of Analysis and Set Theory.

Arithmetic Operations Between the Real Numbers

We have already recalled that addition, subtraction, multiplication and division of rational numbers result in rational numbers. Let us see what happens with irrational numbers and between rational and irrational numbers.

Addition: $\mathbb{I} + \mathbb{I}$. The sum of two irrational numbers can be both irrational and rational. For example, $\sqrt{2} + \sqrt{2} = 2\sqrt{2}$ is irrational number, but $\sqrt{2} + (-\sqrt{2}) = 0$ is rational.

Addition: $\mathbb{Q} + \mathbb{I}$. Let a be rational and b irrational. Let us show that $c = a + b$ is irrational. Suppose for sake of contradiction that c is rational. Then, $b = c + (-a)$ and it is known that the sum of rational numbers is again a rational number, that is, b is rational. However, this contradicts the original condition that b is irrational. Hence, c is irrational.

Remark The same properties are valid for subtraction.

Multiplication: $\mathbb{I} \cdot \mathbb{I}$. The product of two irrational numbers can be both irrational and rational. For example, $\sqrt{2} \cdot \sqrt{3} = \sqrt{6}$ is irrational, but $\sqrt{2} \cdot \sqrt{2} = 2$ is rational.

Multiplication: $\mathbb{Q} \cdot \mathbb{I}$. Let a be non-zero rational and b irrational. Let us show that $c = ab$ is irrational. Arque by contradiction: suppose that c is rational and rewrite the relation among the three numbers in the form $b = \frac{c}{a}$, whence b is rational (since the ratio of two rationals is again a rational). However, this contradicts the hypothesis that b is irrational, which implies that the supposition is false and c is irrational.

Remark Only in the singular case $a = 0$, the product ab is rational.

Division: $\mathbb{I} : \mathbb{I}$. The ratio of two irrational numbers can be both irrational and rational. For example, $\frac{\sqrt{6}}{\sqrt{2}} = \sqrt{3}$ is irrational, but $\frac{\sqrt{8}}{\sqrt{2}} = 2$ is rational.
Division: $\mathbb{Q} : \mathbb{I}$. Let a be non-zero rational and b irrational. Let us show that $c = \frac{a}{b}$ is irrational. Again we apply the proof by contradiction: suppose that c is rational. Then, $b = \frac{a}{c}$ ($c \neq 0$ because $a \neq 0$, and consequently, we can divide by c) and the ratio of two rational numbers is a rational, which implies that b is rational. However, this contradicts the hypothesis that b is irrational. Therefore, c é irrational.

Remark Only in the singular case $a = 0$, the ratio $\frac{a}{b}$ is rational.

Remark For the division $\mathbb{I} : \mathbb{Q}$ the considerations are analogous.

Examples

1. Transform the decimal number $5, 42(376)$ into a common fraction.
   Solution
   First, separate the periodic part by representing the number in the form $5.42(376) = 5.42 + 0.00(376)$. For the first number we have $5.42 = \frac{542}{100}$ and the second we rewrite in the form $0.00(376) = 0.(376) \cdot 10^{-2}$. The number $t = 0.(376)$ can be converted following the standard procedure: $t \cdot 10^3 - t = 376.(376) - 0.(376) = 376$, and therefore, $t = \frac{376}{10^3-1} = \frac{376}{999}$. Joining all the parts, we find $5.42(376) = \frac{542}{100} + \frac{376}{999} \cdot 10^{-2} = \frac{542 \cdot 999 + 376}{999 \cdot 100}$, where the numerator is integer and the denominator is a natural number, that is, we obtain a common fraction.
2. Answer the following questions:

   (a) is a product of a rational number with irrational always irrational?
   (b) are there two irrational numbers whose sum and product are rationals?
   (c) if the powers a^2 and a^3 are irrationals, should the power a^4 also be irrational?

   Solution

   (a) If the rational number is zero, then the product is also rational number 0. In all other cases, the result is an irrational number. Indeed, if the product $c = a \cdot b$, where $a \neq 0$ is rational and b is irrational, was rational, then the division of this relation by a would result in a rational number, but b is irrational by hypothesis.
   (b) There exist such irrational numbers, for instance, $a = \sqrt{2}$ and $b = -\sqrt{2}$.
   (c) Not necessarily, consider, for instance, $a = \sqrt[4]{2}$.

3. Verify if the following numbers are rational or irrational:

   (a) $\sqrt[5]{50}$;
   (b) $\sqrt{2} + \sqrt{6}$;
   (c) $\sqrt{3 - 2\sqrt{2}} - \sqrt{3 + 2\sqrt{2}}$.

Solution

(a) Let us suppose that $\sqrt[5]{50}$ is a rational number, that is, $\sqrt[5]{50} = \frac{m}{n}$, m, $n \in \mathbb{N}$. Without loss of generality we can assume that the last fraction is in lowest terms (that is, m and n have no common factors other than 1). It follows from this relation that $50 = \frac{m^5}{n^5}$, whence $m^5 = 50n^5$. Then, the number m^5 is even, and consequently, m is even too, that is, $m = 2k$, $k \in \mathbb{N}$. Substituting this representation in the relation between m and n, we have $2^5k^5 = 50n^5$ or $25n^5 = 2^4k^5$. Since 25 is odd, n^5 is even, and so is n. Therefore, m and n have the common divisor 2 that contradicts the initial supposition. Hence, the number $\sqrt[5]{50}$ is irrational.

(b) Let us suppose that $\sqrt{2} + \sqrt{6}$ is a rational number. Then its square—$(\sqrt{2} + \sqrt{6})^2 = 2 + 6 + 2\sqrt{12} = 8 + 4\sqrt{3}$—also is rational. It follows from this that $\sqrt{3}$ is rational, which is false (the irrationality of $\sqrt{3}$ can be shown in the same way as that of $\sqrt{2}$). Hence, the given number is irrational.

(c) Notice, first, that the expressions inside the outer roots can be transformed into full squares: $3 - 2\sqrt{2} = (1 - \sqrt{2})^2$ and $3 + 2\sqrt{2} = (1 + \sqrt{2})^2$. Then,

$$\sqrt{3 - 2\sqrt{2}} - \sqrt{3 + 2\sqrt{2}} = \sqrt{(1 - \sqrt{2})^2} - \sqrt{(1 + \sqrt{2})^2}$$
$$= -(1 - \sqrt{2}) - (1 + \sqrt{2}) = -2,$$

which means that the given number is rational.

3.3 Absolute Value

As we will see later, the operation of *absolute value* applied to real numbers and expressions appears naturally and frequently in geometric constructions. Therefore, it is worth to recall the definition and properties of this operation known from middle school.

Definition of Absolute Value The *absolute value* (or *modulus*) of $a \in \mathbb{R}$ is denoted by $|a|$ and defined analytically as follows: $|a| = \begin{cases} a\,, & a \ge 0 \\ -a\,, & a < 0 \end{cases}$.

Elementary Properties of Absolute Value For any $a, b \in \mathbb{R}$ the following properties are satisfied:

1. $|a| \ge 0\,,\ |a| = 0 \leftrightarrow a = 0$;
2. $|a + b| \le |a| + |b|$;
3. $|a - b| \ge ||a| - |b||$;
4. $|ab| = |a||b|$;
5. $|\frac{a}{b}| = \frac{|a|}{|b|}$, $b \ne 0$.

The proofs are straightforward and are left as exercises to the reader.

Elementary Equations and Inequalities with Absolute Value

An *elementary absolute value equation* has the form $|x| = c$. In the case $c > 0$, this equation has two solutions $x = \pm c$. Indeed, if $x \geq 0$, then the equation $|x| = c$ is equivalent to $x = c$ and this is the solution of the original equation, because $c > 0$. If $x < 0$, then the equation $|x| = c$ becomes $-x = c$ or $x = -c$, which is the second solution, because $-c < 0$.

In the case $c = 0$, the unique solution is $x = 0$. Finally, if $c < 0$, there are no solutions, since $|x|$ is a non-negative quantity by the definition.

Elementary absolute value inequalities have one of the following four forms: $|x| < c$, $|x| \leq c$, $|x| > c$ and $|x| \geq c$. We solve the first and the last one and leave two others to the reader (they can be solved using a similar reasoning).

Consider first the inequality $|x| < c$. In the case $c \leq 0$, there are no solutions, since the inequality implies that $|x|$ should be negative, which is impossible according to the definition. Take now $c > 0$ and consider two options for expression of $|x|$. If $x \geq 0$, then the inequality $|x| < c$ is equivalent to $x < c$ and the solution of the original inequality (the first part of the solution) is $0 \leq x < c$. If $x < 0$, then the inequality $|x| < c$ becomes $-x < c$ or $x > -c$ and we have the second part of the solution of the original inequality: $-c < x < 0$. Joining these two parts, we arrive at the solution $-c < x < c$.

For the inequality $|x| \geq c$, in the case $c \leq 0$ any real number is a solution, since the absolute value of an arbitrary number is non-negative. Hence, in this case, the set of solutions is $\mathbb{R}$. If $c > 0$, then we should open the expression of the absolute value to find solutions. If $x \geq 0$, then the inequality $|x| \geq c$ is equivalent to $x \geq c$, and since $c > 0$, the first part of the solution of the original inequality is $x \geq c$. If $x < 0$, then the inequality $|x| \geq c$ becomes $-x \geq c$ or $x \leq -c$, and this is the second part of the solution. Joining the two parts, we obtain the solution in the form $S = \{x : x \geq c\} \cup \{x : x \leq -c\}$.

Examples

1. Solve the absolute value equation $|2x - 1| = 5$ and inequality $|3x - 2| < 4$.
   Solution

   (a) Recalling that $|a| = 5$ is equivalent to $a = 5$ and $a = -5$, we conclude that the equation $|2x - 1| = 5$ is equivalent to $2x - 1 = 5$ and $2x - 1 = -5$. The former has the solution $x_1 = 3$ and the latter—$x_2 = -2$. These are the two solutions of the original equation.

   (b) Recalling that $|a| < 4$ is equivalent to $-4 < a < 4$, we rewrite the inequality $|3x - 2| < 4$ in the form $-4 < 3x - 2 < 4$. This double inequality can be reduced to $-2 < 3x < 6$ or, isolating x, to $-\frac{2}{3} < x < 2$. This is the solution of the original inequality. It also can be written using the set notation: $S = \{x : -\frac{2}{3} < x < 2\}$.

2. Solve the absolute value equation $\left|\frac{x-1}{x+2}\right| = 3$ and inequality $\left|\frac{x-1}{x+2}\right| < 3$.

Solution

(a) Recalling that $|a| = 3$ means $a = 3$ and $a = -3$, we can see that the equation is equivalent to $\frac{x-1}{x+2} = 3$ and $\frac{x-1}{x+2} = -3$. Solving the former we have $x - 1 = 3x + 6$ and then $x_1 = -\frac{7}{2}$. For the latter we obtain $x - 1 = -3x - 6$ and $x_2 = -\frac{5}{4}$. These are the two solutions of the original equation.

(b) Recalling that $|a| < 3$ is equivalent to $-3 < a < 3$, we rewrite the given inequality in the equivalent form $-3 < \frac{x-1}{x+2} < 3$. To solve the right-hand inequality, we have to consider the two cases. First, if $x + 2 > 0$, then this inequality is reduced to $x - 1 < 3x + 6$, whose solution is $x > -\frac{7}{2}$. Since the last solution is true under the condition $x > -2$ (or $x + 2 > 0$), the first part of the solution is $x > -2$. Second, if $x + 2 < 0$, then the right-hand inequality is simplified to $x - 1 > 3x + 6$ with the solution $x < -\frac{7}{2}$. Since this solution satisfies the condition $x < -2$ (that is, $x + 2 < 0$), it is the second part of the solution. Therefore, the solution of the right-hand inequality is $x > -2$ and $x < -\frac{7}{2}$, or expressing in terms of the sets we have $S_r = \{x : x > -2\} \cup \{x : x < -\frac{7}{2}\}$. Next, we switch to the left-hand inequality, which is solved using the same reasoning. If $x + 2 > 0$, then this inequality can be written as $x - 1 > -3x - 6$, leading to the solution $x > -\frac{5}{4}$. This solution satisfies the restriction $x > -2$, and consequently, it is the first part of the solution. If $x + 2 < 0$, then the left-hand inequality is reduced to $x - 1 < -3x - 6$, whose solution is $x < -\frac{5}{4}$. Since the last solution is true under the condition $x < -2$, the second part of the solution is $x < -2$. Hence, the solution of the left-hand inequality is $x > -\frac{5}{4}$ and $x < -2$, or using the set notation, $S_l = \{x : x > -\frac{5}{4}\} \cup \{x : x < -2\}$. Finally, for the original inequality holds, both right-hand and left-hand inequalities should be satisfied. This means that the solution of the original inequality is the intersection of the two solutions: $S = S_r \cap S_l = \{x : x < -\frac{7}{2}\} \cup \{x : x > -\frac{5}{4}\}$.

Notice that in this algorithm, we solve first the right-hand inequality, second the left-hand inequality (considering for them both options of the denominator sign) and then we find the final solution by the intersection of the two one-handed solutions. Naturally, we can change the order of some steps of this procedure. For example, we can first fix the positive sign of the denominator and solve the corresponding double inequality. Second we fix the negative sign of the denominator and solve the second double inequality. Then we find the final solution as the union of the two obtained solutions. Certainly both algorithms provide the same solution. We recommend to the reader to carry out the second procedure and confirm that the found solution is equal to the obtained above.

Remark The method of solution can be represented in more transparent form if we join analytic and geometric approaches in the algorithm called the method of intervals. We explore this possibility in Examples of section 5.2. after introduction of the relation between the real numbers and the points of the real line, which is needed for the method of intervals.

4 Coordinate Line and its Equivalence with the Set of Real Numbers

Definition of a Coordinate Line A *coordinate line* (or *real line*) is a line with the following attributes: the *origin point*, the *positive/negative direction*, and the *unit of measure*. Notice that the origin, labeled O, is chosen arbitrarily, but after been chosen this is a fixed point. The positive direction is a chosen direction of moving along the coordinate line. A part of the line that lies on the side of the origin corresponding to the positive direction is called the positive part, and the remaining part is called negative. The choice of a positive direction is arbitrary, but if the line has the horizontal orientation, it is traditional to consider the rightward direction to be the positive one. Consequently, the part on the right side of the origin is the positive one, and the negative part is located at the left of the origin. If the line has the vertical orientation, then usually the positive direction is upward, with the part of line above the origin being positive and the remaining part negative. The positive direction is frequently labeled by letter x (or another appropriate letter) (see Fig. 1.1). Finally, the unit of measure is of an arbitrary choice and it represents the distance equal to 1 between two points, in particular, between the origin O and the specific point I located on the positive part of the coordinate line (see Fig. 1.1). The coordinate line is also called *coordinate axis* or the x-axis (in the case x is used to label the positive direction).

Equivalence Between a Coordinate Line and the Set of Real Numbers
Let us establish the relation of *equivalence* between the set of real numbers and all the points of a coordinate line, that is, the relation that associates each real number with a point on the line and vice-verse, each point on the coordinate line is associated with a real number (such a relation is also called bijective or one-to-one). Notice that there are different forms to generate an equivalence relation between these two sets. A specification of the relation which we will introduce is that it is based on the measure of the distance from the origin point to the current point of the coordinate line.

Notice first that it is sufficient to settle the equivalence between the positive part of the coordinate line and the set of positive real numbers. In fact, if this part is done, then we associate the origin O with the number 0 and each point P_- of the

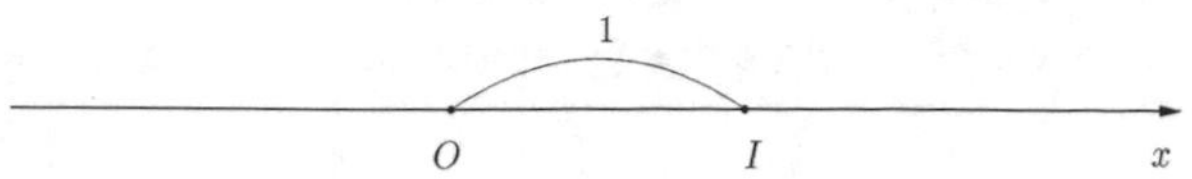

Fig. 1.1 Coordinate line: origin, positive direction and unit of measure

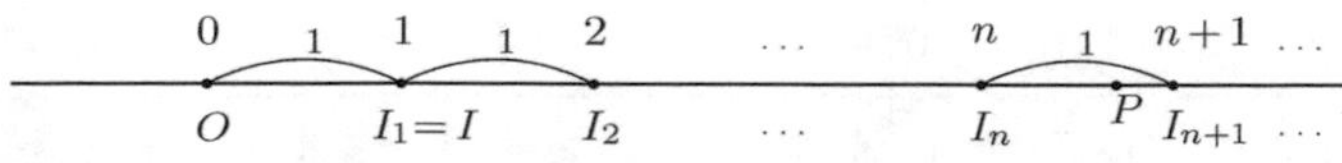

Fig. 1.2 Determining the integer part of x

negative part of the line, which is symmetric to the point P of the positive part with respect to the origin, we associate with the negative number $-x_P$, where x_P is the positive number corresponding to P. The relation between the coordinate line and the real numbers obtained in this way will be the equivalence. Thus, the problem will be solved if we will be able to construct the equivalence relation between the positive part of the coordinate line and the positive real numbers.

Let us specify that the origin O corresponds to 0. We choose an arbitrary point P on the positive part of the line and provide the algorithm that associates P with the unique positive real number. Initially, we mark the points (to the right of O) which corresponds to multiples of the unit of measure: $I_1 \equiv I$ located at a distance of 1 unit from the origin, I_2 at a distance of 2 units from the origin, ..., I_n at a distance of n units from the origin, etc. (see Fig. 1.2). Each of these points we associate with the positive number equal to its distance from the origin: $I_1 \leftrightarrow 1$, $I_2 \leftrightarrow 2$, ..., $I_n \leftrightarrow n, \ldots$ (see Fig. 1.2). In this way we establish the equivalence between all the points, whose distance from O is multiple of the unit of measure, with all the natural numbers. If the chosen point P coincides with one of these point, say with I_n, then the number x corresponding to P is already found: $x = n$, and the algorithm terminates. Otherwise, P is located in between two consecutive points of the constructed sequence I_n, $n = 1, 2, \ldots$. Assume that P lies in between I_n and I_{n+1}. In this case, we determine the integer part of the searched number $x = x_0.x_1x_2\ldots : x_0 = n$, and finalize the step 0 of our algorithm. We start the next step by dividing the segment $[I_n, I_{n+1}]$ (whose length is 1) into ten equal parts using one-tenth of the original unit. The division points we label (from the left to the right) as follows: $A_{1,0} \equiv I_n$, $A_{1,1}$, $A_{1,2}$, $A_{1,3}$, $A_{1,4}$, $A_{1,5}$, $A_{1,6}$, $A_{1,7}$, $A_{1,8}$, $A_{1,9}$,$A_{1,10} \equiv I_{n+1}$ (see Fig. 1.3). If P coincides with one of the points of this division, for instance with $A_{1,7}$, then the desired number is already found: $x = n.7$, and the algorithm terminates. Otherwise, assuming that P lies in between $A_{1,7}$ and $A_{1,8}$, we determine the first digit after the dot $x_1 = 7$ in the decimal representation $x = x_0.x_1x_2\ldots x_i\ldots$ and finalize the step 1.

One can already guess how we proceed on the next step. We divide the interval $[A_{1,7}, A_{1,8}]$, which contains P, into ten equal parts using one-hundredth of the original unit and label the division points by $A_{2,0} \equiv A_{1,7}$, $A_{2,1}$, $A_{2,2}$, $A_{2,3}$, $A_{2,4}$,

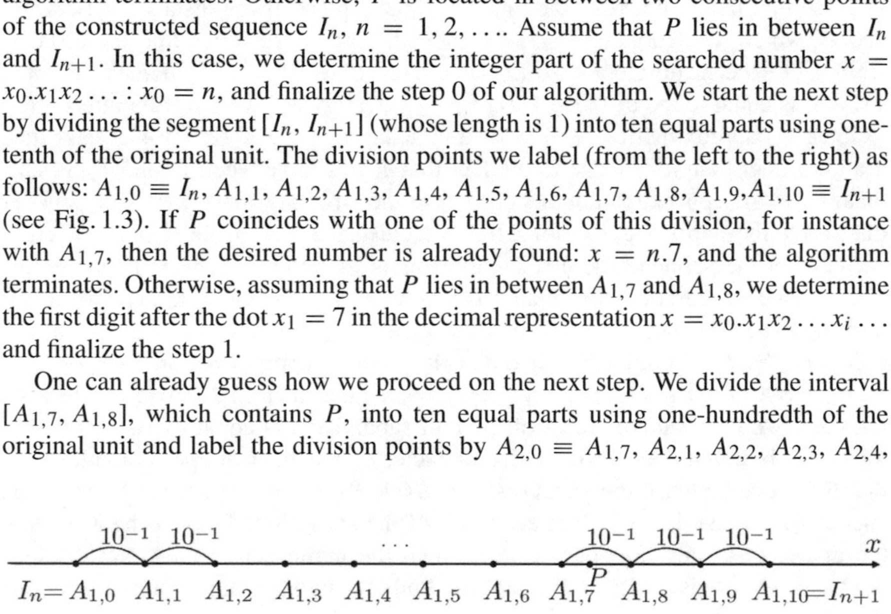

Fig. 1.3 Determining the first decimal digit of x

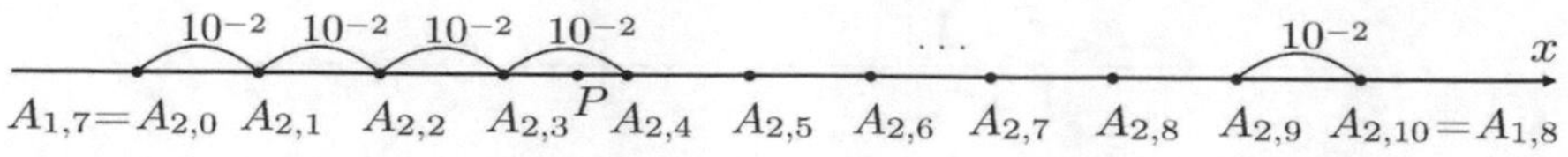

10^{-2} 10^{-2} 10^{-2} 10^{-2} ... 10^{-2} x

$A_{1,7}{=}A_{2,0}$ $A_{2,1}$ $A_{2,2}$ $A_{2,3}$ P $A_{2,4}$ $A_{2,5}$ $A_{2,6}$ $A_{2,7}$ $A_{2,8}$ $A_{2,9}$ $A_{2,10}{=}A_{1,8}$

Fig. 1.4 Determining the second decimal digit of x

$A_{2,5}$, $A_{2,6}$, $A_{2,7}$, $A_{2,8}$, $A_{2,9}$, $A_{2,10} \equiv A_{1,8}$ (see Fig. 1.4). If P coincides with one of these points, for instance with $A_{2,3}$, then the searched number is already found: $x = n.73$, and the algorithm terminates. Otherwise, assuming that P is located in between $A_{2,3}$ and $A_{2,4}$, we determine the second digit after the dot of the number x: $x_2 = 3$, and finalize the step 2. And so on.

At the i-th step, we determine the i-th decimal digit x_i using the i-th subdivision of the current interval of the location of P in the ten equal parts with the length equal to 10^{-i} of the original unit. If P coincides with one of the points of the i-th subdivision, for instance with $A_{i,1}$, then $x_i = 1$ and the wanted number is $x = x_0.x_1x_2\ldots x_i$, that terminates the algorithm. Otherwise, assuming that P lies in between $A_{i,1}$ and $A_{i,2}$, we set $x_i = 1$, finalize the step number i and follow to the next step. In this way, at each step of the algorithm we determine the current decimal digit of x and, depending on location of P, either terminate algorithm (which means that all the remaining digits are 0) or move to the next step, at which we find out the next digit of x, etc. In either of the two cases—in the case of a finite or infinite algorithm–, all the digits of x will be uniquely determined, and therefore, the unique real number associated with a chosen P will be determined.

It is important to note that the found number represents the distance between O and P, which we denote by $d(O, P)$. Indeed, on the step 0, we find the distance from O to the nearest point I_n to the left of P. If $P = I_n$, then the number $x_P = n$ is equal to the distance $d(O, P) = n$. Otherwise, we have the inferior (up to integers) approximation $x_0 = n$ of this distance, since $d(O, P)$ is greater than n and smaller than $n + 1$ due to the fact that P lies in between I_n and I_{n+1}. This means that the approximation of step 0 have the error $\epsilon_0 = d(O, P) - x_0$ smaller than 1. On the step 1, we determine the exact distance $x_0.x_1$ from O if P coincides with one of the points of the first subdivision, or the inferior approximation to this distance with accuracy better than 10^{-1}. And so on. On the i-th step, we either find the exact distance $d(O, P) = x_0.x_1x_2\ldots x_i$ if P coincides with one of the points of the i-th subdivision or obtain its inferior approximation $x_0.x_1x_2\ldots x_i$ with the error $\epsilon_i = d(O, P) - x_0.x_1x_2\ldots x_i$ smaller than 10^{-i} (since the distance between two consecutive points of the i-th subdivision is equal to 10^{-i} and P is located in between them). Notice that at each step the error of approximation ϵ_i is the difference between the exact distance $d(O, P)$ and its inferior approximation $x_0.x_1x_2\ldots x_i$, that is, $\epsilon_i \geq 0$ at each step of the algorithm. Thus, if the algorithm terminates in a finite number of steps, then the number x is the exact distance $d(O, P)$. Otherwise, after i-th step we find the inferior approximation to this distance with accuracy better than 10^{-i}, and on each following step we improve the accuracy 10 times comparing to the preceding step. Hence, after a sufficiently large number of steps, the error of approximation ϵ_i will be smaller than any positive number. Since this error is non-negative and the only non-negative number smaller than any positive number is 0, it follows that at the end of algorithm we find the exact distance between O and P.

Using a similar procedure in the inverse direction, that is, starting from a positive real number and determining the unique corresponding point, we can see that each positive real number can be associated with the only point of the coordinate line (the details of the inverse procedure are left as an exercise for the readers). Thus, each point of the positive part of the coordinate line corresponds to the unique positive real number and vice-verse. This means that between the two sets—the positive part of the coordinate line and the positive real numbers—we have established the equivalence relation.

As it was already mentioned the remaining part is simple: the point O corresponds to the number 0, and each point P_- on the negative part of the coordinate line is associated with the unique negative real number in the following way: first, we find the point P on the positive part of the line, symmetric to P_- about the origin (it has the same distance from O as P_- but belongs to the positive part), then we take the positive number x_P corresponding to P (which exists and is unique according to the first part of the algorithm) and, finally, we choose $-x_P$ as the unique number associated with P_-. Hence, using the proposed procedure, the equivalence between all the points of the coordinate line and all the real numbers is established. The number x_P that corresponds to P is called the *coordinate of the point* P. Due to the equivalence relation, we will frequently use the notation $P = x_P$ and call the points on the coordinate line by numbers and the numbers by points. It does not mean that the two objects (the points and the numbers) are the same, but that they are equivalent.

It is important to note that the algorithm used to show the equivalence between $\mathbb{R}$ and the coordinate line provides such a coordinate of each point that is intimately connected with the distance from the origin: if $P = O$, then the corresponding coordinate $x_P = 0$ is the distance between two coinciding points; if P belongs to the positive part, then x_P is the distance from O to P; finally, if P lies on the negative part, x_P represents the distance from O to P taken with the negative sign. Using as before the notation $d(O, P)$ for distances from the origin, we rewrite these relations in the form: $d(O, P) = 0$ if $x_P = 0$ (or equivalently, if $P = O$); $d(O, P) = x_P$ if $x_P > 0$ (or equivalently, if P is located to the right of O); $d(O, P) = -x_P$ if $x_P < 0$ (or equivalently, if P is located to the left of O). Recalling the definition of an *absolute value*, we immediately conclude that $d(O, P) = |x_P|$.

Remark 1 Due to the equivalence between the sets of real numbers and points of the real line, the real numbers frequently are called points and points of the coordinate line are called the numbers. The coordinate line (coordinate axis) is called *real line* (*real axis*). Following the same analogy, to say that the point A lies to the right of

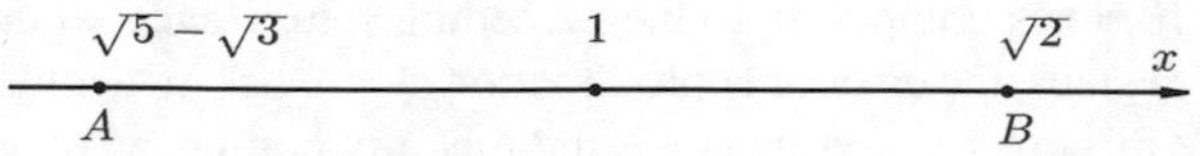

Fig. 1.5 The points $\sqrt{5}-\sqrt{3}$ and $\sqrt{2}$ on the coordinate line

B is the same as to state that $x_A > x_B$ in terms of their coordinates; to say that the point A stays between B and C (with C being rightmost) is the same as to assert that $x_B < x_A < x_C$ in terms of their coordinates. And so on. This means that, from now on, we can use interchangeably the concepts of points and their coordinates without loss of precision of reasoning.

Remark 2 Sometimes it is useful to consider the *extended real line*, including positive $(+\infty)$ and negative $(-\infty)$ infinities. In this case, all the real (geometric) points are called *finite points* and the infinities are called *infinite points*.

Examples

1. Mark (approximately) the points A and B with the coordinates $\sqrt{5}-\sqrt{3}$ and $\sqrt{2}$, respectively, explaining which one is located more to the right.
   Solution
   Use the following elementary evaluations of the irrational numbers: $\sqrt{5} < 2.5$, $\sqrt{3} > 1.5$, $\sqrt{2} > 1$. Then $\sqrt{5}-\sqrt{3} < 2.5-1.5 = 1 < \sqrt{2}$. Hence, the point B lies to the right of A (see Fig. 1.5).
   Another way to compare the locations of A and B is by using the properties of the involved expressions. It can be shown that in general the point A with the coordinate $x_A = \sqrt{a}-\sqrt{b}$ stays to the left of B with the coordinate $x_B = \sqrt{a-b}$ if $a > b \geq 0$. Indeed, since both coordinates are positive, they can be squared without changing their relation: $x_A^2 = a+b-2\sqrt{ab}$ and $x_B^2 = a-b$. Then, we have to compare $b-2\sqrt{ab}$ and $-b$. Since $a > b$, it follows that $b-2\sqrt{ab} < b-2\sqrt{b^2} = b-2b = -b$, that is, x_A is smaller than x_B, or equivalently, A lies to the left of B.
2. Find the coordinates of all the points whose distance from the origin is equal to π.
   Solution
   The distance can be measured both to the right and to the left from the origin O (that is, on the positive and negative parts of the coordinate line). The coordinate of the point A to the right of O is equal to the given distance, that is, $x_A = \pi$. The coordinate of the point B to the left of O is equal to the distance with the negative sign, that is, $x_B = -\pi$.

5 Some Sets of Numbers and Their Properties

5.1 *Distance Between Two Points*

Using the absolute value, we can express the *distance between two points* of the real line in a compact form. First, recall once more that the distance from the origin O to any point P is the absolute value of the coordinate of P: $d(O, P) = |x_P|$. Consider now two arbitrary points A and B. Assume initially that B lies to the right of A, that is, $x_B > x_A$. If both points belong to the positive part of the coordinate line, then $d(A, B) = d(O, B) - d(O, A) = x_B - x_A$ (Fig. 1.6). If the origin is found in between A and B, then $d(A, B) = d(O, B) + d(O, A) = x_B + (-x_A) = x_B - x_A$ (Fig. 1.7). Finally, if A and B stay to the left of the origin, then $d(A, B) = d(O, A) - d(O, B) = -x_A - (-x_B) = x_B - x_A$ (Fig. 1.8). Therefore, in all cases we have $d(A, B) = x_B - x_A$. If the relation between A and B is different—A is located to the right of B –, then it is sufficient to change the naming of the points and their coordinates in the already derived formula to obtain $d(A, B) = x_A - x_B$. Thus, if $x_B > x_A$ (B stays to the right of A), then $d(A, B) = x_B - x_A$; if $x_A > x_B$ (A lies to the right of B), then $d(A, B) = x_A - x_B$; and if $x_B = x_A$ (A coincides with B), then $d(A, B) = 0$. Using the absolute value we obtain

$$d(A, B) = |x_A - x_B| \tag{1.1}$$

for any position of A and B.

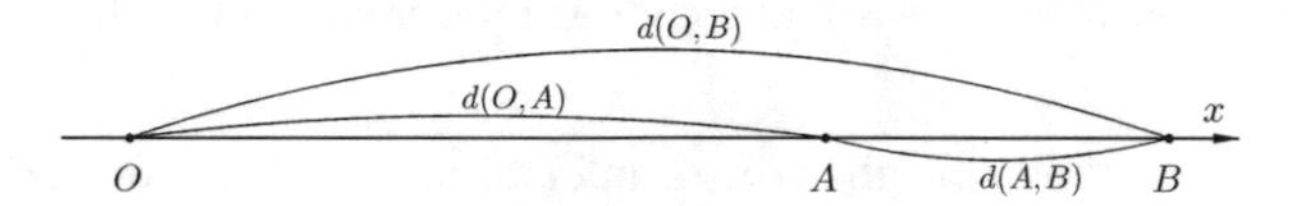

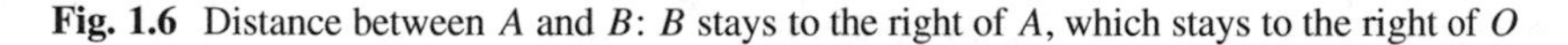

Fig. 1.6 Distance between A and B: B stays to the right of A, which stays to the right of O

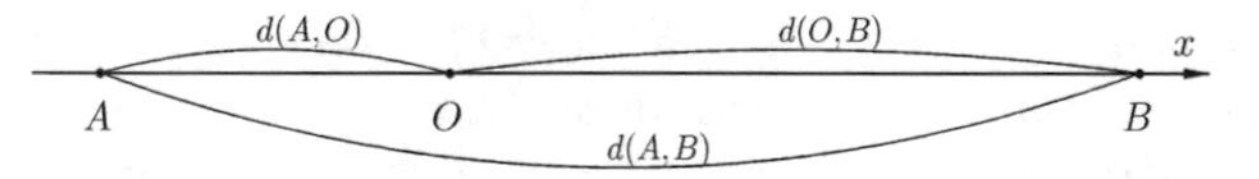

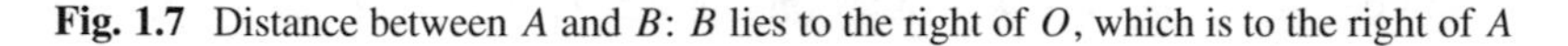

Fig. 1.7 Distance between A and B: B lies to the right of O, which is to the right of A

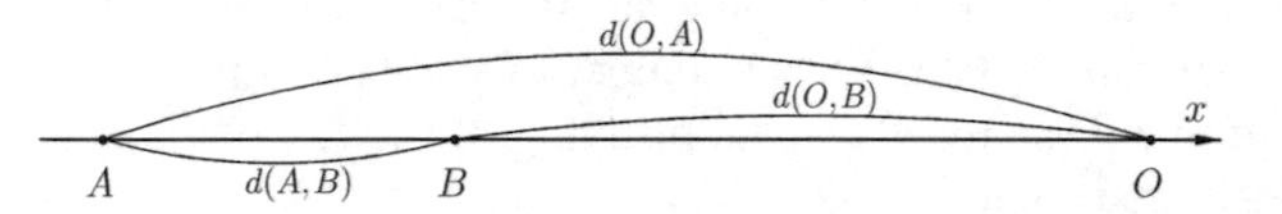

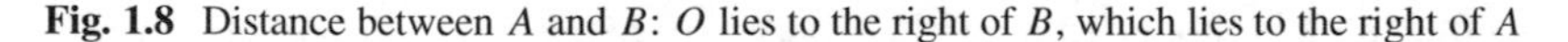

Fig. 1.8 Distance between A and B: O lies to the right of B, which lies to the right of A

5.2 Interval, Midpoint, Symmetric Point, Neighborhood

In what follows, we will use the following natural simplified notations for the coordinates of points: a for coordinate of A, b—for coordinate of B, c—for coordinate of C, etc.

Definition. Interval An *interval (segment)* is a set of all the points of the real line located in between two given points A and B. In the terms of coordinates, the equivalent definition is as follows: an interval is a set of the real numbers in between the given numbers a and b. The points A and B (equivalently, their coordinates a and b), called the endpoints of an interval, can be included in or excluded from the interval. Depending of this, the following four types of the intervals are considered: $a < x < b$ (*open interval*), or $a \leq x \leq b$ (*closed interval*), or $a \leq x < b$, or $a < x \leq b$ (the last two types are called *half-open or half-closed*). The common notations are (a, b), $[a, b]$, $[a, b)$ and $(a, b]$, respectively. Using the points of the coordinate line, the corresponding notations are (A, B), $[A, B]$, $[A, B)$ and $(A, B]$. Unless there is a contrary statement, we will consider "normal" intervals, that is, the intervals whose first endpoint A is located to the left of or coincide with the second endpoint B (or equivalently, $a \leq b$).

Remark 1 Because of the equivalence between the sets of numbers and of points, as it was mentioned above, the usual agreement is to use numerical and geometric notation and terminology interchangeably.

Remark 2 In the case of strict inequality, the point a can be $-\infty$ and the point b can be $+\infty$. In the first case, the formal inequality $-\infty < x < b$ (or $-\infty < x \leq b$) simply means $x < b$ (or $x \leq b$), and similar interpretation is valid in the second case.

Remark 3 In some problems, the type of interval (open, closed or half-open) does not matter. In particular, this is the case of the length of an interval and its midpoint. In such situations we will not specify the type of an interval.

Definition. Bounded Interval An open interval (a, b) is *bounded* when both a and b are finite points (that is, some points of the real axis). If a or b are infinite, then an interval is *unbounded*. A closed interval $[a, b]$ is always bounded. A half-open interval is bounded when an excluded endpoint is finite. Otherwise it is unbounded.

Remark Sometimes a bounded interval is called a *finite interval* and an unbounded—an *infinite*. Clearly, the number of points of any non-singular interval $(a < b)$ is infinite. Usually, a specific meaning of the term "finite/infinite" is clear from the considered context.

The definition of a bounded interval is naturally extended to the definition of a bounded set.

Definition. Bounded Set A set S of real numbers is called *bounded above* (or *bounded on the right*) if there exists a real number M such that for any $x \in S$ it

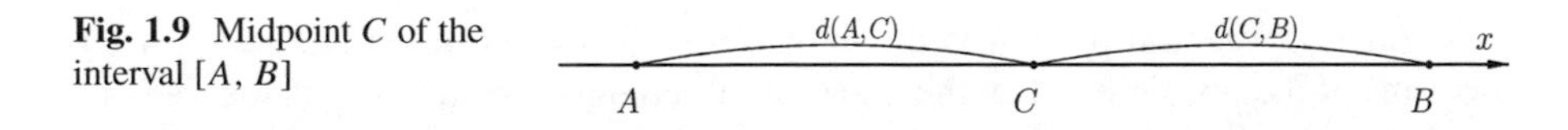

Fig. 1.9 Midpoint C of the interval $[A, B]$

holds $x \leq M$. In the same manner, a set S of real numbers is called *bounded below* (or *bounded on the left*) if there exists a real number m such that for any $x \in S$ it holds $x \geq m$. Finally, a set of real numbers is *bounded* if it is bounded both above and below.

Remark As we can see, a bounded interval (a, b) (or $[a, b]$, or $[a, b)$, or $(a, b]$) is a particular case of a bounded set, since the inequality $a \leq x \leq b$ holds for $\forall x \in (a, b)$ (or $\forall x \in [a, b]$, or $\forall x \in [a, b)$, or $\forall x \in (a, b]$).

Definition. Length of An Interval The *length of an interval* is a distance between its endpoints. (Obviously, the type of interval does not matter in this definition.)

Definition. Midpoint The *midpoint of an interval* is such a point of an interval which is equidistant from its endpoints.

To find the coordinate of a midpoint C of an interval $[A, B]$, we express its geometric definition in the analytic terms: $d(A, C) = d(C, B)$ or $c - a = b - c$ (see Fig. 1.9) or, solving for c:

$$c = \frac{a + b}{2}\,. \tag{1.2}$$

Definition. Symmetric Point The *point B is symmetric* to a given point A about the third point C (called the *center or centerpoint of symmetry*), if B and A are equidistant from C (B and A being different points).

Based on the definition, notice that the condition $d(B, C) = d(A, C)$ means that C is the midpoint of the interval $[A, B]$ (see Fig. 1.9). Then, $c = \frac{a+b}{2}$, and isolating b we have

$$b = 2c - a\,. \tag{1.3}$$

In a particular case when the center of symmetry is the origin O, we have $b = -a$ (this relation was already found before).

Remark The concept of symmetry of a pair of points with respect to the center of symmetry C is naturally extended to a set of points: a set S is symmetric about C if any point of S has a symmetric point also belonging to S. As we will see in the study of even and odd functions, their domains are symmetric about the origin.

Definition. Neighborhood A *neighborhood* of a is an open interval centered at a. To determine the extension of a neighborhood, called the *radius of a neighborhood*, it is used the distance from a to the endpoints. In this way, a neighborhood of a of the radius r is the open interval $(a - r, a + r)$.

Clearly, the central point a is the midpoint of its neighborhood $(a - r, a + r)$. In terms of the absolute value, the equation of a neighborhood of a of the radius r can be written in the form $|x - a| < r$. Indeed, denoting $t = x - a$, we have an elementary inequality $|t| < r$ whose solution is $-r < t < r$, or returning to x we have $-r < x - a < r$. Solving for x, we obtain $a - r < x < a + r$.

Definition. Right/Left Neighborhood A *right neighborhood* of a is an open interval to the right of a: $(a, a + r)$, where r is still called radius. Similarly, a *left neighborhood* of a is an open interval to the left of a: $(a - r, a)$. Clearly, the right neighborhood represents the right half of the "full" neighborhood and the left neighborhood represents its left half. The union of the two one-sided neighborhoods is the entire neighborhood without the central point.

Definition. Interior Point Given a set S, a point c is *interior point* of S if there exists a neighborhood of c contained in S. For example, any point of an open interval (a, b) is its interior point. Of course, an interior point should belong to the considered set, however, it is a necessary, but not sufficient condition: not all the points of a set are necessarily its interior points. For instance, any point $c \in (a, b)$ of the closed interval $[a, b]$ is its interior point, but the endpoints a and b are not the interior points (although they belong to $[a, b]$).

Definition. Exterior Point Given a set S, a point c is *exterior point* of S if there exists a neighborhood of c with no point in S. For example, any point that does not belong to $[a, b]$ is its exterior point. However, if we consider an open interval (a, b), then any point $c < a$ or $c > b$ is it exterior point, but the endpoints a and b are not exterior points (although they do not belong to (a, b)).

Definition. Boundary Point Given a set S, a point c is a *boundary point* of S if any neighborhood of c contains both point of S and points out of S. For example, the endpoints a and b of the open interval (a, b) are its boundary points, and the same is true for the closed interval $[a, b]$.

Examples

1. Give geometric characterization of the points whose coordinates satisfy the following relations:

   (a) $|1 + 3x| \le 2$;
   (b) $x^3 + x^2 - 12x > 0$;
   (c) $\left|\frac{3-x}{x+1}\right| = 2$;
   (d) $\left|\frac{x-1}{x+2}\right| < 3$.

   Solution

   Notice that there are a few cases when given relations admit a direct geometric solution, that is, can be interpreted geometrically without a preliminary analytic solution. The majority of equations and inequalities is treated first analytically and then the found solution is explained geometrically. At this point, the

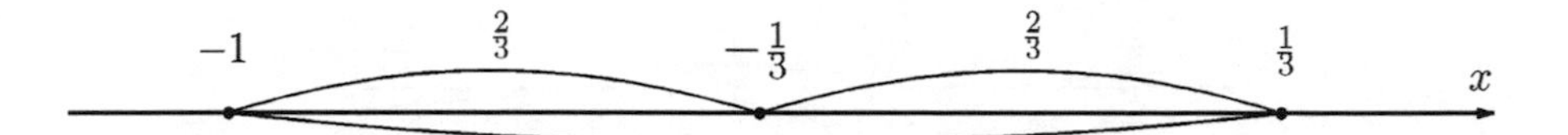

Fig. 1.10 Geometric solution of the inequality $|x - (-\frac{1}{3})| \leq \frac{2}{3}$

importance of the use of coordinates as an indispensable tool becomes to be very clear.

(a) Rewrite the given inequality in the form $-2 \leq 1 + 3x \leq 2$ (recalling that $|a| \leq 2$ is equivalent to $-2 \leq a \leq 2$). Then, it follows that $-3 \leq 3x \leq 1$ or $-1 \leq x \leq \frac{1}{3}$. Hence, the geometric solution is the closed interval $[-1, \frac{1}{3}]$. Notice that in this particular simple case the original inequality can be solved geometrically straightforward applying the characterization of the absolute value. Indeed, the given inequality can be written in the form $|x-(-\frac{1}{3})| \leq \frac{2}{3}$, which means that the desired points stay away from the point $-\frac{1}{3}$ at the distance smaller then or equal to $\frac{2}{3}$ (both to the right and to the left). Then, we mark the centerpoint $-\frac{1}{3}$ of the desired interval and measure the distance $\frac{2}{3}$ from it to both sides, obtaining the point -1 to the left and $\frac{1}{3}$ to the right (see Fig. 1.10). All the points of the interval $[-1, \frac{1}{3}]$ (including its endpoints) represent the (geometric) solutions of the original inequality.

(b) The inequality $x^3 + x^2 - 12x > 0$ cannot be solved geometrically in direct mode. First we need to find its analytic solution. To do this, we use the factorization of the polynomial: $x(x - 3)(x + 4) > 0$ (it is easy to obtain by separating the factor x and solving the quadratic polynomial by standard formula). Notice that the critical points (those where each linear factor change its sign) are -4, 0 and 3 (in increasing order). For convenience, we put the linear factors in the same order: $(x+4)x(x-3) > 0$. If $x < -4$, then all the factors are negative, and consequently, the product is negative, that is, the inequality does not hold; if $-4 < x < 0$, then the first factor is positive, while the other two are negative, resulting in the positive product, which means that the inequality is satisfied; for $0 < x < 3$ the two first factors are positive and the third is negative, providing the negative result, that is, the inequality does not hold; finally, for $x > 3$, all the three factors are positive and so is their product, which means that the inequality is satisfied. Thus, the analytic solution is the union of the two open intervals $(-4, 0)$ and $(3, +\infty)$, and its geometric form is the two corresponding intervals marked on the coordinate line.

We can arrive at the same solution using the analytic-geometric method of intervals. The reasoning follows the above analytic considerations, but for better visualization they are accompanied by geometric representation. We start with the inequality $(x+4)x(x-3) > 0$ and its critical points -4, 0 and 3. Then we mark on the real line the intervals of the positive and negative sign for each linear factor and intersecting these signs we make a conclusion

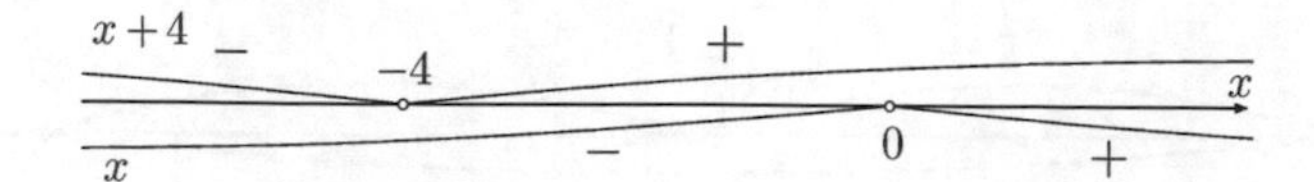

Fig. 1.11 The critical points and signs of the factors $x + 4$ and x

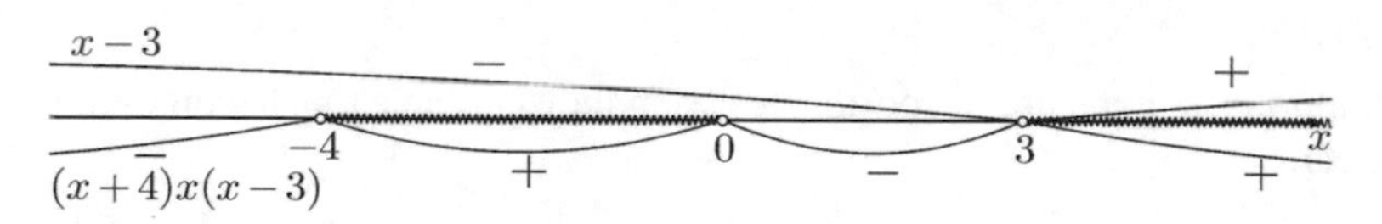

Fig. 1.12 The critical points and signs of the factor $x - 3$ and of the polynomial $(x + 4)x(x - 3)$

about the sign of the product. The intervals of the specific sign for the first two factors are shown in Fig. 1.11 (notations of the first term are made above the line and for the second term below). The sign of the third factor (above the line) and of the resulting product (below the line) are shown in Fig. 1.12. In the last figure, the "waved" part represents the solution of the inequality.

(c) The absolute value equation is equivalent to the two equations without absolute value: $\frac{3-x}{x+1} = 2$ and $\frac{3-x}{x+1} = -2$ (recall that $|a| = 2$ means $a = 2$ and $a = -2$). Solving the former, we have $x_1 = \frac{1}{3}$ and for the latter we obtain $x_2 = -5$. Hence, there are two solutions—points on the real line with the coordinates $\frac{1}{3}$ and -5. Notice that although the analytic solution is straightforward, the equation does not allow us to find directly a geometric solution.

(d) The absolute value inequality $\left|\frac{x-1}{x+2}\right| < 3$ was already solved in Example 2 of Sect. 3.3 using the analytic approach. Let us solve it now applying the method of intervals. First rewrite the given inequality in the form without the absolute value: $-3 < \frac{x-1}{x+2} < 3$. The right-side inequality can be written as $\frac{x-1}{x+2} - 3 = \frac{-2x-7}{x+2} < 0$ or still $\frac{2x+7}{x+2} > 0$. The zeros of the numerator and denominator (the critical points) are $x = -\frac{7}{2}$ and $x = -2$, respectively. Mark these points on the coordinate line and label the intervals of the fixed sign of the numerator and denominator by the symbols "+" and "−": to make no confusion, the notations for the numerator are placed above the line and for the denominator below (see Fig. 1.13). The ratio $\frac{2x+7}{x+2}$ is positive when the signs of the numerator and denominator coincide, and this part of the line is marked as "waved". In the same way, the left inequality can be written in

Fig. 1.13 Critical points and signs of the right inequality

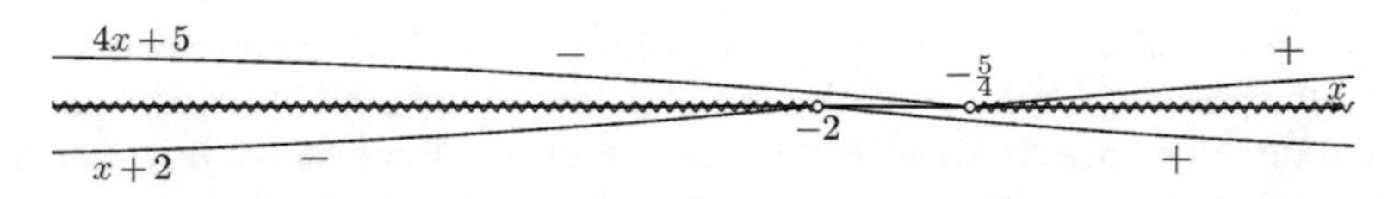

Fig. 1.14 Critical points and signs of the left inequality

the form $\frac{x-1}{x+2} + 3 = \frac{4x+5}{x+2} > 0$. Making similar notations with respect to the critical points $x = -\frac{5}{4}$ and $x = -2$, we immediately conclude that the ratio $\frac{4x+5}{x+2}$ is positive on the "waved" part of the coordinate line (see Fig. 1.14). To satisfy both inequalities at once we should choose the common part (make intersection) of the two "waved" parts. Thus, we arrive at the final solution $S = \{x : x < -\frac{7}{2}\} \cup \{x : x > -\frac{5}{4}\}$.

2. Find the midpoint of the interval AB, where $A = -2$ and $B = 6$.
Solution
From formula (1.2), it follows immediately that the midpoint D has the coordinate $d = \frac{-2+6}{2} = 2$.
3. Find the point B symmetric to A about C:

(a) $A = 5$ and $C = -1$;
(b) $A = -1$ and $d(A, C) = 3$;
(c) $d(A, B) = 6$ and $C = 2$.

Solution

(a) Using formula (1.3), we directly have $B = 2 \cdot (-1) - 5 = -7$.
(b) Under the conditions of this problem we have the two options. If C is located to the right of A, then $C = 2$ and $B = 2 \cdot 2 - (-1) = 5$. If C lies to the left of A, then $C = -4$ and $B = 2 \cdot (-4) - (-1) = -7$.
(c) Under the conditions of this problem we have the two options. If A is located to the right of B, then $A = 2 + 3 = 5$ and $B = 2 - 3 = -1$. Otherwise, $A = -1$ and $B = 5$.

6 Cartesian Coordinates on the Plane

6.1 Definition of the Coordinates

We define the *Cartesian system* on the plane in two steps: first, we introduce the *coordinate axes* and then describe the algorithm that allows us to associate each point on the plane with the unique *ordered pair of the real numbers* and vice-verse. The plane with the introduced Cartesian system is called *Cartesian coordinate plane* or simply *coordinate plane*.

The *coordinate axes* are introduced as follows. Initially, we choose an arbitrary line on the plane and transform it into the coordinate line, that is, we assign to this line the three characteristics—the point of the origin, the positive orientation and the unit of measure. Next, we construct the second coordinate line in such a way that it is perpendicular to the first one, its origin coincides with that of the first line, its unit of measure is the same as of the first line, and finally, its positive direction is found by counterclockwise rotation of the positive direction of the first line by angle $\frac{\pi}{2}$ about the origin. (Usually the counterclockwise rotation is considered a positive rotation on the plane). The first coordinate line is frequently called the x-axis and the second—the y-axis. In this way, both coordinate axes are well defined and we move to the second step.

With the coordinate axes at our disposal, we elaborate an algorithm (one of many possible) which lead to the definition of the *Cartesian coordinates*. Consider an arbitrary point P on the plane and find the corresponding points of its *projection on the coordinate axes*. To find the *projection of* P *on the* x*-axis*, draw the line that passes through P and is perpendicular to the x-axis (parallel to the y-axis) and mark the point P_x of the intersection of this line with the x-axis. In the same manner, the point P_y of the intersection of the y-axis with the line, which passes through P and is perpendicular to the y-axis, is the *projection of* P *on the* y*-axis* (see Fig. 1.15). Since P_x belongs to the coordinate line x, this point is associated with its unique coordinate x. Similarly, the point P_y of the coordinate line y generates the unique coordinate y. The ordered pair (x, y) is called the *Cartesian coordinates of the point* P, where the first coordinate is frequently called *abscissa or* x*-coordinate*

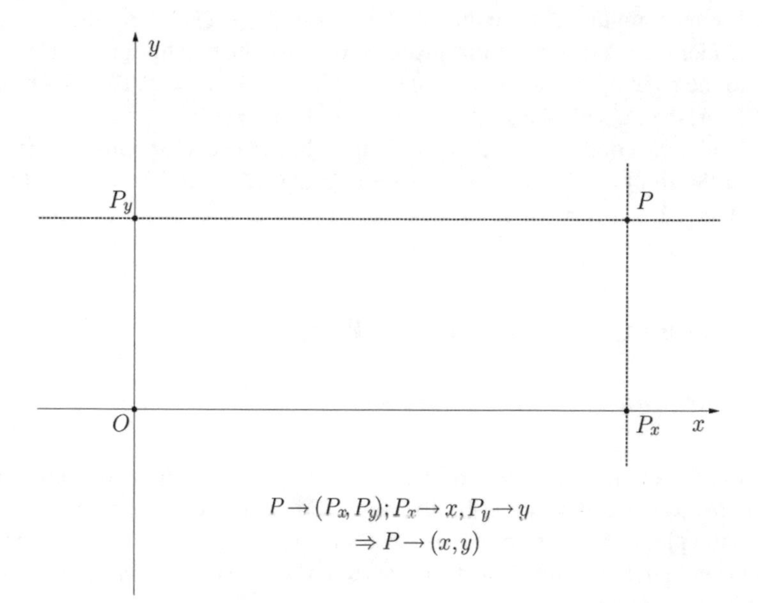

Fig. 1.15 Determining the Cartesian coordinates of a point on the plane

and the second *ordinate or y-coordinate*. Since this algorithm generates the unique projection points P_x and P_y of P, and these projections in turn generates the unique numbers x and y, the ordered pair (x, y) is uniquely defined by the choice of P. Notice that if P has plane coordinates (x, y), its projection points P_x and P_y have plane coordinates $(x, 0)$ and $(0, y)$, respectively.

Using the same procedure in the inverse direction (starting from an arbitrary ordered pair (x, y) and arriving at the unique corresponding point P) it is seen that each ordered pair of real numbers (x, y) defines the unique point on the coordinate plane (the details of this inverse relation are left to the reader). Thus, the proposed algorithm establishes the equivalence relation between all the point of the coordinate plane and all the ordered pairs of real numbers. Due to this equivalence we use the notation $P = (x, y)$ of points and their coordinates, meaning that this two quantities are not the same but equivalent. Since the ordered pairs of the real numbers can be obtained by using the Cartesian product $\mathbb{R} \times \mathbb{R}$, the coordinate plane is frequently denoted by $\mathbb{R}^2$.

6.2 Coordinate Lines

Definition of Coordinate Lines According to the general definition of a coordinate curve, the *Cartesian coordinate lines* are defined as the sets of points that satisfy the equations $x = C_1$ and $y = C_2$, where C_1 and C_2 are constants. In particular, if $y = 0$, we have the set of all the points whose projection on the axis Oy is the origin of the coordinates. Obviously, all these points belong to the line perpendicular to Oy which passes through the origin, that is, these points represent the x-axis. Analogously, the points of the coordinate line $x = 0$ form the y-axis. These are the two *main coordinate lines*. Other coordinate lines can also be simply visualized. The equation $y = C_2$ defines the set of all the points whose projection on the y-axis has the ordinate C_2, that is, all such points lie on the line perpendicular to the y-axis (parallel to the x-axis) which passes through the point $P_y = (0, C_2)$ (see Fig. 1.16). Analogously, $x = C_1$ represents the line perpendicular to the x-axis (parallel to the y-axis) which passes through the point $P_x = (C_1, 0)$. Hence, any Cartesian coordinate line is a line that either coincides with or is parallel to one of the coordinate axis.

For the sake of brevity, we frequently call the main coordinate lines simply coordinate lines when the meaning is clear from the context. If we need to specify that we are talking about the main coordinate lines we use this full name or also the name coordinate axes.

The coordinate axes divide the plane into four parts called *quadrants*: the region between the positive parts of the x-axis and y-axis is called the *first quadrant* (labeled I); the next in the counterclockwise direction part of the plane, which is located between the negative part of the x-axis and the positive part of the y-axis, is the *second quadrant* (II); the third part in the counterclockwise direction is the *third*

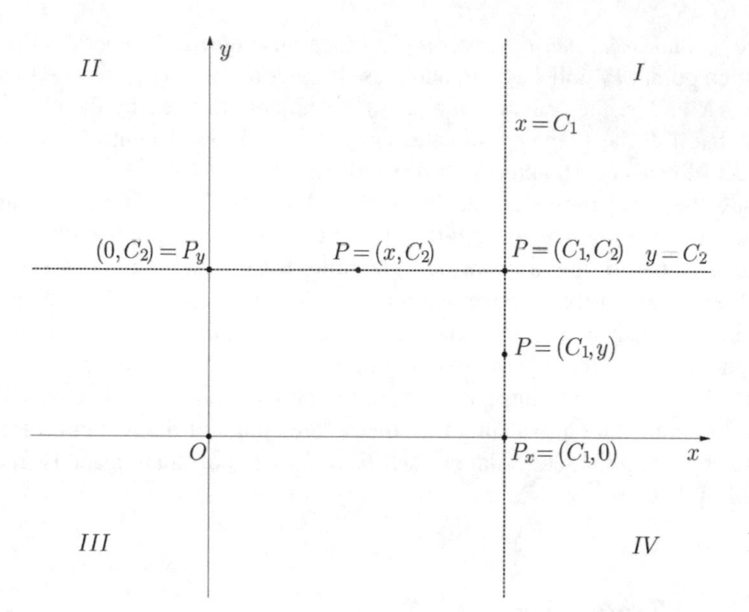

Fig. 1.16 Cartesian coordinate lines

quadrant (III) quadrant; and the last one, between the positive part of the x-axis and the negative part of the y-axis, is the *fourth quadrant* (IV) (see Fig. 1.16).

6.3 Projections on the Coordinate Lines

We have already defined the projection of a point on the coordinate axes. This definition is easily extended to *projections on any coordinate line*.

Definition. Projection of a Point Consider a point $P = (x, y)$ and a coordinate line $y = C$. The projection of P on $y = C$ is the point $P_{y=C}$ of intersection of $y = C$ with the line perpendicular to $y = C$ (and also perpendicular to the x-axis) which passes through P. Clearly, the coordinates of this point are $P_{y=C} = (x, C)$. In the same manner, it is defined the point $P_{x=C}$ of the projection of $P = (x, y)$ on $x = C$ whose coordinates are $P_{x=C} = (C, y)$. In the particular case $C = 0$, we return to the projections on the coordinate axes.

In the next section we will derive a simple formula of the distance between two points. However, the situation is even simpler when the points lie on the same coordinate line, since it is equivalent to the one-dimensional case, and we can present these formulas right now. Indeed, if the points P_1 and P_2 lie on the x-axis, their coordinates are $P_1 = (x_1, 0)$ and $P_2 = (x_2, 0)$, and the distance between them depends only on the x-coordinate like it was on the real line. Therefore, using the known formula for the coordinate line, we promptly have $d(P_2, P_1) = |x_2 - x_1|$. In the case when P_1 and P_2 lie on the coordinate line $y = C$, their coordinates are $P_1 = (x_1, C)$ and $P_2 = (x_2, C)$. Again the second coordinate is the same and we can apply the same formula of the real line: $d(P_2, P_1) = |x_2 - x_1|$. (A bit more strictly and formally: the distance between $P_1 = (x_1, C)$ and $P_2 = (x_2, C)$ is the same as the distance between their projections on the x-axis $P_{1x} = (x_1, 0)$ and $P_{2x} = (x_2, 0)$, that is, $d(P_2, P_1) = d(P_{2x}, P_{1x}) = |x_2 - x_1|$.) The same is true for the points $P_1 = (C, y_1)$ and $P_2 = (C, y_2)$ lying on the same coordinate line $x = C$: the distance between them is $d(P_2, P_1) = |y_2 - y_1|$.

Consider now the *projection of an interval* P_1P_2 on the x-axis. First of all, we need to define (in the geometric form) what is an *interval* (*segment*) on the plane.

Definition (Geometric) of an Interval An *interval* (*segment*) P_1P_2 is a part of the line passing through the points P_1 and P_2 located in between these two points. Certainly, P_2P_1 determines the same interval as P_1P_2. In particular, we will consider the intervals OP where one of the points is the origin O (the order of the points does not matter, but, for convenience we use the origin as the first point in the notation of these intervals).

Similarly to the intervals on the coordinate line, the intervals on the plane can be *closed* (including endpoints), *open* (excluding endpoints) or *half-open/half-closed*. From now on we will usually consider closed intervals (segments), unless it is explicitly stated otherwise.

As shows the next definition, the projection of an interval P_1P_2 on the x-axis is determined by the projection of the endpoints $P_1 = (x_1, y_1)$ and $P_2 = (x_2, y_2)$ on the coordinate axis x.

Definition. Projection of an Interval Let $P_{1x} = (x_1, 0)$ and $P_{2x} = (x_2, 0)$ be projections of the endpoints P_1 and P_2 of the interval P_1P_2 on the x-axis. The interval $P_{1x}P_{2x}$ is called the *projection of the interval* P_1P_2 *on the x-axis* (see Fig. 1.17).

In a similar way it is defined the *projection of the interval on the y-axis* (see Fig. 1.17).

Notice that the projection on the x-axis can result in an "abnormal" interval $P_{1x}P_{2x}$ for which $x_1 > x_2$. The same can happen with the projection $P_{1y}P_{2y}$ of P_1P_2 on the y-axis (see Fig. 1.18). Let us clarify why such situation may occur. Recall that there is a natural (normal) notation of intervals on a coordinate line: an interval is denoted AB if the point A lies to the left of or coincide with the point B (otherwise, the natural notation is BA). The possibility to follow this rule is based on the fact that $\mathbb{R}$ is the ordered field. The situation is different on the plane: if $x_1 < x_2$ but $y_1 > y_2$ (or vice-verse), then it does not exist a natural form to establish the rule whether P_1 precede P_2 or contrary, that is, whether P_1P_2 is a normal notation of the interval or P_2P_1. This is connected with impossibility to establish the order of points in the set $\mathbb{R}^2$ in the same way as it was made in $\mathbb{R}$.

Due to this ambiguity in the notation of intervals on the plane, and consequently, the possibility to find "abnormal" intervals of the projection, sometimes the

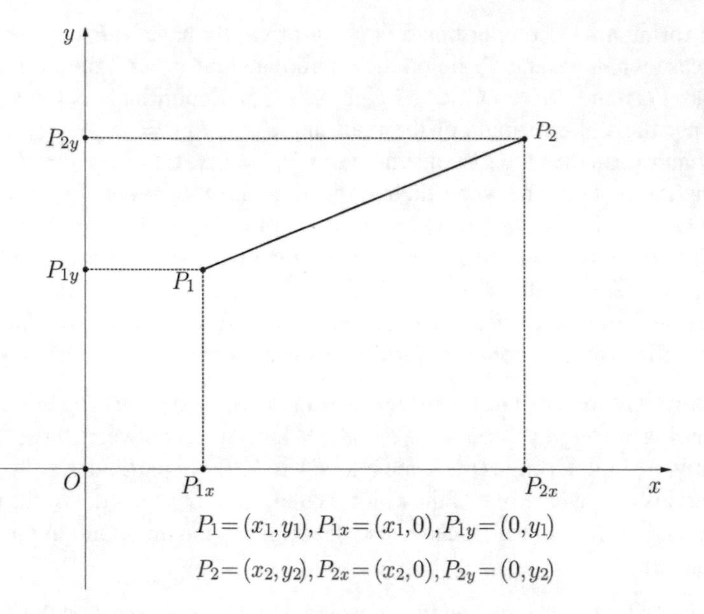

Fig. 1.17 Projection of interval

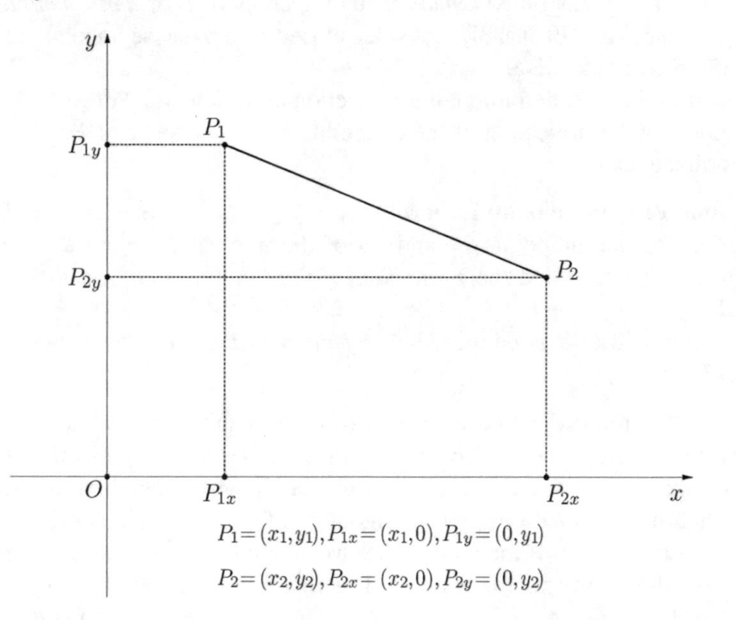

Fig. 1.18 Projection of interval, inverted order

definition of the projection of an interval is given in modified form: the projection of $P_1 P_2$ on the x-axis is the interval $P_{1x} P_{2x}$ if $x_1 \leq x_2$ and $P_{2x} P_{1x}$ if $x_1 > x_2$; and the same is made for the projection on the y-axis. This avoids "abnormal" intervals on the coordinate axes. In the next problems (for example, in division of an interval or finding a symmetric point) it will be more convenient to follow the first definition of the projection, which keeps the order of the points used for the original interval. Anyway, an ambiguity will be eliminated within the context of each problem.

The concept of the projection of an interval on the coordinate axis is naturally extended to the case of the coordinate lines. For a coordinate line $y = C$ the definition is as follows: the interval $P_{1C}P_{2C}$ is the *projection of the interval* P_1P_2 *on the line* $y = C$ if $P_{1C} = (x_1, C)$ and $P_{2C} = (x_2, C)$ are the projections of P_1 and P_2 on the line $y = C$. An analogous definition is used for the projection on $x = C$.

Examples

1. Mark the point whose Cartesian coordinates are $A = (-3, 4)$ and find its projections on the coordinate axes and on the lines $x = 1$ and $y = -2$.
   Solution
   According to the established equivalence relation, to find on the plane a point with given Cartesian coordinates x_0 and y_0 we need, first, to mark the two auxiliary points (projection points) $A_x = (x_0, 0)$ and $A_y = (0, y_0)$ on the x- and y-coordinate axis, respectively. In this specific case $x_0 = -3$, then we have to measure 3 units to the left of the origin on the x-axis and put there the point A_x. Next, we measure 4 units upward from the origin ($y_0 = 4$) along the y-axis and place there the point A_y. Now we trace the line parallel to y-axis and passing through A_x and also the line parallel to x-axis and passing through A_y. These two lines are perpendicular to each other, and consequently, they have the unique intersection point A which is the desired point of the given coordinates (see Fig. 1.19).
   In the course of the above procedure, we have already found the projections $A_x = (-3, 0)$ and $A_y = (0, 4)$ on the coordinate axes. To find the projection

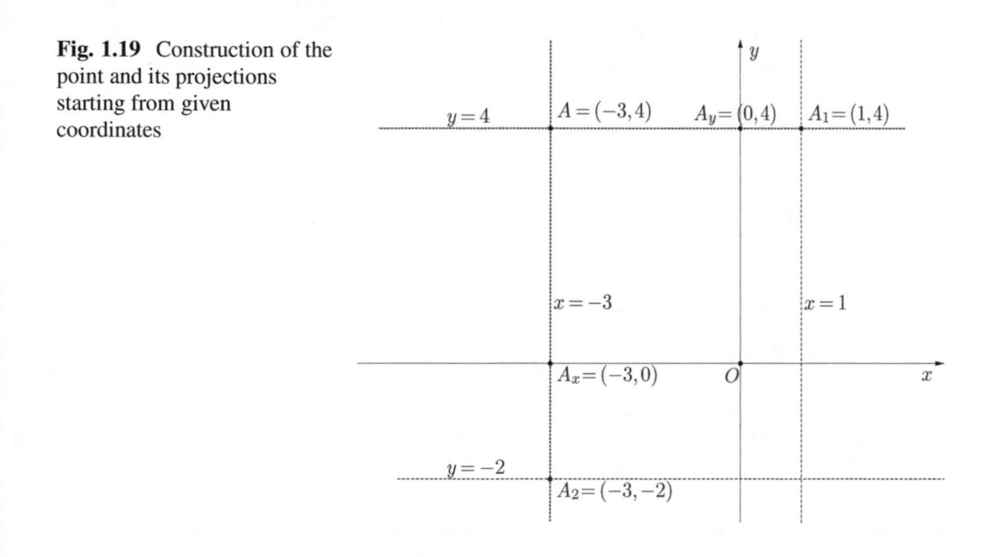

Fig. 1.19 Construction of the point and its projections starting from given coordinates

on the line $x = 1$, we draw the line parallel to the x-axis and passing through A (that is, the coordinate line $y = 4$). The point of the intersection of this line with the line $x = 1$ is the desired projection with the coordinates $A_1 = (1, 4)$. In the same way, drawing the line parallel to the y-axis, which passes through A (that is, the coordinate line $x = -3$), we find the desired projection $A_2 = (-3, -2)$ on $y = -2$ as the point of intersection of the coordinate lines $x = -3$ and $y = -2$. (See the illustration in Fig. 1.19.)

2. Give a geometric description of the points whose coordinates satisfy the inequality $x - y < 0$.

   Solution

   First, consider the corresponding equation $x - y = 0$. Since both coordinates have the same sign ($x = y$), the points are located in quadrants I and III . In quadrant I the point $P_0 = (x_0, x_0)$ has the same distance x_0 from each of the coordinate axes, and these distances are easy to find. Indeed, the distance from P_0 to the x-axis is the distance between P_0 and its projection $P_{0x} = (x_0, 0)$, and both points belong to the same coordinate line $x = x_0$. Analogously, the distance from P_0 to the y-axis is the distance between P_0 and its projection $P_{0y} = (0, x_0)$, and both points belong to the same coordinate line $y = x_0$. In the same way, in quadrant III the distances from P_0 to the coordinate axes are also coincide and are equal to $-x_0$. Hence, we have the set of points equidistant from both coordinate axes, which means that $y = x$ is the equation of the bisector of quadrants I and III. (See the illustration in Fig. 1.20.)

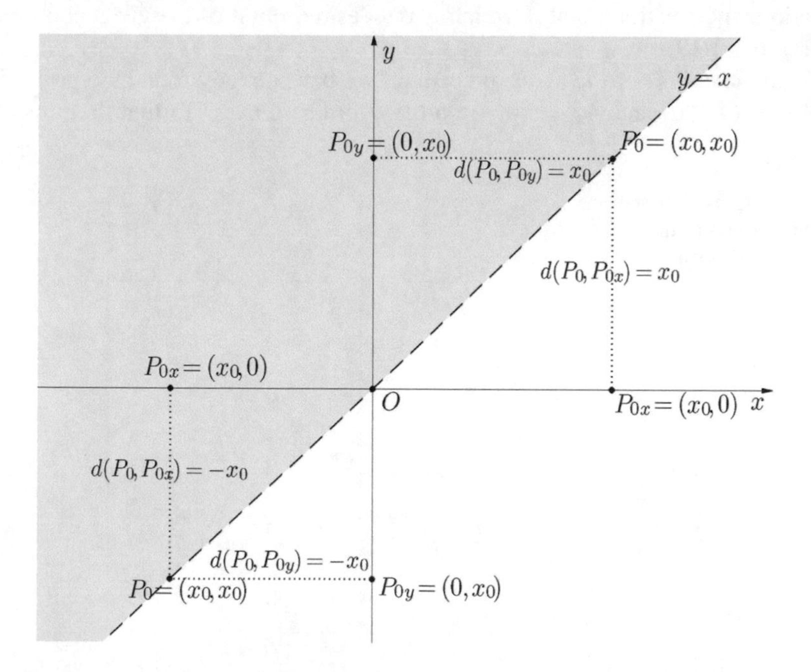

Fig. 1.20 The points satisfying the equation of the bisector $y = x$ and the inequality $y > x$

Returning to the original inequality, we see that the points we need to choose have the second coordinate greater than the first. Since the points with equal coordinates are located on the bisector $y = x$, it follows that the solution of the inequality are all the points located above this bisector. (See the illustration in Fig. 1.20.)

3. Find a geometric description of the points whose coordinates satisfy the inequality $4y^2 - 7y - 2 < 0$.

Solution

We start again from the corresponding equation $4y^2 - 7y - 2 = 0$, whose roots are $y_1 = -\frac{1}{4}$ and $y_2 = 2$. This means that the quadratic expression can be factored as the product of the two linear terms $4y^2 - 7y - 2 = (4y + 1)(y - 2) = 0$ and the two solutions $y = y_1$ and $y = y_2$ represent two coordinate lines parallel to the x-axis.

Returning to the original inequality, we write in the factored form $(4y + 1)(y - 2) < 0$. This inequality holds if the linear factors have opposite signs, which occurs in the two cases: (1) if $4y + 1 > 0$ and $y - 2 < 0$; and (2) if $4y + 1 < 0$ and $y - 2 > 0$. In the first case, we have the solution that can be written in the form $-\frac{1}{4} < y < 2$, but in the second case there is no solution, because one inequality contradicts another. Thus, the wanted points have the ordinates which satisfy the double inequality $-\frac{1}{4} < y < 2$, and consequently, these points are located above the coordinate line $y = -\frac{1}{4}$ and below the coordinate line $y = 2$. In other words, the solution is the infinite horizontal strip bounded by the coordinate lines $y = -\frac{1}{4}$ and $y = 2$ (not including the points of both lines).

7 Some Relations on the Cartesian Plane: Distance, Midpoint, Symmetry

7.1 Distance Between Two Points

Let us derive the well known from high school formula of the *distance between two points* $P_1 = (x_1, y_1)$ and $P_2 = (x_2, y_2)$. From the right triangle $P_1P_2P_3$, where $P_3 = (x_2, y_1)$, it follows that $d^2(P_1, P_2) = d^2(P_1, P_3) + d^2(P_3, P_2)$ (see Fig. 1.21). Since P_1 and P_3 lie on the same coordinate line $y = y_1$, we have $d(P_1, P_3) = |x_2 - x_1|$; also since P_3 and P_2 lie on the same coordinate line $x = x_2$, we have $d(P_3, P_2) = |y_2 - y_1|$. Therefore, $d^2(P_1, P_2) = (x_2 - x_1)^2 + (y_2 - y_1)^2$ or

$$d(P_1, P_2) = \sqrt{(x_2 - x_1)^2 + (y_2 - y_1)^2}. \tag{1.4}$$

The same distance represents the *length of the interval* P_1P_2.

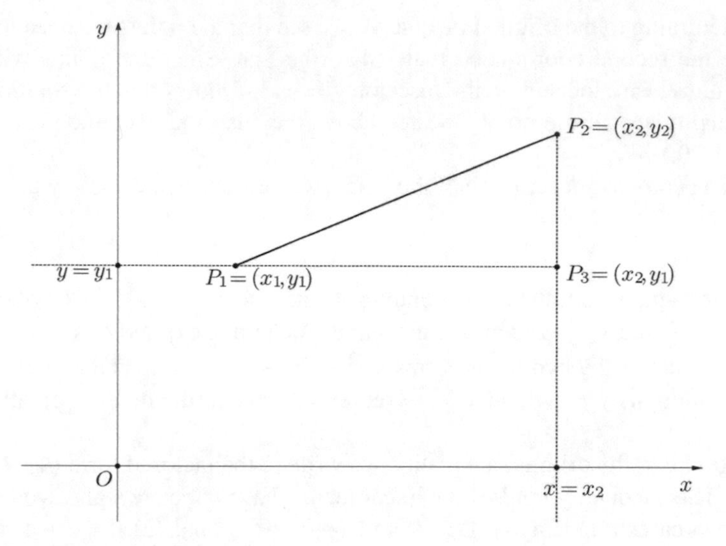

Fig. 1.21 Distance between two points

7.2 *Distance from a Point to a Coordinate Line*

In the case of a general line, the *distance from a point to this line* can be defined as follows.

Definition The *distance* $d(P, R)$ *between a point* P *and a line* R is the distance between P and the point Q of the intersection of R with the line perpendicular to R that passes through P. The point Q is called projection of P on R (see Fig. 1.22). In particular, if P belongs to R, the distance is equal 0.

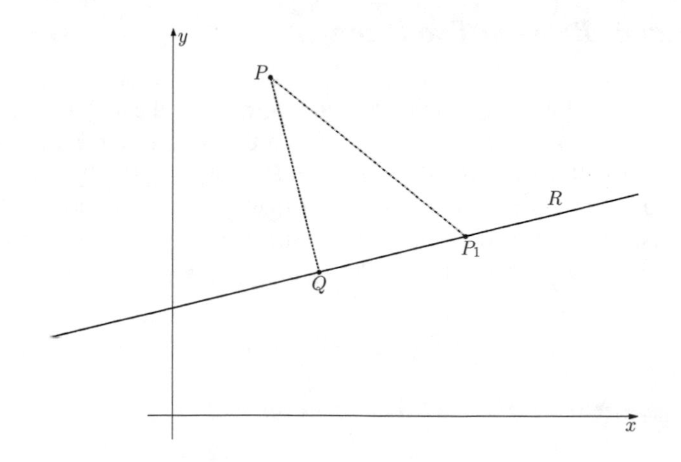

Fig. 1.22 Distance from a point to a line

Remark It can be seen that the distance from a point P to a line R is the smallest distance between P and all the points of R (see Fig. 1.22).

Now we restrict our attention to the coordinate lines. In Sect. 6.3 we have already defined the projections of points on the coordinate lines: the projection $P_{y=C}$ of a point P on the coordinate line $y = C$ is the point of intersection of the line $y = C$ with the line perpendicular to $y = C$ that passes through P. This means that $P_{y=C}$ is the required point Q in the Definition of the distance between P and R. Since the coordinates $P_{y=C}$ are (x, C), applying the formula of the distance between two points, we obtain

$$d(P, y = C) = d(P, P_{y=c}) = \sqrt{(x - x)^2 + (y - C)^2} = |y - C| . \tag{1.5}$$

In particular, the distance from P to the x-axis is equal to $|y|$.

In the same way, the distance from P to a coordinate line $x = C$ is the distance between P and its projection on $x = C$ which is the point $P_{x=C} = (C, y)$. Therefore,

$$d(P, x = C) = d(P, P_{x=c}) = \sqrt{(x - C)^2 + (y - y)^2} = |x - C| . \tag{1.6}$$

In particular, the distance from P to the y-axis is equal to $|x|$.

7.3 Midpoint of an Interval

Definition Given two points $P_1 = (x_1, y_1)$ and $P_2 = (x_2, y_2)$, the *midpoint* P_m of the interval P_1P_2 is the point equidistant from P_1 and P_2: $d(P_1, P_m) = d(P_2, P_m)$.

To find the coordinates (x_m, y_m) of P_m we reduce this problem to the one-dimensional case. To do this we trace the coordinate lines $x = x_1, x = x_2$ and $x = x_m$ through the points P_1, P_2 and P_m (see Fig. 1.23). The projections of these points $P_{1x} = (x_1, 0)$, $P_{2x} = (x_2, 0)$ and $P_{mx} = (x_m, 0)$ lie on the same coordinate line $y = 0$, and consequently, are completely characterized by their first coordinates. Since the lines $x = x_1, x = x_2$ and $x = x_m$ are parallel to each other, the intervals $P_{1x}P_{mx}$ and $P_{mx}P_{2x}$ keep the relation of the original intervals: $d(P_{1x}, P_{mx}) = d(P_{2x}, P_{mx})$, that is, P_{mx} is the midpoint of the interval $P_{1x}P_{2x}$ on the x-axis (see Fig. 1.23). Therefore, the x-coordinate of the midpoint is found by the formula $x_m = \frac{x_1+x_2}{2}$, according to the result of Sect. 5.2 (see formula (1.2)). In the same way, making projections on the y-axis we find the points $P_{1y} = (0, y_1)$, $P_{2y} = (0, y_2)$ and $P_{my} = (0, y_m)$, whose coordinates satisfy the relation $y_m = \frac{y_1+y_2}{2}$. Thus, both coordinates of the midpoint are determined:

$$P_m = (x_m, y_m) : \ x_m = \frac{x_1 + x_2}{2}, \ y_m = \frac{y_1 + y_2}{2} . \tag{1.7}$$

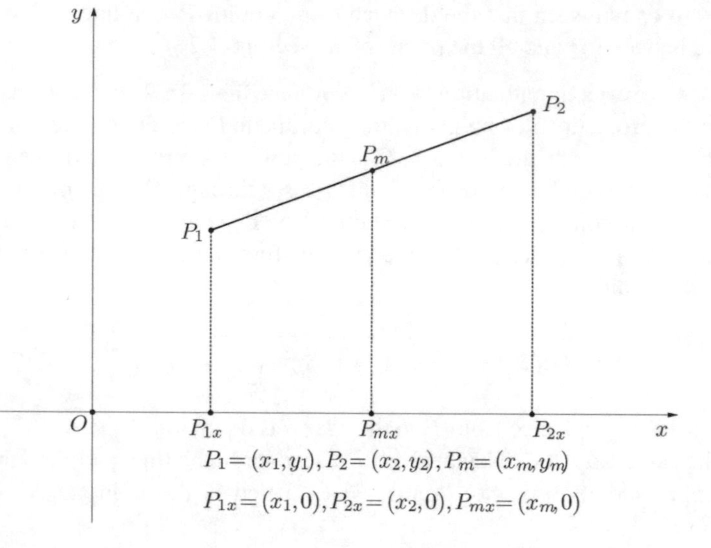

Fig. 1.23 Coordinates of midpoint

7.4 Symmetry with Respect to a Point

Consider now the problem of finding a point P_2 symmetric to a given point P_1 with respect to the centerpoint of the symmetry P_0.

Definition Given a point $P_1 = (x_1, y_1)$, the point $P_2 = (x_2, y_2) \neq P_1$ is *symmetric to* P_1 *with respect to the centerpoint of symmetry* $P_0 = (x_0, y_0)$ if all the three points lie on the same line and P_1 and P_2 are equidistant from P_0, that is, $d(P_1, P_0) = d(P_2, P_0)$.

It follows from the definition that P_0 is the midpoint of the interval P_1P_2 (that is, P_0 coincides with P_m shown in Fig. 1.23). Therefore, $x_0 = \frac{x_1+x_2}{2}$, $y_0 = \frac{y_1+y_2}{2}$, whence

$$P_2 = (x_2, y_2) : x_2 = 2x_0 - x_1\,, \;\; y_2 = 2y_0 - y_1\,. \tag{1.8}$$

In particular, if the centerpoint of symmetry is the origin, that is, $P_0 = O = (0, 0)$, then P_2 symmetric to $P_1 = (x_1, y_1)$ has the coordinates $P_2 = (-x_1, -y_1)$.

Another manner to represent relations (1.8) is by choosing the more symmetric form of involvement of the coordinates of the symmetric points P_1 and P_2:

$$P_1 = (x_0 - x_1, y_0 - y_1),\; P_2 = (x_0 + x_1, y_0 + y_1). \tag{1.9}$$

Remark 1 The concept of symmetry of a pair of points about a centerpoint P_0 is naturally extended to a set of points or a plane figure. A *figure is symmetric with respect to* P_0 if each point P of this figure has the symmetric point P_s about P_0

which belongs to the same figure. In other words, all the point of a figure are separated in pairs symmetric about P_0.

Remark 2 As we will see in the study of functions, the graphs of *odd functions* are symmetric about the origin. Specifying this means that any point $P_1 = (x_1, y_1)$ of the graph of an odd function has a symmetric point (with respect to the origin) $P_2 = (-x_1, -y_1)$ that also belongs to the same graph. This situation is naturally extended to the *symmetry of a graph about a general point* $P_0 = (x_0, y_0)$. However, in Calculus and Analysis the symmetry of graphs about a point is frequently restricted to the case when the center of symmetry is the origin.

7.5 Symmetry with Respect to a Line

Another important type of symmetry is a symmetry with respect to a line.

Definition Given a point P_1 and a line R, the point $P_2 \neq P_1$ is *symmetric to* P_1 *with respect to the line* R *(called the axis of symmetry)*, if P_1 and P_2 lie on the same line S perpendicular to R and are equidistant from the point P_0 of the intersection of R and S.

From this definition it follows that the symmetry of two points about a line is reduced to the symmetry about the specific point—the projection of the original point on the line.

This definition is valid for any axis of symmetry, but the formula for the coordinates of P_2 we will derive only in the case of a coordinate line.

Consider a coordinate line $R : x = C$. In this case, the line perpendicular to R that passes through $P_1 = (x_1, y_1)$ is the coordinate line $S : y = y_1$. The intersection of R and S is the point $P_0 = (C, y_1)$ of the projection of P_1 on R (see Fig. 1.24). The condition of equal distances from P_1 to P_0 and from P_0 to the required point $P_2 = (x_2, y_2)$—$d(P_1, P_0) = d(P_0, P_2)$—implies that the point P_0 is the midpoint of the interval P_1P_2 of the line S, and consequently, its coordinates are $P_0 = (C, y_1) = (\frac{x_1+x_2}{2}, \frac{y_1+y_2}{2})$, from which it follows that

$$P_2 = (x_2, y_2) : \ x_2 = 2C - x_1 \, , \ y_2 = y_1 \, . \tag{1.10}$$

In particular, if the symmetry is performed about the y-axis (that is, $C = 0$), then the point P_2 symmetric to $P_1 = (x_1, y_1)$ has the coordinates $P_2 = (-x_1, y_1)$. A simpler form to remember relations (1.10) is to use more symmetric representation for the coordinates of P_1 and P_2:

$$P_1 = (C - x_1, y_1), \, P_2 = (C + x_1, y_1) \, . \tag{1.11}$$

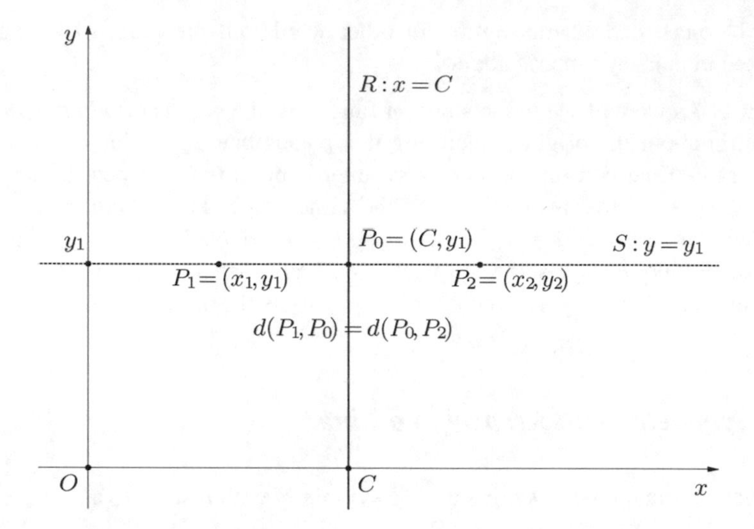

Fig. 1.24 Symmetric point with respect to a coordinate line $x = C$

Analogously, the point P_3 symmetric to $P_1 = (x_1, y_1)$ with respect to the coordinate line $y = C$ has the coordinates

$$P_3 = (x_3, y_3) : \ x_3 = x_1 \, , \ y_3 = 2C - y_1 \, . \tag{1.12}$$

In particular, if the symmetry is performed about the x-axis, then $P_3 = (x_1, -y_1)$.

The two remarks similar to those presented for the symmetry about a point are provided below.

Remark 1 The concept of the symmetry of a pair of points about a line R is naturally extended to a set of points or a figure on the plane. A *figure is symmetric with respect to R* if each point P of this figure has the symmetric point P_s about R which belongs to the same figure. In other words, all the points of a figure are separated in pairs symmetric about R.

Remark 2 In future study of the geometric properties of functions we will see that the graphs of *even functions* are symmetric with respect to the y-axis. This means that any point $P_1 = (x_1, y_1)$ of the graph of an even function has the symmetric point (with respect to the y-axis) $P_3 = (-x_1, y_1)$ that also belongs to the same graph. This situation is naturally extended to the *symmetry of a graph about a general coordinate line $x = x_0$*. However, in Calculus and Analysis the symmetry of graphs about coordinate lines different from the y-axis are rarely explored.

Examples

1. Calculate the distance between the points $A = (-4, 3)$ and $B = (2, -1)$. Find the midpoint of the interval AB.

Solution

Substituting the coordinates of the points in formula (1.4), we find the distance $d(A, B) = \sqrt{(x_A - x_B)^2 + (y_A - y_B)^2} = \sqrt{(-4-2)^2+(3-(-1))^2} = \sqrt{52}$. The coordinates of the midpoint are found by employing formulas (1.7): $x_m = \frac{x_A+x_B}{2} = \frac{-4+2}{2} = -1$, $y_m = \frac{3+(-1)}{2} = 1$.

2. The points $A = (5, 5)$ and $B = (x, y)$ belong to the bisector of the quadrants I and III, and the distance between these points is 8. Find the coordinates of B.

   Solution

   Since B belongs to the bisector of the quadrants I and III, its coordinates are equal: $B = (x, x)$. From the condition of the distance we find $d(A, B) = \sqrt{(5-x)^2 + (5-x)^2} = \sqrt{2}|5 - x| = 8$. The last equation has the two roots: $x-5 = \pm 4\sqrt{2}$, that is, $x_1 = 5-4\sqrt{2}$ and $x_2 = 5+4\sqrt{2}$. Hence, we obtain the two points satisfying the conditions of the problem: the first one $B_1 = (5-4\sqrt{2}, 5-4\sqrt{2})$ stays in the quadrant III, and the second $B_2 = (5+4\sqrt{2}, 5+4\sqrt{2})$ in the quadrant I.

3. Find the points symmetric to $A = (-3, 2)$ with respect to the origin of the coordinates, with respect to the coordinate axes and with respect to the coordinate lines $x = 1$ and $y = -\frac{3}{2}$.

   Solution

   From the formulas of symmetry (1.8), (1.10) and (1.12) it follows immediately that:

   (1) the point symmetric about the origin is $B = (3, -2)$;
   (2) the point symmetric about the x-axis is $C = (-3, -2)$;
   (3) the point symmetric about the y-axis is $D = (3, 2)$;
   (4) the point symmetric about the line $x = 1$ is $E = (5, 2)$;
   (5) the point symmetric about the line $y = -\frac{3}{2}$ is $F = (-3, -5)$.

   (The illustration is shown in Fig. 1.25.)

4. Given the midpoints $D = (7, 8)$, $E = (-4, 5)$ and $F = (1, -4)$ of the three sides of a triangle, find its vertices.

Solution

Denote the required vertices by $A = (x_A, y_A)$, $B = (x_B, y_B)$ and $C = (x_C, y_C)$ in such a way that D is the midpoint of AB, E is the midpoint of AC, and F is the midpoint of BC. According to the formulas of midpoints, we have the following relations: $\frac{x_A+x_B}{2} = 7$, $\frac{y_A+y_B}{2} = 8$, $\frac{x_A+x_C}{2} = -4$, $\frac{y_A+y_C}{2} = 5$ and $\frac{x_B+x_C}{2} = 1$, $\frac{y_B+y_C}{2} = -4$. The odd relations (with x-coordinates) form the decoupled subsystem:

$$x_A + x_B = 14\,, \; x_A + x_C = -8\,, \; x_B + x_C = 2\,.$$

The solution is elementary. For instance, summing the first two equations we have $2x_A + x_B + x_C = 6$ and using the third equation we conclude that $2x_A + 2 = 6$, that is, $x_A = 2$. Then the first two equations give $x_B = 12$ and $x_C = -10$. Another

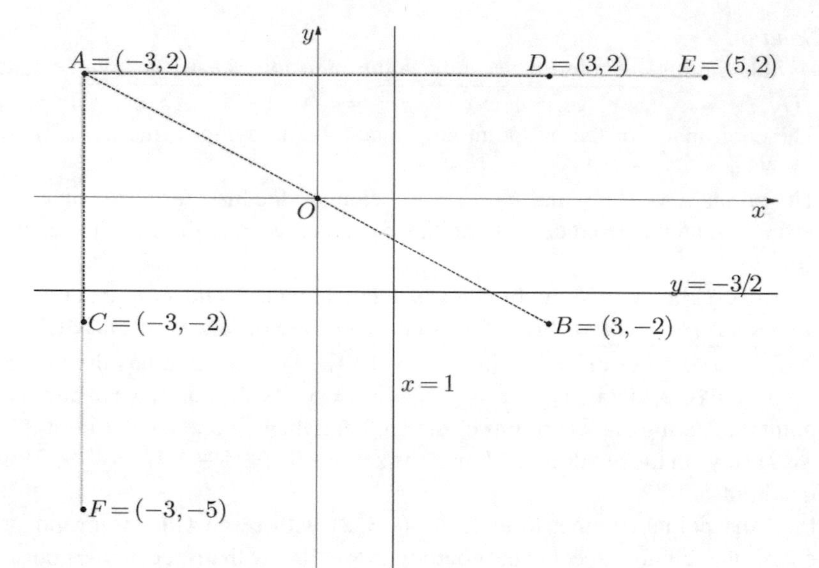

Fig. 1.25 The points symmetric to $A = (-3, 2)$

decoupled system is formed by the even equations (with y-coordinates):

$$y_A + y_B = 16\,, \;\; y_A + y_C = 10\,, \;\; y_B + y_C = -8\,.$$

Solving in a similar manner, we obtain $y_A = 17$, $y_B = -1$ e $y_C = -7$. Hence, the vertices are $A = (2, 17)$, $B = (12, -1)$ and $C = (-10, -7)$.

Problems

General Sets

1. Prove the following properties:

   (a) $A \cap B = B \cap A$;
   (b) $(A \cup B) \cup C = A \cup (B \cup C)$;
   (c) $A \cap (B \cup C) = (A \cap B) \cup (A \cap C)$.

2. Show that $(A \backslash B) \backslash C = A \backslash (B \backslash C)$ if, and only if, $A \cap C = \varnothing$.
3. Find the equal sets among the given ones: $A = \{x : x^2 - 4x + 3 = 0\}$, $B = \{x : x^2 + x - 2 = 0\}$, $C = \{x : \frac{x^2-4x+3}{x-3} = 0\}$, $D = \{1, 3\}$, $E = \{1, -2\}$, $F = \{-2, 3, 1\}$, $G = A \cup B$, $H = A \cap B$.

4. Find $A \cup B$, $A \cap B$, $A \cup C$, $A \cap C$, $B \times C$, $C \times B$, $A \backslash (B \cup C)$ e $(A \backslash B) \cup C$ if $A = \{1, 3, 4, 5\}$, $B = \{3, 7\}$ and $C = \{1, 4, 7\}$.
5. Give an example when $A \cap B = A \cap C$, but $B \neq C$.

Rational Numbers

1. Write the following rational numbers in the form of integers:

   (a) $-\frac{14}{2}$;
   (b) $\frac{-21}{-7}$;
   (c) $\frac{12-60}{6 \cdot 3-22}$.

   What of these numbers are positive and what are negative?
2. Write a rational number whose numerator is the smallest natural number of 5 digits and the denominator is the largest natural number of 3 digits. What is the entire part of this number?
3. Simplify the representation of two rational numbers and compare them:

   (a) $\frac{14+6}{2 \cdot 6-30}$ and $\frac{17-32}{25-5 \cdot 19}$;
   (b) $\frac{1}{3^2}$ and $\frac{1}{2^3}$.

4. Describe a general algorithm of comparison of two rational numbers. (Start with positive numbers with the same denominator, then move to general positive fractions and zero, and finalize with two arbitrary rational numbers.)
5. Find natural n such that $\frac{1}{n}$ is smaller than a given rational number:

   (a) $\frac{1}{99}$;
   (b) $\frac{13}{2778}$.

6. Describe in a general form how to find a natural number n such that $\frac{1}{n}$ is smaller that a positive rational $\frac{p}{q}$. (Start noting that the natural number p is greater than or equal to 1.)
7. The sum $1^2+2^2+\cdots+n^2$, $n \in \mathbb{N}$, which contains the squares of natural numbers and should result in a natural number, is given by fraction $\frac{n(n+1)(2n+1)}{6}$. Explain if there is a contradiction in this result.
8. Verify if the following statement is true or false:

   (a) any integer is a rational number;
   (b) any positive rational number is natural;
   (c) any rational number has an infinite number of representations through fractions $\frac{p}{q}$, where $p \in \mathbb{Z}$, $q \in \mathbb{N}$;
   (d) the set of rational numbers consists of the fractions $\frac{p}{q}$, where $p \in \mathbb{Z}$, $q \in \mathbb{Z} \backslash \{0\}$;
   (e) any rational number $\frac{p}{q}$ with a positive p is greater than 0;

(f) if the numerator p of a rational number $\frac{p}{q}$ is greater that the numerator m of a rational number $\frac{m}{n}$, then the first fraction is greater than the second;
(g) the difference between two negative rational numbers is a negative rational;
(h) if the product of two rational numbers is given, then knowing one of the factors we can always find the second factor.

Real Numbers

1. Transform the given fractions into decimals:
   (a) $\frac{3}{10}$;
   (b) $\frac{10}{3}$;
   (c) $-\frac{15}{8}$;
   (d) $-\frac{19}{12}$;
   (e) $-\frac{20}{7}$.
2. Transform the decimals into common fractions:
   (a) 0.3;
   (b) $0.(3)$;
   (c) $-2.7(25)$;
   (d) $-5.617(23)$;
   (e) $6.01(522)$.
3. Check if a number is rational or irrational:
   (a) $0.111111\ldots$;
   (b) $0.101010\ldots$;
   (c) $0.110111011101\ldots$;
   (d) $0.101101011010110\ldots$;
   (e) $0.01011011101111\ldots$.
4. Verify if a number is rational or irrational:
   (a) $\sqrt{3}$;
   (b) $\sqrt[3]{3}$;
   (c) $\sqrt[3]{49}$;
   (d) $\sqrt[4]{7}$;
   (e) $\sqrt{2}+\sqrt{3}$;
   (f) $\sqrt{3-2\sqrt{2}}+\sqrt{3+2\sqrt{2}}$;
   (g) $\sqrt[3]{\sqrt{5}-2}+\sqrt[3]{\sqrt{5}+2}$;
   (h) $\sqrt[3]{\sqrt{5}-2}-\sqrt[3]{\sqrt{5}+2}$.

5. Let p and q be rational numbers, while a and b be irrational numbers. Verify if the following numbers are rational or irrational:

   (a) $p+q$;
   (b) $a+b$;
   (c) $p+a$;
   (d) $p \cdot q$;
   (e) $a \cdot b$;
   (f) $p \cdot a$;
   (g) p^2;
   (h) a^2;
   (i) $\sqrt{p}$;
   (j) $\sqrt{a}$.

6. Compare the real numbers:

   (a) $\sqrt{2}$ and 1.4;
   (b) $-\frac{2}{3}$ and $-0.(6)$;
   (c) $\sqrt{3}-\sqrt{2}$ and 1;
   (d) $\sqrt{3}+\sqrt{2}$ and $\sqrt{5}$;
   (e) $\sqrt{a}+\sqrt{b}$ and $\sqrt{a+b}$;
   (f) $\sqrt{2ab}$ and $\sqrt{a^2+b^2}$;
   (g) $\sqrt[3]{3}$ and $\sqrt{2}$;
   (h) π^2 and 2^π.

7. Find all real numbers x such that

   (a) $10-x^2<6$;
   (b) $(x-1)^2 \leq 12$;
   (c) $(x-1)(x-5)<0$;
   (d) $x^2-6x+5 \geq 0$;
   (e) $(x-\sqrt{3})(x-2)(x+3)>0$;
   (f) $10^x<\frac{1}{100}$;
   (g) $\frac{2}{x}-\frac{1}{1+x} \leq 0$;
   (h) $x+2^x<3$.

8. Verify if the statement is true or false:

   (a) any rational number is a real number;
   (b) any irrational number is a real number;
   (c) there are numbers which are rationals and irrationals at the same time;
   (d) the set of irrational numbers is the set of all the roots together with the numbers π and e;
   (e) the set of irrational numbers is the set of all the numbers which are not rationals;
   (f) the product of two irrational numbers is an irrational number;

(g) always there exists a rational number in between any two irrational numbers;
(h) if the product of two irrational numbers is given, then knowing one of the factors we can always find another one.

9. Solve the following absolute value equations:

   (a) $|x| = 2$;
   (b) $|2x - 1| = 3$;
   (c) $|2 + x| = 2$;
   (d) $\left|\frac{2-x}{x-1}\right| = 0$;
   (e) $\left|\frac{2x-1}{x-2}\right| = 1$.

10. Solve the following absolute value inequalities:

   (a) $|x| < 2$;
   (b) $|x| \geq 2$;
   (c) $|2x - 1| \leq 3$;
   (d) $|2x - 1| > 3$;
   (e) $\left|\frac{2-x}{x-1}\right| < 2$;
   (f) $\left|\frac{2x-1}{x-2}\right| > 1$.

Coordinate Line

1. Mark the points with the following coordinates:

   (a) $x = 2$;
   (b) $x = -3$;
   (c) $x = \frac{5}{3}$;
   (d) $x = -\frac{3}{5}$;
   (e) $x = \sqrt{2}$;
   (f) $x = -\sqrt{5}$.

2. Mark (approximately) the two points with given coordinates and explain which one lies more to the right:

   (a) $\sqrt{2}$ and 1.4;
   (b) $-\frac{2}{3}$ and $-0.(6)$;
   (c) $\sqrt{2} - \sqrt{3}$ and -1;
   (d) $\sqrt{3} + \sqrt{2}$ and $\sqrt{5}$;
   (e) $-\pi$ and -3.14;
   (f) $(-\pi)^2$ and $(-3.14)^2$.

3. Mark the points whose coordinates are defined as follows:

   (a) three natural numbers smaller than π;
   (b) three integers located in between $-\sqrt{3}$ and $\sqrt{3}$;
   (c) three rational numbers located in between $\sqrt{2}$ and $\sqrt{3}$;
   (d) three irrational numbers multiples of $-\sqrt{5}$;
   (e) three real numbers satisfying the relation $x^3 = 2x$.

Some Sets of Numbers

1. Mark the points whose coordinates satisfy the following equations:

   (a) $|x| = 2$;
   (b) $|2x - 1| = 3$;
   (c) $|2 + x| = 2$;
   (d) $x^2 + x - 2 = 0$;
   (e) $x^2 + x + 2 = 0$;
   (f) $x^2 - 2x + 1 = 0$;
   (g) $x^2 - 8x + 15 = 0$;
   (h) $x^3 + 2x^2 - 3x = 0$;
   (i) $\frac{2-x}{x-1} = 0$;
   (j) $\frac{2x-1}{x-2} = 1$;
   (k) $\left|\frac{2-x}{x-1}\right| = 0$;
   (l) $\left|\frac{2x-1}{x-2}\right| = 1$.

2. Characterize geometrically the points whose coordinates satisfy the following relations:

   (a) $x < 2$;
   (b) $|x| < 2$;
   (c) $|x| \geq 2$;
   (d) $|2x - 1| \leq 3$;
   (e) $|2x - 1| > 3$;
   (f) $x^2 + x - 2 \leq 0$;
   (g) $x^3 + 2x^2 - 3x \geq 0$;
   (h) $x^3 + 2x^2 - 3x < 0$;
   (i) $\frac{2-x}{x-1} < 0$;
   (j) $\frac{2x-1}{x-2} > 1$;
   (k) $\left|\frac{2-x}{x-1}\right| < 2$;
   (l) $\left|\frac{2x-1}{x-2}\right| > 1$.

3. Find the distance between the points A and B whose coordinates are:

   (a) $a = 3$ and $b = 11$;
   (b) $a = 5$ and $b = 2$;
   (c) $a = -1$ and $b = 3$;
   (d) $a = -1$ and $b = -3$;
   (e) $a = 4$ and $b = -6$.

4. Find the midpoint in between the points A and B of Exercise 3.
5. Find the point A if:

   (a) $B = -1, d(A, B) = 5$ and A stays to the right of B;
   (b) $B = -1, d(A, B) = 5$ and A stays to the left of B;
   (c) $B = -2$, and the midpoint of AB is $C = 3$;
   (d) $B = -2$, and the midpoint of AB is $C = -3$.

6. Find the point B symmetric to A about C:

   (a) $A = 2$ and $C = 0$;
   (b) $A = 2$ and $C = 4$;
   (c) $A = -6$ and $C = 3$;
   (d) $A = -3$ and $d(A, C) = 4$;
   (e) $d(A, B) = 5$ and $C = -1$;
   (f) $A = 4$ and $d(C, B) = 3$.

7. Considering that A and B are symmetric about C, answer the following questions:

   (a) may all the three points lie on the negative part of the real line?
   (b) what is the condition which guarantees that A and B stay on the opposite sides of the real line?
   (c) if the coordinate of A is irrational, should the coordinate of B also be irrational?
   (d) if the coordinates of A and C are rationals, should the coordinate of B also be rational?
   (e) if the coordinates of A and B are rationals, should the coordinate of C also be rational?
   (f) if the coordinates of A and C are even integers, should the coordinate of B be the same?
   (g) is there a fourth point D symmetric to both A and B about C?

Cartesian Coordinates on the Plane

1. Mark the points with the following Cartesian coordinates:

   (a) $A = (2, 3)$;
   (b) $B = (-\frac{5}{3}, \frac{2}{3})$;

(c) $C = (-\frac{3}{5}, -2)$;
(d) $D = (1, -\sqrt{2})$.

Find the projections of these points on the coordinate axes and also on the lines $x = -1$ and $y = 2$.

2. Find a geometric location of the points whose coordinates satisfy the following relations:

(a) $x = 3$;
(b) $y = -2$;
(c) $xy = 0$;
(d) $\frac{y}{x} = 1$;
(e) $x + y = 0$;
(f) $xy < 0$;
(g) $x - y \geq 0$;
(h) $x + y < 0$.

3. Find a geometric location of the points whose coordinates satisfy the following equations:

(a) $|x| = 2$;
(b) $|2y - 1| = 3$;
(c) $x^2 + x - 2 = 0$;
(d) $y^2 - 8y + 15 = 0$;
(e) $\frac{2x+1}{y-2} = 0$;
(f) $(2x + 1)(y - 2) = 0$;
(g) $\left|\frac{2x+1}{y-2}\right| = 0$;
(h) $|(2x + 1)(y - 2)| = 0$.

4. Find a geometric location of the points whose coordinates satisfy the following inequalities:

(a) $x > 2$;
(b) $2y - 1 \leq 3$;
(c) $|x| > 2$;
(d) $|2y - 1| \leq 3$;
(e) $y^2 + y - 2 < 0$;
(f) $\frac{y}{x} \geq 1$;
(g) $(2x + 1)(y - 2) > 0$;
(h) $\frac{2x+1}{y-2} < 0$.

Some Relations on the Cartesian Plane

1. Calculate the distance between two points:

   (a) $A = (2, 3)$ and $B = O = (0, 0)$;
   (b) $A = O = (0, 0)$ and $B = (1, -\sqrt{8})$;
   (c) $A = (2, 3)$ and $B = (-2, 6)$;
   (d) $A = (-3, -7)$ and $B = (5, -1)$.

   Find the midpoint of the interval AB.
2. The points $A = (-4, 2)$ and $B = (x, y)$ lie on a line parallel to the x-axis and the distance between them is 2. Find the coordinates of B.
3. The points $A = (-5, 2)$ and $B = (x, y)$ lie on a line parallel to the y-axis and the distance between them is 6. Find the coordinates of B.
4. The points $A = (5, 5)$ and $B = (x, y)$ belong to the bisector of the quadrants I and III and the distance between them is 4. Find the coordinates of B.
5. The points $A = (-3, 3)$ and $B = (x, y)$ belong to the bisector of the quadrants II and IV and the distance between them is 6. Find the coordinates of B.
6. Determine the type of triangle (acute, right or obtuse) if its vertices are:

   (a) $A = (2, 3)$, $B = (6, 7)$, $C = (-7, 2)$;
   (b) $A = (2, -5)$, $B = (-7, -4)$, $C = (-1, 6)$;
   (c) $A = (0, 0)$, $B = (4, 2)$, $C = (-2, 4)$.

7. Find the coordinates of the midpoint of the interval AB:

   (a) $A = (2, 3)$, $B = (6, 7)$;
   (b) $A = (-2, 4)$, $B = (-4, 10)$;
   (c) $A = (-7, 5)$, $B = (11, -9)$.

8. Given the endpoint A and the midpoint C of the interval AB, find the second endpoint:

   (a) $A = (-5, -7)$, $C = (-9, -12)$;
   (b) $A = (-4, 2)$, $C = (-6, 5)$.

9. Find the points symmetric to a given point with respect to the origin and the coordinate axes:

   (a) $A = (2, 3)$;
   (b) $B = (-1, -3)$;
   (c) $C = (-2, 1)$;
   (d) $D = (1, -4)$.

10. Given points $A = (2, -3)$ and $B = (-1, 4)$, find:

    (a) the point C symmetric to A about B;
    (b) the point D symmetric to B about A.

11. Given the midpoints $D = (2, -1)$, $E = (-1, 4)$ and $F = (-2, 2)$ of the three sides of a triangle, find its vertices.
12. Given the three vertices $A = (2, 3)$, $B = (4, -1)$ and $C = (0, 5)$ of a parallelogram $ABCD$, find the fourth vertex.

Chapter 2
Functions and Their Analytic Properties

In this chapter we introduce the most important concept in the branch of analysis, and probably in all mathematics, which is a function. We will present different modes of the definition of a function, and, what is the most important for analysis, the different forms of the definition of a function through a formula. We will consider in a general form the relevant properties of functions which can be treated using rudimentary techniques of a middle and high school (those techniques which are pertinent to Pre-Calculus, available before a study of limits). These properties will be defined and investigated with a mathematical rigor both in general form and in illustrative examples. Some complimentary properties, such as the convergence of functions, whose exact treatment requires a knowledge of more advanced concepts found in Calculus (limits and continuity) will be treated approximately confining to the level accessible for Pre-Calculus.

A study of any function in Analysis (and consequently, in Calculus) consists of investigation of its main properties starting with an analytic definition of a function and using analytic methods, which leads to a possibility of visualization of a geometric form of a given function, that is, plotting its graph. In this text we will follow this standard scheme of the study of elementary function with the only restriction: we will apply only those rudimentary analytic techniques which are available in Pre-Calculus (that is, pre-limit calculus).

1 Function: Definition, Domain, Range

Definition of Function A *function* f from a set X to a set Y is a rule, or mapping, that associates with each element $x \in X$ the only element $y \in Y$. The set X is called the *domain* of f and the set Y *codomain* of f.

A. Bourchtein, L. Bourchtein, *Elementary Functions*,
https://doi.org/10.1007/978-3-031-29075-6_2

The function can be denoted as $f : X \to Y$, or $f(x) : X \to Y$, or $X \xrightarrow{f} Y$, or $y = f(x), x \in X$ without a specification of Y, or still $f(x)$ without specification of both X and Y. The elements x are called *independent variable* or *argument* of a function and the elements y *dependent variable* or *value* of a function. The latter are usually denoted by $y = f(x)$ and the same notation is frequently used to denote a function without specifying its domain and codomain. Although this notation is somewhat ambiguous, because the rule itself f and the result of its action $f(x)$ are different things, but it is convenient to use the notation $y = f(x)$ or $f(x)$ both for values of function and proper function. A specific meaning of this symbol is usually clear from the context. In this text we will consider only the functions whose domain and codomain are subsets of $\mathbb{R}$.

Notice that an indication of a codomain Y in the definition (or notation) of a function is unnecessary since we can always choose $Y = \mathbb{R}$ for any function. Therefore, the codomain is of a little (if any) importance in the definition of a function in Pre-Calculus, since its type is pre-determined: $Y \subset \mathbb{R}$. The following definition corrects this deficiency of a codomain.

Definition of the Range of a Function The *range* of a function $f : X \to Y$ is a part $\tilde{Y}$ of the codomain Y such that for each element $y \in \tilde{Y}$ there exists at least one x of the domain X which the function maps to y: $y = f(x)$. Frequently the range of f is also called the image of a function or the image of the domain.

Notice that by the definition the range is the smallest codomain. Since a codomain usually is out of our interest (because it does not bring any information about a function and can always be $\mathbb{R}$), in the cases when we will be able to determine the range of a function we will use notation Y for the range. It will not give rise to a confusion since the use of the symbol Y as codomain or range will be clear from the context (or will be specified explicitly). In the cases when codomain and range will be considered at the same time, we will keep the notation Y for the codomain and use $\tilde{Y}$ for the range.

Definition of the Image Under a Function Frequently it is important to consider the properties of a function on a part of its domain, say on subset S of X. In particular, it may be important to determine the *image* of S, which is, by the definition, a set $f(S)$ such that for each element $y \in f(S)$ there exists at least one $x \in S$ such that $y = f(x)$. Accordingly, the range of a function is the image of its domain and we will denote it also by $f(X)$.

Examples: Relations Between the Sets

1. Given two sets $X = \{1\}$ and $Y = \{1, 2\}$, the relation $f : X \to Y$ that brings the element $1 \in X$ to $1 \in Y$ and the same element $1 \in X$ to $2 \in Y$ is not a function, because this rule associates to the same element of X the two different elements of Y (see Fig. 2.1).
2. Given two sets $X = \{1, 2\}$ and $Y = \{3\}$, the relation $f : X \to Y$ that associates with the element $1 \in X$ the element $3 \in Y$ is not a function, because there is the element in X (the number 2) which has no corresponding element in Y (see Fig. 2.2).

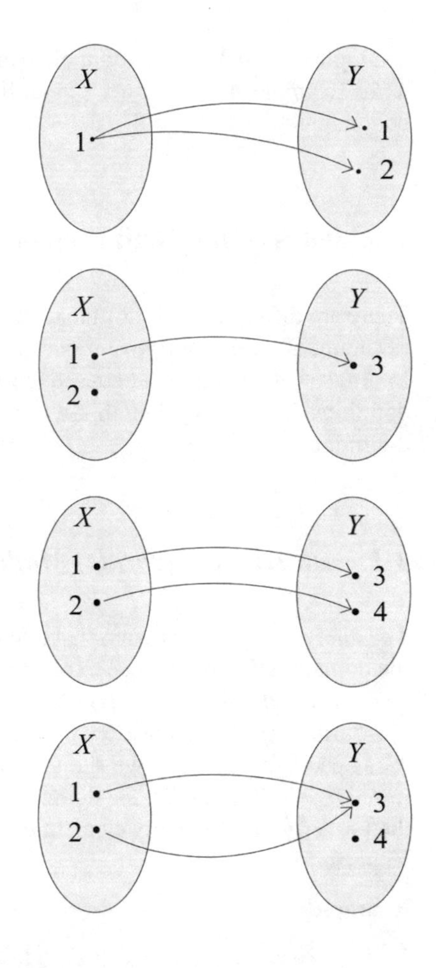

Fig. 2.1 Relation between the sets $X = \{1\}$ and $Y = \{1, 2\}$

Fig. 2.2 Relation between the sets $X = \{1, 2\}$ and $Y = \{3\}$

Fig. 2.3 Relation between the sets $X = \{1, 2\}$ and $Y = \{3, 4\}$

Fig. 2.4 Relation between the sets $X = \{1, 2\}$ and $Y = \{3, 4\}$

3. Given two sets $X = \{1, 2\}$ and $Y = \{3, 4\}$, the relation $f : X \to Y$ that maps the element $1 \in X$ to the element $3 \in Y$ and the element $2 \in X$ to the element $4 \in Y$ is a function according to the definition. The set X is the domain and Y is the range (see Fig. 2.3).
4. Given two sets $X = \{1, 2\}$ and $Y = \{3, 4\}$, the relation $f : X \to Y$ that transforms $1 \in X$ into $3 \in Y$ and $2 \in X$ into the same element $3 \in Y$ is a function according the definition. The set X is the domain, Y is codomain and $\tilde{Y} = \{3\} \subset Y$ is the range (see Fig. 2.4).

Remark By the definition of a function, the roles of the domain X and codomain Y are in general different. First, a function is a mapping of X into Y and not other way. Second, a rule that associates with each element of X an element of Y should be specified (this was missing in Example 2). Third, each element of X may have the only corresponding element of Y (this does not hold in Example 1). These two requirements to the mapping of the domain X are not applied to the codomain Y: the same element of Y can be associated with different elements of X (as in

Example 4) and Y can contain elements not related with any element of X (again as in Example 4). The latter possibility is eliminated if the range is chosen as a codomain.

2 Modes of the Definition of a Function

There are different modes of the definition of a function. Usually analytic, geometric, numeric and descriptive modes are used or some combination of these types. In Analysis/Calculus, and consequently, in Pre-Calculus the primary mode of the definition is analytic. For this reason in what follows we focus on the analytic definition.

2.1 *Analytic (Algebraic) Mode*

The *analytic definition* means that a function is determined through a formula where the domain X is given (in an explicit or implicit form), the codomain Y can always be considered to be $\mathbb{R}$ and the form of the correspondence (rule) f between X and Y is written in a formula $y = f(x)$ that associates the elements of the two sets. If X is specified explicitly we should use an indicated set. Otherwise (in the case of implicit definition of X), the domain X is chosen to be the largest subset of $\mathbb{R}$ with the property that for every element of X all the operations of the given formula can be performed.

Examples

1. $f(x) = x + 2$ (or $y = x + 2$).
   In this case the domain is not determined explicitly, but the formula $x + 2$ has sense for every $x \in \mathbb{R}$, and consequently, $X = \mathbb{R}$. The codomain is $Y = \mathbb{R}$, and the rule that associates with each $x \in \mathbb{R}$ the only $y \in \mathbb{R}$ is the formula $y = x + 2$. In this example the range is $Y = \mathbb{R}$, since for each $y \in \mathbb{R}$ there exists an element of the domain (found by the formula $x = y - 2$, which is the solution of the original formula with respect to x) that the function maps to y.
2. $f(x) = x + 2$, $X = [0, +\infty)$.
   In this case the domain is given explicitly and we have no choice but to use $X = [0, +\infty)$, even though the formula $y = x + 2$ makes sense for any $x \in \mathbb{R}$. The codomain can be $Y = \mathbb{R}$, and the rule that associates with each $x \in [0, +\infty)$ the only element $y \in \mathbb{R}$ is the formula $y = f(x) = x + 2$.
   Clearly in this example the range is not $\mathbb{R}$, since there are real numbers, for instance, $y = 0$, for which there is no corresponding $x \in [0, +\infty)$, because $y = x + 2 \geq 2$ for any $x \geq 0$. However, the last observation allows us to determine the range in the form $\tilde{Y} = [2, +\infty)$. Indeed, for every $y \in [2, +\infty)$ there exists

the corresponding element $x \in [0, +\infty)$ (found by the formula $x = y - 2$) that the function transforms in y.

3. $f(x) = \frac{1}{x}$.
The domain of this function is not defined explicitly, and we have to find the set of all the real numbers x for which the given operation $\frac{1}{x}$ is executable. Clearly, the only restriction goes from vanishing the denominator which occurs when $x = 0$. Then, the formula has sense for any $x \neq 0$, that is, $X = \mathbb{R}\backslash\{0\}$. As usual the codomain can be $Y = \mathbb{R}$. The rule that maps each $x \in X$ in the only value $y \in \mathbb{R}$ is the formula $y = \frac{1}{x}$.
In this case it is easy to find the range: notice that the formula $y = \frac{1}{x}$ can be solved for x in the form $x = \frac{1}{y}$, which means that for any $y \neq 0$ there exists an element x of the domain (found by the formula $x = \frac{1}{y}$) that the function brings to y. Hence, $\tilde{Y} = \mathbb{R}\backslash\{0\}$.
4. $f(x) = \sqrt{x}$.
Since the domain is not specified explicitly we have to find all real x for which the formula $\sqrt{x}$ is meaningful. Recall that the square root (on the field of real numbers) is defined only for non-negative numbers, which means that $X = [0, +\infty)$. The codomain can be $Y = \mathbb{R}$ as usual. The rule that maps each $x \in X$ to the only $y \in \mathbb{R}$ is the formula $y = \sqrt{x}$.
To find the range, recall that by the definition the square root is always non-negative, and consequently, the range $\tilde{Y}$ is contained in $[0, +\infty)$. On the other hand, for any $y \geq 0$ there exists $x = y^2 \in X$ such that $y = \sqrt{x}$ (recall that the square root $\sqrt{x}$, $x \geq 0$ is such real number $y \geq 0$ that $y^2 = x$). This means that every non-negative number belongs to the range, that is, $\tilde{Y} = [0, +\infty)$.
5. $f(x) = |x| = \begin{cases} x, & x \geq 0 \\ -x, & x < 0 \end{cases}$
This function can be defined both by a single formula $f(x) = |x|$ and by a piecewise form using the definition of the absolute value: $f(x) = \begin{cases} x, & x \geq 0 \\ -x, & x < 0 \end{cases}$.
Obviously, the use of a single formula is a relative concept, since it is sufficient to determine a unique symbol of all the operations (in this case the absolute value) applied to the elements of the domains to express rule through a single formula, even when simpler more convenient rules are different in distinct parts of the domain. Recalling the properties of the absolute value, we deduce that this function has the domain $X = \mathbb{R}$ and the range $Y = [0, +\infty)$.
6. $f(x) = \begin{cases} x^2, & x \geq 0 \\ x, & x < 0 \end{cases}$
This function is defined in a piecewise form and there is no a standard symbol to denote this formula (although one can invent such a symbol). The domain of the function is $X = \mathbb{R}$ (there is no restriction for the values of x) and the range is $Y = \mathbb{R}$ (all the non-negative values y are obtained using the first part of the formula and all the negative values are obtained in the second part).

Graph of a Function

Although the analytic definition of a function is the primary form in Pre-Calculus, its geometric representation is very important. This is why one of the main problems solved in Pre-Calculus is a construction of a geometric representation of a function based on its analytic definition. Such a geometric representation is called a graph of a function and its exact definition is as follows.

Definition The *graph* Γ of a function $y = f(x)$ is a set (geometric location) of all the points of Cartesian plane, whose coordinates satisfy the analytic definition of the function, that is, all the points (x, y) such that $x \in X$ and $y = f(x)$.

Remark Notice that this definition of a graph corresponds to the case when an analytic definition is considered to be original. The geometric definition of a function without reference to an analytic form will be given in Sect. 2.2.

A construction of the graph of a function defined by a formula is the result of the study of different properties of this function which determine the main features of its geometric form. At this moment, just for effects of geometric illustration, we show the graphs of the functions of Examples 1–6 without any investigation (see Figs. 2.5, 2.6, 2.7, 2.8, 2.9, and 2.10), noting that these functions are simple and at

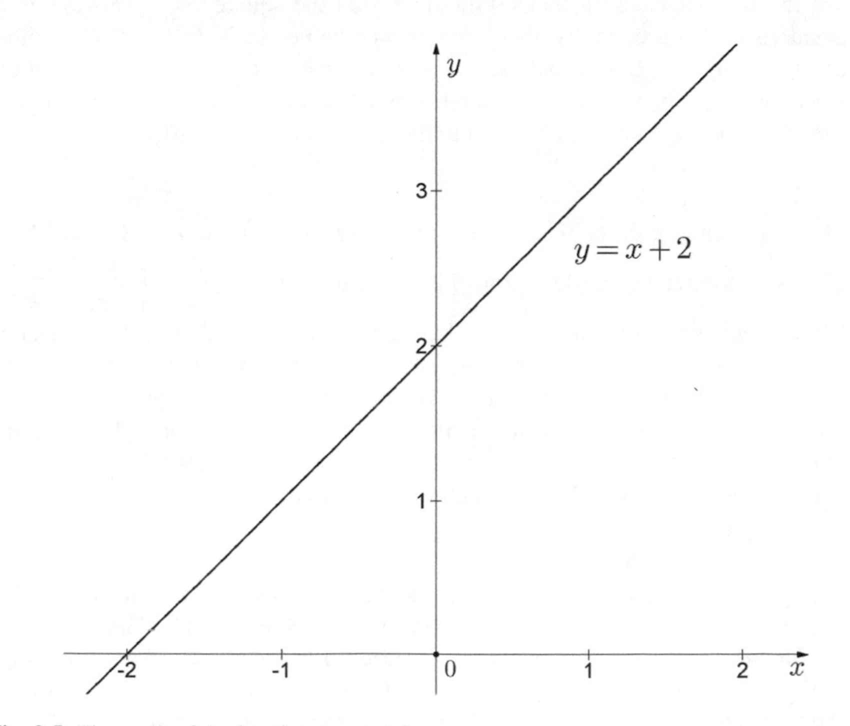

Fig. 2.5 The graph of the function $y = x + 2$

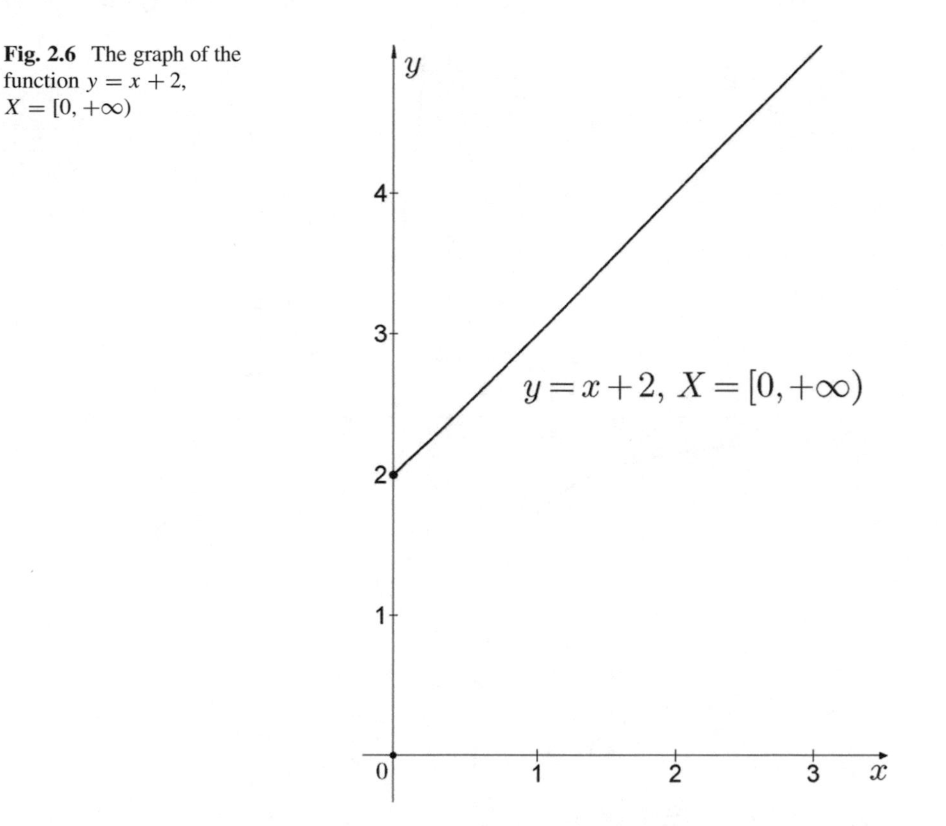

Fig. 2.6 The graph of the function $y = x + 2$, $X = [0, +\infty)$

some degree of precision their graphs can be plotted marking different points on the Cartesian plane and joining these points by a smooth curve.

Notice that there are various important functions in Calculus and Analysis, whose analytic form is simple, but it is not possible to plot their graphs (albeit one can imagine their geometric behavior). An example of such a function is Dirichlet's function $D(x) = \begin{cases} 1, & x \in \mathbb{Q} \\ 0, & x \in \mathbb{I} \end{cases}$. The values of this function are subject to changes from 0 to 1 in no space ("infinitesimal" interval) of the x-axis, because it does not exist a rational number with the smallest distance from a given irrational number and vice-verse. Therefore, the distance between two nearest points x which correspond to different values of the function is zero (if we try to define such a distance). Any attempt to imagine the graph of this function leads to necessity to separate the points where it takes different values by some distance, even very small, which represents only a rough approximation to the desired graph. One of such approximations is shown in Fig. 2.11.

Another important function of a similar type is $f(x) = \sin \frac{1}{x}$, whose oscillations accumulate infinitely near the y-axis. Since in any neighborhood (whatever small

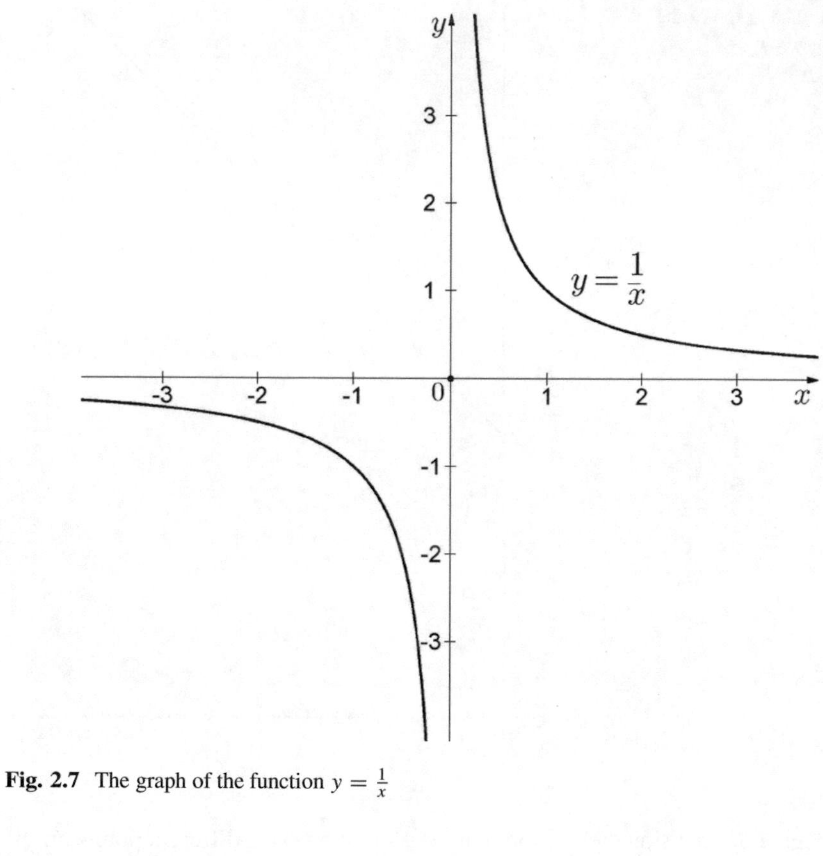

Fig. 2.7 The graph of the function $y = \frac{1}{x}$

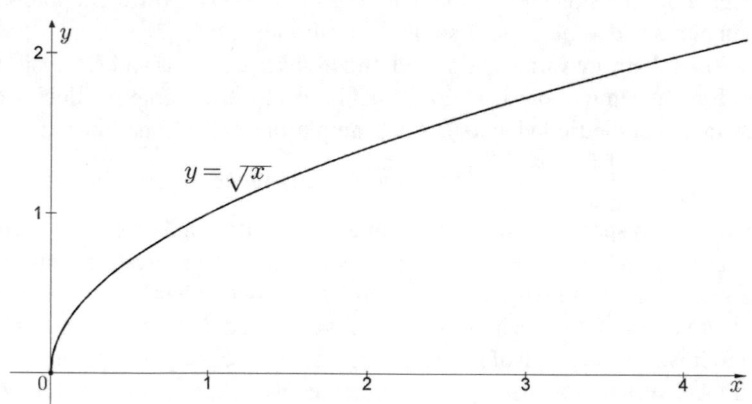

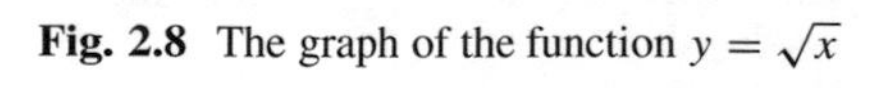
Fig. 2.8 The graph of the function $y = \sqrt{x}$

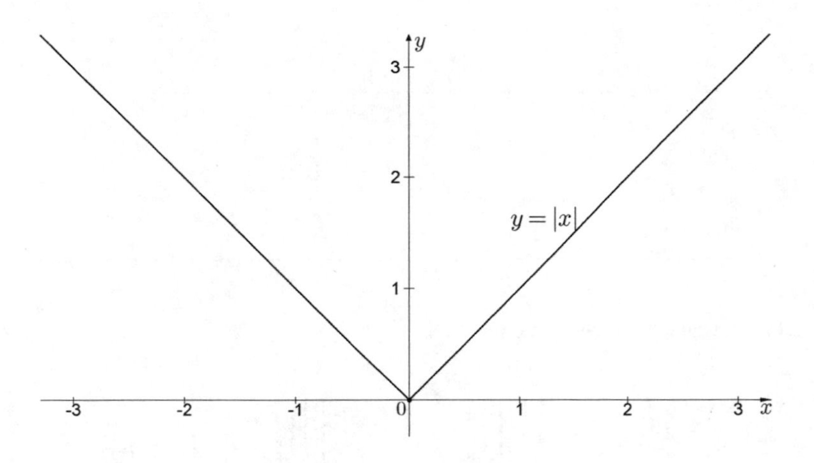

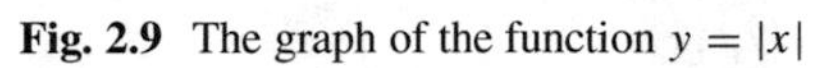

Fig. 2.9 The graph of the function $y = |x|$

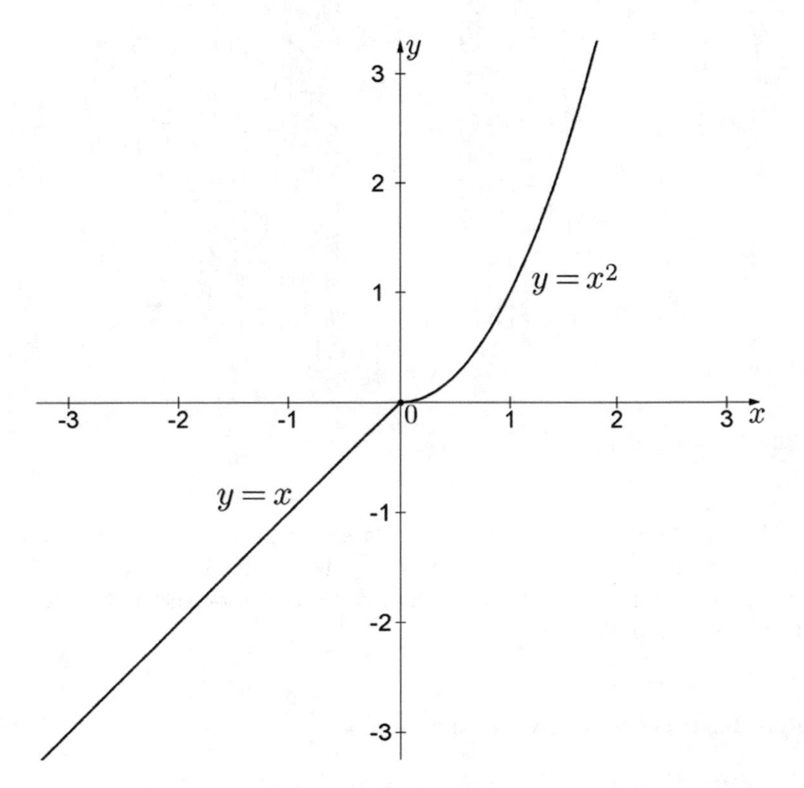

Fig. 2.10 The graph of the function $f(x) = x^2,\ x \geq 0;\, x,\ x < 0$

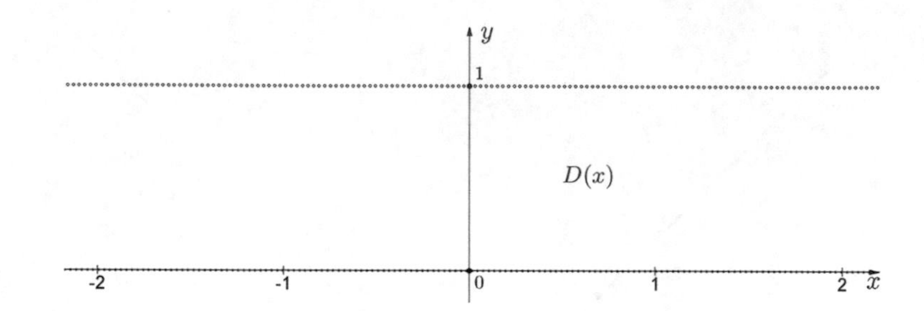

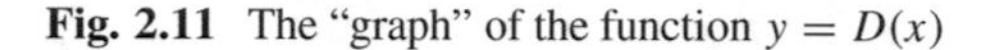

Fig. 2.11 The "graph" of the function $y = D(x)$

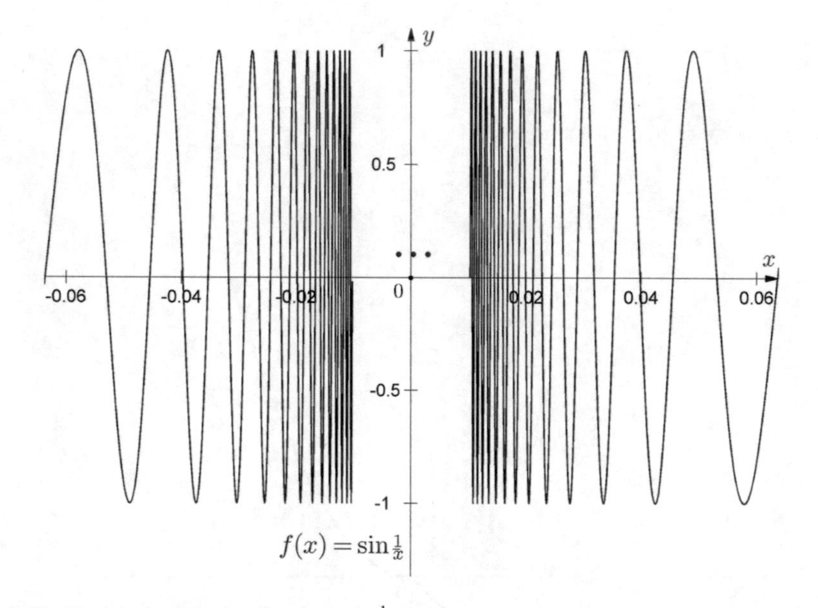

Fig. 2.12 The graph of the function $y = \sin\frac{1}{x}$

it is) of $x = 0$ the function is subject to infinite number of oscillations, there is no way to represent its graph. Again, we can only imagine its rough approximation sketching an accumulation of the oscillations near the y-axis until some chosen point, as is shown in Fig. 2.12.

Analytic Forms of the Definition of a Function

There are three types of general formulas used in analytic definition.

Explicit Form The first form is *explicit*, whose general representation is $y = f(x)$, which means that in this form the variable y is isolated (the formula is solved for y).

This kind of the formula is the simplest one. All the previous six examples of functions have this form.

Implicit Form The second type is *implicit*, where the relation between x and y is not solved for y and has the general form $g(x, y) = 0$.

Formally, any of the previous six examples can be rewritten in this form, joining x and y on the same side of the formula. For instance, Examples 2 and 4, can be rewritten in the form $y - x - 2 = 0$, $X = [0, +\infty)$ and $y - \sqrt{x} = 0$, with $g(x, y) = y - x - 2$ and $g(x, y) = y - \sqrt{x}$, respectively. This is a consequence of the general result that any explicit form $y = f(x)$ can be transformed to implicit one by the formula $g(x, y) = y - f(x) = 0$. The converse is not true: some implicit formulas admit an explicit representation, but many implicit formulas cannot be transformed to explicit form. For instance, the implicit formula $(x + y - 1)^{1/3} = 0$ can be written in the equivalent explicit form $y = 1 - x$; the implicit form $\ln y - x = 0$ can be reduced to the equivalent explicit form $y = e^x$. On the other hand, the implicit formulas $y^5 + 2y^3 + 3y - x = 0$ and $e^y + y + x = 0$ do not admit a transformation to explicit forms, although it can be shown that both formulas define the functions $y = f(x)$ with the domain $X = \mathbb{R}$. Hence, the implicit formulation is more general than explicit. Since many types of problems in mathematics and application areas lead to this kind of functions, we need to have techniques for study of these functions as well, although the treatment of explicit form is usually simpler.

Notice that some implicit formulas define a set of functions, not a single function. For instance, the implicit relation $x^2 + y^2 = 1$ is the equation of the unit circle centered at the origin (the set of all the points equidistant from the origin). Both the equation and its geometric representation indicate that this formula does not define a single function (see Fig. 2.13). Indeed, in the analytic form, we have the two solutions of this equation $y = \pm\sqrt{1 - x^2}$ for each $|x| < 1$ (for $x = \pm 1$ we have the unique solution $y = 0$ and for $|x| > 1$ there is no solution), which means that this implicit relation defines two different functions at once. In geometric terms, this means that any line $x = x_0$, $|x_0| < 1$ crosses the given circle at two points. To determine the only function satisfying the relation $x^2 + y^2 = 1$ we need to use an additional condition which filters out one of the two relations between x and y. For example, in analytic form the function can be defined by two relations $x^2 + y^2 = 1$ and $y \geq 0$, which corresponds geometrically to the choice of the upper half-circle (see Fig. 2.13). Sometimes when an implicit form defines many functions at once, no indication is given what function should be considered. In such cases we have to choose one of possible options or work with all the functions defined by implicit formula. In the example $x^2 + y^2 = 1$, this means that we should consider the two functions $y = \sqrt{1 - x^2}$ and $y = -\sqrt{1 - x^2}$ (upper and lower half-circles), both with the domain $X = [-1, 1]$.

Parametric Form The third type is *parametric*, when the variables x and y are not connected directly, but through an auxiliary variable called parameter. The formula of the relationship has the form $\begin{cases} \varphi(t, x) = 0 \\ \psi(t, y) = 0 \end{cases}$, where t is a parameter.

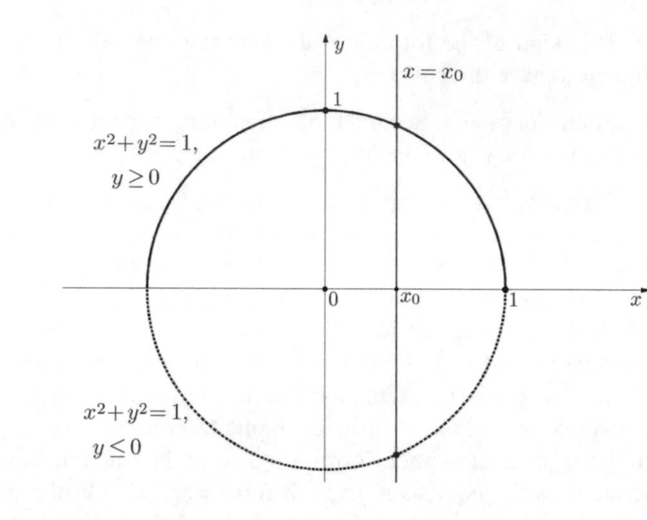

Fig. 2.13 The implicit definition $x^2 + y^2 = 1$

This form is the most general of the three forms and it usually requires the most complicated techniques of investigation. The general implicit form $g(x, y) = 0$ can be easily transformed into the parametric one through an introduction of a parameter $t = x$: $\begin{cases} \varphi(t, x) = t - x = 0 \\ \psi(t, y) = g(t, y) = 0 \end{cases}$. The converse is not true: in general it is not possible to reduce a parametric form to implicit.

If there is a possibility to express t in terms of x (or y) in an explicit form $t = \varphi(x)$ (or $t = \psi(y)$), then the parametric form can be transformed into implicit, and sometimes even into explicit. This happens, for instance, with the following parametric form: $\begin{cases} \varphi(t, x) = t - x = 0 \\ \psi(t, y) = t^2 - y = 0 \end{cases}$. In this case, substituting $t = x$ from the first equation into the second, we obtain $x^2 - y = 0$ and then arrived at the explicit form $y = x^2$.

However, in many cases such a conversion cannot be performed. For instance, the parametric form $\begin{cases} t^5 + t^3 + t + x = 0 \\ t^7 + 2t^3 + 3t + y = 0 \end{cases}$ defines the function $y = f(x)$, but it is not possible to obtain the implicit form of the relationship between x and y (not mentioning explicit form).

2.2 Geometric Mode

A function can be defined in a *geometric form* through a curve in the Cartesian plane which satisfies the condition that no vertical line (parallel to the y-axis) intersects the curve more than once. If this condition holds, a curve determines a function and is called a graph of a function. This verification whether a curve defines a function or not is frequently called the *vertical line test*.

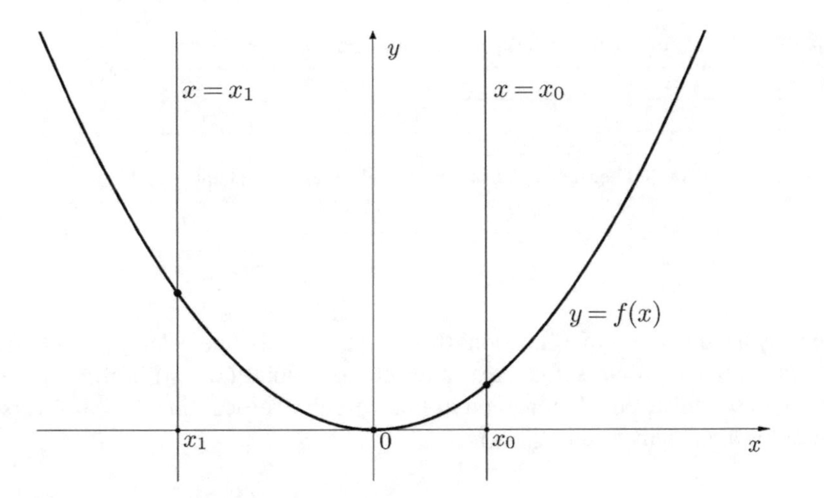

Fig. 2.14 Geometric definition of a function

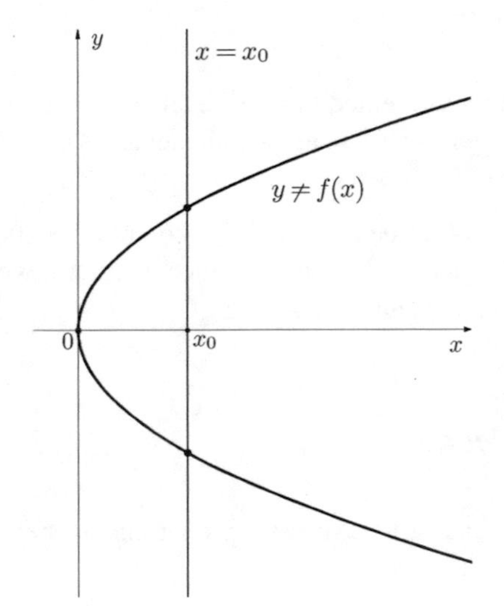

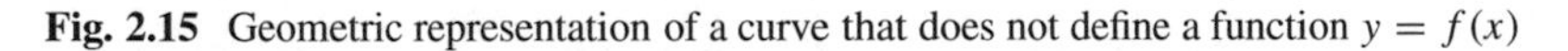

Fig. 2.15 Geometric representation of a curve that does not define a function $y = f(x)$

In the geometric definition a point x_0 belongs to the domain of a function if and only if the vertical line $x = x_0$ intersects a curve of the function. Figure 2.14 shows a geometric definition of a function (with illustration of the vertical line test), while Fig. 2.15 shows a curve that does not define a function $y = f(x)$ (the vertical line test is not satisfied).

Notice that the graph of a function defined analytically satisfies the vertical line test automatically: first, if a point x_0 does not belong to the domain, then no point (x_0, y) (whatever y is) belongs to the graph of the function, that is, the vertical line $x = x_0$ has no intersection with the graph; second, if x_0 belongs to the domain,

Table 2.1 Relationship between independent variable x and dependent y

x	0	1	2	3
y	1	2	3	4

Table 2.2 Relationship between independent variable x and dependent y

x	0	2	2	3
y	1	2	3	4

then (by the definition of a function) there exists exactly one value $y_0 = f(x_0)$ that corresponds to x_0, that is, there exists exactly one point (x_0, y_0) of the graph with the first coordinate equal to x_0, and in this case the vertical line $x = x_0$ intersects the graph at the only point (x_0, y_0).

2.3 Numerical Mode

Numerical form can be represented through a table or a set of ordered pairs of real numbers saved in the memory of an electronic device.

Examples
Table 2.1 defines a function (the general definition of a function is satisfied).

Table 2.2 does not represent a function, since there are two different values of y corresponding to the same value of $x = 2$.

2.4 Descriptive Mode

Descriptive form is a narrative expressing relationship between elements of the domain and codomain in words.

Example
"f is a function that assigns to each point of a real line the distance from this point to the origin."

Transforming this description in the analytic form, we recall about equivalence between all the points of the real line with the set of real numbers. Since the absolute value is the analytic operation on the set of real numbers that represents geometrically the distance between two points, we can deduce that the above sentence defines a function f with the domain $X = \mathbb{R}$ and the formula $y = f(x) = |x|$. The range of this function is the set of non-negative real numbers $\tilde{Y} = [0, +\infty)$, because, on the one hand, the absolute value is always a non-negative number, and on the other hand, for every $y \geq 0$ the formula of the function can be solved for x: $x = \pm y$ indicating such values of the domain which have the image y.

2.5 Relationship Between Different Forms of the Definition of a Function

In Analysis/Calculus, and consequently, in Pre-Calculus, one of the main goals is to investigate the properties of a function defined analytically by using analytic methods. Among the properties we are interested in is a geometric visualization (a sketch of the graph) of a function which frequently allows us to understand better the behavior of a function. A construction of the graph of a function is made based on the analytic properties revealed through analytic study, that is, applying logic reasoning to the analytic formula of a function, without any previous knowledge about the form of the graph. Solution of this problem is very important at the level of Pre-Calculus for at least three reasons. First, this gives a training of logic procedures which are used in Calculus/Analysis albeit involving more advanced tools. Second, this shows relationship between certain analytic formulas (following from the definition of a function) and geometric properties of a function. Finally, this is important didactically, since the graphs of functions shown in middle and high school are usually the curves given without any deduction or, in the best case, constructed approximately marking the values of a function at some points and/or applying a graphic software.

The main point in the logic connection between analytic and geometric properties of functions in the context of Pre-Calculus is that we cannot derive a conclusion about analytic properties based on geometric representation. On the contrary, we should study the analytic properties first and plot the graph of a function as the last step of its investigation. Nevertheless, although a geometric form of a function cannot be used to derive analytic properties, even a rough approximation of the graph can provide a first insight on the properties of a function, which should be confirmed or refuted during analytic study. Besides, sometimes it is appropriate to use a geometric form of definition of some functions for the purposes of illustration of certain concepts in the cases when their analytic properties have not yet been investigated.

3 Bounded Functions

Definition A function $y = f(x)$ is called *bounded above* on a subset S of its domain X, if the image $f(S)$ is bounded above. In other words (expressing in detail the concept of a bounded above set), a function $y = f(x)$ is bounded above on a set $S \subset X$, if there exists a constant M such that for any $x \in S$ it holds $f(x) \leq M$. A constant M is frequently called *upper bound.*

A function $y = f(x)$ is called *bounded below* on a subset S of its domain X, if the image $f(S)$ is bounded below. In other words, a function $y = f(x)$ is bounded below on a set $S \subset X$, if there exists a constant m such that for any $x \in S$ it holds $f(x) \geq m$. A constant m is frequently called *lower bound.*

A function $y = f(x)$ is called *bounded* on a subset S of its domain X if it is bounded both above and below on S, that is, there exist constants m and M such that $m \le f(x) \le M$ for any $x \in S$. Otherwise a function is called *unbounded*.

Notice that the last definition can be equivalently formulated as follows: a function $y = f(x)$ is *bounded* on a subset S of its domain X if there exists constant C such that $|f(x)| \le C$ for any $x \in S$. Indeed, if this definition is satisfied, then the previous holds with $m = -C$ and $M = C$. Conversely, if the previous definition holds, choosing $C = \max\{|m|, |M|\}$ we arrive at the last formulation.

In the most cases, S is the domain X itself.

Geometric Property

In the geometric form, the property of boundedness means that the graph of a function lies between the two coordinate lines $y = m$ and $y = M$. This refers to the entire graph if $S = X$ or to its part related to the set S.

Examples

1a. The function $f(x) = x$, $S = X = \mathbb{R}$ is not bounded above nor below.
1b. The function $f(x) = x$, $S = [0, +\infty)$ is bounded below by the constant $m = 0$ (or any other constant less than 0), but it is not bounded above.
1c. The function $f(x) = x$, $S = [0, 1]$ is bounded above by the constant $M = 1$ (or any other constant greater than 1) and bounded below by the constant $m = 0$ (or any other constant less than 0).
2a. The function $f(x) = x^2$, $S = X = \mathbb{R}$ is bounded below by the constant $m = 0$ (or any other constant less than 0), but it is not bounded above.
2b. The function $f(x) = x^2$, $S = [0, +\infty)$ is bounded below by the constant $m = 0$ (or any other constant less than 0), but it is not bounded above.
2c. The function $f(x) = x^2$, $S = [-1, 2]$ is bounded above by the constant $M = 4$ (or any other constant greater than 4) and bounded below by the constant $m = 0$ (or any other constant less than 0).
3a. The function $f(x) = \cos x$, $S = X = \mathbb{R}$ is bounded above by the constant $M = 1$ (or any other constant greater than 1) and bounded below by the constant $m = -1$ (or any other constant less than -1). Therefore, this function is bounded on the entire domain and on any subset of the domain.

4 Properties of Symmetry

4.1 Even Functions

Definition A function $y = f(x)$ is called *even* if for any x of its domain X the following property holds: $f(-x) = f(x)$, $\forall x \in X$.

Properties

Property of the Domain

It follows immediately from the definition that the domain X of an even function is symmetric about the origin, because if $x \in X$, then $-x \in X$ according to the formula of even functions.

Property of the Graph

Let us show that the graph Γ of an even function is symmetric about the y-axis. Recall that a curve (in particular, a graph of a function) is symmetric about the y-axis if every point $P_1 = (x_1, y_1)$, which belongs to the curve, has the symmetric point $P_2 = (-x_1, y_1)$ also belonging to the curve (see Sect. 7.5 of Chap. 1). Take an arbitrary point $P_1 = (x_1, y_1)$ of the graph Γ of an even function $f(x)$ and show that the symmetric point $P_2 = (x_2, y_2) = (-x_1, y_1))$ also belongs to Γ. The fact that $P_1 = (x_1, y_1) \in \Gamma$ means that $y_1 = f(x_1)$. Since $f(x)$ is even, the condition $x_1 \in X$ implies that $-x_1 \in X$, and taking $x_2 = -x_1$ we get $y_2 = f(x_2) = f(-x_1) = f(x_1) = y_1$. This shows that the point $P_2 = (-x_1, y_1)$ belongs to Γ. Therefore, by the definition of the symmetry, the graph of an even function is symmetric about the y-axis. The graphs of some even functions and the corresponding symmetric points are shown in Figs. 2.16, 2.17, and 2.18.

Examples

1. $y = f(x) = x^2$. This function is even, because $f(-x) = (-x)^2 = x^2 = f(x)$, $\forall x \in X = \mathbb{R}$. The graph of this function is shown in Fig. 2.16.
2. $y = f(x) = |x|$. This function is even, because $f(-x) = |-x| = |x| = f(x)$, $\forall x \in X = \mathbb{R}$. The graph is shown in Fig. 2.17.
3. $y = f(x) = \cos x$. This function is even, because $f(-x) = \cos(-x) = \cos x = f(x)$, $\forall x \in X = \mathbb{R}$. The graph is shown in Fig. 2.18.

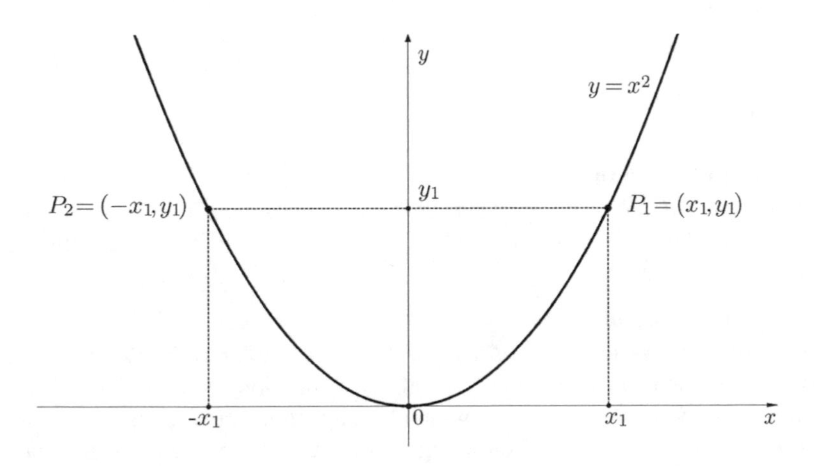

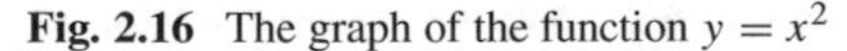
Fig. 2.16 The graph of the function $y = x^2$

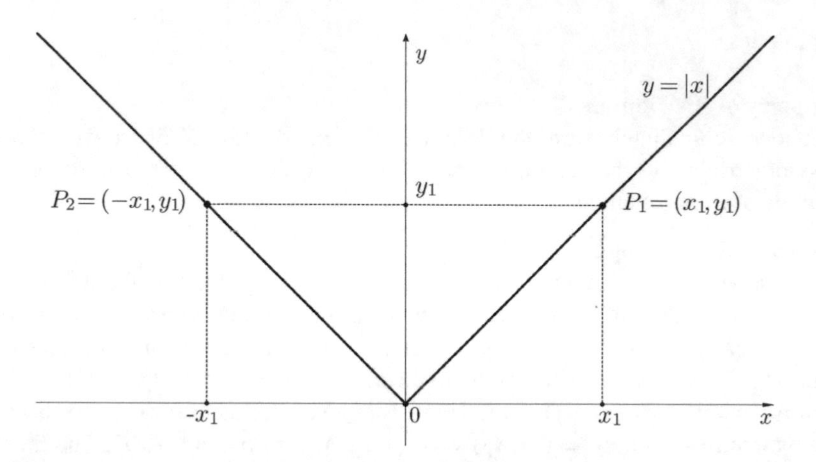

Fig. 2.17 The graph of the function $y = |x|$

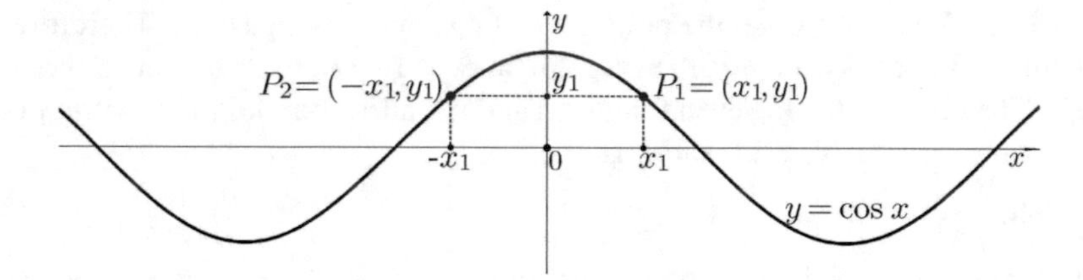

Fig. 2.18 The graph of the function $y = \cos x$

4.2 *Odd Functions*

Definition A function $y = f(x)$ is called *odd* if for any x of its domain X we have $f(-x) = -f(x)$.

Properties

Property of the Domain
It follows immediately from the definition that the domain X *of an odd function is symmetric about the origin*, because if $x \in X$, then $-x \in X$ according to the formula of odd function.

Property of the Graph
Let us show that the graph Γ *of an odd function is symmetric about the origin.* Recall that a curve (in particular, a graph of a function) is symmetric about the origin if every point $P_1 = (x_1, y_1)$, which belongs to the curve, has the symmetric point $P_2 = (-x_1, -y_1)$ also belonging to the curve (see Sect. 7.4 of Chap. 1). Take an arbitrary point $P_1 = (x_1, y_1)$ of the graph Γ of an odd function $f(x)$ and show that the symmetric point $P_2 = (x_2, y_2) = (-x_1, -y_1)$ also belongs to

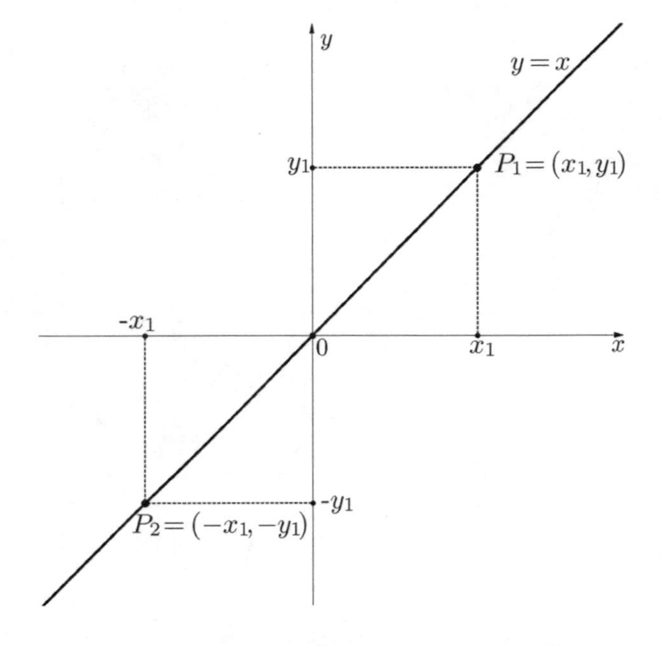

Fig. 2.19 The graph of the function $y = x$

Γ. The fact that $P_1 = (x_1, y_1) \in \Gamma$ means that $y_1 = f(x_1)$. Since $f(x)$ is odd, the condition $x_1 \in X$ implies that $-x_1 \in X$, and taking $x_2 = -x_1$ we get $y_2 = f(x_2) = f(-x_1) = -f(x_1) = -y_1$. This shows that the point $P_2 = (-x_1, -y_1)$ belongs to Γ. Therefore, by the definition of the symmetry, the graph of an odd function is symmetric about the origin. The graphs of some odd functions and the corresponding symmetric points are shown in Figs. 2.19, 2.20, and 2.21.

Examples

1. $y = f(x) = x$. This function is odd, because $f(-x) = -x = -f(x)$, $\forall x \in X = \mathbb{R}$. The graph of this function is shown in Fig. 2.19.
2. $y = f(x) = \frac{1}{x}$. This function is odd, because $f(-x) = \frac{1}{-x} = -\frac{1}{x} = -f(x)$, $\forall x \in X = \mathbb{R}\backslash\{0\}$. The graph of this function is shown in Fig. 2.20.
3. $y = f(x) = \sin x$. This function is odd, because $f(-x) = \sin(-x) = -\sin x = -f(x)$, $\forall x \in X = \mathbb{R}$. The graph of this function is shown in Fig. 2.21.

Definition The property of a function to be even or odd is called *parity*.

Examples of Parity

1. $f(x) = x^2$, $X = (-1, 1)$. This function is even, because $f(-x) = f(x)$ for each $x \in X = (-1, 1)$.
2. $f(x) = \frac{x^3-x^2}{x-1}$. This function is not even nor odd, because its domain $X = (-\infty, 1) \cup (1, +\infty)$ is not symmetric about the origin. Even though $f(x) = \frac{x^3-x^2}{x-1} = x^2$, $\forall x \neq 1$, and consequently, $f(-x) = f(x)$ for $\forall x \neq \pm 1$, the fact that the function is defined at $x = -1$ and not defined at $x = 1$ implies that for the pair of points $x = \pm 1$ the property of parity is not satisfied, although one of

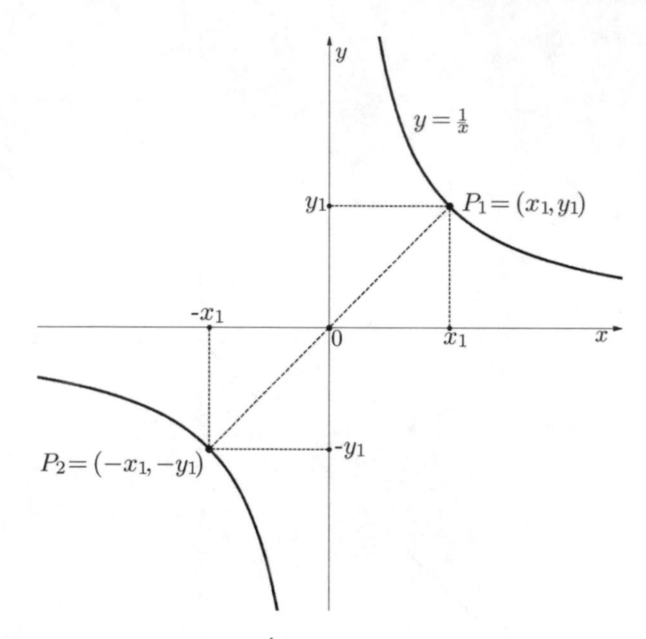

Fig. 2.20 The graph of the function $y = \frac{1}{x}$

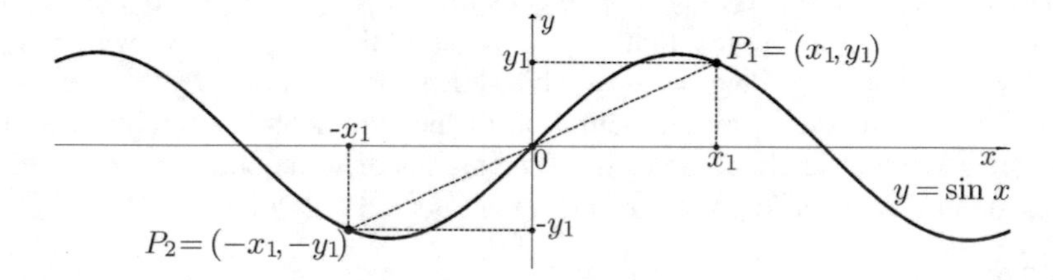

Fig. 2.21 The graph of the function $y = \sin x$

them belongs to the domain. To restore the evenness of the function, we have to use a symmetric domain, such as $X = (-\infty, -1) \cup (-1, 1) \cup (1, +\infty)$.

3. $f(x) = \sqrt{x}$. This function is not even nor odd, because its domain $X = [0, +\infty)$ is not symmetric about the origin. There is no way to restore a symmetry of the domain unless by choosing $X = \{0\}$, that leads to a singular case of a function defined at the only point, which is out of interest.

4.3 Arithmetic Operations with Even and Odd Functions

Remark If the domains of the functions involved in arithmetic operations are different, it is necessary to consider these functions on the intersection of their domains, and this can result in a situation when an operation is not defined because the intersection is the empty set. To simplify considerations, we assume below that all the functions used in a specific operation are defined on the same domain. We also assume that in the case of the division the domain includes all the points where the denominator does not vanish.

1. Even Functions

Proposition *Let $f(x) : X \to \mathbb{R}$ and $g(x) : X \to \mathbb{R}$ be even functions. Then $f(x) + g(x) : X \to \mathbb{R}$, $f(x) - g(x) : X \to \mathbb{R}$, $f(x) \cdot g(x) : X \to \mathbb{R}$ and $\frac{f(x)}{g(x)} : \tilde{X} \to \mathbb{R}$ are also even functions, the last one on the set $\tilde{X}$ where $g(x) \neq 0$.*

Proof The proofs are trivial and we perform them in the case of the sum and quotient, leaving other two operations to the reader.

Denote $h(x) = f(x) + g(x)$ and notice that $h(x)$ is defined on X, which is symmetric about the origin. Then, $h(-x) = f(-x)+g(-x) = f(x)+g(x) = h(x)$, $\forall x \in X$, that is, $h(x)$ is an even function.

In the same manner, denote $h(x) = \frac{f(x)}{g(x)}$ and notice that $h(x)$ is defined on $\tilde{X}$ where $g(x) \neq 0$. Since $g(x)$ is even, its zeros are symmetric about the origin, and together with the symmetry of X this guarantees that $\tilde{X}$ is symmetric about the origin. Then, $h(-x) = \frac{f(-x)}{g(-x)} = \frac{f(x)}{g(x)} = h(x)$, $\forall x \in \tilde{X}$, that is, $h(x)$ is an even function. □

2. Odd Functions

Proposition *Let $f(x) : X \to \mathbb{R}$ and $g(x) : X \to \mathbb{R}$ be odd functions. Then $f(x) + g(x) : X \to \mathbb{R}$ and $f(x) - g(x) : X \to \mathbb{R}$ are also odd functions, while $f(x) \cdot g(x) : X \to \mathbb{R}$ and $\frac{f(x)}{g(x)} : \tilde{X} \to \mathbb{R}$ are even functions, the last one on the set $\tilde{X}$ where $g(x) \neq 0$.*

Proof The proofs are straightforward and we perform them for the difference and product, leaving the other two operations for the readers.

Denote $h(x) = f(x) - g(x)$ and notice that $h(x)$ is defined on X, which is symmetric about the origin. Then, $h(-x) = f(-x)-g(-x) = -f(x)-(-g(x)) = -(f(x) - g(x)) = -h(x)$, $\forall x \in X$, that is, $h(x)$ is an odd function.

In a similar way, denote $h(x) = f(x) \cdot g(x)$ and notice that $h(x)$ is defined on X, which is symmetric about the origin. Then, $h(-x) = f(-x) \cdot g(-x) = -f(x) \cdot (-g(x)) = h(x)$, $\forall x \in X$, that is, $h(x)$ is an even function. □

3. Even and Odd Functions

Proposition *Let $f(x) : X \to \mathbb{R}$ be an even function and $g(x) : X \to \mathbb{R}$ be an odd function. Then $f(x) + g(x) : X \to \mathbb{R}$ and $f(x) - g(x) : X \to \mathbb{R}$ are neither even nor odd, unless one of these functions is zero. At the same time, $f(x)\cdot g(x) : X \to \mathbb{R}$ and $\frac{f(x)}{g(x)} : \tilde{X} \to \mathbb{R}$ are odd, the last being defined on the set $\tilde{X}$ where $g(x) \neq 0$.*

Proof We provide elementary proofs for the sum and product and leave other two operations to the reader.

Denote $h(x) = f(x) + g(x)$ and notice that $h(x)$ is defined on X, which is symmetric about the origin. Suppose, for the contradiction, that $h(x)$ is even. Then, by the previous property, $g(x) = h(x) - f(x)$ should be even, that contradicts the condition of $g(x)$ (unless $g(x) \equiv 0$, in which case it is both even and odd). In the same way, we arrive at a contradiction if we suppose that $h(x)$ is odd. Therefore, $h(x)$ is neither even nor odd (unless one of the functions $f(x)$ or $g(x)$ is zero).

Denote $h(x) = f(x) \cdot g(x)$ and notice that $h(x)$ is defined on X, which is symmetric about the origin. Then, $h(-x) = f(-x) \cdot g(-x) = f(x) \cdot (-g(x)) = -h(x)$, $\forall x \in X$, that is, $h(x)$ is an odd function. □

Examples

1. $f(x) = x^2 + \cos x$. This is an even function on $\mathbb{R}$ as the sum of two even functions on $\mathbb{R}$.
2. $f(x) = x \cos x$. The product of an odd x and even $\cos x$ functions on $\mathbb{R}$ results in an odd function on $\mathbb{R}$.
3. $f(x) = \frac{\sin x}{x^3}$. This function is even on $\mathbb{R}\backslash\{0\}$ as the quotient of two odd functions with the denominator vanishing only at the origin.
4. $f(x) = x - |x|$. The difference of an odd x and even $|x|$ functions is neither even nor odd. Another way to show this is by choosing an appropriate pair of symmetric points: for instance, for $x = -1$ we have $f(-1) = -2$, while for $x = 1$ we get $f(1) = 0$.

4.4 Periodic Functions

Definition A function $y = f(x)$ is called *periodic* if there exists a non-zero constant T such that for any x of the domain X the relation $f(x + T) = f(x)$ holds. The constant T is called *period*.

If there exists a minimum positive number T such that this property holds, this number is called *minimum period* or *fundamental period*.

Properties

Properties of the Domain

Before to formulate and demonstrate the properties of the domain, we make some important preliminary remarks.

First, it follows directly from the definition that if x belongs to the domain X, then $x + T$ also belongs to X, because both points are involved in the main formula of periodicity. In the same manner, if $x_1 = x + T$ is a point of the domain, then $x_2 = x_1 + T = x + 2T$ is also a point of the domain. Proceeding in this way, we can see that if $x \in X$, then $(x + nT) \in X$ for $\forall n \in \mathbb{N}$.

Analogously, $x - T$ belongs to X if $x \in X$, since using the notation $\tilde{x} = x + T$, the main formula can be written as $f(\tilde{x}) = f(\tilde{x} - T)$. Consequently, $(x - nT) \in X$, $\forall n \in \mathbb{N}$ if $x \in X$. Hence, if $x \in X$, then $(x + nT) \in X$, $\forall n \in \mathbb{Z}$.

Notice also that the formula $f(\tilde{x}) = f(\tilde{x} - T)$ shows that if T is a period of $f(x)$, then $-T$ also is a period.

Let us now recall the definitions of bounded sets (see Sect. 5.2 of Chap. 1). A set S is called *bounded on the right* if there exists a real number M such that for any x of S we have $x \le M$. In a similar manner, a set S is called *bounded on the left* if there exists a real number m such that for any x of S we have $x \ge m$. Finally, a set is called *bounded* if it is bounded both on the right and on the left.

Let us formulate now the first property of the domain.

Property 1 of the Domain *The domain of a periodic function cannot be bounded either on the right or on the left.*

Proof First, we show that the domain is not bounded on the right. Without loss of generality we can assume that the period T is positive (if T is negative we take $-T$). We can argue by contradiction. Suppose that there exists a constant M such that $x \le M$, $\forall x \in X$. Take some $x_0 \in X$ and calculate the distance between x_0 and M: $d(x_0, M) = M - x_0$. Compare this distance with the period T using the quantity $a = \frac{M-x_0}{T}$ and take any natural number n greater than a, say $n = [a] + 1$, where $[a]$ denotes the entire part of a. In this case, from $n > a = \frac{M-x_0}{T}$ it follows that $nT > M - x_0$ and $x_0 + nT > M$. However, as it was discussed previously, $x_0 + nT$ is a point of the domain X, that contradicts the supposition. Therefore, X is not bounded on the right.

Similarly, it can be proved that X is not bounded on the left. The reader can provide details of this proof. □

As was discussed above, roughly speaking, any property observed for $x \in X$ should be held also for $x + nT \in X$, $\forall n \in \mathbb{Z}$. In particular, if X is the domain, then $X + nT = \{x + nT, \forall x \in X\}$, $\forall n \in \mathbb{Z}$ is the same domain. This leads to the following property useful in practice.

Property 2 of the Domain *A periodic function cannot be undefined at a finite number of points.*

Proof Let T be a positive period of function. Suppose, for contradiction, that the domain is $X = \mathbb{R} \backslash \{x_1, \ldots, x_n\}$. Then we can always find the largest number among $x_1, \ldots, x_n$. Suppose, without loss of generality, that this number is x_n. By the property of periodic functions, the number $x_0 = x_n + T$ should be out of the domain, for if $x_0 \in X$ it would imply that $x_0 - T = x_n \in X$, which is not true. But x_n, not x_0, is the largest point out of the domain, which makes the contradiction. □

Property of the Range *A periodic function cannot take some value at a finite number of points. In particular, a periodic functions cannot have a finite number of zeros.*

Proof If $f(x_0) = a$, then the same value a the function takes on the infinite set of points $\{x_0 + kT, \forall k \in \mathbb{Z}\}$, where T is a period of function. □

Property of a Period *If T is a period of a function $f(x)$, then any number nT, $\forall n \in \mathbb{Z}\backslash\{0\}$ is also a period of this function.*

Proof This property is intimately connected to the property of the domain and we will take advantage of some previous considerations in its proof.

First, notice that if $x \in X$, then $(x + nT) \in X$ for $\forall n \in \mathbb{Z}$, as was shown before.

Let us start with the proof that if T is a period, then $2T$ is also a period. This part will reveal how we can proceed in a general case. If T is a period of $f(x) : X \to \mathbb{R}$, then $f(x + T) = f(x), \forall x \in X$. Consequently, denoting $x_1 = x + T$, we can write $f(x + 2T) = f(x_1 + T) = f(x_1) = f(x + T) = f(x), \forall x \in X$, that is, $2T$ is a period of the same function. Now we extend the same reasoning to the case of nT, $\forall n \in \mathbb{N}$:

$$f(x + nT) = f(x + (n-1)T + T) = f(x + (n-1)T) = f(x + (n-2)T + T)$$

$$= f(x + (n-2)T) = \ldots = f(x + T) = f(x),$$

which shows that nT is a period. Notice additionally that $-T$ is also a period (as was discussed before): denoting $x_1 = x - T$, we have $f(x - T) = f(x_1) = f(x_1 + T) = f(x)$. Therefore, from the fact that nT, $\forall n \in \mathbb{N}$ is a period it follows that $-nT$, $\forall n \in \mathbb{N}$ is also a period. This completes the proof. □

Property of the Graph *Let $f(x)$ be a periodic function with a period T. If the graph of $f(x)$ is known on an interval with the length T, then the entire graph can be found by repetition (horizontal translation) of the original part to the right and to the left infinitely many times.*

Remark The original interval of the length T should be half-open or closed if the function is defined at the endpoints of this interval. (Clearly, this interval is open if the function is not defined at its endpoints.)

Proof To specify considerations, let us suppose that $f(x)$ is defined at the endpoints of the original interval $[a, a + T]$. Then the graph of the function $f(x)$ is known on the interval $[a, a + T]$ and we intend to construct its graph on the entire domain of $f(x)$. Without loss of generality we can assume that $T > 0$.

To extend the graph to the right, consider first the interval $[a + T, a + 2T]$. Take any point $P_0 = (x_0, f(x_0))$ of the graph Γ of $f(x)$ and show that the point $P_1 = (x_0 + T, f(x_0))$, $x_0 \in [a, a + T]$, obtained by horizontal translation of P_0 by T units to the right, also belongs to Γ. Indeed, if $x_0 \in X$, then $(x_0 + T) \in X$ and, by the definition of periodicity, $f(x_0 + T) = f(x_0)$. Consequently, $P_1 = (x_0 + T, f(x_0)) \in \Gamma$. Since this property holds for any P_0 of the original part of the graph, the part of the graph on the interval $[a + T, a + 2T]$ is obtained by shifting the original part T units to the right.

In the same way, the graph of $f(x)$ on the interval $[a + 2T, a + 3T]$ is obtained by horizontal translation of the original part $2T$ units to the right, etc. In general, the graph of $f(x)$ on the interval $[a + nT, a + (n + 1)T]$, $\forall n \in \mathbb{N}$ is obtained by shifting horizontally the original part nT units to the right.

Similarly, it can be demonstrated that the graph of $f(x)$ on the interval $[a - nT, a - (n - 1)T]$, $\forall n \in \mathbb{N}$ is obtained by shifting horizontally the original part nT units to the left. The reader is supposed to complete details of this part of the proof. □

Various periodic functions can be seen in Figs. 2.22, 2.24, 2.25, and 2.26, which illustrate the following examples.

Examples

1. A constant function $f(x) = 1$ is periodic, because $f(x+T) = f(x) = 1, \forall T \neq 0$. This function does not have a minimum period, since any number $T > 0$ is a period and a smaller number $\frac{T}{2}$ is also a period of this function. See the graph of this function in Fig. 2.22.
2. $f(x) = [x] = n, \forall x \in [n, n + 1), \forall n \in \mathbb{Z}$ (the integer part of x).
   In other words, the function $[x]$ transforms every real number x into the greatest integer less than or equal to x (geometrically speaking, the rule associates with each x the first integer point located at the left or coinciding with x). For example, $[1, 234] = 1$, $[-4, 67] = -5$, $\left[\sqrt{2}\right] = 1$, $[\pi] = 3$. The range of this function is $\mathbb{Z}$.
   To better understand the definition of this function see its graph in Fig. 2.23.
   This function is not periodic, because it has distinct values on any two different intervals $[n, n + 1)$ and $[m, m + 1)$, $n, m \in \mathbb{Z}, n \neq m$.
3. $f(x) = x-[x]$. This function is periodic and its minimum period is 1: $f(x+1) = x + 1 - [x + 1] = x - [x] = f(x)$. Its graph is shown in Fig. 2.24.
4. $f(x) = \cos x$. The function is periodic with the minimum period equal to 2π: $f(x + 2\pi) = \cos(x + 2\pi) = \cos x = f(x)$. Its graph is shown in Fig. 2.25.
5. $f(x) = \sin x$. The function is periodic with the minimum period equal to 2π: $f(x + 2\pi) = \sin(x + 2\pi) = \sin x = f(x)$. Its graph is shown in Fig. 2.26.

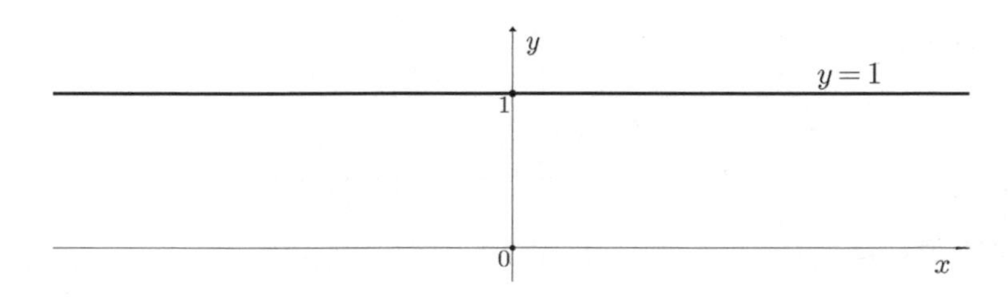

Fig. 2.22 The graph of the function $y = 1$

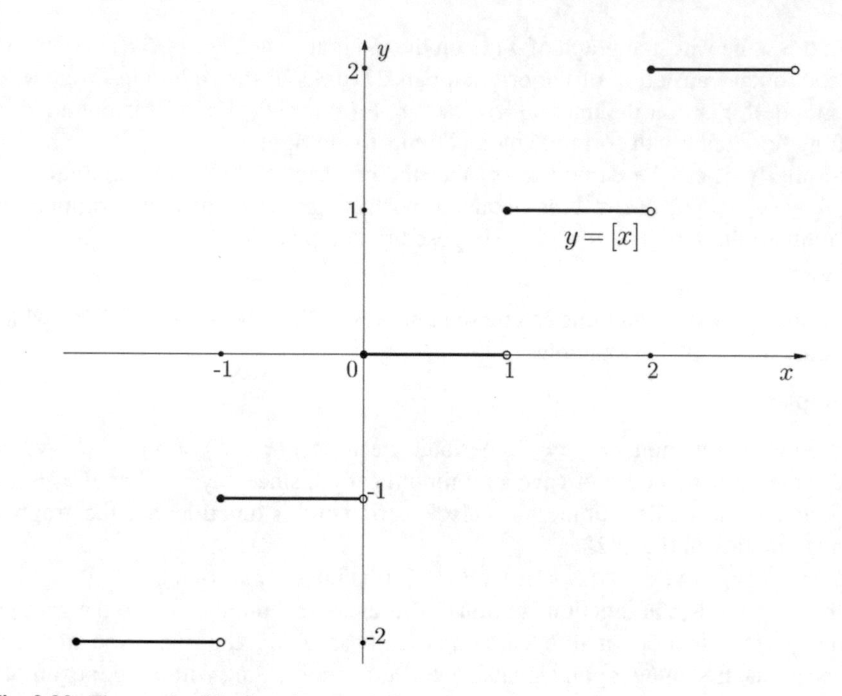

Fig. 2.23 The graph of the function $y = [x]$

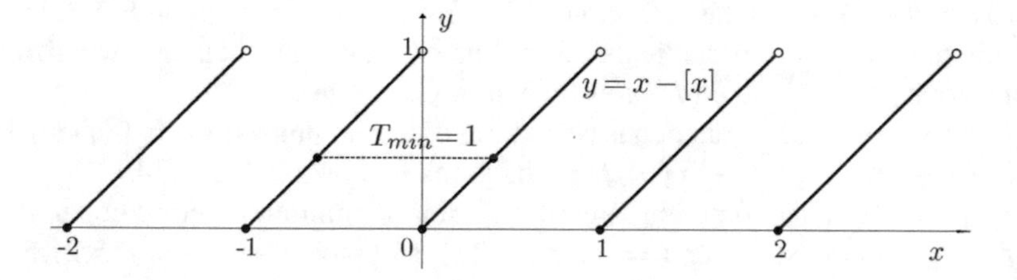

Fig. 2.24 The graph of the function $y = x - [x]$

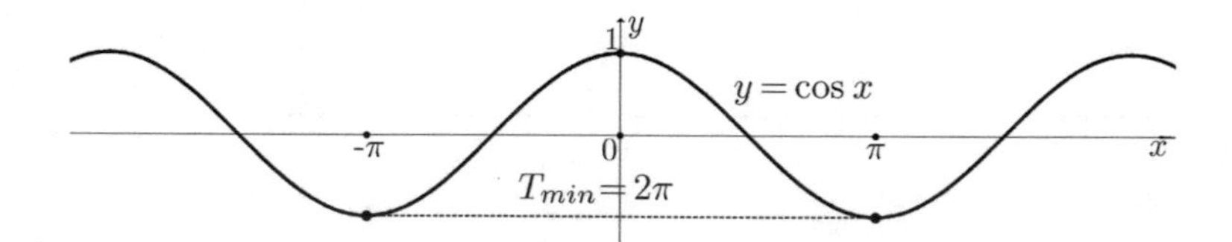

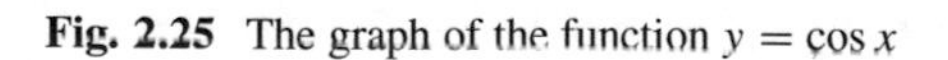

Fig. 2.25 The graph of the function $y = \cos x$

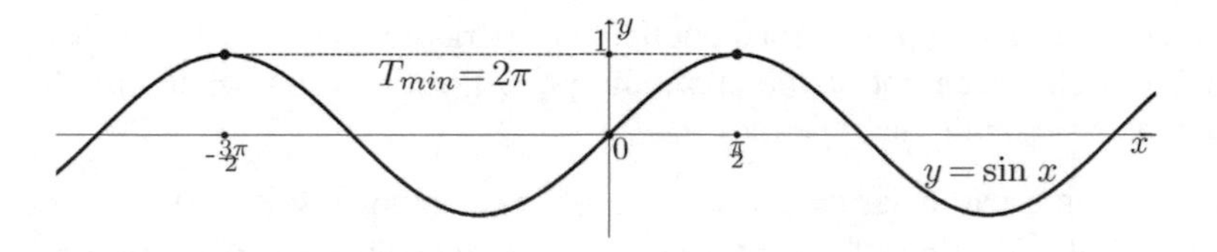

Fig. 2.26 The graph of the function $y = \sin x$

4.5 Elementary Operations with Periodic Functions

Arithmetic Operations with Periodic Functions

Remark If the domains of the functions involved in arithmetic operations are different, it is necessary to consider these functions on the intersection of their domains, and this can result in a situation when an operation is not defined because the intersection is the empty set. To simplify considerations, we assume below that all the functions used in a specific operation are defined on the same domain. We also assume that in the case of the division the domain includes all the points where the denominator does not vanish.

Arithmetic Properties *Let $f(x) : X \to \mathbb{R}$ and $g(x) : X \to \mathbb{R}$ be periodic functions with the period T. Then $f(x) + g(x) : X \to \mathbb{R}$, $f(x) - g(x) : X \to \mathbb{R}$, $f(x) \cdot g(x) : X \to \mathbb{R}$ and $\frac{f(x)}{g(x)} : \tilde{X} \to \mathbb{R}$ are periodic functions with the same period T, the last be defined on $\tilde{X}$ where $g(x) \neq 0$. (We assume here that $g(x)$ is not identically* 0.*)*

Proof The proofs are trivial and we perform them only for the sum and product, leaving the remaining two operations to the readers.

Denote $h(x) = f(x) + g(x)$ and notice that $h(x)$ is defined on X. Then $h(x + T) = f(x + T) + g(x + T) = f(x) + g(x) = h(x)$, $\forall x \in X$, that is, $h(x)$ is a periodic function with the period T.

Similarly, denote $h(x) = f(x) \cdot g(x)$ and notice that $h(x)$ is defined on X. Then $h(x + T) = f(x + T) \cdot g(x + T) = f(x) \cdot g(x) = h(x)$, $\forall x \in X$, that is, $h(x)$ is a periodic function with the period T. □

Remark 1 The same statement for the minimum period is not true, even if both functions has the same minimum period. An elementary example is $f(x) = \sin x$, $g(x) = -\sin x$, both functions defined on $X = \mathbb{R}$ and having the minimum period 2π, but their sum is the constant function $h(x) = f(x) + g(x) \equiv 0$ which does not have a minimum period (any real number is its period). A less trivial example is the functions $f(x) = \sin x$, $g(x) = \cos x$ with the minimum period 2π and the operation of multiplication: the function $h(x) = f(x) \cdot g(x) = \frac{1}{2} \sin 2x$ has the minimum period π.

In general, the following can be said about the minimum period: if $f(x)$ and $g(x)$ are periodic functions with the same domain and the minimum period T, then their sum, difference, product and quotient is a periodic function with the period T, which can be or can not be the minimum period of the result (including the case when there is no minimum period).

Remark 2 In a particular case when one of the functions, say $g(x)$, is a constant $g(x) = C$, the functions $f(x) + C$, $f(x) - C$, $Cf(x)$ and $\frac{f(x)}{C}$, $C \neq 0$ are periodic with the period T (according to the above statement), and moreover, they keep the same minimum period of the function $f(x)$.

Operations with the Argument of Periodic Functions

Remark To simplify considerations we assume below that all the involved functions are defined on $\mathbb{R}$.

Argument Properties *If $f(x) : \mathbb{R} \to \mathbb{R}$ is a periodic function with the period T, then $f(x + c)$, $c = constant$ is periodic with the period T and $f(cx)$, $c \neq 0$ is periodic with the period $\frac{T}{c}$. Moreover, if $T > 0$ is the minimum period of $f(x)$, then T is the minimum period of $f(x + c)$ and $\frac{T}{|c|}$ is the minimum period of $f(cx)$.*

Proof Denote $g(x) = f(x+c)$ and using the periodicity of $f(x)$ obtain $g(x+T) = f(x+c+T) = f(x+c) = g(x)$, that is, T is a period of $g(x)$. To show that $T > 0$ is the minimum period of $g(x)$, let us suppose, for contradiction, that there exists a smaller positive period $0 < \tilde{T} < T$. Then, $f(x+\tilde{T}) = g(x-c+\tilde{T}) = g(x-c) = f(x)$, which means that $\tilde{T}$ is also a period of $f(x)$ smaller than the minimum period T, that contradicts the definition of the minimum period.

Analogously, denote $g(x) = f(cx)$ and using the periodicity of $f(x)$ obtain $g(x + \frac{T}{c}) = f(cx + T) = f(cx) = g(x)$, that is, $\frac{T}{c}$ is a period of $g(x)$. To show that $\frac{T}{|c|} > 0$ is the minimum period of $g(x)$ suppose, for contradiction, that there exists a smaller positive period $0 < \frac{\tilde{T}}{|c|} < \frac{T}{|c|}$. Notice that in this case both $\frac{\tilde{T}}{|c|}$ and $\frac{\tilde{T}}{c}$ are periods of $g(x)$. Then, $f(x + \tilde{T}) = g(\frac{x+\tilde{T}}{c}) = g(\frac{x}{c} + \frac{\tilde{T}}{c}) = g(\frac{x}{c}) = f(x)$, which means that $\tilde{T}$ is a period of $f(x)$ smaller than the minimum period T, that contradicts the definition of the minimum period. □

Examples

1. The function $f(x) = \cos x + \sin x$ is periodic with the period $T = 2\pi$ as a linear combination of the two periodic functions with the period 2π. Although in general a linear combination does not keep the minimum period of the involved functions, but in this case $T = 2\pi$ is the minimum period of $f(x)$. Indeed, using trigonometric formulas, we can write the formula of the function in the form $f(x) = \cos x + \sin x = \sqrt{2}\left(\frac{\sqrt{2}}{2}\cos x + \frac{\sqrt{2}}{2}\sin x\right) = \sqrt{2}\left(\cos\frac{\pi}{4}\cos x + \sin\frac{\pi}{4}\sin x\right) = \sqrt{2}\cos\left(x - \frac{\pi}{4}\right)$. Since $\cos\left(x - \frac{\pi}{4}\right)$ has the minimum period 2π (according to the properties of the argument of periodic functions), the function $f(x)$ also has the minimum period 2π.

2. The function $f(x) = \sin x \cos x$ is periodic with the period $T = 2\pi$ as a product of the two functions with the period 2π. In this case, the minimum period 2π of the original functions is not preserved, and $f(x)$ has the minimum period π. This follows directly from the application of the trigonometric formula $f(x) = \sin x \cos x = \frac{1}{2} \sin 2x$ and the properties of the argument of periodic functions.
3. The function $f(x) = \frac{\sin x}{\cos x} = \tan x$ (called tangent) is periodic with the period $T = 2\pi$ on the domain $X = \mathbb{R}\backslash\{\frac{\pi}{2}+k\pi, k \in \mathbb{Z}\}$ (where the denominator does not vanish) as the quotient of the two functions with the period 2π. In this case, the minimum period 2π of both original functions is not preserved, and the minimum period of $f(x)$ is equal to π. Indeed, on the one side, $f(x+\pi) = \frac{\sin(x+\pi)}{\cos(x+\pi)} = \frac{-\sin x}{-\cos x} = f(x)$, that is, $\tilde{T} = \pi$ is a period of $f(x)$, and on the other side, the function takes the value 0 only at the points $k\pi, k \in \mathbb{Z}$ (where the numerator $\sin x$ vanishes), with the minimum distance between these points equal to π, and consequently, it cannot exist a period smaller than $\tilde{T} = \pi$.
4. The function $f(x) = \cos\left(\frac{x}{3} - 2\right)$ is periodic with the minimum period $T = \frac{2\pi}{1/3} = 6\pi$ according to the properties of the argument of periodic functions.

5 Monotonicity of a Function

5.1 Definitions and Examples

Increase and Decrease on a Set

Definition A function $f(x) : X \to \mathbb{R}$ is called *increasing* (*non-strictly increasing*) on a subset S of its domain X if for any $x_1, x_2 \in S$, $x_1 < x_2$ it follows that $f(x_1) < f(x_2)$ ($f(x_1) \leq f(x_2)$). Similarly, a function $f(x) : X \to \mathbb{R}$ is called *decreasing* (*non-strictly decreasing*) on a subset S of its domain X if for any $x_1, x_2 \in S, x_1 < x_2$ it follows that $f(x_1) > f(x_2)$ ($f(x_1) \geq f(x_2)$).

Remark The set S in this definition can be a part of the domain X or the entire domain. If S is not specified, it is usually considered the entire domain X.

The particular case when S is an interval is the most important for our study. For this reason, we formulate below once more the same definition for an interval.

Definition A function $f(x) : X \to \mathbb{R}$ is called *increasing* (*non-strictly increasing*) on an interval $I \subset X$ if for any $x_1, x_2 \in I$, $x_1 < x_2$ it follows that $f(x_1) < f(x_2)$ ($f(x_1) \leq f(x_2)$). Analogously, a function $f(x) : X \to \mathbb{R}$ is called *decreasing* (*non-strictly decreasing*) on an interval $I \subset X$ if for any $x_1, x_2 \in I$, $x_1 < x_2$ it follows that $f(x_1) > f(x_2)$ ($f(x_1) \geq f(x_2)$).

Remark An interval I in this definition can be of any type (open, half-open, or closed, finite or infinite).

Increasing and Decreasing at a Point

Definition A function $f(x)$ is called *increasing* (*non-strictly increasing*) at a point of its domain, if there exists a neighborhood of this point where the function is increasing (non-strictly increasing). Similarly, a function $f(x)$ is called *decreasing* (*non-strictly decreasing*) at a point of its domain if there exists a neighborhood of this point where the function is decreasing (non-strictly decreasing).

Remark Notice that according to the definition, a point of increasing/decreasing should belong to the domain together with a neighborhood.

Definition A function is called *monotonic* on a set, interval or at a point, if it is increasing or decreasing on this set, interval or at this point. This terminology is used both for *general monotonicity* and *strict monotonicity*.

Remark 1 For the sake of briefness we will refer to the points where the function is increasing/decreasing/monotonic as to increasing/decreasing/monotonic points.

Remark 2 In some sources increasing/decreasing function is called strictly increasing/decreasing, while non-strictly increasing/decreasing function is called increasing/decreasing. In this text we will follow the terminology given in the above definitions, unless the contrary is stated explicitly. If we will need to specify the type of considered monotonicity we will indicate whether it is strict or non-strict.

Examples

1. $f(x) = x$. This function is increasing on its domain $X = \mathbb{R}$: for any $x_1, x_2 \in \mathbb{R}$, $x_1 < x_2$ it follows that $f(x_1) < f(x_2)$. Notice that the function has the same property on any part of its domain. Notice also that each point of the domain is an increasing point.

   The graph of the function is the bisector of the odd quadrants (see Fig. 2.27).
2. $f(x) = |x|$. According to the definition of the absolute value, we consider separately the behavior of the function on the intervals $(-\infty, 0]$ and $[0, +\infty)$.

   Take two arbitrary points $x_1 < x_2$ in the interval $(-\infty, 0]$, that is, $x_1 < x_2 \le 0$ and compare the corresponding values of the function: $f(x_1) = -x_1 > -x_2 = f(x_2)$. Then, by the definition, the function $f(x) = |x|$ is decreasing on $(-\infty, 0]$.

   Using similar considerations on the interval $[0, +\infty)$, we take two arbitrary points such that $0 \le x_1 < x_2$ and compare the corresponding values of the function: $f(x_1) = x_1 < x_2 = f(x_2)$. This means that $f(x) = |x|$ is increasing on $[0, +\infty)$.

   From these results it follows that each point of the interval $(-\infty, 0)$ is decreasing (since it belongs to the decreasing interval $(-\infty, 0)$) and each point of the interval $(0, +\infty)$ is increasing (since it lies in the increasing interval $(0, +\infty)$). The point $x = 0$ is not increasing or decreasing (it does not belong to $(-\infty, 0)$ or to $(0, +\infty)$). In any neighborhood of this point for $x < 0$ we have $f(0) < f(x)$ (due to decreasing property), and for $x > 0$ we also get $f(0) < f(x)$ (due to increasing property).

   The graph of the function is shown in Fig. 2.28.

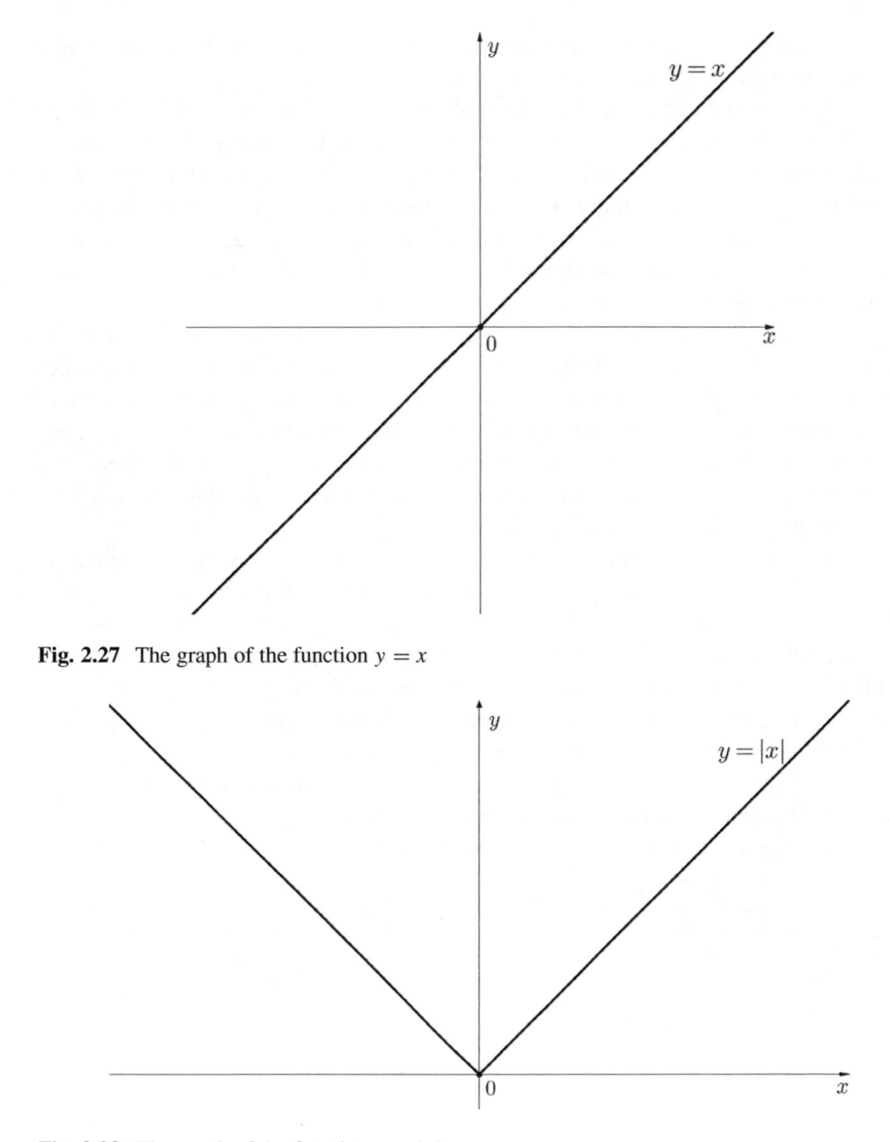

Fig. 2.27 The graph of the function $y = x$

Fig. 2.28 The graph of the function $y = |x|$

3. $f(x) = x^2$. Recalling the graph of this function, known from high school, we can note that the function is decreasing on the interval $(-\infty, 0]$ and increasing on the interval $[0, +\infty)$. Although in the course of Elementary Functions the graph is usually not used to derive analytic properties, its approximated form can be used to give a first impression about the properties of a function, which should be confirmed or rejected in the course of an analytic investigation. Under this approach, we can only suppose that $f(x) = x^2$ decreases on the interval

$(-\infty, 0]$ and increases on the interval $[0, +\infty)$, but this intuitive assumption should be verified using an analytic method.

Let us consider first the interval $(-\infty, 0]$. Take two arbitrary points $x_1 < x_2$ in this interval, that is, $x_1 < x_2 \le 0$ and evaluate the difference between the corresponding values of the function: $f(x_2) - f(x_1) = x_2^2 - x_1^2 = (x_2 - x_1)(x_2 + x_1)$. The first factor is positive, because $x_1 < x_2$, but the second is negative, because $x_1 < x_2 \le 0$. Therefore, the product is negative, that is, $f(x_2) - f(x_1) < 0$, and consequently, $f(x_2) < f(x_1)$. Hence, by the definition, the function $f(x) = x^2$ is decreasing on $(-\infty, 0]$.

Now apply the same reasoning on the interval $[0, +\infty)$. Take two arbitrary points $0 \le x_1 < x_2$ and evaluate the difference between the corresponding values of the function: $f(x_2) - f(x_1) = x_2^2 - x_1^2 = (x_2 - x_1)(x_2 + x_1)$. The first factor is positive, since $x_1 < x_2$, and the second is also positive, since $0 \le x_1 < x_2$. Therefore, $f(x_2) - f(x_1) > 0$, and consequently, $f(x_2) > f(x_1)$, that is, the function $f(x) = x^2$ is increasing on $[0, +\infty)$ according to the definition.

The graph of the function is shown in Fig. 2.29.

Notice that by the definition each point of the interval $(-\infty, 0)$ is decreasing and each point of the interval $(0, +\infty)$ is increasing. At the same time, the point 0, included in both intervals of monotonicity, is not monotonic, because there is no neighborhood of the origin where the function is increasing or decreasing: to the left of 0 the function decreases, while to the right it increases.

4. $f(x) = [x]$. For any pair $x_1 < x_2$ we have the two options: if $x_1, x_2 \in [n, n+1), n \in \mathbb{Z}$, then $f(x_1) = f(x_2)$; if $x_1 \in [n, n+1), n \in \mathbb{Z}$ and $x_2 \in [m, m+1), m \in \mathbb{Z}, m > n$, then $f(x_1) < f(x_2)$. Therefore, by the definition, $f(x)$ is non-strictly increasing on the whole domain $X = \mathbb{R}$.

   The graph of the function is shown in Fig. 2.30.

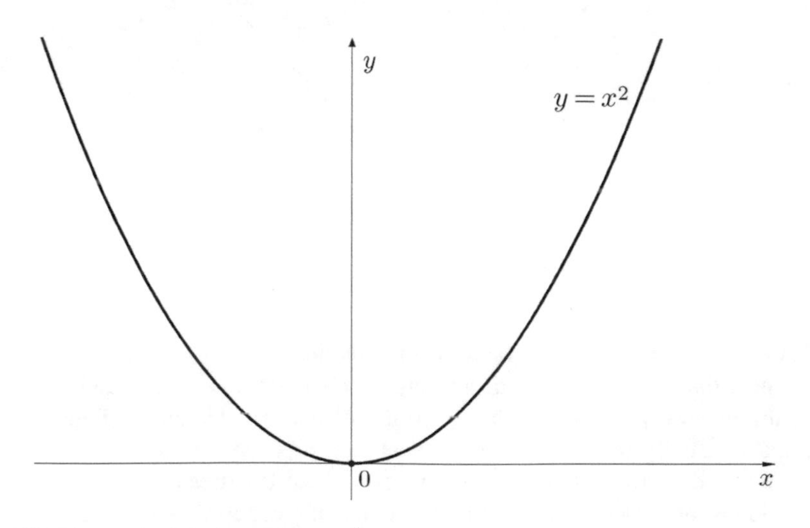

Fig. 2.29 The graph of the function $y = x^2$

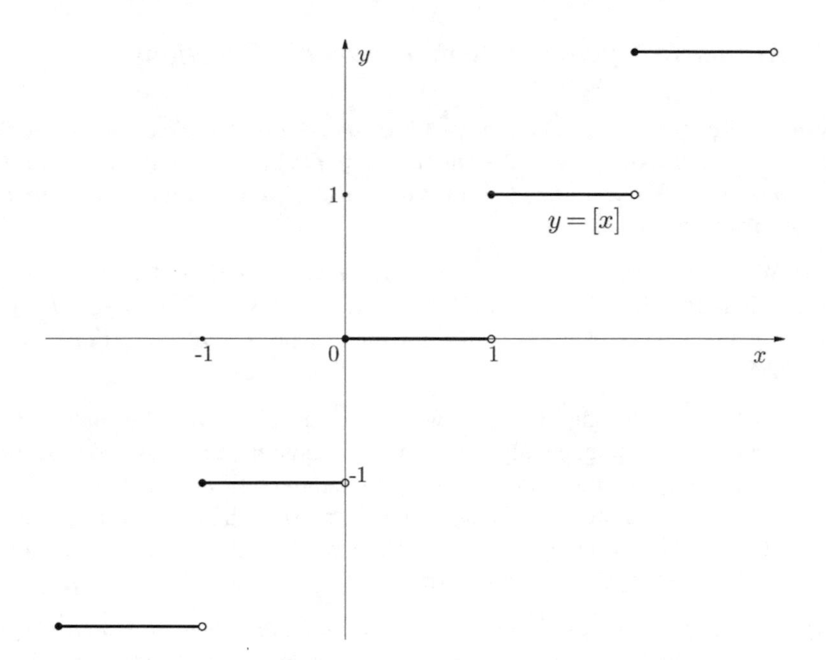

Fig. 2.30 The graph of the function $y = [x]$

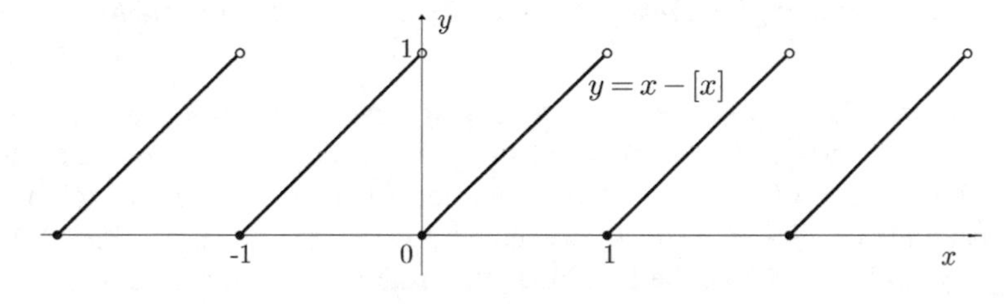

Fig. 2.31 The graph of the function $y = x - [x]$

5. $f(x) = x - [x]$. Choosing a specific interval $[n, n+1), n \in \mathbb{Z}$, we notice that for any x in this interval we have $f(x) = x - [x] = x - n$, where n is a constant. Therefore, for any pair x_1, x_2 such that $n \leq x_1 < x_2 < n+1$ we have $f(x_1) = x_1 - n < x_2 - n = f(x_2)$, that is, $f(x)$ is increasing on any interval $[n, n+1), n \in \mathbb{Z}$.

   If we consider any interval of the length greater than 1 (in particular, the whole domain $X = \mathbb{R}$), then the function is not monotonic. Indeed, any such interval has the integer point n where $f(n) = 0$ and also the points to the left and to the right of n where $f(x) > 0 = f(n)$. Therefore, no type of monotonicity is observed in any interval whose length is greater than 1. For the same reason the function is not monotonic in any interval that has an integer point inside, even when its length is smaller than 1.

   The graph of the function is shown in Fig. 2.31.

5.2 Arithmetic Operations with Monotonic Functions

Property 1 *If $f(x)$ and $g(x)$ are increasing functions on the same domain X, then $f(x) + g(x)$ is increasing on X. Analogously, if $f(x)$ and $g(x)$ are decreasing on the same domain X, then $f(x) + g(x)$ is decreasing on X. Similar results are true for non-strict monotonicity.*

Proof We provide the proof only for increasing functions (all other cases are handled similarly). Taking two arbitrary points $x_1, x_2 \in X, x_1 < x_2$, we get $h(x_1) = f(x_1) + g(x_1) < f(x_2) + g(x_2) = h(x_2)$, which means that $h(x) = f(x) + g(x)$ is increasing on X. □

Remark Obviously, the difference between two functions with the same type of monotonicity does not generally result in a monotonic function. For example, $f(x) = x$ and $g(x) = 2x$ are two increasing functions on $\mathbb{R}$, but $f(x) - g(x) = -x$ is decreasing on $\mathbb{R}$, while $g(x) - f(x) = x$ is increasing on $\mathbb{R}$. Furthermore, taking the functions $f(x) = x$ and $g(x) = x^2$, both increasing on $[0, +\infty)$, we obtain the function $f(x) - g(x) = x - x^2$ increasing on $[0, \frac{1}{2}]$ and decreasing on $[\frac{1}{2}, +\infty)$.

Property 2 *If $f(x)$ and $g(x)$ are increasing on the same domain X, and also are positive on X, then $f(x)g(x)$ is increasing (and positive) on X. Analogously, if $f(x)$ and $g(x)$ are increasing on the same domain X, and also are negative on X, then $f(x)g(x)$ is decreasing (and positive) on X. Similar results are true for non-strict monotonicity.*

Proof We provide the proof only for increasing positive functions (all other cases are handled similarly). The positivity of the product $f(x)g(x)$ is obvious, and we focus on the proof of increase of the function $h(x) = f(x)g(x)$. Taking two arbitrary points $x_1, x_2 \in X, x_1 < x_2$, we have $h(x_1) = f(x_1)g(x_1) < f(x_2)g(x_1) < f(x_2)g(x_2) = h(x_2)$, where the first inequality follows from increase of $f(x)$ (and positivity of $g(x)$) and the second from increase of $g(x)$ (and positivity of $f(x)$). Therefore, $h(x) = f(x)g(x)$ is increasing on X. □

Remark If the functions do not preserve the required sign the property is not valid. For example, $f(x) = x$ and $g(x) = x$ are two increasing functions on $\mathbb{R}$, but $f(x)g(x) = x^2$ is decreasing on $(-\infty, 0]$ and increasing on $[0, +\infty)$.

Property 3 *If $f(x)$ and $g(x)$ are decreasing on the same domain X, and also are positive on X, then $f(x)g(x)$ is decreasing (and positive) on X. Analogously, if $f(x)$ and $g(x)$ are decreasing on the same domain X, and also are negative on X, then $f(x)g(x)$ is increasing (and positive) on X. Similar results are valid for non-strict monotonicity.*

Proof The formulation and the proof of this property is completely analogous to those of Property 2. □

Remark A mixture of an increasing and decreasing functions in the product generally does not result in a monotonic function, even when both functions keep the same sign. For example, $f(x) = x$ is increasing and positive on $(0, 1)$, and $g(x) = 1 - x$ is decreasing and positive on $(0, 1)$, but their product $f(x)g(x) = x - x^2$ is increasing on $(0, \frac{1}{2}]$ and decreasing on $[\frac{1}{2}, 1)$.

6 Extrema of a Function

6.1 Global Extrema

Definition A point x_0 is called *global maximum* (*strict global maximum*) of a function $y = f(x)$ on a set S of its domain if for each x of S different from x_0 the following inequality holds: $f(x) \le f(x_0)$ $(f(x) < f(x_0))$.

In the same manner, a point x_0 is called *global minimum* (*strict global minimum*) of a function $y = f(x)$ on a set S of its domain if for each x of S different from x_0 the following inequality holds: $f(x) \ge f(x_0)$ $(f(x) > f(x_0))$.

A (strict) global maximum or minimum is called *(strict) global extremum.*

Remark A global extremum can also be called absolute extremum.

Relationship Between Extrema and Monotonicity

Sometimes it is convenient to take advantage of the properties of monotonicity (if it was already investigated) to find extrema of a function. More specifically, it is quite evident that if a point x_0 is monotonic, then it cannot be an extremum. Indeed, suppose that x_0 is an increasing point of $f(x)$. Then there exists its neighborhood $(x_0 - r, x_0 + r)$ where the function $f(x)$ increases. Therefore, for the midpoint $x_1 = x_0 + \frac{r}{2}$ between x_0 and $x_0 + r$, we have $f(x_1) > f(x_0)$ which shows that x_0 is not a maximum. On the other hand, for the midpoint $x_2 = x_0 - \frac{r}{2}$ between x_0 and $x_0 - r$ we get $f(x_2) < f(x_0)$ which shows that x_0 is not a minimum. Hence, x_0 is not an extremum. The same reasoning can be applied to a decreasing point.

Notice that a global extremum can belong to an interval of monotonicity, like in the case of the function $f(x) = x$ on the set $S = [0, 1]$. In this example, the entire interval $S = [0, 1]$ is the interval where $f(x) = x$ is increasing (the proof is trivial), but the endpoints of this interval are global extrema: $f(x) = x > 0 = f(0)$, $\forall x \in (0, 1]$ and $f(x) = x < 1 = f(1)$, $\forall x \in [0, 1)$, that is, $x_1 = 0$ is a global minimum and $x_2 = 1$ is a global maximum on the set $S = [0, 1]$ (see Fig. 2.32). This does not contradict the considerations made in the previous paragraph, because the points 0 and 1 are not increasing: they belong to the interval of increase $S = [0, 1]$, but do not belong to this interval together with some neighborhood.

When a monotonicity is non-strict, the corresponding points still can be global extrema (albeit non-strict). Consider as example the function $f(x) =$

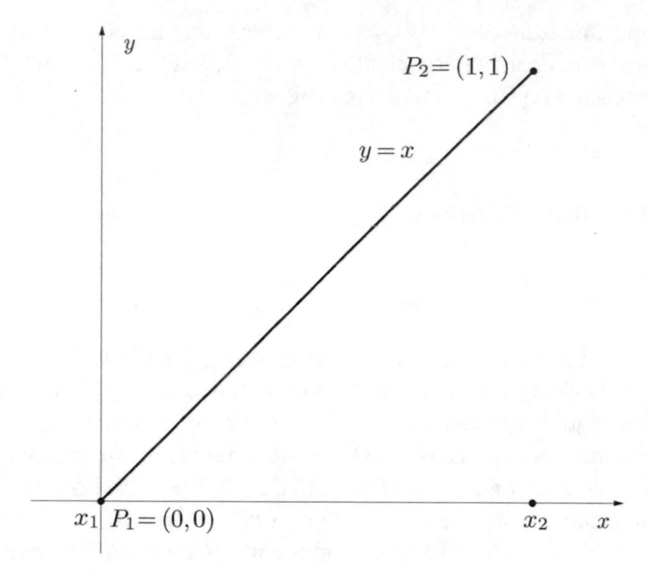

Fig. 2.32 Function $y = x, x \in [0, 1]$ and its global extrema

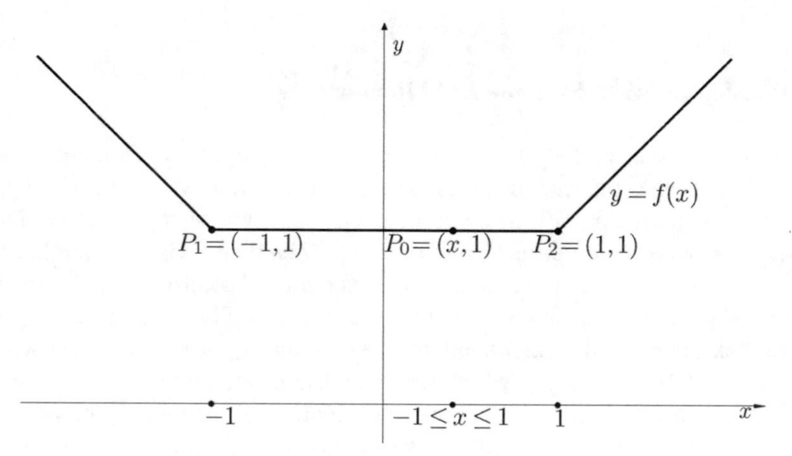

Fig. 2.33 The function $y = f(x)$ and its non-strict global minima

$\begin{cases} -x, \ x \leq -1 \\ 1, \ -1 < x < 1 \\ x, \ x \geq 1 \end{cases}$. A construction of its graph is elementary (it consists of three line parts) and it makes possible to better visualize the behavior of this function (see Fig. 2.33). Clearly, the entire interval $(-\infty, 1]$ is the interval of non-strict decrease of $f(x)$, and at the same time, every point of the interval $[-1, 1]$ is the point of non-strict global minimum on the domain $\mathbb{R}$ of this function.

Fig. 2.34 The function $y = h(x)$ that has no global extrema

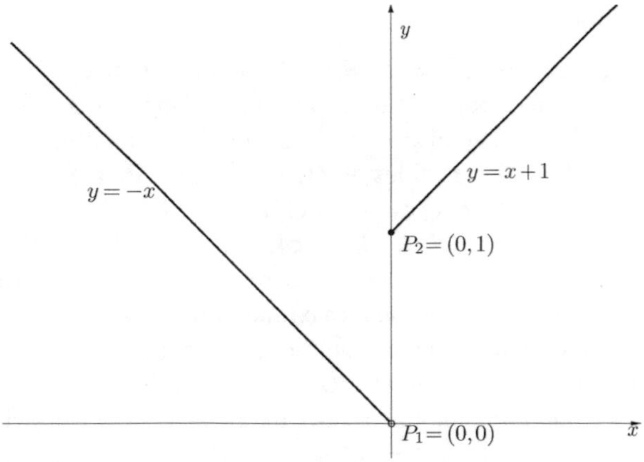

One more relationship between monotonicity and extrema is not so evident as it may seem at first glance. If we consider a function that is decreasing on the interval $(-\infty, x_0)$ and increasing on the interval $(x_0, +\infty)$, then it may appear that x_0 should be a global minimum. This indeed can happen for some functions, like, for instance, in the case of $f(x) = |x|$ with the global minimum $x_0 = 0$, but generally it is not true. First, if a function is not defined at x_0, then this point cannot be its minimum, like in the case of the function $g(x) = \frac{x^2}{|x|}$ equal to $f(x) = |x|$ at all the points, except for the origin, where $g(x)$ is not defined. Moreover, even if a function is defined at x_0 there is no guarantee that this is a minimum. Indeed, consider the function $h(x) = \begin{cases} -x, \ x < 0 \\ x+1, \ x \geq 0 \end{cases}$. Its graph has a simple form (it consists of the two line parts) and is useful to reveal what happens near the point $x_0 = 0$ (see Fig. 2.34). Like $f(x) = |x|$, the function $h(x)$ decreases on $(-\infty, 0)$ and increases on $(0, +\infty)$, but $x_0 = 0$ is not its minimum (even in the non-strict sense). It can be seen just comparing the values at the points 0 and $-\frac{1}{2}$: $h(0) = 1 > \frac{1}{2} = h(-\frac{1}{2})$. In general, $h(x)$ has no global minimum, although its behavior of monotonicity is very similar to that of $f(x) = |x|$. In fact, any point $x_0 \geq 0$ is not a minimum due to the comparison $h(x_0) \geq 1 > \frac{1}{2} = h(-\frac{1}{2})$. At the same time, every point $x_0 < 0$ is not a minimum, since choosing $x_1 = \frac{x_0}{2}$ we get $h(x_0) > h(x_1)$.

Naturally, analogous considerations are applied to the relationship between monotonicity and global maxima.

We recommend to the reader to analyze monotonicity and extrema of the following functions: $f(x) = \begin{cases} |x|, \ x \neq 0 \\ 1, \ x = 0 \end{cases}$ and $g(x) = \begin{cases} |x|, \ x \neq 0 \\ -1, \ x = 0 \end{cases}$.

Remark There is certain ambiguity in the terminology of extrema. An extremum point x_0, the value of a function at this point $f(x_0)$ and the point of the graph of the function $P_0 = (x_0, f(x_0))$ can be called extremum. Usually it does not cause any confusion since a specific meaning of this term becomes clear in the considered context.

Examples

1a. $f(x) = x$, $S = X = \mathbb{R}$. Since $f(x)$ increases over the entire domain $\mathbb{R}$, there is no extremum: every point x_0 of the domain is increasing. For this function we can use also a more "general" justification: for every point x_0, to the right of x_0 all the values of $f(x)$ are greater than $f(x_0)$, while to the left of x_0 all the values of $f(x)$ are smaller than $f(x_0)$.

1b. $f(x) = x$, $S = [0, +\infty)$. On the set $S = [0, +\infty)$, the function $f(x)$ has the strict global minimum $x_0 = 0$, because $f(x) = x > 0 = f(0)$, $\forall x > 0$. However, the global maximum does not exist, since for every point $x_0 \in S$, to the right of x_0 the function takes greater values than $f(x_0)$.

1c. $f(x) = x$, $S = [0, 10]$. On the set $S = [0, 10]$, the function has both global minimum and maximum: the point $x_1 = 0$ is a strict global minimum, because $f(x) = x > 0 = f(0)$, $\forall x \in (0, 10]$, and the point $x_2 = 10$ is a strict global maximum, because $f(x) = x < 10 = f(10)$, $\forall x \in [0, 10)$.

1d. $f(x) = x$, $S = (0, 10)$. On the set $S = (0, 10)$, the function has no extremum. Indeed, choosing any $x_0 \in (0, 10)$, we can always find a point, for instance, the midpoint $x_1 = \frac{x_0}{2} \in (0, 10)$ between 0 and x_0 where $f(x_1) < f(x_0)$, that is, x_0 cannot be a global minimum. For the same reason, choosing any $x_0 \in (0, 10)$, we can always find a point, for instance, the midpoint $x_2 = \frac{x_0+10}{2} \in (0, 10)$ between x_0 and 10, where $f(x_2) > f(x_0)$, that is, x_0 cannot be a global maximum. Despite that, the function is bounded on S, since $0 < f(x) = x < 10$, $\forall x \in (0, 10)$.

2a. $f(x) = |x|$, $S = X = \mathbb{R}$. On the set $S = \mathbb{R}$ the function $f(x) = |x|$ has only a strict global minimum at the point $x_0 = 0$. Indeed, by the definition of the absolute value, $f(x) = |x| > 0 = f(0)$, $\forall x \neq 0$.
To show that there is no global maximum, we notice that $x_0 = 0$ cannot be a maximum (since it is a strict minimum), and any other point x_0 is monotonic (decreasing for $x_0 < 0$ and increasing for $x_0 > 0$), and consequently, it cannot be a maximum.
Another way to show the absence of a maximum is by the definition. Taking any point $x_0 \neq 0$, we can choose $x_1 = 2x_0$ and get $f(x_1) = 2|x_0| > |x_0| = f(x_0)$ which contradicts the definition of the maximum at the point x_0. For $x_0 = 0$ we take any $x_1 \neq 0$, for instance, $x_1 = 1$, and have $f(x_1) = |x_1| = 1 > 0 = f(0)$.

2b. $f(x) = |x|$, $S = (-3, 2]$. On the interval $(-3, 2]$, the function $f(x) = |x|$ has a strict global minimum at the point $x_0 = 0$, but it does not have a global maximum. Indeed, for any $x \in (-3, 2], x \neq 0$ we have $f(x) = |x| > 0 = f(0)$, and consequently, $x_0 = 0$ is a strict global minimum. To show that there is no maximum, notice first that any $x_0 \in [-2, 2]$ cannot be a maximum, because $f(x_0) \leq f(2) = 2 < f(-2.5) = 2.5$, where the point $x_1 = -2.5$ belongs to the interval $(-3, 2]$. Take now any $x_0 \in (-3, -2)$ and choose the midpoint $x_1 = \frac{-3+x_0}{2} \in (-3, -2)$ between -3 and x_0. Then $f(x_0) = -x_0 < \frac{3-x_0}{2} = f(x_1)$, which means that x_0 is not a maximum.

2c. $f(x) = |x|$, $S = [-3, 2)$. On the interval $[-3, 2)$, the function $f(x) = |x|$ has a strict global minimum at the point $x_0 = 0$ and a strict global maximum at the point $x_1 = -3$. Indeed, for any $x \in [-3, 2), x \neq 0$ we have $f(x) = |x| > 0 = f(0)$, and therefore, $x_0 = 0$ is a strict global minimum. On the other hand, for any $x \in (-3, 2)$ we have $f(x) = |x| < 3 = f(-3)$, which shows that $x_1 = -3$ is a strict global maximum.

2d. $f(x) = |x|$, $S = (1, 2)$. On the interval $(1, 2)$, the function $f(x) = |x|$ has no global extremum. Indeed, for any $x_0 \in (1, 2)$ we can choose the midpoint $x_1 = \frac{1+x_0}{2} \in (1, 2)$ between 1 and x_0, and get $f(x_0) = x_0 > \frac{1+x_0}{2} = f(x_1)$. This shows that x_0 is not a minimum. Similarly, for any $x_0 \in (1, 2)$ we can choose the midpoint $x_1 = \frac{x_0+2}{2} \in (1, 2)$ between x_0 and 2, and obtain $f(x_0) = x_0 < \frac{x_0+2}{2} = f(x_1)$. This means that x_0 is not a maximum.

3a. $f(x) = -x^2$, $S = X = \mathbb{R}$. Considering all the real numbers, the function $f(x) = -x^2$ has only a strict global maximum at the point $x_0 = 0$. Indeed, according to the definition of the maximum, $f(x) = -x^2 < 0 = f(0), \forall x \neq 0$. To show that there is no global minimum, we notice that $x_0 = 0$ cannot be a minimum (since it is a strict maximum) and any other point x_0 is monotonic (increasing for $x_0 < 0$ and decreasing for $x_0 > 0$), and consequently, cannot also be a minimum.

Another way to show the absence of a global minimum is by the definition. Taking any point $x_0 \neq 0$, we can choose $x_1 = 2x_0$ and obtain $f(x_1) = -4x_0^2 < -x_0^2 = f(x_0)$, that contradicts the definition of minimum at the point x_0. For $x_0 = 0$ we take any $x_1 \neq 0$, for instance, $x_1 = 1$, and have $f(x_1) = -x_1^2 = -1 < 0 = f(0)$.

3b. $f(x) = -x^2$, $S = [-1, 2]$. On the interval $[-1, 2]$, the function $f(x) = -x^2$ has a strict global maximum at the point $x_0 = 0$ and a strict global minimum at the point $x_1 = 2$. Indeed, for any $x \neq 0$ we have $f(x) = -x^2 < 0 = f(0)$, and for any $-1 \leq x < 2$ we have $f(x) = -x^2 > -4 = f(2)$.

3c. $f(x) = -x^2$, $S = [-1, 1]$. On the interval $[-1, 1]$, the function $f(x) = -x^2$ has a strict global maximum at the point $x_0 = 0$ and a non-strict global minimum at the two points $x_1 = -1$ and $x_2 = 1$. Indeed, for any $x \neq 0$ we have $f(x) = -x^2 < 0 = f(0)$, and for any $-1 < x < 1$ we get $f(x) = -x^2 > -1 = f(-1) = f(1)$.

3d. $f(x) = -x^2$, $S = (0, 1]$. On the interval $(0, 1]$, the function $f(x) = -x^2$ has a strict global minimum at the point $x_1 = 1$ and does not have a global maximum. Indeed, for any $0 < x < 1$ we have $f(x) = -x^2 > -1 = f(1)$. On the other hand, for any $0 < x_0 \leq 1$ we can choose $x_2 = \frac{x_0}{2} \in (0, 1]$ such that $f(x_0) = -x_0^2 < -\frac{x_0^2}{4} = f(\frac{x_0}{2})$, which means that x_0 is not a global maximum.

6.2 *Local Extrema*

Definition A point x_0 is called *local maximum* (*strict local maximum*) of a function $f(x)$, if there exists a neighborhood of this point, contained in the domain of $f(x)$, in which x_0 is a global maximum (strict global maximum). Recalling the definition of a global maximum, we can specify this definition as follows: a point x_0 is called *local maximum* (*strict local maximum*) of a function $f(x)$, if there exists a neighborhood of this point, contained in the domain of $f(x)$, such that for any x of this neighborhood different from x_0, the inequality $f(x) \leq f(x_0)$ ($f(x) < f(x_0)$) holds.

In the same manner, a point x_0 is called *local minimum* (*strict local minimum*) of a function $f(x)$, if there exists a neighborhood of this point, contained in the domain of $f(x)$, in which x_0 is a global minimum (strict global minimum). Using the definition of a global minimum, we can formulate this definition in a more detailed form: a point x_0 is called *local minimum* (*strict local minimum*) of a function $f(x)$, if there exists a neighborhood of this point, contained in the domain of $f(x)$, such that for any x of this neighborhood different from x_0, the inequality $f(x) \geq f(x_0)$ ($f(x) > f(x_0)$) holds.

A (strict) local maximum or minimum is called *(strict) local extremum.*

Remark 1 Notice that, according to the definition, a local minimum/maximum should belong to the domain together with its neighborhood.

Remark 2 A local extremum is sometimes called relative extremum.

Remark 3 The same type of ambiguity of the terminology, that has been noted for global extrema, also takes place for local extrema: the x-coordinate of extremum x_0, the value of a function at this point $f(x_0)$ and the point of the graph $P_0 = (x_0, f(x_0))$ can be called local extremum.

Relationship Between Local Extrema and Monotonicity

The same type of relationship between extrema and monotonicity, that was already discussed in the case of global extrema, is valid for local extrema, keeping the same arguments and examples. In a specified form:

1. If a point x_0 is monotonic, then it cannot be a local extremum.
2. If a monotonicity is non-strict, then the corresponding points still can be (non-strict) local extrema.
3. A function decreasing on (x_0-r, x_0) and increasing on (x_0, x_0+r) (or vice-verse) does not necessarily have a local minimum (local maximum) at x_0.

The unique exception is a possibility of occurrence of a global extremum at the endpoint of an interval, which is not possible for a local extremum.

Relationship Between Local and Global Extrema

Naturally, a local extremum can be or not a global extremum. For example, the function $f(x) = x^2$ (with the domain $\mathbb{R}$) has a local minimum $x_0 = 0$ which is also a global minimum. On the other hand, recalling the graph of the function

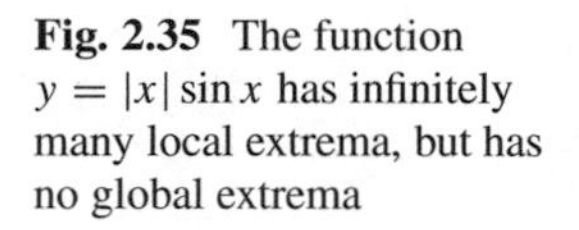

Fig. 2.35 The function $y = |x| \sin x$ has infinitely many local extrema, but has no global extrema

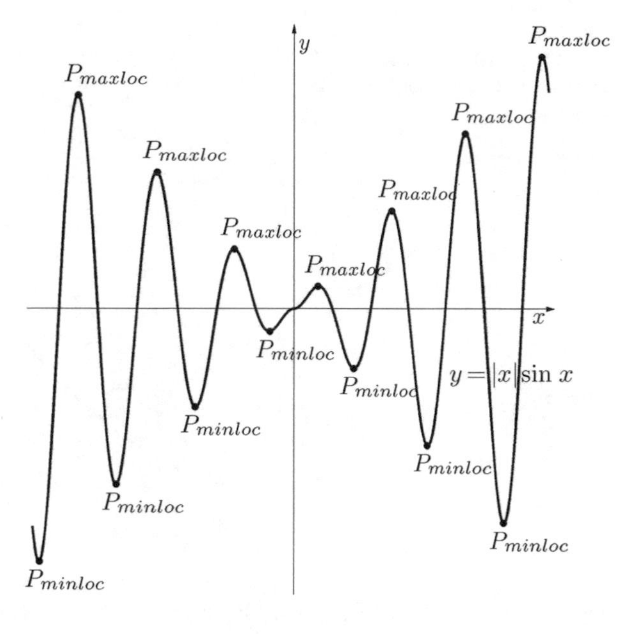

$f(x) = \sin x$ (with the domain $\mathbb{R}$), we notice that every point $x_k = \frac{\pi}{2} + 2k\pi$, $\forall k \in \mathbb{Z}$ is a strict local maximum and every point $x_n = -\frac{\pi}{2} + 2n\pi$, $\forall n \in \mathbb{Z}$ is a strict local minimum. At the same time, all these points are non-strict global extrema (the function $f(x) = \sin x$ has no strict global extremum on $\mathbb{R}$). The graphs of these two function are shown in Figs. 2.25 and 2.26.

We can also provide a more radical example of the function whose graph is shown in Fig. 2.35 (the analytic form of this function is $f(x) = |x| \sin x$). Strict local extrema of the function $f(x) = |x| \sin x$ occur near the extrema of $\sin x$, but no one of them is a global extremum of $f(x)$, since the values of the function at local minima decrease even more and at local maxima increase even more at the points located farther from the origin.

It happens that a global extremum also is not necessarily a local extremum. For example, the function $f(x) = x^2$ has a global minimum $x_0 = 0$ on $\mathbb{R}$ and this point is also a local minimum. At the same time, the function $f(x) = x^2$ on the domain $[0, 1]$ has both a global minimum $x_0 = 0$ and a global maximum $x_1 = 1$, but it does not have a local extremum on this interval: by the definition, a local extremum cannot be find at the endpoints of an interval and the remaining points of $[0, 1]$ are increasing (see Fig. 2.36). In general, if a global extremum occurs at a point belonging to the domain together with a neighborhood, then by the definition this point is also local extremum. However, if global extremum coincides with a endpoint of an interval of consideration, then it cannot be a local extremum.

Examples

1a. $f(x) = x$, $S = X = \mathbb{R}$. Since $f(x)$ increases over the entire domain $\mathbb{R}$, each point $x_0 \in \mathbb{R}$ is increasing, which implies that there is no local extremum.

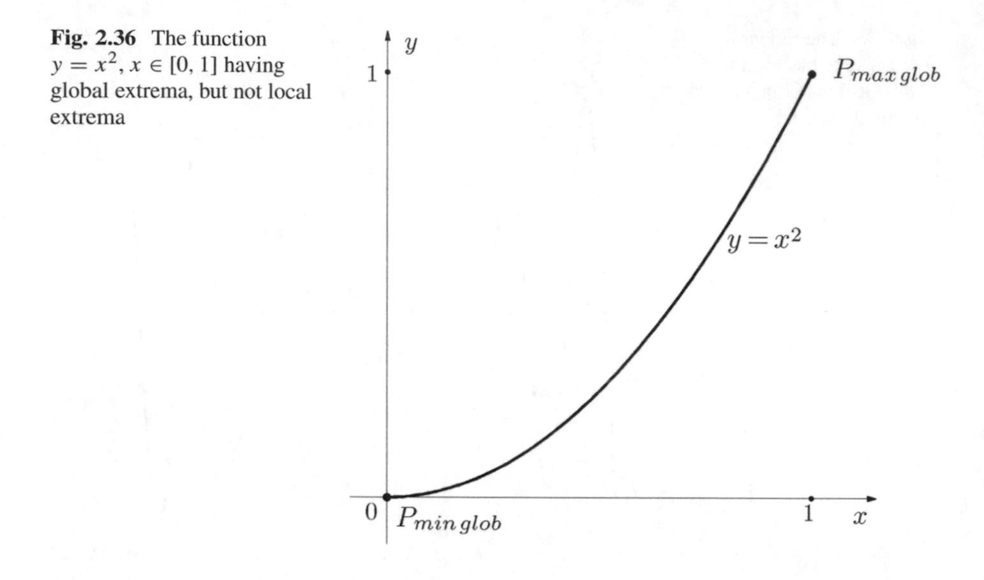

Fig. 2.36 The function $y = x^2$, $x \in [0, 1]$ having global extrema, but not local extrema

1b. $f(x) = x$, $S = [0, +\infty)$. On the set $S = [0, +\infty)$, the function $f(x)$ has no local extremum, since each point of the open interval $(0, +\infty)$ is increasing and the endpoint 0 cannot be a local extremum (it is not contained in S with a neighborhood). At the same time, the endpoint $x = 0$ is a global minimum, because $f(0) = 0 < x = f(x)$, $\forall x > 0$.

1c. $f(x) = x$, $S = (a, b)$. Each point of the open interval (a, b) is increasing, and consequently, there is no local extremum (nor global).

1d. $f(x) = x$, $S = [a, b]$. On the closed interval $S = [a, b]$ the function has no local extremum, since each point of the open interval (a, b) is increasing and no endpoint can be a local extremum. At the same time, the left endpoint $x = a$ is a global minimum and the right endpoint $x = b$ is a global maximum: $f(a) = a < x = f(x) < b = f(b)$, $\forall x \in (a, b)$.

2a. $f(x) = |x|$, $S = X = \mathbb{R}$. It was already shown that the function $f(x) = |x|$ has a strict global minimum at the point $x_0 = 0$ and has no global maximum. Since $x_0 = 0$ belongs to $\mathbb{R}$ together with a neighborhood, this point is also a local minimum.

Let us show that there is no other local extremum. Indeed, each point $x_0 \in (-\infty, 0)$ is decreasing, and each point $x_0 \in (0, +\infty)$ is increasing, which implies that none of these points can be a local extremum.

2b. $f(x) = |x|$, $S = (a, b)$, $0 \leq a < b$. On the interval (a, b), $a \geq 0$ the function $f(x) = |x| = x$ increases, and consequently, it has no local extremum. It has no global extremum either, because each point of (a, b) is increasing and interior.

2c. $f(x) = |x|$, $S = [a, b]$, $0 \leq a < b$. On the interval $[a, b]$, $a \geq 0$ the function $f(x) = |x| = x$ increases, which means that each interior point $x_0 \in (a, b)$ is increasing, and consequently, it cannot be a local extremum or a global extremum. At the remaining two points (the endpoints) a local extremum cannot occur, since there is no a neighborhood of any of these points contained in $[a, b]$. At the same time, the function has a global minimum at the endpoint $x = a$ and global maximum at the endpoint $x = b$: $f(a) = a < x = f(x) < b = f(b)$, $\forall x \in (a, b)$, $a \geq 0$.

2d. $f(x) = |x|$, $S = (a, b)$, $a < 0 < b$. It was already shown that $x_0 = 0$ is a global minimum in this case. Since this point lies inside the interval (a, b), it is also a local minimum. No other point can be a local extremum, since the function decreases on the interval $(a, 0)$ and increases on the interval $(0, b)$. The same is true for any interval $S = [a, b]$, $a < 0 < b$.
Notice that the situation with global extrema is different on the intervals (a, b) and $[a, b]$, $a < 0 < b$. On the former, the function has the only global extremum which is the global minimum $x_0 = 0$, while on the latter the function has also a global maximum which occurs at one of the endpoints of the interval: if $|a| > b$, the global maximum occurs at $x = a$, if $|a| < b$ the global maximum occurs at $x = b$, and if $|a| = b$ the global maximum is attained at both endpoints.

3a. $f(x) = -x^2$, $S = X = \mathbb{R}$. It was shown before that the function $f(x) = -x^2$ has a strict global maximum at the point $x_0 = 0$. Since this point belongs to $\mathbb{R}$ together with a neighborhood, it is also a strict local maximum.
Let us see that there is no other local extremum. In fact, each point $x_0 \in (-\infty, 0)$ is increasing, and each point $x_0 \in (0, +\infty)$ is decreasing, which means that none of these points can be a local extremum.

3b. $f(x) = -x^2$, $S = (a, b)$, $0 \leq a < b$. On the interval (a, b), $a \geq 0$ the function $f(x) = -x^2$ decreases, and consequently, it has no local extremum. The function also has no global extremum, since each point in (a, b) is an interior point of this interval.

3c. $f(x) = -x^2$, $S = [a, b]$, $0 \leq a < b$. On the interval $[a, b]$, $a \geq 0$ the function $f(x) = -x^2$ decreases, which means that each interior point $x_0 \in (a, b)$ is increasing, and consequently, it cannot be a local extremum. At the two remaining points (endpoints) a local extremum cannot occur, because these points are not interior (none of them has a neighborhood contained in $[a, b]$). At the same time, the function attains a global maximum at the point $x = a$ and a global minimum at $x = b$: $f(a) = -a^2 > -x^2 = f(x) > -b^2 = f(b)$, $\forall x \in (a, b)$, $a \geq 0$.

3d. $f(x) = -x^2$, $S = (a, b)$, $a < 0 < b$. Since the global maximum $x_0 = 0$ lies inside the interval (a, b), it is also a local maximum. No other point can be a local extremum, because on the interval $(a, 0)$ the function increases and on the interval $(0, b)$ it decreases. The same can be said about the interval $S = [a, b]$, $a < 0 < b$.
Notice that the situation with global extrema is different on the intervals (a, b) and $[a, b]$, $a < 0 < b$. On the former the function has the only global extremum, which is the global maximum $x_0 = 0$, while on the latter the function has also a global minimum attained at one of the endpoints of the interval: if $|a| > b$, the global minimum is attained at $x = a$, if $|a| < b$ the global minimum is at $x = b$, and if $|a| = b$ the global minimum is attained at both endpoints.

6.3 *Monotonicity, Extrema and Symmetry*

Property 1 *If $f(x)$ is an even function, increasing on an interval $[a, b]$, $0 \le a < b$, then it is decreasing on the interval $[-b, -a]$.*

Proof Clearly the interval $[-b, -a]$ is symmetric to $[a, b]$ about the origin, and consequently, $f(x)$ is defined on $[-b, -a]$ since it is even and defined on $[a, b]$. Take two arbitrary points $x_1 < x_2$ of $[-b, -a]$ and notice that the corresponding symmetric points $-x_1 > -x_2$ lie in $[a, b]$. Then, from the evenness and increase on $[a, b]$ it follows that $f(x_1) = f(-x_1) > f(-x_2) = f(x_2)$, which shows that $f(x)$ decreases on $[-b, -a]$. □

Remark 1 The same statement (and proof) is valid for any kind of an interval (open, half-open or closed, bounded or unbounded). It can be formulated in the following general form: if $f(x)$ is even and increasing on an interval $I \subset [0, +\infty)$, then it is decreasing on the interval $\tilde{I}$ symmetric to I about the origin.

In particular, if the original interval is $[0, b)$ (where b can be $+\infty$), then $f(x)$ decreases on $(-b, 0]$, and for this reason 0 is a strict minimum of $f(x)$ on $(-b, b)$ (if $b = +\infty$, then 0 is a strict local and global minimum on $\mathbb{R}$).

Remark 2 In the case of decrease we have a similar statement: if $f(x)$ is an even function, decreasing on an interval $I \subset [0, +\infty)$, then it is increasing on the interval $\tilde{I}$ symmetric to I about the origin. In particular, if the original interval is $[0, b)$ (where b can be $+\infty$), then $f(x)$ increases on $(-b, 0]$, and consequently, 0 is a strict maximum of $f(x)$ on $(-b, b)$ (if $b = +\infty$, then 0 is a strict local and global maximum on $\mathbb{R}$).

Property 2 *If $f(x)$ is an odd function, increasing on an interval $I \subset [0, +\infty)$, then it is increasing on the interval $\tilde{I}$ symmetric to I about the origin.*

Proof The demonstration of this statement is completely analogous to that of Property 1 and it is left to the reader. □

Remark 1 If the original interval is $[0, b)$ (where b can be $+\infty$), then $f(x)$ increases on the entire interval $(-b, b)$ (if $b = +\infty$, then $f(x)$ increases on $\mathbb{R}$).

Remark 2 In the case of decrease we have a similar statement: if $f(x)$ is an odd function, decreasing on an interval $I \subset [0, +\infty)$, then it is decreasing on the interval $\tilde{I}$ symmetric to I about the origin. In particular, if the original interval is $[0, b)$ (where b can be $+\infty$), then $f(x)$ decreases on the entire interval $(-b, b)$ (if $b = +\infty$, then $f(x)$ decreases on $\mathbb{R}$).

Remark 1g Naturally, the role of the intervals on the positive and negative parts of the real axis can be interchanged in these properties.

Remark 2g Similar statements are true for non-strict monotonicity.

Property 3 *If a function $f(x)$ is periodic with a period T and increasing on an interval $[a, b]$, then it is increasing on the interval $[a + T, b + T]$.*

Proof Clearly the interval $[a+T, b+T]$ is a part of the domain of $f(x)$, since $f(x)$ is periodic and $[a, b]$ is contained in the domain. Then, taking two arbitrary points $x_1 < x_2$ in $[a+T, b+T]$, we notice that $x_1 - T$ and $x_2 - T$ belong to $[a, b]$ and obtain $f(x_1) = f(x_1 - T) < f(x_2 - T) = f(x_2)$, which shows that $f(x)$ increases on $[a+T, b+T]$. □

Remark 1 Analogous statement (and proof) is valid for any type of the interval (open, half-open or closed).

Remark 2 A similar statement is true in the case of decreasing function.

Remark 3 Similar statements are true for non-strict monotonicity.

7 Concavity and Inflection

7.1 General Concavity

In this section our goal is to define one more (and more subtle) characteristic of functions which makes available more precise specification of the behavior of a function and the form of its graph. Let us establish both geometric and analytic conditions which determine this characteristic. The former is natural and represents a distinction between graphs of functions, while the latter is relevant for investigation of this characteristic in different functions.

Let us start with a geometric illustration. Consider the graphs of the two functions $f(x) = x^2$ and $g(x) = \sqrt{|x|}$, which we consider at the moment as known or constructed using computer graphic tools, but shortly thereafter we will derive all the properties of these functions analytically, which make possible to draw their graphs without any additional information. These two graphs are shown in Figs. 2.37 and 2.38.

We can notice that both graphs have different common properties. First, each of the graphs is symmetric about the y-axis, which reflects the evenness of the functions: $f(-x) = (-x)^2 = x^2 = f(x)$ and $g(-x) = \sqrt{|-x|} = \sqrt{|x|} = g(x)$, $\forall x \in \mathbb{R}$.

Second, both graphs show decrease of the functions on the interval $(-\infty, 0]$ and increase on $[0, +\infty)$. This is a consequence of the analytic properties of these functions. Indeed, taking $x_1 < x_2 \leq 0$, for $f(x) = x^2$ we have $f(x_2) - f(x_1) = x_2^2 - x_1^2 = (x_2 - x_1)(x_2 + x_1) < 0$, because the first factor is positive and the second is negative. Similarly, for the second function we have

$$g(x_2) - g(x_1) = \sqrt{|x_2|} - \sqrt{|x_1|} = \frac{(\sqrt{|x_2|} - \sqrt{|x_1|})(\sqrt{|x_2|} + \sqrt{|x_1|})}{\sqrt{|x_2|} + \sqrt{|x_1|}}$$

$$= \frac{|x_2| - |x_1|}{\sqrt{|x_2|} + \sqrt{|x_1|}} = \frac{x_1 - x_2}{\sqrt{|x_2|} + \sqrt{|x_1|}} < 0,$$

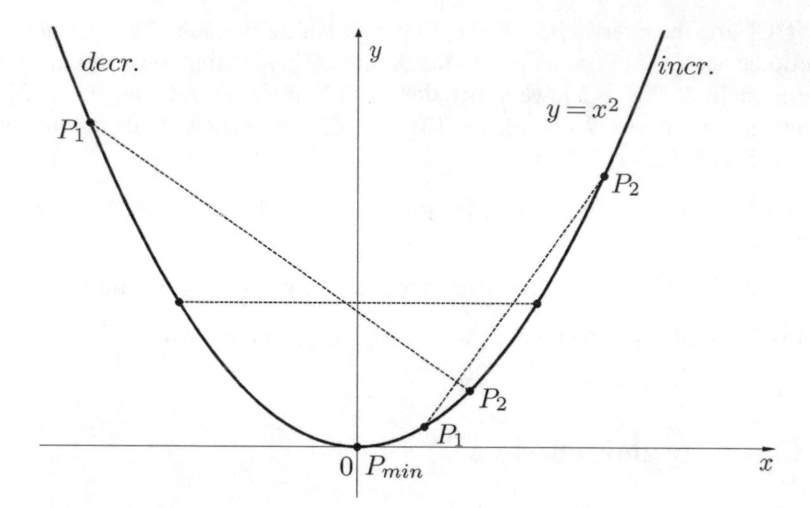

Fig. 2.37 The graph and concavity of the function $y = x^2$

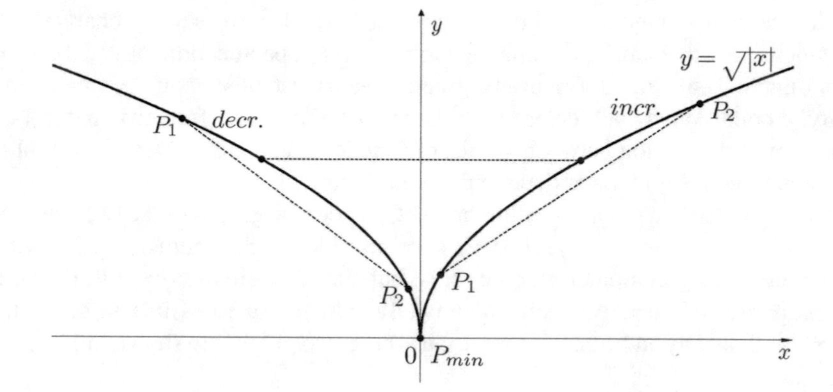

Fig. 2.38 The graph and concavity of the function $y = \sqrt{|x|}$

since the numerator is negative and denominator is positive. This means that both functions are decreasing on $(-\infty, 0]$. In the same way, taking $0 \le x_1 < x_2$, we get $f(x_2) - f(x_1) = x_2^2 - x_1^2 = (x_2 - x_1)(x_2 + x_1) > 0$, since both factors are positive. And for $g(x)$ we get

$$g(x_2) - g(x_1) = \sqrt{|x_2|} - \sqrt{|x_1|} = \frac{(\sqrt{|x_2|} - \sqrt{|x_1|})(\sqrt{|x_2|} + \sqrt{|x_1|})}{\sqrt{|x_2|} + \sqrt{|x_1|}}$$

$$= \frac{|x_2| - |x_1|}{\sqrt{|x_2|} + \sqrt{|x_1|}} = \frac{x_2 - x_1}{\sqrt{|x_2|} + \sqrt{|x_1|}} > 0,$$

since the numerator and denominator are positive. This shows that both functions are increasing on $[0, +\infty)$.

Third, both graphs have a strict global minimum at the origin (and do not have any other extremum, local or global, due to their properties of monotonicity). In fact, $f(x) = x^2 > 0 = f(0)$ and $g(x) = \sqrt{|x|} > 0 = g(0)$, $\forall x \neq 0$. From the last property it follows that both functions are bounded below (for example, by the constant 0). Besides, it can be shown that both functions are unbounded above.

Thus, both graphs, which reflect the demonstrated analytic properties, have many common characteristics. However, they are visibly different. What makes these graphs clearly distinct is a property of functions which we will now reveal. Roughly speaking the first graph (parabola x^2) has the form of a upward cup, that maintain a water inside it, while the second is composed of the two inverted half-cups, which do not keep the water. In a slightly more precise form, the graph of x^2 is bent up, while that of $\sqrt{|x|}$ consists of the two parts, each of which is bent down. This characterization is far from being exact, but we can improve it noting that the type of bending can be specified in terms of the position of some lines connected with the graph. To do this, we introduce secant lines of a graph, which is a quite simple concept.

Definition Given two (different) points $P_1 = (x_1, f(x_1))$ and $P_2 = (x_2, f(x_2))$ of the graph of a function $f(x)$, the *secant line* P_1P_2 is the line passing through these two points.

Notice now that for a bent up graph (like that of x^2) each secant line P_1P_2 has the segment between the points $P_1 = (x_1, f(x_1))$ and $P_2 = (x_2, f(x_2))$ located above the graph. On the contrary, for a bent down graph (like the left or right part of $\sqrt{|x|}$) each secant line P_1P_2 has the segment between the points $P_1 = (x_1, f(x_1))$ and $P_2 = (x_2, f(x_2))$ located below the graph. This is an exact geometric characterization of the type of bending and it is called concavity. In the first case (x^2) we have the upward concavity and in the second ($\sqrt{|x|}$)—downward concavity. Formalizing and generalizing this description we arrive at the following definition.

Definition (Geometric) Let I be an arbitrary interval (open, half-open or closed, finite or infinite) contained in the domain X of a function $f(x)$. We say that $f(x)$ is *concave upward* on I if for any two points $x_1 \neq x_2$ of this interval, the line segment joining the points $P_1 = (x_1, f(x_1))$ and $P_2 = (x_2, f(x_2))$ lies above the graph of $f(x)$.

Analogously, $f(x)$ is *concave downward* on an interval I, if for any two points $x_1, x_2 \in I$, $x_1 \neq x_2$, the line segment joining the points $P_1 = (x_1, f(x_1))$ and $P_2 = (x_2, f(x_2))$ lies below the graph of $f(x)$.

If every line segment P_1P_2 does not lie below (above) the graph of a function, it is said that the function has non-strict upward (downward) concavity.

Remark 1 The line segment P_1P_2 is a part of the secant line P_1P_2 between the points $P_1 = (x_1, f(x_1))$ and $P_2 = (x_2, f(x_2))$ of intersection with the graph of the function $f(x)$.

Remark 2 Sometimes the concavity defined above is called strict concavity, while non-strict concavity is called simply concavity. In this text we will follow the terminology set up in the above definitions. If we will need to specify the type of considered concavity we will indicate whether it is strict or non-strict.

Remark 3 The presented terminology is standard in Calculus. However, in Analysis another terminology is more popular: a concave upward function is called convex and concave downward—simply concave. Besides, only convex functions are usually studied in Analysis, since the properties of any concave function $f(x)$ are directly obtained from the properties of the associated convex function $-f(x)$. In this text we follow the approach used in Calculus.

Let us translate the geometric definitions into the analytic form, which is important for verification of this property in the case of non-trivial functions. First, we write the equation of a secant line P_1P_2: $\frac{\tilde{y}-y_1}{y_2-y_1} = \frac{x-x_1}{x_2-x_1}$ or $\tilde{y} = f(x_1) + \frac{f(x_2)-f(x_1)}{x_2-x_1}(x - x_1)$. The segment between the points P_1 and P_2 can be defined in the form $\tilde{y} = f(x_1) + \frac{f(x_2)-f(x_1)}{x_2-x_1}(x - x_1)$, $\forall x \in (x_1, x_2)$ (we take only the interior points of this segment, since at the endpoints the values of the function and the secant line coincide). Hence, the condition of a upward concavity takes the following form.

Definition (Analytic) We say that $f(x)$ is *concave upward* on an interval $I \subset X$ if for any two different points $x_1, x_2 \in I$ the following inequality holds:

$$f(x) < f(x_1) + \frac{f(x_2) - f(x_1)}{x_2 - x_1}(x - x_1), \forall x \in (x_1, x_2). \tag{2.1}$$

In the same way, $f(x)$ is *concave downward* on an interval $I \subset X$, if for any two different points $x_1, x_2 \in I$, the following inequality holds:

$$f(x) > f(x_1) + \frac{f(x_2) - f(x_1)}{x_2 - x_1}(x - x_1), \forall x \in (x_1, x_2). \tag{2.2}$$

If the strict inequality is changed to non-strict one in these definitions, then we obtain the definitions of non-strict concavity.

Let us show that the two functions $f(x) = x^2$ and $g(x) = \sqrt{|x|}$ have the expected type of concavity according to analytic definitions.

Indeed, for $f(x) = x^2$ we take $I = X = \mathbb{R}$ and to specify considerations we can choose $x_1 < x < x_2$, $\forall x_1, x_2 \in \mathbb{R}$ without loss of generality. Then, evaluating the difference between the values of the secant $\tilde{y} = f(x_1) + \frac{f(x_2)-f(x_1)}{x_2-x_1}(x - x_1)$ and the corresponding values of the function $y = f(x) = x^2$ on the interval (x_1, x_2), we have:

$$y - \tilde{y} = x^2 - \left[x_1^2 + \frac{x_2^2 - x_1^2}{x_2 - x_1}(x - x_1) \right] = x^2 - x_1^2 - (x_2 + x_1)(x - x_1)$$

$$= (x - x_1)(x - x_2) < 0.$$

The last inequality follows from the fact that the first factor is positive and the second is negative. According to the analytic definition, this shows that the function $f(x) = x^2$ has upward concavity over the entire domain. This is what we have expected looking at the graph of this function.

To evaluate $g(x) = \sqrt{|x|}$ we have to consider the two different parts of its domain. Let us start with the interval $I_1 = [0, +\infty)$. Again, without loss of generality we can choose $0 \le x_1 < x < x_2$, and evaluating the difference between the values of the secant $\tilde{y} = g(x_1) + \frac{g(x_2) - g(x_1)}{x_2 - x_1}(x - x_1)$ and the corresponding values of the function $y = g(x) = \sqrt{|x|}$ on the interval (x_1, x_2), we obtain:

$$y - \tilde{y} = \sqrt{x} - \left[\sqrt{x_1} + \frac{\sqrt{x_2} - \sqrt{x_1}}{x_2 - x_1}(x - x_1) \right]$$

$$= \sqrt{x} - \sqrt{x_1} - \frac{\sqrt{x_2} - \sqrt{x_1}}{(\sqrt{x_2} - \sqrt{x_1})(\sqrt{x_2} + \sqrt{x_1})}(x - x_1)$$

$$= \sqrt{x} - \sqrt{x_1} - \frac{1}{\sqrt{x_2} + \sqrt{x_1}}(\sqrt{x} - \sqrt{x_1})(\sqrt{x} + \sqrt{x_1})$$

$$= \frac{\sqrt{x} - \sqrt{x_1}}{\sqrt{x_2} + \sqrt{x_1}}(\sqrt{x_2} - \sqrt{x}) > 0.$$

The last inequality follows from the positivity of all the three factors. Then, by the definition, the function $g(x) = \sqrt{|x|}$ has downward concavity on the entire interval $I_1 = [0, +\infty)$. This reflects the form of bending of its graph.

In the same way, it can be shown that $g(x) = \sqrt{|x|}$ is concave downward on the entire interval $I_2 = (-\infty, 0]$. We leave this task to the reader.

Clearly, we cannot conclude that $g(x) = \sqrt{|x|}$ is concave downward on its domain $X = \mathbb{R}$. Indeed, considering the interval $I_3 = [-1, 1]$ and taking the points $x_1 = -1$ and $x_2 = 1$, we find the secant segment $\tilde{y} = 1$, $x \in (-1, 1)$ lying above the corresponding points of the graph, since $y = \sqrt{|x|} < 1$ when $x \in (-1, 1)$.

7.2 *Midpoint Concavity*

Although the conditions (2.1) and (2.2) are direct translations of geometric features into analytic form, its application can be technically problematic already in the cases of relatively simple functions, not mentioning complex ones. A simple function $g(x) = \sqrt{|x|}$ has required some algebra manipulations to achieve a conclusive result about the concavity. What can complicate a comparison between the values of a secant and the original function is a necessity to compare these two functions at all the points of the interval (x_1, x_2). If it was possible to reduce this comparison to only one point of this interval and if such point had a specific location within the interval, then it would made an investigation of concavity much simpler. It happens that such a simplification is available for all the elementary functions and it consists of comparison between the values of a secant and a function only at the midpoint of the interval (x_1, x_2) (which corresponds to the midpoint of the segment of secant). The corresponding concept is called midpoint concavity and it can be defined as follows.

Definition Let us say that $f(x)$ is *midpoint concave upward* on an interval $I \subset X$, if for any two different points $x_1, x_2 \in I$, the following inequality is satisfied:

$$f\left(\frac{x_1+x_2}{2}\right) < \frac{f(x_1)+f(x_2)}{2}. \tag{2.3}$$

In the same manner, $f(x)$ is *midpoint concave downward* on an interval $I \subset X$, if for any two different points $x_1, x_2 \in I$, the following inequality is satisfied:

$$f\left(\frac{x_1+x_2}{2}\right) > \frac{f(x_1)+f(x_2)}{2}. \tag{2.4}$$

If the sign of the strict inequality in both definitions is changed to the non-strict inequality, then the midpoint concavity is called non-strict.

Remark Sometimes it is more convenient to write the formulas (2.3) and (2.4) in the form

$$2f(x) < f(x-h) + f(x+h) \tag{2.5}$$

and

$$2f(x) > f(x-h) + f(x+h)], \tag{2.6}$$

respectively, where x is an arbitrary point of I and h is an arbitrary increment such that both $x - h$ and $x + h$ belong to I.

Clearly, the inequalities (2.3) and (2.4) follow from (2.1) and (2.2), respectively. In general, the converse is not true. However, it holds for all the elementary functions, which is sufficient for our study. A proof of the converse for elementary functions is complex and is way beyond the scope of this text (and even beyond the scope of Calculus). Nevertheless, we will systematically use the result of the equivalence of general concavity and midpoint concavity (for the elementary functions) to simplify technical manipulations required to investigate the concavity.

Even for simple functions like $f(x) = x^2$ and $g(x) = \sqrt{|x|}$ the inequalities of the midpoint concavity (2.3) and (2.4) can reduce a technical work. For the function $f(x) = x^2$ we have

$$\left(\frac{x_1 + x_2}{2}\right)^2 - \frac{x_1^2 + x_2^2}{2} = \frac{2x_1x_2 - x_1^2 - x_2^2}{4} = -\frac{(x_1 - x_2)^2}{4} < 0,$$

which corresponds to the general upward concavity of $f(x) = x^2$ over the entire domain. For the function $g(x) = \sqrt{|x|}$ on the interval $I_1 = [0, +\infty)$ the condition of the midpoint concavity can be used as follows. A comparison between $\sqrt{\frac{x_1+x_2}{2}}$ and $\frac{\sqrt{x_1}+\sqrt{x_2}}{2}$ is equivalent to a comparison between the squares of these quantities $\frac{x_1+x_2}{2}$ and $\frac{x_1+x_2+2\sqrt{x_1x_2}}{4}$, since all the values are positive. Then, we obtain:

$$\begin{aligned}\frac{x_1 + x_2}{2} - \frac{x_1 + x_2 + 2\sqrt{x_1x_2}}{4} &= \frac{(\sqrt{x_1})^2 + (\sqrt{x_2})^2 - 2\sqrt{x_1x_2}}{4} \\ &= \frac{(\sqrt{x_1} - \sqrt{x_2})^2}{4} > 0,\end{aligned}$$

that indicates downward concavity on $I_1 = [0, +\infty)$. The evaluation on $I_2 = (-\infty, 0]$ is analogous.

Examples

Besides the two solved cases of the functions $f(x) = x^2$ and $g(x) = \sqrt{|x|}$, we consider the two more examples of investigation of concavity.

1. $f(x) = |x|$. We start with an application of the definition of the general concavity. Let us consider $I = X = \mathbb{R}$ and without loss of generality choose $x_1 < x < x_2$, $\forall x_1, x_2 \in \mathbb{R}$. Then, evaluating the difference between the values of the secant $\tilde{y} = f(x_1) + \frac{f(x_2)-f(x_1)}{x_2-x_1}(x - x_1)$ and the values of the function $y = f(x) = |x|$ on the interval (x_1, x_2), we have:

$$D \equiv y - \tilde{y} = |x| - \left[|x_1| + \frac{|x_2| - |x_1|}{x_2 - x_1}(x - x_1)\right].$$

To determine the sign of D, we evaluate the four different cases which can occur. First, if $x_1 < x_2 \le 0$, then

$$D = -x + x_1 - \frac{-x_2 + x_1}{x_2 - x_1}(x - x_1) = -x + x_1 + (x - x_1) = 0.$$

Second, if $0 \le x_1 < x_2$, then

$$D = x - x_1 - \frac{x_2 - x_1}{x_2 - x_1}(x - x_1) = x - x_1 - (x - x_1) = 0.$$

Third, for $x_1 < 0 \leq x < x_2$, we get

$$D = x + x_1 - \frac{x_2 + x_1}{x_2 - x_1}(x - x_1) = \frac{(x + x_1)(x_2 - x_1) - (x_2 + x_1)(x - x_1)}{x_2 - x_1}$$
$$= \frac{2x_1(x_2 - x)}{x_2 - x_1} < 0.$$

Finally, for $x_1 < x \leq 0 < x_2$, we obtain

$$D = -x + x_1 - \frac{x_2 + x_1}{x_2 - x_1}(x - x_1) = \frac{x - x_1}{x_2 - x_1}(-(x_2 - x_1) - (x_2 + x_1))$$
$$= \frac{x - x_1}{x_2 - x_1}(-2x_2) < 0.$$

Hence, for $\forall x_1, x_2 \in \mathbb{R}$ we have $D \equiv y - \tilde{y} \leq 0$, that is, the function is non-strictly concave upward over the domain $\mathbb{R}$, which was expected from the form of its graph.

Next we investigate the midpoint concavity. (According to the general statement on the equivalence of the two types of concavity we expect the same result.) On an arbitrary interval (x_1, x_2) we have to compare the value of the function at the midpoint $f\left(\frac{x_1+x_2}{2}\right)$ with the mean value of the secant $\frac{f(x_1)+f(x_2)}{2}$ (which is the same as the value of the secant at the midpoint). In the case of $f(x) = |x|$ we need to show that $\left|\frac{x_1+x_2}{2}\right| \leq \frac{|x_1|+|x_2|}{2}$, which is a direct consequence of the well-known property of absolute values: $|x_1 + x_2| \leq |x_1| + |x_2|$, $\forall x_1, x_2 \in \mathbb{R}$.

A geometric illustration is given in Fig. 2.39.

2. $f(x) = 3+2x-x^2$. First we use the definition of the general concavity. Consider $I = X = \mathbb{R}$ and choose, without loss of generality, $x_1 < x < x_2$, $\forall x_1, x_2 \in \mathbb{R}$. Then, evaluating the difference between the values of the secant $\tilde{y} = f(x_1) +$

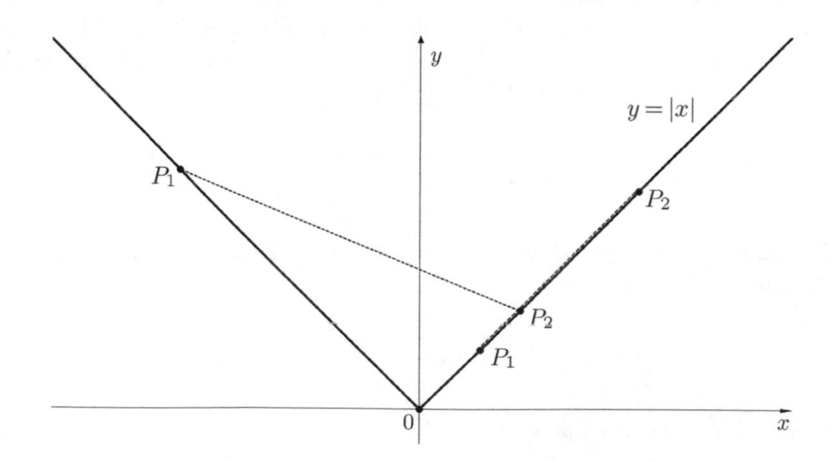

Fig. 2.39 The graph and concavity of the function $y = |x|$

$\frac{f(x_2)-f(x_1)}{x_2-x_1}(x-x_1)$ and the values of the function $y = f(x) = 3 + 2x - x^2$ on the interval (x_1, x_2), we obtain:

$$\begin{aligned} y - \tilde{y} &= 3 + 2x - x^2 \\ &\quad - \left[3 + 2x_1 - x_1^2 + \frac{(3 + 2x_2 - x_2^2) - (3 + 2x_1 - x_1^2)}{x_2 - x_1}(x - x_1) \right] \\ &= 2(x - x_1) - (x^2 - x_1^2) - \frac{2(x_2 - x_1) - (x_2^2 - x_1^2)}{x_2 - x_1}(x - x_1) \\ &= (x - x_1)\left[2 - (x + x_1) - (2 - (x_2 + x_1))\right] = (x - x_1)(x_2 - x) > 0. \end{aligned}$$

The last inequality follows from the positivity of both factors. Therefore, the function $f(x) = 3 + 2x - x^2$ has downward concavity over the entire domain.

Now we apply the definition of the midpoint concavity by comparing the value of the function at the midpoint $f\left(\frac{x_1+x_2}{2}\right)$ and the mean value of the secant $\frac{f(x_1)+f(x_2)}{2}$. In the case of $f(x) = 3 + 2x - x^2$ we have to evaluate the difference

$$\begin{aligned} &3 + 2\frac{x_1 + x_2}{2} - \left(\frac{x_1 + x_2}{2}\right)^2 - \frac{1}{2}\left[3 + 2x_1 - x_1^2 + (3 + 2x_2 - x_2^2)\right] \\ &= -\frac{(x_1 + x_2)^2}{4} + \frac{x_1^2 + x_2^2}{2} = \frac{(x_1 - x_2)^2}{4} > 0, \forall x_1, x_2 \in \mathbb{R}. \end{aligned}$$

As was expected, this confirms the result that $f(x) = 3 + 2x - x^2$ is concave downward on $\mathbb{R}$.

A geometric illustration is given in Fig. 2.40.

7.3 *Elementary Properties of Concave Functions*

In this section we formulate and prove some properties of functions concave upward. The same task in the case of functions concave downward is left to the reader. In the most proofs we use the concept of general concavity.

Property 1 *If $f(x)$ and $g(x)$ are concave upward on an interval I, then $f(x) + g(x)$ is also concave upward on I.*

Proof By the definition, for any distinct $x_1, x_2 \in I$ we have the following two inequalities:

$$f(x) < f(x_1) + \frac{f(x_2) - f(x_1)}{x_2 - x_1}(x - x_1), \forall x \in (x_1, x_2)$$

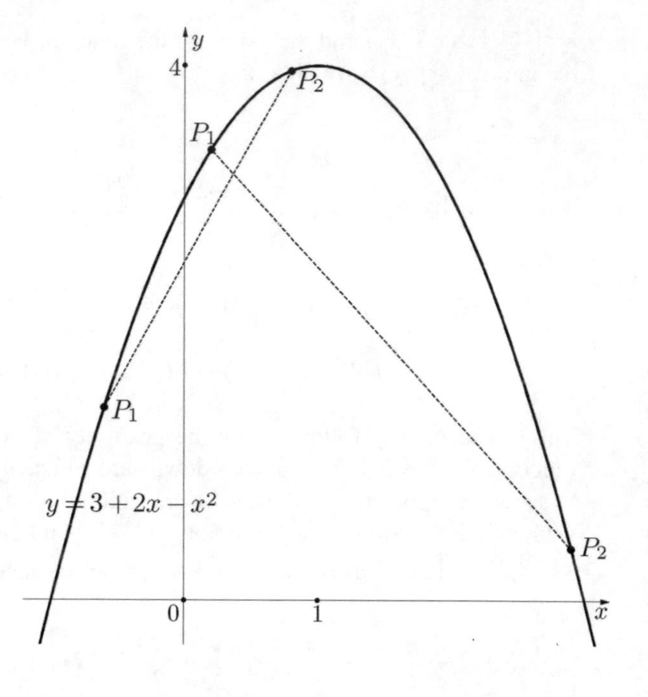

Fig. 2.40 The graph and concavity of the function $y = 3 + 2x - x^2$

and

$$g(x) < g(x_1) + \frac{g(x_2) - g(x_1)}{x_2 - x_1}(x - x_1), \forall x \in (x_1, x_2).$$

Then, for the function $h(x) = f(x) + g(x)$ we get

$$\begin{aligned} h(x) &< f(x_1) + \frac{f(x_2) - f(x_1)}{x_2 - x_1}(x - x_1) + g(x_1) + \frac{g(x_2) - g(x_1)}{x_2 - x_1}(x - x_1) \\ &= f(x_1) + g(x_1) + \frac{f(x_2) + g(x_2) - (f(x_1) + g(x_1))}{x_2 - x_1}(x - x_1) \\ &= h(x_1) + \frac{h(x_2) - h(x_1)}{x_2 - x_1}(x - x_1), \forall x \in (x_1, x_2), \end{aligned}$$

which is the definition of upward concavity of $h(x)$ on the interval I. □

Remark Analogous property and elementary proof are true for non-strict concavity.

Property 2 *If $f(x)$ is concave upward on an interval I, then:*

(1) $\alpha f(x)$ is concave upward on I if $\alpha > 0$;
(2) $\alpha f(x)$ is concave downward on I if $\alpha < 0$.

Proof The proof is elementary and shown only in the case $\alpha > 0$. By the definition, for any distinct $x_1, x_2 \in I$ we have

$$f(x) < f(x_1) + \frac{f(x_2) - f(x_1)}{x_2 - x_1}(x - x_1), \forall x \in (x_1, x_2).$$

Therefore,

$$\alpha f(x) < \alpha f(x_1) + \frac{\alpha f(x_2) - \alpha f(x_1)}{x_2 - x_1}(x - x_1), \forall x \in (x_1, x_2),$$

which means that $\alpha f(x)$ is concave upward on I. □

Remark Analogous property and elementary proof are true for non-strict concavity.

Property 3 *If $f(x)$ is concave upward on an interval I and has a local minimum at a point $x_0 \in I$, then this point is a global minimum in I.*

Proof The idea of the demonstration follows the geometric illustration of concavity. Suppose, for contradiction, that there exists $x_1 \in I$ such that $f(x_1) < f(x_0)$. To specify a situation assume that $x_1 > x_0$. Consider the interval $[x_0, x_1] \in I$ and notice first that all the points of the secant segment P_0P_1, $P_0 = (x_0, f(x_0))$, $P_1 = (x_1, f(x_1))$ lie below the point P_0 (since $f(x_1) < f(x_0)$ by the supposition), and second that the corresponding part of the graph of $f(x)$ (between P_0 and P_1) lies below the secant segment P_0P_1 (due to upward concavity of $f(x)$) (see Fig. 2.41).

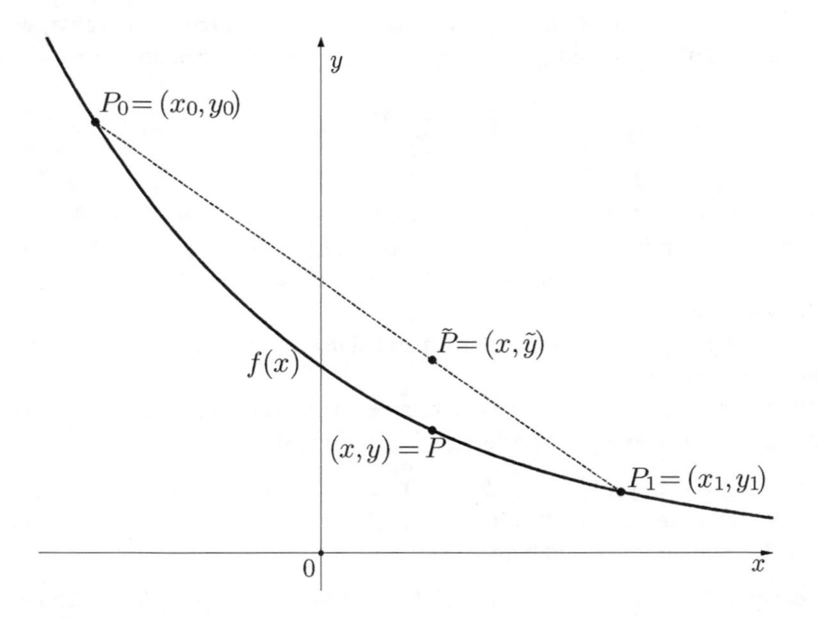

Fig. 2.41 False supposition of a global minimum at $x_1 > x_0$

In particular, in any neighborhood of P_0 there are points of the graph lying below P_0, which contradicts the fact that P_0 is a local minimum.

Now let us represent these arguments analytically. On the interval $(x_0, x_1) \in I$, by the definition of upward concavity (2.1), we have

$$f(x) < f(x_0) + \frac{f(x_1) - f(x_0)}{x_1 - x_0}(x - x_0), \forall x \in (x_0, x_1).$$

Since $x - x_0 > 0$, $x_1 - x_0 > 0$ and $f(x_1) - f(x_0) < 0$ (by the supposition), then $\frac{f(x_1)-f(x_0)}{x_1-x_0}(x - x_0) < 0$, and consequently, $f(x) < f(x_0)$, $\forall x \in (x_0, x_1)$. This is true, in particular, for the points of any right neighborhood of x_0, that contradicts the definition of a local minimum.

In the case $x_1 < x_0$, the same arguments are applied to the interval (x_1, x_0). □

Remark 1 From the provided proof it follows that if a function $f(x)$, concave upward on an interval I, has a local minimum in this interval, then this local minimum is unique in I.

Remark 2 A similar property can be formulated for functions concave downward: if $f(x)$ is concave downward on an interval I and has a local maximum at a point $x_0 \in I$, then this point is a global maximum in I and there is no other local maximum in I. The proof follows the same reasoning.

Property 4 *If $f(x)$ is concave upward on an interval $I = [a, b]$, then $f(x)$ attains its global maximum on I at the point a or b, or at both.*

Proof To specify a situation, let us suppose that $f(a) \geq f(b)$. In this case, we use formula (2.1) of the definition of upward concavity with $x_1 = a$ and $x_2 = b$:

$$f(x) < f(a) + \frac{f(b) - f(a)}{b - a}(x - a), \forall x \in (a, b).$$

Noting that $x-a > 0$, $b-a > 0$ and $f(b)-f(a) \leq 0$, we obtain $\frac{f(b)-f(a)}{b-a}(x-a) \leq 0$, and consequently, $f(x) < f(a)$, $\forall x \in (a, b)$. Therefore, $x = a$ is a global maximum of $f(x)$ on the interval $[a, b]$. (If $f(a) = f(b)$, then $x = b$ is also a global maximum.)

The case $f(a) < f(b)$ can be demonstrated in a similar way using formula (2.1) with $x_1 = b$ and $x_2 = a$.

We recommend to the reader to make a geometric illustration which may clarify better the provided analytic reasoning, like in Property 3. □

Remark A similar property holds for functions concave downward: if $f(x)$ is concave downward on an interval $I = [a, b]$, then $f(x)$ attains its global minimum on I at the point a or b, or at both.

Property 5 *If $f(x)$ is concave upward on an interval I, then its variations (on subintervals of equal length) increase on this interval. In particular, if $f(x)$ is increasing on I, then it increases faster with the increase of x; if $f(x)$ is decreasing* *on a part of I, then it will decrease slower with increase of x or will become increasing on the next part of I.*

Proof This property follows directly from the midpoint concavity: formula (2.5) can be written in the form $f(x)-f(x-h) < f(x+h)-f(x)$, where $x-h, x, x+h \in I$, $h > 0$, which makes clear that on the subinterval $[x, x+h]$ the variation of the function is greater than on the preceding subinterval $[x-h, x]$ (notice that both subintervals has the length h). □

The same result can be expressed in the terms of relative variations: $\frac{f(x)-f(x-h)}{h} < \frac{f(x+h)-f(x)}{h}$. Both quotients in this formula represent the slopes of the secant lines: the secant line with the slope equal to the first quotient passes through the points $P_1 = (x-h, f(x-h))$ and $P = (x, f(x))$, and the secant line with the slope equal to the second quotient passes through the points $P = (x, f(x))$ and $P_2 = (x+h, f(x+h))$. In this form of relative variations the result can be generalized to the case of subintervals of arbitrary lengths, using the general concavity. Indeed, by the analytic definition, the function $y = f(x)$ is concave upward on the interval I if for any two distinct points $x_1, x_2 \in I$ the inequality $y = f(x) < \tilde{f}(x) = \tilde{y}$ is satisfied, where x is an arbitrary point between x_1, x_2 and $\tilde{y} = \tilde{f}(x)$ is the secant line passing through the points $P_1 = (x_1, y_1)$ and $P_2 = (x_2, y_2)$, where $y_1 = f(x_1)$ and $y_2 = f(x_2)$ (see Fig. 2.42). To precise the following reasoning, we can always assume without loss of generality that $x_1 < x_2$ (if it is not so, we just renumber the points).

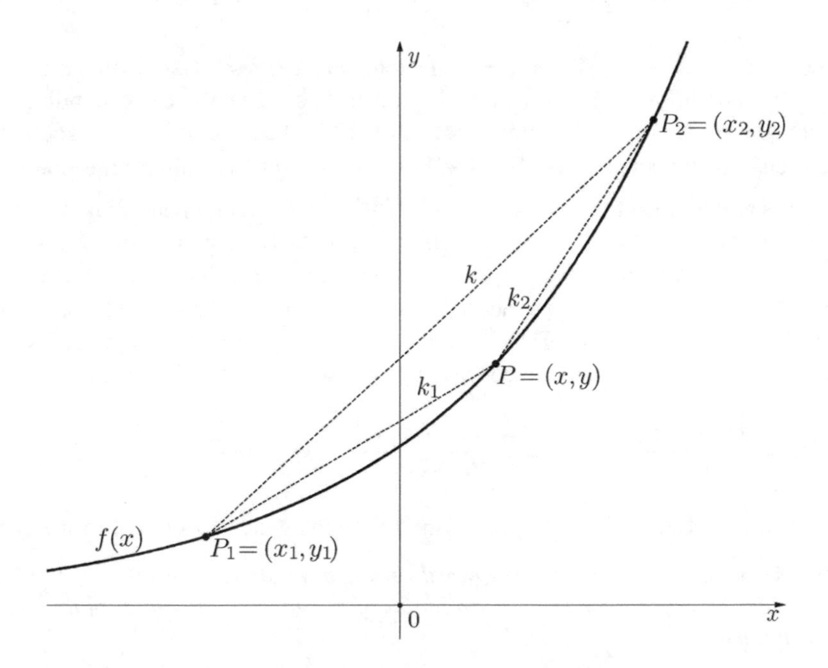

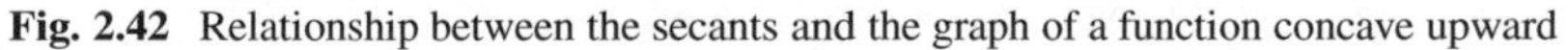
Fig. 2.42 Relationship between the secants and the graph of a function concave upward

Let us write the condition of the general concavity in the two equivalent forms, using the forms of the secant line. The equation of the secant line can be written as $\tilde{f}(x) = f(x_1) + \frac{f(x_2)-f(x_1)}{x_2-x_1}(x - x_1)$, or equivalently $\tilde{f}(x) = f(x_2) + \frac{f(x_2)-f(x_1)}{x_2-x_1}(x - x_2)$. Using the first formula, we arrive at condition (2.1) in the definition of the upward concavity (Sect. 7.1):

$$f(x) < f(x_1) + \frac{f(x_2) - f(x_1)}{x_2 - x_1}(x - x_1), \forall x \in (x_1, x_2),$$

which can also be written in the terms of the relative variations as follows:

$$\frac{f(x) - f(x_1)}{x - x_1} < \frac{f(x_2) - f(x_1)}{x_2 - x_1}, \forall x \in (x_1, x_2). \tag{2.7}$$

Using the second formula of the secant line, we arrive at the condition of the upward concavity in the form

$$f(x) < f(x_2) + \frac{f(x_2) - f(x_1)}{x_2 - x_1}(x - x_2), \forall x \in (x_1, x_2),$$

or equivalently, in the terms of the relative variations

$$\frac{f(x) - f(x_2)}{x - x_2} > \frac{f(x_2) - f(x_1)}{x_2 - x_1}, \forall x \in (x_1, x_2). \tag{2.8}$$

Notice that these inequalities are equivalent and express analytically the same geometric condition: any segment of the secant lies above the corresponding part of the graph. At the same time, inequality (2.7) has the following geometric interpretation: the slope $k = \frac{f(x_2)-f(x_1)}{x_2-x_1}$ of the secant line joining the points P_1 and P_2 is greater than the slope $k_1 = \frac{f(x)-f(x_1)}{x-x_1}$ of the secant line joining the points P_1 and P, where $P = (x, y)$ is an arbitrary point of the graph between P_1 and P_2. Analogously, inequality (2.8) represents the geometric property that the slope $k = \frac{f(x_2)-f(x_1)}{x_2-x_1}$ is smaller than the slope $k_2 = \frac{f(x_2)-f(x)}{x_2-x}$ of the secant line passing through the points P and P_2. Hence, for the slopes of the secant lines (that is, for the relative variations) of a function concave upward we have the following inequalities

$$k_1 = \frac{f(x) - f(x_1)}{x - x_1} < k = \frac{f(x_2) - f(x_1)}{x_2 - x_1} < k_2 = \frac{f(x_2) - f(x)}{x_2 - x} \tag{2.9}$$

(see Fig. 2.42). Hence, we have proved the following generalization of Property 5:

Property 5′ *If $f(x)$ is concave upward on an interval I, then the slopes of the secant lines increase when the x-coordinates of the intersection points of the secant lines increase.*

In a similar manner, the downward concavity can be expressed by the following equivalent inequalities:

$$\frac{f(x) - f(x_1)}{x - x_1} > \frac{f(x_2) - f(x_1)}{x_2 - x_1}, \forall x \in (x_1, x_2) \tag{2.10}$$

or

$$\frac{f(x) - f(x_2)}{x - x_2} < \frac{f(x_2) - f(x_1)}{x_2 - x_1}, \forall x \in (x_1, x_2). \tag{2.11}$$

We leave to the reader the task to specify a geometric meaning of the last two conditions in the terms of the slopes of secant lines.

7.4 Inflection Points

Until now we have seen examples of functions which keep the same type of concavity over the entire domain or over the different parts of the domain. Naturally, there are functions which change the type of concavity moving form one part of the domain to another. A typical example of such a function is $y = f(x) = x^3$, whose graph is shown in Fig. 2.43.

Like in many other examples of this section, at the first moment we take this graph as given to make some geometric considerations. Later we confirm the

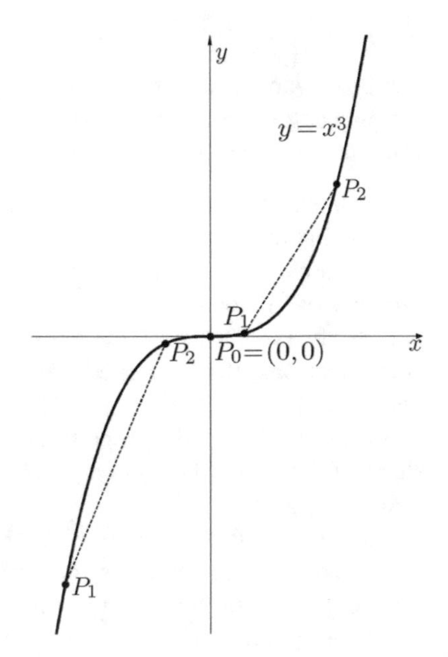

Fig. 2.43 The graph, concavity and inflection point P_0 of the function $y = x^3$

geometric arguments by the analytic study, including a specification of the graph. We can see from the shape of the graph that on the interval $(-\infty, 0]$ the function has strict downward concavity, while on the interval $[0, +\infty)$ it has strict upward concavity. The point 0 where the graph change the type of concavity is called inflection point. This specific example of an inflection point is easily generalized in the following definition.

Definition A point x_0 of the domain of a function $f(x)$ is called an *inflection point* if $f(x)$ has different types of concavity on the intervals $[a, x_0]$ and $[x_0, b]$.

Remark 1 The intervals in this definition can be open at the end points a and b, but should be closed at the point x_0.

Remark 2 In this definition, the concavity is usually considered in a strict sense on both sides of inflection point. Hereinafter we follow this standard agreement.

Remark 3 The order in which concavity appears at different sides of an inflection point does not matter in this definition: a function can have downward concavity to the left of an inflection point and upward concavity to the right, or vice-verse.

Remark 4 Like in the case of extrema, the terminology of inflection point is somewhat ambiguous: an inflection point can be x_0 (as stated in the definition) or also the corresponding point of the graph $P_0 = (x_0, f(x_0))$.

Remark 5 It is tempting to formulate the definition as follows: a point x_0 of the domain of a function $f(x)$ is called an inflection point if $f(x)$ changes the type of concavity at this point. Although this formulation is shorter and simpler, it is not quite accurate, since it admits the cases like the point $x_0 = 0$ of the functions $f(x) = \begin{cases} x^3, x \le 0 \\ x^3 + 1, x > 0 \end{cases}$ (see illustration in Fig. 2.44). Clearly, this function has downward concavity on $(-\infty, 0]$ and upward concavity on $(0, +\infty)$, so the simplified "definition" holds. However, there is no interval $[0, b)$, $b > 0$ where the function is concave upward, and consequently, the original definition is not satisfied.

It happens that some classes of functions possess the property (called continuity in Calculus and Analysis) that avoid such jumps or other kinds of abrupt changes of functions at the points of the domain. (We cannot show not even define this property in this text, since it is based on the concept of limit studied in Calculus and Analysis.) One of these classes is the set of the elementary functions. For such functions, the simplified "definition" is acceptable, since it is equivalent to the original.

Let us verify now the properties of the function $y = f(x) = x^3$ which justify the shape of the graph shown in Fig. 2.43. In particular, we will show that $x_0 = 0$ is an inflection point.

First, notice that $f(x) = x^3$ is an odd function: $f(-x) = (-x)^3 = -x^3 = -f(x)$, which implies the symmetry of the graph about the origin. Second, the function is increasing over the entire domain $X = \mathbb{R}$ that can be shown by the following evaluation for any pair of points $x_1 < x_2$: $f(x_2) - f(x_1) = x_2^3 - x_1^3 =$

Fig. 2.44 Absence of an inflection point of the function $f(x)$ in Remark 5

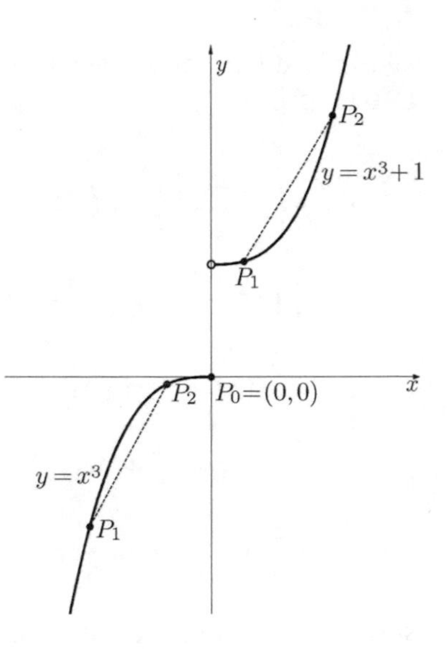

$(x_2-x_1)(x_2^2+x_2x_1+x_1^2) = (x_2-x_1)\left((x_2+\frac{1}{2}x_1)^2+\frac{3}{4}x_1^2\right) > 0$. Since the function is increasing on $\mathbb{R}$, there is no extremum of any type. Finally, let us investigate the concavity of $f(x)$ using the concept of the midpoint concavity. We start with the following evaluation of the difference between the value of the function at the midpoint $f\left(\frac{x_1+x_2}{2}\right)$ and the mean value of the secant $\frac{f(x_1)+f(x_2)}{2}$:

$$
\begin{aligned}
D &\equiv \left(\frac{x_1+x_2}{2}\right)^3 - \frac{x_1^3+x_2^3}{2} = \frac{1}{8}\left(x_1^3+x_2^3+3x_1^2x_2+3x_1x_2^2-4x_1^3-4x_2^3\right) \\
&= \frac{3}{8}\left(-x_1^3-x_2^3+x_1^2x_2+x_1x_2^2\right) = \frac{3}{8}\left(x_1^2(x_2-x_1)-x_2^2(x_2-x_1)\right) \\
&= -\frac{3}{8}(x_2-x_1)^2(x_2+x_1).
\end{aligned}
$$

Now we specify the choice of the points x_1 and x_2. Taking the points in the interval $(-\infty, 0]$, that is, $x_1 < x_2 \le 0$, we have $D = -\frac{3}{8}(x_2-x_1)^2(x_2+x_1) > 0$, since the third factor is negative. This means that $f(x) = x^3$ is concave downward on $(-\infty, 0]$. For the values $0 \le x_1 < x_2$, we get $D = -\frac{3}{8}(x_2-x_1)^2(x_2+x_1) < 0$, since the third factor is positive. Therefore, $f(x) = x^3$ is concave upward on $[0, +\infty)$. Since the function changes the type of concavity at the point $x_0 = 0$, this point is an inflection point.

Let us consider one more example of an inflection point in the case of the function $f(x) = 3x^2 - x^3$. To determine the concavity we use a simpler formula of the midpoint concavity. We start with the evaluation of the difference between the value of the function at the midpoint $f\left(\frac{x_1+x_2}{2}\right)$ and the mean value of the secant $\frac{f(x_1)+f(x_2)}{2}$:

$$
\begin{aligned}
D &\equiv 3\left(\frac{x_1+x_2}{2}\right)^2 - \left(\frac{x_1+x_2}{2}\right)^3 - \frac{3x_1^2 - x_1^3 + 3x_2^2 - x_2^3}{2} \\
&= \frac{3}{4}\left(x_1^2 + 2x_1x_2 + x_2^2 - 2x_1^2 - 2x_2^2\right) \\
&\quad - \frac{1}{8}\left(x_1^3 + x_2^3 + 3x_1^2x_2 + 3x_1x_2^2 - 4x_1^3 - 4x_2^3\right) \\
&= \frac{3}{4}\left(-x_1^2 - x_2^2 + 2x_1x_2\right) + \frac{3}{8}\left(x_1^3 + x_2^3 - x_1^2x_2 - x_1x_2^2\right) \\
&= -\frac{3}{4}(x_2 - x_1)^2 + \frac{3}{8}(x_2 - x_1)^2(x_2 + x_1) = \frac{3}{8}(x_2 - x_1)^2(x_2 + x_1 - 2).
\end{aligned}
$$

Now we specify the location of the points x_1 and x_2. Choosing two arbitrary points $x_1 < x_2 \le 1$, we have $D = \frac{3}{8}(x_2 - x_1)^2(x_2 + x_1 - 2) < 0$, since the third factor is negative. This means that $f(x)$ is concave upward on the interval $(-\infty, 1]$. For the values $1 \le x_1 < x_2$, we get $D = \frac{3}{8}(x_2 - x_1)^2(x_2 + x_1 - 2) > 0$, since the third factor is positive. Consequently, $f(x)$ is concave downward on the interval $[1, +\infty)$. Since the graph changes the type of concavity passing through the point $P_0 = (1, 2)$, this point is an inflection point (see Fig. 2.45).

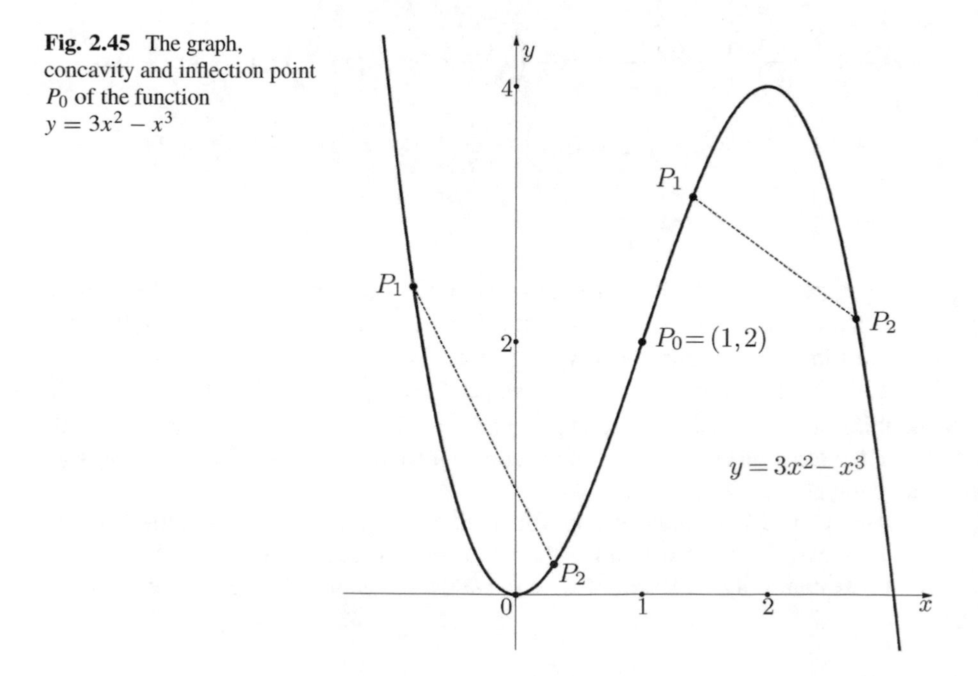

Fig. 2.45 The graph, concavity and inflection point P_0 of the function $y = 3x^2 - x^3$

Finally we present an example when a function has different types of concavity to the left and right sides of a point x_0, but this point is not an inflection. Consider the function $f(x) = \frac{1}{x}$. The evaluation of the difference between the value of the function at the midpoint $f\left(\frac{x_1+x_2}{2}\right)$ and the mean value of the secant $\frac{f(x_1)+f(x_2)}{2}$ gives the following result:

$$D \equiv \frac{1}{\frac{x_1+x_2}{2}} - \frac{\frac{1}{x_1}+\frac{1}{x_2}}{2} = \frac{2}{x_1+x_2} - \frac{1}{2x_1} - \frac{1}{2x_2}$$

$$= \frac{4x_1x_2 - (x_1+x_2)x_2 - (x_1+x_2)x_1}{2x_1x_2(x_1+x_2)} = -\frac{(x_1-x_2)^2}{2x_1x_2(x_1+x_2)}.$$

Choosing now $x_1 < x_2 < 0$, we have $D = -\frac{(x_1-x_2)^2}{2x_1x_2(x_1+x_2)} > 0$, since the denominator is negative. This means that $f(x)$ is concave downward on the interval $(-\infty, 0)$. At the same time, for $0 < x_1 < x_2$, we obtain $D = -\frac{(x_1-x_2)^2}{2x_1x_2(x_1+x_2)} < 0$, since the denominator is positive. This indicates that $f(x)$ is concave upward on the interval $(0, +\infty)$. Nevertheless, $x_0 = 0$ is not an inflection point, simply because the function is not defined at the origin (see the graph of $f(x) = \frac{1}{x}$ in Fig. 2.46).

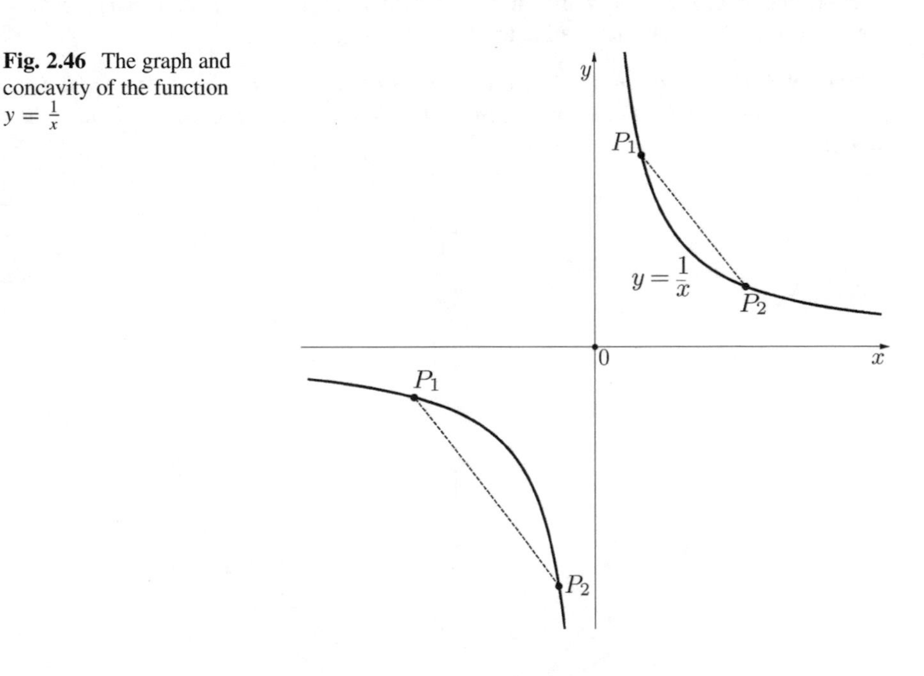

Fig. 2.46 The graph and concavity of the function $y = \frac{1}{x}$

7.5 Concavity, Inflection and Symmetry

Property 1 *If a function $f(x)$ is even and concave upward on an interval $I \subset [0, +\infty)$, then it is concave upward on the interval $\tilde{I}$ symmetric to I about the origin.*

Proof Obviously, $f(x)$ is defined on $\tilde{I}$, since it is even and defined on I. Take two arbitrary points x_1, x_2 of $\tilde{I}$ and notice that the corresponding points $-x_1, -x_2$ belong to I. Then, from the evenness and concavity on I it follows that $f\left(\frac{x_1+x_2}{2}\right) = f\left(\frac{-x_1-x_2}{2}\right) < \frac{f(-x_1)+f(-x_2)}{2} = \frac{f(x_1)+f(x_2)}{2}$, that demonstrates the upward concavity on $\tilde{I}$. □

Remark 1 If $I = [0, b)$ (where b can be $+\infty$), then $f(x)$ is concave upward on $\tilde{I} = (-b, 0]$. However, in general, the function is not concave upward on $(-b, b)$. A counterexample can be given by the function $f(x) = x^2 - |x|$, which is concave upward on the intervals $(-\infty, 0]$ and $[0, +\infty)$, but is not concave upward (even in a non-strict sense) on the interval $(-\infty, +\infty)$ (or any interval (a, b) containing the origin). See illustration in Fig. 2.47.

Remark 2 There is a similar statement for downward concavity: if a function $f(x)$ is even and concave downward on an interval $I \subset [0, +\infty)$, then it is concave downward on the interval $\tilde{I}$ symmetric to I about the origin.

Property 2 *If a function $f(x)$ is odd and concave upward on an interval $I \subset [0, +\infty)$, then it is concave downward on the interval $\tilde{I}$ symmetric to I about the origin.*

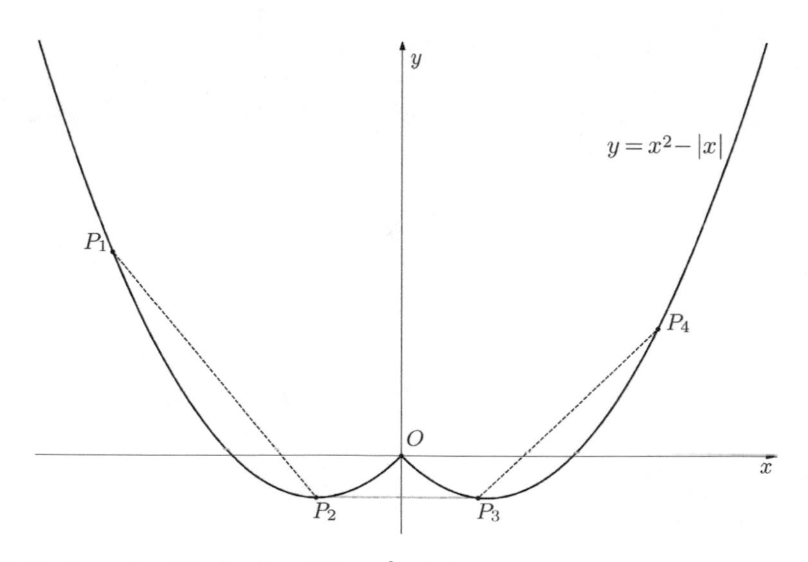

Fig. 2.47 Concavity of the function $f(x) = x^2 - |x|$

Proof The proof follows the same reasoning as in Property 1. Notice that $f(x)$ is defined in $\tilde{I}$, since it is odd and defined in I. Take two arbitrary points $x_1, x_2 \in \tilde{I}$ with the corresponding points $-x_1, -x_2 \in I$ and obtain $f\left(\frac{x_1+x_2}{2}\right) = -f\left(\frac{-x_1-x_2}{2}\right) > -\frac{f(-x_1)+f(-x_2)}{2} = \frac{f(x_1)+f(x_2)}{2}$, that shows the downward concavity on $\tilde{I}$. □

Remark 1 If $I = [0, b)$ (where b can be $+\infty$), then $f(x)$ is concave downward on $\tilde{I} = (-b, 0]$. Therefore, $f(x)$ changes the type of concavity at the point 0 that belongs to the domain, and consequently, 0 is an inflection point.

Remark 2 There is a similar statement in the case of downward concavity: if a function $f(x)$ is odd and concave downward on an interval $I \subset [0, +\infty)$, then it is concave upward on the interval $\tilde{I}$ symmetric to I about the origin.

Remark 1g Naturally, the role of the intervals on the positive and negative parts can be interchanged in the two properties.

Property 3 *If a function $f(x)$ is concave upward on an interval I, then the function $g(x) = f(-x)$ is concave upward on the interval $\tilde{I}$ symmetric to I about the origin.*

Proof The reasoning follows the line of the proof of Property 1. Notice first that $g(x) = f(-x)$ is defined on $\tilde{I}$, since $-x \in \tilde{I}$ for any $x \in I$. Then, using the concavity of $f(x)$ on I, we have $g\left(\frac{x_1+x_2}{2}\right) = f\left(\frac{-x_1-x_2}{2}\right) < \frac{f(-x_1)+f(-x_2)}{2} = \frac{g(x_1)+g(x_2)}{2}$, which shows the upward concavity of $g(x)$ on $\tilde{I}$. □

Remark A similar statement for downward concavity is as follows: if a function $f(x)$ is concave downward on an interval I, then the function $g(x) = f(-x)$ is concave downward on the interval $\tilde{I}$ symmetric to I about the origin.

Property 4 *If a function $f(x)$ is concave upward on an interval I, then the function $g(x) = -f(x)$ is concave downward on the same interval.*

Proof This is a particular case of Property 2 in Sect. 7.3. □

Remark A similar statement for downward concavity is as follows: if a function $f(x)$ is concave downward on an interval I, then the function $g(x) = -f(x)$ is concave upward on the same interval.

Remark 2g Similar statements of the four properties are valid for non-strict concavity.

8 Complimentary Properties of Functions

8.1 Behavior at Infinity and Convergence to Infinity

In this section we consider some complimentary but important properties of functions. They are called complimentary in these text because we have no tools (they will be introduced in Calculus) to define exactly these properties, not saying about their study. Instead we provide an intuitive approximate approach which is a reasonable alternative at this stage of knowledge.

All these properties are related to the *convergence* of a function, that is, the behavior of a function when x approaches some specific point or increases/decreases without restriction. This type of behavior is connected with the notion of limit studied rigorously in Calculus. In this text we can handle this type of behavior only approximately, but will use the terminology of convergence and limit, albeit without a precise meaning of these concepts.

In the first place we will focus on the two situations called the *behavior at infinity* and *convergence to infinity*. The first situation occurs when the values of x increase in the absolute value without restriction, as it is said "x goes to $+\infty$ or to $-\infty$". In this case, it may happen that a function $y = f(x)$ also goes to infinity (positive or negative) or it approaches a certain (finite) value, or still does not have any specific convergence.

In slightly more precise terms, if a convergence of x to $+\infty$ results in a convergence of $f(x)$ to $+\infty$, this means that the values of the function $y = f(x)$ become greater than any positive constant chosen in advance, if we choose sufficiently large values of x. A notation used in this case is $\lim_{x \to +\infty} f(x) = +\infty$ (the abbreviation "lim" comes from the term limit). If the convergence of x to $-\infty$ results in the convergence of $f(x)$ to $+\infty$, this means that the values of the function $y = f(x)$ become greater than any positive constant chosen in advance, if we choose negative x sufficiently large in the absolute value. The corresponding notation is $\lim_{x \to -\infty} f(x) = +\infty$. Other two situation—$f(x)$ converges to $-\infty$ when x goes to $+\infty$ or to $-\infty$ –, have a similar description and are denoted by the symbols $\lim_{x \to +\infty} f(x) = -\infty$ and $\lim_{x \to -\infty} f(x) = -\infty$, respectively.

If the convergence of x to $+\infty$ implies that $f(x)$ approaches a specific number A, then it means that the values of $f(x)$ stay as close to A as it is desired when sufficiently large values of x are chosen (greater than a certain constant). A similar description can be done in the case when x goes to $-\infty$. The corresponding symbols are $\lim_{x \to +\infty} f(x) = A$ and $\lim_{x \to -\infty} f(x) = A$.

For example, the values of the function $y = x$ increase without any restriction when the argument x goes to $+\infty$ (takes increasing values). In more precise terms this means that the values of the function $y = x$ become greater than any constant chosen in advance, if we choose sufficiently large values of x. Indeed, for any constant $M > 0$, choosing $x > M$ we guarantee that $y = x > M$, that is, the values of the function become arbitrary large, going to $+\infty$. In a similar manner, when x goes to $-\infty$ (that is, the values of $|x|$ increase without restriction but x is negative), then $y = x$ also converges to $-\infty$. In more precise terms this means that for any constant $M > 0$, we can choose the points $x < -M$ such that the corresponding values of the function are smaller than $-M$: $y = x < -M$. For the function $y = x$ this behavior at the infinity is obvious, since the values of the function coincide with the values of the independent variable. Using the symbols of limits we can write that $\lim\limits_{x\to+\infty} x = +\infty$ and $\lim\limits_{x\to-\infty} x = -\infty$.

Consider one more example of the behavior at infinity using the function $y = x^2$. The graph of this function (shown in Fig. 2.16 of Sect. 4.1) suggests that the function goes to $+\infty$ when $|x|$ increases without restriction. Since the function is quite simple, this behavior is obvious. Even so, let us justify it using an approximate description of the convergence to infinity. Indeed, if $x > 1$ then $x^2 > x$, and consequently, choosing $x > M$ (where $M > 1$ is an arbitrary constant chosen in advance) we guarantee that $x^2 > M$, that is, the values of the function also becomes sufficiently large, that is, the function converges to $+\infty$. Due to the evenness of this function, the same is true when the values of $|x|$ increase without restriction but in the negative direction (x goes to $-\infty$). Using the symbols of limits we have $\lim\limits_{x\to+\infty} x^2 = +\infty$ and $\lim\limits_{x\to-\infty} x^2 = +\infty$.

In both considered examples the functions $y = x$ and $y = x^2$ have the convergence to infinity when $|x|$ goes to infinity, in other words, the functions become infinite on infinity. In the next example we take the function which approaches a specific value on infinity. Consider the behavior of the function $\frac{1}{x}$ when $|x|$ goes to infinity. The graph of the function (Fig. 2.46 in Sect. 7.4) suggests that this function approaches 0 when $|x|$ increases without restriction, that is, the values $\frac{1}{|x|}$ becomes smaller than any chosen in advance constant $m > 0$ if we choose sufficiently large values of $|x|$. Indeed, given an arbitrary constant $m > 0$, for the values $|x| > M = \frac{1}{m}$ we guarantee that $\frac{1}{|x|} < m$. For example, if we choose $m = 10^{-l}, l \in \mathbb{N}$, then for $|x| > 10^l$ the values of the function satisfy the inequality $\frac{1}{|x|} < 10^{-l}$. Obviously, when x goes to $+\infty$, then the function $\frac{1}{x}$ approaches 0 by the positive values, and when x goes to $-\infty$, then 0 is approached by the negative values of the function. In this case we write $\lim\limits_{x\to+\infty} \frac{1}{x} = 0$ and $\lim\limits_{x\to-\infty} \frac{1}{x} = 0$. This justify, albeit in an approximate form, the behavior of the function shown on its graph.

The second situation, *convergence to infinity at a finite point*, occurs when convergence of x to a specific (finite) point leads to unrestricted increase/decrease of the values of function. We say that a function goes to $+\infty$ as x approaches a point a, if the values of the function becomes greater than any chosen in advance constant when the values of x are chosen in close proximity to a, without considering the point a itself. The symbol used for this behavior of a function is $\lim\limits_{x\to a} f(x) = +\infty$. Here, "close proximity" means, in more precise form, that the distance from x to a is smaller than a chosen quantity $\delta > 0$, that is, $|x - a| < \delta$. In other words, x stay in a δ-neighborhood of a, where δ is sufficiently small. The condition that we do not care what happens with the function at the point a means that an information about the function at this point has no effect on the behavior of the function when x approaches a (that is, near a): a function can be defined at a in any way and even not defined. This condition is due to the necessity to consider the functions which are defined around a, but not at a (the domain has a "hole" in a). In analytic terms it is usually expressed as $|x - a| > 0$. Therefore, when x approaches a we are working with the points in a δ-neighborhood of a different from a, that is, $0 < |x - a| < \delta$.

In a similar way we can define the situation when a function goes to $-\infty$ as x approaches a. The corresponding notation is $\lim_{x \to a} f(x) = -\infty$.

Sometimes, it is necessary to investigate what happens with a function separately to the left and to the right of a, that is, when x approaches a only from the left (by the values less than a) and only from the right (by the values greater than a). In the former case, the proximity of x to a is defined by the inequality $a - \delta < x < a$, and in the latter by $a < x < a + \delta$. The meaning of the convergence of a function is still the same, but it is related to one-sided convergence of x to a. If $f(x)$ goes to infinity as x approaches a from one of the sides, then the following symbols are used: $\lim_{x \to a^-} f(x) = +\infty$ and $\lim_{x \to a^-} f(x) = -\infty$ for the left-hand approximation and $\lim_{x \to a^+} f(x) = +\infty$ and $\lim_{x \to a^+} f(x) = -\infty$ for the right-hand approximation.

To exemplify such behavior, let us consider again the function $\frac{1}{x}$ and choose the point $a = 0$. Although this point does not belong to the domain, but it does not impede to consider convergence of x to this point and behavior of the function under this convergence. First, let us consider approximation of a from the right. It is clear intuitively that in this case the function $\frac{1}{x}$ increases its values unrestrictedly, in other words, goes to $+\infty$. To justify this, we use a more precise description: for any given in advance constant $M > 0$ we are able to find $\delta = \frac{1}{M}$ such that for all x in the interval $0 < x < \delta = \frac{1}{M}$ (that is, for all positive x sufficiently close to 0) it follows that $\frac{1}{x} > M$. For the level of precision we can use, this is the proof that $\lim_{x \to 0^+} \frac{1}{x} = +\infty$. In a similar manner we can show that $\frac{1}{x}$ goes to $-\infty$ as x approaches 0 from the left: for any $M > 0$, we are able to find $\delta = \frac{1}{M}$ such that for all x in the interval $-\frac{1}{M} = -\delta < x < 0$ (that is, for all negative x sufficiently close to 0) it follows that $\frac{1}{x} < -M$. Under the level of precision we can achieve, this means that $\lim_{x \to 0^-} \frac{1}{x} = -\infty$. See illustration in Fig. 2.48 where the graph of $y = \frac{1}{x}$ is reproduced together with additional characteristics.

Notice that the same kind of the convergence at the origin we have for the function $g(x) = \begin{cases} 1/x, x \neq 0 \\ 0, x = 0 \end{cases}$, which is defined at 0, since the functions $g(x)$ and $f(x)$ differ only at the point 0. Just recall that the value of the function at 0 (or the information that the function is or is not defined at 0) does not affect the evaluation of the behavior of the function as x approaches 0. Hence, $\lim_{x \to 0^+} g(x) = +\infty$ and $\lim_{x \to 0^-} g(x) = -\infty$.

Finally, it can also occur that a function $f(x)$ *approaches a specific number A* when x approaches a. The simplest (intuitively) situation is when this number A is the value of $f(x)$ at a. In this case, a function is called *continuous* at a and the used notation is $\lim_{x \to a} f(x) = f(a)$. Notice that this type of convergence (as well as other considered types) can be discussed only approximately in this text, since its exact treatment requires the concepts and results of Calculus and Analysis.

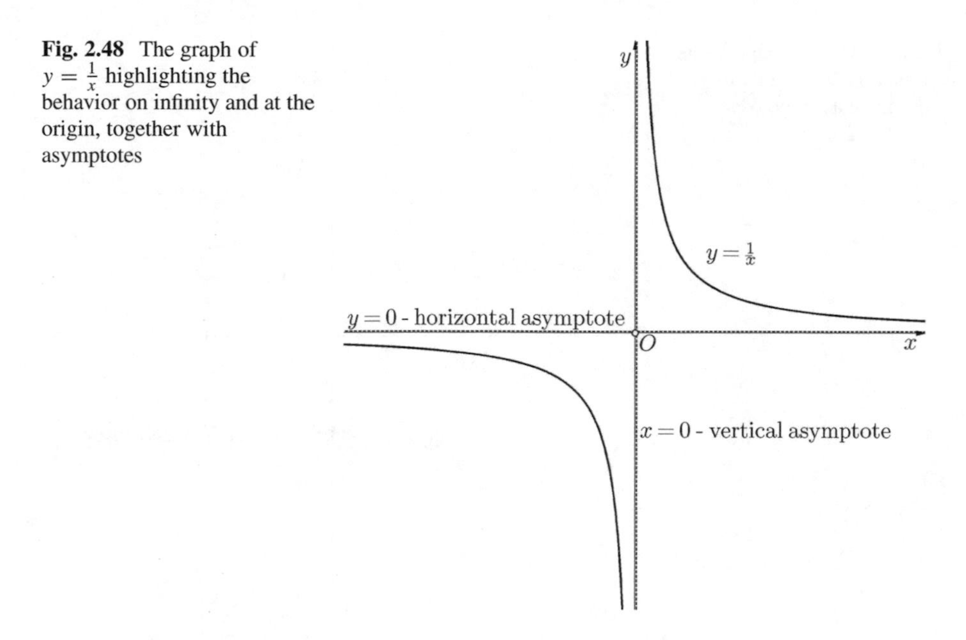

Fig. 2.48 The graph of $y = \frac{1}{x}$ highlighting the behavior on infinity and at the origin, together with asymptotes

8.2 *Horizontal and Vertical Asymptotes*

In study of behavior of a function at infinity (when x goes to $+\infty$ or to $-\infty$) we have noted that if a function $f(x)$ approaches a constant A, then geometrically this means that its graph approaches the line $y = A$ when $|x|$ increases. In other words, the distance between the graph of $f(x)$ and the horizontal line $y = A$ goes to 0 when $|x|$ goes to infinity. In this case, the line $y = A$ is called horizontal asymptote. Let us formalize this in the following definition.

Definition of Horizontal Asymptote The line $y = A$ is called a *horizontal asymptote* of the graph of a function $y = f(x)$ if at least one of the two conditions is satisfied: $\lim\limits_{x\to+\infty} f(x) = A$ or/and $\lim\limits_{x\to-\infty} f(x) = A$.

Notice that the number of horizontal asymptotes of a function can vary from zero to two. If $f(x)$ approaches A as x goes to $+\infty$, but does not show a convergence to a specific number when x goes to $-\infty$, then there is the only one asymptote $y = A$. The same situation with the only asymptote $y = A$ occurs if $f(x)$ approaches the same constant $A = B$ both when x goes to $+\infty$ and when it goes to $-\infty$. An example of this case is the function $y = \frac{1}{x}$ with the limits $\lim\limits_{x\to+\infty} \frac{1}{x} = \lim\limits_{x\to-\infty} \frac{1}{x} = 0$ (see Fig. 2.48). It may also happen that a function has two different horizontal asymptotes if $\lim\limits_{x\to+\infty} f(x) = A \neq B = \lim\limits_{x\to-\infty} f(x)$. An example of such a function is $f(x) = \begin{cases} 1/x, x < 0 \\ 1 + 1/x, x > 0 \end{cases}$ which has the two asymptotes $y = 0$ and $y = 1$: the former is related to the limit $\lim\limits_{x\to-\infty} f(x) = 0$ and the latter to $\lim\limits_{x\to+\infty} f(x) = 1$.

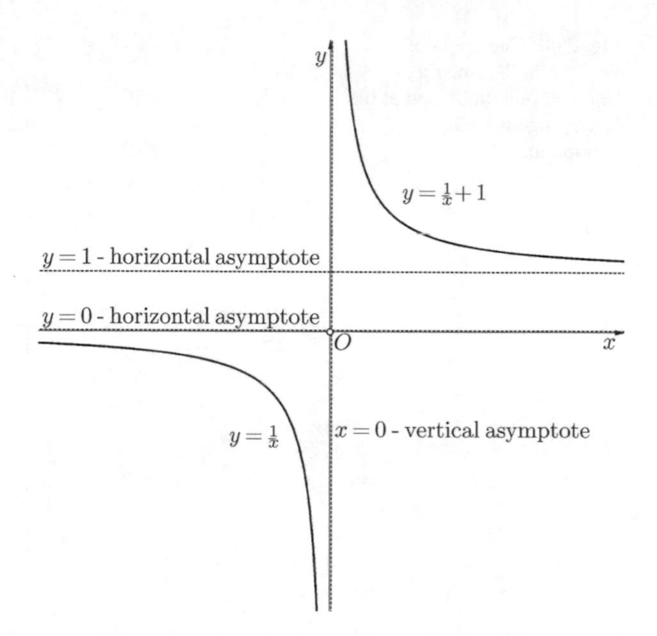

Fig. 2.49 The graph of the function with the two horizontal asymptotes

This example is illustrated in Fig. 2.49. Finally, it may happen that a function has no horizontal asymptote, like the function $y = x^2$.

If a function $f(x)$ converges to infinity when x approaches a finite point a, we have noted that geometrically this means that the graph of $f(x)$ stay closer and closer to the vertical line $x = a$ as x approaches a. In other words, the distance between the graph of $f(x)$ and the line $x = a$ converges to 0 when $|x - a|$ approaches 0. In this case, the line $x = a$ is called a vertical asymptote and its formal definition goes as follows.

Definition of Vertical Asymptote A line $x = a$ is called a *vertical asymptote* of the graph of $y = f(x)$ if at least one of the following conditions is satisfied: $\lim\limits_{x \to a^+} f(x) = +\infty$ or $\lim\limits_{x \to a^+} f(x) = -\infty$ or $\lim\limits_{x \to a^-} f(x) = +\infty$ or $\lim\limits_{x \to a^-} f(x) = -\infty$.

An example of a vertical asymptote is the line $x = 0$ for the function $y = \frac{1}{x}$ (see Fig. 2.48). Notice that, by the definition, it is not required that the convergence to infinity occurs from the both sides of a. For example, the function $f(x) = \begin{cases} 0, x \leq 0 \\ 1/x, x > 0 \end{cases}$ has the convergence to $+\infty$ only when x approaches 0 from the right (see Fig. 2.50). Even so, the line $x = 0$ is a vertical asymptote of this function.

Notice that the number of the vertical asymptotes of a function can vary from zero to infinity: for example, the function $y = x^2$ has no vertical asymptote, the function $y = \frac{1}{x}$ has the only vertical asymptote $x = 0$ and the function $y = \tan x$ (it will thoroughly studied in Chap. 4) has infinite number of the vertical asymptotes $x = \frac{\pi}{2} + n\pi, \forall n \in \mathbb{Z}$.

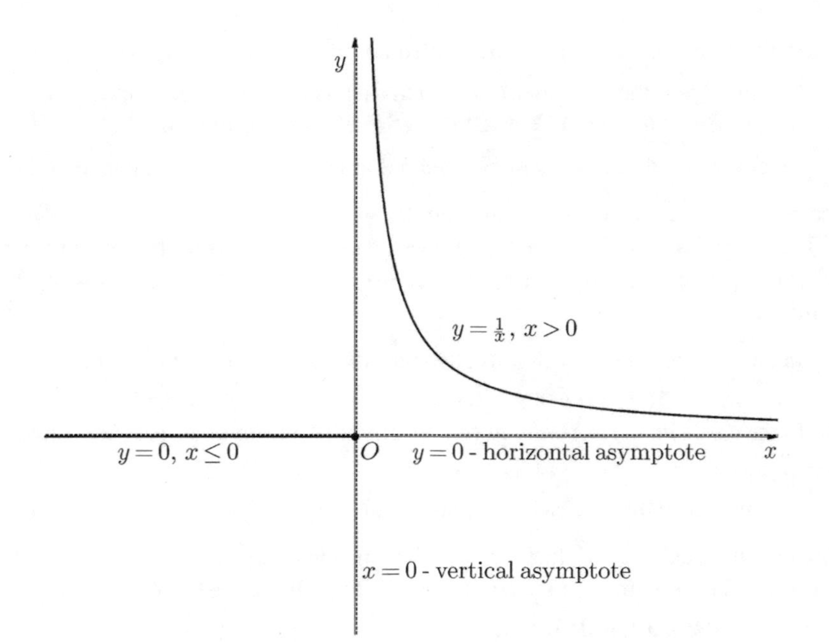

Fig. 2.50 The graph of the function with the convergence to the vertical asymptote only from the right

Without any proof, we state that any elementary function $f(x)$ is continuous at each point of its domain X, that is, $\lim_{x\to a} f(x) = f(a)$, $\forall a \in X$. This implies that infinite decrease/increase cannot happen at any point of the domain of an elementary function, and consequently, such a function has no vertical asymptote at any point of the domain. However, vertical asymptotes can be found at other points, which can be approached through the points of the domain. The typical examples of this situation are the functions $f(x) = \frac{1}{x}$ with the vertical asymptote $x = 0$ and the function $f(x) = \tan x$ with infinite number of the vertical asymptotes $x = \frac{\pi}{2} + n\pi$, $\forall n \in \mathbb{Z}$. Of course, it may happen that there is no vertical asymptote at a point a that does not belong to the domain of an elementary function. First, this may occur when such a point a is not "approachable" by the points of the domain, like in the case of $f(x) = \sqrt{x}$ and $a = -1$. Second, this happens when the function approaches some finite value, like in the case of $f(x) = \frac{x^2-1}{x+1}$ and $a = -1$. In the latter case, the function can be simplified to the form $f(x) = x - 1$, $x \neq -1$, which shows (intuitively) that $\lim_{x\to a} f(x) = -2$, that is, an infinite behavior does not observed near the point $a = -1$.

For the reader interested in more advanced concepts we can add that the first situation, behavior at infinity, is related to the limits at infinity ($+\infty$ or $-\infty$) whose exact definitions are as follows.

(1) In the case $\lim_{x\to+\infty} f(x) = +\infty$ (called an infinite limit at infinity):
for any $E > 0$ there exists $M > 0$ such that $f(x) > E$ whenever $x > M$.
(2) In the case $\lim_{x\to+\infty} f(x) = A$ (called a finite limit at infinity):
for any $\varepsilon > 0$ there exists $M > 0$ such that $|f(x) - A| < \varepsilon$ whenever $x > M$.

The remaining cases are defined in a similar way.

The second situation, when the convergence of x to some finite point causes the convergence of a function to infinity ($+\infty$ or $-\infty$) has the following exact definitions.

(1) In the case $\lim_{x\to a} f(x) = +\infty$ (called an infinite limit at a finite point):
for any $E > 0$ there exists $\delta > 0$ such that $f(x) > E$ whenever $0 < |x-a| < \delta$.
(2) In the case $\lim_{x\to a^+} f(x) = +\infty$ (called an infinite limit at a finite point from the right):
for any $E > 0$ there exists $\delta > 0$ such that $f(x) > E$ whenever $0 < x - a < \delta$.

The remaining cases are defined in a similar manner.

Finally, the case $\lim_{x\to a} f(x) = A$, that is, the finite limit at a finite point (belonging or not to the domain), is defined as follows:

for any $\varepsilon > 0$ there exists $\delta > 0$ such that $|f(x) - A| < \varepsilon$ whenever $0 < |x - a| < \delta$.

In a special case when $A = f(a)$, that is, under additional conditions that the function is defined at a and its value $f(a)$ coincides with the value of the limit A, the function is called continuous at a.

In this text we do not use these exact notions, which are important concepts of more advanced courses such as Calculus and Analysis. Here, we restrict our consideration to approximate descriptions of the limit and convergence provided in this section.

9 Composite Functions

9.1 Composition of Functions

Let us suppose that we have two functions $y = f(x) : X \to Y$ and $z = g(y) : Y \to Z$, where the range Y of the former is the domain of the latter. The first function maps the elements of X into Y and the second send the elements of Y in Z. In this case, instead of performing a mapping of elements $x \in X$ in $z \in Z$ into the two steps (the two functions) we can construct a direct, one-step transformation from X into Z keeping the same relationship between the elements of the two sets. For this purpose we need to define a composite function.

Definition Given two functions $y = f(x) : X \to Y$ and $z = g(y) : Y \to Z$, their *composition*, called a *composite function* $z = h(x) : X \to Z$, is such the function that takes each element $x \in X$ and associates with it the element $z \in Z$, which is obtained under the successive application of the function $y = f(x)$ to the elements x and the function $z = g(y)$ to the corresponding elements y. According to this definition, the formula of a composite function is $z = h(x) = g(f(x)) : X \to Z$.

There are two standard forms of notation of a composite function: the first is the formula $g(f(x))$ of the definition itself, and the second is $g \circ f(x)$. In this text we use the first notation.

Notice that, by the definition, the domain of a composition is the domain X of the first function $y = f(x)$. The definition also impose the condition that Y is the range of the first function $y = f(x)$ and the domain of the second function $z = g(y)$. If additionally Z is the range of the second function, then the range of the composite function is also Z. Indeed, if $Z = g(Y)$ (Z is the range of Y under $z = g(y)$), then for every $z_0 \in Z$ there exists (at least one) $y_0 \in Y$ such that $z_0 = g(y_0)$. In turn, since $Y = f(X)$ (Y is the range of X under $y = f(x)$), for this $y_0 \in Y$ there exists (at least one) $x_0 \in X$ such that $y_0 = f(x_0)$. Therefore, for every $z_0 \in Z$ we have found (at least one) $x_0 \in X$ such that $y_0 = f(x_0)$ and $z_0 = g(y_0)$, that is, $h(x_0) = g(f(x_0)) = g(y_0) = z_0$, which shows that Z is the range of $h(x)$.

See an illustration of this definition in Fig. 2.51.

Examples

1. $y = f(x) = x : X = \mathbb{R} \to Y = \mathbb{R}$ and $z = g(y) = 2y : Y = \mathbb{R} \to Z = \mathbb{R}$.
   The composite function is defined by the formula $h(x) = g(f(x)) = g(x) = 2x : X = \mathbb{R} \to Z = \mathbb{R}$. Since $Z = \mathbb{R}$ is the range of the second function, it is also the range of the composition.
2. $y = f(x) = x + 1 : X = \mathbb{R} \to Y = \mathbb{R}$ and $z = g(y) = y^2 : Y = \mathbb{R} \to Z = [0, +\infty)$.
   The composite function is defined by the formula $h(x) = g(f(x)) = g(x+1) = (x + 1)^2 : X = \mathbb{R} \to Z = [0, +\infty)$. Since $Z = [0, +\infty)$ is the range of the second function, it is also the range of the composition.
3. $y = f(x) = x^2 : X = \mathbb{R} \to Y = [0, +\infty)$ and $z = g(y) = \sqrt{y} : Y = [0, +\infty) \to Z = [0, +\infty)$.

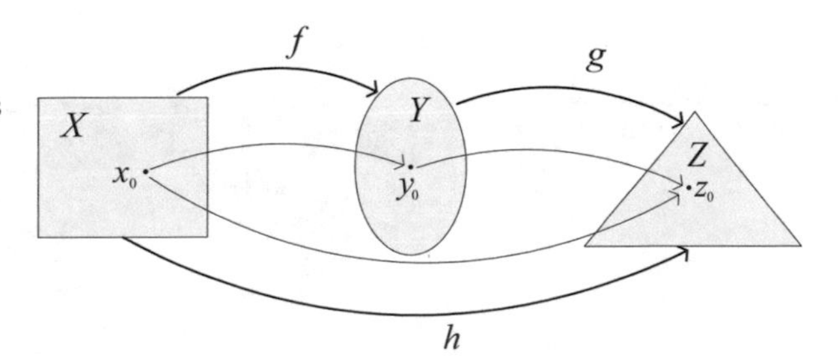

Fig. 2.51 Composite function, the case when the range Y of the first function is the domain of the second one

The composite function is defined by the formula $h(x) = g(f(x)) = g(x^2) = \sqrt{x^2} = |x| : X = \mathbb{R} \to Z = [0, +\infty)$. Since $Z = [0, +\infty)$ is the range of the second function, it is also the range of the composition.

4. $y = f(x) = x^2 : X = (-\infty, 0] \to Y = [0, +\infty)$ and $z = g(y) = \sqrt{y} : Y = [0, +\infty) \to Z = [0, +\infty)$.
   The composite function is defined by the formula $h(x) = g(f(x)) = g(x^2) = \sqrt{x^2} = -x : X = (-\infty, 0] \to Z = [0, +\infty)$. Since $Z = [0, +\infty)$ is the range of the second function, it is also the range of the composition.

In all these examples, according to the definition, the range Y of the first function coincide with the domain of the second. Now let us extend the definition of a composition to the cases when it does not happen.

The first case is simple. If $y = f(x) : X \to Y$ and $z = g(y) : \tilde{Y} \to Z$, where $Y \subset \tilde{Y}$ (here Y can be a codomain, not necessarily the range of the first function), then the composite function can be defined by the same rule as before: $z = h(x) = g(f(x)) : X \to Z$. The only difference is that the relationship between the range of the second function and that of the composition is not so simple as in the original definition: even when Z is the range of the second function, this does not mean that Z is necessarily the range of the composition, because the composition is associated only with the part Y of the domain $\tilde{Y}$ of the second function. Therefore, in general, Z is a codomain of the composition even when Z is the range of the second function.

See illustration of this case in Fig. 2.52.

Examples

1. $y = f(x) = x^2 : X = \mathbb{R} \to Y = [0, +\infty)$ and $z = g(y) = 2y : \tilde{Y} = \mathbb{R} \to Z = \mathbb{R}$.
   The composite function is defined by the formula $h(x) = g(f(x)) = g(x^2) = 2x^2 : X = \mathbb{R} \to Z = \mathbb{R}$. Although $Z = \mathbb{R}$ is the range of the second function, but it is only a codomain of the composition (the range of the composition is $\tilde{Z} = [0, +\infty)$).
2. $y = f(x) = \sqrt{x} : X = [0, +\infty) \to Y = [0, +\infty)$ and $z = g(y) = y^2 : \tilde{Y} = \mathbb{R} \to Z = [0, +\infty)$.
   The composite function is defined by the formula $h(x) = g(f(x)) = g(\sqrt{x}) = (\sqrt{x})^2 = x : X = [0, +\infty) \to Z = [0, +\infty)$. In this case $Z = [0, +\infty)$ is

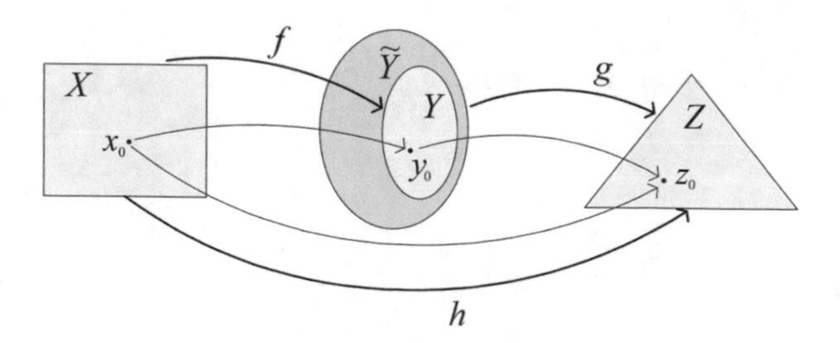

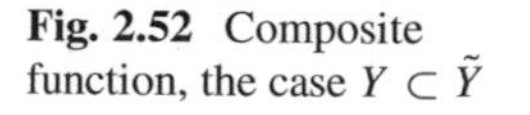
Fig. 2.52 Composite function, the case $Y \subset \tilde{Y}$

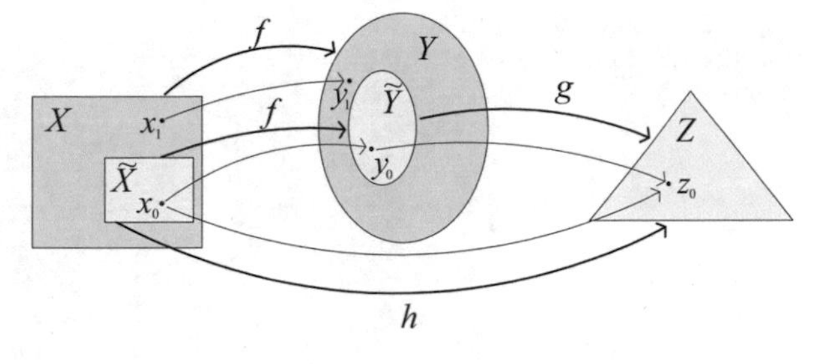

Fig. 2.53 Composite function, the case $Y \supset \tilde{Y}$

the range of the second function and also the range of the composition. Notice that the composition is not a complete function $u(x) = x$, since its domain is restricted to $X = [0, +\infty)$.

The second extension is related to the case when $y = f(x) : X \to Y$ and $z = g(y) : \tilde{Y} \to Z$, where $Y \supset \tilde{Y}$ (here Y is the range of the first function). This situation is more complicated. Since the second function is not defined on a part of the range of the first one, there are values $x_1 \in X$ such that $y_1 = f(x_1) \in Y \backslash \tilde{Y}$ (that is, these $x \in X$ are sent by $f(x)$ outside the domain $\tilde{Y}$ of $g(y)$), and consequently, the composition is not defined at these points. Usually we have no possibility to enlarge the domain of the second function, but the problem can be circumvented by reducing the domain X of the first function to such $\tilde{X}$ whose range under $f(x)$ is $\tilde{Y}$: $y = f(x) : \tilde{X} \to \tilde{Y}$. After the first function has been modified in this way, we can return to the original definition of the composition $z = h(x) = g(f(x)) : \tilde{X} \to Z$.

See illustration of this extension in Fig. 2.53.

Examples

1. $y = f(x) = 2x : X = \mathbb{R} \to Y = \mathbb{R}$ and $z = g(y) = \sqrt{y} : \tilde{Y} = [0, +\infty) \to Z = [0, +\infty)$.
   In this case $Y = \mathbb{R} \supset \tilde{Y} = [0, +\infty)$, and consequently, there is no possibility to define the composition on the entire domain $X = \mathbb{R}$ of the first function. Since we cannot enlarge the domain of the second function, we need to modify the first function (keeping the same formula) in such a way that its range would be $\tilde{Y} = [0, +\infty)$. Clearly it is achieved by using the domain $\tilde{X} = [0, +\infty)$ instead of $X = \mathbb{R}$. Then the first function takes the form $y = f(x) = 2x : \tilde{X} = [0, +\infty) \to \tilde{Y} = [0, +\infty)$ (the second remains the same) and the composition is $z = h(x) = g(f(x)) = g(2x) = \sqrt{2x} : \tilde{X} = [0, +\infty) \to Z = [0, +\infty)$.
2. $y = f(x) = x^2 - 1 : X = \mathbb{R} \to Y = [-1, +\infty)$ and $z = g(y) = \sqrt{y} : \tilde{Y} = [0, +\infty) \to Z = [0, +\infty)$.
   Since $Y = [-1, +\infty) \supset \tilde{Y} = [0, +\infty)$, the composition cannot be defined on the entire domain $X = \mathbb{R}$ of the first function. We need to modify the first function (keeping the same formula) in such a way that its range would be $\tilde{Y} = [0, +\infty)$. To do this we have to solve the inequality $y = x^2 - 1 \ge 0$, which is equivalent to $|x| \ge 1$, whose solutions are $x \le -1$ and $x \ge 1$, that is, $\tilde{X} = (-\infty, -1] \cup [1, +\infty)$. Then the function $f(x) = x^2 - 1$ with the domain

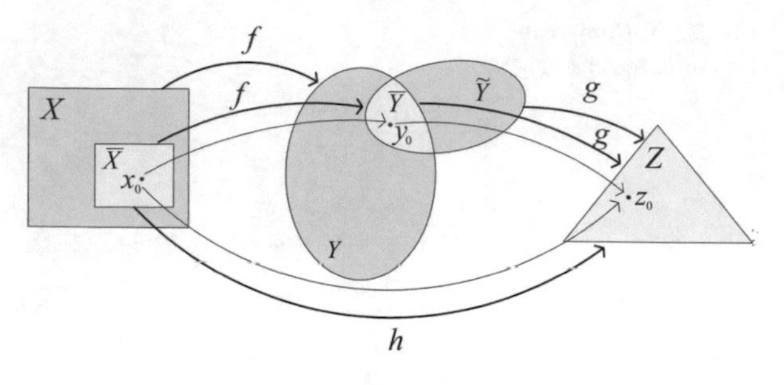

Fig. 2.54 Composite function, the case $Y \cap \tilde{Y} \neq \varnothing$, $Y \not\subset \tilde{Y}, Y \not\supset \tilde{Y}$

$\tilde{X} = (-\infty, -1] \cup [1, +\infty)$ has the required image $\tilde{Y} = [0, +\infty)$, and using this function as the first one, we find the composition in the form: $z = h(x) = g(f(x)) = g(x^2-1) = \sqrt{x^2-1} : \tilde{X} = (-\infty, -1] \cup [1, +\infty) \to Z = [0, +\infty)$.

Finally, the third, more general, extension occurs when $y = f(x) : X \to Y$, $z = g(y) : \tilde{Y} \to Z$ and $Y \cap \tilde{Y} \neq \varnothing$, but $Y \not\subset \tilde{Y}, Y \not\supset \tilde{Y}$. (Obviously, if $Y \cap \tilde{Y} = \varnothing$, there is no way to define a composition.) In this case, we reduce the domain X of $f(x)$ to the set $\overline{X} \subset X$ whose range under $f(x)$ is $\overline{Y} = Y \cap \tilde{Y}$ (the common part of Y and $\tilde{Y}$): $y = f(x) : \overline{X} \to \overline{Y}$. After the first function has been modified in this way, the functions satisfy the conditions of the first extension and we can define their composition in the form $z = h(x) = g(f(x)) : \overline{X} \to Z$.

The illustration of the general extension is given in Fig. 2.54.

Examples

1. $y = f(x) = 1 - x^2 : X = \mathbb{R} \to Y = (-\infty, 1]$ and $z = g(y) = \sqrt{y} : \tilde{Y} = [0, +\infty) \to Z = [0, +\infty)$.
   Since $Y \cap \tilde{Y} = (-\infty, 1] \cap [0, +\infty) = [0, 1]$ and none of the sets $Y = (-\infty, 1]$ and $\tilde{Y} = [0, +\infty)$ is a subset of another one, we have a general case. Therefore, we need to reduce the domain of the first function in such a way that its range would coincide with the set $\overline{Y} = Y \cap \tilde{Y} = (-\infty, 1] \cap [0, +\infty) = [0, 1]$. To find this new domain $\overline{X}$, we have to solve the inequality $y = 1 - x^2 \geq 0$, which is equivalent to $|x| \leq 1$, whose solution is $-1 \leq x \leq 1$, that is, $\overline{X} = [-1, 1]$. Then the function $f(x) = 1 - x^2$ with the domain $\overline{X} = [-1, 1]$ has the required range $\overline{Y} = [0, 1]$ and using this function as the first one in the composition, we obtain $z = h(x) = g(f(x)) = g(1 - x^2) = \sqrt{1 - x^2} : \overline{X} = [-1, 1] \to Z = [0, +\infty)$. Obviously, $Z = [0, +\infty)$ is a codomain and not the range of the composite function (although Z is the range of $g(y)$). The range of the composition is $\overline{Z} = [0, 1]$.
2. $y = f(x) = \cos x : X = \mathbb{R} \to Y = [-1, 1]$ and $z = g(y) = \sqrt{y} : \tilde{Y} = [0, +\infty) \to Z = [0, +\infty)$.
   In this case $Y \cap \tilde{Y} = [-1, 1] \cap [0, +\infty) = [0, 1]$, and none of the sets $Y = [-1, 1]$ and $\tilde{Y} = [0, +\infty)$ is a subset of another one. Therefore, we need to reduce the domain of the first function to the set whose range would be $\overline{Y} = [0, 1]$. To find this new domain $\overline{X}$, we have to solve the inequality $y = \cos x \geq 0$. Recalling the properties of $\cos x$ (and/or using its graph), we conclude that the inequality

$\cos x \geq 0$ holds when $x \in [-\frac{\pi}{2}, \frac{\pi}{2}] + 2n\pi$, $\forall n \in \mathbb{Z}$. Then the function $f(x) = \cos x$ with the domain $\overline{X} = \cup_{n=-\infty}^{+\infty}[-\frac{\pi}{2} + 2n\pi, \frac{\pi}{2} + 2n\pi]$ has the required range $\overline{Y} = [0, 1]$. Using this function as the first one (and keeping the same second function), we find the composition in the form $z = h(x) = g(f(x)) = g(\cos x) = \sqrt{\cos x} : \overline{X} = \cup_{n=-\infty}^{+\infty}[-\frac{\pi}{2} + 2n\pi, \frac{\pi}{2} + 2n\pi] \to Z = [0, +\infty)$. The image of the composition is $\overline{Z} = [0, 1]$.

9.2 Decomposition of Complicated Functions

There are many situations in Calculus/Analysis when it is convenient or even required to represent a more complicated function as a composition of simpler functions. This procedure is called *decomposition*. Let us solve some examples of this type.

Examples

1. Given a function $h(x) = \cos 2x$, find the functions $y = f(x)$ and $g(y)$ such that $g(f(x)) = h(x)$.
   The formula of $h(x)$ itself suggests a natural decomposition. Choosing $f(x) = 2x$ and $g(y) = \cos y$, we obtain the required formula of the composition $h(x) = g(f(x))$. It remains to specify the domain and range of each function. The original function has the domain $X = \mathbb{R}$ and the range $Z = [-1, 1]$. Then the domain of $f(x)$ should be $X = \mathbb{R}$ that gives the range $Y = \mathbb{R}$. Taking the same Y as the domain of the second function $g(y)$, we obtain its range $Z = [-1, 1]$. Hence, the given function $h(x) = \cos 2x : X = \mathbb{R} \to Z = [-1, 1]$ can be considered as the composition of the following two functions: $y = f(x) = 2x : X = \mathbb{R} \to Y = \mathbb{R}$ and $z = g(y) = \cos y : Y = \mathbb{R} \to Z = [-1, 1]$.
2. Given a function $h(x) = \sqrt{1 - x^2}$, fund the functions $y = f(x)$ and $g(y)$, such that $g(f(x)) = h(x)$.
   The formula of $h(x)$ itself suggests a natural decomposition. Choosing $f(x) = 1 - x^2$ and $g(y) = \sqrt{y}$, we obtain the required formula of the composition $h(x) = g(f(x))$. It remains to specify the domain and range of each function. The original function is defined for $1 - x^2 \geq 0$, that is, its domain is $X = [-1, 1]$. Then the domain of $f(x)$ should be $X = [-1, 1]$, and consequently, the image is $Y = [0, 1]$. Taking the same Y as the domain of the second function $g(y)$, we obtain its range $Z = [0, 1]$. Hence, the given function $h(x) = \sqrt{1 - x^2} : X = [-1, 1] \to Z = [0, 1]$ can be decomposed as follows: $h(x) = g(f(x))$, where $y = f(x) = 1 - x^2 : X = [-1, 1] \to Y = [0, 1]$ and $z = g(y) = \sqrt{y} : Y = [0, 1] \to Z = [0, 1]$.

9.3 Compositions of Specific Types of Functions

Composition of Even/Odd Functions

Property 1 *If $y = f(x) : X_f \to Y_f$ is an even function and $g(y) : Y_f \to \mathbb{R}$ is an arbitrary function, then $g(f(x)) : X_f \to \mathbb{R}$ is even.*

Proof For any $x \in X_f$ we have: $h(-x) = g(f(-x)) = g(f(x)) = h(x)$, which means that $h(x)$ is even. □

Property 2 *If $y = f(x) : X_f \to Y_f$ is odd, $g(y) : Y_f \to \mathbb{R}$ is even and $u(y) : Y_f \to \mathbb{R}$ is odd, then $g(f(x)) : X_f \to \mathbb{R}$ is even and $u(f(x)) : X_f \to \mathbb{R}$ is odd.*

Proof We demonstrate only the case of $h(x) = u(f(x))$. Another case is equally simple. For any $x \in X_f$ we have: $h(-x) = u(f(-x)) = u(-f(x)) = -u(f(x)) = -h(x)$, which means that $h(x)$ is odd. Notice that the symmetry of the domain Y_f of the functions $g(y)$ and $u(y)$ is guaranteed by the fact that Y_f is the range of $f(x)$ with the property $f(-x) = -f(x)$, whence both $y = f(x)$ and $-y = -f(x) = f(-x)$ belong to Y_f. □

The shown results on the parity of the composition are collected in Table 2.3.

Composition of Periodic Functions

Property *If $y = f(x) : X_f \to Y_f$ is periodic with a period T and $g(y) : Y_f \to \mathbb{R}$ is an arbitrary function, then $h(x) = g(f(x)) : X_f \to \mathbb{R}$ is periodic with the period T.*

Proof For any $x \in X_f$ we have $h(x+T) = g(f(x+T)) = g(f(x)) = h(x)$, which means that $h(x)$ is periodic with the period T. □

Remark 1 The composition in the inverted order does not provide, in general, a periodic function, that is, the function $g(f(x)) : X_f \to \mathbb{R}$ where $y = f(x) : X_f \to Y_f$ is an arbitrary function and $g(y) : Y_f \to \mathbb{R}$ is periodic with a period T, can be non-periodic. A counterexample is given by $h(x) = g(f(x)) = \cos x^3 : \mathbb{R} \to \mathbb{R}$ with $y = f(x) = x^3 : \mathbb{R} \to \mathbb{R}$ and $g(y) = \cos y : \mathbb{R} \to \mathbb{R}$. The function $f(x)$ is increasing (and consequently, non-periodic) and $g(y)$ is periodic with the minimum period 2π. The composition $h(x)$ is non-periodic, which can be shown by the contradiction. Indeed, let us suppose that $h(x)$ is periodic, that is, $h(x+T) = \cos(x+T)^3 = \cos x^3 = h(x)$ for some $T > 0$ and for all real x. Then,

Table 2.3 Parity of the composition

$f(x) : X_f \to Y_f$	$g(y) : Y_f \to \mathbb{R}$	$g(f(x)) : X_f \to \mathbb{R}$
Even	Arbitrary	Even
Odd	Even	Even
Odd	Odd	Odd

it follows from the 2π-periodicity of $\cos y$ that $(x+T)^3 = \pm x^3 + 2k\pi$ for some $k \in \mathbb{N}$ and for all $x \in \mathbb{R}$. Consider first the equation with a positive sign, in which case it reduces to $3x^2T + 3xT^2 + T^3 = 2k\pi$. In particular, taking in the last equation $x_1 = 2x$, we obtain $3 \cdot (2x)^2T + 3 \cdot (2x)T^2 + T^3 = 2k\pi$ for the same k. Then the difference between the two equations gives $9x^2T + 3xT^2 = 0$ or $3Tx(3x+T) = 0$. Since T is a non-zero constant, the last equation has no solution for T. In a similar way it can be shown that the equation with a negative sign also has no solution for a non-zero constant T. Therefore, the assumption about periodicity of $h(x)$ is false.

Remark 2 The composition $h(x) = g(f(x))$ does not inherit the fundamental period of $f(x)$, that is, the function $h(x) = g(f(x)) : X_f \to \mathbb{R}$, where $y = f(x) : X_f \to Y_f$ is periodic with a fundamental period T and $g(y) : Y_f \to \mathbb{R}$ is an arbitrary function, has the period T, but this period is not necessarily fundamental for $h(x)$. Indeed, consider the functions $y = f(x) = \cos x : \mathbb{R} \to [-1, 1]$ and $g(y) = y^2 : [-1, 1] \to \mathbb{R}$. The former has the fundamental period 2π and the latter is not periodic. Their composition $h(x) = g(f(x)) = \cos^2 x : \mathbb{R} \to \mathbb{R}$ has period 2π, but its fundamental period is π.

Composition of Monotonic Functions

Property 1 *If $y = f(x) : X_f \to Y_f$ is increasing and $g(y) : Y_f \to \mathbb{R}$ is increasing, then $g(f(x)) : X_f \to \mathbb{R}$ is also increasing. Analogously, if $y = f(x) : X_f \to Y_f$ is non-strictly increasing and $g(y) : Y_f \to \mathbb{R}$ is non-strictly increasing, then $g(f(x)) : X_f \to \mathbb{R}$ is non-strictly increasing. In the same way, decrease of two functions results in increase of their composition.*

Proof We provide the proof only in the case of increase, because the other cases are handled similarly. Take two arbitrary points $x_1, x_2 \in X_f$, $x_1 < x_2$ and notice that increase of $f(x)$ means that $y_1 = f(x_1) < f(x_2) = y_2$. Then, from increase of $g(y)$ it follows that $h(x_1) = g(f(x_1)) = g(y_1) < g(y_2) = g(f(x_2)) = h(x_2)$, $\forall x_1, x_2 \in X_f$, $x_1 < x_2$, which means that $h(x)$ is increasing on X_f. □

Property 2 *If $y = f(x) : X_f \to Y_f$ is increasing and $g(y) : Y_f \to \mathbb{R}$ is decreasing, then $g(f(x)) : X_f \to \mathbb{R}$ is decreasing. Analogously, if $y = f(x) : X_f \to Y_f$ is non-strictly increasing and $g(y) : Y_f \to \mathbb{R}$ is non-strictly decreasing, then $g(f(x)) : X_f \to \mathbb{R}$ is non-strictly decreasing. The same results are true if the order of increasing and decreasing functions is interchanged.*

Proof We demonstrate only the case of increasing $f(x)$ and decreasing $g(x)$, since the other cases are proved in a similar way. Take two arbitrary points $x_1, x_2 \in X_f$, $x_1 < x_2$ and notice that increase of $f(x)$ means that $y_1 = f(x_1) < f(x_2) = y_2$. Then, from decrease of $g(y)$ it follows that $h(x_1) = g(f(x_1)) = g(y_1) > g(y_2) = g(f(x_2)) = h(x_2)$, $\forall x_1, x_2 \in X_f$, $x_1 < x_2$, which means that $h(x)$ is decreasing on X_f. □

Table 2.4 Monotonicity of the composition

$f(x): X_f \to Y_f$	$g(y): Y_f \to \mathbb{R}$	$g(f(x)): X_f \to \mathbb{R}$
Increasing	Increasing	Increasing
Increasing	Decreasing	Decreasing
Decreasing	Increasing	Decreasing
Decreasing	Decreasing	Increasing

The results on the monotonicity of the composition are collected in Table 2.4.

Composition of Concave Functions

We formulate and prove the property of the composition in the case when the inner (first) function in the composition is concave upward, leaving to the reader a similar task for functions concave downward.

Property 1 *If $y = f(x)$ is concave upward (at least non-strictly) on an interval I_f whose image is an interval I_g, and $g(y)$ is concave upward (at least non-strictly) and increasing on I_g, then $g(f(x))$ is non-strictly concave upward on I_f. Additionally, if at least one of the functions $f(x)$ or $g(y)$ is strictly concave, then $g(f(x))$ is also strictly concave.*

Proof We provide the proof in the case of a strict upward concavity of $g(y)$ (the other cases are demonstrated similarly). According to the definition of concavity, we have the following inequalities for the functions $f(x)$ and $g(y)$:

$$f(x) \leq f(x_1) + \frac{f(x_2) - f(x_1)}{x_2 - x_1}(x - x_1), \forall x \in (x_1, x_2), \forall x_1, x_2 \in I_f$$

and

$$g(y) < g(y_1) + \frac{g(y_2) - g(y_1)}{y_2 - y_1}(y - y_1), \forall y \in (y_1, y_2), \forall y_1, y_2 \in I_g.$$

The concavity of $g(y)$ implies the following inequality for the function $h(x) = g(f(x))$:

$$\begin{aligned} h(x) = g(f(x)) = g(y) < g(y_1) + \frac{g(y_2) - g(y_1)}{y_2 - y_1}(y - y_1) \\ = g(f(x_1)) + \frac{g(f(x_2)) - g(f(x_1))}{f(x_2) - f(x_1)}(f(x) - f(x_1)) \\ = h(x_1) + \frac{h(x_2) - h(x_1)}{f(x_2) - f(x_1)}(f(x) - f(x_1)). \end{aligned}$$

Taking into account the positivity of the quotient $\frac{g(y_2)-g(y_1)}{y_2-y_1} = \frac{h(x_2)-h(x_1)}{f(x_2)-f(x_1)}$ (due to strict increasing of $g(y)$) and applying the concavity of $f(x)$, we finalize the evaluation of $h(x)$ in the following manner:

$$\begin{aligned} h(x) &< h(x_1) + \frac{h(x_2)-h(x_1)}{f(x_2)-f(x_1)}(f(x)-f(x_1)) \\ &\le h(x_1) + \frac{h(x_2)-h(x_1)}{f(x_2)-f(x_1)} \cdot \frac{f(x_2)-f(x_1)}{x_2-x_1}(x-x_1) \\ &= h(x_1) + \frac{h(x_2)-h(x_1)}{x_2-x_1}(x-x_1), \forall x \in (x_1,x_2), \forall x_1,x_2 \in I_f. \end{aligned}$$

Hence, we arrive to the definition of (strict) upward concavity of $h(x)$ on the interval I_f. □

Remark 1 If $y = f(x)$ is concave upward (at least non-strictly) on the interval I_f whose image is an interval I_g, and $g(y)$ is concave downward (at least non-strictly) and decreasing on I_g, then $g(f(x))$ is non-strictly concave downward on I_f. Additionally, if at least one of the functions $f(x)$ or $g(y)$ has strict concavity, then $g(f(x))$ is also strictly concave. The proof is similar to that of Property 1.

Remark 2 If $y = f(x)$ is concave upward on an interval I_f whose image is an interval I_g, and $g(y)$ is concave upward and decreasing on I_g, then $g(f(x))$ has no determined concavity. Indeed, the function $f(x) = x^2 : I_f = [0, +\infty) \to I_g = [0, +\infty)$ is concave upward on I_f. The function $g(y) = -y^{1/3} : I_g = [0, +\infty) \to I_h = (-\infty, 0]$ is concave upward and decreasing on I_g. The composition $h(x) = g(f(x)) = -x^{2/3} : I_f = [0, +\infty) \to I_h = (-\infty, 0]$ is concave upward on I_f. Keeping the same $f(x)$, choose now $g(y) = -y^{2/3} : I_g = [0, +\infty) \to I_h = (-\infty, 0]$ which again is concave upward and decreasing on I_g. However, this time the composition $h(x) = g(f(x)) = -x^{4/3} : I_f = [0, +\infty) \to I_h = (-\infty, 0]$ is concave downward on I_f. Moreover, combining the functions of these two examples, we can construct the composition that changes the type of concavity on its domain. Indeed, keeping the same $f(x) = x^2 : I_f = [0, +\infty) \to I_g = [0, +\infty)$ and choosing $g(y) = \left\{ \begin{matrix} -y^{1/3}, y \in [0, 1] \\ -y^{2/3}, y \in [1, +\infty) \end{matrix} \right\} : I_g = [0, +\infty) \to I_h = (-\infty, 0]$, we get the composite function $h(x) = g(f(x)) = \left\{ \begin{matrix} -x^{2/3}, x \in [0, 1] \\ -x^{4/3}, x \in [1, +\infty) \end{matrix} \right\} : I_f = [0, +\infty) \to I_h = (-\infty, 0]$, which is concave upward on the interval $[0, 1]$ and concave downward on the interval $[1, +\infty)$.

Finally, we formulate the corresponding property in the case of concave downward function $f(x)$, leaving to the reader the task to prove this statement.

Property 2 *If $y = f(x)$ is concave downward (at least non-strictly) on an interval I_f whose image is an interval I_g, and $g(y)$ is concave downward (at least non-strictly) and increasing on I_g, then $g(f(x))$ is non-strictly concave downward on*

Table 2.5 Concavity of the composition

$f(x) : I_f \to I_g$	$g(y) : I_g \to \mathbb{R}$	$g(f(x)) : I_f \to \mathbb{R}$
Concave upward	Concave upward and increasing	Concave upward
Concave upward	Concave downward and decreasing	Concave downward
Concave downward	Concave downward and increasing	Concave downward
Concave downward	Concave upward and decreasing	Concave upward

I_f. Additionally, if at least one of the functions $f(x)$ or $g(y)$ has strict concavity, then $g(f(x))$ is also strictly concave.

Remark If $y = f(x)$ is concave downward (at least non-strictly) on an interval I_f whose image is an interval I_g, and $g(y)$ is concave upward (at least non-strictly) and decreasing on I_g, then $g(f(x))$ is non-strictly concave upward on I_f. Additionally, if at least one of the functions $f(x)$ or $g(y)$ has strict concavity, then $g(f(x))$ is also strictly concave.

The obtained results on the concavity of the composition are collected in Table 2.5.

Remark About Notations of the Composite Functions For the purposes of the definition and initial illustration it is convenient to use independent variable x for the first function and independent variable y for the second, in order to emphasize the relationship $y = f(x)$ between the two variables in the construction of a composite function. However, in order to compare different functions it is more appropriate to use the same notation x of an independent variable for all the functions. Hereinafter, in the most cases, we will follow this more usual notation.

10 Elementary Transformations of Functions and Their Graphs

In this section we consider the transformations of the graph of an original function $y = f(x)$ caused by linear variations of both the independent variable $x \to ax + b$ and values of function $y \to cy + d$, where a, b, c, d are constant and $a \neq 0$, $c \neq 0$.

10.1 Vertical and Horizontal Translations

Vertical Translation

Given the graph of a function $f(x)$, the graph of the function $f(x) + c$ is obtained by *vertical translation (shift)* of the graph of $f(x)$ by c units upward. Hereinafter we understand that for negative c the translation is made $|c|$ units downward. In other words, if $c > 0$ then the graph of $f(x)$ is moved c units upward to obtain the graph of $f(x) + c$, and c units downward to obtain the graph of $f(x) - c$.

Indeed, consider an arbitrary point $P_0 = (x_0, y_0) = (x_0, f(x_0))$ of the graph Γ of the function $f(x)$. Then the corresponding point $\tilde{P}_0 = (x_0, \tilde{y}_0) = (x_0, f(x_0)+c)$ belongs to the graph $\tilde{\Gamma}$ of the function $f(x) + c$ just because its coordinates satisfy the relation $\tilde{y}_0 = f(x_0) + c$. Geometrically this means that the point $\tilde{P}_0$ is obtained moving the point P_0 a distance of c units upward (parallelly to the y-axis). Notice that in the case of negative c, the vertical shift is made $|c|$ units downward. Since this translation is applied to every point of Γ, the entire graph $\tilde{\Gamma}$ is obtained by shifting Γ a distance of c units upward.

Consequently, in analytic terms, the function $f(x) + c$ has the same domain as $f(x)$ and the range shifted c units. In particular, if the range of $f(x)$ is an interval (a, b), the function $f(x) + c$ has the range $(a + c, b + c)$ (the same is true for half-open and closed intervals). It is understood that if $a = -\infty$ in the interval (a, b), then $a + c = -\infty$, and if $b = +\infty$, then $b + c = +\infty$. A function $f(x) + c$ has the same type of monotonicity and concavity as $f(x)$, and the same points of extremum and inflection (considering here only the x-coordinates of these points, since the y-coordinates are shifted c units).

See an illustration of the vertical translation in Fig. 2.55.

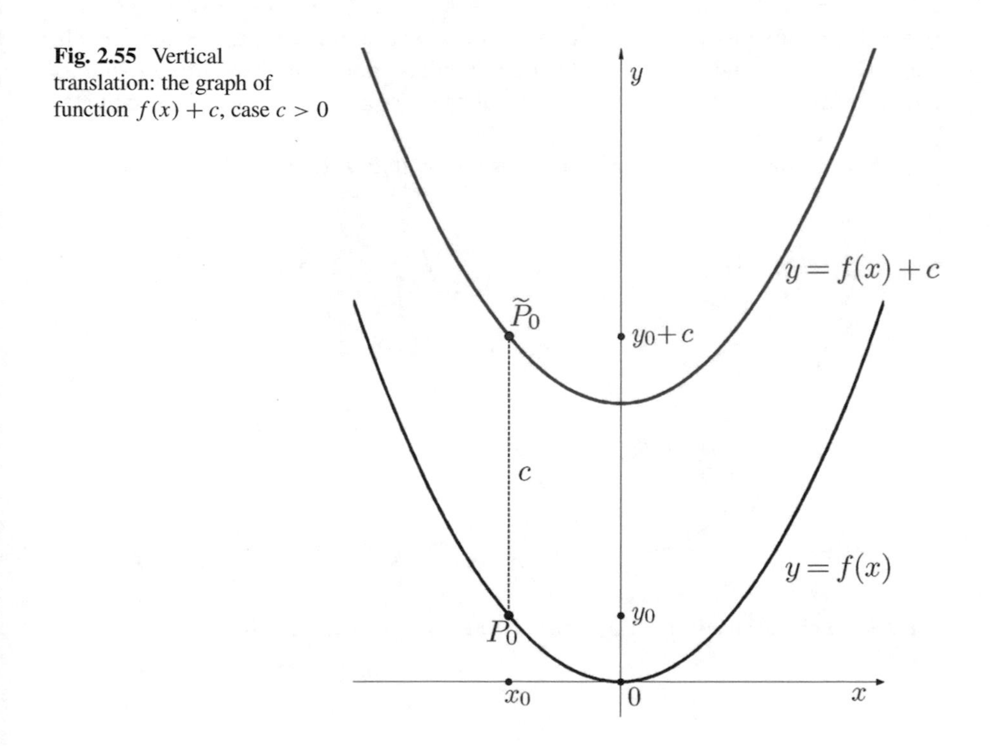

Fig. 2.55 Vertical translation: the graph of function $f(x) + c$, case $c > 0$

Horizontal Translation

Given the graph of a function $f(x)$, the graph of the function $f(x+c)$ is obtained by *horizontal translation (shift)* of the graph of $f(x)$ by c units to the left. Hereinafter we understand that for negative c the translation is made $|c|$ units to the right. In other words, if $c > 0$ then the graph of $f(x)$ is moved c units to the left to obtain the graph of $f(x+c)$, and c units to the right to obtain the graph of $f(x-c)$.

Indeed, consider an arbitrary point $P_0 = (x_0, y_0) = (x_0, f(x_0))$ of the graph Γ of the function $f(x)$. Then the corresponding point $\tilde{P}_0 = (\tilde{x}_0, y_0) = (x_0 - c, f(x_0))$ belongs to the graph $\tilde{\Gamma}$ of the function $f(x+c)$ just because its coordinates satisfy the relation $y_0 = f(\tilde{x}_0 + c) = f(x_0)$. Geometrically this means that the point $\tilde{P}_0$ is obtained by moving P_0 a distance of c units to the left (parallelly to the x-axis). Notice that if c is negative, the horizontal shift is made $|c|$ units to the right. Since this translation is applied to every point of Γ, the entire graph $\tilde{\Gamma}$ is obtained by shifting Γ a distance of c units to the left.

Consequently, in analytic terms, the function $f(x+c)$ has the same range as $f(x)$ and its domain is shifted $-c$ units. For example, the domain (a, b) of $f(x)$ will become the domain $(a-c, b-c)$ of $f(x+c)$ (the same is true for half-open and closed intervals). It is understood that if $a = -\infty$ in the interval (a, b), then $a+c = -\infty$, and if $b = +\infty$, then $b+c = +\infty$. A function $f(x+c)$ keeps the same type of monotonicity and concavity as $f(x)$, albeit on the intervals of the domain shifted $-c$ units, and the extremum and inflection points are also shifted $-c$ units (considering here only the x-coordinates of these points, since their y-coordinates coincide with those of $f(x)$).

See an illustration of the horizontal translation in Fig. 2.56.

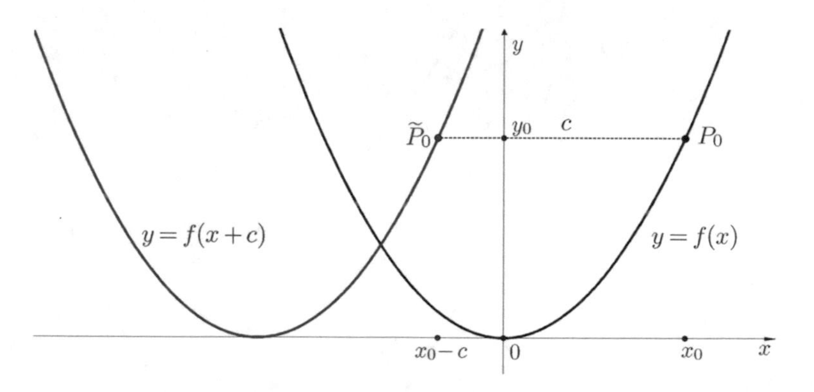

Fig. 2.56 Horizontal translation: the graph of function $f(x+c)$, case $c > 0$

10.2 *Vertical and Horizontal Reflection*

Vertical Reflection

Given the graph of a function $f(x)$, the graph of $-f(x)$ is obtained by *vertical reflection (symmetry)* of the graph of $f(x)$ about the x-axis.

Indeed, consider an arbitrary point $P_0 = (x_0, y_0) = (x_0, f(x_0))$ of the graph Γ of the function $f(x)$. Then the corresponding point $\tilde{P}_0 = (x_0, \tilde{y}_0) = (x_0, -f(x_0))$ belongs to the graph $\tilde{\Gamma}$ of the function $-f(x)$ simply because its coordinates satisfy the formula of $-f(x)$: $\tilde{y}_0 = -f(x_0)$. Geometrically this means that the point $\tilde{P}_0$ is obtained by reflecting P_0 with respect to the x-axis. Since this is applied to every point of Γ, the entire graph $\tilde{\Gamma}$ is obtained by reflecting Γ about the x-axis.

Consequently, in analytic terms, the function $-f(x)$ has the same domain as $f(x)$ and if the range of $f(x)$ is (a, b) then the range of $-f(x)$ will be $(-b, -a)$ (the same is true for half-open and closed intervals). It is understood that if $a = -\infty$ in the interval (a, b), then $-a = +\infty$, and if $b = +\infty$, then $-b = -\infty$. The function $-f(x)$ will have the inverted type of monotonicity (on the sets where $f(x)$ is increasing $-f(x)$ will be decreasing and vice-verse) and the inverted type of concavity (the intervals of upward concavity of $f(x)$ will become the intervals of downward concavity of $-f(x)$ and vice-verse). The function $-f(x)$ will keep the extremum and inflection points of $f(x)$, but changing the type of the extrema (the maximum of $f(x)$ will be the minimum of $-f(x)$ and vice-verse), considering only the x-coordinates of these points, since the y-coordinates will have the opposite sign.

See an illustration of the vertical reflection in Fig. 2.57.

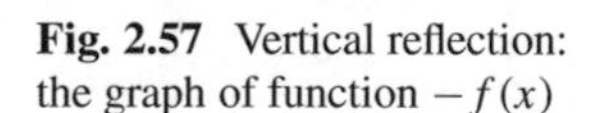

Fig. 2.57 Vertical reflection: the graph of function $-f(x)$

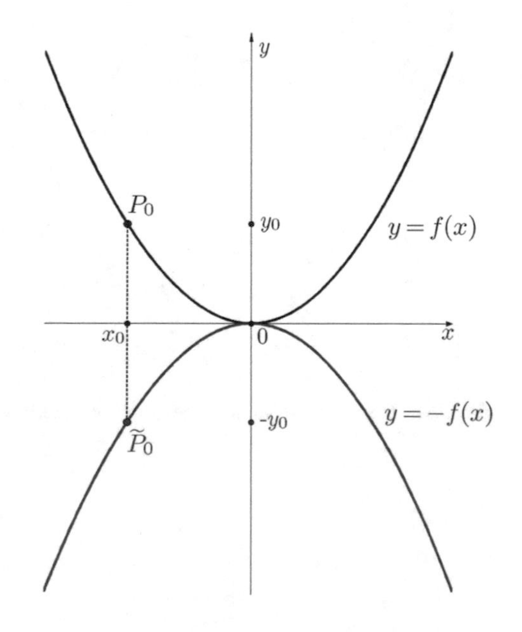

Horizontal Reflection

Given the graph of a function $f(x)$, the graph of $f(-x)$ is obtained by *horizontal reflection (symmetry)* of the graph of $f(x)$ about the y-axis.

Indeed, consider an arbitrary point $P_0 = (x_0, y_0) = (x_0, f(x_0))$ of the graph Γ of the function $f(x)$. Then the corresponding point $\tilde{P}_0 = (\tilde{x}_0, y_0) = (-x_0, f(x_0))$ belongs to the graph $\tilde{\Gamma}$ of the function $f(-x)$ simply because its coordinates satisfy the formula of $f(-x)$: $y_0 = f(-(-x_0)) = f(x_0)$. Geometrically this means that the point $\tilde{P}_0$ is obtained by reflecting P_0 with respect to the y-axis. Since this is applied to every point of Γ, the entire graph $\tilde{\Gamma}$ is obtained by reflecting Γ about the y-axis.

Consequently, in analytic terms, the function $f(-x)$ has the same range as $f(x)$, but the domain of $f(-x)$ is symmetric to the domain of $f(x)$ about the origin. In particular, if the domain of $f(x)$ is (a, b) then the domain of $f(-x)$ will be $(-b, -a)$ (the same is true for half-open and closed intervals). It is understood that if $a = -\infty$ in the interval (a, b), then $-a = +\infty$, and if $b = +\infty$, then $-b = -\infty$. The function $f(-x)$ has the inverted type of monotonicity (if $f(x)$ is increasing on a set S, then $f(-x)$ will decrease on $\tilde{S}$ symmetric to S about the origin and vice-verse), but it keeps the same type of concavity on the corresponding intervals (if $f(x)$ is concave upward on an interval (a, b), then $f(-x)$ will be concave upward on the interval $(-b, -a)$ and vice-verse). The relationship between the extremum and inflection points of the two functions is as follows: if $f(x)$ has minimum at the point $P_1 = (a, f(a))$, then $f(-x)$ has minimum at the point $P_2 = (-a, f(a))$ (the same is true for maximum), and if $f(x)$ has the inflection point $P_1 = (a, f(a))$, then $f(-x)$ will have the inflection point $P_2 = (-a, f(a))$.

See an illustration of the horizontal reflection in Fig. 2.58.

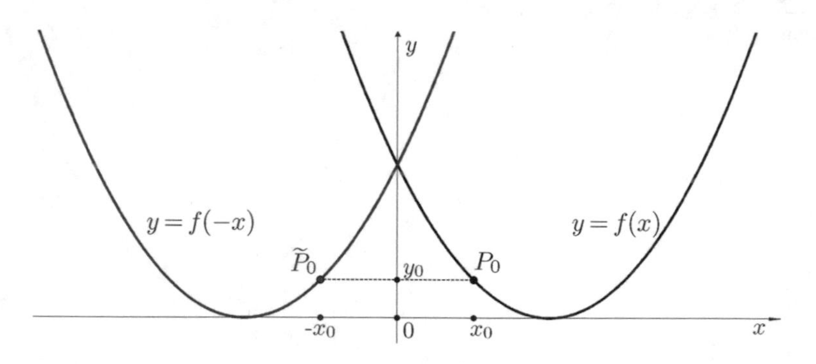

Fig. 2.58 Horizontal reflection: the graph of function $f(-x)$

10.3 Vertical and Horizontal Stretching/Shrinking

Vertical Stretching/Shrinking

Given the graph of a function $f(x)$, the graph of the function $cf(x)$, $c > 0$ is obtained by *stretching (expanding)* the graph of $f(x)$ *vertically* by a factor of c. Hereinafter for the values $0 < c < 1$ the stretching is understood as *shrinking* by a factor of $\frac{1}{c}$. In other words, if $c > 1$, then the graph of $f(x)$ is expanded c times away from the x-axis, while for $0 < c < 1$ the graph of $f(x)$ is contracted $\frac{1}{c}$ times toward the x-axis.

Indeed, consider an arbitrary point $P_0 = (x_0, y_0) = (x_0, f(x_0))$ of the graph Γ of the function $f(x)$. Then the corresponding point $\tilde{P}_0 = (x_0, \tilde{y}_0) = (x_0, cf(x_0))$ belongs to the graph $\tilde{\Gamma}$ of the function $cf(x)$ because its coordinates satisfy the formula of $cf(x)$: $\tilde{y}_0 = cf(x_0)$. This means that the y-coordinate of the point $\tilde{P}_0$ is obtained by increasing c times the y-coordinate of P_0 if $c > 1$ or decreasing $\frac{1}{c}$ times the y-coordinate of P_0 if $0 < c < 1$. Geometrically the point $\tilde{P}_0$ is obtained by moving P_0 away from the x-axis c times if $c > 1$ or moving toward the x-axis $\frac{1}{c}$ times if $0 < c < 1$. Since this is applied to every point of Γ, the entire graph $\tilde{\Gamma}$ is obtained by stretching Γ vertically by a factor of c if $c > 1$ or shrinking vertically by a factor of $\frac{1}{c}$ if $0 < c < 1$.

Consequently, in analytic terms, the function $cf(x)$ has the same domain as $f(x)$, but the range of $cf(x)$ grows c times comparing with $f(x)$. In particular, if the range of $f(x)$ is (a, b), then the range of $cf(x)$ will be (ca, cb) (the same is true for half-open and closed intervals). It is understood that if $a = -\infty$ in the interval (a, b), then $ca = -\infty$, and if $b = +\infty$, then $cb = +\infty$. The function $cf(x)$ keeps the type of monotonicity and concavity of the function $f(x)$, and it also keeps the extremum and inflection points (considering here only the x-coordinates of these points, since the y-coordinates will increase c times).

See an illustration of the vertical stretching in Fig. 2.59.

If we need to find the graph of the function $h(x) = cf(x)$, $c < 0$, then we represent $h(x)$ as the following composition: $h(x) = -g(x)$, $g(x) = -cf(x)$. The first transformation from $f(x)$ to $g(x)$ is a vertical expansion/contraction by a factor of $-c > 0$, and the second transformation is symmetric reflection about the x-axis. These transformations have already been considered.

Horizontal Stretching/Shrinking

Given the graph of a function $f(x)$, the graph of the function $f(cx)$, $c > 0$ is obtained by *shrinking (contracting)* the graph of $f(x)$ *horizontally* by a factor of c. Hereinafter for the values $0 < c < 1$ the shrinking is understood as *stretching* by a factor of $\frac{1}{c}$. In other words, if $c > 1$ then the graph of $f(x)$ is compressed c times toward the y-axis, while for $0 < c < 1$ the graph of $f(x)$ is expanded $\frac{1}{c}$ times away from the y-axis.

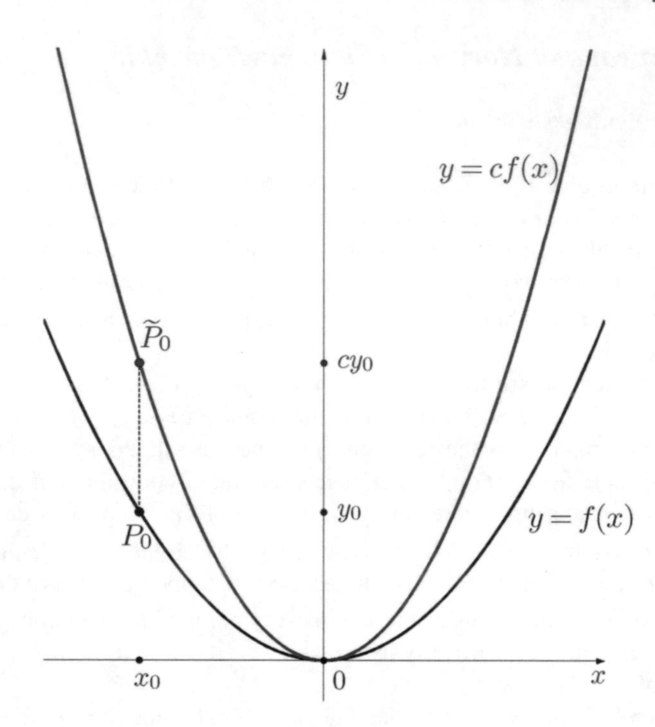

Fig. 2.59 Vertical stretching: the graph of function $cf(x)$, case $c > 1$

Indeed, consider an arbitrary point $P_0 = (x_0, y_0) = (x_0, f(x_0))$ of the graph Γ of the function $f(x)$. Then the corresponding point $\tilde{P}_0 = (\tilde{x}_0, y_0) = (\frac{1}{c}x_0, f(x_0))$ belongs to the graph $\tilde{\Gamma}$ of the function $f(cx)$ because its coordinates satisfy the formula of $f(cx)$: $y_0 = f(c\tilde{x}_0) = f(x_0)$. This means that the x-coordinate of the point $\tilde{P}_0$ is obtained by decreasing c times the x-coordinate of P_0 if $c > 1$ or increasing $\frac{1}{c}$ times the x-coordinate of P_0 if $0 < c < 1$. Geometrically the point $\tilde{P}_0$ is obtained by approximating P_0 to the y-axis c times if $c > 1$ or taking away from the y-axis $\frac{1}{c}$ times if $0 < c < 1$. Since this is applied to every point of Γ, the entire graph $\tilde{\Gamma}$ is obtained by shrinking Γ horizontally by a factor of c if $c > 1$ or stretching horizontally by a factor of $\frac{1}{c}$ if $0 < c < 1$.

Consequently, in analytic terms, the function $f(cx)$ has the same range as $f(x)$, but the domain of $f(cx)$ contracts c times comparing with $f(x)$. In particular, if the domain of $f(x)$ is (a, b), then the domain of $f(cx)$ will be $(\frac{a}{c}, \frac{b}{c})$ (the same is true for half-open and closed intervals). It is understood that if $a = -\infty$ in the interval (a, b), then $\frac{a}{c} = -\infty$, and if $b = +\infty$, then $\frac{b}{c} = +\infty$. The function $f(cx)$ keeps the type of monotonicity and concavity of the function $f(x)$ on the corresponding intervals: for instance, if $f(x)$ is increasing on (a, b), then $f(cx)$ will be increasing on $(\frac{a}{c}, \frac{b}{c})$. Similarly, $f(cx)$ keeps the extremum and inflection points with the due change of the x-coordinate: for instance, if $f(x)$ has a minimum at a point $P_1 = (a, f(a))$, then $f(cx)$ will have a minimum at the point $P_2 = (\frac{a}{c}, f(a))$.

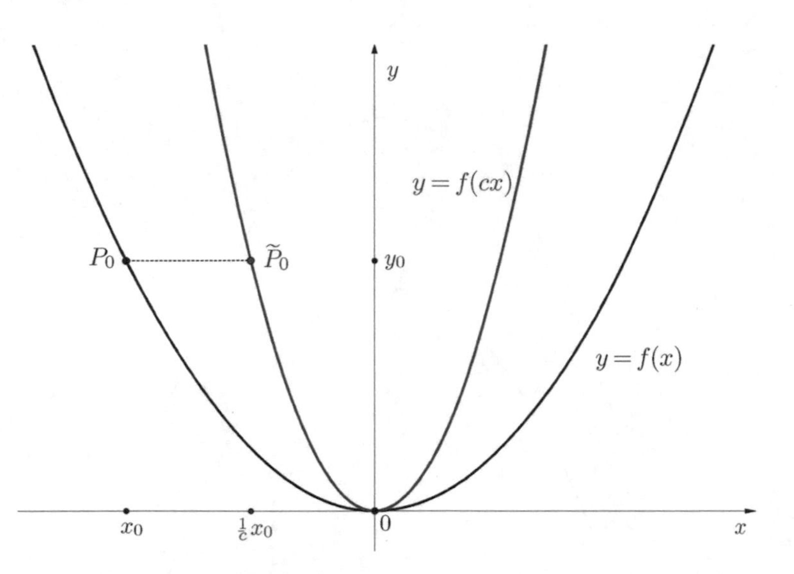

Fig. 2.60 Horizontal shrinking: the graph of function $f(cx)$, case $c > 1$

See an illustration of the horizontal shrinking in Fig. 2.60.

If we need to find the graph of the function $h(x) = f(cx)$, $c < 0$, then we represent $h(x)$ as the following composition $h(x) = g(-x)$, $g(x) = f(-cx)$. The first transformation from $f(x)$ to $g(x)$ is a horizontal expansion/contraction by a factor of $-c > 0$, and the second transformation is symmetric reflection about the y-axis. These transformations have already been considered.

Examples

1. $f(x) = -x^2 \to g(x) = 1 - x^2$
   Recall that the graph of $f(x) = -x^2$ is a parabola with the vertex at the origin and concave downward. This graph is original (given). Writing the second function in the form $g(x) = 1 + f(x)$, we conclude that the graph of $g(x)$ is obtained by vertical translation of the original graph 1 unit upward (see Fig. 2.61).
2. $f(x) = -x^2 \to g(x) = -(x-3)^2$
   Again the graph of $f(x) = -x^2$ is original (given). Representing the second function in the form $g(x) = f(x-3)$, we conclude that the graph of $g(x)$ is obtained by horizontal translation of the original graph 3 units to the right (see Fig. 2.62).
3. $f(x) = \cos x \to g(x) = \frac{1}{3}\cos x$
   The graph of $f(x) = \cos x$ has already been shown and it is the original graph in this example. Writing the second function in the form $g(x) = \frac{1}{3}f(x)$, we conclude that the graph of $g(x)$ is obtained by the vertical shrinking of the original graph 3 times toward the x-axis (see Fig. 2.63).

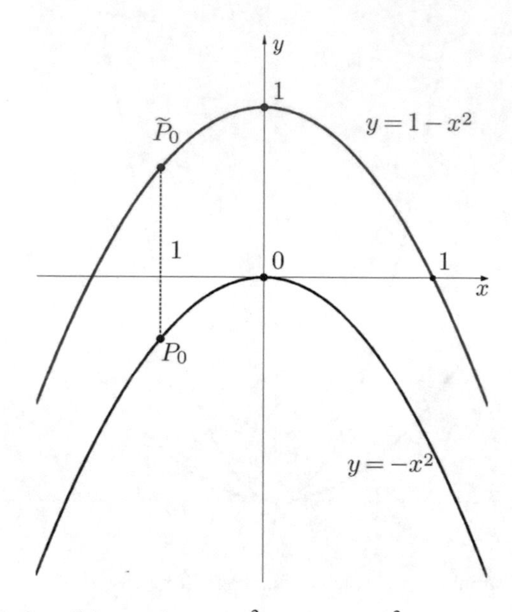

Fig. 2.61 Transformation of the graph $y = -x^2 \to y = 1 - x^2$

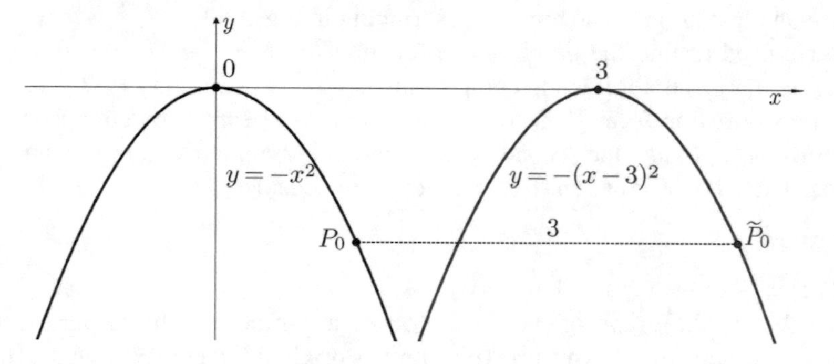

Fig. 2.62 Transformation of the graph $y = -x^2 \to y = -(x-3)^2$

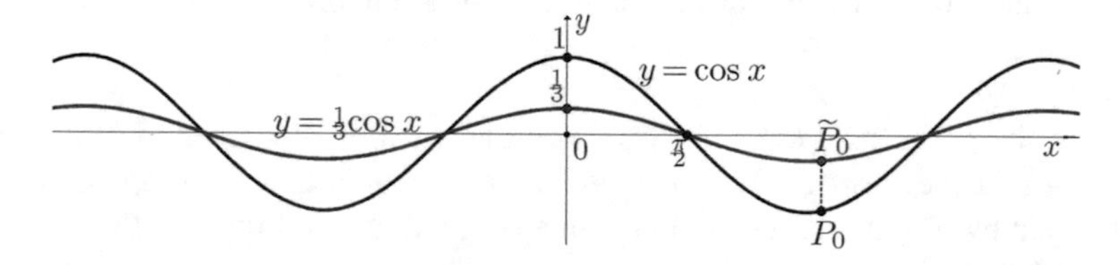

Fig. 2.63 Transformation of the graph $y = \cos x \to y = \frac{1}{3}\cos x$

4. $f(x) = \cos x \to g(x) = \cos 2x$
   The graph of $f(x) = \cos x$ is given in this example. Representing the second function in the form $g(x) = f(2x)$, we conclude that the graph of $g(x)$ is obtained by contracting the original graph 2 times to the y-axis (see Fig. 2.64).

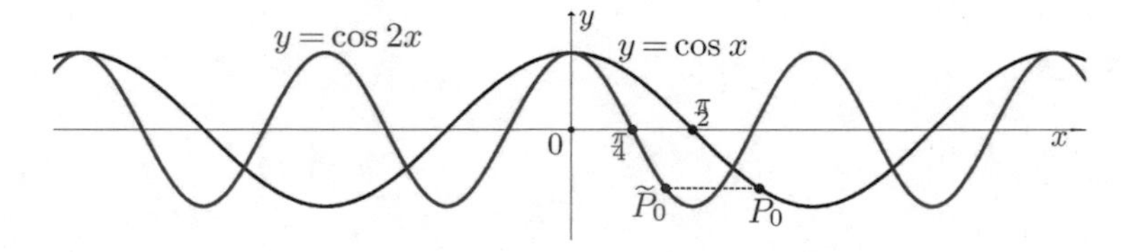

Fig. 2.64 Transformation of the graph $y = \cos x \to y = \cos 2x$

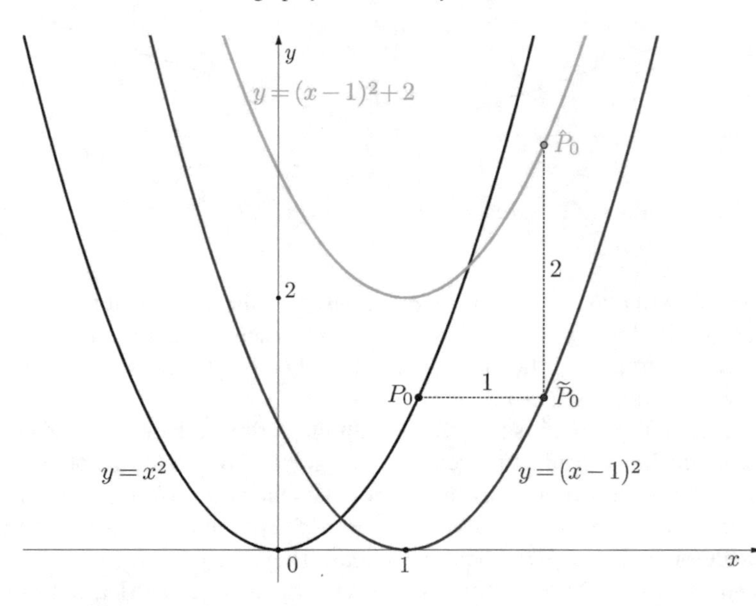

Fig. 2.65 Transformation of the graph $y = x^2 \to y = (x-1)^2 + 2$

5. $f(x) = x^2 \to g(x) = (x-1)^2 + 2$
   Recall that the graph of $f(x) = x^2$ is a parabola with the vertex at the origin and concave upward. This parabola is given in this example. From the graph of $f(x) = x^2$ we obtain first the graph of $\tilde{f}(x) = (x-1)^2 = f(x-1)$ by translating the original graph 1 unit to the right. Next, considering the graph of $f(x-1)$ as given, we find the graph of $g(x) = (x-1)^2 + 2$. Comparing $\tilde{f}(x)$ and $g(x)$, we see that $g(x) = \tilde{f}(x) + 2$, which represents geometrically the shifting upward of the graph of $\tilde{f}(x)$ a distance of 2 units. Hence, the graph of $g(x) = (x-1)^2 + 2$ is obtained from the original graph of $f(x) = x^2$ by translating the latter 1 unit to the right and 2 units upward (see Fig. 2.65).
6. $f(x) = |2x| \to g(x) = |\frac{1}{3}x|$
   Recall that the graph of $f(x) = |2x|$ consists of the two line parts $y = -2x$, for $x \le 0$ and $y = 2x$ for $x \ge 0$ which meet at the origin (vertex of this graph). This graph is original in this example. To obtain the graph of $g(x) = |\frac{1}{3}x|$ we can stretch the original graph horizontally 6 times, since $g(x) = |\frac{1}{6}2x| = f(\frac{1}{6}x)$.

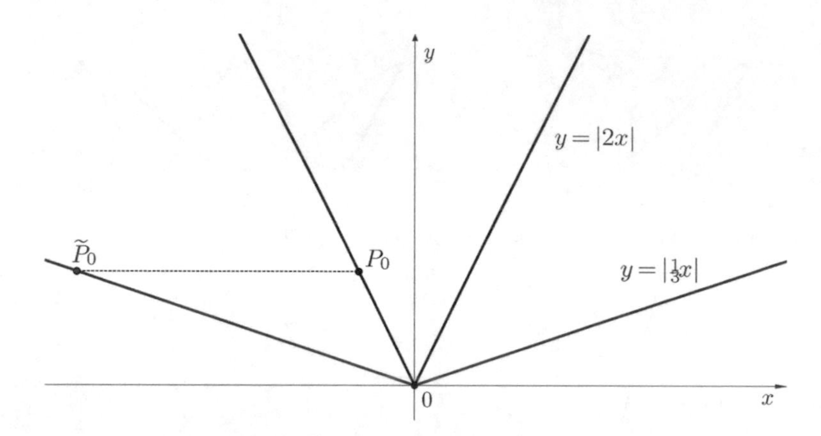

Fig. 2.66 Transformation of the graph $y = |2x| \to y = |\frac{1}{3}x|$

Another way to obtain the same graph is based on the equivalent representation of $g(x)$ in the form $g(x) = \frac{1}{6}|2x| = \frac{1}{6}f(x)$. Then the graph of $g(x)$ can be obtained by shrinking of the original graph vertically 6 times (see Fig. 2.66).

7. $f(x) = x^2 \to g(x) = 3 - (2x + 1)^2$.
   The graph of $f(x) = x^2$ is a parabola with the vertex at the origin and concave upward, which is considered given in this example. To find the graph of $g(x) = 3-(2x+1)^2$ we have to perform a chain of elementary transformations. First, we transform $f(x) = x^2$ into $f_1(x) = -x^2 = -f(x)$, which means geometrically to reflect the graph of $f(x)$ about the x-axis. Then, we transform $f_1(x) = -x^2$ into $f_2(x) = -(2x)^2 = f_1(2x)$, which corresponds geometrically to making the graph of $f_1(x)$ two times closer to the y-axis. Next, we move from $f_2(x) = -(2x)^2$ to $f_3(x) = -(2x + 1)^2 = -(2(x + \frac{1}{2}))^2 = f_2(x + \frac{1}{2})$, which consists geometrically in translation of the graph of $f_2(x)$ a distance of $\frac{1}{2}$ to the left. Finally, from $f_3(x) = -(2x + 1)^2$ we arrive to $g(x) = 3 - (2x + 1)^2 = 3 + f_3(x)$, by shifting the graph of $f_3(x)$ three units upward. Thus, this chain of transformations leads to the following consecutive geometric steps: the graph of $f(x)$ is reflected about the x-axis, then the obtained graph is shrunk twice horizontally, the result is shifted half a unit to the left and finally the last graph is shifted three unites upward.
   An illustration of these transformations is shown in Fig. 2.67.
   Notice that there are other ways to arrive to the same graph through elementary transformations.
8. $f(x) = x^2 \to g(x) = 3(\frac{1}{2}x - 1)^2 - 2$.
   Reasoning used in this example is quite similar to the previous one. For this reason we represent the chain of transformations in a concise form. The graph of $g(x) = 3(\frac{1}{2}x - 1)^2 - 2$ can be obtained from the graph of $f(x) = x^2$ using the following sequence of elementary geometric transformations: first, the graph of $f(x)$ is stretched twice horizontally, then the result is translated two units to the right, next the obtained graph is stretched thrice vertically, and finally the last

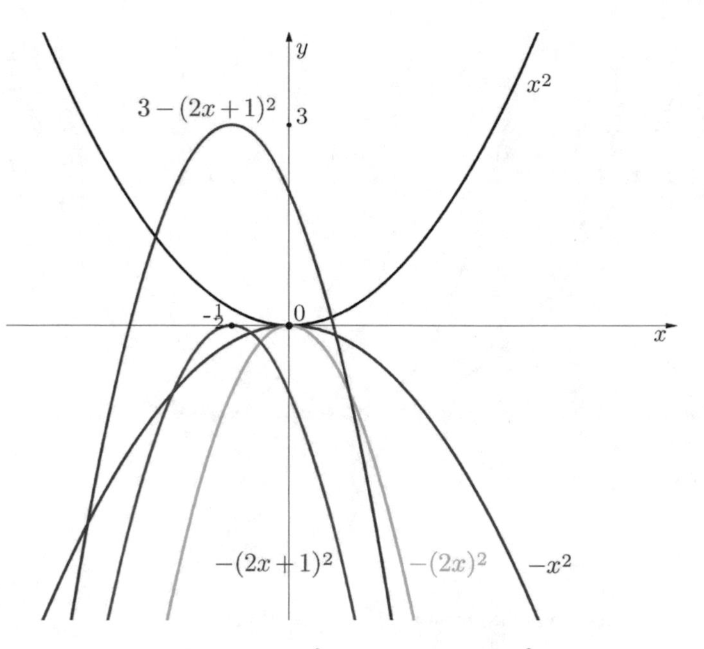

Fig. 2.67 Transformation of the graph $y = x^2 \to y = 3 - (2x+1)^2$

graph is moved two units down. These geometric transformations are based on the following chain of the analytic transformations:

$$f(x) = x^2 \to \left(\frac{1}{2}x\right)^2 \to \left(\frac{1}{2}(x-2)\right)^2$$
$$= \left(\frac{1}{2}x - 1\right)^2 \to 3\left(\frac{1}{2}x - 1\right)^2 \to 3\left(\frac{1}{2}x - 1\right)^2 - 2 = g(x)\,.$$

The detailed explanations are left to the reader.

Figure 2.68 illustrates the described geometric transformations.

Notice that there are other ways to arrive to the same graph through elementary transformations.

11 Injective, Surjective and Bijective Functions

Definition A function $y = f(x) : X \to Y$ is *injective* (*one-to-one*) if $f(x_1) \neq f(x_2)$ whenever $x_1 \neq x_2$, $x_1, x_2 \in X$, that is, an injective function maps different elements of the domain to different elements of the range.

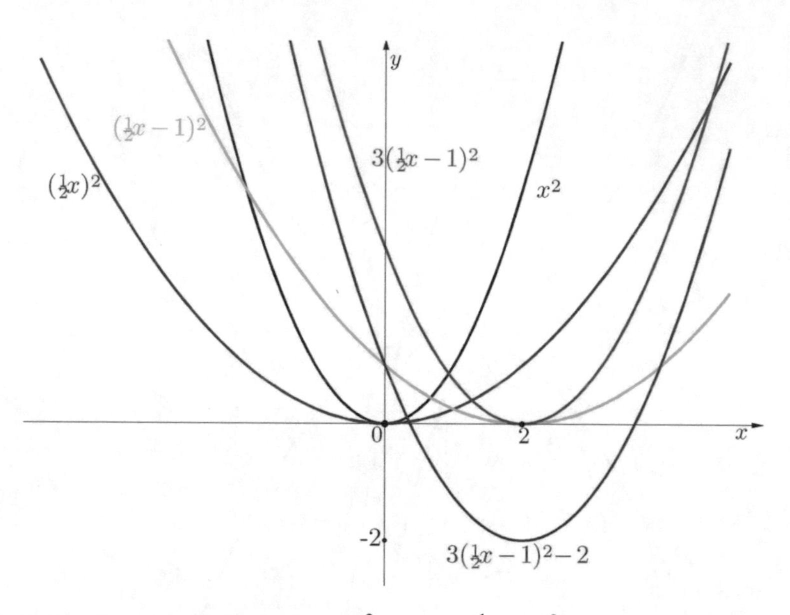

Fig. 2.68 Transformation of the graph $y = x^2 \to y = 3(\frac{1}{2}x - 1)^2 - 2$

Geometrically, the condition of an injective function can be verified by so-called *horizontal line test*: a function is injective if no horizontal line (parallel to the x-axis) intersects its graph more than once.

An injective function is also called *injection*.

Definition A function $y = f(x) : X \to Y$ is *surjective* (*onto*) if its codomain Y coincides with the range: $Y = f(X)$. In other words, for every element $y \in Y$ of a surjective function there exists at least one element $x \in X$ such that $f(x) = y$.

In geometric terms it can be formulated as follows: a function with the codomain Y is surjective if any horizontal line $y = y_0$, $y_0 \in Y$ intersects its graph at least once.

A surjective function is also called *surjection*.

Definition A function is *bijective* if it is both injective and surjective. A bijective function is also called *bijection* or *one-to-one correspondence*.

Geometrically, a bijection is characterized by the following property: a function with the codomain Y is bijective if any horizontal line $y = y_0$, $y_0 \in Y$ intersects its graph exactly one time.

Notice that a *one-to-one correspondence* is a particular case of a *one-to-one function*: the former is both injective and surjective, while the latter is injective, but can be non-surjective.

Composition of Injections *If $y = f(x) : X \to Y$ is injective and $g(y) : Y \to Z$ is injective, then their composition $g(f(x)) : X \to Z$ is also injective.*

Proof Since $f(x)$ is injective, for any pair of points $x_1, x_2 \in X$, $x_1 \neq x_2$ we have $y_1 = f(x_1) \neq f(x_2) = y_2$, $y_1, y_2 \in \tilde{Y} \subset Y$. Next, notice that $g(y)$ is injective on $\tilde{Y}$, because it is injective on Y which contains the range $\tilde{Y}$ of $f(x)$. Therefore, for any pair of points $y_1, y_2 \in \tilde{Y}$, $y_1 \neq y_2$ it follows that $z_1 = g(y_1) \neq g(y_2) = z_2$. Hence, the property of injection is satisfied for the composition $h(x) = g(f(x))$ defined on X: $h(x_1) = g(f(x_1)) = g(y_1) \neq g(y_2) = g(f(x_2)) = h(x_2)$ whenever $\forall x_1 \neq x_2, x_1, x_2 \in X$. □

Composition of Surjections *If $y = f(x) : X \to Y$ is surjective and $g(y) : Y \to Z$ is surjective, then their composition $g(f(x)) : X \to Z$ is also surjective.*

Proof Take an arbitrary element $z_0 \in Z$. Since $g(y)$ is surjective, there exists at least one element $y_0 \in Y$ such that $g(y_0) = z_0$. In turn, $f(x)$ is subjective, which means that there exists at least one element $x_0 \in X$ such that $f(x_0) = y_0$. Therefore, for any element $z_0 \in Z$ there exists at least one element $x_0 \in X$ such that $h(x_0) = g(f(x_0)) = g(y_0) = z_0$, that is, $h(x) : X \to Z$ is surjective. □

Composition of Bijections *If $y = f(x) : X \to Y$ is bijective and $g(y) : Y \to Z$ is bijective, then their composition $g(f(x)) : X \to Z$ is also bijective.*

Proof This result follows directly from composition of injections and surjections. □

Examples

1a. $y = f(x) = x : X = \mathbb{R} \to Y = \mathbb{R}$.
The codomain $Y = \mathbb{R}$ of this function is its range, because for each $y \in Y = \mathbb{R}$ there exists $x = y \in X = \mathbb{R}$ such that $f(x) = f(y) = y$. This means that the function is surjective. Besides, the value of x, which corresponds to the chosen element y, is unique. In other words, if we take two different elements of the domain $x_1 \neq x_2$, then $y_1 = f(x_1) = x_1 \neq x_2 = f(x_2) = y_2$, which means that $f(x)$ is injective. Therefore, the function is bijective.

1b. $y = f(x) = x : X = [0, +\infty) \to Y = \mathbb{R}$.
The codomain $Y = \mathbb{R}$ of this function does not coincide with its range, since there is no $x \in X = [0, +\infty)$ which the function maps to a negative y. This means that the function is not surjective (and consequently, it is not bijective). At the same time, any two different elements of the domain $x_1 \neq x_2$ the function sends to the different elements of the codomain: $y_1 = f(x_1) = x_1 \neq x_2 = f(x_2) = y_2$, which means that $f(x)$ is injective.

1c. $y = f(x) = x : X = [0, +\infty) \to Y = [0, +\infty)$.
The codomain $Y = [0, +\infty)$ of this function is its range, because for each $y \geq 0$ there exists $x = y \geq 0$ such that $f(x) = f(y) = y$. This means that the function is surjective. Besides, this element x, which corresponds to the chosen value of y, is unique. In other words, if we take two different elements of the domain $x_1 \neq x_2$, then $y_1 = f(x_1) = x_1 \neq x_2 = f(x_2) = y_2$, which means that $f(x)$ is injective. Consequently, the function is bijective.

Remark Notice that if a function is injective on a domain X, it is also injective on any subset of X, whatever set is chosen for its codomain. This property, observed in the above examples, is a general rule. Indeed, if $f(x)$ is injective on X, then for $\forall x_1, x_2 \in \tilde{X} \subset X$, $x_1 \neq x_2$ it follows that $f(x_1) \neq f(x_2)$, that is, $f(x)$ is injective on $\tilde{X}$, and this property does not depend on the choice of a codomain.

Notice also that the property of surjection depends both on the choice of the domain and codomain. If a function is not surjective for a specific pair of domain and codomain, it is always possible to make it surjective keeping the same domain and reducing its codomain to the range.

2a. $y = f(x) = x^2 : X = \mathbb{R} \to Y = \mathbb{R}$.

The codomain $Y = \mathbb{R}$ of this function is different from its range, since for any $x \in \mathbb{R}$ we have $y = f(x) = x^2 \geq 0$, that is, no negative value is found in the range. This means that the function is not surjective (and consequently, it is not bijective). Besides, two different values of the domain $x_2 = -x_1 \neq 0$ the function sends to the same element of the codomain: $y_1 = f(x_1) = x_1^2 = (-x_1)^2 = f(x_2) = y_2$, which means that $f(x)$ is not injective.

2b. $y = f(x) = x^2 : X = [0, +\infty) \to Y = \mathbb{R}$.

The function is injective, since two different elements of the domain $x_1 \neq x_2$ are mapped to two different elements of the range: it follows directly from increase of $f(x)$ on the interval $[0, +\infty)$. At the same time, the function is not surjective, because any element of the range is non-negative.

2c. $y = f(x) = x^2 : X = \mathbb{R} \to Y = [0, +\infty)$.

The codomain $Y = [0, +\infty)$ of this function coincides with its range, because for each $y \in [0, +\infty)$ there exists the corresponding point of the domain $x = \sqrt{y} \in \mathbb{R}$ which the function maps to the chosen y: $f(x) = f(\sqrt{y}) = (\sqrt{y})^2 = y$. This means that the function is surjective. At the same time, two different values of the domain $x_2 = -x_1 \neq 0$ the function sends to the same element of the codomain: $y_1 = f(x_1) = x_1^2 = (-x_1)^2 = f(x_2) = y_2$, which means that $f(x)$ is not injective (and consequently, it is not bijective).

2d. $y = f(x) = x^2 : X = [0, +\infty) \to Y = [0, +\infty)$.

The codomain $Y = [0, +\infty)$ of this function coincides with its range, because for each $y \in [0, +\infty)$ there exists the corresponding point of the domain $x = \sqrt{y} \in [0, +\infty)$ which the function maps to the chosen y: $f(x) = f(\sqrt{y}) = (\sqrt{y})^2 = y$. This means that the function is surjective. At the same time, two different values of the domain $x_1 \neq x_2$ the function sends to the two different elements of the range, since the function is increasing on the interval $[0, +\infty)$. This means that $f(x)$ is injective. Consequently, the function is bijective.

Remark Notice a general rule observed in this example: a modification of a codomain does not affect the fact that a function is not injective, since this property is related only to the type of mapping and its domain. Indeed, if a function is

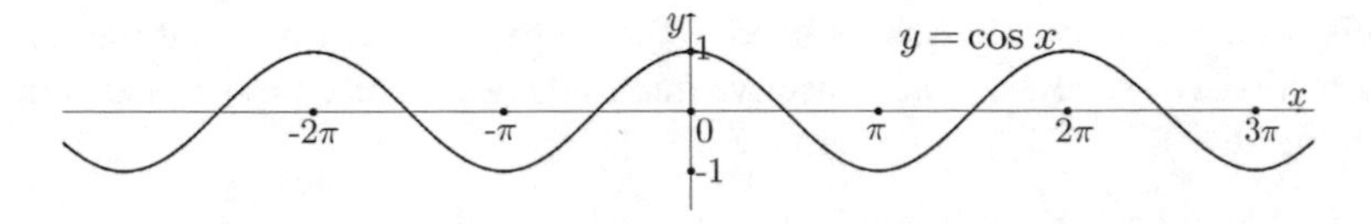

Fig. 2.69 The graph of the function $f(x) = \cos x$

not injective, then there exists at least one pair $x_1 \neq x_2$ in its domain such that $y_1 = f(x_1) = f(x_2)$, where y_1 is an element of the range, and any change of the codomain keeps the range as its subset, and consequently, the function will continue to be non-injective with the modified codomain (even if the codomain will be reduced to the range).

This result, along with the Remark to the previous example, leads to the conclusion that any change of a codomain have no influence on the fact that the function is or is not injective.

On the other side, reducing in a appropriate way the domain of a function, we can arrive to an injective function. Clearly, in some special cases, like in the case of the constant function $f(x) = c$, this decreasing of the domain can necessarily lead to the domain containing the only point. Thus, there is always a possibility to create an injection from a function that is not originally injective, by reducing the domain of this function and keeping its formula (rule).

3a. $f(x) = \cos x : X = \mathbb{R} \to Y = \mathbb{R}$.

As before, we appeal at this moment to the graph of this function shown in Fig. 2.69 and to its properties known from the high school.

The function $f(x) = \cos x$ has all its values contained in the interval $[-1, 1]$, more precisely, its range is $[-1, 1]$. For this reason, if the codomain is $Y = \mathbb{R}$, then the function is not surjective. We also know that $f(x) = \cos x$ is a periodic function with the minimum period 2π. This means that the function $f(x) = \cos x$ is not injective on the domain $X = \mathbb{R}$.

Using the two general rules deduced during the solution of the previous two examples—to turn a function surjective it is enough to choose its range as the codomain (keeping the same domain), and to turn a function injective its domain should be reduced correspondingly—we can transform $f(x) = \cos x$ into a bijective function. First we eliminate the repetitive values of function by reducing its domain.

3b. $f(x) = \cos x : X = [0, \pi] \to Y = \mathbb{R}$.

Looking at the graph, we can see that the function is decreasing on the interval $[0, \pi]$, and consequently, it does not repeat any of its values, that is, $f(x) = \cos x$ is injective on the domain $[0, \pi]$. However, its range is still $[-1, 1]$, and it is not surjective with the codomain $Y = \mathbb{R}$ (and consequently, it is not bijective).

3c. $f(x) = \cos x : X = \mathbb{R} \to Y = [-1, 1]$.

In this case, the codomain was reduced to the range, which means that the function is surjective. However, it is not injective due to its periodicity (and consequently, it is not bijective).

3d. $f(x) = \cos x : X = [0, \pi] \to Y = [-1, 1]$.

Finally, choosing the domain $X = [0, \pi]$ as in Example 3b, and the codomain $Y = [-1, 1]$ as in Example 3c, we obtain the injective and surjective function, that is, a bijection.

12 Inverse Function

12.1 One-Sided Inverses

Property of Injection *A function $y = f(x) : X \to Y$ is injective if and only if there exists a function $g(y) : Y \to X$ such that $g(f(x)) = x$, $\forall x \in X$.*

Proof First, we consider an injective function $y = f(x) : X \to Y$ and construct $g(y) : Y \to X$ with the required property. To do this, we define $g(y)$ on the set Y as follows:

(1) for each y which belongs to the range $\tilde{Y} \subset Y$ we set $g(y) = x \in X$ where x is such that $f(x) = y$; this x exists and is unique because the function $f(x)$ is injective, that is, every y of its range is associated with the only x of the domain X;

(2) for each y which belongs to $Y \backslash \tilde{Y}$ (if this difference is not empty) we set $g(y) = x_0 \in X$, where x_0 is any fixed element of X.
In this way, we obtain $g(y) : Y \to X$ with the property $g(f(x)) = g(y) = x$, $\forall x \in X$, because each element $x \in X$ the function $f(x)$ sends to $y \in \tilde{Y}$ and this y the function $g(y)$ returns to x, according to the construction of $g(y)$.

Now let us prove the converse: if $g(y) : Y \to X$ is such that $g(f(x)) = x$, $\forall x \in X$, then $f(x) : X \to Y$ is injective. Take any $x_1, x_2 \in X$, $x_1 \neq x_2$ and show that $y_1 = f(x_1) \neq f(x_2) = y_2$. Indeed, assuming that $y_1 = y_2$, we have $x_1 = g(y_1) = g(y_2) = x_2$ that contradicts the supposition $x_1 \neq x_2$. □

Remark Notice that, in accordance with the proof, if $f(x)$ is not surjective, then there are infinitely many functions $g(y)$ which satisfy the conditions of the statement: an element $y \in Y \backslash \tilde{Y}$ can be sent to any $x_0 \in X$, preserving the property $g(f(x)) = x$, $\forall x \in X$. For example, the function $f(x) = x : X = (0, 1] \to Y = [0, 1]$ is injective, but not surjective. Then, the element $y = 0 \in Y$, which does not belong to the range of $f(x)$, can be sent to X in infinitely many ways, generating each time a new function $g(y)$, each of which has the property $g(f(x)) = x$, $\forall x \in X$. Specifying, for each $x_0 \in X = (0, 1]$ the function $g(y) = \begin{cases} y, & y \in (0, 1] \\ x_0, & y = 0 \end{cases}$ satisfies the required condition.

Definition Given a function $y = f(x) : X \to Y$, a function $g(y) : Y \to X$ such that $g(f(x)) = x, \forall x \in X$ is called *left inverse* of $f(x)$.

Property of injection. Corollary *The property of injection can be reformulated as follows: a function* $y = f(x) : X \to Y$ *is injective if and only if it has a left inverse.*

Property of Surjection *A function* $y = f(x) : X \to Y$ *is surjective if and only if there exists a function* $g(y) : Y \to X$ *such that* $f(g(y)) = y, \forall y \in Y$.

Proof For the first implication, we recall that a function $y = f(x) : X \to Y$ is surjective if for each $y \in \tilde{Y} = Y$ there exists at least one $x \in X$ such that $f(x) = y$. Then, we define $g(y)$ on the set Y as follows: for each $y \in Y$ we set $g(y) = x \in X$, where x is one of the elements (if there are many) that $f(x)$ sends to y, that is, $f(x) = y$ for a chosen x. In this case, we obtain $g(y) : Y \to X$ with the property $f(g(y)) = f(x) = y, \forall y \in Y$.

For the converse, if $g(y) : Y \to X$ is such that $f(g(y)) = y, \forall y \in Y$, then the values of $f(x)$ cover all the set Y even when we count only the elements x which belong to the range of $g(y)$, that is, such x that $x \in \tilde{X} \subset X$. Consequently, Y is the range of $f(x) : X \to Y$, that is, $f(x)$ is surjective. □

Remark Notice that, in accordance with the proof, if $f(x)$ is not injective, then there are many functions $g(y)$ which satisfy the conditions of the statement: we have multiple choices to determine the element $x_0 = g(y_0) \in X$ if the function $f(x)$ sends different elements $x \in X$ to the same point y_0. For example, the function $f(x) = \begin{cases} x, x \in [0,1) \\ 0, x = 1 \end{cases} : X = [0,1] \to Y = [0,1)$ is surjective, but not injective. Then, the element $y = 0 \in Y$ can be returned by the function $g(y) : Y \to X$ both to 0 and to 1. Therefore, both functions $g(y) = \begin{cases} y, y \in (0,1) \\ 0, y = 0 \end{cases}$ and $h(y) = \begin{cases} y, y \in (0,1) \\ 1, y = 0 \end{cases}$ satisfy the required condition: $f(g(y)) = y, \forall y \in Y$ and $f(h(y)) = y, \forall y \in Y$. If the property of injection is violated in an infinite number of points, there are infinitely many functions $g(y)$. For example, the surjective function $f(x) = \begin{cases} x, x \in [0,1) \\ 0, x \in [1,2] \end{cases} : X = [0,2] \to Y = [0,1)$ takes the value 0 in all the points of the interval $[1,2]$ and also at the origin. Then, choosing any $x_0 \in [1,2] \cup \{0\}$, we obtain the function $g(y) = \begin{cases} y, y \in (0,1) \\ x_0, y = 0 \end{cases}$ that satisfies the relevant property: $f(g(y)) = y, \forall y \in Y$. Hence, we construct infinitely many functions $g(y)$.

Definition Given a function $y = f(x) : X \to Y$, a function $g(y) : Y \to X$ such that $f(g(y)) = y, \forall y \in Y$ is called *right inverse* of $f(x)$.

Property of Surjection. Corollary *The property of surjection can be reformulated as follows: a function $y = f(x) : X \to Y$ is surjective if and only if it has a right inverse.*

Remark 1 According to the property of injection, the original function $y = f(x) : X \to Y$ represents a right inverse for its left inverse $x = g(y) : Y \to X$, which means (by the property of surjection) that $g(y)$ is surjective.

Remark 2 According to the property of surjection, the original function $y = f(x) : X \to Y$ represents a left inverse for its right inverse $x = g(y) : Y \to X$, which means (by the property of injection) that $g(y)$ is injective.

Let us consider a few examples, taking advantage of the functions studied in the previous section.

Examples

1b. $y = f(x) = x : X = [0, +\infty) \to Y = \mathbb{R}$.
In Example 1b of the previous section we have shown that this function is injective, but not surjective. Then, by the property of injection, it has different left inverses. A "natural" left inverse is $g(y) = \left\{ \begin{array}{l} y,\ y \in [0, +\infty) \\ -y,\ y \in (-\infty, 0) \end{array} \right\} = |y| : Y = \mathbb{R} \to X = [0, +\infty)$. Another left inverse can be $h(y) = \begin{cases} y,\ y \in [0, +\infty) \\ x_0,\ y \in (-\infty, 0) \end{cases}, \forall x_0 \geq 0 : Y = \mathbb{R} \to X = [0, +\infty)$. In this manner, preserving the same first (obligatory) formula in $[0, +\infty)$, we can invent infinitely many ways to map the interval $(-\infty, 0)$ onto $[0, +\infty)$. One of them involves Dirichlet's function: $u(y) = \begin{cases} y,\ y \in [0, +\infty) \\ D(y),\ y \in (-\infty, 0) \end{cases} : Y = \mathbb{R} \to X = [0, +\infty)$.

2a. $y = f(x) = x^2 : X = \mathbb{R} \to Y = \mathbb{R}$.
In Example 2a of the previous section it was shown that this function is not injective nor surjective. Consequently, it does not have left inverse nor right.

2b. $y = f(x) = x^2 : X = [0, +\infty) \to Y = \mathbb{R}$.
In Example 2b of the previous section it was shown that this function is injective, but not surjective. Therefore, by the property of injection, it has different left inverses. The "first" left inverse is constructed using non-negative roots of the equations $y = x^2$ and $-y = x^2$ for unknown x: $g(y) = \left\{ \begin{array}{l} \sqrt{y},\ y \in [0, +\infty) \\ \sqrt{-y},\ y \in (-\infty, 0) \end{array} \right\} = \sqrt{|y|} : Y = \mathbb{R} \to X = [0, +\infty)$. Clearly, the property of the left inverse is satisfied: $g(f(x)) = \sqrt{|x^2|} = \sqrt{x^2} = |x| = x$, $\forall x \in X = [0, +\infty)$.
Notice that the construction of this left inverse consists of the two parts, reflected in the two formulas of the function in different parts of the domain. First, we have to return all the elements y to their original points, which demands to specify the range $\tilde{Y} = [0, +\infty)$ of the original function (in other words, we need to identify the bijective part of the original function, which is

$y = f(x) = x^2 : X = [0, +\infty) \to \tilde{Y} = [0, +\infty)$) and find the original element of each $y \in \tilde{Y} = [0, +\infty)$. Technically, this means to solve the equation $y = x^2$ for every given y in $[0, +\infty)$, searching for the root x which belongs to $X = [0, +\infty)$. The result of this operation is the first formula of the function $g(y)$ which is obligatory and uniquely determined. The second step consists in completing the first formula by the second one defined for the remaining values of $y \in Y$, that is, for $y \in (-\infty, 0)$. At this step, we can use any rule under the only condition that the corresponding points $x = g(y)$ lie in the set $X = [0, +\infty)$ (the domain of the original function and the range of a left inverse). This can be made in various manners, one of them is to use the square root of $-y \in (0, +\infty)$, the same already used for non-negative y of the first formula. In this way, we obtain the second formula of $g(y)$.

Keeping the same obligatory (first) formula on $[0, +\infty)$, we can now create infinitely many other left inverses. For instance, another left inverse can be $h(y) = \begin{cases} \sqrt{y}, y \in [0, +\infty) \\ c^2, y \in (-\infty, 0) \end{cases}$, $\forall c \in \mathbb{R} : Y = \mathbb{R} \to X = [0, +\infty)$ and one more has the form $h(y) = \begin{cases} \sqrt{y}, y \in [0, +\infty) \\ e^y, y \in (-\infty, 0) \end{cases} : Y = \mathbb{R} \to X = [0, +\infty)$. It is easy to show that the property of a left inverse is satisfied for these functions.

2c. $y = f(x) = x^2 : X = \mathbb{R} \to Y = [0, +\infty)$.

We have demonstrated in Example 2c of the previous section that this function is surjective and not injective. Then, by the property of surjection, it has various right inverses. One of them algebraically is the positive root of the equation $y = f(x) = x^2$ for unknown x, which generates the function $x = g(y) = \sqrt{y} : Y = [0, +\infty) \to X = [0, +\infty)$. Another one is the negative root of the same equation, that is, the function $h(y) = -\sqrt{y} : Y = [0, +\infty) \to X = (-\infty, 0]$. Clearly, both functions satisfy the property of the right inverse: $f(g(y)) = (\sqrt{y})^2 = y$, $\forall y \in Y = [0, +\infty)$ and $f(h(y)) = (-\sqrt{y})^2 = y$, $\forall y \in Y = [0, +\infty)$.

Notice that in the construction of a right inverse somehow we deal with a bijective part of an original function. This part can be established in different ways, choosing an appropriate part of the original domain $X = \mathbb{R}$. Setting the bijective part as $y = f(x) = x^2 : X_g = [0, +\infty) \to Y = [0, +\infty)$ (that is, restricting the original domain $X = \mathbb{R}$ to the interval $X_g = [0, +\infty)$), we arrive at the right inverse $g(y)$, while choosing the bijective part in the form $y = f(x) = x^2 : X_h = (-\infty, 0] \to Y = [0, +\infty)$ (that is, reducing the original domain $X = \mathbb{R}$ to the interval $X_h = (-\infty, 0]$), we obtain the right inverse $h(y)$. Choosing another part of the original domain, whose image is $Y = [0, +\infty)$, we construct another right inverse. Since there are infinitely many ways to choose such parts of the original domain, there are infinitely many right inverses.

Using the two "natural" right inverses $g(y)$ and $h(y)$ (which correspond to the two roots of the equation $y = x^2$), we can construct infinitely many right inverses combining the values of these "natural" inverses. For instance,

$u(y) = \begin{cases} \sqrt{y}, y \in [0, n] \\ -\sqrt{y}, y \in (n, +\infty) \end{cases}$ for any $n \in \mathbb{N}$ is a right inverse of $f(x)$, as well as $v(y) = \begin{cases} \sqrt{y}, y \in [2n-2, 2n-1) \\ -\sqrt{y}, y \in [2n-1, 2n) \end{cases}$, $\forall n \in \mathbb{N}$ and $w(y) = \begin{cases} \sqrt{y}, y \in \mathbb{Q}, y \geq 0 \\ -\sqrt{y}, y \in \mathbb{I}, y > 0 \end{cases}$.

2d. $y = f(x) = x^2 : X = [0, +\infty) \to Y = [0, +\infty)$.
It was shown that this function is bijective, and consequently, it has both left and right inverse. It happens that these two inverses coincide: $g(y) = \sqrt{y} : Y = [0, +\infty) \to X = [0, +\infty)$. In the next section we will see that this does not happen by chance, but it is a general rule. Actually, the proofs of the properties of injection and surjection already indicate this.

3b. $f(x) = \cos x : X = [0, \pi] \to Y = \mathbb{R}$.
We have already demonstrated that this function is injective, but not surjective. Therefore, it admits different left inverses, but none right inverse.
A left inverse can be constructed using the roots of the equation $y = \cos x$ for unknown $x \in [0, \pi]$. Since, at the moment, we appeal to the geometric definition of the cosine function, we will solve this equation geometrically using the graph of the function. Any line $y = y_0 \in [-1, 1]$ intersects $y = \cos x$, $x \in [0, \pi]$ exactly at one point, which means that $y_0 = \cos x$ has the only root x_0 in $[0, \pi]$ for any $y_0 \in [-1, 1]$ (see Fig. 2.70). This root is called arccosine and denoted by $x_0 = \arccos y_0$. Using these roots on the interval $x \in [0, \pi]$, we obtain the function defined on $[-1, 1]$ whose range is $[0, \pi]$, which is called arccosine (the same name as that of the roots): $\tilde{g}(y) = \arccos y : \tilde{Y} = [-1, 1] \to X = [0, \pi]$. This function represents the principal (obligatory) part

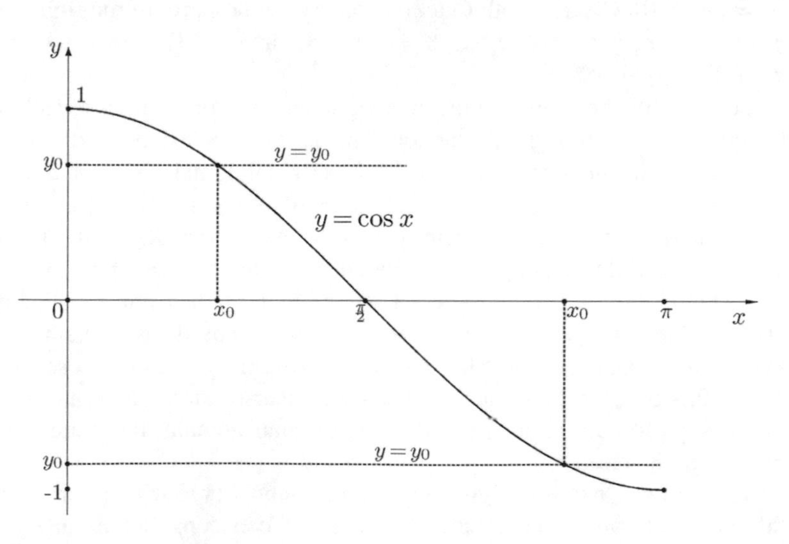

Fig. 2.70 Solution of the equation $y = \cos x$, $x \in [0, \pi]$

of the left inverse. Now we just need to complete this part with the definition of the function at the remaining values of y (that is, on the set $Y \backslash \tilde{Y}$) in such a way that $g(Y \backslash \tilde{Y}) \subset [0, \pi]$ (the image of the remaining values is contained in $[0, \pi]$). It can be made in infinitely many ways, for instance, setting a set of the left inverses by the formula: $g(y) = \begin{cases} \arccos y, \ y \in [-1, 1] \\ x_0, \ y \in \mathbb{R} \backslash [-1, 1] \end{cases}, \forall x_0 \in [0, \pi] :$ $Y = \mathbb{R} \to X = [0, \pi]$. Verifying the property of the left inverse, we have: $g(f(x)) = g(\cos x) = \arccos(\cos x) = x, \forall x \in X = [0, \pi]$ (the second formula of $g(y)$ is not acting here because $y = f(x) = \cos x \in [-1, 1]$).

3d. $f(x) = \cos x : X = [0, \pi] \to Y = [-1, 1]$.
As we already know, this function is bijective, and consequently, it admits the uniquely defined one-sided inverses, which coincide with each other. This general inverse is the arccosine function $\tilde{g}(y) = \arccos y : Y = [-1, 1] \to X = [0, \pi]$ constructed in Example 3b.

12.2 General Inverse. Definition and Elementary Examples

Naturally, an injective function is not necessarily surjective, and vice-verse. Therefore, a left inverse is not necessarily a right inverse, and vice-verse. The situation is simpler if a function is bijective. Joining the two properties of the previous section we arrive to the following result.

Property of Bijection *A function $y = f(x) : X \to Y$ is bijective if and only if there exist two functions $g(y) : Y \to X$ such that $g(f(x)) = x, \forall x \in X$ and $h(y) : Y \to X$ such that $f(h(y)) = y, \forall y \in Y$. In the terms of one-sided inverses this means that a function $y = f(x) : X \to Y$ is bijective if, and only if, it has left and right inverses.*

We can even strengthen this result with the statement that both left inverse $g(y)$ and right inverse $h(y)$ are unique and coincide: $g(y) = h(y), \forall y \in Y$. Indeed, consider one of left inverses $g(y)$ and one of right inverses $h(y)$. Take an arbitrary $y \in Y$ and compare the values of $g(y)$ and $h(y)$. Notice that $g(y)$ is defined for any $\forall y \in Y$ since $f(x)$ is surjective and $h(y)$ is also defined for any $\forall y \in Y$ due to its description in the formulation of the property of surjection. For the first function we have $g(y) = x$ such that $f(x) = y$, and for the second $h(y) = \bar{x}$ such that $f(\bar{x}) = y$. Since $f(x)$ is injective, the relations $f(x) = y$ and $f(\bar{x}) = y$ imply that $x = \bar{x}$. Since this is valid for $\forall y \in Y$, then $g(y) = x = h(y), \forall y \in Y$, that is, $g(y)$ and $h(y)$ are the same function. Due to arbitrariness of the choice of the inverses, any left inverse $g(y)$ coincides with the right inverse $h(y)$, and any right inverse $h(y)$ coincides with the left inverse $g(y)$. Therefore, there exists the only left inverse and the only right inverse, which coincide with each other and this unique function is called the inverse (or general inverse).

The property of bijection in a strengthened form, which we have just demonstrated, can be stated as follows.

Property of Bijection. Corollary *A function* $y = f(x) : X \to Y$ *is bijective if and only if there exists a unique function* $g(y) : Y \to X$ *such that* $g(f(x)) = x, \forall x \in X$ *and* $f(g(y)) = y, \forall y \in Y$*. For obvious reasons, this function* $x = g(y) : Y \to X$ *is called general inverse, or simply inverse of* $y = f(x)$ *and the notation* $g(y) \equiv f^{-1}(y)$ *is used.*

Remark The same property of bijection can be derived without appealing to left and right inverses, using direct reasoning for a bijective function. Indeed, notice that for a bijective function the codomain Y has the same properties as the domain X: for any $y \in Y$ there exists at least one $x \in X$ (because Y is the range) and this x is unique (because the function is injective). In this way, together with the original function $y = f(x) : X \to Y$, it is uniquely defined the function $x = g(y) : Y \to X$ which acts in the inverse direction: $g(y)$ takes each element $y \in Y$ and associates with it such a single element $x \in X$ that $f(x) = y$. That is, the function $x = g(y)$ maps each $y \in Y$ to such specific x that the original function transforms into $f(x) = y$. In other words, applying the function $y = f(x)$ to a specific element $x_0 \in X$, we get $f(x_0) = y_0 \in Y$, and then, applying the function $x = g(y)$ to the obtained element y_0, we return to the same point x_0 from which we started $x_0 = g(y_0)$. This means that the following property (the property of a left inverse) is satisfied: $g(f(x)) = x$, $\forall x \in X$. On the other hand, if $g(y)$ is the inverse of $f(x)$, then $f(x)$, in turn, is the inverse of $g(y)$, and consequently, the following property (the property of a right inverse) holds: $f(g(y)) = y, \forall y \in Y$ (see Fig. 2.71).

In the majority of cases, in Analysis and Calculus, one-sided inverses are not used, and for this reason a general inverse is usually called simply inverse without any ambiguity. Hereinafter we will follow this short name. Let us formalize this concept in the next definition.

Definition of the Inverse Given a function $y = f(x) : X \to Y$, the function $g(y) : Y \to X$ such that $g(f(x)) = x, \forall x \in X$ and $f(g(y)) = y, \forall y \in Y$ is called the *inverse* of $f(x)$. A standard notation of the inverse function is $g(y) = f^{-1}(y)$. The function that admits the inverse is called *invertible*.

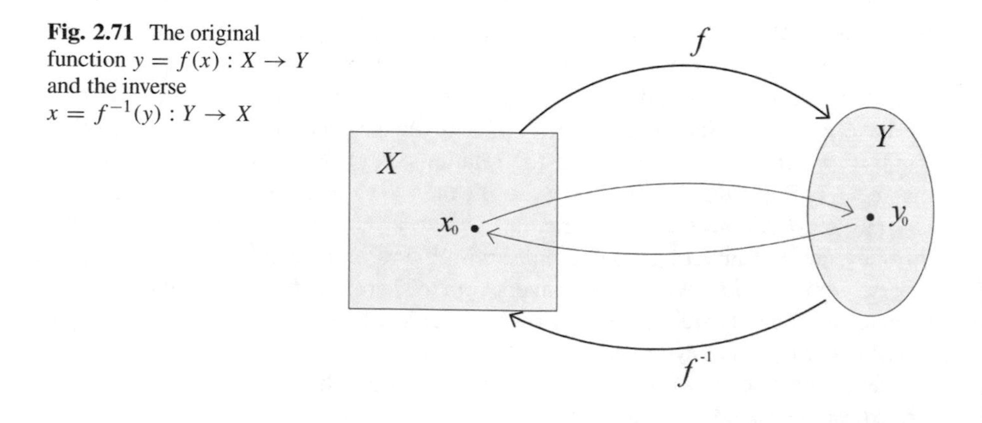

Fig. 2.71 The original function $y = f(x) : X \to Y$ and the inverse $x = f^{-1}(y) : Y \to X$

The proved results show that a function $y = f(x) : X \to Y$ admits the inverse if and only if it is bijective. That is why the following alternative definition of the inverse is frequently used.

Alternative Definition of the Inverse Given a bijective function $y = f(x) : X \to Y$, the function $x = g(y) : Y \to X$ is the *inverse* of $f(x)$ if it associates with each element $y \in Y$ such element $x \in X$ which the original function maps to $y \in Y$.

Notice that the range of the original function becomes the domain of the inverse, and the domain of the original turns into the range of the inverse (see Fig. 2.71). The formula of the inverse is the same as that of the original, $y = f(x)$, but acting in the opposite direction, from y to x. Therefore, if the formula of the original function $y = f(x) : X \to Y$ is given in the explicit form, then the same formula used for the inverse $f^{-1}(y) : Y \to X$ will be implicit. In this case, to find an explicit formula of the inverse, we need to solve the equation $y = f(x)$ with respect to unknown x for all $y \in Y$. Frequently it is not possible to solve this equation algebraically even when the inverse exists. One of examples of this type is the function $y = f(x) = x + e^x : X = \mathbb{R} \to Y = \mathbb{R}$, which is bijective, and consequently, admits the inverse, but the equation $y = x + e^x$ can not be solved algebraically for x, and we can only use the implicit formula for the inverse function. The details of this example are provided in the next section.

The fact that a function is not bijective, and consequently, does not admits an inverse over the entire domain, does not prevent this function from being bijective on a part of the domain and admits the inverse there. As we have seen before, for any function $f(x)$ it is possible to restrict its domain and codomain in such a way that the modified function becomes bijective which makes possible to define its inverse. The reduction of a codomain to the range is necessarily to make a function surjective, while a restriction of the domain is made to turn the function injective.

Let us consider a few examples, using the functions whose properties of injection and surjection were studied in the previous sections.

Examples

1. $y = f(x) = x : X = \mathbb{R} \to Y = \mathbb{R}$
   This function is bijective and its inverse is $x = f^{-1}(y) = y : Y = \mathbb{R} \to X = \mathbb{R}$.
2. $y = f(x) = x^2 : X = \mathbb{R} \to Y = \mathbb{R}$
   This function is not bijective (it is not surjective nor injective), and consequently, it does not admit the inverse. However, it is easy to transform this function (keeping its formula) into a bijection. To do this, it is sufficient to reduce its domain and codomain as follows: $y = f(x) = x^2 : X = [0, +\infty) \to Y = [0, +\infty)$. The last function has the inverse $x = f^{-1}(y) = \sqrt{y} : Y = [0, +\infty) \to X = [0, +\infty)$. Another way to contract its domain is $y = f(x) = x^2 : X = (-\infty, 0] \to Y = [0, +\infty)$. In this case, the inverse is different $x = f^{-1}(y) = -\sqrt{y} : Y = [0, +\infty) \to X = (-\infty, 0]$.
3. $y = f(x) = |x| : X = \mathbb{R} \to Y = [0, +\infty)$
   This function is not bijective (it is surjective, but not injective), and consequently, it does not have the inverse. However, it is easy to modify this function

(preserving the formula), by shrinking its domain, to make this function bijective. There are different ways to do this, one of them is $y = f(x) = |x| = x : X = [0, +\infty) \to Y = [0, +\infty)$. The last function has the inverse $x = f^{-1}(y) = y : Y = [0, +\infty) \to X = [0, +\infty)$. Another choice of the reduced domain is as follows: $y = f(x) = |x| = -x : X = (-\infty, 0] \to Y = [0, +\infty)$. In this case, the inverse is different: $x = f^{-1}(y) = -y : Y = [0, +\infty) \to X = (-\infty, 0]$.

4. $y = f(x) = \cos x : X = \mathbb{R} \to Y = \mathbb{R}$
   This function is not bijective (it is not surjective nor injective), and consequently, it has no inverse. However, reducing the domain and codomain (but preserving the formula) we can easily transform this function into bijection. One of options of such reducing is as follows: $y = f(x) = \cos x : X = [0, \pi] \to Y = [-1, 1]$. The last function is bijective and admits the inverse called arccosine $x = f^{-1}(y) = \arccos y : Y = [-1, 1] \to X = [0, \pi]$.

12.3 Conditions of the Existence of the Inverse

A necessary and sufficient condition of the existence of the inverse was proved in the preceding subsection as well as the uniqueness of the inverse (see the property of bijection and its Corollary). For completeness of exposition of this section these results are reproduced below.

Necessary and Sufficient Condition of the Inverse *A function is invertible if and only if it is bijective.*

Uniqueness of the Inverse *If the inverse exists, then it is unique.*

One of the conditions that guarantees the existence of the inverse is monotonicity of an original function. This result is stated in the following theorem.

Sufficient Condition of the Inverse *If a function $y = f(x)$ is monotonic on a set X whose image is Y, then there exists the inverse of $f(x)$ defined on the set Y with the image X. Moreover, the inverse is monotonic on Y with the same type of monotonicity as the original function.*

Proof We demonstrate the statement only for the case of increasing function on X, since the case of decreasing function is shown analogously. Since $Y = f(X)$, the function $f(x)$ is surjective on X. To show that $f(x)$ is injective, take two arbitrary points $x_1, x_2 \in X$, $x_1 < x_2$ and obtain $f(x_1) < f(x_2)$ due to strict increasing on X, that is, two different elements of X are mapped to the different elements of Y. Therefore, $f(x) : X \to Y$ is bijective, and consequently, invertible.

To show that the inverse $f^{-1}(y)$ is increasing on Y, suppose, for contradiction, that there exists a pair of $y_1 < y_2$ in Y such that $x_1 = f^{-1}(y_1) \geq f^{-1}(y_2) = x_2$. Then, using the increase of $f(x)$, we obtain $y_1 = f(x_1) \geq f(x_2) = y_2$ that contradicts the supposition. □

Remark 1 The most used form of this Theorem in Calculus is when X is an interval: if a function $y = f(x)$ is monotonic on an interval X (whatever type of this interval) whose image is Y, then there exists the inverse of $f(x)$ defined on the set Y with the image X. Moreover, the inverse is monotonic on Y with the same type of monotonicity as the original function.

Remark 2 The theoretic existence of the inverse ensured by the Theorem does not mean that the analytic formula of the inverse can be found in the explicit form. For example, the function $y = f(x) = x + e^x : X = \mathbb{R} \to Y = \mathbb{R}$ is increasing over the entire domain $\mathbb{R}$, since both summands (x and e^x) are increasing functions on $\mathbb{R}$. Therefore, by the Theorem, there exists the inverse function. However, an explicit expression for this inverse is not accessible, because the equation $y = x + e^x$ for the unknown x cannot be solved algebraically.

Remark 3 Monotonicity of a function is sufficient but not necessarily condition of the existence of the inverse. There are many non-monotonic functions which admit the inverse. For example, the function $f(x) = \begin{cases} x, x \in [0,1] \\ 1-x, x \in (1,2] \end{cases}$ is not monotonic on its domain $X = [0, 2]$, but it has the inverse. Indeed, $f(x)$ is increasing on the interval $[0, 1]$ and decreasing on the interval $(1, 2]$, which means that it does not monotonic on the entire interval $[0, 2]$. According to the first formula $f(x) = x, x \in [0, 1]$, the function transforms the interval $X_1 = [0, 1]$ into $Y_1 = [0, 1]$, where Y_1 is the image of X_1: $Y_1 = f(X_1)$. According to the second formula $f(x) = 1 - x, x \in (1, 2]$, the function maps the interval $X_2 = (1, 2]$ onto $Y_2 = [-1, 0)$: $Y_2 = f(X_2)$. Since $X = X_1 \cup X_2$, the entire image is $Y = f(X) = f(X_1) \cup f(X_2) = [0, 1] \cup [-1, 0) = [-1, 1]$. Notice that each element of Y is associated with the only element of the domain, which means that $f(x)$ is injective. In fact, on the part $X_1 = [0, 1]$ the function is increasing, and consequently, none of its values is repeated. On the second part $X_2 = (1, 2]$ the function is decreasing, and again none of its values is repeated. To conclude that $f(x)$ is injective on X, it remains to observe that the parts $Y_1 = f(X_1)$ and $Y_2 = f(X_2)$ of the image have no common element. Hence, the given function with the range $Y = [-1, 1]$ is bijective, and consequently, invertible.

We can also specify the formula of this inverse. Notice that the inverse of the first part is $f^{-1}(y) = y : Y_1 = [0, 1] \to X_1 = [0, 1]$, and the inverse of the second part is $f^{-1}(y) = 1 - y : Y_2 = [-1, 0) \to X_2 = (1, 2]$. Joining these two results, we obtain the inverse of the original function: $f^{-1}(y) = \begin{cases} y, y \in [0,1] \\ 1-y, y \in [-1,0) \end{cases}$.

See the graphs of the two functions in Fig. 2.72.

Another, more radical, example is the function $f(x) = x + D(x) = \begin{cases} x+1, x \in \mathbb{Q} \\ x, x \in \mathbb{I} \end{cases} : X = \mathbb{R} \to Y = \mathbb{R}$ which is not monotonic on any interval of its domain $\mathbb{R}$. Indeed, whatever interval is chosen, it contains both rational and irrational points, and $f(x)$ permanently "jumps" between the points of the line $y = x + 1$ (when x is rational) and the points of the line $y = x$ (when x is irrational). Even so, $f(x)$ is invertible. In fact, $f(x)$ is injective, because it takes

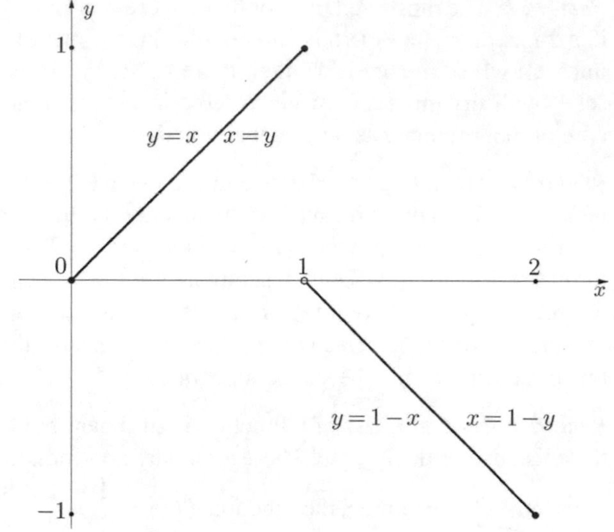

Fig. 2.72 Function $y = f(x)$ of Remark 3 and its inverse

two different values $f(x_1) = x_1 + 1 \neq x_2 + 1 = f(x_2)$ at any two distinct rational points $x_1 \neq x_2$, it also takes different values $f(x_1) = x_1 \neq x_2 = f(x_2)$ at two distinct irrational points $x_1 \neq x_2$, and finally, its value $f(x_1)$ at any rational x_1 differs from a value $f(x_2)$ at any irrational x_2 since the former is rational and the latter is irrational. Besides, $f(x)$ is surjective, since for each $y \in \mathbb{Q}$ there exists a point $x = (y - 1) \in \mathbb{Q}$ of the domain such that $f(x) = y$, and for each $y \in \mathbb{I}$ there exists a point $x = y \in \mathbb{I}$ of the domain such that $f(x) = y$. Hence, $f(x)$ is bijective, and consequently, it admits the inverse given by the formula $f^{-1}(y) = y - D(y) = \begin{cases} y - 1, \ y \in \mathbb{Q} \\ y, \ y \in \mathbb{I} \end{cases} : Y = \mathbb{R} \to X = \mathbb{R}$.

Examples

In Examples 1–4 of the preceding section, the original functions have satisfied the conditions of the last Theorem or were changed in such a way that the modified function would satisfy the conditions of this Theorem.

1. The original function $y = f(x) = x : X = \mathbb{R} \to Y = \mathbb{R}$ is increasing over the entire domain and $Y = \mathbb{R}$ is its range.
2. The function $y = f(x) = x^2 : X = \mathbb{R} \to Y = \mathbb{R}$ was transformed restricting its domain in such a way that the modified function $y = f(x) = x^2 : X = [0, +\infty) \to Y = [0, +\infty)$ would be increasing on $X = [0, +\infty)$ with the image $Y = [0, +\infty)$.
3. The function $y = f(x) = |x| : X = \mathbb{R} \to Y = [0, +\infty)$ was changed to the function $y = f(x) = |x| = x : X = [0, +\infty) \to Y = [0, +\infty)$, which is increasing on its domain and has the image $Y = [0, +\infty)$. Another modification of the original function leaded to the function $y = f(x) = |x| = -x : X = (-\infty, 0] \to Y = [0, +\infty)$ decreasing on its domain with the image

$Y = [0, +\infty)$. Both modifications were made using the parts (the intervals) of the domain where the original function is monotonic.

4. The original function $y = f(x) = \cos x : X = \mathbb{R} \to Y = \mathbb{R}$ has infinitely many intervals of increase and decrease. Choosing the interval of decrease $[0, \pi]$, we transform the original function into the function $y = f(x) = \cos x : X = [0, \pi] \to Y = [-1, 1]$ decreasing on the restricted domain with the image $Y = [-1, 1]$. Again, the original function was transformed to the invertible one by restricting the original domain to the interval of monotonicity.

12.4 Analytic Properties of the Inverse

The inverse function preserves different important properties of the original function.

First, it follows directly from the definition of the inverse function that the range of the original function becomes the domain of the inverse and the domain of the original function becomes the range of the inverse.

Considering the main symmetries of functions, we can notice that a periodic or even functions has no inverse, because these functions are not injective (the former repeats the same value at infinitely many points of the domain, and the latter has the same value at least at two points symmetric about the origin). Clearly there are also odd functions which do not have the inverse, like $y = \sin x$ defined on $X = \mathbb{R}$. However, the property of odd function does not prevent from being an invertible function, like the function $y = x^3$ on the domain $X = \mathbb{R}$ or the function $y = \sin x$ on $X = [-\frac{\pi}{2}, \frac{\pi}{2}]$. If an odd function admits the inverse, then this inverse inherits the property of being odd. This is quite obvious and elementary proof is provided in the following statement.

Theorem About Oddness of the Inverse *If an odd function is invertible, then its inverse is also odd.*

Proof Consider an odd function $y = f(x) : X \to Y$ and its inverse $x = f^{-1}(y) : Y \to X$. Since $y = f(x)$ is odd, each point y belongs to the domain Y of the inverse together with the symmetric point $-y$. Taking two arbitrary points y and $-y$ of Y (that is, such points that $y = f(x)$ and $-y = f(-x)$) we have $f^{-1}(-y) = -x = -f^{-1}(y)$. Due to arbitrariness of $y, -y \in Y$, we can conclude that $f^{-1}(y)$ is an odd function. □

Another property inherited by the inverse is monotonicity (of course, it should be a strict monotonicity which guarantees that the values of a function cannot be repeated). It was already seen in the previous section that monotonicity of the original function guarantees the existence of the inverse and its monotonicity. Let us formulate that result once again for completeness.

Theorem About Monotonicity of the Inverse *If $y = f(x)$ is monotonic on a set X whose image is Y, then it admits the inverse, which is monotonic on its domain Y keeping the type of monotonicity of the original function.*

Remark The most used form of this Theorem in Calculus is when X is an interval (of any type): if a function $y = f(x)$ is monotonic on an interval X whose image is Y, then there exists the inverse of $f(x)$ defined on Y with the image X, and this inverse is monotonic on its domain Y, preserving the type of monotonicity of the original function. Notice that the set Y is not necessarily an interval even when X is an interval.

One more property transferable from the original function to the inverse is the concavity, which is shown in the next theorem in the case of strict concavity. Hereinafter, to specify the considerations, we will assume that $x_1 < x_2$ without loss of generality.

Theorem About Concavity of the Inverse *Let $y = f(x)$ be a monotonic function on an interval X whose image is an interval Y. If $y = f(x)$ is concave upward and increasing on X, then its inverse is concave downward on Y; if $y = f(x)$ is concave upward and decreasing on X, then its inverse is concave upward on Y. Analogously, if $y = f(x)$ is concave downward and increasing on X, then its inverse is concave upward on Y; if $y = f(x)$ is concave downward and decreasing on X, then its inverse is concave downward on Y. Summarizing, the inverse preserves the type of concavity of the original decreasing function, and the inverse reverses the type of concavity of the original increasing function.*

Proof First of all, recall that the hypotheses of the theorem guarantee the existence of the inverse, according to the previous Theorem. Moreover, the same Theorem ensures that the inverse has the same type of the monotonicity as the original function. Let us demonstrate the statement about concavity of the inverse in the case when the original function is concave upward.

We start with the case when $y = f(x)$ is increasing on X. We use condition (2.7) (in Sect. 7.3) of the concavity of $f(x)$:

$$\frac{f(x) - f(x_1)}{x - x_1} < \frac{f(x_2) - f(x_1)}{x_2 - x_1}, \forall x \in (x_1, x_2) \subset X.$$

Take now two arbitrary points $y_1 < y_2$ of the interval Y (the domain of the inverse) and evaluate the quotient $\frac{f^{-1}(y)-f^{-1}(y_1)}{y-y_1}$. Notice that increasing of $f(x)$ implies in increasing of $f^{-1}(y)$, and therefore, the condition $y_1 < y_2$ is equivalent to $x_1 = f^{-1}(y_1) < f^{-1}(y_2) = x_2$. Using the relationship between the original and inverse functions and also the concavity of the original function in the form (2.7), we obtain:

$$\begin{aligned}\frac{f^{-1}(y) - f^{-1}(y_1)}{y - y_1} &= \frac{x - x_1}{f(x) - f(x_1)} > \frac{x_2 - x_1}{f(x_2) - f(x_1)} \\ &= \frac{f^{-1}(y_2) - f^{-1}(y_1)}{y_2 - y_1}, \forall y \in (y_1, y_2) \subset Y.\end{aligned}$$

Hence, we arrive to the definition of downward concavity of the inverse in the form (2.10) (in Sect. 7.3).

Consider now the case when $y = f(x)$ is decreasing on X. Take again two arbitrary points $y_1 < y_2$ of the interval Y and evaluate the quotient $\frac{f^{-1}(y)-f^{-1}(y_1)}{y-y_1}$. Notice that decreasing of $f(x)$ implies in decreasing of $f^{-1}(y)$, and consequently, the condition $y_1 < y_2$ leads to $x_1 = f^{-1}(y_1) > f^{-1}(y_2) = x_2$. Therefore, in the formulas of concavity of $f(x)$ the meaning of x_1 and x_2 should be reversed. In particular, the condition of the upward concavity (2.8) (in Sect. 7.3) takes the form

$$\frac{f(x) - f(x_1)}{x - x_1} > \frac{f(x_1) - f(x_2)}{x_1 - x_2}, \forall x \in (x_2, x_1).$$

Using the relationship between the original and inverse functions and also the concavity of the original function, we obtain:

$$\frac{f^{-1}(y)-f^{-1}(y_1)}{y - y_1} = \frac{x - x_1}{f(x)-f(x_1)} < \frac{x_1 - x_2}{f(x_1)-f(x_2)} = \frac{x_2 - x_1}{f(x_2)-f(x_1)}$$
$$= \frac{f^{-1}(y_2)-f^{-1}(y_1)}{y_2 - y_1}, \forall y \in (y_1, y_2) \subset Y.$$

This is the definition of the upward concavity of the inverse function in the form (2.7) (in Sect. 7.3).

The case of downward concavity of the original function is proved similarly. This task is left to the reader. □

Remark An analogous result is valid for non-strict concavity of monotonic functions which map an interval X onto an interval Y. The proof follows the same reasoning applied in the above Theorem about strict concavity. For this reason we only formulate the corresponding statement. If $y = f(x)$ is concave upward non-strictly and increasing on X, then its inverse is concave downward non-strictly on Y; if $y = f(x)$ is concave upward non-strictly and decreasing on X, then its inverse is concave upward non-strictly on Y. Analogously, if $y = f(x)$ is concave downward non-strictly and increasing on X, then its inverse is concave upward non-strictly on Y; if $y = f(x)$ is concave downward non-strictly and decreasing on X, then its inverse is concave downward non-strictly on Y.

The last result of this section is about the inverse of a composite function.

Inverse of a Composition *If $y = f(x) : X \to Y$ has the inverse $x = f^{-1}(y) : Y \to X$ and $z = g(y) : Y \to Z$ has the inverse $y = g^{-1}(z) : Z \to Y$, then the composition $h(x) = g(f(x)) : X \to Z$ is also invertible and its inverse is defined by the formula $h^{-1}(z) = f^{-1}(g^{-1}(z)) : Z \to X$.*

Proof Since $f(x)$ and $g(y)$ are invertible, both functions are bijective, and by the Theorem about composition of bijections, the function $h(x) = g(f(x))$ is also bijective, and consequently, invertible. To verify the formula of the inverse, notice that $g^{-1}(z)$ sends each $z_0 \in Z$ to the element $y_0 \in Y$ such that $g(y_0) = z_0$, and the function $f^{-1}(y)$ maps y_0 to the element $x_0 \in X$ such that $f(x_0) = y_0$. Hence, $h^{-1}(z) = f^{-1}(g^{-1}(z))$ takes each $z_0 \in Z$ and associates with it the element x_0 such that $h(x_0) = z_0$, which corresponds to the definition of the inverse. □

12.5 Geometric Property of the Inverse

Finally, let us show that a shape of the graph of the inverse is obtained directly from the graph of the original function.

Like in the case of composite functions, the notation $x = f^{-1}(y)$ of the inverse is very useful for illustration of the definition and proofs of some properties. But when we intend to compare a behavior of different functions, it is natural to consider that all of them has the domain contained in the x-axis and the range in the y-axis. Hence, in the theorem about comparison of the graphs of the original and inverse functions we will use the independent variable x and the value of function y for both functions, that is, the inverse of $y = f(x)$ will be denoted by $y = f^{-1}(x)$.

Notice first that the using the notation $y = f(x) : X \to Y$ for the original function and $x = f^{-1}(y) : Y \to X$ for its inverse, we have the same graph for both functions. Indeed, any pair of the coordinates $P_0 = (x_0, y_0)$ that satisfies the formula $y_0 = f(x_0)$ of the original function (that is, P_0 is a point of the graph of $y = f(x)$), also satisfies the relation $f^{-1}(y_0) = x_0$ of the inverse, which means that the same point P_0 belongs to the graph of the inverse function $x = f^{-1}(y)$.

The situation changes if we use the notation $y = f^{-1}(x) : X_i \to Y_i$ for the inverse, where $X_i = Y$ and $Y_i = X$. In this case we have the following result about the graphs of the original and inverse functions.

Theorem About the Graph of the Inverse *The graph of the inverse function $y = f^{-1}(x) : Y \to X$ is symmetric to the graph of the original function $y = f(x) : X \to Y$ with respect to the line $y = x$ (the bisector of the first and third quadrants).*

Proof Recall once more that if we use the notation $x = f^{-1}(y) : Y \to X$ for the inverse function, then the graphs of the two functions coincide. Changing the notation of the inverse to $y = f^{-1}(x) : Y \to X$ we simply interchange the notation of the independent and dependent variables. Therefore, if a point $P_0 = (x_0, y_0)$ belongs to the graph of the original function, then $\tilde{P}_0 = (y_0, x_0)$ is the point of the graph of the inverse.

Notice that if $y_0 = x_0$ then both points have the same coordinates: $P_0 = \tilde{P}_0 = (x_0, x_0)$ and they belong to the line $y = x$. Therefore, in this case we have a singular situation of the symmetry of the points P_0 and $\tilde{P}_0$ with respect to the line $y = x$.

Let us suppose now that $y_0 \neq x_0$. Consider the midpoint between $P_0 = (x_0, y_0)$ and $\tilde{P}_0 = (y_0, x_0)$ (the midpoint of the segment $P_0\tilde{P}_0$): $\overline{P}_0 = (\frac{x_0+y_0}{2}, \frac{y_0+x_0}{2})$. By the definition, this point is equidistant from the points P_0 and $\tilde{P}_0$ and it belongs to the line $y = x$ because its coordinates coincide (see Fig. 2.73). To show that P_0 and

Fig. 2.73 Relationship between the graphs of the original function $y = f(x)$ and the inverse function $y = f^{-1}(x)$

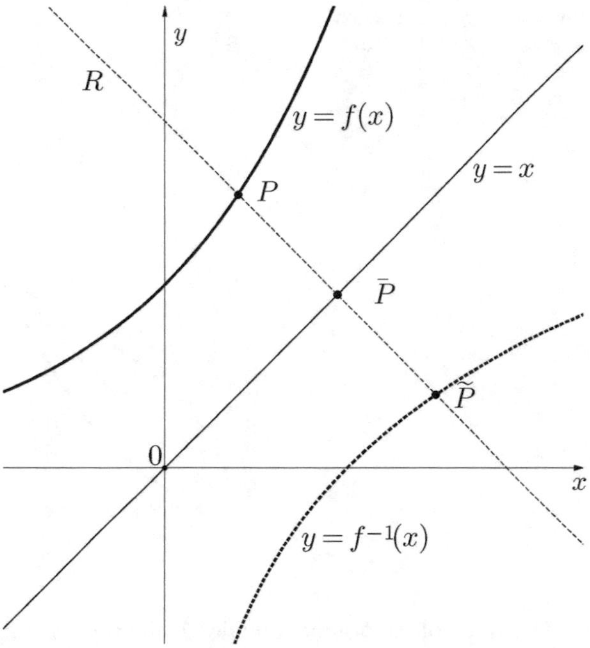

$\tilde{P}_0$ are symmetric about $y = x$ we need to verify that the line R passing through these two points is perpendicular to the line $y = x$. Recalling the equation of the line joining two given points, we have $\frac{y-x_0}{y_0-x_0} = \frac{x-y_0}{x_0-y_0}$ or $y = x_0 + \frac{y_0-x_0}{x_0-y_0}(x - y_0)$. Simplifying we get $y = -x + x_0 + y_0$. This shows that the line R is parallel to the bisector of the second and fourth quadrants $y = -x$, which is perpendicular to the bisector of the first and third quadrants $y = x$. Therefore, the line R is perpendicular to $y = x$ (see Fig. 2.73). Hence, the points $P_0 = (x_0, y_0)$ and $\tilde{P}_0 = (y_0, x_0)$ are symmetric about the line $y = x$. Since this is true for any pair of the points of the two graphs, we can conclude that the graph of the inverse is symmetric to the graph of the original function with respect to the line $y = x$.

□

Examples

Let us draw the graphs of the inverse functions considered earlier in this section applying the last theorem.

1. $y = f(x) = x : X = \mathbb{R} \to Y = \mathbb{R}$. This function is bijective and its inverse is $x = f^{-1}(y) = y : Y = \mathbb{R} \to X = \mathbb{R}$. The graphs of $y = f(x)$ and $x = f^{-1}(y)$ coincide. Rewriting the inverse in the form with the independent variable x, we have $y = f^{-1}(x) = x : X_i = Y = \mathbb{R} \to Y_i = X = \mathbb{R}$, but the graph of the inverse is still the same, because the original and inverse functions coincide.
2. $y = f(x) = x^2 : X = [0, +\infty) \to Y = [0, +\infty)$. This function is invertible: $x = f^{-1}(y) = \sqrt{y} : Y = [0, +\infty) \to X = [0, +\infty)$, or interchanging x and y, we get $y = f^{-1}(x) = \sqrt{x} : X_i = Y = [0, +\infty) \to Y_i = X = [0, +\infty)$.

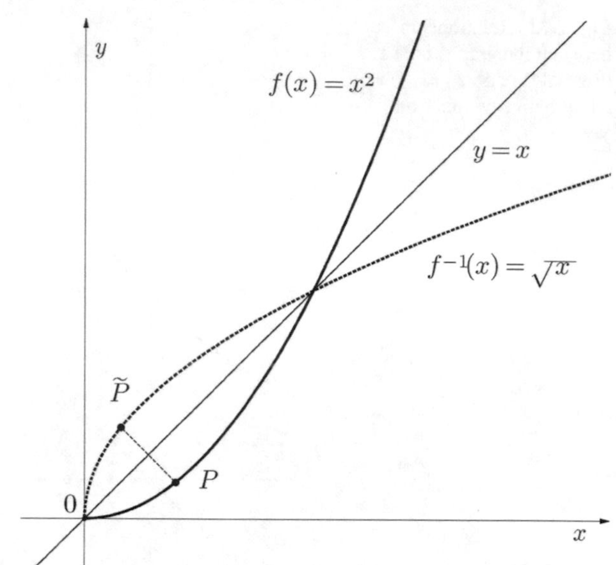

Fig. 2.74 Original function $f(x) = x^2$, $X = [0, +\infty)$ and its inverse $f^{-1}(x) = \sqrt{x}$, $X = [0, +\infty)$

The graph of the original function is the right part of the parabola and the graph of the inverse is symmetric to this half-parabola about the bisector $y = x$ (see Fig. 2.74). From the relationship between the graphs, we can see geometrically that increasing of the original function implies increasing of the inverse, and that the upward concavity of the original (together with its increasing) implies downward concavity of the inverse.

3. $y = f(x) = |x| = -x : X = (-\infty, 0] \to Y = [0, +\infty)$. The inverse of this function is $x = f^{-1}(y) = -y : Y = [0, +\infty) \to X = (-\infty, 0]$, or, interchanging x and y, $y = f^{-1}(x) = -x : X_i = Y = [0, +\infty) \to Y_i = X = (-\infty, 0]$. The graphs of the two functions are shown in Fig. 2.75.
4. $y = f(x) = \cos x : X = [0, \pi] \to Y = [-1, 1]$. The inverse of this function is $x = f^{-1}(y) = \arccos y : Y = [-1, 1] \to X = [0, \pi]$. Interchanging the notation of independent variable and values of function, we get $y = f^{-1}(x) = \arccos x : X_i = Y = [-1, 1] \to Y_i = X = [0, \pi]$. The graphs of the two functions are shown in Fig. 2.76. The relationship between the two graphs reveals geometrically that decreasing of the original function implies decreasing of the inverse. Additionally, under the condition of decreasing, the downward concavity of $f(x)$ on the interval $[0, \frac{\pi}{2}]$ results in downward concavity of $f^{-1}(x)$ on the part $[0, 1]$ of its domain, and the upward concavity of $f(x)$ on $[\frac{\pi}{2}, \pi]$ implies the upward concavity of $f^{-1}(x)$ on the part $[-1, 0]$ of the domain.

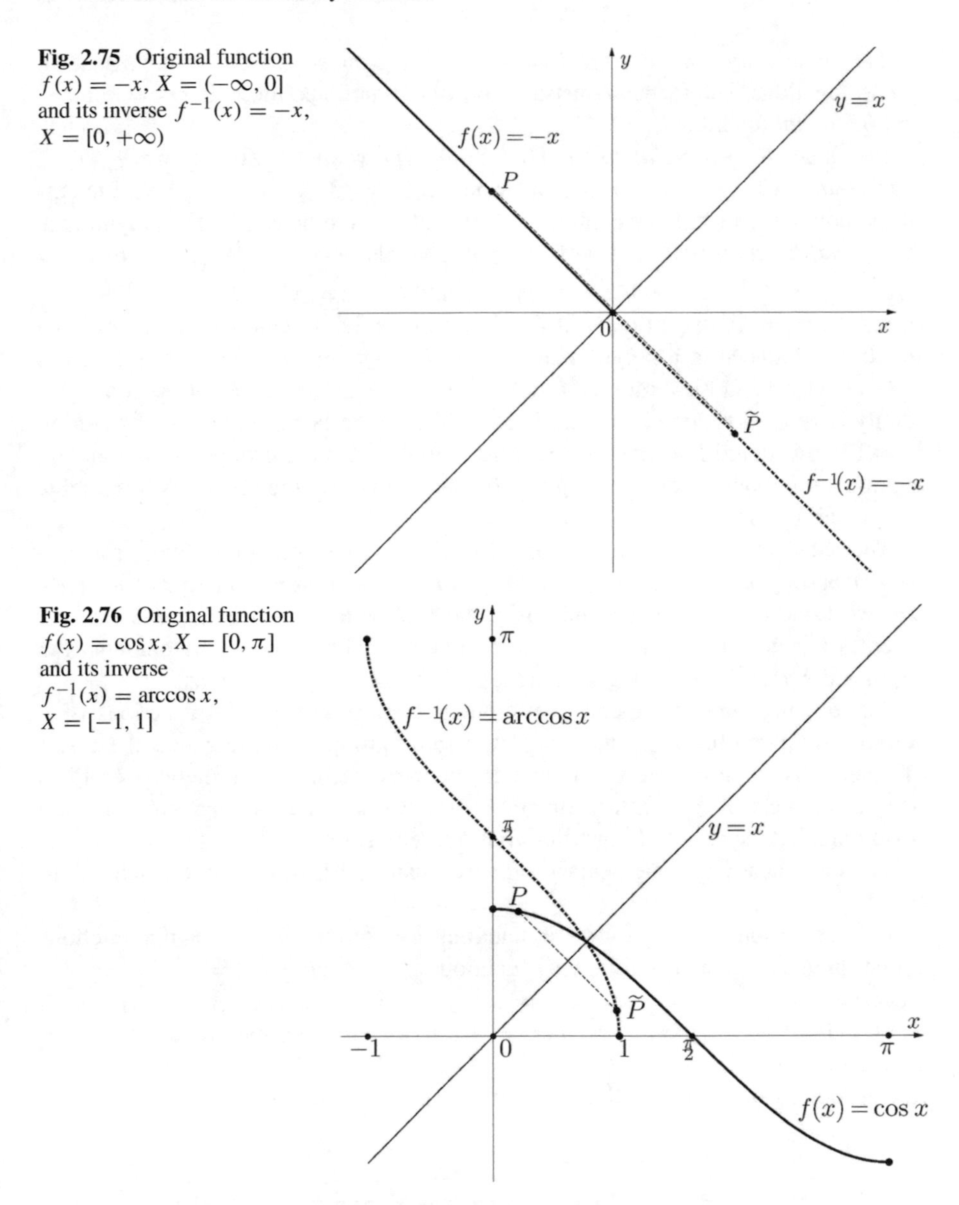

Fig. 2.75 Original function $f(x) = -x$, $X = (-\infty, 0]$ and its inverse $f^{-1}(x) = -x$, $X = [0, +\infty)$

Fig. 2.76 Original function $f(x) = \cos x$, $X = [0, \pi]$ and its inverse $f^{-1}(x) = \arccos x$, $X = [-1, 1]$

13 Classification of Elementary Functions

At this point, we give an overview of the class of functions called *elementary functions*, without specification of the definitions of many involved functions and characterization of their properties, appealing just to a preliminary knowledge of these functions from the high school.

The elementary functions are divided into two groups—*algebraic* and *transcendental* functions. The former consists of *rational* functions (including polynomials) and *irrational* functions.

Recall that a *polynomial function* (or simply a *polynomial*) has the form $P_n(x) = a_n x^n + a_{n-1}x^{n-1} + \ldots + a_1 x + a_0$, where $a_i, i = 0, \ldots, n$ are the coefficients of the polynomial (real constants), $a_n \neq 0$, and n is the degree of the polynomial. A *rational function* is the ratio of two polynomials: $R(x) = \frac{P_n(x)}{Q_m(x)}$, where $P_n = a_n x^n + a_{n-1}x^{n-1} + \ldots + a_1 x + a_0$, $a_n \neq 0$ and $Q_m = b_m x^m + b_{m-1}x^{m-1} + \ldots + b_1 x + b_0$, $b_m \neq 0$. In particular, if Q_m is a polynomial of degree 0, then a rational function is reduced to a polynomial. An *irrational function* is defined in the form $\sqrt[n]{R(x)}$, where $R(x)$ is a rational function and $n \in \mathbb{N}$ is the degree of the root. This family is generalization of rational functions since the last are obtained by setting $n = 1$ in this definition. Even so, it is usual to separate the groups of rational and irrational functions, understanding that the latter contains the functions irreducible to rational functions.

The class of *transcendental functions* is composed of the two groups—*trigonometric functions* and *exponential/logarithmic functions*. The former includes the two basic functions $\sin x$ and $\cos x$, their ratios $\tan x$ and $\cot x$, and also the inverses $\arcsin x$, $\arccos x$, $\arctan x$ and $\text{arccot}\, x$. The second part contains the exponential a^x, $a > 0$ and logarithmic $\log_a x$, $a > 0, a \neq 1$ functions.

Additionally, any arithmetic operation or composition of the functions of a certain group results in a function of the same group (assuming that the result of operations has non-empty domain). In the same manner, arithmetic operations or compositions of elementary functions produce again an elementary function (assuming that the result of operations has non-empty domain).

The classification of elementary functions can be illustrated by the scheme in Fig. 2.77.

Let us give some examples of the functions of different types using the functions studies in this chapter. The following functions are rational: $f(x) = x$, $f(x) = -x$, $f(x) = x + 2$, $f(x) = 2x$, $f(x) = x^2$, $f(x) = -x^2$, $f(x) = (x-1)^2 + 2$, $f(x) = 3 + 2x - x^2$, $f(x) = x^3$, $f(x) = \frac{1}{x}$ (all of them are polynomials, except for the last one). The functions $f(x) = \sqrt{x}$, $f(x) = |x|$ and $f(x) = \sqrt{|x|}$ are irrational (the absolute value is an irrational function since it can be represented in the form

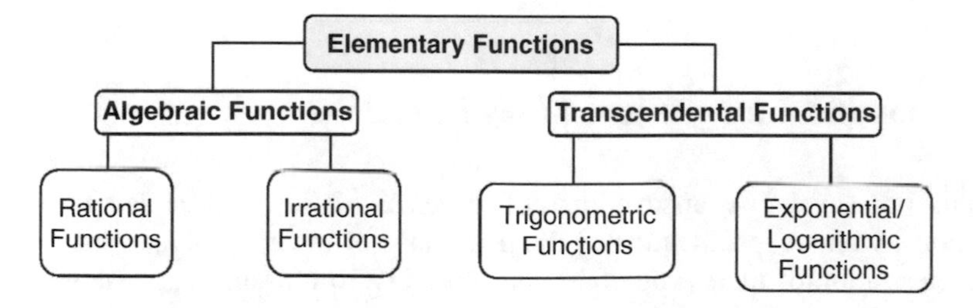

Fig. 2.77 Classification of elementary functions

$|x| = \sqrt{x^2}$). The functions $f(x) = \sin x$, $f(x) = \cos x$, $f(x) = \frac{1}{3}\cos x$ and $f(x) = \cos 2x$ are trigonometric. The functions $f(x) = x^2 + \cos x$, $f(x) = \frac{\sin x}{x^3}$, $f(x) = |x| \sin x$ and $f(x) = \sin \frac{1}{x}$ are elementary, but not of a specific group. The functions $f(x) = \begin{cases} x^2, & x \geq 0 \\ x, & x < 0 \end{cases}$, $f(x) = \begin{cases} -x, & x < 0 \\ x+1, & x \geq 0 \end{cases}$ and $D(x) = \begin{cases} 1, & x \in \mathbb{Q} \\ 0, & x \in \mathbb{I} \end{cases}$ are not elementary.

14 Solved Exercises

14.1 Domain and Range

1. Given a function $f(x)$, find its values at the indicated points:
   (a) $f(x) = \sqrt{1+x^2}$, find $f(0)$, $f\left(-\frac{3}{4}\right)$, $f(-x)$, $f\left(\frac{1}{x}\right)$;
   (b) $f(x) = \sqrt{x^2-9}$, find $f(0)$, $f(-3)$, $f(x+3)$, $f(\frac{x}{2})$, $f(\frac{1}{x^2})$.

   Solution.

   (a) $f(0) = \sqrt{1+0^2} = \sqrt{1} = 1$, $\quad f\left(-\frac{3}{4}\right) = \sqrt{1+\left(-\frac{3}{4}\right)^2} = \sqrt{1+\frac{9}{16}} = \sqrt{\frac{25}{16}} = \frac{5}{4}$,
   $f(-x) = \sqrt{1+(-x)^2} = \sqrt{1+x^2}$, $\quad f\left(\frac{1}{x}\right) = \sqrt{1+\left(\frac{1}{x}\right)^2} = \sqrt{1+\frac{1}{x^2}} = \sqrt{\frac{x^2+1}{x^2}}$, $x \neq 0$;
   (b) the function is not defined at 0, $\quad f(-3) = \sqrt{(-3)^2-9} = \sqrt{9-9} = 0$,
   $f(x+3) = \sqrt{(x+3)^2-9} = \sqrt{x^2+6x}$, $\quad f(\frac{x}{2}) = \sqrt{\left(\frac{x}{2}\right)^2 - 9} = \sqrt{\frac{x^2}{4} - 9} = \frac{1}{2}\sqrt{x^2-36}$,
   $f(\frac{1}{x^2}) = \sqrt{\left(\frac{1}{x^2}\right)^2 - 9} = \sqrt{\frac{1}{x^4} - 9} = \frac{1}{x^2}\sqrt{1-9x^4}$, $x \neq 0$.

2. Determine the domain of the function:
   (a) $f(x) = \sqrt{x+1}$;
   (b) $f(x) = \sqrt{9-x^2}$;
   (c) $f(x) = \frac{\sqrt{x^2-2}}{x}$;
   (d) $f(x) = \sqrt{x-x^3}$;
   (e) $f(x) = \frac{1}{4-x^2}$;
   (f) $f(x) = \frac{\sqrt{1-x}}{\sqrt{x^2-1}}$.

Solution.

In all these exercises the domain X is not specified explicitly, which means, by agreement, that the domain is the largest set of real numbers such that the formula of the function is defined for each point of this set.

(a) The expression inside the square root cannot be negative, that is, $x + 1 \geq 0$ or $x \geq -1$. Therefore, $X = [-1, +\infty)$.
(b) The definition of the square root leads to the restriction $9-x^2 \geq 0$ or $x^2 \leq 9$, whose solution is $-3 \leq x \leq 3$. There is no other restriction, which means that the domain is $X = [-3, 3]$.
(c) The numerator is defined only for $x^2 - 2 \geq 0$ or $x^2 \geq 2$, which is equivalent to $|x| \geq \sqrt{2}$. The solution of this inequality is $x \geq \sqrt{2}$ and $x \leq -\sqrt{2}$. The denominator gives the restriction $x \neq 0$, but this point is already excluded because of the numerator. Therefore, $X = (-\infty, -\sqrt{2}] \cup [\sqrt{2}, +\infty)$.
(d) For the square root to be defined, the inequality $x - x^3 = x(1 - x^2) \geq 0$ should be satisfied. This happens when both factors x and $1 - x^2$ have the same sign. If both are non-negative, then $1 - x^2 \geq 0$ implies that $x^2 \leq 1$ or $|x| \leq 1$. The solution of the last inequality is $-1 \leq x \leq 1$, and together with the restriction $x \geq 0$ this gives the solution $X_1 = [0, 1]$. If both factors are non-positive, then $1 - x^2 \leq 0$, whence $x^2 \geq 1$ or $|x| \geq 1$ whose solution is $x \leq -1$ and $x \geq 1$. Adding the condition $x \leq 0$, we get the second part of the solution $X_2 = (-\infty, -1]$. Hence, the domain is $X = X_1 \cup X_2 = (-\infty, -1] \cup [0, 1]$.
(e) The only restriction is non-vanishing of the denominator: $4 - x^2 \neq 0$ or $x^2 \neq 4$, which results in $x \neq \pm 2$. Then, $X = \mathbb{R}\backslash\{-2, 2\}$.
(f) The expression inside the square root in the numerator should be non-negative: $1 - x \geq 0$ or $x \leq 1$. The expression inside the square root in the denominator should be positive $x^2 - 1 > 0$. The last inequality has the solutions $x > 1$ and $x < -1$. The intersection of $x \leq 1$ and $x \in (-\infty - 1) \cup (1, +\infty)$ gives the domain $X = (-\infty, -1)$.

3. Determine the domain and the range of the following functions:

(a) $f(x) = 1 - 2x$;
(b) $f(x) = x^2$;
(c) $f(x) = 10 - x^2$;
(d) $f(x) = \frac{x^2}{1-x^2}$;
(e) $f(x) = \sqrt{4 - x^2}$.

Solution.

In what follows the domain is denoted by X and the range by Y.

(a) There is no restriction for the formula, and consequently, the domain is $X = \mathbb{R}$. To find the range, we need to solve the equation $y = 1 - 2x$ for x: $x = \frac{1-y}{2}$. This means that for every real y there exists the corresponding element x of the domain found by the last formula. Therefore, the range is $Y = \mathbb{R}$.

(b) The formula is applicable for any real x, which means that $X = \mathbb{R}$. To find the range, notice that the equation $y = x^2$ with respect to x has a solution for every $y \geq 0$ (if $y > 0$ there are two solutions). This means that for every $y \geq 0$ there exists at least one corresponding x in the domain that the function sends to y. At the same time, the equation $y = x^2$ has no (real) solution if $y < 0$. Therefore, $Y = [0, +\infty)$.

(c) Clearly the domain is $X = \mathbb{R}$ because the formula does not impose any restriction. Since $x^2 \geq 0, \forall x$ and $x^2 > 0, \forall x \neq 0$, we have $y = 10 - x^2 \leq 10$, $\forall x$ and $10 - x^2 < 10$, $\forall x \neq 0$. Therefore, the range Y is contained in $(-\infty, 10]$. Let us show that each number of the last interval is the element of Y. By the definition, we should take an arbitrary $y \in (-\infty, 10]$ and prove that it is associated with at least one element of the domain. It is so, because the equation $y = 10 - x^2$ with respect to x has at least one solution $x = \pm\sqrt{10 - y}$ for every $y \leq 10$. Therefore, $Y = (-\infty, 10]$.

(d) The domain contains all the points that do not vanish denominator, that is, $x \neq \pm 1$. To find out the range, we need to verify for what values of y the equation $y = \frac{x^2}{1-x^2}$ for x has at least one solution. To solve this equation, rewrite it in the form $(1 - x^2)y = x^2$ and then $x^2(1 + y) = y$. Assuming, at the moment, that $y \neq -1$, we can still represent this equation equivalently as $x^2 = \frac{y}{1+y}$. The last equation has solution when the right-hand side is non-negative: $\frac{y}{1+y} \geq 0$, which happens when $y \geq 0$ and $1 + y > 0$, or when $y \leq 0$ and $1 + y < 0$. From the first two conditions we have $y \geq 0$, and other two conditions lead to $y < -1$. It remains to verify if $y = -1$ belongs to the range: in this case we have $\frac{x^2}{1-x^2} = -1$ or $x^2 = x^2 - 1$ or $0 = -1$ which is false statement. Therefore, $y = -1$ is out of the range. Hence, the range is $Y = (-\infty, -1) \cup [0, +\infty)$.

(e) The expression inside the square root should be non-negative: $4 - x^2 \geq 0$, whence $-2 \leq x \leq 2$, that is, the domain is $X = [-2, 2]$. To find the range, notice first that, by the definition of the square root, y is always non-negative. Next, under the restriction $y \geq 0$, we solve the equation $y = \sqrt{4 - x^2}$ with respect to x: $y^2 = 4 - x^2$ or $x^2 = 4 - y^2$. The last equation has solutions x only when $4 - y^2 \geq 0$, that is, $-2 \leq y \leq 2$. Since the negative values y were already discarded, the only values remained $0 \leq y \leq 2$. Hence, $Y = [0, 2]$.

4. Determine the domain and sketch the graph of the function:

(a) $f(x) = \begin{cases} 0, x < 2 \\ 1, x \geq 2 \end{cases};$

(b) $f(x) = \begin{cases} -1, x < 0 \\ 0, x = 0 \\ 1, x > 0 \end{cases};$

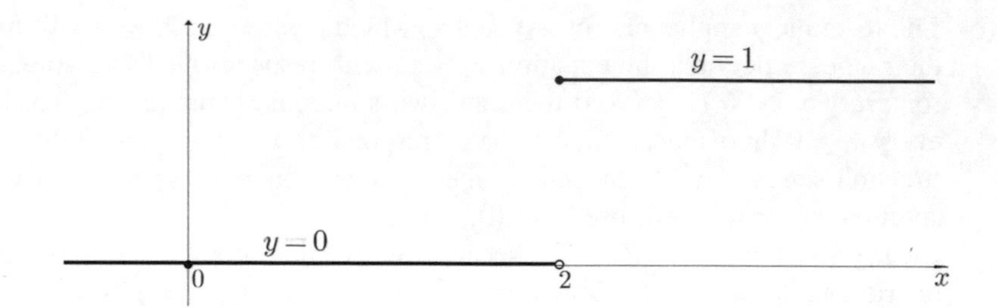

Fig. 2.78 The graph of the function of exercise 4(a)

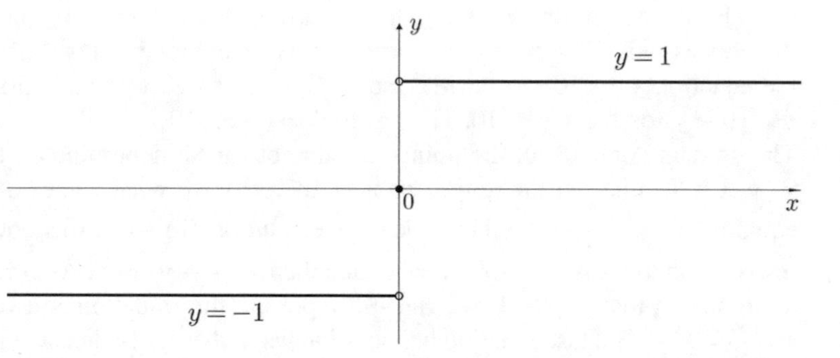

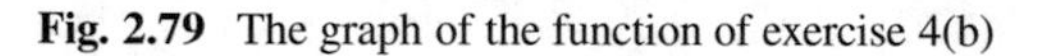
Fig. 2.79 The graph of the function of exercise 4(b)

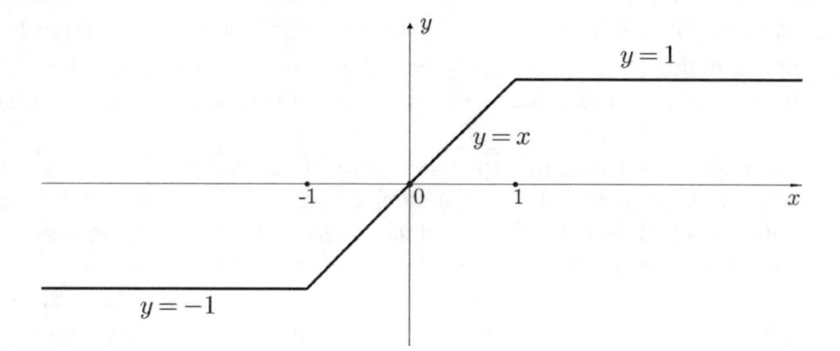

Fig. 2.80 The graph of the function of exercise 4(c)

(c) $f(x) = \begin{cases} -1, x < -1 \\ x, -1 \leq x \leq 1 \\ 1, x > 1 \end{cases}.$

Solution.

(a) The domain is $X = \mathbb{R}$. The graph is shown in Fig. 2.78.
(b) The domain is $X = \mathbb{R}$. The graph is shown in Fig. 2.79.
(c) The domain is $X = \mathbb{R}$. The graph is shown in Fig. 2.80.

14.2 Bounded Functions

1. Verify if the function is bounded, bounded above, bounded below or unbounded on the set S:

   (a) $f(x) = 4x - x^2 - 3$, $S = \mathbb{R}$;
   (b) $f(x) = 4x - x^2 - 3$, $S = [-1, 1]$;
   (c) $f(x) = \sqrt{x+1}$, $S = [-1, +\infty)$.

   Solution.

   (a,b) To simplify the reasoning let us represent the quadratic function in the canonical form of a parabola: $f(x) = 1 - (x-2)^2$. This form makes clear that any value of the function is less than or equal to 1: $f(x) < f(2) = 1, \forall x \neq 2$. Therefore, the function is bounded above on any set S and one of the upper bounds is $M = 1 = f(2)$. On the set $\mathbb{R}$ the function is unbounded below, because for any constant m there are points x such that $f(x) < m$. Indeed, consider an arbitrary negative constant m (for non-negative m it is sufficient to take $x = 0$) and solve the inequality $f(x) = 1-(x-2)^2 < m$. This inequality is equivalent to $(x-2)^2 > 1-m$, and for $x > 2$ we have $x > 2 + \sqrt{1-m}$. Therefore, the required points x are found. On the other hand, on the set $S = [-1, 1]$ the function is limited below, since $f(x) \geq f(-1), \forall x \in [-1, 1]$, and one of the lower bounds can be $m = f(-1) = -8$. Therefore, the function is bounded on $S = [-1, 1]$.

   (c) By the definition of the square root, $f(x) = \sqrt{x+1} \geq 0$ over the entire domain, which is the given set $S = [-1, +\infty)$. Therefore, $f(x)$ is bounded below, and one of the lower bounds is $m = f(-1) = 0$. However, $f(x)$ is not bounded above, since the inequality $f(x) = \sqrt{x+1} > M$ has solutions for any constant M. Indeed, considering $M > 0$ (for $M \leq 0$ we can take any $x > -1$), we transform the inequality for $f(x)$ into the equivalent form $x > M^2 - 1$ that shows an infinite set of the elements x such that $f(x) > M$. Hence, the function is not bounded above.

14.3 Even, Odd and Periodic Functions

1. Verify if the functions are even, odd or none of these:

   (a) $f(x) = x^4 - 2x^2$;
   (b) $f(x) = \sqrt{1+x+x^2} - \sqrt{1-x+x^2}$;
   (c) $f(x) = \frac{1}{x}$;
   (d) $f(x) = \frac{1}{2x-1}$;
   (e) $f(x) = x^{-2} + \sqrt{x}$;
   (f) $f(x) = \frac{1}{x} - 2x^3$;

(g) $f(x) = 3x^3 - \frac{1}{x^2}$;
(h) $f(x) = \ln(\sqrt{x^2+1} + x)$.

Solution.

(a) The domain $\mathbb{R}$ of the function is symmetric about the origin and $f(-x) = (-x)^4 - 2(-x)^2 = x^4 - 2x^2 = f(x)$, $\forall x \in \mathbb{R}$. Therefore, the function is even.

(b) Since $1 + x + x^2 = (x + \frac{1}{2})^2 + \frac{3}{4}$ and $1 - x + x^2 = (x - \frac{1}{2})^2 + \frac{3}{4}$, the square roots are defined for all the real x, and consequently, the domain is $\mathbb{R}$, symmetric about the origin. Besides,

$$\begin{aligned} f(-x) &= \sqrt{1 + (-x) + (-x)^2} - \sqrt{1-(-x)+(-x)^2} \\ &= \sqrt{1-x+x^2} - \sqrt{1+x+x^2} \\ &= -\left(\sqrt{1+x+x^2} - \sqrt{1-x+x^2}\right) = -f(x)\,, \ \forall x \in \mathbb{R}, \end{aligned}$$

which shows that the function is odd.

(c) The domain $X = \mathbb{R}\backslash\{0\}$ is symmetric about the origin and $f(-x) = \frac{1}{-x} = -\frac{1}{x} = -f(x)$, $\forall x \in X$. Therefore, the function is odd.

(d) The domain $X = \mathbb{R}\backslash\{\frac{1}{2}\}$ is not symmetric about the origin, which implies that the function cannot be even or odd.

(e) The domain $X = (0, +\infty)$ is not symmetric about the origin, which implies that the function cannot be even or odd.

(f) The domain $X = \mathbb{R}\backslash\{0\}$ is symmetric about the origin and

$$f(-x) = \frac{1}{-x} - 2(-x)^3 = -\frac{1}{x} + 2x^3 = -\left(\frac{1}{x} - 2x^3\right) = -f(x), \ \forall x \in X.$$

Therefore, the function is odd.

(g) The domain $X = \mathbb{R}\backslash\{0\}$ is symmetric about the origin, but $f(-x) = 3(-x)^3 - \frac{1}{(-x)^2} = -3x^3 - \frac{1}{x^2} \neq f(x), \neq -f(x)$. Therefore, the function is not even or odd. Another manner to show this is by choosing a specific pair of the symmetric points of the domain: $f(1) = 2$, $f(-1) = -4$ $\rightarrow f(1) \neq \pm f(-1)$.

(h) Since $x^2 + 1 > 0$ and $\sqrt{x^2+1} + x > 0$, $\forall x \in \mathbb{R}$, the formula is defined for all the real x, and consequently, the domain is $\mathbb{R}$, symmetric about the origin. Besides,

$$\begin{aligned} f(-x) &= \ln(\sqrt{(-x)^2+1} - x) = \ln \frac{x^2 + 1 - x^2}{\sqrt{x^2+1} + x} \\ &= -\ln(\sqrt{x^2+1} + x) = -f(x), \ \forall x \in \mathbb{R}, \end{aligned}$$

which shows that the function is odd.

2. Let $f(x)$ be defined on a symmetric set about the origin. Show that $f(x)$ can be represented in the form $f(x) = g(x) + h(x)$, where $g(x)$ is even and $h(x)$ is odd.

   Solution.

   Consider the functions $g(x) = \frac{1}{2}(f(x) + f(-x))$ and $h(x) = \frac{1}{2}(f(x) - f(-x))$ defined on the domain X of $f(x)$. By the definition, $g(-x) = g(x)$ and $h(-x) = -h(x)$ for every $x \in X$, that is, $g(x)$ is even and $h(x)$ is odd. At the same time, $g(x) + h(x) = f(x)$ for each $x \in X$.

3. Verify if the following functions are periodic; if so, find a period and, if possible, the minimum period:

   (a) $f(x) = x + 1 - [x + 1]$;
   (b) $f(x) = x + 1 - [x]$;
   (c) $f(x) = 2x - [2x]$;
   (d) $f(x) = x^2 + 3$;
   (e) $f(x) = \sqrt{x + 1}$;
   (f) $f(x) = 2 \cos x$;
   (g) $f(x) = 3 - 4 \cos x$;
   (h) $f(x) = \cos 3x - 2$.

   Solution.

   In the solution of these exercises we will apply the properties of periodicity of some simple functions studied before. Usually this is the simplest way to prove the periodicity. Another approach is a direct verification of the definition, without references to the previous results.

   (a) By the definition of the integer part, $[x + 1] = [x] + 1$, and consequently, the function can written in a simpler form $f(x) = x + 1 - [x + 1] = x - [x]$. The last function was already studied in Sect. 4.4, where it was shown that the function is periodic with the minimum period 1.
   (b) The function $h(x) = x - [x]$ is periodic with the minimum period 1 and $f(x) = x + 1 - [x] = h(x) + g(x)$, where $g(x) = 1$ is periodic with an arbitrary period, including the period 1. Then, $f(x)$ is periodic with the period 1. To show that this period is minimum, let us suppose, for contradiction, that $f(x)$ has a positive period $T < 1$. Then, $h(x) = f(x) - 1$ has the same period, which contradicts the fact that the minimum period of $h(x)$ is 1.
   (c) Notice that $[2x] \neq 2[x]$ (for example, $[2 \cdot 1.6] = [3.2] = 3$, while $2 \cdot [1.6] = 2 \cdot 1 = 2$), and consequently, we cannot put the factor 2 as a common factor of the two terms. However, we can use the property of the argument of a periodic function, noting that $f(x) = 2x - [2x] = h(2x)$, $h(x) = x - [x]$. Since $h(x)$ is periodic with the minimum period 1, then $f(x)$ is periodic with the minimum period $\frac{1}{2}$.
   (d) We use the method of contradiction: suppose that $f(x)$ is a periodic, and consequently, there exists $T \neq 0$ such that the relation $f(x + T) = f(x)$ is satisfied for $\forall x \in \mathbb{R}$. Specifying the equation of the periodic function, we get $f(x + T) = (x + T)^2 + 3 = x^2 + 3 = f(x)$, or simplifying, $2xT + T^2 = 0$,

or still $T(2x+T)=0$. Since $T\neq 0$, the only remaining option is $T=-2x$, but T is a constant, independent of x by the definition. In this way we arrive at the contradiction, which means that assumption about periodicity of $f(x)$ is false.

(e) The simplest way to solve this exercise is to recall that the domain of a periodic function should be unbounded both on the left and on the right. The domain of the function $f(x)=\sqrt{x+1}$ is bounded on the left: $x\geq -1$, which means that the function is not periodic.

(f) The function $g(x)=\cos x$ is periodic with the minimum period 2π. Then, by the arithmetic properties of periodic functions, the function $f(x)=2g(x)$ is periodic with the same minimum period.

(g) The function $g(x)=\cos x$ is periodic with the minimum period 2π. Then, by the arithmetic properties of periodic functions, the function $f(x)=3-4g(x)$ is periodic with the same minimum period.

(h) The function $g(x)=\cos x$ is periodic with the minimum period 2π. Then, by the properties of the argument, $h(x)=\cos 3x=g(3x)$ is periodic with the minimum period $\frac{2\pi}{3}$. Using also the arithmetic properties, we conclude that $f(x)=\cos 3x-2=h(x)-2$ is periodic with the minimum period $\frac{2\pi}{3}$.

4. Find a non-constant function that has both rational and irrational periods.

Solution.

Consider the following set of real numbers: $S=\{x=p+\sqrt{2}q,\ p,q\in\mathbb{Q}\}$, and define the function $f_S(x)$ as follows: $f_S(x)=\begin{cases}1, x\in S\\ 0, x\notin S\end{cases}$. Let us show that $T_1=1$ and $T_2=\sqrt{2}$ are periods of this function. Indeed, if $x\in S$, then $(x+1)\in S$ and $(x+\sqrt{2})\in S$, and therefore, $f_S(x)=f_S(x+1)=f_S(x+\sqrt{2})=1$; if $x\notin S$, then $(x+1)\notin S$ and $(x+\sqrt{2})\notin S$, and consequently, $f_S(x)=f_S(x+1)=f_S(x+\sqrt{2})=0$. Hence, for each $x\in\mathbb{R}$ we have $f_S(x)=f_S(x+1)=f_S(x+\sqrt{2})$.

14.4 Monotonicity

1. Find the largest set of monotonicity of a given function:

(a) $f(x)=x^2-2x+3$;
(b) $f(x)=\sqrt{x+1}$;
(c) $f(x)=\cos x$.

Solution.

(a) To have a first intuitive impression about the behavior of this function, let us represent it in the canonical form of a parabola: $f(x)=(x-1)^2+2$. Then, we can see that the function is decreasing when x increases up to 1, at the point 1 it takes the minimum value $f(1)=2$, and after this it starts to increase. Now we verify this intuitive suggestions in a strong analytic manner. Initially, we take an arbitrary pair of points $x_1 < x_2$ and evaluate the difference between the values of the function:

$$\begin{aligned} f(x_2) - f(x_1) &= x_2^2 - 2x_2 + 3 - (x_1^2 - 2x_1 + 3) \\ &= (x_2 - x_1)(x_2 + x_1) - 2(x_2 - x_1) \\ &= (x_2 - x_1)(x_2 + x_1 - 2). \end{aligned}$$

Notice that the first factor is positive, which means that the result depends on the second factor.

Next, we consider separately the two parts of the domain: the intervals $(-\infty, 1]$ and $[1, +\infty)$. If we choose $x_1 < x_2 \le 1$, then $x_2 + x_1 - 2 < 0$, which implies that $f(x_2) < f(x_1)$, and consequently, the function decreases on $(-\infty, 1]$. If we choose $1 \le x_1 < x_2$, then $x_2 + x_1 - 2 > 0$, and consequently, $f(x_2) > f(x_1)$, which means that the function increases on $[1, +\infty)$. Hence, $(-\infty, 1]$ is the largest set of decrease and $[1, +\infty)$ is the largest set of increase (both in a strict sense). Clearly, any subset of $(-\infty, 1]$ is also a decreasing set, and any subset of $[1, +\infty)$ is an increasing set.

(b) The domain of $f(x) = \sqrt{x+1}$ is $[-1, +\infty)$ and an increase of x on this set implies the increase of the argument $x + 1$ of the square root, which causes, by the properties of the square root, the increase of $f(x)$. In the terms of formulas we have: for any pair $-1 \le x_1 < x_2$ it follows that $x_1 + 1 < x_2 + 1$ and the last inequality results in $f(x_1) = \sqrt{x_1+1} < \sqrt{x_2+1} = f(x_2)$, that is, $f(x)$ is increasing on the entire domain.

(c) Recalling the analytic definition of $\cos x$ (or even using its graphic representations as a definition), we conclude that on the interval $[0, \pi]$ the function $f(x) = \cos x$ decreases from 1 to -1 (the x-coordinate of the point of the unit circle decreases when the angle varies from 0 to π). On the next interval, $[\pi, 2\pi]$, the function $f(x) = \cos x$ increases from -1 to 1 (the x-coordinate of the point of the unit circle increases when the angle varies from π to 2π). Using the 2π-periodicity of the function, we conclude that $f(x) = \cos x$ decreases on any interval $[2k\pi, \pi + 2k\pi]$, $k \in \mathbb{Z}$, and increases on any interval $[\pi + 2n\pi, 2\pi + 2n\pi]$, $n \in \mathbb{Z}$. Therefore, each of the intervals of the first type is the largest interval of decrease, and each of the intervals of the second type is the largest interval of increase.

2. Answer if an increasing on $\mathbb{R}$ function can be bounded.

Solution.

There are many simple functions, especially monomials, such as x, x^3 and in general x^{2k+1}, $k \in \mathbb{N}$, which are more known and are increasing on $\mathbb{R}$ and at the same time unbounded. There are also monomials of even degree, such as x^2, x^4 and in general x^{2k}, $k \in \mathbb{N}$, which are increasing only on a part of the domain, but even so all of them are unbounded. For this reason, there is a natural tendency to generalize this situation to the cases of other elementary functions. However, this is not the case, as is seen already for algebraic functions. For instance, the function $f(x) = \frac{x}{|x|+1}$ is increasing on $\mathbb{R}$. Indeed, the function is odd, and consequently, we can verify its monotonicity only on the non-negative part of the real axis: taking $0 \le x_1 < x_2$ we get

$$f(x_2) - f(x_1) = \frac{x_2}{x_2+1} - \frac{x_1}{x_1+1} = \frac{x_2(x_1+1) - x_1(x_2+1)}{(x_2+1)(x_1+1)}$$
$$= \frac{x_2 - x_1}{(x_2+1)(x_1+1)}.$$

The numerator and denominator are positive, which indicates that $f(x)$ increases on $[0, +\infty)$. Then, due to the oddness, the function is increasing on $\mathbb{R}$. Even so, $f(x)$ is bounded, keeping its values in $(-1, 1)$, that follows from the evaluation $|f(x)| = \frac{|x|}{|x|+1} < 1, \forall x \in \mathbb{R}$.

Another classical example of an increasing and bounded on $\mathbb{R}$ function is $f(x) = \arctan x$.

14.5 Global and Local Extrema

1. Verify if a function $f(x)$ has global and local extrema on a given set S; if so, find these extrema:

   (a) $f(x) = 1 - 2x$, $S = [-4, 2)$;
   (b) $f(x) = |1 - 2x|$, $S = [-4, 2)$;
   (c) $f(x) = 2 - |x + 1|$, $S = \mathbb{R}$;
   (d) $f(x) = \begin{cases} 1-2x, x \le 0 \\ 1+4x-4x^2, \ x \in (0,1) \\ 2x-1, x \ge 1 \end{cases}$, $S = \mathbb{R}$;
   (e) $f(x) = x - [x]$, $S = \mathbb{R}$;
   (f) $f(x) = \cos x$, $S = \mathbb{R}$.

   Solution.

   (a) The function $f(x) = 1 - 2x$ is decreasing over the entire domain $\mathbb{R}$ and, in particular, on $S = [-4, 2)$: $f(x_2) - f(x_1) = 1 - 2x_2 - (1 - 2x_1) = 2(x_1 - x_2) < 0, \forall x_1 < x_2$. Then, $f(x)$ has no local extremum: the endpoint -4 cannot be a local extremum and any interior point x_0 of S is decreasing, which means that $f(x) > f(x_0)$ for $x < x_0$ and $f(x) < f(x_0)$ for $x > x_0$, which shows that x_0 also cannot be a local extremum. The same property of decrease implies that any interior point x_0 of S cannot be a global extremum, and consequently, a global extremum can occur only at the endpoint -4, which is a global maximum: $f(-4) = 9 > 1 - 2x = f(x), \forall x \in (-4, 2)$. The right endpoint does not belong to S, and consequently, it is not a global extremum. If, on the other hand, we try to take another point x_0, a bit smaller than 2, as a global minimum, then we can always find the point $x_1 = \frac{x_0+2}{2}$ belonging to S and such that $f(x_1) < f(x_0)$ because $x_1 > x_0$, which shows that x_0 is not a global minimum.

(b) Due to the definition of the absolute value, it is clear that $f(\frac{1}{2}) = 0$ is a global minimum of the function, because at any other point $1 - 2x \neq 0$, and consequently, $f(x) > 0, \forall x \neq \frac{1}{2}$. Opening the absolute value $f(x) = |1 - 2x| = \begin{cases} 1 - 2x, x < \frac{1}{2} \\ 2x - 1, x \geq \frac{1}{2} \end{cases}$ we can see that the function is decreasing on the part $x \leq \frac{1}{2}$ of its domain $\mathbb{R}$, and increasing on the part $x \geq \frac{1}{2}$. Therefore, $f(x)$ has no other global or local extremum on $\mathbb{R}$ but $x = \frac{1}{2}$. However, additional extrema can occur on a finite interval like $S = [-4, 2)$. First notice that $x_0 = \frac{1}{2}$ is an interior point of S, and for this reason x_0 is both global and local minimum of $f(x)$ on S (in the same way as on $\mathbb{R}$). Since $f(x)$ decreases on $[-4, \frac{1}{2}]$, the greatest value on this part of S the function has at the left endpoint $f(-4) = 9$. On the part $[\frac{1}{2}, 2)$ the function increases, and therefore, $f(x) < f(2) = 3$. Although $f(x)$ does not attain the value $f(2) = 3$ on S, but this value is smaller than $f(-4)$, which indicates that $f(x) < f(-4), \forall x \in (-4, 2)$. Therefore, $x_1 = -4$ is a global maximum of $f(x)$ on S (but it is not a local maximum, because -4 is not an interior point of S).

(c) Recall that by the definition of the absolute value, the function definition has the two formulas: $f(x) = 2 - |x + 1| = \begin{cases} 2 + (x + 1), & x < -1 \\ 2 - (x + 1), & x \geq -1 \end{cases} = \begin{cases} 3 + x, & x < -1 \\ 1 - x, & x \geq -1 \end{cases}$. So it is clear intuitively that the function has only a global maximum at the point $x_0 = -1$. Let us verify this analytically.
First, let us show that the function is increasing on the interval $(-\infty, -1]$ and decreasing on the interval $[-1, +\infty)$. Indeed, by the definition, for any pair $x_1 < x_2 \leq -1$ we have $f(x_1) = 3 + x_1 < 3 + x_2 = f(x_2)$, and for any pair $-1 \leq x_1 < x_2$ we get $f(x_1) = 1 - x_1 > 1 - x_2 = f(x_2)$. Therefore, any point in $(-\infty, -1)$ is increasing and any point in $(-1, +\infty)$ is decreasing, which means that none of these points can be an extremum of any type. The only remaining point $x_0 = -1$ is strict global and local maximum, because $f(x) = 2 - |x + 1| < 2 = f(-1), \forall x \neq -1$ and this point belongs to the domain together with its neighborhood.

(d) Consider separately three parts of the domain $\mathbb{R}$. On the part $x \leq 0$ the function $1 - 2x$ is decreasing, attains the global minimum value $f(0) = 1$ and does not have a maximum, since for negative x with sufficiently large absolute value the function becomes greater than any positive constant chosen in advance. On the part $0 < x < 1$ the quadratic formula $p(x) = 1 + 4x - 4x^2$ can be written in the canonical form $p(x) = 2 - (2x - 1)^2$ which evidences that this parabola goes downward and has the vertex at

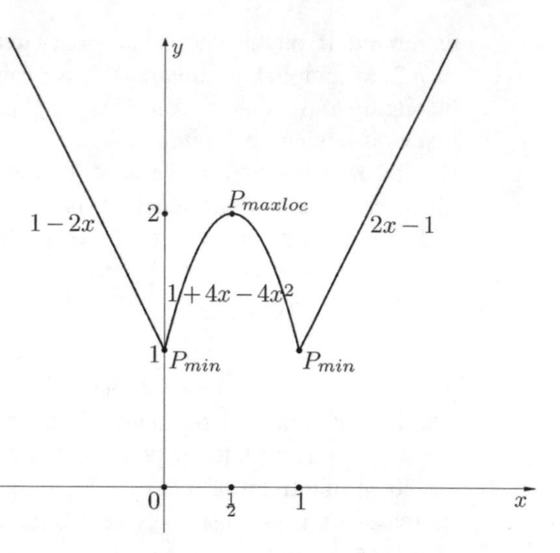

Fig. 2.81 Extrema of the function $y = f(x)$ of exercise 1(d)

the point $(\frac{1}{2}, 2)$, which implies that on the subinterval $(0, \frac{1}{2})$ the function $2-(2x-1)^2$ increases and on the next subinterval $(\frac{1}{2}, 1)$ it decreases (clearly, these results can be deduced analytically, without appealing to the properties of a parabola). Notice that at the endpoints of $(0, 1)$ the parabola takes the values $p(0) = p(1) = 1$. Finally, on the part $x \geq 1$ the function $2x - 1$ increases starting from its minimum $f(1) = 1$ and it is unbounded taking the values greater than any constant given in advance for sufficiently large x (see Fig. 2.81).

Using these observations, we can conclude that there is no global or local extremum on the intervals $(-\infty, 0)$ and $(1, +\infty)$ (monotonicity of the linear parts) and on the intervals $(0, \frac{1}{2})$ and $(\frac{1}{2}, 0)$ (monotonicity of the parabolic part), since every point of these intervals is monotonic and interior in the corresponding interval. Then, there are the three remaining points 0, $\frac{1}{2}$ and 1. Among them there are two global and local minima—the points 0 and 1: $f(0) = f(1) = 1 < f(x), \forall x \neq 0,\ 1$. Still, by the properties of the parabola, there is a local maximum $x = \frac{1}{2}$, but it is not a global maximum, because on the intervals $(-\infty, 0)$ and $(1, +\infty)$ the function takes the values greater than any constant, in particular, greater than $f(\frac{1}{2}) = 2$.

(e) To facilitate the reasoning, recall the graph of the function $f(x) = x - [x]$ shown in Fig. 2.82.

We start with global extrema. All the integers $x_n = n$, $n \in \mathbb{Z}$ are global non-strict minima:

$$\forall x \in \mathbb{R}, x \neq x_n = n \in \mathbb{Z} : f(x) = x - [x] > 0 = x_n - [x_n] = f(x_n).$$

Let us show that there is no global maximum. Let us suppose, for contradiction, that x_1 is a global maximum, $x_1 \in [n, n + 1), n \in \mathbb{Z}$. Then we

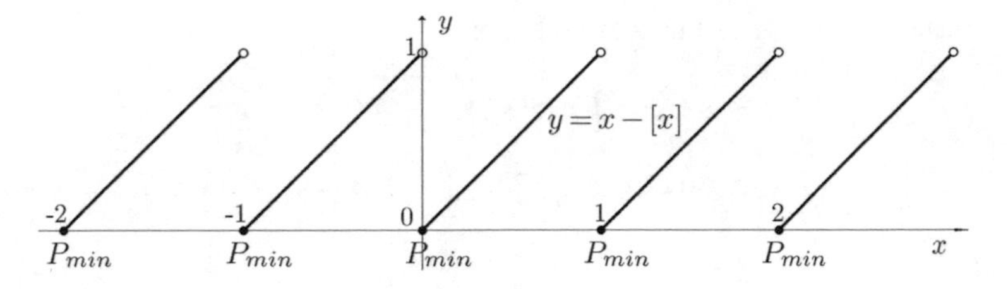

Fig. 2.82 Extrema of the function $y = x - [x]$

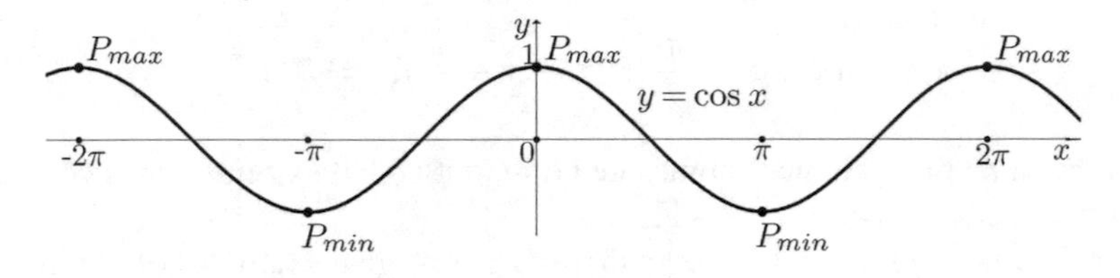

Fig. 2.83 Extrema of the function $y = \cos x$

take $x_2 = \frac{x_1+n+1}{2}$, $x_2 \in [n, n+1)$ and obtain $f(x_1) = x_1 - [x_1] < \frac{x_1+n+1}{2} - [x_1] = \frac{x_1+n+1}{2} - \left[\frac{x_1+n+1}{2}\right] = f(x_2)$. This means that x_1 is not a global maximum.
Consider now local extrema. Each of the points $x_n = n, n \in \mathbb{Z}$ is a global minimum contained in $\mathbb{R}$ together with its neighborhood, which means that these points are also local minima. If we choose a neighborhood of n with the radius smaller than 1, for instance with the radius $\frac{1}{2}$, then all other integers will stay out of this neighborhood, and consequently, $f(x) = x - [x] > 0 = n - [n] = f(n), \forall x \in (n - \frac{1}{2}, n + \frac{1}{2})$, which means that each $x_n = n, n \in \mathbb{Z}$ is a strict local minimum. At the same time, any point of the interval $(n, n+1)$, $n \in \mathbb{Z}$ is increasing and interior for this interval, which shows that none of these points can be a local extremum.

(f) Although our principal line of an investigation is to study the properties of a function and sketch its graph starting from the analytic definition, in this exercise we appeal to the graph of the function, since the analytic properties of trigonometric functions were not investigated yet. In the cases like this, we appeal to an intuitive knowledge of a function learned in high school or consider the graph as the original definition of a function in geometric form. The graph of $y = \cos x$ is shown in Fig. 2.83.

According to the graph of $f(x) = \cos x$, notice that each point $x_k = 2k\pi, \forall k \in \mathbb{Z}$ is a strict local maximum (where the function takes the value 1), and each point $x_n = \pi + 2n\pi, \forall n \in \mathbb{Z}$ is a strict local minimum (where the function takes the value -1). It is sufficient to choose a neighborhood of the radius smaller than π, for instance with the radius $\frac{\pi}{2}$, to see that

$$f(x) = \cos x < 1 = \cos 2k\pi = f(x_k),$$
$$\forall x \in (2k\pi - \frac{\pi}{2}, 2k\pi + \frac{\pi}{2}), x \neq 2k\pi, \forall k \in \mathbb{Z}$$

and

$$f(x) = \cos x > -1 = \cos(2n\pi + \pi) = f(x_n),$$
$$\forall x \in (2n\pi + \pi - \frac{\pi}{2}, 2n\pi + \pi + \frac{\pi}{2}), x \neq 2n\pi + \pi, \forall n \in \mathbb{Z}.$$

At the same time, all these points are non-strict global extrema, since $\cos x \leq 1$, $\forall x \in \mathbb{R}$ and $\cos x \geq -1$, $\forall x \in \mathbb{R}$.

2. For the functions $f(x) = -|x|$ and $f(x) = x^2$, give examples of the intervals such that $f(x)$:

   (a) has neither maximum nor minimum;
   (b) has a maximum, but does not have a minimum;
   (c) has a minimum, but does not have a maximum;
   (d) has both maximum and minimum.

   Solution.

   (a) For both functions one of such intervals is $(-2, 0)$, and another one is $(1, 4)$.
   (b) For $f(x) = -|x|$ one of such intervals is $(-2, 1)$ or $[1, 4)$. For $f(x) = x^2$ one of such intervals is $(0, 1]$ or $[-2, -1)$.
   (c) For $f(x) = -|x|$ one of such intervals is $(0, 1]$ or $[-2, -1)$. For $f(x) = x^2$ one of such intervals is $(-2, 1)$ or $[1, 4)$.
   (d) For both functions one of such intervals is $[-2, 0]$, and another one is $[1, 4]$.

   Prove all these statements.

3. Show that if $y = f(x)$ has strict global minimum at x_0 and $g(y)$ is monotonic on $\mathbb{R}$, then $g(f(x))$ has strict global extremum at x_0.

   Solution.

   Notice first that the composite function is defined for each x of the domain X_f of $f(x)$, because $g(y)$ is defined on $\mathbb{R}$.

   If $g(y)$ increases on $\mathbb{R}$, then $g(y) > g(y_0)$ for any pair $y > y_0$. Since x_0 is a strict global minimum of $f(x)$, we have $y = f(x) > y_0 = f(x_0)$ for any $x \in X_f, x \neq x_0$. Then, $g(f(x)) = g(y) > g(y_0) = g(f(x_0))$ for any $x \in X_f$, $x \neq x_0$, which means that x_0 is a strict global minimum of $g(f(x))$ on X_f.

   In the same manner, if $g(y)$ decreases on $\mathbb{R}$, then $g(y) < g(y_0)$ for any pair $y > y_0$. Since x_0 is a strict global minimum of $f(x)$, we have $y = f(x) > y_0 = f(x_0)$ for any $x \in X_f, x \neq x_0$. Then, $g(f(x)) = g(y) < g(y_0) = g(f(x_0))$ for any $x \in X_f, x \neq x_0$, which means that x_0 is a strict global maximum of $g(f(x))$ on X_f.

   Formulate and prove a similar property for a maximum at x_0.

14.6 Concavity and Inflection

1. Investigate concavity and inflection of the following functions:
   (a) $f(x) = \frac{1}{x}$;
   (b) $f(x) = \frac{1}{x^2}$;
   (c) $f(x) = \frac{1}{x^3}$;
   (d) $f(x) = \sqrt[3]{x}$;
   (e) $f(x) = \frac{1}{\sqrt[3]{x}}$;
   (f) $f(x) = e^x$;
   (g) $f(x) = \ln x$;
   (h) $f(x) = \cos x$.

   Solution.

   (a) Since concavity is defined on an interval contained in the domain, which is $X = \mathbb{R}\backslash\{0\}$ for the function $f(x) = \frac{1}{x}$, we have the two options: to choose the interval $(-\infty, 0)$ or $(0, +\infty)$. Under this restriction, for any pair $x_1, x_2 \in X$ we have the following evaluation of the difference between the mean value of the secant $\frac{f(x_1)+f(x_2)}{2}$ and the value of the function at the midpoint $f\left(\frac{x_1+x_2}{2}\right)$:

$$D_2 \equiv \frac{f(x_1) + f(x_2)}{2} - f\left(\frac{x_1 + x_2}{2}\right) = \frac{\frac{1}{x_1} + \frac{1}{x_2}}{2} - \frac{1}{\frac{x_1+x_2}{2}}$$

$$= \frac{x_1 + x_2}{2x_1x_2} - \frac{2}{x_1 + x_2}$$

$$= \frac{(x_1 + x_2)^2 - 4x_1x_2}{2x_1x_2(x_1 + x_2)} = \frac{(x_1 - x_2)^2}{2x_1x_2(x_1 + x_2)}.$$

   First, if $x_1, x_2 < 0$, then $x_1x_2(x_1 + x_2) < 0$, and consequently, $D_2 < 0$, which means that the function is concave downward on the interval $(-\infty, 0)$. Second, if $x_1, x_2 > 0$, then $x_1x_2(x_1 + x_2) > 0$, and consequently, $D_2 > 0$, which means that the function is concave upward on the interval $(0, +\infty)$. Although the function changes the type of concavity at the point $x_0 = 0$, but this point does not belong to the domain, so it is not an inflection point. Since the function keeps the same concavity over the entire interval $(-\infty, 0)$ and also over the entire interval $(0, +\infty)$, there is no inflection point.
   Notice that the function $f(x) = \frac{1}{x}$ is odd, and consequently, the study of concavity can be initially performed only on the interval $(0, +\infty)$ (or $(-\infty, 0)$) and then the obtained result can be extended to the remaining part of the domain using the oddness according Property 2 in Sect. 7.5.

   (b) Since the function $f(x) = \frac{1}{x^2}$ is even: $f(-x) = \frac{1}{(-x)^2} = \frac{1}{x^2} = f(x)$, $\forall x \in X = \mathbb{R}\backslash\{0\}$ (where X is the domain), by Property 2 of Sect. 7.5, we can perform investigation of its concavity initially only on the interval $(0, +\infty)$ and then extend the results to the remaining part of the domain using the evenness.

Choosing two arbitrary points x_1, x_2 of the interval $(0, +\infty)$, we obtain the following evaluation of the difference between the mean value of the secant $\frac{f(x_1)+f(x_2)}{2}$ and the value of the function at the midpoint $f\left(\frac{x_1+x_2}{2}\right)$:

$$D_2 \equiv \frac{f(x_1)+f(x_2)}{2} - f\left(\frac{x_1+x_2}{2}\right) = \frac{\frac{1}{x_1^2}+\frac{1}{x_2^2}}{2} - \frac{1}{(\frac{x_1+x_2}{2})^2}$$

$$= \frac{x_1^2+x_2^2}{2x_1^2x_2^2} - \frac{4}{(x_1+x_2)^2}$$

$$= \frac{(x_1^2+x_2^2)(x_1+x_2)^2 - 8x_1^2x_2^2}{2x_1^2x_2^2(x_1+x_2)^2}$$

$$= \frac{(x_1^4+x_2^4-2x_1^2x_2^2) + \left(2x_1x_2(x_1^2+x_2^2) - 4x_1^2x_2^2\right)}{2x_1^2x_2^2(x_1+x_2)^2}$$

$$= \frac{(x_1^2-x_2^2)^2 + 2x_1x_2(x_1-x_2)^2}{2x_1^2x_2^2(x_1+x_2)^2}.$$

Obviously, both numerator and denominator are positive, and consequently, $D_2 > 0$, which means that $f(x)$ is concave upward on $(0, +\infty)$.

Applying now Property 1 of Sect. 7.5, we conclude that $f(x)$ is also concave upward on $(-\infty, 0)$. Therefore, there is no inflection point.

(c) Since the function $f(x) = \frac{1}{x^3}$ is odd: $f(-x) = \frac{1}{(-x)^3} = -\frac{1}{x^3} = f(x)$, $\forall x \in X = \mathbb{R}\backslash\{0\}$ (where X is the domain), by Property 2 of Sect. 7.5, we can perform investigation of its concavity initially only on the interval $(0, +\infty)$ and then extend the results to the remaining part of the domain using the oddness.

Choosing two arbitrary points x_1, x_2 of the interval $(0, +\infty)$, we denote the midpoint $x_0 = \frac{x_1+x_2}{2}$ and the increment $h = x_2 - x_0 = x_0 - x_1$, and evaluate the difference between the mean value of the secant $\frac{f(x_1)+f(x_2)}{2}$ and the value of the function at the midpoint $f(x_0)$ as follows:

$$D_2 \equiv \frac{f(x_1)+f(x_2)}{2} - f(x_0) = \frac{\frac{1}{x_1^3}+\frac{1}{x_2^3}}{2} - \frac{1}{x_0^3} = \frac{x_1^3+x_2^3}{2x_1^3x_2^3} - \frac{1}{x_0^3}$$

$$= \frac{(x_0-h)^3+(x_0+h)^3}{2x_1^3x_2^3} - \frac{1}{x_0^3} = \frac{x_0^3+3x_0h^2}{x_1^3x_2^3} - \frac{1}{x_0^3}$$

$$= \frac{(x_0^3 + 3x_0h^2)x_0^3 - (x_0^2 - h^2)^3}{x_1^3 x_2^3 x_0^3}$$

$$= \frac{h^2(6x_0^4 - 3x_0^2h^2 + h^4)}{x_1^3 x_2^3 x_0^3} = \frac{h^2\left((\frac{3}{2}x_0^2 - h^2)^2 + \frac{15}{4}x_0^4\right)}{x_1^3 x_2^3 x_0^3}.$$

Obviously, the numerator and denominator are positive, and consequently, $D_2 > 0$, which means that the function is concave upward on the interval $(0, +\infty)$.
Applying Property 2 of Sect. 7.5, we conclude that $f(x)$ is concave downward on the interval $(-\infty, 0)$.
Although the function changes the type of concavity passing the point $x_0 = 0$, but this point does not belong to the domain, and so it is not an inflection point. Since the function is concave on both intervals of the domain, there is no inflection point.

(d) Take two arbitrary points such that $x_1 < x_2$, denote the midpoint $x_0 = \frac{x_1+x_2}{2}$ and increment $h = x_2 - x_0 = x_0 - x_1 > 0$, and find the following evaluation of the difference between the mean value of the secant $\frac{f(x_1)+f(x_2)}{2}$ and the value of the function at the midpoint $f(x_0)$:

$$2D_2 \equiv f(x_1) + f(x_2) - 2f(x_0) = x_1^{1/3} + x_2^{1/3} - 2x_0^{1/3}$$

$$= (x_2^{1/3} - x_0^{1/3}) - (x_0^{1/3} - x_1^{1/3})$$

$$= \frac{x_2 - x_0}{x_2^{2/3} + x_2^{1/3}x_0^{1/3} + x_0^{2/3}} - \frac{x_0 - x_1}{x_0^{2/3} + x_0^{1/3}x_1^{1/3} + x_1^{2/3}}$$

$$= h\frac{(x_0^{2/3} + x_0^{1/3}x_1^{1/3} + x_1^{2/3}) - (x_2^{2/3} + x_2^{1/3}x_0^{1/3} + x_0^{2/3})}{(x_2^{2/3} + x_2^{1/3}x_0^{1/3} + x_0^{2/3})(x_0^{2/3} + x_0^{1/3}x_1^{1/3} + x_1^{2/3})}$$

$$= h\frac{x_0^{1/3}(x_1^{1/3} - x_2^{1/3}) + (x_1^{2/3} - x_2^{2/3})}{(x_2^{2/3} + x_2^{1/3}x_0^{1/3} + x_0^{2/3})(x_0^{2/3} + x_0^{1/3}x_1^{1/3} + x_1^{2/3})}$$

$$= \frac{h(x_1^{1/3} - x_2^{1/3})}{(x_2^{2/3} + x_2^{1/3}x_0^{1/3} + x_0^{2/3})(x_0^{2/3} + x_0^{1/3}x_1^{1/3} + x_1^{2/3})} \cdot (x_0^{1/3} + x_1^{1/3} + x_2^{1/3}).$$

Since $h > 0$, $x_2^{2/3} + x_2^{1/3}x_0^{1/3} + x_0^{2/3} = (x_2^{1/3} + \frac{1}{2}x_0^{1/3})^2 + \frac{3}{4}x_0^{2/3} > 0$, $x_0^{2/3} + x_0^{1/3}x_1^{1/3} + x_1^{2/3} = (x_0^{1/3} + \frac{1}{2}x_1^{1/3})^2 + \frac{3}{4}x_1^{2/3} > 0$ and $x_1^{1/3} - x_2^{1/3} < 0$, the first factor is negative and the sign of D_2 is determined by the second factor. If $x_2 \le 0$, then $x_0^{1/3} + x_1^{1/3} + x_2^{1/3} < 0$, and consequently, $D_2 > 0$, which means that the functions is concave upward on $(-\infty, 0]$. If $0 \le x_1$, then $x_0^{1/3} + x_1^{1/3} + x_2^{1/3} > 0$, and therefore, $D_2 < 0$, which means that the function is concave downward on $[0, +\infty)$.

Since the function changes the type of concavity at the point 0 (and it belongs to the domain), this point is inflection.
Notice that the function $f(x) = x^{1/3}$ is odd, which allows us to perform a study of concavity initially only on the interval $(0, +\infty)$ (or $(-\infty, 0)$) with posterior extension of the obtained results to the remaining part of the domain according to Property 2 of Sect. 7.5.

(e) Since the function $f(x) = \frac{1}{x^{1/3}}$ is odd: $f(-x) = \frac{1}{(-x)^{1/3}} = -\frac{1}{x^{1/3}} = f(x)$, $\forall x \in X = \mathbb{R}\backslash\{0\}$ (where X is the domain), by Property 2 of Sect. 7.5, we can perform investigation of its concavity initially only on the interval $(0, +\infty)$ and then extend the results to the remaining part of the domain using the oddness.
Choosing two arbitrary points $x_1 < x_2$ of the interval $(0, +\infty)$ and denoting the midpoint and increment by $x_0 = \frac{x_1+x_2}{2}$ and $h = x_2 - x_0 = x_0 - x_1$, respectively, we obtain the following evaluation of the difference between the mean value of the secant $\frac{f(x_1)+f(x_2)}{2}$ and the value of the function at the midpoint $f(x_0)$:

$$2D_2 \equiv f(x_1) + f(x_2) - 2f(x_0) = \frac{1}{x_1^{1/3}} + \frac{1}{x_2^{1/3}} - 2\frac{1}{x_0^{1/3}}$$

$$= \left(\frac{1}{x_2^{1/3}} - \frac{1}{x_0^{1/3}}\right) - \left(\frac{1}{x_0^{1/3}} - \frac{1}{x_1^{1/3}}\right)$$

$$= \frac{x_0 - x_2}{x_2^{1/3} x_0^{1/3} (x_2^{2/3} + x_2^{1/3} x_0^{1/3} + x_0^{2/3})} - \frac{x_1 - x_0}{x_0^{1/3} x_1^{1/3} (x_0^{2/3} + x_0^{1/3} x_1^{1/3} + x_1^{2/3})}$$

$$= -h \frac{x_1^{1/3}(x_0^{2/3} + x_0^{1/3} x_1^{1/3} + x_1^{2/3}) - x_2^{1/3}(x_2^{2/3} + x_2^{1/3} x_0^{1/3} + x_0^{2/3})}{x_1^{1/3} x_0^{1/3} x_2^{1/3} (x_2^{2/3} + x_2^{1/3} x_0^{1/3} + x_0^{2/3})(x_0^{2/3} + x_0^{1/3} x_1^{1/3} + x_1^{2/3})}$$

$$= -h \frac{x_1 - x_2 + x_0^{2/3}(x_1^{1/3} - x_2^{1/3}) + x_0^{1/3}(x_1^{2/3} - x_2^{2/3})}{x_1^{1/3} x_0^{1/3} x_2^{1/3} (x_2^{2/3} + x_2^{1/3} x_0^{1/3} + x_0^{2/3})(x_0^{2/3} + x_0^{1/3} x_1^{1/3} + x_1^{2/3})} .$$

Since $h > 0$, the numerator of the quotient is negative and the denominator is positive, which guarantees that $D_2 > 0$, and so the function is concave upward on $(0, +\infty)$.
Using Property 2 of Sect. 7.5, we conclude that $f(x)$ is concave downward on $(-\infty, 0)$.
Although the function changes the type of concavity passing $x = 0$, but this point is out of the domain, and consequently, it is not an inflection point. On both intervals of the domain the function is concave, which implies that there is no inflection point.

(f) For any pair of points $x_1, x_2 \in \mathbb{R}$, we have the evaluation

$$f(x_1)+f(x_2)-2f\left(\frac{x_1+x_2}{2}\right)=e^{x_1}+e^{x_2}-2e^{\frac{x_1+x_2}{2}}=\left(e^{\frac{x_1}{2}}-e^{\frac{x_2}{2}}\right)^2>0,$$

which reveals the upward concavity over the entire domain. From this result it follows immediately that there is no inflection point.

(g) For any pair of points $x_1, x_2 \in X = (0, +\infty)$, we have the following evaluation:

$$\begin{aligned}f(x_1)+f(x_2)-2f\left(\frac{x_1+x_2}{2}\right)&=\ln x_1+\ln x_2-2\ln\frac{x_1+x_2}{2}\\&=\ln\frac{4x_1x_2}{(x_1+x_2)^2}<0.\end{aligned}$$

The last inequality follows from the formula $(x_1+x_2)^2-4x_1x_2=(x_1-x_2)^2>0$. Then, the function is concave downward over the entire domain X. From this result it follows immediately that there is no inflection point.

(h) Using periodicity of the function $f(x)=\cos x$, we can initially perform our investigation on the interval $[-\frac{\pi}{2}, \frac{3\pi}{2}]$. Taking two arbitrary points $x_1 < x_2$ in $[-\frac{\pi}{2}, \frac{3\pi}{2}]$, we have

$$\begin{aligned}2D_2 \equiv f(x_1)+f(x_2)-2f\left(\frac{x_1+x_2}{2}\right)&=\cos x_1+\cos x_2-2\cos\frac{x_1+x_2}{2}\\&=2\cos\frac{x_1+x_2}{2}\cos\frac{x_1-x_2}{2}-2\cos\frac{x_1+x_2}{2}\\&=2\cos\frac{x_1+x_2}{2}\left(\cos\frac{x_1-x_2}{2}-1\right).\end{aligned}$$

To determine the sign of D_2, consider separately the two subintervals. First, choose the interval $-\frac{\pi}{2} \leq x_1 < x_2 \leq \frac{\pi}{2}$. In this case, $-\frac{\pi}{2} < \frac{x_1+x_2}{2} < \frac{\pi}{2}$, which assures that $\cos\frac{x_1+x_2}{2} > 0$, and at the same time $0 < \frac{x_2-x_1}{2} \leq \frac{\pi}{2}$, which guarantees that the second factor is negative. Therefore, $D_2 < 0$ and the function has downward concavity on $[-\frac{\pi}{2}, \frac{\pi}{2}]$. In a similar manner, on the second subinterval $\frac{\pi}{2} \leq x_1 < x_2 \leq \frac{3\pi}{2}$, we have $\frac{\pi}{2} < \frac{x_1+x_2}{2} < \frac{3\pi}{2}$ and $0 < \frac{x_2-x_1}{2} \leq \frac{\pi}{2}$, from which it follows that both factors are negative, and then $D_2 > 0$. So, the function has upward concavity on $[\frac{\pi}{2}, \frac{3\pi}{2}]$.

Using the 2π-periodicity, we can conclude that $\cos x$ is concave downward on the intervals $[-\frac{\pi}{2}, \frac{\pi}{2}] + 2k\pi$, $\forall k \in \mathbb{Z}$ and it is concave upward on the intervals $[\frac{\pi}{2}, \frac{3\pi}{2}] + 2n\pi$, $\forall n \in \mathbb{Z}$.

Since the function changes the type of concavity at the points $x_k = \frac{\pi}{2} + k\pi$, $\forall k \in \mathbb{Z}$ (all of which belong to the domain), each of these points is an inflection point.

2. Answer if a concave upward on $\mathbb{R}$ function can be bounded above.

   Solution.

   The answer is negative. Let us show even stronger result: if a non-constant function is concave upward, even non-strictly, on $\mathbb{R}$, then it is unbounded above. Indeed, since the function is non-constant, there are two points x_0 and x_1 such that $f(x_0) \neq f(x_1)$. Without loss of generality we can assume that $x_1 < x_0$ and consider first the case $f(x_1) < f(x_0)$. Take now an arbitrary point $x_2 > x_0$ and consider the general non-strict concavity of $f(x)$ on the interval $[x_1, x_2]$ with the point x_0 lying inside: $f(x_0) \leq f(x_1) + \frac{f(x_2)-f(x_1)}{x_2-x_1}(x_0 - x_1)$. The last inequality can be written in the form $\frac{f(x_0)-f(x_1)}{x_0-x_1}(x_2 - x_1) + f(x_1) \leq f(x_2)$. Since this is true for any $x_2 > x_0$, then considering variable x_2, we have on the left-hand side a linear function of x_2 with the positive main coefficient, and consequently, this linear function increases without restriction as x_2 increases. Therefore, according to the inequality, $f(x_2)$ also increases without restriction, which implies that $f(x)$ is unbounded above.

   Next, consider the case $x_1 < x_0$ and $f(x_1) > f(x_0)$. It can be reduced to the previous situation by considering the function $g(x) = f(-x)$. Indeed, by Property 3 of Sect. 7.5, $g(x)$ is concave upward on $\mathbb{R}$. Making notations $\tilde{x}_0 = -x_0$ and $\tilde{x}_1 = -x_1$, we have $\tilde{x}_0 < \tilde{x}_1$ and $g(\tilde{x}_0) = f(x_0) < f(x_1) = g(\tilde{x}_1)$, that is, we arrive at the first case for $g(x)$ with indices of the points interchanged. Therefore, $g(x)$ is unbounded above, which means that $f(x)$ is unbounded above.

   Since a concave upward on $\mathbb{R}$ function cannot be constant, it follows from the proved result that such a function is unbounded above.

3. Verify if concavity of $f(x)$ on a set has an implication for concavity of $\frac{1}{f(x)}$ on the intervals where the latter is defined.

   Solution.

   Some elementary examples show that $\frac{1}{f(x)}$ can preserve the type of concavity of $f(x)$. For example, $f(x) = e^x$ (see exercise 1(f)) and $\frac{1}{f(x)} = e^{-x}$ have upward concavity on $\mathbb{R}$ (the concavity of e^{-x} can be proved in the same way as for e^x); $f(x) = x^2$ (see Sects. 7.1 and 7.2) has upward concavity on $\mathbb{R}$ and $\frac{1}{f(x)} = \frac{1}{x^2}$ has upward concavity on the intervals $(-\infty, 0)$ and $(0, +\infty)$ (see exercise 1(b)). However, in general, such a rule does not exist. An elementary example, which shows that such a rule looks suspicious, is $f(x) = x$: this function has a neutral concavity (non-strict concavity of both types) on $\mathbb{R}$, while $\frac{1}{f(x)} = \frac{1}{x}$ has downward concavity on $(-\infty, 0)$ and upward concavity on $(0, +\infty)$ (see exercise 1(a)). A simple example, which disproves supposed relationship between concavities of $f(x)$ and $\frac{1}{f(x)}$, is the function $f(x) = x^{1/3}$. This function has upward concavity on $(-\infty, 0]$ and downward concavity on $[0, +\infty)$ (see exercise 1(d)), while the function $\frac{1}{f(x)} = \frac{1}{x^{1/3}}$ has downward concavity on $(-\infty, 0)$ and upward concavity on $(0, +\infty)$ (see exercise 1(e)).

Another example is $f(x) = e^x - 1$ that has upward concavity on $(-\infty, 0)$, while $\frac{1}{f(x)} = \frac{1}{e^x-1}$ has downward concavity on $(-\infty, 0)$.

4. Show that if $f(x)$ is concave upward (even non-strictly) on an interval I of its domain, then it satisfies Jensen's inequality: $f\left(\sum_{i=1}^{n} \lambda_i x_i\right) \le \sum_{i=1}^{n} \lambda_i f(x_i)$ for any set of the points $x_1, \ldots, x_n \in I$ and a set of non-negative parameters $\lambda_i, i = 1, \ldots, n$ such that $\lambda_1 + \ldots + \lambda_n = 1$.

   Solution.

   We prove by induction. For $n = 1$ the inequality is a tautology and for $n = 2$ it expresses the condition of non-strict concavity of $f(x)$. Let us assume that the inequality is true for some n and show its validity for $n + 1$. If $\lambda_{n+1} = 1$, then $\lambda_1 = \ldots = \lambda_n = 0$ and we come back to the case of the one parameter. If $\lambda_{n+1} < 1$, then we separate the last parameter on the left-hand side and apply there the inequality for $n = 2$:

$$f\left(\sum_{i=1}^{n+1} \lambda_i x_i\right) = f\left(\lambda_{n+1} x_{n+1} + (1-\lambda_{n+1}) \sum_{i=1}^{n} \frac{\lambda_i}{1-\lambda_{n+1}} x_i\right)$$

$$\le \lambda_{n+1} f(x_{n+1}) + (1 - \lambda_{n+1}) f\left(\sum_{i=1}^{n} \frac{\lambda_i}{1-\lambda_{n+1}} x_i\right) \equiv A\,.$$

Now, noting that the parameters $\mu_i = \frac{\lambda_i}{1-\lambda_{n+1}}$ are not negative and that $\mu_1 + \ldots + \mu_n = 1$, we apply the hypothesis of induction to the last term and obtain the required result:

$$A \le \lambda_{n+1} f(x_{n+1}) + (1 - \lambda_{n+1}) \sum_{i=1}^{n} \frac{\lambda_i}{1 - \lambda_{n+1}} f(x_i) = \sum_{i=1}^{n+1} \lambda_i f(x_i).$$

5. Using the concavity of e^x show Young's inequality $ab \le \frac{a^p}{p} + \frac{b^q}{q}$, where a, b are non-negative numbers and parameters p, q are positive and related by the formula $\frac{1}{p} + \frac{1}{q} = 1$.

   Solution.

   If one of the numbers a or b is zero in Young's inequality, then the result is trivial. Let us consider the case when $ab > 0$. Recall that e^x is concave upward on $\mathbb{R}$ (see exercise 1(f)), which means, by the general definition, that for any $x_1, x_2 \in \mathbb{R}$ and $t \in (0, 1)$ the inequality $e^{(1-t)x_1+tx_2} < (1-t)e^{x_1} + te^{x_2}$ is true. Then represent ab in the following form:

$$ab = e^{\ln ab} = e^{\ln a+\ln b} = \exp\left[(1-t)\frac{\ln a}{1-t} + t\frac{\ln b}{t}\right] \equiv A,$$

where $t \in (0, 1)$. Applying the concavity of e^x with $x_1 = \frac{\ln a}{1-t}$ and $x_2 = \frac{\ln b}{t}$, we obtain Young's inequality:

$$A < (1-t)\exp\left[\frac{\ln a}{1-t}\right] + t\exp\left[\frac{\ln b}{t}\right] = (1-t)a^{\frac{1}{1-t}} + tb^{\frac{1}{t}},$$

where $\frac{1}{1-t} = p > 0$ and $\frac{1}{t} = q > 0$, with the relationship $\frac{1}{p}+\frac{1}{q} = (1-t)+t = 1$.

6. Using the concavity of $\ln x$ and Jensen's inequality, prove the inequality between geometric and arithmetic means: $\left(\prod_{i=1}^{n} x_i\right)^{\frac{1}{n}} \le \frac{1}{n}\sum_{i=1}^{n} x_i$, where $x_i \ge 0, \forall i$.

   Solution.

   If one of the numbers x_i is zero, the result is trivial. Let us consider the case when $x_i > 0, \forall i$. Recall that $\ln x$ is concave downward on $(0, +\infty)$ (see exercise 1(g)), and consequently, $-\ln x$ is concave upward on $(0, +\infty)$. Applying $-\ln x$ to the product $\left(\prod_{i=1}^{n} x_i\right)^{\frac{1}{n}}$ and using Jensen's inequality with the parameters $\lambda_i = \frac{1}{n}, \forall i$, we obtain

$$-\ln\left(\prod_{i=1}^{n} x_i\right)^{\frac{1}{n}} = -\sum_{i=1}^{n}\frac{1}{n}\ln x_i \ge -\ln\left(\sum_{i=1}^{n}\frac{1}{n}x_i\right).$$

   Multiplying the last inequality by -1 and applying the exponential function to both sides, we arrive to the required result:

$$\left(\prod_{i=1}^{n} x_i\right)^{\frac{1}{n}} = \exp\left[\ln\left(\prod_{i=1}^{n} x_i\right)^{\frac{1}{n}}\right] \le \exp\left[\ln\left(\sum_{i=1}^{n}\frac{1}{n}x_i\right)\right] = \frac{1}{n}\sum_{i=1}^{n} x_i.$$

14.7 Composite Functions

1. Find the composite functions $g(f(x))$ and $f(g(x))$:

   (a) $f(x) = x^2 - 1$, $g(x) = \frac{1}{x-1}$;
   (b) $f(x) = \frac{1}{x+1}$, $g(x) = \frac{1}{x-1}$.

   Solution.

   (a) $f(x) = x^2 - 1$, $g(x) = \frac{1}{x-1}$
   First, we determine the domain and range of each function separately. The domain of $f(x)$ is $D_f = \mathbb{R}$ and the range is $I_f = [-1, +\infty)$; the domain of $g(x)$ is $D_g = \mathbb{R}\backslash\{1\}$ and the range is $I_g = \mathbb{R}\backslash\{0\}$. In abbreviated form it is denoted as follows: $f(x) : D_f = \mathbb{R} \to I_f = [-1, +\infty)$ and $g(x) : D_g = \mathbb{R}\backslash\{1\} \to I_g = \mathbb{R}\backslash\{0\}$.

Now consider the composite function $h(x) = g(f(x)) : D_h \to I_h$. Notice that $I_f \not\subset D_g$, and therefore, we need to restrict (shrink) I_f that can be made by reducing D_f. To guarantee the condition $I_f \subset D_g$ we have to eliminate from D_f the elements whose image is 1, that is, such x that satisfy the relation $f(x) = x^2 - 1 = 1$, or equivalently, $x = \pm\sqrt{2}$. Then, we get the new domain $\bar{D}_f = \mathbb{R}\backslash\{\pm\sqrt{2}\}$ of $f(x)$ which, under the same formula $f(x) = x^2 - 1$, has the image contained in D_g: $\bar{I}_f = [-1, +\infty)\backslash\{1\} \subset D_g$. Strictly speaking, we have the new function $\bar{f}(x) : \bar{D}_f = \mathbb{R}\backslash\{\pm\sqrt{2}\} \to \bar{I}_f = [-1, +\infty)\backslash\{1\}$, which substitutes the function $f(x)$ and is frequently denoted by the same letter f. After this adjustment of the domain of $f(x)$, the conditions of the composition $g(\bar{f}(x))$ are satisfied and we obtain the following composite function:

$$h(x) = g(\bar{f}(x)) = g(x^2 - 1) = \frac{1}{x^2 - 2} : D_h = \bar{D}_f = \mathbb{R}\backslash\{\pm\sqrt{2}\} \to I_h \subset I_g$$
$$= \mathbb{R}\backslash\{0\}.$$

Although in this composition the modified function $\bar{f}$ was used instead of f, for the sake of simplicity, the notation used for this modified function is frequently just f. In what follows we will use such simplified notation in many cases.
The range I_h of $h(x) = g(\bar{f}(x))$ may not contain all the elements of I_g, because the range of $\bar{f}(x)$ does not cover all the domain of $g(x)$. If it is required, we can specify the range I_h knowing the formula of $h(x)$ and its domain D_h. Indeed, denoting the values of the range by y, we have the following formula of the composition $\frac{1}{x^2-2} = y$. To determine the range we have to find all y for which the last relation has at least one corresponding element of the domain. In other words, we need to find all y such that the equation $\frac{1}{x^2-2} = y$ has at least one solution x in D_h. Notice first that for $y = 0$ there is no solution. Then, we can divide by y without loss of solutions and rewrite the equation in the form $x^2 = \frac{1}{y} + 2$. Since $x^2 \ge 0$, the admissible values of y should satisfy the condition $\frac{1}{y} + 2 \ge 0$, that is, $\frac{1}{y} \ge -2$ (for other values of y there is no corresponding x). For any $y > 0$ this condition holds. If $y < 0$, then the condition can be written as $y \le -\frac{1}{2}$. Hence, the values y, for which the quadratic equation $x^2 = \frac{1}{y} + 2$ has solutions in D_h, are $y > 0$ and $y \le -\frac{1}{2}$. Therefore, the range of the function $h(x) = g(f(x))$ is $I_h = (-\infty, -\frac{1}{2}] \cup (0, +\infty)$.

Let us consider now the composition $p(x) = f(g(x)) : D_p \to I_p$. Notice that $I_g \subset D_f$, which makes possible to construct $p(x)$ directly, with no adjustments:

$$f(g(x)) = f(\frac{1}{x-1}) = \frac{1}{(x-1)^2} - 1 : D_p = D_g = \mathbb{R}\backslash\{1\} \to I_p \subset I_f$$
$$= [-1, +\infty).$$

Although we didn't modify f or g, we can only state that the range of the composition is contained in I_f, because the range of $g(x)$ does not cover all the domain of $f(x)$. If it is required, we can specify the range of I_p knowing the formula of $p(x)$ and its domain D_p. Another way to determine I_p, simpler in this example, is to find the image of the point $0 \in D_f$ under $f(x)$ (the only point of D_f that does not belong to I_g) and exclude it from the range of f. Obviously, $f(0) = -1$, and therefore, the range of $p(x) = f(g(x))$ is $I_p = I_f\backslash\{-1\} = (-1, +\infty)$.

(b) $f(x) = \frac{1}{x+1}$, $g(x) = \frac{1}{x-1}$
First, we determine the domain and range of each function separately. The domain of $f(x)$ is $D_f = \mathbb{R}\backslash\{-1\}$ and the range is $I_f = \mathbb{R}\backslash\{0\}$. The domain of $g(x)$ is $D_g = \mathbb{R}\backslash\{1\}$ and the range is $I_g = \mathbb{R}\backslash\{0\}$. In the abbreviated form we have: $f(x) : D_f = \mathbb{R}\backslash\{-1\} \to I_f = \mathbb{R}\backslash\{0\}$ and $g(x) : D_g = \mathbb{R}\backslash\{1\} \to I_g = \mathbb{R}\backslash\{0\}$.
Now consider the composite function $h(x) = g(f(x)) : D_h \to I_h$. Notice that $I_f \not\subset D_g$, which means that we need to shrink I_f that can be made by reducing D_f. To guarantee the condition $I_f \subset D_g$ we have to eliminate the elements of D_f whose image is 1, that is, those x that satisfy the relation $f(x) = \frac{1}{x+1} = 1$. The only solution of the last equation is $x = 0$, and the new domain is $\bar{D}_f = D_f\backslash\{0\} = \mathbb{R}\backslash\{-1, 0\}$. Under the same formula $f(x) = \frac{1}{x+1}$ the domain $\bar{D}_f$ has the image contained in D_g: $\bar{I}_f = \mathbb{R}\backslash\{0, 1\} \subset D_g$. Thus, we construct the new function $\bar{f}(x) : \bar{D}_f = \mathbb{R}\backslash\{-1, 0\} \to \bar{I}_f = \mathbb{R}\backslash\{0, 1\}$, but, due to simplicity, we denote this function by the same letter f. After this modification of f, the conditions of the composition $g(f(x))$ are satisfied and we obtain the following function:

$$h(x) = g(f(x)) = g(\frac{1}{x+1})$$
$$= \frac{x+1}{-x} : D_h = \bar{D}_f = \mathbb{R}\backslash\{-1, 0\} \to I_h \subset I_g = \mathbb{R}\backslash\{0\}.$$

For now we can only state that the range of the composition I_h is contained in I_g, since the range of $f(x)$ does not cover the entire domain of $g(x)$. If we need to specify I_h, the simplest way, in this case, is to find the image of the point $0 \in D_g$ under $g(x)$ (the only point of D_g that does not belong to $\bar{I}_f$)

and exclude it from the range I_g. Obviously, $g(0) = -1$, and therefore, the range of $h(x) = g(f(x))$ is $I_h = I_g \backslash \{-1\} = \mathbb{R} \backslash \{-1, 0\}$.
Let us switch to the second composition $p(x) = f(g(x)) : D_p \to I_p$. Notice that $I_g \not\subset D_f$, which implies that we need to reduce I_g which is achieved by restricting D_g. To ensure the condition $I_g \subset D_f$ we have to exclude from D_g the elements whose image under g is -1, that is, such x that satisfy the relation $g(x) = \frac{1}{x-1} = -1$, whence $x = 0$. Then, we get the new domain $\bar{D}_g = \mathbb{R} \backslash \{0, 1\}$ whose image under the same rule $g(x) = \frac{1}{x-1}$ is contained in D_f: $\bar{I}_g = \mathbb{R} \backslash \{-1, 0\} \subset D_f$. Thus, we obtain the modified function $\bar{g}(x) : \bar{D}_g = \mathbb{R} \backslash \{0, 1\} \to \bar{I}_g = \mathbb{R} \backslash \{-1, 0\}$ that we denote again by the same letter g. After this adjustment, the conditions of the composition $f(g(x))$ are satisfied and we obtain the following function:

$$p(x) = f(g(x)) = f(\frac{1}{x-1})$$
$$= \frac{x-1}{x} : D_p = \bar{D}_g = \mathbb{R} \backslash \{0, 1\} \to I_p \subset I_f = \mathbb{R} \backslash \{0\}.$$

For now we can only state that the range of the composition I_p is contained in I_f, since the range of $g(x)$ does not cover the domain of $f(x)$. If we want to determine I_p, the simplest way is to find the image of the point $0 \in D_f$ under f (this is the only point of D_f that does not belong to $\bar{I}_g$) and eliminate it from the range of f. Obviously, $f(0) = 1$, and therefore, the range of $p(x) = f(g(x))$ is $I_p = I_f \backslash \{1\} = \mathbb{R} \backslash \{0, 1\}$.

14.8 *Injection, Surjection, Bijection, Inverse*

1. Verify if a function is bijective; if necessary, make minimum modifications of a function (without changing its formula) to turn it into a bijection; find the inverse of the modified function:
   (a) $y = f(x) = \frac{1}{x+1} : X = \mathbb{R} \backslash \{-1\} \to Y = \mathbb{R} \backslash \{0\}$;
   (b) $y = f(x) = \frac{7x+4}{x} : X = \mathbb{R} \backslash \{0\} \to Y = \mathbb{R} \backslash \{7\}$;
   (c) $y = |x + 2| : X = \mathbb{R} \to Y = [0, +\infty)$;
   (d) $y = x^2 - 1$;
   (e) $y = \sqrt{x + 1}$;
   (f) $y = \frac{1}{2x-1}$;
   (g) $y = 1 + 12x - 4x^2$.

   Solution.

   (a) $y = f(x) = \frac{1}{x+1} : X = \mathbb{R} \backslash \{-1\} \to Y = \mathbb{R} \backslash \{0\}$

To find the range of the function, we need to solve the equation $\frac{1}{x+1} = y$ for x: $x = \frac{1}{y} - 1$. Therefore, y can takes any real value except 0, which gives the range Y already provided in the definition of the function, that is, the given function is surjective. To show that $f(x)$ is injective, take two arbitrary values in the domain $x_1, x_2 \in X$ and evaluate $y_2 - y_1 = f(x_2) - f(x_1) = \frac{1}{x_2+1} - \frac{1}{x_1+1} = \frac{x_1-x_2}{(x_2+1)(x_1+1)}$. It follows that $y_1 = y_2$ if and only if $x_1 = x_2$, which means that the function is injective. Hence, $f(x)$ is bijective and its inverse is $x = g(y) = \frac{1}{y} - 1 : \mathbb{R}\backslash\{0\} \to \mathbb{R}\backslash\{-1\}$. We can also verify the properties $g(f(x)) = x, \forall x \in X$ and $f(g(y)) = y, \forall y \in Y$. For the former we have $g(f(x)) = g\left(\frac{1}{x+1}\right) = \frac{1}{\frac{1}{x+1}} - 1 = x + 1 - 1 = x, \forall x \in X$ and the latter is verified with the same readiness.

(b) $y = f(x) = \frac{7x+4}{x} : X = \mathbb{R}\backslash\{0\} \to Y = \mathbb{R}\backslash\{7\}$
To find the range of the function, we solve the given formula for unknown x: $x = \frac{4}{y-7}$. Therefore, y can take any real value except 7, which gives the range Y already provided in the definition of the function, that is, the given function is surjective. To show that $f(x)$ is injective, take two arbitrary values in the domain $x_1, x_2 \in X$ and evaluate $y_2 - y_1 = f(x_2) - f(x_1) = 7 + \frac{4}{x_2} - \left(7 + \frac{4}{x_1}\right) = \frac{4(x_1-x_2)}{x_2x_1}$. It is clear that $y_1 = y_2$ if and only if $x_1 = x_2$, which means that the function is injective. Hence, $f(x)$ is bijective and its inverse is $x = g(y) = \frac{4}{y-7} : \mathbb{R}\backslash\{7\} \to \mathbb{R}\backslash\{0\}$. We can also verify the properties $g(f(x)) = x, \forall x \in X$ and $f(g(y)) = y, \forall y \in Y$. For the former we have $g(f(x)) = g\left(\frac{7x+4}{x}\right) = \frac{4}{\left(\frac{7x+4}{x}\right)-7} = \frac{4}{\frac{7x+4-7x}{x}} = \frac{4x}{4} = x, \forall x \in X$ and the latter is verified with the same readiness.

(c) $y = |x+2| : X = \mathbb{R} \to Y = [0, +\infty)$
To find the range of the function, we open the absolute value by the definition: if $x \geq -2$ then $y = |x+2| = x+2$ and the image of this part is $[0, +\infty)$; if $x \leq -2$ then $y = |x+2| = -x-2$ and the image of the second part is also $[0, +\infty)$. Hence, the range of the function is $Y = [0, +\infty)$, the set given in the definition, which means that the function is surjective. However, it is not injective, because the points x and $-x-4$ have the same image: $f(-x-4) = |-x-4+2| = |x+2| = f(x)$. On the other hand, in each of the two intervals $(-\infty, -2]$ and $[-2, +\infty)$ the function takes the form of a linear function (decreasing on the former interval and increasing on the latter), and consequently, it is injective on each of these intervals. Restricting, for example, the domain to the interval $[-2, +\infty)$, the function becomes bijective: $y = f(x) = |x+2| = x+2 : \tilde{X} = [-2, +\infty) \to Y = [0, +\infty)$ with the inverse $x = g(y) = y - 2 : Y = [0, +\infty) \to \tilde{X} = [-2, +\infty)$. Verifying the property $g(f(x)) = x, \forall x \in \tilde{X}$, we get $g(f(x)) = g(|x+2|) = g(x+2) = x + 2 - 2 = x, \forall x \in \tilde{X}$. The property $f(g(y)) = y, \forall y \in Y$ is verified with the same readiness.
Another option to obtain a bijective function is by choosing the domain $(-\infty, -2]$: $y = f(x) = |x+2| = -x-2 : \bar{X} = (-\infty, -2] \to Y =$

$[0, +\infty)$ with the inverse $x = g(y) = -y - 2 : Y = [0, +\infty) \to \bar{X} = (-\infty, -2]$.

(d) $y = x^2 - 1$
Notice first that the domain of this function is $X = \mathbb{R}$. Next find its range Y. The minimum value that the function can take is $f(0) = -1$, when the quadratic term vanishes. If $x \neq 0$, then $f(x) > -1$ (the constant -1 is added with a positive quantity). So the range is contained in the interval $[-1, +\infty)$. To show that all the points of this interval belong to the range, we have to solve the equation $y = x^2 - 1$ with unknown x for any given $y \geq -1$. Obviously, the solution is found in the form $x = \pm\sqrt{y+1}$ that confirms that the range is $Y = [-1, +\infty)$. Therefore, the function $y = x^2 - 1 : X = \mathbb{R} \to Y = [-1, +\infty)$ is surjective. However, it is not injective, because the function takes the same value at x and $-x$: $f(-x) = (-x)^2 - 1 = x^2 - 1 = f(x)$. Since x and $-x$ are symmetric about 0, choosing one of the parts of the domain, $(-\infty, 0]$ or $[0, +\infty)$, we eliminate the symmetry about the origin. Since the function is monotonic on each of these two intervals (decreasing on $(-\infty, 0]$ and increasing on $[0, +\infty)$), it is injective on both the former and the latter. If we choose, for instance, $\tilde{X} = (-\infty, 0]$, then the function $f(x) = x^2 - 1 : \tilde{X} = (-\infty, 0] \to Y = [-1, +\infty)$ is bijective and has the inverse $x = g(y) = -\sqrt{y+1} : Y = [-1, +\infty) \to \tilde{X} = (-\infty, 0]$.
The second option is to choose $\bar{X} = [0, +\infty)$, which gives the bijection $f(x) = x^2 - 1 : \bar{X} = [0, +\infty) \to Y = [-1, +\infty)$ with the inverse $x = g(y) = \sqrt{y+1} : Y = [-1, +\infty) \to \bar{X} = [0, +\infty)$.

(e) $y = \sqrt{x+1}$
First determine the domain—$X = [-1, +\infty)$. Next, evaluate the range Y. Since the square root has only non-negative values, the range is contained in $[0, +\infty)$ and the value 0 is attained at the point $x = -1$. Let us show that any point of the interval $[0, +\infty)$ belongs to the range of the function. Indeed, taking an arbitrary $y \geq 0$ and solving the equation $y = \sqrt{x+1}$ with respect to x, we obtain the solution $x = y^2 - 1 \in X = [-1, +\infty)$. Therefore, $Y = [0, +\infty)$ is the range of the function, which is surjective with the domain and the range specified. To verify whether the function is injective, take two arbitrary points $x_1, x_2 \in X$, and notice that $x_1 \neq x_2$ implies that $x_1 + 1 \neq x_2 + 1$, and consequently, that $\sqrt{x_1+1} \neq \sqrt{x_2+1}$. Hence, the function $y = \sqrt{x+1} : X = [-1, +\infty) \to Y = [0, +\infty)$ is a bijection and admits the inverse $x = g(y) = y^2 - 1 : Y = [0, +\infty) \to X = [-1, +\infty)$.

(f) $y = \frac{1}{2x-1}$
Notice first that the domain is $X = \mathbb{R}\backslash\{\frac{1}{2}\}$. The range does not contain 0, since the numerator never vanishes. Let us see if other points are included in the range by solving the given formula $y = \frac{1}{2x-1}$ with respect to x: $x = \frac{1}{2y} + \frac{1}{2}$, $\forall y \neq 0$. Therefore, the range is $Y = \mathbb{R}\backslash\{0\}$. Let us verify if the function $y = \frac{1}{2x-1} : X = \mathbb{R}\backslash\{\frac{1}{2}\} \to Y = \mathbb{R}\backslash\{0\}$ is injective. Take two different elements of the domain $x_1, x_2 \in X$ and compare the corresponding

values of the function $y_1 = \frac{1}{2x_1-1} = \frac{1}{2x_2-1} = y_2$. From this equality follows that $2x_1 - 1 = 2x_2 - 1$ and then $x_1 = x_2$, that is, $y_1 = y_2$ if and only if $x_1 = x_2$, which means that the function is injective. Hence, $f(x) = \frac{1}{2x-1} : X = \mathbb{R}\backslash\{\frac{1}{2}\} \to Y = \mathbb{R}\backslash\{0\}$ is bijective and admits the inverse $x = g(y) = \frac{1}{2y} + \frac{1}{2} : Y = \mathbb{R}\backslash\{0\} \to X = \mathbb{R}\backslash\{\frac{1}{2}\}$.

(g) $y = 1 + 12x - 4x^2$
The domain of this function includes all the real values, $X = \mathbb{R}$. The range is simpler to determine by using the canonical form of a parabola $y = 10 - (2x - 3)^2$. Then it becomes clear that the function takes the greatest value at the point $x = \frac{3}{2}$: $y(\frac{3}{2}) = 10$. The values of the function at all other points are smaller than this, since a positive quantity is subtracted from 10. Therefore, the range is contained in $(-\infty, 10]$. To show that this interval is actually the range of the function, take an arbitrary $y \in (-\infty, 10]$ and check that the equation $y = 10 - (2x - 3)^2$ for unknown x has the solution $x = \frac{3}{2} \pm \frac{1}{2}\sqrt{10 - y}$. Notice that the function $f(x) = 1 + 12x - 4x^2 : X = \mathbb{R} \to Y = (-\infty, 10]$ is not injective, because it takes the same value at the points $\frac{3}{2} - x$ and $\frac{3}{2} + x$: $f(\frac{3}{2} - x) = 10 - (2 \cdot (\frac{3}{2} - x) - 3)^2 = 10 - (-2x)^2 = 10 - (2x)^2 = 10 - (2 \cdot (\frac{3}{2} + x) - 3)^2 = f(\frac{3}{2} + x)$. Since these points are symmetric about $\frac{3}{2}$, we can expect that $f(x)$ is injective on one of the two intervals $(-\infty, \frac{3}{2}]$ or $[\frac{3}{2}, +\infty)$. Indeed, $f(x)$ is monotonic on each of these intervals (increasing on $(-\infty, \frac{3}{2}]$ and decreasing on $[\frac{3}{2}, +\infty)$), and consequently, it is injective on either of the two intervals. Choosing, for instance, $\tilde{X} = (-\infty, \frac{3}{2}]$ we obtain the bijective, and consequently, invertible function $f(x) = 1 + 12x - 4x^2 : \tilde{X} = (-\infty, \frac{3}{2}] \to Y = (-\infty, 10]$. To find the inverse, solve the formula $y = 10 - (2x - 3)^2$ with respect to x: $x = \frac{3}{2} \pm \frac{1}{2}\sqrt{10 - y}$, and choose the solution contained in $(-\infty, \frac{3}{2}]$. Hence, the inverse is $x = g(y) = \frac{3}{2} - \frac{1}{2}\sqrt{10 - y} : Y = (-\infty, 10] \to \tilde{X} = (-\infty, \frac{3}{2}]$.
Another option to obtain a bijection is by choosing $\bar{X} = [\frac{3}{2}, +\infty)$ for the domain, which gives the invertible function $f(x) = 1 + 12x - 4x^2 : \bar{X} = [\frac{3}{2}, +\infty) \to Y = (-\infty, 10]$. In this case, the inverse is $x = g(y) = \frac{3}{2} + \frac{1}{2}\sqrt{10 - y} : Y = (-\infty, 10] \to \bar{X} = [\frac{3}{2}, +\infty)$.

2. Determine the smallest value of b in $B = \{y \in \mathbb{R} : y \geq b\}$ such that the function $f(x) = x^2 - 4x + 6 : \mathbb{R} \to B$ is surjective.
Solution.
Representing the function in the form $y = f(x) = (x - 2)^2 + 2$, we notice that $y \geq 2$ for $\forall x \in \mathbb{R}$ and that any $y \geq 2$ belongs to the range: $(x - 2)^2 + 2 = y$ is cquivalent to $(x - 2)^2 = y - 2$ and this equation has solution x for every $y \geq 2$. Therefore, $b = 2$.
3. Determine the smallest value of a in $A = \{x \in \mathbb{R} : x \geq a\}$ such that the function $f(x) = 2x^2 - 6x + 7 : A \to \mathbb{R}$ is injective.
Solution.

Representing the function in the form $y = f(x) = 2(x - \frac{3}{2})^2 + \frac{5}{2}$, we notice that for any $x = \frac{3}{2} + t$, $t \neq 0$ there exists the point $\bar{x} = \frac{3}{2} - t$ symmetric to x about $\frac{3}{2}$ such that the function takes the same value at these two points: $f(x) = 2(\frac{3}{2}+t-\frac{3}{2})^2+\frac{5}{2} = 2t^2+\frac{5}{2} = 2(-t)^2+\frac{5}{2} = 2(\frac{3}{2}-t-\frac{3}{2})^2+\frac{5}{2} = f(\bar{x})$. Therefore, to eliminate these repetitive values, we have to choose the interval $[\frac{3}{2}, +\infty)$ or $(-\infty, \frac{3}{2}]$ of the domain of the function. Since we should satisfy the condition $x \geq a$, the only option we have is to choose $[\frac{3}{2}, +\infty)$ with the value $a = \frac{3}{2}$ in the definition of A. Let us verify that the function $f(x) = 2x^2 - 6x + 7$ is injective on this interval. Take two different points $x_1, x_2 \in [\frac{3}{2}, +\infty)$ and solve the equation $f(x_1) = 2x_1^2 - 6x_1 + 7 = 2x_2^2 - 6x_2 + 7 = f(x_2)$. Simplifying, we get $2(x_2^2 - x_1^2) - 6(x_2 - x_1) = 0$ or $(x_2 - x_1)(2(x_2 + x_1) - 6) = 0$. Since $x_1 \neq x_2$, it should be $x_2 + x_1 = 3$. But $x_1, x_2 \geq \frac{3}{2}$ and $x_1 \neq x_2$, which shows that the last equation has no solutions. Therefore, there are no points in $[\frac{3}{2}, +\infty)$ at which the function takes the same value, which means that it is injective. (This result can also be derived using the property of decreasing of $f(x)$ on $(-\infty, \frac{3}{2}]$ and increasing on $[\frac{3}{2}, +\infty)$, which is easy to verify.)

14.9 Study of Functions

In this group of exercises we apply different properties of functions studied so far to analyze the behavior of a given function and sketch its graph. The scheme of the study follows the patterns of Calculus and Analysis, though in a simplified form. Given an analytic form (formula) of a function, we carry out the following steps:

(1) identify its domain and range,
(2) find the points of intersection with the coordinate axes and the regions where the function maintains the same sign,
(3) determine if it is bounded or not,
(4) verify the properties of symmetry (parity and periodicity),
(5) investigate the monotonicity and extrema,
(6) investigate the concavity and inflection,
(7) plot the graph based on the studied analytic properties.

It is worth noting that not all of these steps are obligatory and executable for any elementary function: for some simple functions the results can be obvious, while for others the investigation of some properties can lead to hard technical problems, especially when we are allowed to use only rudimentary tools of Pre-Calculus like in this text. Even when all the steps can be performed, their order is not strict and depending on the type of function it may be convenient to change the sequence of the studied properties, since some more advanced properties (like monotonicity and extremes) can immediately provide information about preceding properties, dismissing their separate investigation, which may cause technical difficulties.

Problem A Analyze the function $f(x) = x^4 - 4x^2 + 3$ and use the revealed properties to sketch its graph.

Solution

1. Domain and range.
   We start with finding the domain. Since the domain is not specified explicitly, it consists of all the points for which the formula of the function has a sense. Any polynomial function, including the given one, involves the operations executable for any real number, which means that the function is defined on the entire real axis, that is, $X = \mathbb{R}$.
   To determine the range, it is suitable to use a canonical representation of the bi-quadratic function $f(x) = (x^2 - 2)^2 - 1$. This makes clear that $f(x) \geq -1$ and the minimum value -1 the function takes at the points $x = \pm\sqrt{2}$: $f(\pm\sqrt{2}) = -1$. Then, the range Y is contained in the interval $[-1, +\infty)$. To show that any point of this interval belongs to the range, we need to solve the equation $y = (x^2 - 2)^2 - 1$ for x, where y is any given number in $[-1, +\infty)$. This equation can be written in the form $(x^2 - 2)^2 = y + 1$ and $x^2 - 2 = \pm\sqrt{y+1}$ and still $x^2 = 2 \pm \sqrt{y+1}$. To avoid the situations without solutions when the right-hand side is negative, we choose only the relation with positive square root and have the roots $x = \pm\sqrt{2 + \sqrt{y+1}}$. Therefore, any $y \geq -1$ is associated with at least one x of the domain (actually, there are at least two roots) under the given function. Hence, the range is $Y = [-1, +\infty)$.
2. Intersections with the coordinate axes and the sign of the function.
   The point of intersection with the y-axis is $P_0 = (0, 3)$. To find the points of intersection with the x-axis we have to solve the equation $(x^2 - 2)^2 - 1 = 0$. Using the same approach as for determining the range, we find first $x^2 = 2 \pm 1$. The right-hand side is positive in both cases, and consequently, we have the four roots $x = \pm\sqrt{2 \pm 1}$, that is, $x = -\sqrt{3}, -1, 1, \sqrt{3}$.
   Using the canonical form $f(x) = (x^2 - 2)^2 - 1$ it is easy to determine the sign of the polynomial:

   (1) if $x < -\sqrt{3}$, then $f(x) > 0$;
   (2) if $-\sqrt{3} < x < -1$, then $f(x) < 0$;
   (3) if $-1 < x < 1$, then $f(x) > 0$;
   (4) if $1 < x < \sqrt{3}$, then $f(x) < 0$;
   (5) if $x > \sqrt{3}$, then $f(x) > 0$.

3. Boundedness.
   Since the range is $Y = [-1, +\infty)$, the function is bounded below and unbounded above.
4. Symmetries.
   Since $f(-x) = (-x)^4 - 4(-x)^2 + 3 = x^4 - 4x^2 + 3 = f(x)$, $\forall x \in \mathbb{R}$, the function is even and its graph is symmetric about the y-axis. For this reason we can initially investigate the properties for the values $x \geq 0$ and subsequently extend the obtained result using the symmetry of the function.

We postpone a verification of periodicity until after the study of monotonicity.

5. Monotonicity and extrema.
   Take $0 \leq x_1 < x_2$ and evaluate the difference $D_1 = f(x_2) - f(x_1)$:

$$D_1 = f(x_2) - f(x_1) = (x_2^4 - 4x_2^2 + 3) - \left(x_1^4 - 4x_1^2 + 3\right) = (x_2^4 - x_1^4) - 4(x_2^2 - x_1^2)$$

$$= (x_2^2 - x_1^2)(x_2^2 + x_1^2) - 4(x_2^2 - x_1^2) = (x_2^2 - x_1^2)(x_2^2 + x_1^2 - 4).$$

   In the last expression, the first factor is positive, and consequently, the sign of D_1 depends only on the second factor. The form of this factor makes clear that its sign is well determined on the two regions: on the interval $[0, \sqrt{2}]$ and on the interval $[\sqrt{2}, +\infty)$. On the former, $x_2 \leq \sqrt{2}$, and consequently, $x_2^2 + x_1^2 - 4 < 0$, while on the latter, $x_1 \geq \sqrt{2}$ and $x_2^2 + x_1^2 - 4 > 0$. Therefore, on $[0, \sqrt{2}]$ we have $D_1 < 0$ and $f(x)$ is decreasing, while on $[\sqrt{2}, +\infty)$ we have $D_1 > 0$ and $f(x)$ is increasing.
   According to the evenness of the function, we can extend the properties of monotonicity established on $[0, +\infty)$ to $(-\infty, 0]$ as follows:

   (1) $f(x)$ is decreasing on $(-\infty, -\sqrt{2}]$;
   (2) $f(x)$ is increasing on $[-\sqrt{2}, 0]$;
   (3) $f(x)$ is decreasing on $[0, \sqrt{2}]$;
   (4) $f(x)$ is increasing on $[\sqrt{2}, +\infty)$.

   Therefore, the points $x = \pm\sqrt{2}$ are local minima and the point $x = 0$ is a local maximum. As was shown in the study of the range, the points $x = \pm\sqrt{2}$ are also global minima. The point $x = 0$ is not a global maximum since the range is not bounded above (for instance, compare $f(10)$ and $f(0)$).
   Another important consequence of monotonicity: the increase on the infinite interval $[\sqrt{2}, +\infty)$ ensures that $f(x)$ is not periodic.

6. Concavity and inflection.
   As usual, take two arbitrary points such that $0 \leq x_1 < x_2$, define the midpoint $x_0 = \frac{x_1+x_2}{2}$ and increment $h = x_2 - x_0 = x_0 - x_1 > 0$, and use the quantity $2D_2 \equiv f(x_1) + f(x_2) - 2f(x_0)$ to evaluate the difference between the mean value of the secant $\frac{f(x_1)+f(x_2)}{2}$ and the value of the function at the midpoint $f(x_0)$. Taking advantage of the expression found for D_1, the evaluation of D_2 can be made as follows:

$$2D_2 = f(x_1) + f(x_2) - 2f(x_0) = (f(x_2) - f(x_0)) - (f(x_0) - f(x_1))$$

$$= (x_2^2 - x_0^2)\left(x_2^2 + x_0^2 - 4\right) - (x_0^2 - x_1^2)\left(x_0^2 + x_1^2 - 4\right)$$

$$= h(x_2 + x_0)\left(x_2^2 + x_0^2 - 4\right) - h(x_0 + x_1)\left(x_0^2 + x_1^2 - 4\right)$$

$$= h\left[(2x_0 + h)\left((x_0 + h)^2 + x_0^2 - 4\right) - (2x_0 - h)\left(x_0^2 + (x_0 - h)^2 - 4\right)\right]$$

$$= h\left[2x_0\left((x_0+h)^2 + x_0^2 - 4 - x_0^2 - (x_0-h)^2 + 4\right)\right.$$
$$\left.+h\left((x_0+h)^2 + x_0^2 - 4 + x_0^2 + (x_0-h)^2 - 4\right)\right]$$
$$= h\left[2x_0 \cdot 4x_0h + h\left(4x_0^2 + 2h^2 - 8\right)\right] = 2h^2(6x_0^2 + h^2 - 4).$$

The choice of x_0 determines the sign of the second factor, and the division between negative and positive values occurs at the points $6x_0^2 = 4$, that is, $x_0 = \pm\sqrt{\frac{2}{3}}$. Consider first non-negative values of x. Then, there is the only point of the division $x_0 = \sqrt{\frac{2}{3}}$. On the interval $[0, \sqrt{\frac{2}{3}}]$ we have $x_2 \le \sqrt{\frac{2}{3}}$ and $x_0 + h \le \sqrt{\frac{2}{3}}$, which implies that $(x_0+h)^2 \le \frac{2}{3}$ or $x_0^2 + h^2 \le \frac{2}{3} - 2x_0h$, and consequently, $6x_0^2 + h^2 < 4$. In this case, the second factor is negative and so $D_2 < 0$, which means that $f(x)$ is concave downward on $[0, \sqrt{\frac{2}{3}}]$. On the interval $[\sqrt{\frac{2}{3}}, +\infty)$ we have $x_1 \ge \sqrt{\frac{2}{3}}$, and consequently, $x_0 > \sqrt{\frac{2}{3}}$ and $6x_0^2 > 4$. Therefore, the second factor is positive and $D_2 > 0$, which means that $f(x)$ is concave upward on $[\sqrt{\frac{2}{3}}, +\infty)$.

Extending these results according to the evenness of the function, we can conclude that:

(1) $f(x)$ is concave upward on $(-\infty, -\sqrt{\frac{2}{3}}]$;

(2) $f(x)$ is concave downward on $[-\sqrt{\frac{2}{3}}, 0]$;

(3) $f(x)$ is concave downward on $[0, \sqrt{\frac{2}{3}}]$;

(4) $f(x)$ is concave upward on $[\sqrt{\frac{2}{3}}, +\infty)$.

Since the function changes the type of concavity at the points $x = \pm\sqrt{\frac{2}{3}}$, these are inflection points.

As is shown in Remark 1 to Property 1 in Sect. 7.5, in general, a function with the same concavity on intervals $[-b, 0]$ and $[0, b]$ does not necessarily preserve this concavity on the entire interval $[-b, b]$. Therefore, we cannot deduce directly that $f(x)$ is concave downward on $[-\sqrt{\frac{2}{3}}, \sqrt{\frac{2}{3}}]$. However, the given function keeps the same concavity on the entire interval $[-\sqrt{\frac{2}{3}}, \sqrt{\frac{2}{3}}]$, but to show this, we need to make additional considerations. Return to the obtained expression $D_2 = h^2(6x_0^2 + h^2 - 4)$ and consider now the points such that $-\sqrt{\frac{2}{3}} \le x_1 = x_0 - h < x_0 + h = x_2 \le \sqrt{\frac{2}{3}}$. The left-hand and right-hand inequalities imply that $(x_0-h)^2 \le \frac{2}{3}$ and $(x_0+h)^2 \le \frac{2}{3}$, respectively. Therefore, $(x_0-h)^2 + (x_0+h)^2 \le \frac{4}{3}$ or $2x_0^2 + 2h^2 \le \frac{4}{3}$ or still $6x_0^2 + 6h^2 \le 4$. The last inequality guarantees that

$6x_0^2 + h^2 - 4 < 0$, and consequently $D_2 < 0$. This means that $f(x)$ is concave downward on $[-\sqrt{\frac{2}{3}}, \sqrt{\frac{2}{3}}]$.

The same result about the concavity on $[-\sqrt{\frac{2}{3}}, \sqrt{\frac{2}{3}}]$ can be derived from the following general statement: if $f(x)$ is concave downward on the intervals $[a, b]$ and $[b, d]$, and the point $x = b$ is a local maximum, then $f(x)$ is concave downward on the entire interval $[a, d]$. However, the proof of this statement is provided only in Sect. 4.1 of Chap. 3.

7. Using the established properties and marking some important points, such as $P_0 = (0, 3)$ (intersection with the y-axis and local maximum), $P_1 = (-\sqrt{3}, 0)$, $P_2 = (-1, 0)$, $P_3 = (1, 0)$ and $P_4 = (\sqrt{3}, 0)$ (intersections with the x-axis), $P_5 = (-\sqrt{2}, -1)$ and $P_6 = (\sqrt{2}, -1)$ (local and global minima), $P_7 = (-\sqrt{\frac{2}{3}}, \frac{7}{9})$ and $P_8 = (\sqrt{\frac{2}{3}}, \frac{7}{9})$ (inflection points), we can plot the graph shown in Fig. 2.84.

Problem B Analyze the function $f(x) = \frac{x}{1+x^2}$ and use the revealed properties to sketch its graph.

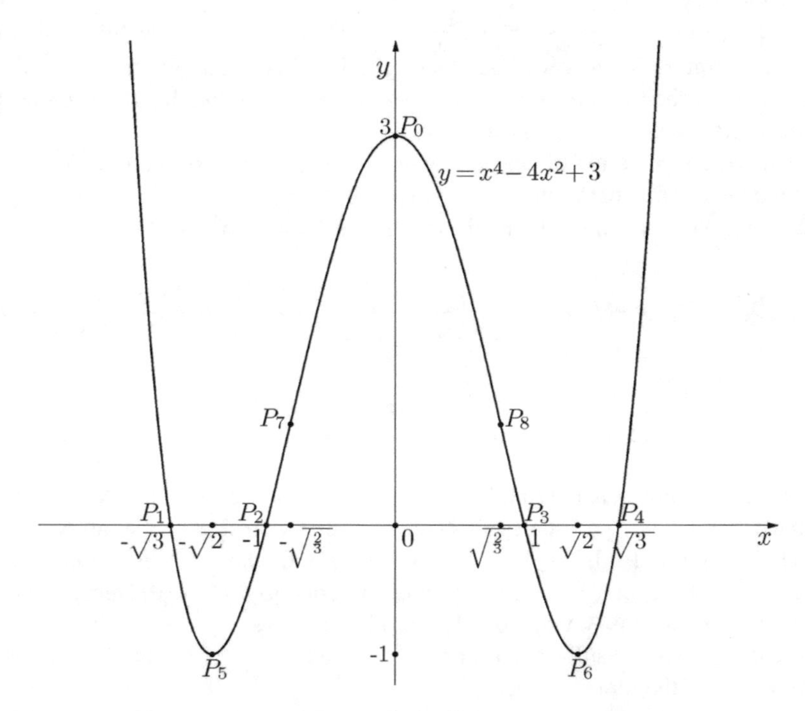

Fig. 2.84 The graph of the function $f(x) = x^4 - 4x^2 + 3$

Solution

1. Domain and range.
Any rational function is defined at all points where its denominator does not vanish. The denominator $1+x^2$ of the given rational function is different from 0 at all real points, which means that the function is defined on $X=\mathbb{R}$.
To determine the range, we need to verify for which y the equation $\frac{x}{1+x^2}=y$ for unknown x has solutions. Writing this equation in the canonical form of a quadratic equation $yx^2-x+y=0$, we find solutions $x=\frac{1}{2}(1\pm\sqrt{1-4y^2})$. Therefore, the real roots exist if and only if $y\in[-\frac{1}{2},\frac{1}{2}]$. Hence, the range is $Y=[-\frac{1}{2},\frac{1}{2}]$.
2. Intersections with the coordinate axes and the sign of the function.
The point of intersection with the y-axis is $P_0=(0,0)$ and it is also the only point of intersection with the x-axis.
The sign of the function is defined by the numerator. Obviously, $f(x)<0$ when $x<0$ and $f(x)>0$ when $x>0$.
3. Boundedness.
Since the range is $Y=[-\frac{1}{2},\frac{1}{2}]$, the function is bounded.
4. Symmetries.
Since $f(-x)=\frac{-x}{1+(-x)^2}=-\frac{x}{1+x^2}=-f(x)$, $\forall x\in\mathbb{R}$, the function is odd and its graph is symmetric about the origin. For this reason we can proceed with the study of the properties on the interval $[0,+\infty)$, completing the negative part afterward according to the symmetry.
We leave a verification of periodicity until after the study of monotonicity.
5. Monotonicity and extrema
Take $0\le x_1<x_2$ and evaluate the difference $D_1=f(x_2)-f(x_1)$:

$$D_1=f(x_2)-f(x_1)=\frac{x_2}{1+x_2^2}-\frac{x_1}{1+x_1^2}=\frac{x_2(1+x_1^2)-x_1(1+x_2^2)}{(1+x_2^2)(1+x_1^2)}$$

$$=\frac{(x_2-x_1)+x_2x_1(x_1-x_2)}{(1+x_2^2)(1+x_1^2)}=\frac{x_2-x_1}{(1+x_2^2)(1+x_1^2)}\cdot(1-x_2x_1).$$

In the last expression, the first factor is positive, and the sign of D_1 depends only on the factor $1-x_2x_1$. Clearly, this factor has different sign on the intervals $[0,1]$ and $[1,+\infty)$. Indeed, if $x_2\le 1$, then $1-x_2x_1>0$, and so $D_1>0$. On the other hand, if $x_1\ge 1$, then $1-x_2x_1<0$, and consequently, $D_1<0$. Hence, on $[0,1]$ the function increases, while on $[1,+\infty)$ it decreases.
Extending these results according to the oddness, we obtain the following properties of the monotonicity:

(1) $f(x)$ is decreasing on $(-\infty,-1]$;
(2) $f(x)$ is increasing on $[-1,1]$;
(3) $f(x)$ is decreasing on $[1,+\infty)$.

It follows from these results that the point $x = -1$ is a local minimum and $x = 1$ is a local maximum. Since $f(-1) = -\frac{1}{2}$ and $f(1) = \frac{1}{2}$, the same points are also global extrema according to the specification of the range $Y = [-\frac{1}{2}, \frac{1}{2}]$. Another direct implication of the monotonicity is that the function cannot be periodic, since it is decreasing on the infinite interval $[1, +\infty)$.

6. Concavity and inflection.
   As usual, take two arbitrary points such that $0 \leq x_1 < x_2$, define the midpoint $x_0 = \frac{x_1+x_2}{2}$ and increment $h = x_2 - x_0 = x_0 - x_1 > 0$, and use the quantity $2D_2 \equiv f(x_1) + f(x_2) - 2f(x_0)$ to evaluate the difference between the mean value of the secant $\frac{f(x_1)+f(x_2)}{2}$ and the value of the function at the midpoint $f(x_0)$. Taking advantage of the expression found for D_1, the evaluation of D_2 can be made as follows:

$$
\begin{aligned}
2D_2 &= f(x_1) + f(x_2) - 2f(x_0) = (f(x_2) - f(x_0)) - (f(x_0) - f(x_1)) \\
&= \frac{(x_2 - x_0)(1 - x_2 x_0)}{(1 + x_2^2)(1 + x_0^2)} - \frac{(x_0 - x_1)(1 - x_0 x_1)}{(1 + x_0^2)(1 + x_1^2)} \\
&= \frac{h}{1 + x_0^2} \left(\frac{1 - x_2 x_0}{1 + x_2^2} - \frac{1 - x_0 x_1}{1 + x_1^2} \right) \\
&= \frac{h}{1 + x_0^2} \cdot \frac{(1 - x_2 x_0)(1 + x_1^2) - (1 - x_0 x_1)(1 + x_2^2)}{(1 + x_2^2)(1 + x_1^2)} \\
&\equiv \frac{h}{1 + x_0^2} \cdot \frac{A}{(1 + x_2^2)(1 + x_1^2)}.
\end{aligned}
$$

The expression $A = (1 - x_2 x_0)(1 + x_1^2) - (1 - x_0 x_1)(1 + x_2^2)$ can be transformed in the following way:

$$
\begin{aligned}
A &= (1 - x_2 x_0 + x_1^2 - x_2 x_0 x_1^2) - (1 - x_0 x_1 + x_2^2 - x_0 x_1 x_2^2) \\
&= x_0(x_1 - x_2) + (x_1 - x_2)(x_1 + x_2) + x_0 x_1 x_2 (x_2 - x_1)
\end{aligned}
$$

$$
= (x_2 - x_1)(x_0 x_1 x_2 - x_0 - x_1 - x_2) = 2h(x_0 x_1 x_2 - 3x_0) = 2hx_0(x_1 x_2 - 3).
$$

The sign of D_2 depends only on A, because all other terms are positive. In turn, the sign of A depends on the factor $x_1 x_2 - 3$ that has different sign to the left and to the right of $\sqrt{3}$. Indeed, if $x_2 \leq \sqrt{3}$, then $x_1 x_2 < 3$ whence $A < 0$, and consequently, $D_2 < 0$. If $x_1 \geq \sqrt{3}$, then $x_1 x_2 > 3$ whence $A > 0$, and consequently, $D_2 > 0$. Hence, on the interval $[0, \sqrt{3}]$ the function is concave downward, while on the interval $[\sqrt{3}, +\infty)$ it is concave upward.

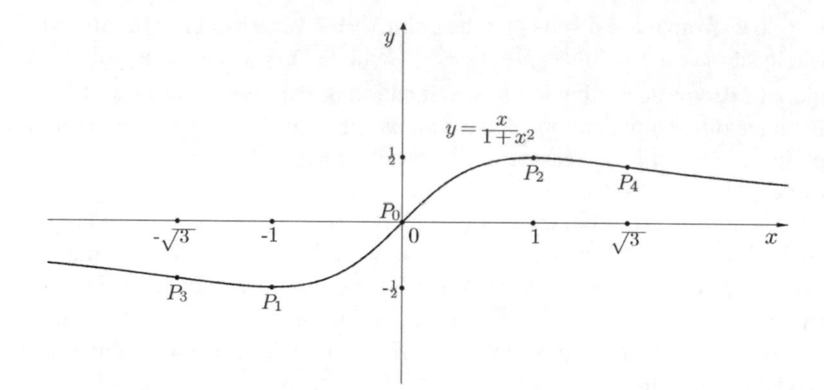

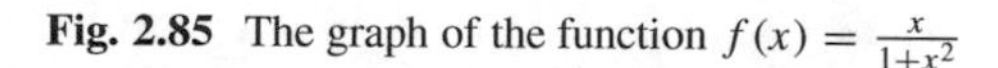
Fig. 2.85 The graph of the function $f(x) = \frac{x}{1+x^2}$

Extending these results to the entire domain according to the oddness, we obtain that:

(1) $f(x)$ is concave downward on $(-\infty, -\sqrt{3}]$;
(2) $f(x)$ is concave upward on $[-\sqrt{3}, 0]$;
(3) $f(x)$ is concave downward on $[0, \sqrt{3}]$;
(4) $f(x)$ is concave upward on $[\sqrt{3}, +\infty)$.

Since the function changes the type of the concavity at the points $x = \pm\sqrt{3}$ and $x = 0$, these are inflection points.

7. Using the established properties and marking some important points, such as $P_0 = (0, 0)$ (intersection with the coordinate axes and one of the inflection points), $P_1 = (-1, -\frac{1}{2})$ and $P_2 = (1, \frac{1}{2})$ (local and global minimum and maximum, respectively), $P_3 = (-\sqrt{3}, -\frac{\sqrt{3}}{4})$ and $P_4 = (\sqrt{3}, \frac{\sqrt{3}}{4})$ (inflection points), we plot the graph of the function shown in Fig. 2.85.

Problem C Analyze the function $f(x) = \sqrt{4 - 2x}$ and use the revealed properties to sketch its graph.

Solution

1. Domain and range.
   The restrictions on the domain and range are caused by the square root. On the one hand, the square root is defined only for non-negative values, that leads to the restriction $x \le 2$ and to the domain $X = (-\infty, 2]$. On the other hand, the square root is always a non-negative number, which means that the range is contained in $[0, +\infty)$. To show that each y of this interval belongs to the range, we solve the equation $\sqrt{4 - 2x} = y$, $\forall y \ge 0$ and obtain the corresponding values x: $x = 2 - \frac{y^2}{2}$, which belong to the domain. Therefore, the range is $Y = [0, +\infty)$.
2. Intersections with axes and the sign of the function.

The point of intersection with the y-axis is $P_0 = (0, 2)$. Since the values of the range are greater than or equal to 0, the only point of intersection with the x-axis is $P_1 = (2, 0)$.
According to the specified range, the function is positive at all points of the domain, but $x = 2$ where it vanishes.

3. Boundedness.
Since the range is $Y = [0, +\infty)$, the function is bounded below and unbounded above.
4. Symmetries.
The domain $X = (-\infty, 2]$ is not symmetric about the origin, which implies that the function is not even nor odd.
Besides, the domain is bounded to the right, which implies that the function is not periodic.
5. Monotonicity and extrema.
For $x_1 < x_2 \leq 2$ we have $4 - 2x_1 > 4 - 2x_2$, which implies that $f(x_1) = \sqrt{4 - 2x_1} > \sqrt{4 - 2x_2} = f(x_2)$. Therefore, the function is decreasing over the entire domain. Another way to arrive at the same conclusion is by applying the property of composition of monotonic functions (Sect. 9.3): the function $\tilde{f}(x) = 4 - 2x$ is decreasing and $\tilde{g}(y) = \sqrt{y}$ is increasing, that gives the decreasing composition $f(x) = \tilde{g}(\tilde{f}(x)) = \sqrt{4 - 2x}$.
The immediate consequence of the monotonicity is that each point $x \in (-\infty, 2)$ is decreasing, and consequently, it cannot be a local or global extremum. However, the endpoint $x = 2$ is a global minimum (although it is not a local minimum since it is not contained in the domain together with a neighborhood.)
6. Concavity and inflection.
Take two arbitrary points such that $x_1 < x_2 \leq 2$, define the midpoint $x_0 = \frac{x_1+x_2}{2}$ and increment $h = x_2 - x_0 = x_0 - x_1 > 0$, and use the quantity $2D_2 \equiv f(x_1) + f(x_2) - 2f(x_0)$ to evaluate the difference between the mean value of the secant $\frac{f(x_1)+f(x_2)}{2}$ and the value of the function at the midpoint $f(x_0)$:

$$2D_2 = f(x_2) + f(x_1) - 2f(x_0) = \sqrt{4 - 2x_2} + \sqrt{4 - 2x_1} - 2\sqrt{4 - 2x_0}$$

$$= \frac{(\sqrt{4 - 2x_2} - \sqrt{4 - 2x_0})(\sqrt{4 - 2x_2} + \sqrt{4 - 2x_0})}{\sqrt{4 - 2x_2} + \sqrt{4 - 2x_0}}$$

$$+ \frac{(\sqrt{4 - 2x_1} - \sqrt{4 - 2x_0})(\sqrt{4 - 2x_1} + \sqrt{4 - 2x_0})}{\sqrt{4 - 2x_1} + \sqrt{4 - 2x_0}}$$

$$= \frac{2(x_0 - x_2)}{\sqrt{4 - 2x_2} + \sqrt{4 - 2x_0}} + \frac{2(x_0 - x_1)}{\sqrt{4 - 2x_1} + \sqrt{4 - 2x_0}}$$

$$= 2h \frac{\sqrt{4 - 2x_2} - \sqrt{4 - 2x_1}}{(\sqrt{4 - 2x_2} + \sqrt{4 - 2x_0})(\sqrt{4 - 2x_1} + \sqrt{4 - 2x_0})}$$

$$= 2h \frac{(\sqrt{4-2x_2} - \sqrt{4-2x_1})(\sqrt{4-2x_2} + \sqrt{4-2x_1})}{(\sqrt{4-2x_2} + \sqrt{4-2x_0})(\sqrt{4-2x_1} + \sqrt{4-2x_0})(\sqrt{4-2x_2} + \sqrt{4-2x_1})}$$

$$= \frac{-8h^2}{(\sqrt{4-2x_2} + \sqrt{4-2x_0})(\sqrt{4-2x_1} + \sqrt{4-2x_0})(\sqrt{4-2x_2} + \sqrt{4-2x_1})}.$$

The numerator is negative and the denominator is positive for any x_1, x_2, which shows that $D_2 < 0$ on the entire domain, and consequently, $f(x)$ has downward concavity on the entire domain. Therefore, there is no inflection point.
Another method of investigation of concavity consists of application of the property of composition of concave functions (Sect. 9.3). Indeed, the function $\tilde{f}(x) = 4 - 2x$ is concave downward (non-strictly) and $\tilde{g}(y) = \sqrt{y}$ is concave downward and increasing. Therefore, the composition $f(x) = \tilde{g}(\tilde{f}(x)) = \sqrt{4-2x}$ is concave downward.

7. Using the established properties and marking some important points, such as $P_0 = (0, 2)$ (intersection with the y-axis), $P_1 = (2, 0)$ (intersection with the x-axis and global minimum) and $P_2 = (-2, \sqrt{8})$ (a point to the left of the y-axis), we construct the graph of the function shown in Fig. 2.86.
The properties of this function can also be derived through the geometric transformations of the simpler function $\sqrt{x}$ whose properties were already studied in this chapter. The transformation of $f_0(x) = \sqrt{x}$ into $f(x) = \sqrt{4-2x}$

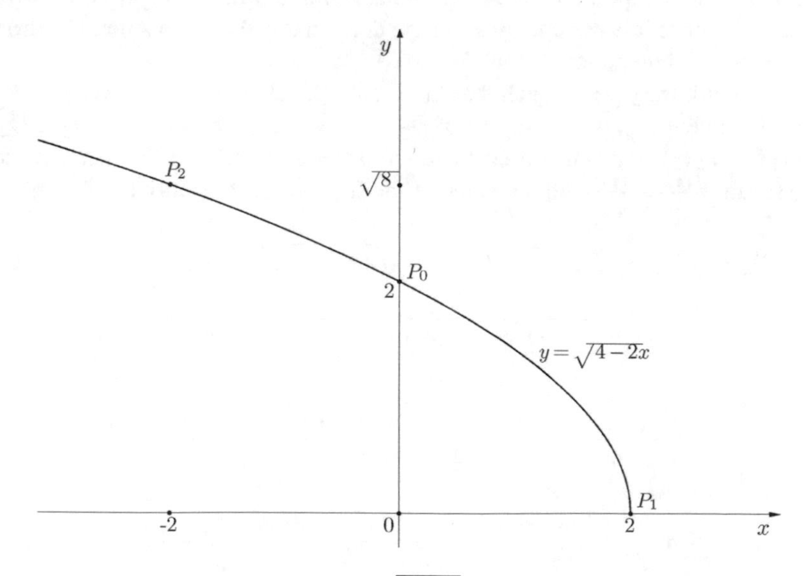

Fig. 2.86 The graph of the function $f(x) = \sqrt{4-2x}$

can be made by the following chain of elementary transformations:

$$f_0(x) = \sqrt{x} \to f_1(x) = f_0(2x) = \sqrt{2x} \to f_2(x)$$
$$= f_1(-x) = \sqrt{-2x} \to f(x) = f_2(x-2) = \sqrt{4-2x}.$$

This geometric approach and comparison of its results with the analytic study presented above is left to the reader.

Problems

Domain, Range and Graph

1. Given a function $f(x)$, find its values at the indicated points:

   (a) $f(x) = \frac{x}{1-x}$, find $f(0)$, $f(1)$, $f(2)$, $f(-x)$, $f(x^2)$, $f\left(\frac{1}{x^2}\right)$;
   (b) $f(x) = \sqrt{x^2+5}$, find $f(0)$, $f(2)$, $f(-2)$, $f\left(\frac{1}{x}\right)$, $f(x^2+1)$, $f\left(\frac{x}{1+x}\right)$;
   (c) $f(x) = \sqrt{16-x^2}$, find $f(0)$, $f(-3)$, $f(4)$, $f(5)$, $f\left(\frac{1}{x}\right)$, $f\left(\frac{x+1}{x}\right)$.

2. Determine the domain of the function:

   (a) $f(x) = \sqrt{2x+3}$;
   (b) $f(x) = \sqrt[3]{2x+3}$;
   (c) $f(x) = \sqrt{16-x^2}$;
   (d) $f(x) = \sqrt{x^2-16}$;
   (e) $f(x) = \sqrt{9x-x^3}$;
   (f) $f(x) = \frac{x}{x^2-1}$;
   (g) $f(x) = \frac{\sqrt{16-x^2}}{x^2-4}$;
   (h) $f(x) = \frac{\sqrt{x+4}}{\sqrt{16-x^2}}$;
   (i) $f(x) = \ln\frac{2-x}{1-x}$;
   (j) $f(x) = \ln(2-x) - \ln(1-x)$;
   (k) $f(x) = \frac{1}{\sin x}$;
   (l) $f(x) = \sqrt{\sin x}$.

3. Determine the domain and the range of the following functions:

   (a) $f(x) = 5-3x$;
   (b) $f(x) = x^2-2$;
   (c) $f(x) = 6-2x^2$;
   (d) $f(x) = \frac{x^2+1}{4-x^2}$;
   (e) $f(x) = \sqrt{16-x^2}$;
   (f) $f(x) = \sqrt{x^2-25}$;

(g) $f(x) = \frac{\sqrt{25-x^2}}{x}$.

4. Determine the domain and range, and sketch the graph of the function:

(a) $f(x) = \begin{cases} 0, x < 0 \\ 2x, x \geq 0 \end{cases}$;

(b) $f(x) = \begin{cases} 1, |x| > 1 \\ |x|, |x| \leq 1 \end{cases}$;

(c) $f(x) = \begin{cases} -2, x < -1 \\ x, -1 \leq x \leq 1 \\ 2, x > 1 \end{cases}$;

(d) $f(x) = \begin{cases} 2, x < -1 \\ -2x, -1 \leq x \leq 1 \\ -2, x > 1 \end{cases}$.

Bounded Functions

1. Verify if the function is bounded, bounded above, bounded below or unbounded on the set S:

   (a) $f(x) = x^2 - 3x + 2$, $S_1 = \mathbb{R}$, $S_2 = [-5, +\infty)$, $S_3 = [-5, 5]$;
   (b) $f(x) = 3 - 2x$, $S_1 = \mathbb{R}$, $S_2 = [-5, +\infty)$, $S_3 = [-5, 5]$;
   (c) $f(x) = \sqrt{2-x}$, $S_1 = (-\infty, 2]$, $S_2 = (-7, 1]$.

2. Show that the sum, difference and product of two bounded functions (considered on the same domain) is again a bounded function. Verify if the same is true for the quotient.
3. Verify if the sum of two unbounded functions is also an unbounded function.
4. Verify if a bounded function attains its global extrema.

*5. Construct a function defined on $\mathbb{R}$ such that it does not bounded in any neighborhood of any point. (Hint: consider the function $f(x) = \begin{cases} n, x \in \mathbb{Q}, x = \frac{m}{n} \\ 0, x \in \mathbb{I} \text{ and } x = 0 \end{cases}$, where m is an integer, n is a natural number, and the fraction $\frac{m}{n}$ is in lowest terms.)

Symmetries: Even, Odd and Periodic Functions

1. Verify whether the functions are even, odd or none of these:

   (a) $f(x) = 5x - 2x^3$;
   (b) $f(x) = x^6 - 2x^4 + 3$;
   (c) $f(x) = \frac{1}{2x+3}$;
   (d) $f(x) = \sqrt[3]{(2-x)^2} - \sqrt[3]{(2+x)^2}$;
   (e) $f(x) = \frac{1}{\sqrt{x}} + 2x$;
   (f) $f(x) = \sqrt{x^2 - 2x + 3} - \sqrt{x^2 + 2x + 3}$;
   (g) $f(x) = \sqrt[3]{1-x} - \sqrt[3]{1+x}$;
   (h) $f(x) = \sqrt[5]{x - x^3} - \sqrt[5]{x + x^3}$;
   (i) $f(x) = \sqrt[3]{1 - x^3} + \sqrt[3]{1 + x^3}$;

(j) $f(x) = 2^x + 2^{-x}$;
(k) $f(x) = \ln \frac{1-x}{1+x}$;
(l) $f(x) = \cos(x + \frac{3\pi}{2})$.

2. Give an example of two non-even functions whose sum is even. The same task for the product.
3. Answer if a function can be even and odd at the same time.
4. Suppose $f(x)$ is an even function and $g(x)$ is an odd function, both defined on the same domain. What you can say about the following functions:

   (a) $|g(x)|$;
   (b) $f(-x) + g(|x|)|$;
   (c) $xf(x) + x^2 g(x)$;
   (d) $f(x|x|)$.

5. Find a centerpoint of symmetry of the graph of a function $f(x)$:

   (a) $f(x) = x^3 + 3x^2 + 3x$;
   (b) $f(x) = \cos(x - \frac{\pi}{4})$.

6. Find an axis of symmetry of the graph of a function $f(x)$:

   (a) $f(x) = 2x - x^2$;
   (b) $f(x) = \cos(x - \frac{\pi}{4})$.

7. Let $f(x)$ be a non-constant function. Can a graph of $f(x)$ have a few centerpoints of symmetry? Can a graph of $f(x)$ have a few axes of symmetry? Can a graph of $f(x)$ have a centerpoint and an axis of symmetry at the same time?
8. Verify if the following functions are periodic; if so, find a period and, if possible, the minimum period:

   (a) $f(x) = x + 3 - [x + 3]$;
   (b) $f(x) = x + 3 - [x + 1]$;
   (c) $f(x) = 3x - [3x]$;
   (d) $f(x) = \frac{1}{x-[x]}$;
   (e) $f(x) = x^2 + x - 1$;
   (f) $f(x) = \sqrt{2x - 3}$;
   (g) $f(x) = \sin^2 x$;
   (*h) $f(x) = \sin x^2$;
   (i) $f(x) = \sqrt{\sin x}$;
   (j) $f(x) = \sin \sqrt{x}$;
   (k) $f(x) = \cos x + 2\cos 2x + 3\cos 3x$.

9. Find a non-constant function for which any rational number is a period.
10. Prove that any of the following conditions determines a periodic function:

    (a) $f(x + T) = -f(x), \forall x \in \mathbb{R}, T \neq 0$;
    (b) $f(x + T) = \frac{1}{f(x)}, \forall x \in \mathbb{R}, T \neq 0$;

(*c) $f(x+T) = \frac{1}{1-f(x)}$, $\forall x \in \mathbb{R}$, $T \neq 0$;

(d) $f(x+T) = \frac{f(x)+1}{f(x)-1}$, $\forall x \in \mathbb{R}$, $T \neq 0$;

(*e) $f(x+T) = \frac{1}{2} + \sqrt{f(x) - f^2(x)}$, $\forall x \in \mathbb{R}$, $T \neq 0$.

11. Give an example of two periodic functions with the same minimum period T, whose product has the minimum period $\frac{T}{4}$.
12. Give an example of two periodic functions with the same minimum period T, whose product has no minimum period.

*13. Construct an example of two periodic functions whose sum is not periodic. (Hint: consider two sine functions, one with a rational period and another with irrational period.)

14. Can you assert that $f(x)$ is periodic if:

(a) $f^2(x)$ is periodic;
(b) $f^3(x)$ is periodic.

*15. Suppose $f(x)$ is defined on $\mathbb{R}$ and its graph is symmetric about the two lines $x = a$ and $x = b$. Prove that $f(x)$ is periodic.

*16. Suppose $f(x)$ is defined on $\mathbb{R}$ and its graph is symmetric about the line $x = a$ and about the point $x = x_0$. Prove that $f(x)$ is periodic.

Monotonicity

1. Find the sets of monotonicity of a given function:

(a) $f(x) = 5 + 4x - x^2$;
(b) $f(x) = \sqrt{5 + 4x - x^2}$;
(c) $f(x) = \sqrt{x + 5}$;
(d) $f(x) = \sqrt[3]{x + 5}$;
(e) $f(x) = \log_2(x - 1)$;
(f) $f(x) = \sin x$.

2. Answer if the square of an increasing on $\mathbb{R}$ function is also increasing. Is the converse true?
3. Give an example of two non-monotonic functions whose sum is monotonic. The same task for the product.

*4. Give an example of a function that has infinitely many intervals of increase and decrease within the interval $[0, 1]$.

5. Give an example of a function defined on $\mathbb{R}$ that has no interval of monotonicity.

Global and Local Extrema

1. Verify if a function $f(x)$ has global and local extrema on a given set S; if so, find these extrema:

(a) $f(x) = 5x - 2$, $S = (-1, 1]$;
(b) $f(x) = |5x - 2|$, $S = (-1, 1]$;
(c) $f(x) = 3 - |5x - 2|$, $S = (-1, 1]$;
(d) $f(x) = 3 - |5x - 2|$, $S = \mathbb{R}$;
(e) $f(x) = [x] - x$, $S = \mathbb{R}$;

(f) $f(x) = 2\sin x$, $S = \mathbb{R}$;

(g) $f(x) = \begin{cases} 3x - 3, x < 1 \\ x^2 - 3x + 2, \ 1 \le x \le 2 \\ 6 - 3x, x > 2 \end{cases}$, $S = \mathbb{R}$.

2. For the functions $f(x) = |x - 2|$ and $f(x) = -x^4$, give examples of the intervals such that:

   (a) $f(x)$ has neither maximum nor minimum;
   (b) $f(x)$ has a maximum, but does not have a minimum;
   (c) $f(x)$ has a minimum, but does not have a maximum;
   (d) $f(x)$ has both maximum and minimum.

3. Suppose that $y = f(x)$ has strict global maximum at x_0. Give example of a function $g(y)$ defined on $\mathbb{R}$ such that:

   (a) $g(f(x))$ has strict global maximum at x_0;
   (b) $g(f(x))$ has strict global minimum at x_0;
   (c) $g(f(x))$ has no extremum at x_0.

*4. Give an example of a function that has infinitely many strict local maxima and only one global maximum.

*5. Verify whether the following statement is true: if x_0 is a minimum (local or global) of $f(x)$, then x_0 is a maximum (local or global) of $\frac{1}{f(x)}$.

*6. Verify whether the following statement is true: if x_0 is a maximum (local or global) of $f(x)$ and $g(x)$, then x_0 is a maximum (local or global) of $f(x) \cdot g(x)$.

Concavity and Inflection

1. Investigate concavity and inflection of the following functions:

   (a) $f(x) = \frac{1}{x-2}$;
   (b) $f(x) = x^2 - 3x + 2$;
   (c) $f(x) = x^3 - 2x + 5$;
   (d) $f(x) = x^3 - 3x^2 + 2$;
   (e) $f(x) = \sqrt{x - 2}$;
   (f) $f(x) = \sqrt[3]{x - 2}$;
   (g) $f(x) = \frac{1}{\sqrt{x-2}}$;
   (h) $f(x) = \frac{1}{\sqrt[3]{x-2}}$;
   (i) $f(x) = \frac{1}{x^2-1}$;
   (j) $f(x) = 2^x$;
   (k) $f(x) = \log_2(x - 3)$;
   (l) $f(x) = 2\sin x$.

2. Show that a concave downward on $\mathbb{R}$ function cannot be bounded below.

3. Suppose that $y = f(x)$ has inflection at x_0. Give examples of a function $g(y)$ defined on $\mathbb{R}$ such that:

   (a) $g(f(x))$ has inflection point at x_0, keeping the types of concavity of $f(x)$ to the left and right of x_0;
   (b) $g(f(x))$ has inflection point at x_0, changing the types of concavity of $f(x)$ to the left and right of x_0;
   (c) $g(f(x))$ does not have an inflection point at x_0.

Composite Functions

1. Find the composite functions $f(f(x))$, $g(g(x))$, $g(f(x))$ and $f(g(x))$:

   (a) $f(x) = x^2 - 4$, $g(x) = \frac{1}{x+1}$;
   (b) $f(x) = x^2 + 1$, $g(x) = 2^x$;
   (c) $f(x) = \frac{1}{x+2}$, $g(x) = \frac{1}{x-3}$;
   (d) $f(x) = \sqrt{x - 10}$, $g(x) = \frac{1}{x}$;
   (e) $f(x) = \begin{cases} 0, x < 0 \\ -x, \ x \geq 0 \end{cases}$, $g(x) = \begin{cases} 0, x < 0 \\ x^2, \ x \geq 0 \end{cases}$.

2. Find the composite functions $f(f(x))$ and $f(f(f(x)))$:

   (a) $f(x) = \frac{1}{1-x}$;
   (b) $f(x) = \frac{x}{\sqrt{1+x^2}}$.

3. Find $f(x)$ if:

   (a) $f(\frac{x}{x+1}) = x^2$;
   (b) $f(\frac{1}{x}) = x + \sqrt{1 + x^2}$;
   (c) $f(x + \frac{1}{x}) = x^2 + \frac{1}{x^2}$.

4. Show that composition of two non-periodic functions can be a periodic function.
5. Show that composition of two non-monotonic functions can be a monotonic function.

Injection, Surjection, Bijection, Inverse

1. Verify if a function is bijective; if necessary, make minimum modifications of a function (without changing its formula) to turn it into a bijection; find the inverse of the modified function:

   (a) $y = f(x) = \frac{1}{x-3} : X = \mathbb{R}\backslash\{3\} \to Y = \mathbb{R}\backslash\{0\}$;
   (b) $y = f(x) = x^2 - 4 : X = \mathbb{R} \to Y = [-4, +\infty)$;
   (c) $y = f(x) = \frac{5x-2}{x+1}$;
   (d) $y = x^2 - 5x + 4$;
   (e) $y = \sqrt{2x + 3}$;
   (f) $y = \sqrt[3]{x^2 - 1}$;
   (g) $y = |2x - 7|$;
   (h) $y = \sqrt{x^2 - 4}$.

2. Find the largest value of b in $B = \{y \in \mathbb{R} : y \leq b\}$ such that the function $f(x) = 3 - |x + 1| : \mathbb{R} \to B$ is surjective.
3. Find the smallest value of a in $A = \{x \in \mathbb{R} : x \geq a\}$ such that the function $f(x) = x^2 - 5x + 4 : A \to \mathbb{R}$ is injective.
4. Find the largest value of a in $A = \{x \in \mathbb{R} : x \leq a\}$ such that the function $f(x) = 2 - 3x - x^2 : A \to \mathbb{R}$ is injective.
5. Give an example of functions $f(x) : \mathbb{R} \to \mathbb{R}$ and $g(x) : \mathbb{R} \to \mathbb{R}$ such that $f(x)$ is injective, $g(x)$ is non-injective, and $g(f(x))$ is injective. Can you do the same with surjection property?

Study of Functions

1. Determine the properties of a function $f(x)$ and sketch its graph by using a chain of elementary transformations starting from a known function $g(x)$:

   (a) $f(x) = 12x - 9x^2 - 2$ $(g(x) = x^2)$;
   (b) $f(x) = \sqrt{5 - 3x}$ $(g(x) = \sqrt{x})$;
   (c) $f(x) = 8x^3 - 12x^2 + 6x + 2$ $(g(x) = x^3)$;
   (d) $f(x) = 1 - |2x - 3|$ $(g(x) = |x|)$.

2. Perform analytic investigation of a given function and sketch its graph:

   (a) $f(x) = x^4 - 6x^2 + 5$;
   (b) $f(x) = x^3 - 3x - 2$;
   (c) $f(x) = x^3 - 3x^2 + 4$;
   (d) $f(x) = 8x^3 - 12x^2 + 6x + 2$;
   (e) $f(x) = \frac{1}{x^2+3}$;
   (f) $f(x) = \frac{x}{x^2+3}$;
   (g) $f(x) = \sqrt{5 - 3x}$;
   (h) $f(x) = \sqrt[3]{x - 2}$.

Chapter 3
Algebraic Functions: Polynomial, Rational and Irrational

The preceding sections have prepared a basis for detailed analysis of different specific functions and construction of their graphs. A few examples of such investigation were already presented at the end of Chap. 2, but the focus of that chapter was on properties themselves: what are the important characteristics of functions and how they can be revealed. From now on we start to study more deeply the properties of different elementary functions.

In this chapter we consider the first group of the elementary functions, algebraic functions, starting with the simpler polynomial functions, moving to more general rational functions and finalizing with the most complicated functions of this group which are the irrational functions. The scheme of the study follows the patterns of Calculus and Analysis, though in a simplified form. Given an analytic form (formula) of a function, we carry out the following steps of examination:

(1) identify its domain and range,
(2) determine if it is bounded or not,
(3) verify the properties of symmetry (parity and periodicity),
(4) find the points of intersection with the coordinate axes and the regions where the function maintains the same sign,
(5) investigate the monotonicity and extrema,
(6) investigate the concavity and inflection,
(7) analyze complimentary properties, such as behavior at infinity and convergence to infinity, vertical and horizontal asymptotes,
(8) plot a sketch of the graph based on the studied analytic properties.

It is worth noting that not all of these steps are obligatory and executable for any elementary function: for some simple functions the results can be obvious, while for others the investigation of some properties can lead to hard technical problems, especially when we are allowed to use only rudimentary tools of Pre-Calculus like in this text. Even when all the steps can be performed, their order is not strict and, depending on the type of function, it may be convenient to change the sequence

A. Bourchtein, L. Bourchtein, *Elementary Functions*,
https://doi.org/10.1007/978-3-031-29075-6_3

of the studied properties, since some more advanced properties (like monotonicity and extremes) can immediately provide information about preceding properties, dismissing their separate investigation, which may cause technical difficulties.

According to the agreement mentioned in Chap. 2, in the study of concavity we will systematically use the simpler concept of the midpoint concavity without mentioning it again.

1 Polynomial Functions

Let us start with the definition of a polynomial function.

Definition of a Polynomial Function A *polynomial function* (a *polynomial*) has the form $y = P_n(x) = a_n x^n + a_{n-1}x^{n-1} + \ldots + a_1 x + a_0$, where $a_i, i = 0, \ldots, n$ are real constants (called *coefficients*) and $a_n \neq 0$. The highest power n is called the *degree of a polynomial*, the term with the highest power $a_n x^n$ is called the *leading term* and a_n is the *leading coefficient*.

In this section we consider some particular cases of polynomial functions, starting with the simplest linear function ($n = 0$ and $n = 1$), then moving to quadratic functions ($n = 2$) and finalize with monomials $y = x^n$, $n \in \mathbb{N}$.

1.1 Linear Function

Definition of a Linear Function A function of the form $y = ax + b$, where a and b are the real constants (*coefficients*), is a *linear function*. The coefficient a is called a *slope coefficient* or *angular coefficient*.

As we will soon see, this terminology follows from the geometric form of a linear function, whose graph is a line with the slope a with respect to x-axis. (Recall that a slope of a line is the tangent of the angle that forms this line with a positive direction of the x-axis.)

Let us start the study of these functions with the two particular simplest cases: $y = b$ $(a = 0)$ and $y = x$ $(a = 1, b = 0)$.

Study of $y = b$ $(a = 0)$

This function is called *constant*.

1. Domain and range. The domain is $X = \mathbb{R}$ (the formula is defined for $\forall x \in \mathbb{R}$) and the range is $Y = \{b\}$ because the function takes the same value b for any x.
2. Boundedness. The function is bounded: for instance, the upper and lower bounds can be chosen equal to b.

3. Parity. Since $f(-x) = b = f(x), \forall x \in X$, the function is even.
4. Periodicity. The relation $f(x+T) = b = f(x), \forall x \in X, \forall T > 0$ shows that the function is periodic, but it does not have the minimum period, because any $T \neq 0$ is its period.
5. Monotonicity and extrema. For any pair $x_1 < x_2$ we have $f(x_1) = b = f(x_2)$ which means that the function is non-strictly increasing and decreasing at the same time over the entire domain. For any point x_0 we have $f(x_0) \geq f(x)$, for $\forall x \in X$, and also $f(x_0) \leq f(x)$ for $\forall x \in X$. This means that any point of the domain is the (global and local) maximum and minimum at the same time, but only in a non-strict sense.
6. Concavity and inflection. Like the monotonicity and extrema, the concavity (of both types) is observed on the entire domain for this function but only in a non-strict sense. For this reason there is no inflection point.
7. Intersections and sign. For all points of the domain the function has the same sign: positive if $b > 0$, zero if $b = 0$ or negative if $b < 0$. Consequently, it has no intersections with the x-axis if $b \neq 0$ or it coincides with the x-axis if $b = 0$. The point of intersection with the y-axis is $(0, b)$.
8. Inverse. The function $f(x) = b : X = \mathbb{R} \to Y = \{b\}$ is surjective but not injective, since different values of x correspond to the same value $y = b$. Consequently, the function is not bijective and does not admit an inverse.

   To make the function injective (and consequently, invertible) we need to restrict its domain to the only point, which makes the function very singular, without any interest for study.
9. The graph of the function is the line parallel to the x-axis with the distance of b units between the line and this axis (it is shown in Fig. 3.1).

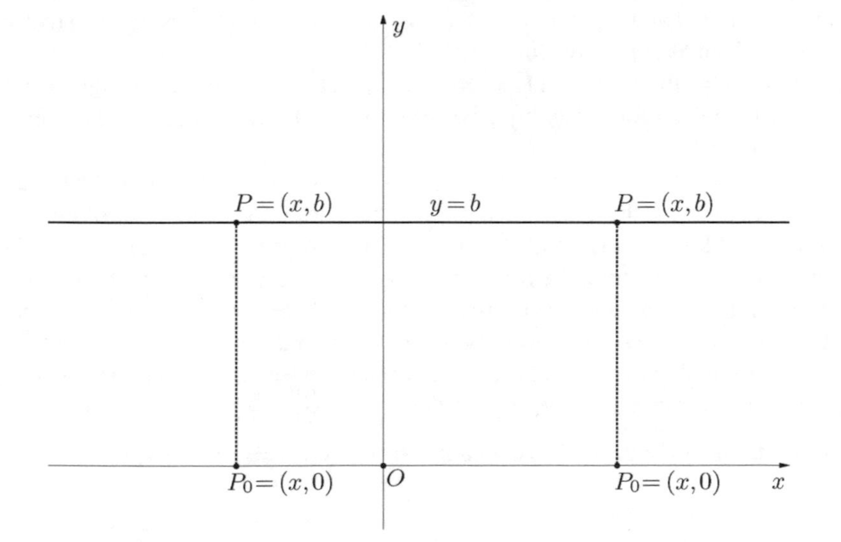

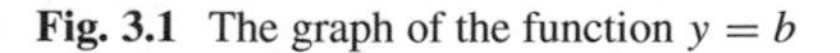
Fig. 3.1 The graph of the function $y = b$

Study of $y = x$ ($a = 1, b = 0$)

This function is called *identity*.

1. Domain and range. The domain is $X = \mathbb{R}$ (the formula is defined for $\forall x \in \mathbb{R}$) and the range is $Y = \mathbb{R}$, since for each $y \in \mathbb{R}$ there exists (the unique) value $x = y$ of the domain such that $f(x) = f(y) = y$.
2. Boundedness. According to the determined range, the function is unbounded both above and below.
3. Parity. The property $f(-x) = -x = -f(x), \forall x \in X$ shows that this function is odd.
4. Periodicity. The condition $f(x + T) = x + T = x = f(x), \forall x \in X$ leads to the conclusion that it is possible only if $T = 0$ that contradicts the definition of periodicity. Therefore, the function is not periodic.
5. Monotonicity and extrema. For any pair $x_1 < x_2$ we have $f(x_1) = x_1 < x_2 = f(x_2)$, which shows that the function is increasing on its domain. This property guarantees the absence of the extrema both local and global, strict or not. This follows from the general relation between extrema and monotonicity (Sect. 6.1 of Chap. 2). The same statement can be directly shown in the specific case of $f(x) = x$: considering an arbitrary point x_0, we notice that any neighborhood of this point contains both the points such that $f(x) < f(x_0)$ and those where $f(x) > f(x_0)$: $\forall x_1 < x_0 \to f(x_1) = x_1 < x_0 = f(x_0)$ and $\forall x_2 > x_0 \to f(x_2) = x_2 > x_0 = f(x_0)$.
6. Concavity and inflection. The function has concavity of both types in non-strict sense. For this reason there is no inflection point.
7. Intersections and sign. Intersection with the y-axis occurs at the origin and the very same point is the only intersection with the x-axis. Solving the inequalities $f(x) = x < 0$ and $f(x) = x > 0$, we see that the function has the negative sign for $x < 0$ and the positive for $x > 0$.
8. Inverse. The function $f(x) = x : X = \mathbb{R} \to Y = \mathbb{R}$ is injective and surjective. Therefore, it is bijective and has the inverse, which is the function itself: $f^{-1}(x) = x : \mathbb{R} \to \mathbb{R}$.
9. The graph of the function is the line passing through the origin and having the slope of 45° with the x-axis, that is, this line represents the bisector of quadrants I and III. Indeed, any point $P = (x, x)$ of the graph has the same values of both coordinates. Therefore, the triangle PP_0O, where $P_0 = (x, 0)$ is the projection of P on the x-axis, is isosceles triangle with $|PP_0| = |OP_0| = |x|$, which means that the angle POP_0 is measuring 45° (see Fig. 3.2). Notice that the coefficient $a = 1$ is equal to the ratio of the length of the two legs of the triangle PP_0O that expresses the tangent of the angle POP_0: $a = \frac{|PP_0|}{|OP_0|} = \tan POP_0 = \tan \alpha = 1$.

Consider now the general case of a linear function $y = ax + b$ with $a \neq 0$.

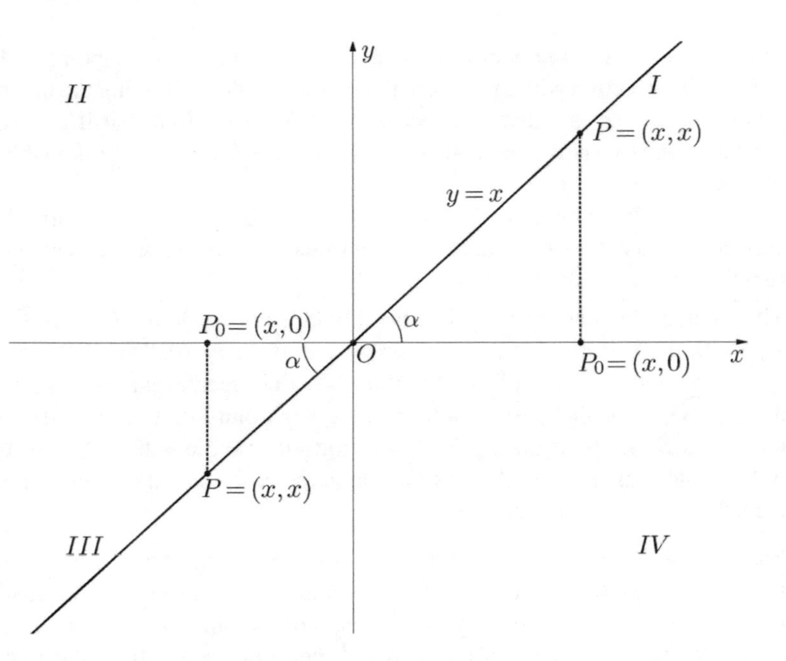

Fig. 3.2 The graph of the function $y = x$

Study of $y = ax + b, a \neq 0$

1. Domain and range. The domain is $X = \mathbb{R}$ (the formula is defined for $\forall x \in \mathbb{R}$) and the range is $Y = \mathbb{R}$ since for any $y \in \mathbb{R}$ there exists (the unique) value $x = \frac{y-b}{a}$ of the domain such that $f(x) = f(\frac{y-b}{a}) = a\frac{y-b}{a} + b = y$.
2. Boundedness. According to the specification of the range, the function is unbounded both above and below.
3. Parity. If $b = 0$, then the function is odd since $f(-x) = -ax = -f(x), \forall x \in X$. If $b \neq 0$, then the function is neither even nor odd: $f(-x) = -ax+b \neq ax+b = f(x)$ and $f(-x) = -ax + b \neq -ax - b = -f(x), \forall x \in X$. For instance, $f(-1) = -a + b \neq a + b = f(1)$ and $f(-1) = -a + b \neq -a - b = -f(1)$.
4. Periodicity. The condition $f(x+T) = a(x+T)+b = ax+b+aT = ax+b = f(x), \forall x \in X$ leads to the conclusion that this is possible only if $T = 0$, which contradicts the definition of the periodicity. Hence, the function is not periodic.
5. Monotonicity and extrema. For any pair $x_1 < x_2$ we have $f(x_2) - f(x_1) = a(x_2 - x_1)$. Since $x_2 - x_1 > 0$, the function is increasing on its domain if $a > 0$, and decreasing if $a < 0$.

   Since the function is monotonic in $\mathbb{R}$, it has no extrema of any type, local or global, strict or not. This follows immediately from the general relation between monotonicity and extrema (Sect. 6.1 of Chap. 2).
6. Concavity and inflection. The function has concavity of both types in non-strict sense. For this reason there is no inflection point.

7. Intersections and sign. Intersection with the y-axis occurs at the point $(0, b)$, and the only intersection with the x-axis at the point $(-\frac{b}{a}, 0)$. Solving the inequalities $f(x) = ax + b < 0$ and $f(x) = ax + b > 0$, we find that if $a > 0$, then the function is negative for $x < -\frac{b}{a}$ and positive for $x > -\frac{b}{a}$; the situation is opposite if $a < 0$.
8. Inverse. The function $f(x) = ax + b : X = \mathbb{R} \to Y = \mathbb{R}$ is injective and surjective. Therefore, it is bijective and admits the inverse, which is also a linear function: $f^{-1}(x) = \frac{x-b}{a} : \mathbb{R} \to \mathbb{R}$.
9. The graph of the function is the line passing through the points $P_1 = (-\frac{b}{a}, 0)$ and $P_2 = (0, b)$ (if $b = 0$ we can choose the points $P_1 = (0, 0)$ and $P_2 = (1, a)$). To verify that the graph of this function is a line, recall that, by the geometric description, a line that passes through two given points P_1 and P_2 is the set of all the points P whose segments P_1P form the same angle with a chosen segment or line. The illustration of this description in the case when the angle is measured from the x-axis is provided in Fig. 3.3.

Based on this description, we can show that any point $P = (x, y)$ of the line that passes through the points $P_1 = (-\frac{b}{a}, 0)$ and $P_2 = (0, b)$ $(b \neq 0)$, satisfies the relation $y = ax + b$, that is, belongs to the graph of the function $y = ax + b$. Indeed, using the geometric property of this line and measuring angle from the x-axis, we take an arbitrary $P = (x, y)$ on this line, find its projection $P_0 = (x, 0)$ on the x-axis and note that the right triangles OP_1P_2 and P_0P_1P are similar (see Fig. 3.3). Therefore, $\frac{d(P,P_0)}{d(P_1,P_0)} = \frac{d(P_2,O)}{d(P_1,O)}$, or expressing in coordinates, $\frac{|y|}{|x+b/a|} = \frac{|b|}{|b/a|} = |a|$, and consequently, $|y| = |ax + b|$. From the two options $y = \pm(ax + b)$ we choose only that with the positive sign $y = ax + b$, because the condition that P_2 belongs to

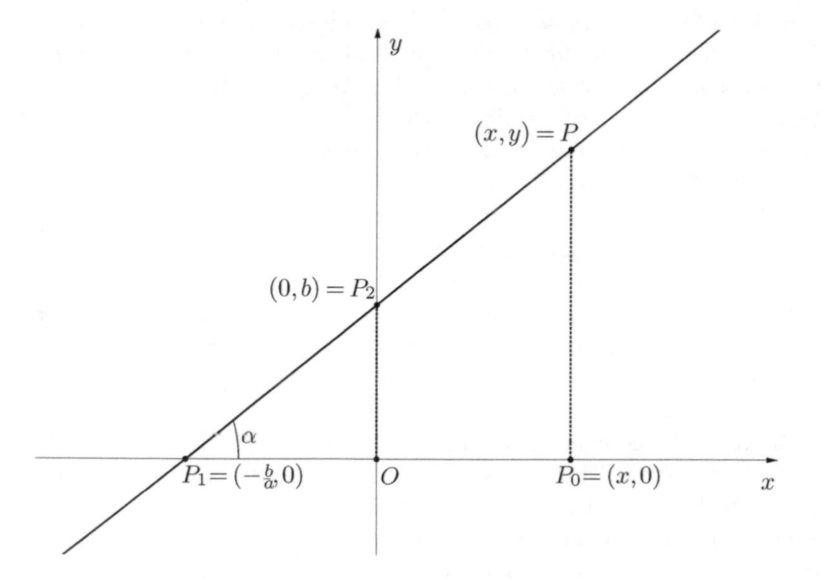

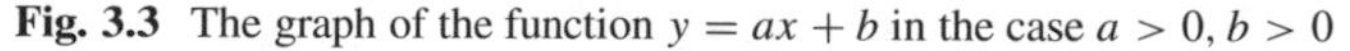

Fig. 3.3 The graph of the function $y = ax + b$ in the case $a > 0, b > 0$

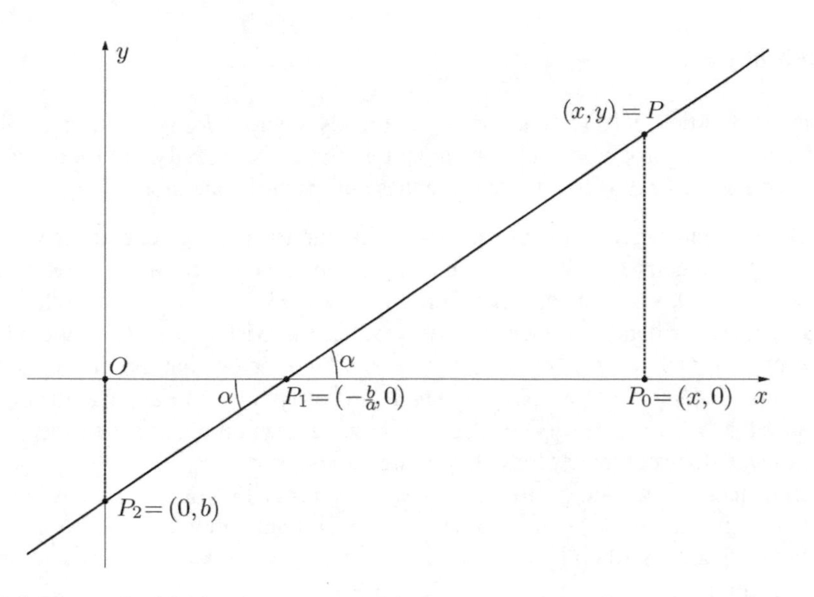

Fig. 3.4 The graph of the function $y = ax + b$ in the case $a > 0, b < 0$

the line implies that $y(0) = b$. The illustrations of the two cases of the coefficients $a > 0, b > 0$ and $a > 0, b < 0$ are given in Figs. 3.3 and 3.4. The formulas are the same for other positions of the points P_1 and P_2 with respect to the coordinate axes, which correspond to the cases when $a < 0, b > 0$ and $a < 0, b < 0$.

Notice that the coefficient a is associated with the angle α between the line (graph of the linear function) and the x-axis, namely: $\tan \alpha = \frac{d(P_2, O)}{d(P_1, O)} = |a|$. If $a > 0$, then the angle α is measured counterclockwise (in the positive direction) from the x-axis and it is considered in the interval $(0, \frac{\pi}{2})$, where $\tan \alpha$ is also positive, and therefore, $\tan \alpha = a$ (see Figs. 3.3 and 3.4). If $a < 0$, then the angle is measured clockwise (in the negative direction) from the x-axis and it is considered that $\alpha \in (-\frac{\pi}{2}, 0)$; for these angles, $\tan \alpha$ is negative and again $\tan \alpha = a$. In this last case, the angle α between the line and the x-axis is not the same angle of the acute triangle OP_1P_2, but its opposite (the angle OP_1P_2 with the negative sign).

The same results can be derived by using the established properties of the function $y = x$ and applying the elementary transformations which allow us to start from the graph of $y = x$ and arrive at the graph of $y = ax + b$.

1.2 Quadratic Function

Definition of a Quadratic Function A *quadratic function* has the form $y = ax^2 + bx + c$, where $a \neq 0, b, c$ are real constants (*coefficients*). The constant a is called the *leading coefficient*.

Let us start with the simplest form of quadratic functions: $y = x^2$.

Study of $y = x^2$

Some properties of this function have already been investigated previously in examples of the properties of parity, monotonicity and concavity, but for the sake of completeness of exposition we analyze these properties once more.

1. Domain and range. The domain is $X = \mathbb{R}$ (the formula is defined for $\forall x \in \mathbb{R}$) and the range is $Y = [0, +\infty)$. Indeed, on the one hand for any $x \in \mathbb{R}$ we have $y = x^2 \geq 0$, which means that all negative values are excluded from the range. On the other hand, for each $y \in [0, +\infty)$ there exists $x = \sqrt{y}$ in the domain such that $f(x) = f(\sqrt{y}) = (\sqrt{y})^2 = y$, which means that the image contains all $y \geq 0$. (Actually for each $y > 0$ there are two points $x = \pm\sqrt{y}$ in the domain such that $f(x) = f(\pm\sqrt{y}) = (\pm\sqrt{y})^2 = y$, but the number of the corresponding points x does not matter for determining the range.)
2. Boundedness. It follows from the form of the range that the function is bounded below (for example, by the constant $m = 0$) and unbounded above.
3. Parity. The formula $f(-x) = (-x)^2 = x^2 = f(x), \forall x \in X$ shows that the function is even.
4. Periodicity. The condition $f(x + T) = (x + T)^2 = x^2 = f(x), \forall x \in X$, is simplified to $T(2x + T) = 0$ that results in $T = 0$ and $T = -2x$, which is impossible for a periodic function, since a period should be a constant different from 0. Hence, the function is not periodic.
5. Monotonicity and extrema. Taking $x_1 < x_2$, we have $f(x_2) - f(x_1) = x_2^2 - x_1^2 = (x_2 - x_1)(x_2 + x_1)$, where the first factor is positive, and consequently, the sign of $f(x_2) - f(x_1)$ depends only on the second factor. If $x_1 < x_2 \leq 0$, then $f(x_2) - f(x_1) < 0$, that is, the function is decreasing on $(-\infty, 0]$. On the other hand, if $0 \leq x_1 < x_2$, then $f(x_2) - f(x_1) > 0$, which means that the function is increasing on $[0, +\infty)$. Therefore, each point $x_0 < 0$ is decreasing, since there exists its neighborhood (for example, $(2x_0, 0)$) in which the function is decreasing. Analogously, each point $x_0 > 0$ is increasing.

   Since each point $x \neq 0$ is monotonic, none of these points is extremum of any kind, global or local, strict or not. The unique remaining point is $x = 0$, which evidently is a strict global and local minimum, since $f(x) = x^2 > 0 = f(0)$, $\forall x \neq 0$.
6. Concavity and inflection. For any pair of points $x_1 \neq x_2 \in \mathbb{R}$, the following inequality holds: $\frac{f(x_1)+f(x_2)}{2} - f\left(\frac{x_1+x_2}{2}\right) = \frac{x_1^2+x_2^2}{2} - \left(\frac{x_1+x_2}{2}\right)^2 = \frac{x_1^2+x_2^2-2x_1x_2}{4} = \frac{(x_1-x_2)^2}{4} > 0$. This shows upward concavity over the entire domain. This also implies that there are no inflection points.
7. Intersection and sign. Intersection with the y-axis occurs at the origin and the same is true for the only intersection with the x-axis. As it was already indicated in the study of the range, $f(x) = x^2 \geq 0, \forall x$ and $f(x) = x^2 > 0, \forall x \neq 0$.
8. Inverse. The function $f(x) = x^2 : X = \mathbb{R} \to Y = [0, +\infty)$ is surjective, but not injective. Indeed, there are two points of the domain $x_\pm = \pm\sqrt{y}$, which correspond to each $y > 0$: $f(x_\pm) = (\pm\sqrt{y})^2 = y$. Therefore, the function

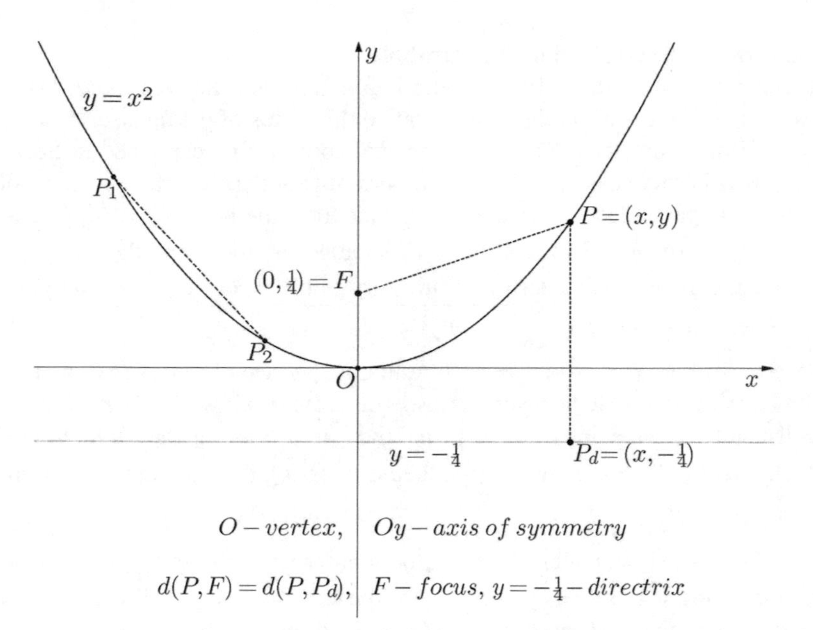

Fig. 3.5 The graph of the function $y = x^2$ and the geometric characterization of the parabola

is not bijective. However, if we restrict the domain of the function to be the interval $[0, +\infty)$, then it becomes bijective and has the inverse $f^{-1}(x) = \sqrt{x} : [0, +\infty) \to [0, +\infty)$. Analogously, the restriction of the domain to the interval $(-\infty, 0]$ also provides the bijective function whose inverse is $f^{-1}(x) = -\sqrt{x} : [0, +\infty) \to (-\infty, 0]$.

9. Based on the revealed properties we can plot the graph of the function, called *parabola*, as it is shown in Fig. 3.5.

Complementary Properties of $y = x^2$

The graph of parabola suggests that the function $y = x^2$ increases without restriction when $|x|$ approaches $+\infty$ (that is, x itself approaches $+\infty$ or $-\infty$). Indeed, if $x > 1$ then $x^2 > x$, and consequently, taking $x > M$ (where $M > 1$ is an arbitrarily large constant) we guarantee that $x^2 > M$, that is, the values of the function also become arbitrarily great, in other words, the function approaches $+\infty$. Since the function is even, the same is true when $|x|$ increases without restriction keeping negative values (x approaches $-\infty$). This implies, in particular, that the function $y = x^2$ has no horizontal asymptotes. In this way, we confirm the tendency of the graph of the function at infinity which are shown in Fig. 3.5. Since the function $y = x^2$ has no infinite limit at the finite points of its domain, its graph has no vertical asymptotes.

Geometric Characterization of a Parabola

The graph of $y = x^2$ (and of any quadratic function) is called *parabola*. The point $(0, 0)$, which is the global minimum, is called the vertex of parabola, and the y-axis is its axis of symmetry (recall the symmetry about a line explained in Sect. 7.5, Chap. 1). This curve has the following important geometric characterization: all the points of the parabola $y = x^2$ are equidistant from the point $F = (0, \frac{1}{4})$ and the line $y = -\frac{1}{4}$ (see Fig. 3.5). The point F is called the focus and the line $y = -\frac{1}{4}$ the directrix. Indeed, considering an arbitrary point $P = (x, x^2)$ of the graph of $y = x^2$, we have: $d(P, F) = \sqrt{(x-0)^2 + (x^2 - \frac{1}{4})^2} = \sqrt{(x^2 + \frac{1}{4})^2} = x^2 + \frac{1}{4}$ and $d(P, y = -\frac{1}{4}) = |x^2 - (-\frac{1}{4})| = x^2 + \frac{1}{4}$, that is, the two distances are equal.

It is easy to show that the reverse is also true: a curve whose points $P = (x, y)$ are equidistant from the point $F = (0, \frac{1}{4})$ and the line $y = -\frac{1}{4}$ is the parabola $y = x^2$. In fact, if $d(P, F) = d(P, y = -\frac{1}{4})$, then expressing this equality in coordinates, we have: $d(P, F) = \sqrt{x^2 + (y - \frac{1}{4})^2} = |y + \frac{1}{4}| = d(P, y = -\frac{1}{4})$, or, squaring both sides, $x^2 + (y - \frac{1}{4})^2 = (y + \frac{1}{4})^2$. Opening the squares $x^2 + y^2 - \frac{1}{2}y + \frac{1}{16} = y^2 + \frac{1}{2}y + \frac{1}{16}$ and simplifying $x^2 - \frac{1}{2}y = \frac{1}{2}y$, we arrive at the equation $y = x^2$. Thus, the parabola is uniquely defined by its geometric characterization.

Study of $y = -x^2$

The properties of the function $y = -x^2$ can be investigate in the same way as those of $y = x^2$. However, it is simpler to use the property of the transformation of graphs which says the multiplication of the function by -1 makes its graph to be reflected about the x-axis. Based on this geometric property we can deduce all the results about $y = -x^2$.

1. The domain of $y = -x^2$ is $X = \mathbb{R}$ and the range is $Y = (-\infty, 0]$.
2. The function is bounded above (for example, by the constant $M = 0$) and unbounded below.
3. The function is even.
4. The function is not periodic.
5. Monotonicity and extrema. The function increases on $(-\infty, 0]$ and decreases on $[0, +\infty)$. Then, each point $x_0 < 0$ is increasing, and each point $x_0 > 0$ is decreasing.

   The unique extreme $x = 0$ is the strict global and local maximum. The point $(0, 0)$ is the vertex of the parabola $y = -x^2$.
6. Concavity and inflection. The graph is concave downward on all the domain. There are no inflection points.
7. The only intersection with the coordinate axes occurs at the origin.
8. Inverse. The function $f(x) = -x^2 : X = \mathbb{R} \to Y = (-\infty, 0]$ is surjective, but not injective. If its domain is restricted to the interval $[0, +\infty)$, then it becomes bijective and has the inverse $f^{-1}(x) = \sqrt{-x} : (-\infty, 0] \to [0, +\infty)$.

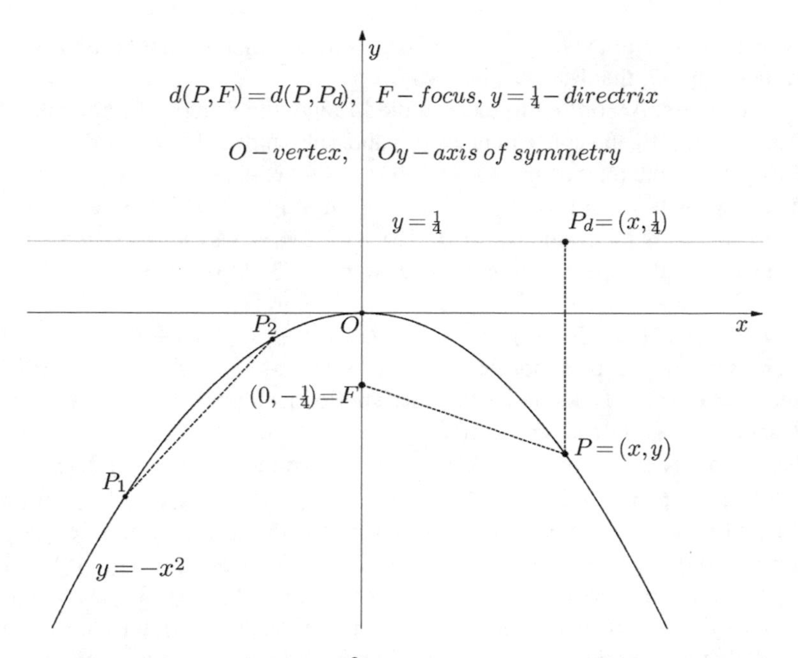

Fig. 3.6 The graph of the function $y = -x^2$ and the geometric characterization of the parabola

Analogously, the restriction of the domain to the interval $(-\infty, 0]$ also provides the bijective function whose inverse is $f^{-1}(x) = -\sqrt{-x} : (-\infty, 0] \to (-\infty, 0]$.

9. The graph of the function is shown in Fig. 3.6.

The geometric characterization of the parabola $y = -x^2$ is as follows: all its points are equidistant from the focus $F = (0, -\frac{1}{4})$ and the directrix $y = \frac{1}{4}$ (see Fig. 3.6).

Before to analyze the quadratic function in the general form, let us consider an example which will serve as a prototype for the investigation of the general function.

Study of $y = x^2 - 6x + 2$

Notice first that the formula of the function can be written as $y = (x - 3)^2 - 7$, which will be frequently used in the following investigation.

1. Domain and range. The domain is $X = \mathbb{R}$ (the formula is defined for $\forall x \in \mathbb{R}$) and the range is $Y = [-7, +\infty)$. To show the latter, notice first that for any $x \in \mathbb{R}$ we have $y = (x - 3)^2 - 7 \geq -7$, which means that the values smaller than -7 are excluded from the range, that is, $Y \subset [-7, +\infty)$. Second, for each $y \geq -7$ there exists the real number $x = 3 + \sqrt{y + 7}$ such that $f(x) = f(3 + \sqrt{y + 7}) =$

$(\sqrt{y+7})^2 - 7 = (y+7) - 7 = y$, which means that the range contains all the values $y \geq -7$, that is, $Y = [-7, +\infty)$.

2. Boundedness. According to the specified range, the function is bounded below (for instance, by the constant $m = -7$) and unbounded above.
3. Parity. From the relation $f(-x) = (-x)^2 - 6(-x) + 2 = x^2 + 6x + 2, \forall x \in X$, it follows that $f(-x) \neq \pm f(x)$, that is, the function is neither even nor odd. However, using the formula $y = (x-3)^2 - 7$ it is easy to see that there exists the symmetry with respect to the coordinate line $x = 3$: $f(3-x) = (3-x-3)^2 - 7 = x^2 - 7 = (3+x-3)^2 - 7 = f(3+x)$.
4. Periodicity. The condition $f(x+T) = (x+T)^2 - 6(x+T) + 2 = x^2 - 6x + 2 = f(x), \forall x \in X$ can be simplified to $T(2x + T - 6) = 0$, which results in $T = 0$ and $T = 6 - 2x$. This implies that the function is not periodic, because a period T should be a non-zero constant.
5. Monotonicity and extrema. Taking $x_1 < x_2$, we have $f(x_2) - f(x_1) = (x_2^2 - 6x_2 + 2) - (x_1^2 - 6x_1 + 2) = (x_2^2 - x_1^2) - 6(x_2 - x_1) = (x_2 - x_1)(x_2 + x_1 - 6)$, where the first factor is always positive, which means that the sign of $f(x_2) - f(x_1)$ depends only on the second factor. If $x_1 < x_2 \leq 3$, then $f(x_2) - f(x_1) < 0$, that is, the function decreases on $(-\infty, 3]$. On the other hand, if $3 \leq x_1 < x_2$, then $f(x_2) - f(x_1) > 0$, that is, the function increases on $[3, +\infty)$. It follows from this result that each point $x_0 < 3$ is decreasing, because there exists its neighborhood (for instance, $(2x_0 - 3, 3)$) where the function decreases; and similarly, each point $x_0 > 3$ is increasing.

   Since each point $x \neq 3$ is monotonic, none of these points can be extremum of any type, global or local, strict or not. The only remaining point is $x = 3$, which clearly is the strict global and local minimum, since $f(x) = (x-3)^2 - 7 > -7 = f(3), \forall x \neq 3$.
6. Concavity and inflection. For any two points $x_1 \neq x_2 \in \mathbb{R}$, the following inequality holds $\frac{f(x_1)+f(x_2)}{2} - f\left(\frac{x_1+x_2}{2}\right) = \left[\frac{x_1^2+x_2^2}{2} - 6\frac{x_1+x_2}{2} + 2\right] - \left[\left(\frac{x_1+x_2}{2}\right)^2 - 6\frac{x_1+x_2}{2} + 2\right] = \frac{x_1^2+x_2^2-2x_1x_2}{4} = \frac{(x_1-x_2)^2}{4} > 0$. This shows that the function is concave upward over the entire domain. Consequently, there are no inflection points.
7. Intersection and sign. Intersection with the y-axis occurs at the point $(0, 2)$ and with the x-axis at the points $(3 \pm \sqrt{7}, 0)$. The function has positive values when $x < 3 - \sqrt{7}$ and $x > 3 + \sqrt{7}$, and negative values on the interval $(3 - \sqrt{7}, 3 + \sqrt{7})$.
8. Inverse. The function $f(x) = (x-3)^2 - 7 : X = \mathbb{R} \to Y = [-7, +\infty)$ is surjective, but not injective. Indeed, each number $y > -7$ is associated with the two points of the domain $x_\pm = 3 \pm \sqrt{y+7}$: $f(x_\pm) = (\pm\sqrt{y+7})^2 - 7 = y + 7 - 7 = y$. Therefore, the function is not bijective.

   However, restricting the domain to the interval $[3, +\infty)$, we obtain a bijective function whose inverse is $f^{-1}(x) = 3 + \sqrt{x+7} : [-7, +\infty) \to [3, +\infty)$. Similarly, choosing for the domain the interval $(-\infty, 3]$ we also get the bijective function with the inverse $f^{-1}(x) = 3 - \sqrt{x+7} : [-7, +\infty) \to (-\infty, 3]$.

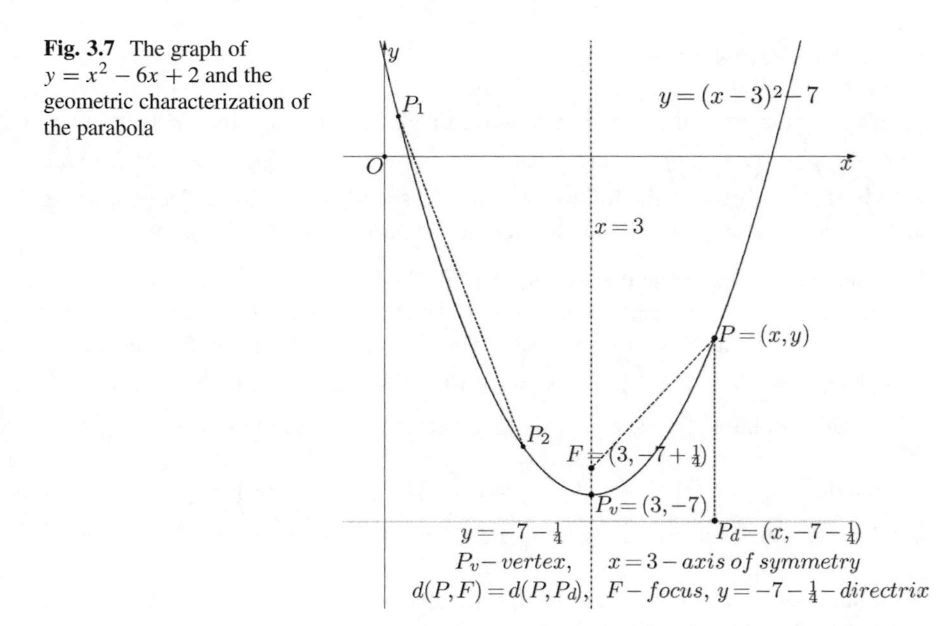

Fig. 3.7 The graph of $y = x^2 - 6x + 2$ and the geometric characterization of the parabola

9. Based on the investigated properties, we can plot the graph of the function shown in Fig. 3.7.

The parabola $y = (x - 3)^2 - 7$ has the vertex at the point $(3, -7)$ and the axis of symmetry $x = 3$. This curve has the following geometric characterization: all the point of the parabola $y = (x - 3)^2 - 7$ are equidistant from the focus $F = (3, -7 + \frac{1}{4})$ and the directrix $y = -7 - \frac{1}{4}$ (see Fig. 3.7). In fact, for an arbitrary point $P = (x, (x - 3)^2 - 7)$ of this parabola we have: $d(P, F) = \sqrt{(x-3)^2 + \left((x-3)^2 - 7 - (-7 + \frac{1}{4})\right)^2} = \sqrt{\left((x-3)^2 + \frac{1}{4}\right)^2} = (x - 3)^2 + \frac{1}{4}$ and $d(P, y = -7 - \frac{1}{4}) = |(x - 3)^2 - 7 - (-7 - \frac{1}{4})| = (x - 3)^2 + \frac{1}{4}$, that is, the two distances coincide. It is simple to prove that the reverse also holds: the curve whose points $P = (x, y)$ are equidistant from the point $F = (3, -7 + \frac{1}{4})$ and the line $y = -7 - \frac{1}{4}$ is the parabola $y = (x - 3)^2 - 7$ (a proof of this fact is left to the reader).

Another manner to investigate the properties of the function $y = (x - 3)^2 - 7$ and sketch its graph is to find geometric transformations that bring from $y = x^2$ to $y = (x - 3)^2 - 7$. This task is left to the reader.

Using the pattern of the study of the functions $y = x^2$ and $y = x^2 - 6x + 2$ we proceed to the analysis of the general quadratic function $y = ax^2 + bx + c$.

Study of $y = ax^2 + bx + c$

Notice that the formula of the function can also be written in the form $y = a\left(x + \frac{b}{2a}\right)^2 + \left(c - \frac{b^2}{4a}\right) = a\,(x - x_0)^2 + y_0$, $x_0 = -\frac{b}{2a}$, $y_0 = c - \frac{b^2}{4a}$, which we will frequently use in the following study. To specify the analysis, let us suppose that $a > 0$. In the case $a < 0$ the investigation is performed in a similar way.

1. Domain and range. The domain is $X = \mathbb{R}$ (the formula is defined for $\forall x \in \mathbb{R}$) and the range is $Y = [y_0, +\infty)$. To prove the latter statement, notice first that for any $x \in \mathbb{R}$ we have $y = a\,(x - x_0)^2 + y_0 \geq y_0$, which means that the values smaller than $y_0 = c - \frac{b^2}{4a}$ are excluded from the range, that is, $Y \subset [y_0, +\infty)$. On the other hand, for each $y \geq y_0$ there exists the real number $x = x_0 + \sqrt{\frac{y-y_0}{a}}$ such that $f(x) = f\left(x_0 + \sqrt{\frac{y-y_0}{a}}\right) = a\left(x_0 + \sqrt{\frac{y-y_0}{a}} - x_0\right)^2 + y_0 = a\frac{y-y_0}{a} + y_0 = y$, which means that the range contains each $y \geq y_0$, and consequently, $Y = [y_0, +\infty)$.
2. Boundedness. It follows from the specification of the range that the function is bounded below (for instance, by the constant $m = y_0$) and unbounded above.
3. Parity. From the condition $f(-x) = a(-x)^2 + b(-x) + c = ax^2 + bx + c, \forall x \in X$ it follows that $f(-x) = f(x)$ if, and only if, $b = 0$, in which case the function is even. If $b \neq 0$, then $f(-x) \neq \pm f(x)$ and the function is neither even nor odd. However, using the form $y = a\,(x - x_0)^2 + y_0$ it is easy to see that the function has symmetry with respect to the coordinate line $x = x_0 = -\frac{b}{2a}$: $f\,(x_0 - x) = a(-x)^2 + y_0 = ax^2 + y_0 = f\,(x_0 + x)$ (recall the relations of symmetry (1.11) in Sect. 7.5, Chap. 1).
4. Periodicity. The condition $f(x+T) = a(x+T)^2 + b(x+T) + c = ax^2 + bx + c = f(x), \forall x \in X$ can be simplified to $T(2ax + aT + b) = 0$, which has the two solutions for T: $T = 0$ and $T = -\frac{2ax+b}{a}$. This shows that the function is not periodic, since a period T should be a non-zero constant.
5. Monotonicity and extrema. Taking $x_1 < x_2$, we have

$$f(x_2) - f(x_1) = (ax_2^2 + bx_2 + c) - (ax_1^2 + bx_1 + c) = a(x_2^2 - x_1^2) + b(x_2 - x_1)$$

$$= (x_2 - x_1)(a(x_2 + x_1) + b) = a(x_2 - x_1)(x_2 + x_1 + \frac{b}{a}),$$

where the first two factors are positive, meaning that the sign of $f(x_2) - f(x_1)$ depends only on the last factor. If $x_1 < x_2 \leq -\frac{b}{2a}$, then $f(x_2) - f(x_1) < 0$, that is, the function is decreasing on $(-\infty, -\frac{b}{2a}]$. On the other hand, if $-\frac{b}{2a} \leq x_1 < x_2$, then $f(x_2) - f(x_1) > 0$, that is, the function is increasing on $[-\frac{b}{2a}, +\infty)$. It follows from this result that each point $x_0 < -\frac{b}{2a}$ is decreasing since there exists its neighborhood (for instance, $(2x_0 + \frac{b}{2a}, -\frac{b}{2a})$) in which the function is decreasing. For the same reasons, each point $x_0 > -\frac{b}{2a}$ is increasing.

Since each point $x \neq -\frac{b}{2a}$ is monotonic, none of these points can be an extreme of any type, global or local, strict or not. The only remaining point is $x_0 = -\frac{b}{2a}$, which is clearly a strict global and local minimum, because $f(x) = a(x - x_0)^2 + y_0 > y_0 = f(x_0), \forall x \neq x_0$.

6. Concavity and inflection. For any two points $x_1 \neq x_2 \in \mathbb{R}$ we obtain the inequality

$$\frac{f(x_1) + f(x_2)}{2} - f\left(\frac{x_1 + x_2}{2}\right) = \left[a\frac{x_1^2 + x_2^2}{2} + b\frac{x_1 + x_2}{2} + c\right] - \left[a\left(\frac{x_1 + x_2}{2}\right)^2 + b\frac{x_1 + x_2}{2} + c\right]$$

$$= a\frac{x_1^2 + x_2^2 - 2x_1x_2}{4} = a\frac{(x_1 - x_2)^2}{4} > 0,$$

which shows upward concavity over the entire domain. Consequently, there are no inflection points.

7. Intersection and sign. Intersection with the y-axis occurs at the point $(0, c)$, while intersection with the x-axis and the sign of the function depend on the discriminant $D = b^2 - 4ac$ of the quadratic equation $ax^2 + bx + c = 0$. If $D > 0$, then there are two points of intersection $x_{1,2} = \frac{-b \pm \sqrt{D}}{2a} = x_0 \pm \frac{\sqrt{D}}{2a}$ (symmetric about x_0), and consequently, the function has positive values when $x < x_1 = x_0 - \frac{\sqrt{D}}{2a}$ and $x > x_2 = x_0 + \frac{\sqrt{D}}{2a}$, and negative values on the interval (x_1, x_2). If $D = 0$, there is the only point of intersection $x_0 = \frac{-b}{2a}$ with the x-axis and the function is positive at each point, except for x_0 where the function is zero. Finally, if $D < 0$, then there is no intersection with the x-axis and the function is positive over the entire domain. Notice that since $D = -4ay_0$, the same classification can be done in the terms of y_0, which is directly connected with the second form of the function $y = a(x - x_0)^2 + y_0$.
8. The function $f(x) = a(x - x_0)^2 + y_0 : X = \mathbb{R} \to Y = [y_0, +\infty)$ is surjective, but not injective. In fact, each number $y > y_0$ is associated with the points of the domain $x_{\pm} = x_0 \pm \sqrt{\frac{y-y_0}{a}}$: $f(x_{\pm}) = f\left(x_0 \pm \sqrt{\frac{y-y_0}{a}}\right) = a\left(x_0 \pm \sqrt{\frac{y-y_0}{a}} - x_0\right)^2 + y_0 = a\frac{y-y_0}{a} + y_0 = y$. Therefore, the function is not bijective. However, restricting the domain to the interval $[x_0, +\infty)$, we turn the function into bijective with the inverse $f^{-1}(x) = x_0 + \sqrt{\frac{x-y_0}{a}} = -\frac{b}{2a} + \frac{\sqrt{b^2+4a(x-c)}}{2a} : [y_0, +\infty) \to [x_0, +\infty)$. Similarly, restricting the domain to the interval $(-\infty, x_0]$ we obtain the bijective function whose inverse is $f^{-1}(x) = x_0 - \sqrt{\frac{x-y_0}{a}} = -\frac{b}{2a} - \frac{\sqrt{b^2+4a(x-c)}}{2a} : [y_0, +\infty) \to (-\infty, x_0]$.

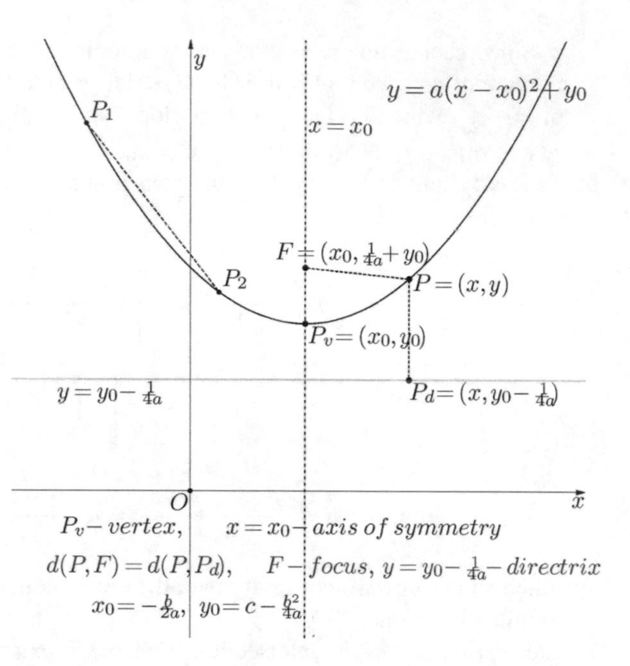

Fig. 3.8 The graph of the function $y = ax^2 + bx + c = a(x - x_0)^2 + y_0$ and the geometric characterization of the parabola

9. Based on the investigated properties, we can plot the graph of the functions as is shown in Fig. 3.8.

The parabola $y = a(x - x_0)^2 + y_0$ has the vertex $P_v = (x_0, y_0)$ and the axis of symmetry $x = x_0$ (parallel to the axis Oy). This curve is characterized by the following geometric feature: all the points of the parabola $y = a\,(x - x_0)^2 + y_0$ are equidistant from the focus $F = (-\frac{b}{2a}, y_0 + \frac{1}{4a})$ and the directrix $y = y_0 - \frac{1}{4a}$ (see Fig. 3.8). Indeed, considering an arbitrary point $P = (x, a\,(x - x_0)^2 + y_0)$ of the parabola, we obtain: $d(P, F) = \sqrt{(x - x_0)^2 + \left[a\,(x - x_0)^2 - \frac{1}{4a}\right]^2} = \sqrt{\left[a\,(x - x_0)^2 + \frac{1}{4a}\right]^2} = a\,(x - x_0)^2 + \frac{1}{4a}$ and $d(P, y = y_0 - \frac{1}{4a}) = \left|a\,(x - x_0)^2 + y_0 - (y_0 - \frac{1}{4a})\right| = a\,(x - x_0)^2 + \frac{1}{4a}$, that is, the two distances are equal. It is easy to show that the reverse is also true: the curve whose points $P = (x, y)$ are equidistant from the point $F = (x_0, y_0 + \frac{1}{4a})$ and the line $y = y_0 - \frac{1}{4a}$ is the parabola $y = a\,(x - x_0)^2 + y_0$. In fact, from the equality of the two distances we have: $(x - x_0)^2 + \left(y - (y_0 + \frac{1}{4a})\right)^2 = \left(y - (y_0 - \frac{1}{4a})\right)^2$, or simplifying, $(x - x_0)^2 = 4(y - y_0)\frac{1}{4a}$, whence $y = a\,(x - x_0)^2 + y_0$. Notice that the axis of symmetry of the parabola is the line which passes through the vertex P_v and focus F (it is also perpendicular to the directrix).

Another way to determine the properties of the function $y = a\,(x - x_0)^2 + y_0$, $x_0 = -\frac{b}{2a}$, $y_0 = c - \frac{b^2}{4a}$, and to plot its graph consists of application of elementary geometric transformations, starting with $y = x^2$ and arriving at $y = a\,(x - x_0)^2 +$

y_0. The following chain of transformations can be used:

$$g(x) = x^2 \to g_1(x) = (x - x_0)^2 = g(x - x_0) \to g_2(x) = a\,(x - x_0)^2 = ag_1(x)$$

$$\to f(x) = a\,(x - x_0)^2 + y_0 = g_2(x) + y_0.$$

Geometrically, the first transformation is the horizontal translation (shift) of the graph of x^2 by x_0 units (recall that a shift is made to the left if $x_0 < 0$ and to the right if $x_0 > 0$). Under this translation the vertex $P_0 = (0, 0)$ of the original parabola $y = x^2$ moves to $P_1 = (x_0, 0)$, the symmetry axis moves from $x = 0$ to $x = x_0$, the focus $F_0 = (0, \frac{1}{4})$ becomes $F_1 = (x_0, \frac{1}{4})$ and the directrix $y = -\frac{1}{4}$ stay on the same place. The second transformation is the vertical extension by the factor a of the graph of $g_1(x)$ (recall that this extension represents a stretching a times from the x-axis if $a > 1$ and compressing $\frac{1}{a}$ times if $0 < a < 1$). Under this, the vertex $P_1 = (x_0, 0)$ of the parabola $g_1(x)$ remains in the same place, $P_2 = P_1$, the symmetry axis $x = x_0$ also remains the same, the focus $F_1 = (x_0, \frac{1}{4})$ moves to $F_2 = (x_0, \frac{1}{4a})$ and the directrix $y = -\frac{1}{4}$ to $y = -\frac{1}{4a}$. The last transformation represents the vertical translation of the graph of $g_2(x)$ by y_0 units (recall that a vertical shift is made upward if $y_0 > 0$ and downward if $y_0 < 0$). Under this transformation, the vertex $P_2 = (x_0, 0)$ of the parabola $g_2(x)$ moves to $P_v = (x_0, y_0)$, the symmetry axis $x = x_0$ stays the same, the focus $F_2 = (x_0, \frac{1}{4a})$ moves to $F = (x_0, \frac{1}{4a} + y_0)$ and the directrix $y = -\frac{1}{4a}$ to $y = y_0 - \frac{1}{4a}$. See illustration of these transformations in Fig. 3.9.

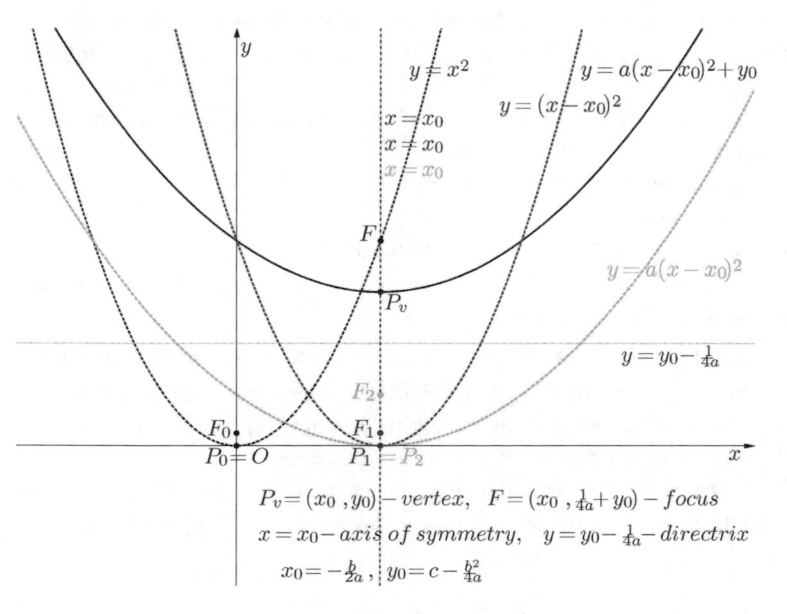

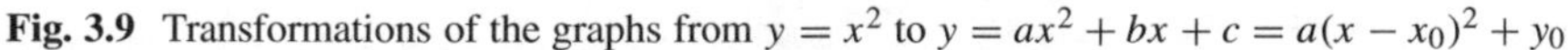
Fig. 3.9 Transformations of the graphs from $y = x^2$ to $y = ax^2 + bx + c = a(x - x_0)^2 + y_0$

The case $a < 0$ can be investigated using similar reasoning both in analytic and geometric form. This problem is left to the reader.

1.3 Monomials

Let us recall the general definition of a polynomial function.

Definition of a Polynomial Function A *polynomial function* (a *polynomial*) has the form $y = P_n(x) = a_n x^n + a_{n-1}x^{n-1} + \ldots + a_1 x + a_0$, where $a_i, i = 0, \ldots, n$ are real constants (called *coefficients*) and $a_n \neq 0$ is the *leading coefficient*.

In this section we consider a particular case of *monomials*, that is, the polynomials that contain only the *leading term*: $P_n(x) = a_n x^n$. Let us start with $y = x^n$, $n = 2k - 1$, $k \in \mathbb{N}$. The special case of $k = 1$ when we have a linear function $y = x$ was already studied. We consider now the function $y = x^3$ and then generalize this study to an arbitrary odd power.

Study of $y = x^3$

Some properties of this function have already been analyzed in Sect. 7 of Chap. 2, but for the sake of completeness of the presentation we prove them once more.

1. Domain and range. The domain is $X = \mathbb{R}$ (the formula is defined for $\forall x \in \mathbb{R}$) and the range is $Y = \mathbb{R}$. Indeed, for each y there exists $x = \sqrt[3]{y}$ such that $f(x) = f(\sqrt[3]{y}) = (\sqrt[3]{y})^3 = y$, which means that the range contains all real numbers.
2. Boundedness. The specification of the range shows that the function is unbounded both below and above.
3. Parity. The relation $f(-x) = (-x)^3 = -x^3 = -f(x), \forall x \in X$ shows that the function is odd.
4. Monotonicity and extrema. The function is increasing on the entire domain $X = \mathbb{R}$ due to the following evaluation for an arbitrary pair of the points $x_1 < x_2$: $f(x_2) - f(x_1) = x_2^3 - x_1^3 = (x_2 - x_1)(x_2^2 + x_2 x_1 + x_1^2) = (x_2 - x_1)\left((x_2 + \frac{1}{2}x_1)^2 + \frac{3}{4}x_1^2\right) > 0$. Since the function is increasing on $\mathbb{R}$, it has no extrema, neither local nor global, neither strict nor non-strict.
5. Periodicity. The increase on the entire domain implies that the function is not periodic, since it does not repeat any of its values.
6. Concavity and inflection. For any pair of points x_1, x_2 we have the following evaluation of the difference between the mean value of the secant $\frac{f(x_1)+f(x_2)}{2}$

and the value of the function at the midpoint $f\left(\frac{x_1+x_2}{2}\right)$:

$$D_2 \equiv \frac{x_1^3+x_2^3}{2} - \left(\frac{x_1+x_2}{2}\right)^3 = \frac{1}{8}\left(4x_1^3+4x_2^3-x_1^3-x_2^3-3x_1^2x_2-3x_1x_2^2\right)$$

$$= \frac{3}{8}\left(x_1^3+x_2^3-x_1^2x_2-x_1x_2^2\right) = \frac{3}{8}\left(x_2^2(x_2-x_1)-x_1^2(x_2-x_1)\right)$$

$$= \frac{3}{8}(x_2-x_1)^2(x_2+x_1).$$

Choosing two different points such that $x_1, x_2 \le 0$, we obtain $D_2 = \frac{3}{8}(x_2 - x_1)^2(x_2+x_1) < 0$, because the last factor is negative. This means that $f(x) = x^3$ is concave downward on the interval $(-\infty, 0]$. For the points $x_1, x_2 \ge 0$, we get $D_2 = \frac{3}{8}(x_2 - x_1)^2(x_2 + x_1) > 0$, because the last factor is positive. This means that $f(x) = x^3$ is concave upward on the interval $[0, +\infty)$. Since the function changes the type of concavity passing through the point $x_0 = 0$, this point is the inflection point.

7. Intersection and sign. Intersection with the y-axis occurs at the origin and the same point is the only point of intersection with the x-axis. It follows from the monotonicity properties that $f(x) = x^3 < 0$, $\forall x < 0$ and $f(x) = x^3 > 0$, $\forall x > 0$.
8. Inverse. The function $f(x) = x^3 : X = \mathbb{R} \to Y = \mathbb{R}$ is surjective and, due to the increase over the entire domain, it is also injective. Therefore, the function is bijective and its inverse is $f^{-1}(x) = \sqrt[3]{x} : \mathbb{R} \to \mathbb{R}$.
9. Based on the investigated properties, we can sketch the graph of the function as it is shown in Fig. 3.10.

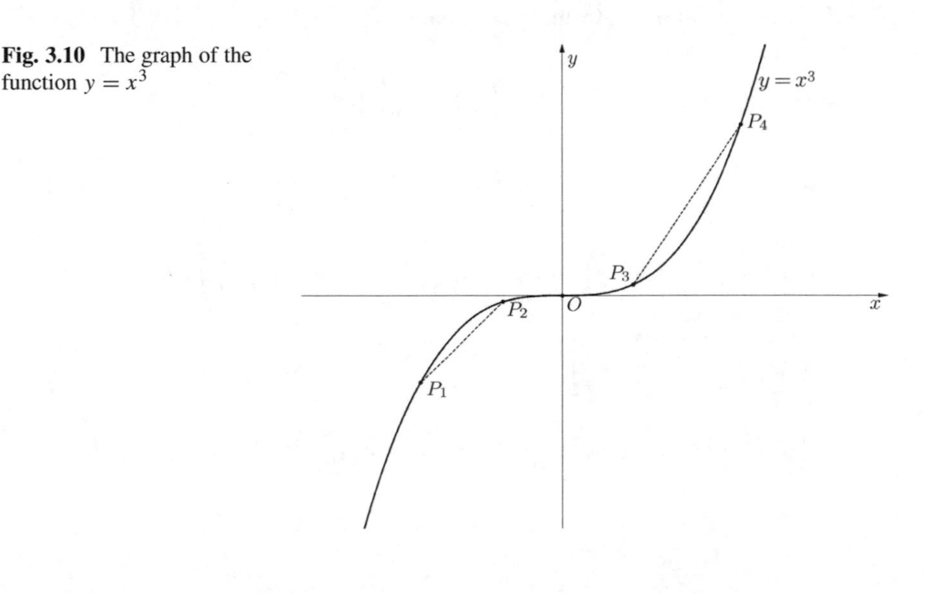

Fig. 3.10 The graph of the function $y = x^3$

The graph of $y = x^3$ (and of any function $y = ax^3 + bx^2 + cx + d, a \neq 0$) is called *cubic parabola*.

Complimentary Properties of $y = x^3$
The graph of the cubic parabola suggests that the function $y = x^3$ approaches $+\infty$ (increases without restriction) as x approaches $+\infty$ and it approaches $-\infty$ (decreases without restriction) as x approaches $-\infty$. This can be verified in more precise form. Indeed, if $x > 1$ then $x^3 > x$, and consequently, taking $x > M$ (where $M > 1$ is an arbitrarily large constant) we guarantee that $x^3 > M$, that is, the values of the function also become arbitrarily large, in other words, approach $+\infty$. Since the function is odd, this implies that increase of $|x|$ without restriction with negative values (that is, sending x to $-\infty$) results in the convergence of x^3 to $-\infty$. This implies, in particular, that the function $y = x^3$ has no horizontal asymptote. Since the function $y = x^3$ does not have an infinite limit at any finite point, the graph has no vertical asymptote. The absence of any kind of asymptote can also be noted in the graph of the function.

It happens that the behavior of monomials x^n, $n \in \mathbb{N}$ is similar to the one of the already analyzed functions x^2 or x^3. In the case of the odd power the function is similar to x^3, while the even power is similar to x^2.

Before to show the properties of the general monomials we will prove an auxiliary technical result which will be used in the study of concavity of monomials and also in future investigations of other functions.

Lemma *For an arbitrary real parameter $\varepsilon \neq 0$ and any $n \in \mathbb{N}$, $n \geq 2$ the following evaluation holds:*

$$E_n \equiv (1-\varepsilon)^n + (1+\varepsilon)^n > 2. \tag{3.1}$$

Proof Using the binomial formula, we obtain

$$\begin{aligned}
E_n &= (1-\varepsilon)^n + (1+\varepsilon)^n = \Big[1 - \binom{n}{1}\varepsilon + \binom{n}{2}\varepsilon^2 + \ldots \\
&\quad - \binom{n}{2k-1}\varepsilon^{2k-1} + \binom{n}{2k}\varepsilon^{2k} + \ldots + (-1)^{n-1}\binom{n}{n-1}\varepsilon^{n-1} + (-1)^n\varepsilon^n\Big] \\
&\quad + \Big[1 + \binom{n}{1}\varepsilon + \binom{n}{2}\varepsilon^2 + \ldots + \binom{n}{2k-1}\varepsilon^{2k-1} + \binom{n}{2k}\varepsilon^{2k} + \ldots \\
&\quad + \binom{n}{n-1}\varepsilon^{n-1} + \varepsilon^n\Big] \\
&= 2 + 2\binom{n}{2}\varepsilon^2 + \ldots + 2\binom{n}{2k}\varepsilon^{2k} + \ldots + 2\binom{n}{2m}\varepsilon^{2m} > 2,
\end{aligned}$$

where $2m = n - 1$ if n is odd or $2m = n$ if n is even. In other words, all the terms with odd powers are canceled, while those with even powers are doubled. Since all the remaining terms of ε have even powers, it follows that $E_n > 2$. □

Now let us show that the study of the function $y = x^3$ can be generalized to the monomials with odd powers $y = x^{2k+1}$, $k \in \mathbb{N}$.

Study of $y = x^{2k+1}$, $k \in \mathbb{N}$

1. Domain and range. The domain is $X = \mathbb{R}$ (the formula is defined for $\forall x \in \mathbb{R}$) and the range is $Y = \mathbb{R}$. Indeed, for each y there exists $x = \sqrt[2k+1]{y}$ such that $f(x) = f(\sqrt[2k+1]{y}) = (\sqrt[2k+1]{y})^{2k+1} = y$, which means that the range contains all the real numbers.
2. Boundedness. From the specification of the range it follows that the function is unbounded both below and above.
3. Parity. The relation $f(-x) = (-x)^{2k+1} = -x^{2k+1} = -f(x), \forall x \in X$ shows that the function is odd.
4. Monotonicity and extrema. Let us show that the function is increasing on the entire domain $X = \mathbb{R}$. Indeed, for an arbitrary pair of the points $x_1 < x_2$ the following evaluation holds:

$$\begin{aligned} f(x_2) - f(x_1) &= x_2^{2k+1} - x_1^{2k+1} \\ &= (x_2 - x_1)(x_2^{2k} + x_2^{2k-1}x_1 + x_2^{2k-2}x_1^2 + \ldots + x_2x_1^{2k-1} + x_1^{2k}) \\ &\equiv (x_2 - x_1)F_0. \end{aligned}$$

   The first factor is positive and the second represents the sum of the terms each of which has the even combined power. Due to this property, if x_1 and x_2 has the same sign, then $F_0 > 0$. More specifically, if $x_1 < x_2 \le 0$, then $F_0 > 0$ since there is at least one positive summand (x_1^{2k}) and others are non-negative. If $0 \le x_1 < x_2$, then again $F_0 > 0$ since there is at least one positive summand (x_2^{2k}) and others are non-negative. Therefore, in both cases $f(x_2) > f(x_1)$. It remains to consider the situation when $x_1 < 0 < x_2$. In this case it is sufficient to introduce the intermediate point $x_0 = 0$ and recall that (according to the already demonstrated results) the inequality $x_1 < x_0 = 0$ implies that $f(x_0) > f(x_1)$ and $0 = x_0 < x_2$ implies that $f(x_2) > f(x_0)$. Joining these two inequalities, we obtain $f(x_2) > f(x_0) > f(x_1)$. Hence, for any pair $x_1 < x_2$ we have $f(x_1) < f(x_2)$.

   Since the function is increasing on $\mathbb{R}$, it has no extrema, neither local nor global, neither strict nor not strict.
5. Periodicity. The increase over the entire domain implies that the function is not periodic, since it does not repeat any of its values.
6. Concavity and inflection. Take two arbitrary points x_1, x_2 and, for convenience, use such numbering that $x_1 < x_2$ (it can always be made without loss of

generality). The difference between the mean value of the secant $\frac{f(x_1)+f(x_2)}{2}$ and the value of the function at the midpoint $f\left(\frac{x_1+x_2}{2}\right)$ takes a more convenient form if we use the midpoint $x_0 = \frac{x_1+x_2}{2}$ together with the absolute and relative deviations $\delta = x_2 - x_0 = x_0 - x_1$ and $\varepsilon = \frac{\delta}{x_0}$, $x_0 \neq 0$:

$$D_2 \equiv \frac{f(x_1)+f(x_2)}{2} - f\left(\frac{x_1+x_2}{2}\right) = \frac{x_1^{2k+1}+x_2^{2k+1}}{2} - \left(\frac{x_1+x_2}{2}\right)^{2k+1}$$

$$= \frac{1}{2}\left((x_0-\delta)^{2k+1} + (x_0+\delta)^{2k+1}\right) - x_0^{2k+1}$$

$$= x_0^{2k+1}\left[\frac{1}{2}\left((1-\varepsilon)^{2k+1} + (1+\varepsilon)^{2k+1}\right) - 1\right]$$

$$= x_0^{2k+1}\left[\frac{1}{2}E_{2k+1} - 1\right].$$

According to the Lemma $E_{2k+1} > 2$. Therefore, if $x_1 < x_2 \leq 0$, then $x_0 < 0$, and consequently,

$$D_2 = x_0^{2k+1}\left[\frac{1}{2}E_{2k+1} - 1\right] < 0,$$

that is, the function is concave downward on the interval $(-\infty, 0]$. On the other hand, if $0 \leq x_1 < x_2$, then $x_0 > 0$, and consequently,

$$D_2 = x_0^{2k+1}\left[\frac{1}{2}E_{2k+1} - 1\right] > 0,$$

that is, the function is concave upward on the interval $[0, +\infty)$. Since the function changes the type of concavity passing through the point $x_0 = 0$, this point is the inflection point.

7. Intersection and sign. Intersection with the y-axis occurs at the origin and the same point is the only point of intersection with the x-axis. It follows from the monotonicity properties that $f(x) = x^{2k+1} < 0, \forall x < 0$ and $f(x) = x^{2k+1} > 0$, $\forall x > 0$.
8. Inverse. It was already shown that the function $f(x) = x^{2k+1} : X = \mathbb{R} \to Y = \mathbb{R}$ is surjective, and, due to the increase over the entire domain, it is also injective. Therefore, the function is bijective and its inverse is $f^{-1}(x) = \sqrt[2k+1]{x} : \mathbb{R} \to \mathbb{R}$.
9. Based on the investigated properties, we can sketch the graph of the function which has qualitatively the same shape as the graph of $y = x^3$.

Complimentary properties of $y = x^{2k+1}$

It follows from similarity of the properties of $y = x^3$ and $y = x^{2k+1}$, $k \in \mathbb{N}$ that when x approaches $+\infty$, the function $y = x^{2k+1}$ also approaches $+\infty$ and when x approaches $-\infty$, the function goes to $-\infty$. This shows that the function $y = x^{2k+1}$

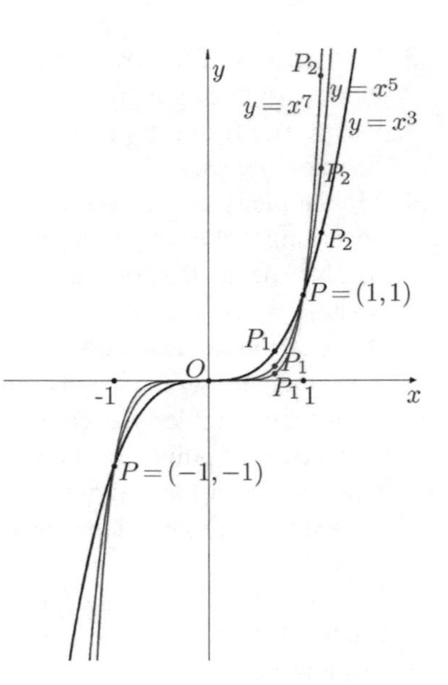

Fig. 3.11 The graphs of $y = x^3$, $y = x^5$ and $y = x^7$

has no horizontal asymptote. Since the function $y = x^{2k+1}$ does not have an infinite limit at any finite point, the graph has no vertical asymptote.

Comparison Between the Functions $y = x^{2k+1}$, $k \in \mathbb{N}$
It follows from the properties of powers that on the interval $(1, +\infty)$ the inequality $x^{2k+1} > x^{2m+1}$ holds when $k > m$. Besides, $\left(\frac{x_2}{x_1}\right)^{2k+1} > \left(\frac{x_2}{x_1}\right)^{2m+1}$ for any pair $x_2 > x_1 \geq 1$ and $k > m$. This shows that on the interval $(1, +\infty)$ the function $y_k = x^{2k+1}$ has greater values and increases faster than the function $y_m = x^{2m+1}$ if $k > m$. On the interval $(0, 1)$ we have $x^{2k+1} < x^{2m+1}$ when $k > m$, that is, $y_k = x^{2k+1}$ has smaller values than $y_m = x^{2m+1}$. Since these functions are odd, a similar relationship is observed on the intervals $(-\infty, -1)$ and $(-1, 0)$. At the points $x = -1, 0, 1$ all the functions $y = x^{2k+1}$ have the same values $y = -1, 0, 1$, respectively. This comparison is illustrated in Fig. 3.11.

Now let us switch to the even power: $y = x^{2k}$, $k \in \mathbb{N}$. In this case, we will generalize the study of the function $y = x^2$.

Study of $y = x^{2k}$, $k \in \mathbb{N}$

1. Domain and range. The domain is $X = \mathbb{R}$ (the formula is defined for $\forall x \in \mathbb{R}$) and the range is $Y = [0, +\infty)$. To prove the latter, notice first that for any $x \in \mathbb{R}$ we have $y = x^{2k} \geq 0$, which means that all negative values are excluded from the range, that is, $Y \subset [0, +\infty)$. On the other hand, for each $y \in [0, +\infty)$ there exists $x = \sqrt[2k]{y}$ such that $f(x) = f(\sqrt[2k]{y}) = (\sqrt[2k]{y})^{2k} = y$, which means that the image contains all $y \geq 0$. Hence, $Y = [0, +\infty)$.

2. Boundedness. It follows from the form of the range that the function is bounded below (for example, by the constant $m = 0$) and unbounded above.
3. Parity. The formula $f(-x) = (-x)^{2k} = x^{2k} = f(x), \forall x \in X$, shows that the function is even.
4. Monotonicity and extrema. Taking $x_1 < x_2 \leq 0$, we have $|x_2| < |x_1|$ and according to the power properties $f(x_2) - f(x_1) = x_2^{2k} - x_1^{2k} = |x_2|^{2k} - |x_1|^{2k} < 0$. This means that the function decreases on the interval $(-\infty, 0]$. Similarly, taking $0 \leq x_1 < x_2$, we have $f(x_2) - f(x_1) = x_2^{2k} - x_1^{2k} > 0$, which means that the function increases on $[0, +\infty)$. Therefore, each point $x_0 < 0$ is decreasing, since there exists its neighborhood (for instance, $(2x_0, 0)$) where the function is decreasing. Similarly, each point $x_0 > 0$ is increasing.

   Since each point $x \neq 0$ is monotonic, none of these points is extremum of any kind, global or local, strict or not. The unique remaining point is $x = 0$, which obviously is a strict global and local minimum since $f(x) = x^{2k} > 0 = f(0)$, $\forall x \neq 0$.
5. Periodicity. The strict increase of the function on the infinite interval $[0, +\infty)$ implies that the function is not periodic, since none of its values is repeated in this interval.
6. Concavity and inflection. For any pair of points $x_1 < x_2$ the difference between the mean value of the secant $\frac{f(x_1)+f(x_2)}{2}$ and the value of the function at the midpoint $f\left(\frac{x_1+x_2}{2}\right)$ takes a more convenient form if we use the midpoint $x_0 = \frac{x_1+x_2}{2}$ together with the absolute and relative deviations $\delta = x_2 - x_0 = x_0 - x_1$ and $\varepsilon = \frac{\delta}{x_0}$, $x_0 \neq 0$:

$$D_2 \equiv \frac{f(x_1) + f(x_2)}{2} - f\left(\frac{x_1 + x_2}{2}\right) = \frac{x_1^{2k} + x_2^{2k}}{2} - \left(\frac{x_1 + x_2}{2}\right)^{2k}$$

$$= \frac{1}{2}\left((x_0 - \delta)^{2k} + (x_0 + \delta)^{2k}\right) - x_0^{2k}$$

$$= x_0^{2k}\left[\frac{1}{2}\left((1 - \varepsilon)^{2k} + (1 + \varepsilon)^{2k}\right) - 1\right] = x_0^{2k}\left[\frac{1}{2}E_{2k} - 1\right].$$

According to the Lemma $E_{2k} > 2$. Therefore, for any pair such that $x_1 \neq -x_2$ (that is, $x_0 \neq 0$) we obtain

$$D_2 = x_0^{2k}\left(\frac{1}{2}E_{2k} - 1\right) > 0.$$

If $x_1 = -x_2$ (that is, $x_0 = 0$), then

$$D_2 \equiv \frac{f(x_1) + f(x_2)}{2} - f\left(\frac{x_1 + x_2}{2}\right) = \frac{1}{2}(f(x_1) + f(x_2)) - f(0) = x_1^{2k} > 0.$$

Hence, the function is concave upward on the entire domain. Therefore, the function has no inflection point.

7. Intersection and sign. Intersection with the y-axis occurs at the origin and the same point is the only intersection with the x-axis. As it was already indicated in the study of the range, $f(x) = x^{2k} > 0, \forall x \neq 0$.
8. Inverse. The function $f(x) = x^{2k} : X = \mathbb{R} \to Y = [0, +\infty)$ is surjective but not injective. In fact, each number $y > 0$ is associated with the two points of the domain $x_{\pm} = \pm \sqrt[2k]{y}$: $f(x_{\pm}) = (\pm \sqrt[2k]{y})^{2k} = y$. Consequently, the function is not bijective. If we restrict the domain to the interval $[0, +\infty)$, then the function becomes bijective and has the inverse $f^{-1}(x) = \sqrt[2k]{x} : [0, +\infty) \to [0, +\infty)$. Analogously, restricting the domain to the interval $(-\infty, 0]$ we obtain a bijective function with the inverse $f^{-1}(x) = -\sqrt[2k]{x} : [0, +\infty) \to (-\infty, 0]$.
9. Based on the studied properties we can plot the graph of the function with a shape similar to the graph of $y = x^2$.

Complimentary Properties of $y = x^{2k}$

Due to similarity of the properties between $y = x^2$ and $y = x^{2k}$, $k \in \mathbb{N}$, we can deduce that when $|x|$ approaches $+\infty$, the function goes to $+\infty$. This implies that the function $y = x^{2k}$ has no horizontal asymptote. Since the function $y = x^{2k}$ does not have an infinite limit at any finite point, the graph has no vertical asymptote.

Comparison Between the Functions $y = x^{2k}$

It follows from the properties of powers that on the interval $(1, +\infty)$ the inequality $x^{2k} > x^{2m}$ holds when $k > m$. Besides, $\left(\frac{x_2}{x_1}\right)^{2k} > \left(\frac{x_2}{x_1}\right)^{2m}$ for any pair $x_2 > x_1 \geq 1$ and $k > m$. This shows that on the interval $(1, +\infty)$, the function $y_k = x^{2k}$ has greater values and increases faster than the function $y_m = x^{2m}$ if $k > m$. On the interval $(0, 1)$ we have $x^{2k} < x^{2m}$ when $k > m$, that is, $y_k = x^{2k}$ has smaller values than $y_m = x^{2m}$. Since these functions are even, similar relations are observed on the intervals $(-\infty, -1)$ and $(-1, 0)$. At the points $x = -1, 0, 1$ all the functions $y = x^{2k}$ have the same values $y = 1, 0, 1$, respectively. This comparison is illustrated in Fig. 3.12.

The functions $y = ax^n$, $a > 0$, $n \in \mathbb{N}$ behave in the same way as $y = x^n$ and their graphs can be obtained by means of the vertical extension by the factor a of the graph of x^n (recall that this extension is understood as stretching a times from the x-axis if $a > 1$ and compressing $\frac{1}{a}$ times if $0 < a < 1$). If $a < 0$, we should add the reflection of the graph about the x-axis with the corresponding change of the properties. To exemplify we list below the properties and show the graph of the function $y = -2x^7$, employing the investigated properties of $y = x^{2k+1}$, $k \in \mathbb{N}$ as the reference.

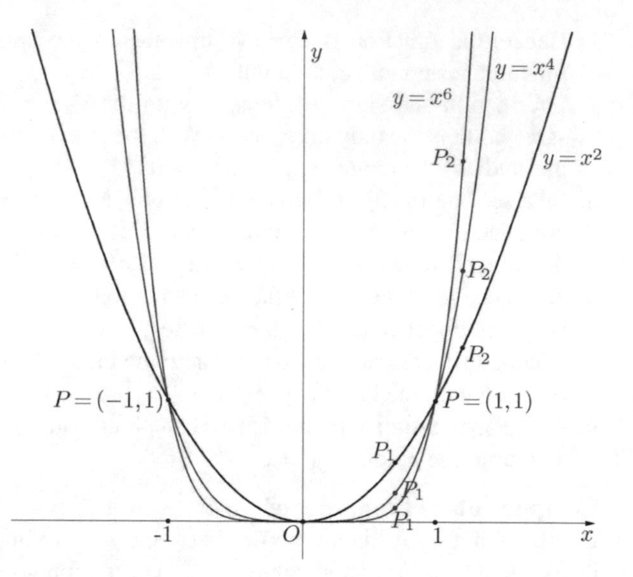

Fig. 3.12 The graphs of $y = x^2$, $y = x^4$ and $y = x^6$

Study of $y = -2x^7$

1. Domain and range. The domain is $X = \mathbb{R}$ and the range is $Y = \mathbb{R}$.
2. Boundedness. The function is unbounded both below and above.
3. Parity. The function is odd.
4. Periodicity. The function is not periodic.
5. Monotonicity and extrema. The function is decreasing over the entire domain $X = \mathbb{R}$. It has no extreme, neither local nor global, neither strict nor non-strict.
6. Concavity and inflection. The function is concave upward on the interval $(-\infty, 0]$ and downward on the interval $[0, +\infty)$. The only inflection point is $x_0 = 0$.
7. Intersection and sign. Intersection with the y-axis occurs at the origin and the same point is the only intersection with the x-axis. The sign of the function is as follows: $f(x) > 0, \forall x < 0$ and $f(x) < 0, \forall x > 0$.
8. Inverse. The function $f(x) = -2x^7 : X = \mathbb{R} \to Y = \mathbb{R}$ is surjective and injective. Consequently, it is bijective and admits the inverse $f^{-1}(x) = -\sqrt[7]{\frac{x}{2}}$: $\mathbb{R} \to \mathbb{R}$.
9. Based on the presented properties we can sketch the graph of the function shown in Fig. 3.13.

Complimentary Properties of Polynomials

The behavior of a general polynomial at infinity can be revealed by representing it in the form $P_n(x) = a_n x_n \left(1 + \frac{a_{n-1}}{a_n}\frac{1}{x} + \frac{a_{n-2}}{a_n}\frac{1}{x^2} + \ldots + \frac{a_0}{a_n}\frac{1}{x^n}\right)$. Notice that each term $\frac{1}{|x|^k}$, $k \in \mathbb{N}$ becomes very small, almost negligible, when the values of $|x|$

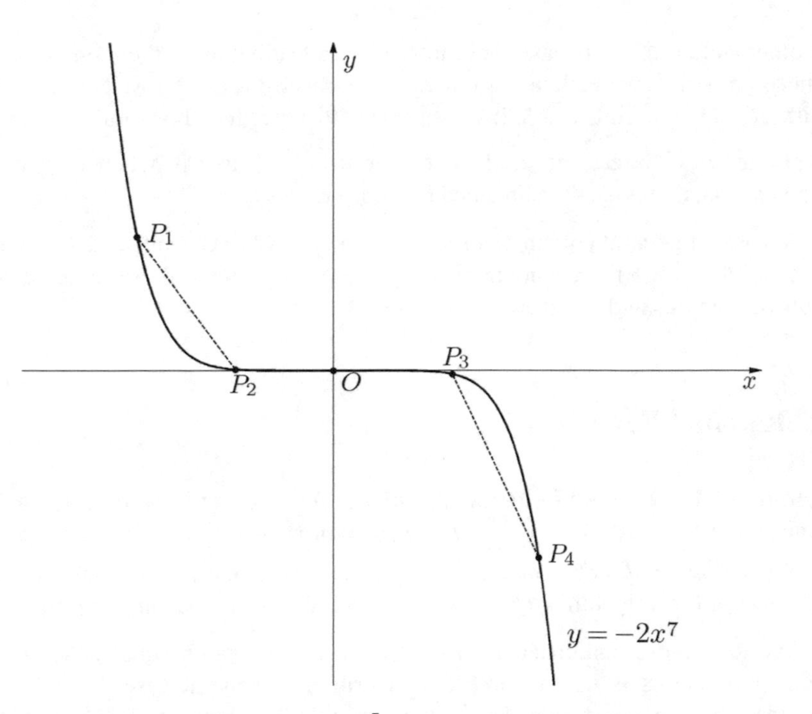

Fig. 3.13 The graph of the function $y = -2x^7$

are sufficiently large, that is, $\lim_{|x|\to\infty} \frac{1}{|x|^k} = 0, \forall k \in \mathbb{N}$. Specifying this a bit more, we can make $\frac{1}{|x|^k}$ smaller than any positive constant chosen beforehand, if we consider sufficiently large values of $|x|$. Indeed, if we choose an arbitrary constant $m > 0$, then taking $|x| > M = \max\{\frac{1}{m}, 1\}$ we guarantee that $|x|^k > |x| > M$, whence $\frac{1}{|x|^k} < m$. For instance, if we choose $m = 10^{-l}$, $l \in \mathbb{N}$, then it is sufficient to consider $|x| > 10^l$ in order to guarantee that $\frac{1}{|x|^k} < 10^{-l}$. This means that for any value of the coefficient $\frac{a_k}{a_n}$, if $|x|$ is sufficiently large, then the term $\left|\frac{a_k}{a_n}\frac{1}{x^k}\right|$ is very small. Reformulating with in a more precise form, we can always choose the values of $|x|$ large enough to make this term smaller than any constant $m > 0$ chosen beforehand. In this case it is said that the term $\left|\frac{a_k}{a_n}\frac{1}{x^k}\right|$ approaches 0 as $|x|$ approaches infinity, or using symbols: $\lim_{|x|\to\infty} \left|\frac{a_k}{a_n}\frac{1}{x^k}\right| = 0$. Using these considerations, we can conclude that among all the terms inside the parentheses in the expression $P_n(x) = a_n x_n \left(1 + \frac{a_{n-1}}{a_n}\frac{1}{x} + \frac{a_{n-2}}{a_n}\frac{1}{x^2} + \ldots + \frac{a_0}{a_n}\frac{1}{x^n}\right)$ the only term which does not approach 0 when $|x|$ is sufficiently large is the constant 1. Therefore, this constant is the dominant term at infinity (meaning that all other terms are negligible comparing with 1 when $|x|$ is sufficiently large). Hence, the behavior of $P_n(x)$ at infinity is uniquely determined by the first term.

In other words, $P_n(x)$ behaves at infinity like its leading term, the monomial $a_n x^n$, whose properties were already studied. Symbolically it can be written as follows: $\lim\limits_{|x|\to\infty} P_n(x) = \lim\limits_{|x|\to\infty} a_n x^n$. It is important to note that this similarity between $P_n(x)$ and $a_n x^n$ refers only to the behavior at infinity, that is, when x approaches $+\infty$ or $-\infty$, and it is not maintained for finite values of x.

A study of general polynomial functions can be rather complicated and in many cases all the properties cannot be investigated even if one will use more advanced tools of Calculus and Analysis.

2 Rational Functions

Definition of a Rational Function A *rational function* is the ratio of two polynomials, that is: $y = R(x) = \frac{P_n(x)}{Q_m(x)}$, where $P_n = a_n x^n + a_{n-1}x^{n-1} + \ldots + a_1 x + a_0$, $a_n \neq 0$ and $Q_m = b_m x^m + b_{m-1}x^{m-1} + \ldots + b_1 x + b_0$, $b_m \neq 0$. Notice that if the degree m of the polynomial Q_m is zero, a rational function becomes a polynomial.

Due to a more complicated form comparing with polynomials, the study of rational functions is not possible to perform in a general case. Therefore, we consider only some particular forms, starting with the functions $y = \frac{1}{x^n}$, $n \in \mathbb{N}$.

2.1 Functions $y = \frac{1}{x^n}$, $n \in \mathbb{N}$

In this class of functions, we consider first the two specific functions $y = \frac{1}{x}$ and $y = \frac{1}{x^2}$, and then we generalize the techniques of the analysis to the functions $y = \frac{1}{x^{2k+1}}$ and $y = \frac{1}{x^{2k}}$, $k \in \mathbb{N}$.

Let us start with the function $y = \frac{1}{x}$ (here $P_0(x) = 1$ and $Q_1(x) = x$).

Study of $y = \frac{1}{x}$

Some properties of this function have been already investigated in examples of Chap. 2, but for the sake of completeness we show these properties once more.

1. Domain and range. The domain is $X = \mathbb{R}\backslash\{0\}$ (the formula is defined for $\forall x \neq 0$) and the range is $Y = \mathbb{R}\backslash\{0\}$. Indeed, for each $y \neq 0$ there exists $x = \frac{1}{y}$ in the domain such that $f(x) = f(\frac{1}{y}) = \frac{1}{\frac{1}{y}} = y$. On the other hand, there is no x such that $\frac{1}{x} = 0$, which means that $y = 0$ is excluded from the range.

2. Boundedness. The specification of the range implies that the function is unbounded both below and above.
3. Parity. The relation $f(-x) = \frac{1}{-x} = -\frac{1}{x} = -f(x), \forall x \in X$ shows that the function is odd.
4. Monotonicity and extrema. Evaluating the difference between the values of the function at the pair of points $x_1 < x_2$, $x_1x_2 \neq 0$, we have: $f(x_2) - f(x_1) = \frac{1}{x_2} - \frac{1}{x_1} = \frac{x_1-x_2}{x_1x_2}$. The numerator is always negative, and consequently, the sign of the ratio depends only on denominator. If $x_1 < x_2 < 0$, then the denominator is positive and $f(x_2) - f(x_1) < 0$, that is, the function decreases on $(-\infty, 0)$. Analogously, if $0 < x_1 < x_2$, then the denominator is also positive and $f(x_2) - f(x_1) < 0$, that is, the function decreases on $(0, +\infty)$. Hence, each point $x_0 \neq 0$ is a decreasing point, because there exists its neighborhood where the function is decreasing. Notice that the function does not decrease on the entire domain, despite be decreasing on both intervals whose union form the entire domain. This follows from the obtained evaluation (taking $x_1 < 0 < x_2$) and also from an elementary specification of points (for instance, choosing the points $x_1 = -1$ and $x_2 = 1$ we get $f(x_1) = \frac{1}{x_1} = -1 < 1 = \frac{1}{x_2} = f(x_2)$).

   Since each point of the domain ($x \neq 0$) is decreasing, it cannot be an extremum of any type, global or local, strict or non-strict. Therefore, the function has no extrema.
5. Periodicity. The decrease of the function on the unbounded interval $(0, +\infty)$ indicates that the function cannot be periodic.
6. Concavity and inflection. Since the concavity is defined on an interval of the domain, we can choose two points x_1, x_2 either in the interval $(-\infty, 0)$ or in the interval $(0, +\infty)$ (there is no sense to take one point in $(-\infty, 0)$ and another in $(0, +\infty)$). Under this restriction, for any pair $x_1, x_2 \in X$, we have the following evaluation of the difference between the mean value of the secant $\frac{f(x_1)+f(x_2)}{2}$ and the value of the function at the midpoint $f\left(\frac{x_1+x_2}{2}\right)$:

$$D_2 \equiv \frac{f(x_1) + f(x_2)}{2} - f\left(\frac{x_1 + x_2}{2}\right) = \frac{\frac{1}{x_1} + \frac{1}{x_2}}{2} - \frac{1}{\frac{x_1+x_2}{2}} = \frac{x_1 + x_2}{2x_1x_2} - \frac{2}{x_1 + x_2}$$

$$= \frac{(x_1 + x_2)^2 - 4x_1x_2}{2x_1x_2(x_1 + x_2)} = \frac{(x_1 - x_2)^2}{2x_1x_2(x_1 + x_2)}.$$

If $x_1, x_2 < 0$, then $x_1x_2(x_1 + x_2) < 0$, and consequently, $D_2 < 0$, which means that the function is concave downward on the interval $(-\infty, 0)$. If $x_1, x_2 > 0$, then $x_1x_2(x_1+x_2) > 0$, and consequently, $D_2 > 0$, which means that the function is concave upward on the interval $(0, +\infty)$. Although the function changes the type of the concavity passing the point $x_0 = 0$, but this point is out of the domain and it cannot be an inflection point. Since the function keeps the same type of concavity on each of the two intervals which compose the domain, there are no inflection points.

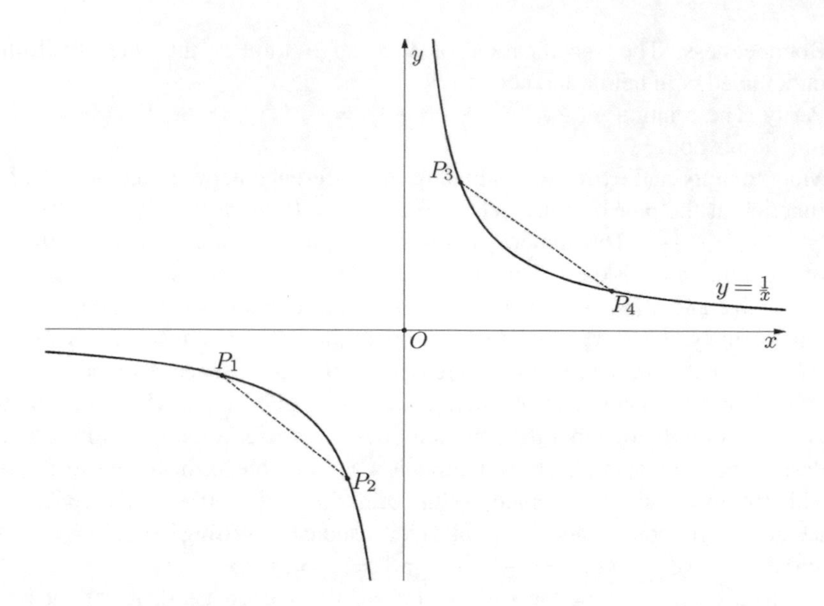

Fig. 3.14 The graph of the function $y = \frac{1}{x}$

7. Intersection and sign. There is no intersection with the y-axis, because the function is not defined at $x = 0$. Also there is no intersection with the x-axis, because the value $y = 0$ is out of the range. The sign of the function is defined by the following obvious inequalities: $f(x) = \frac{1}{x} < 0$, $\forall x < 0$ and $f(x) = \frac{1}{x} > 0$, $\forall x > 0$.
8. Inverse. The function $f(x) = \frac{1}{x} : X = \mathbb{R}\backslash\{0\} \to Y = \mathbb{R}\backslash\{0\}$ is surjective and injective, since the function is monotonic on the intervals $(-\infty, 0)$ and $(0, +\infty)$ and also recalling that it is negative on the former and positive on the latter. Therefore, the function is bijective and has the inverse $f^{-1}(x) = \frac{1}{x} : \mathbb{R}\backslash\{0\} \to \mathbb{R}\backslash\{0\}$.
9. Based on the investigated properties, we can sketch the graph of the function as shown in Fig. 3.14.

Complimentary Properties of the Function $y = \frac{1}{x}$

1. Behavior at infinity

   The function $\frac{1}{x}$ converges to 0 as $|x|$ goes to infinity, that is, the values of $\frac{1}{|x|}$ becomes smaller than any constant $m > 0$ chosen beforehand if $|x|$ is sufficiently large. Indeed, given any constant $m > 0$, choosing the points x such that $|x| > M = \frac{1}{m}$ we ensure that $\frac{1}{|x|} < m$. For instance, for a constant $m = 10^{-l}$, $l \in \mathbb{N}$, choosing $|x| > 10^l$ we guarantee that $\frac{1}{|x|} < 10^{-l}$. This means that $\frac{1}{|x|}$ approaches 0 as $|x|$ goes to $+\infty$, or in symbols $\lim\limits_{x\to+\infty} \frac{1}{x} = 0$ and $\lim\limits_{x\to-\infty} \frac{1}{x} = 0$. Returning to the variable x itself, we notice that when x goes to $+\infty$ then $\frac{1}{x}$ approaches 0 by positive values, while when x goes to $-\infty$ the convergence to 0 occurs by negative values (see Fig. 3.14). This implies that the line $y = 0$ is a horizontal asymptote of the graph of $y = \frac{1}{x}$.

2. Behavior near the origin

   When x approaches 0 from the right (by the positive values), the values of the function $\frac{1}{x}$ increase without restriction, in other words, $\frac{1}{x}$ goes to $+\infty$. In fact, for a given constant $M > 1$, choosing the values $0 < x < \frac{1}{M}$, we guarantee that $\frac{1}{x} > M$ for all these x. This shows that the function approaches the vertical line $x = 0$ as x approaches 0 from the right, meaning that the distance between the corresponding points of the graph of $f(x)$ and the line converges to 0. Consequently, the line $x = 0$ is a vertical asymptote of the graph of $f(x)$. In the same way, when x approaches 0 from the left (by the negative values), the values of $\frac{1}{x}$ decrease without restriction (the values of $\frac{1}{|x|}$ increase without restriction and $\frac{1}{x}$ keeps negative values), in other words, $f(x)$ goes to $-\infty$. Indeed, for a given constant $M > 1$, we can choose the values $0 > x > -\frac{1}{M}$ to ensure that $\frac{1}{x} < -M$ for all these x. Therefore, the line $x = 0$ is also a vertical asymptote of the graph of $f(x)$ from the left.

Geometric Characterization of Hyperbola

The graph of the function $y = \frac{1}{x}$ is called *hyperbola*. It has the following geometric characterization: the absolute value of the difference between the distances from any point of this curve to the two points $F_1 = (\sqrt{2}, \sqrt{2})$ and $F_2 = (-\sqrt{2}, -\sqrt{2})$ is equal $2\sqrt{2}$. The points F_1 and F_2 are called the foci and the bisector of the odd quadrants $y = x$, which contains these foci, is called focal axis (or transverse axis). To show this property, take an arbitrary point $P = (x, \frac{1}{x})$ of the hyperbola $y = \frac{1}{x}$ and calculate its distances to F_1 and F_2:

$$d(P, F_1) = \sqrt{(x - \sqrt{2})^2 + (\frac{1}{x} - \sqrt{2})^2} = \sqrt{x^2 + \frac{1}{x^2} - 2\sqrt{2}(x + \frac{1}{x}) + 4}$$

$$= \sqrt{(x + \frac{1}{x} - \sqrt{2})^2} = |x + \frac{1}{x} - \sqrt{2}|$$

and

$$d(P, F_2) = \sqrt{(x + \sqrt{2})^2 + (\frac{1}{x} + \sqrt{2})^2} = \sqrt{x^2 + \frac{1}{x^2} + 2\sqrt{2}(x + \frac{1}{x}) + 4}$$

$$= \sqrt{(x + \frac{1}{x} + \sqrt{2})^2} = |x + \frac{1}{x} + \sqrt{2}|.$$

Notice that if $x > 0$, then $x + \frac{1}{x} \geq 2$, and if $x < 0$, then $x + \frac{1}{x} \leq -2$. Therefore, for the points $P = (x, \frac{1}{x})$ with positive x-coordinates we have

$$\begin{aligned} d(P, F_1) - d(P, F_2) &= |x + \frac{1}{x} - \sqrt{2}| - |x + \frac{1}{x} + \sqrt{2}| \\ &= x + \frac{1}{x} - \sqrt{2} - (x + \frac{1}{x} + \sqrt{2}) = -2\sqrt{2}, \end{aligned}$$

while for the points with negative x-coordinates we get

$$\begin{aligned} d(P, F_1) - d(P, F_2) &= |x + \frac{1}{x} - \sqrt{2}| - |x + \frac{1}{x} + \sqrt{2}| \\ &= -(x + \frac{1}{x} - \sqrt{2}) + (x + \frac{1}{x} + \sqrt{2}) = 2\sqrt{2}. \end{aligned}$$

Joining the two results, we arrive at the geometric characterization:

$$|d(P, F_1) - d(P, F_2)| = 2\sqrt{2}.$$

It can be demonstrated that the reverse is also true: the geometric set of all points satisfying the property that the absolute value of the difference of the distances from a point to the two given points (foci) $F_1 = (\sqrt{2}, \sqrt{2})$ and $F_2 = (-\sqrt{2}, -\sqrt{2})$ is equal to $2\sqrt{2}$ represents the hyperbola $y = \frac{1}{x}$ (see Fig. 3.15). Indeed, take an arbitrary point $P(x, y)$ that satisfies the above property of the distances and write

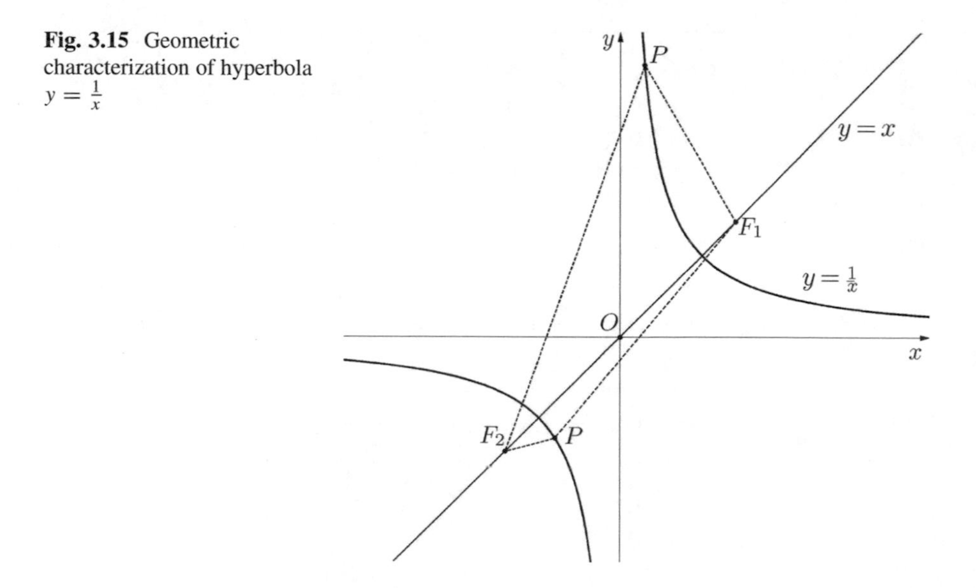

Fig. 3.15 Geometric characterization of hyperbola $y = \frac{1}{x}$

this property using coordinates:

$$|d(P,F_1)-d(P,F_2)| = |\sqrt{(x-\sqrt{2})^2+(y-\sqrt{2})^2}-\sqrt{(x+\sqrt{2})^2+(y+\sqrt{2})^2}|$$
$$= 2\sqrt{2}$$

or opening the absolute value

$$\sqrt{(x-\sqrt{2})^2+(y-\sqrt{2})^2}-\sqrt{(x+\sqrt{2})^2+(y+\sqrt{2})^2} = \pm 2\sqrt{2}.$$

Transferring one of the roots to the right-hand side and squaring both sides, we have

$$\left(\sqrt{(x-\sqrt{2})^2+(y-\sqrt{2})^2}\right)^2 = \left(\sqrt{(x+\sqrt{2})^2+(y+\sqrt{2})^2} \pm 2\sqrt{2}\right)^2$$

or

$$(x-\sqrt{2})^2+(y-\sqrt{2})^2 = (x+\sqrt{2})^2+(y+\sqrt{2})^2+8\pm 4\sqrt{2}\sqrt{(x+\sqrt{2})^2+(y+\sqrt{2})^2}$$

or simplifying

$$-x-y = \pm\sqrt{(x+\sqrt{2})^2+(y+\sqrt{2})^2}+\sqrt{2}.$$

Moving the remaining root with x, y to the left-hand side and joining all the other terms on the right-hand side, we get

$$\mp\sqrt{(x+\sqrt{2})^2+(y+\sqrt{2})^2} = x+y+\sqrt{2}.$$

Squaring once more both sides

$$(x+\sqrt{2})^2+(y+\sqrt{2})^2 = (x+y+\sqrt{2})^2$$

and simplifying, we arrive at the equation

$$xy = 1,$$

which is equivalent to the formula of hyperbola $y = \frac{1}{x}$.

In a more general form, any function $y = \frac{c}{x}$, where $c \neq 0$ is a constant, is called *hyperbola*. If $c > 0$, the shape of hyperbola is similar to that of $y = \frac{1}{x}$ and it has the following geometric feature: the absolute value of the difference between the distances from any point of the hyperbola to the two points (called foci) $F_1 = (\sqrt{2c}, \sqrt{2c})$ and $F_2 = (-\sqrt{2c}, -\sqrt{2c})$ is equal to $2\sqrt{2c}$. The bisector of the odd quadrants $y = x$, which contains the foci, is called focal axis. In the case $c < 0$, the

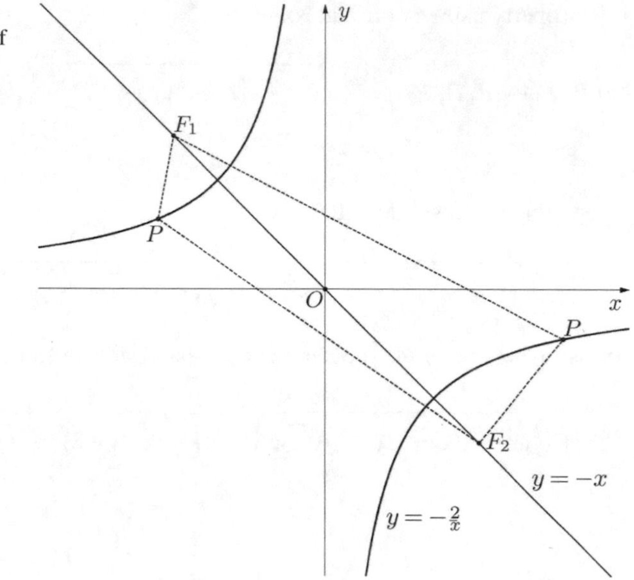

Fig. 3.16 The graph and geometric characterization of the function $y = -\frac{2}{x}$

graph of the hyperbola is obtained by reflecting the graph of $y = \frac{-c}{x}$ about the y-axis and it has the following geometric feature: the absolute value of the difference between the distances from any point of the hyperbola to the two points (foci) $F_1 = (-\sqrt{-2c}, \sqrt{-2c})$ and $F_2 = (\sqrt{-2c}, -\sqrt{-2c})$ is equal to $2\sqrt{-2c}$. The bisector of the even quadrants $y = -x$, which contains the foci, is called focal axis.

The proof of this geometric property of the hyperbola follows the same reasoning as in the case of $y = \frac{1}{x}$ and it is left to the reader. The illustration of the hyperbola with $c < 0$ is provided in Fig. 3.16.

Another important and simple rational function is $y = \frac{1}{x^2}$ (here $P_0(x) = 1$ and $Q_2(x) = x^2$).

Study of $y = \frac{1}{x^2}$

Some properties of this function have been already investigated in examples of Chap. 2, but for the sake of completeness we analyze these properties once more.

1. Domain and range. The domain is $X = \mathbb{R}\backslash\{0\}$ (the formula is defined for $\forall x \neq 0$) and the range is $Y = (0, +\infty)$. Indeed, any value of the function is positive: $f(x) = \frac{1}{x^2} > 0$, $\forall x \in X$. On the other hand, for each $y > 0$ there exists $x = \frac{1}{\sqrt{y}}$ in the domain such that $f(x) = f(\frac{1}{\sqrt{y}}) = \frac{1}{(\frac{1}{\sqrt{y}})^2} = y$.
2. Boundedness. The specification of the range implies that the function is bounded below (for instance, by $m = 0$), but unbounded above.

3. Parity. The relation $f(-x) = \frac{1}{(-x)^2} = \frac{1}{x^2} = f(x), \forall x \in X$ shows that the function is even.
4. Monotonicity and extrema. Evaluating the difference between the values of the function at the pair of points $x_1 < x_2$, $x_1 x_2 \neq 0$, we have: $f(x_2) - f(x_1) = \frac{1}{x_2^2} - \frac{1}{x_1^2} = \frac{(x_1-x_2)(x_1+x_2)}{x_1^2 x_2^2}$. The denominator is always positive, and consequently, the sign of the ratio depends only on the second factor of the numerator. If $x_1 < x_2 < 0$, then the second factor is negative and $f(x_2) - f(x_1) > 0$, that is, the function increases on $(-\infty, 0)$. If $0 < x_1 < x_2$, then the second factor is positive and $f(x_2) - f(x_1) < 0$, which means that the function decreases on $(0, +\infty)$. Hence, each point $x_0 < 0$ is increasing and each point $x_0 > 0$ is decreasing. Since each point $x \neq 0$ is monotonic, it cannot be extremum of any type, either global or local, either strict or not. Although the function change the type of monotonicity at the point $x = 0$, this point is not extremum since it does not belong to the domain. Therefore, the function has no extrema of either type.
5. Periodicity. The decrease on the unbounded interval $(0, +\infty)$ implies that the function is not periodic.
6. Concavity and inflection. Since the concavity is defined on an interval of the domain, we can choose two points x_1, x_2 either in the interval $(-\infty, 0)$ or in the interval $(0, +\infty)$ (there is no sense to take one point in $(-\infty, 0)$ and another in $(0, +\infty)$). Under this restriction, for any pair $x_1, x_2 \in X$, we have the following evaluation of the difference between the mean value of the secant $\frac{f(x_1)+f(x_2)}{2}$ and the value of the function at the midpoint $f\left(\frac{x_1+x_2}{2}\right)$:

$$D_2 \equiv \frac{f(x_1) + f(x_2)}{2} - f\left(\frac{x_1 + x_2}{2}\right) = \frac{\frac{1}{x_1^2} + \frac{1}{x_2^2}}{2} - \frac{1}{(\frac{x_1+x_2}{2})^2}$$

$$= \frac{x_1^2 + x_2^2}{2x_1^2 x_2^2} - \frac{4}{(x_1 + x_2)^2}$$

$$= \frac{(x_1^2 + x_2^2)(x_1 + x_2)^2 - 8x_1^2 x_2^2}{2x_1^2 x_2^2 (x_1 + x_2)^2}$$

$$= \frac{(x_1^4 + x_2^4 - 2x_1^2 x_2^2) + \left(2x_1 x_2 (x_1^2 + x_2^2) - 4x_1^2 x_2^2\right)}{2x_1^2 x_2^2 (x_1 + x_2)^2}$$

$$= \frac{(x_1^2 - x_2^2)^2 + 2x_1 x_2 (x_1 - x_2)^2}{2x_1^2 x_2^2 (x_1 + x_2)^2}.$$

This clearly shows that $D_2 > 0$ for any pair x_1, x_2, either in the interval $(-\infty, 0)$ or in $(0, +\infty)$. Therefore, the function is concave upward both on the interval $(-\infty, 0)$ and on $(0, +\infty)$. This implies that there are no points of inflection.
7. Intersection and sign. There is no intersection with the y-axis, because the function is not defined at $x = 0$. There is no intersection with the x-axis either

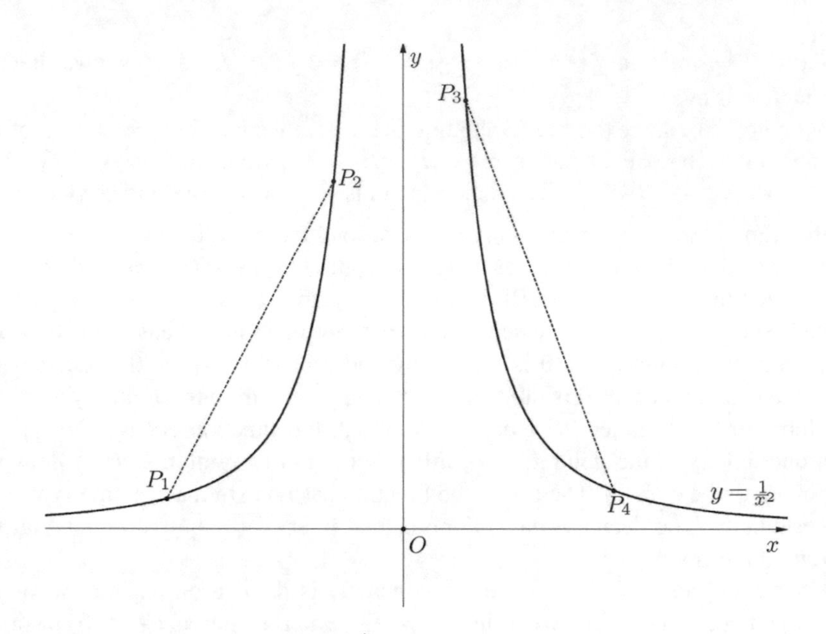

Fig. 3.17 The graph of the function $y = \frac{1}{x^2}$

because the value $y = 0$ is out of the range. The function is positive, which follows from the obvious inequality: $f(x) = \frac{1}{x^2} > 0, \forall x \in X$.

8. Inverse. The function $f(x) = \frac{1}{x^2} : X = \mathbb{R}\backslash\{0\} \to Y = (0, +\infty)$ is surjective, but not injective, since each $y > 0$ is associated with the two different points of the domain: $x_\pm = \pm\frac{1}{\sqrt{y}}$. To obtain a bijective function we should restrict the domain to the interval $(0, +\infty)$ or to $(-\infty, 0)$. In the first case, we have the bijective function $f(x) = \frac{1}{x^2} : X = (0, +\infty) \to Y = (0, +\infty)$ with the inverse $f^{-1}(x) = \frac{1}{\sqrt{x}} : (0, +\infty) \to (0, +\infty)$. In the second case, we get the bijective function $f(x) = \frac{1}{x^2} : X = (-\infty, 0) \to Y = (0, +\infty)$ with the inverse $f^{-1}(x) = -\frac{1}{\sqrt{x}} : (0, +\infty) \to (-\infty, 0)$.
9. Based on the investigated properties, we can sketch the graph of the function as shown in Fig. 3.17.

Complimentary Properties of $y = \frac{1}{x^2}$

1. Behavior at infinity

The function $\frac{1}{x^2}$ converges to 0 as $|x|$ goes to infinity, that is, the values of $\frac{1}{x^2}$ becomes smaller than any constant $m > 0$ chosen beforehand if $|x|$ is sufficiently large. Indeed, given an arbitrary constant $m > 0$, we can choose the points x such that $|x| > M = \max\{\frac{1}{m}, 1\}$ to ensure that $\frac{1}{x^2} < \frac{1}{|x|} < m$. For instance, for a constant $m = 10^{-l}$, $l \in \mathbb{N}$, choosing $|x| > 10^l$ we guarantee that $\frac{1}{x^2} < 10^{-l}$ (actually we have the values even smaller: $\frac{1}{x^2} < 10^{-2l}$). Since the function is even, the behavior is same when x goes to $+\infty$ and to $-\infty$. Therefore, the line $y = 0$ is a horizontal asymptote of the graph of $y = \frac{1}{x^2}$.

2. Behavior near the origin

   When x approaches 0 (both from the right and the left), the values of the function $\frac{1}{x^2}$ increase without restriction, that is, converges to $+\infty$. In fact, for a given constant $M > 1$, we can choose x such that $0 < x < \frac{1}{M}$ to guarantee that the corresponding values of the function are greater than M: $\frac{1}{x^2} > \frac{1}{x} > M$. This means that the graph of the function approaches the vertical line $x = 0$ when x approaches 0 from the right and from the left, that is, the distance between the corresponding points of the graph and the line $x = 0$ converges to 0. Therefore, the line $x = 0$ is a vertical asymptote of the graph of the function.

In a similar manner as the properties of the functions x^3 and x^2 were generalized to the cases of x^{2k+1} and x^{2k}, the qualitative behavior of $\frac{1}{x}$ and $\frac{1}{x^2}$ can be extended to the functions $\frac{1}{x^{2k+1}}$ and $\frac{1}{x^{2k}}$, $k \in \mathbb{N}$, respectively.

Study of $y = \frac{1}{x^{2k+1}}$

1. Domain and range. The domain is $X = \mathbb{R}\backslash\{0\}$ (the formula is defined for $\forall x \neq 0$) and the range is $Y = \mathbb{R}\backslash\{0\}$. Indeed, for each $y \neq 0$ there exists $x = \frac{1}{\sqrt[2k+1]{y}}$ in the domain such that $f(x) = f(\frac{1}{\sqrt[2k+1]{y}}) = \frac{1}{(\frac{1}{\sqrt[2k+1]{y}})^{2k+1}} = y$. On the other hand, there is no x such that $\frac{1}{x^{2k+1}} = 0$, which means that $y = 0$ is excluded from the range.
2. Boundedness. The specification of the range implies that the function is unbounded both below and above.
3. Parity. The relation $f(-x) = \frac{1}{(-x)^{2k+1}} = -\frac{1}{x^{2k+1}} = -f(x)$, $\forall x \in X$ shows that the function is odd.
4. Monotonicity and extrema. Evaluating the difference between the values of the function at the pair of points $x_1 < x_2$, $x_1 x_2 \neq 0$, we have: $f(x_2) - f(x_1) = \frac{1}{x_2^{2k+1}} - \frac{1}{x_1^{2k+1}} = \frac{x_1^{2k+1} - x_2^{2k+1}}{x_1^{2k+1} x_2^{2k+1}}$. If $x_1 < x_2 < 0$, then $x_1 x_2 > 0$ and $x_1^{2k+1} - x_2^{2k+1} < 0$, whence $f(x_2) - f(x_1) < 0$, that is, the function is decreasing on $(-\infty, 0)$. If $0 < x_1 < x_2$, then again $x_1 x_2 > 0$ and $x_1^{2k+1} - x_2^{2k+1} < 0$, whence $f(x_2) - f(x_1) < 0$, which means that the function is also decreasing on $(0, +\infty)$. This implies that each point $x_0 \neq 0$ is monotonic, since there exists its neighborhood where the function is decreasing. Notice that the function is not decreasing over the entire domain, though it is decreasing in both intervals that compose the domain. This follows from the above obtained evaluation (take $x_1 < 0 < x_2$ there) and also from an elementary specification of the points (for instance, using the points $x_1 = -1$ and $x_2 = 1$ we have $f(-1) = -1 < 1 = f(1)$).

Since each point $x \neq 0$ is decreasing, none of these points can be extremum of any type. Additionally, $x = 0$ is not an extremum because it does not belong to the domain. Hence, the function has no extremum, either global or local, either strict or non-strict.

5. Periodicity. The decrease of the function on the unbounded interval $(0, +\infty)$ indicates that the function is not periodic.
6. Concavity and inflection. Since the concavity is defined on an interval of the domain, we can choose two points x_1, x_2 either in the interval $(-\infty, 0)$ or in the interval $(0, +\infty)$ (there is no sense to take one point in $(-\infty, 0)$ and another in $(0, +\infty)$). Under this restriction, we choose two arbitrary points $x_1, x_2 \in X$ and, without a loss of generality, number them in such a way that $x_1 < x_2$. The difference between the mean value of the secant $\frac{f(x_1)+f(x_2)}{2}$ and the value of the function at the midpoint $f\left(\frac{x_1+x_2}{2}\right)$ is more convenient to represent using the midpoint $x_0 = \frac{x_1+x_2}{2}$ and the absolute and relative deviations $\delta = x_2 - x_0 = x_0 - x_1$ and $\varepsilon = \frac{\delta}{x_0}$. Notice that both in the case $x_1 < x_2 < 0$ and $0 < x_1 < x_2$ the absolute deviation satisfies the inequality $\delta < |x_0|$, and consequently, $|\varepsilon| < 1$. Hence, we obtain

$$D_2 \equiv \frac{f(x_1) + f(x_2)}{2} - f\left(\frac{x_1 + x_2}{2}\right) = \frac{\frac{1}{x_1^{2k+1}} + \frac{1}{x_2^{2k+1}}}{2} - \frac{1}{\left(\frac{x_1+x_2}{2}\right)^{2k+1}}$$

$$= \frac{1}{2}\left(\frac{1}{(x_0 - \delta)^{2k+1}} + \frac{1}{(x_0 + \delta)^{2k+1}}\right) - \frac{1}{x_0^{2k+1}}$$

$$= \frac{1}{x_0^{2k+1}}\left[\frac{1}{2}\left(\frac{1}{(1-\varepsilon)^{2k+1}} + \frac{1}{(1+\varepsilon)^{2k+1}}\right) - 1\right]$$

$$= \frac{1}{x_0^{2k+1}}\left[\frac{1}{2}\frac{(1+\varepsilon)^{2k+1} + (1-\varepsilon)^{2k+1}}{(1-\varepsilon^2)^{2k+1}} - 1\right] = \frac{1}{x_0^{2k+1}}\left[\frac{1}{2}\frac{E_{2k+1}}{(1-\varepsilon^2)^{2k+1}} - 1\right].$$

Since $E_{2k+1} > 2$ (see formula (3.1) of Lemma in Sect. 1.3) and $(1-\varepsilon^2)^{2k+1} < 1$ (because $|\varepsilon| < 1$), we get $\frac{E_{2k+1}}{(1-\varepsilon^2)^{2k+1}} > 2$, and consequently, $\frac{1}{2}\frac{E_{2k+1}}{(1-\varepsilon^2)^{2k+1}} - 1 > 0$. Therefore, if $x_1 < x_2 < 0$, then $x_0 < 0$ and

$$D_2 = \frac{1}{x_0^{2k+1}}\left[\frac{1}{2}\frac{E_{2k+1}}{(1-\varepsilon^2)^{2k+1}} - 1\right] < 0,$$

that is, on the interval $(-\infty, 0)$ the concavity is downward. On the other hand, if $0 < x_1 < x_2$, then $x_0 > 0$ and

$$D_2 = \frac{1}{x_0^{2k+1}}\left[\frac{1}{2}\frac{E_{2k+1}}{(1-\varepsilon^2)^{2k+1}} - 1\right] > 0,$$

which means that on the interval $(0, +\infty)$ the concavity is upward.

Although the function changes the type of concavity passing through the point $x_0 = 0$, this point is not an inflection point because it does not belong to the domain. Besides, on each of the intervals $(-\infty, 0)$ and $(0, +\infty)$ the function keeps the same concavity. Hence, the function has no points of inflection.

7. Intersection and sign. There is no intersection with the y-axis, because the function is not defined at $x = 0$. Also there is no intersection with the x-axis, because the value $y = 0$ is out of range. The sign of the function is defined by the following obvious inequalities: $f(x) = \frac{1}{x^{2k+1}} < 0, \forall x < 0$ and $f(x) = \frac{1}{x^{2k+1}} > 0, \forall x > 0$.
8. Inverse. The function $f(x) = \frac{1}{x^{2k+1}} : X = \mathbb{R}\backslash\{0\} \to Y = \mathbb{R}\backslash\{0\}$ is surjective and injective. The latter is the consequence of monotonicity on the intervals $(-\infty, 0)$ and $(0, +\infty)$ and of the fact that on the first interval the function is negative and on the second is positive. Therefore, the function is bijective and admits the inverse $f^{-1}(x) = \frac{1}{\sqrt[2k+1]{x}} : \mathbb{R}\backslash\{0\} \to \mathbb{R}\backslash\{0\}$.
9. Based on the investigated properties we can sketch the graph of the function similar to that of $\frac{1}{x}$.

Complimentary Properties of the Function $y = \frac{1}{x^{2k+1}}$

Behavior at infinity and near the origin.

Similarly to $\frac{1}{x}$, the function $\frac{1}{x^{2k+1}}$ converges to 0 as $|x|$ goes to infinity, that is, the values $\frac{1}{|x|^{2k+1}}$ becomes smaller than any constant $m > 0$ chosen beforehand for sufficiently large values of $|x|$. This implies that the line $y = 0$ is a horizontal asymptote of the function $y = \frac{1}{x^{2k+1}}$.

When x approaches 0 from the right (by the positive values), the values of the function $\frac{1}{x^{2k+1}}$ increase without restriction, that is, go to $+\infty$. In consequence of this, the line $x = 0$ is a vertical asymptote of the graph of the function. In the same way, when x approaches 0 from the left (by the negative values), the values of the function $\frac{1}{x^{2k+1}}$ decrease without restriction (the values of $\frac{1}{|x|^{2k+1}}$ increase without restriction and the function $\frac{1}{x^{2k+1}}$ keeps negative values). In other words, the function goes to $-\infty$. Hence, the line $x = 0$ is a vertical asymptote also for negative values.

Comparison Between Functions $y = \frac{1}{x^{2k+1}}$

It follows from the properties of powers that on the interval $(1, +\infty)$ we have $x^{2k+1} > x^{2m+1}$ when $k > m$, whence $y_k = \frac{1}{x^{2k+1}} < \frac{1}{x^{2m+1}} = y_m$. Besides, $\left(\frac{x_2}{x_1}\right)^{2k+1} > \left(\frac{x_2}{x_1}\right)^{2m+1}$ for any $x_2 > x_1 \geq 1$ and $k > m$, and consequently, $\frac{y_k(x_1)}{y_k(x_2)} = \left(\frac{x_2}{x_1}\right)^{2k+1} > \left(\frac{x_2}{x_1}\right)^{2m+1} = \frac{y_m(x_1)}{y_m(x_2)}$. This shows that on the interval $(1, +\infty)$ the function $y_k = \frac{1}{x^{2k+1}}$ has smaller values and decreases faster than $y_m = \frac{1}{x^{2m+1}}$ if $k > m$. On the interval $(0, 1)$ we have $x^{2k+1} < x^{2m+1}$ when $k > m$, that is, $y_k = \frac{1}{x^{2k+1}}$ has greater values than $y_m = \frac{1}{x^{2m+1}}$. Since the functions are odd, similar relationships hold on the intervals $(-\infty, -1)$ and $(-1, 0)$. At the points $x = -1, 1$ the values of all the functions of this type coincide and are equal to $y = -1, 1$, respectively. This comparison is illustrated in Fig. 3.18.

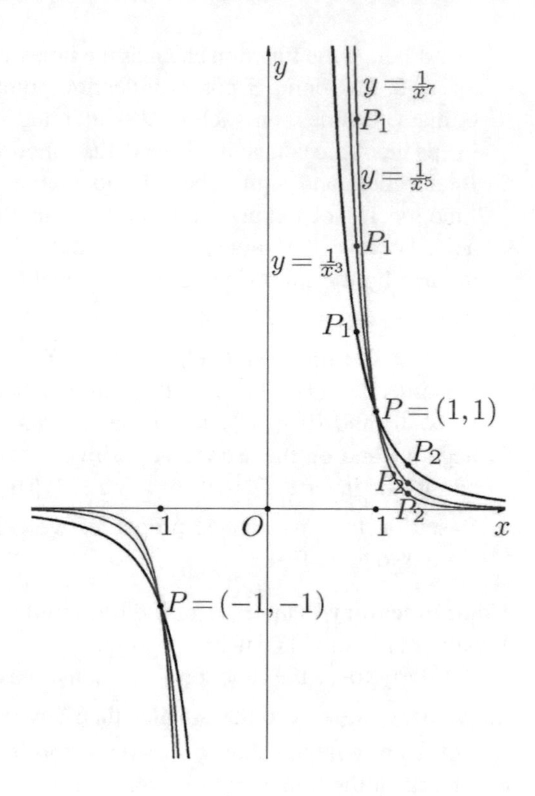

Fig. 3.18 The graphs of the functions $y = \frac{1}{x^3}$, $y = \frac{1}{x^5}$ and $y = \frac{1}{x^7}$

In a similar way we generalize next the behavior of the function $\frac{1}{x^2}$ to the case of the functions $\frac{1}{x^{2k}}$, $k \in \mathbb{N}$.

Study of $y = \frac{1}{x^{2k}}$

1. Domain and range. The domain is $X = \mathbb{R}\backslash\{0\}$ (the formula is defined for $\forall x \neq 0$) and the range is $Y = (0, +\infty)$. Indeed, any value of the function is positive: $f(x) = \frac{1}{x^{2k}} > 0$, $\forall x \in X$. Besides, for each $y > 0$ there exists $x = \frac{1}{\sqrt[2k]{y}}$ in the domain such that $f(x) = f(\frac{1}{\sqrt[2k]{y}}) = \frac{1}{(\frac{1}{\sqrt[2k]{y}})^{2k}} = y$.
2. Boundedness. The specification of the range implies that the function is bounded below (for instance, by $m = 0$), but unbounded above.
3. Parity. The relation $f(-x) = \frac{1}{(-x)^{2k}} = \frac{1}{x^{2k}} = f(x)$, $\forall x \in X$ shows that the function is even.
4. Monotonicity and extrema. Evaluating the difference between the values of the function at the pair of points $x_1 < x_2$, $x_1 x_2 \neq 0$, we have: $f(x_2) - f(x_1) = \frac{1}{x_2^{2k}} - \frac{1}{x_1^{2k}} = \frac{x_1^{2k} - x_2^{2k}}{x_1^{2k} x_2^{2k}}$. The denominator is always positive, and consequently, the sign of the ratio depends only on the numerator. If $x_1 < x_2 < 0$, then $|x_1| > |x_2|$

and $x_1^{2k} - x_2^{2k} > 0$, whence $f(x_2) - f(x_1) > 0$, that is, the function increases on $(-\infty, 0)$. If $0 < x_1 < x_2$, then $x_1^{2k} - x_2^{2k} < 0$, whence $f(x_2) - f(x_1) < 0$, that is, the function decreases on $(0, +\infty)$. Hence, each point $x_0 < 0$ is increasing and each point $x_0 > 0$ is decreasing. Since each point $x \neq 0$ is monotonic, it cannot be extremum of any type, either global or local, either strict or not. Although the function changes the type of monotonicity at the point $x = 0$, this point is not extremum since it does not belong to the domain. Therefore, the function has no extrema of either type.

5. Periodicity. The decrease on the unbounded interval $(0, +\infty)$ implies that the function is not periodic.
6. Concavity and inflection. Since the concavity is defined on an interval of the domain, we can choose two points x_1, x_2 either in the interval $(-\infty, 0)$ or in the interval $(0, +\infty)$ (there is no sense to take one point in $(-\infty, 0)$ and another in $(0, +\infty)$). Under this restriction, we choose two arbitrary points $x_1, x_2 \in X$ and, without a loss of generality, number them in such a way that $x_1 < x_2$. The difference between the mean value $\frac{f(x_1)+f(x_2)}{2}$ and the value of the function at the midpoint $f\left(\frac{x_1+x_2}{2}\right)$ is more convenient to represent using the midpoint $x_0 = \frac{x_1+x_2}{2}$ and the absolute and relative deviations $\delta = x_2 - x_0 = x_0 - x_1$ and $\varepsilon = \frac{\delta}{x_0}$. Notice that both in the case $x_1 < x_2 < 0$ and $0 < x_1 < x_2$ the absolute deviation satisfies the inequality $\delta < |x_0|$, and consequently, $|\varepsilon| < 1$. Hence, we obtain

$$D_2 \equiv \frac{f(x_1) + f(x_2)}{2} - f\left(\frac{x_1 + x_2}{2}\right) = \frac{\frac{1}{x_1^{2k}} + \frac{1}{x_2^{2k}}}{2} - \frac{1}{\left(\frac{x_1+x_2}{2}\right)^{2k}}$$

$$= \frac{1}{2}\left(\frac{1}{(x_0 - \delta)^{2k}} + \frac{1}{(x_0 + \delta)^{2k}}\right) - \frac{1}{x_0^{2k}}$$

$$= \frac{1}{x_0^{2k}}\left[\frac{1}{2}\left(\frac{1}{(1-\varepsilon)^{2k}} + \frac{1}{(1+\varepsilon)^{2k}}\right) - 1\right]$$

$$= \frac{1}{x_0^{2k}}\left[\frac{1}{2}\frac{(1+\varepsilon)^{2k} + (1-\varepsilon)^{2k}}{(1-\varepsilon^2)^{2k}} - 1\right] = \frac{1}{x_0^{2k}}\left[\frac{1}{2}\frac{E_{2k}}{(1-\varepsilon^2)^{2k}} - 1\right].$$

Since $E_{2k} > 2$ (see formula (3.1) of Lemma in Sect. 1.3) and $(1 - \varepsilon^2)^{2k} < 1$ (because $|\varepsilon| < 1$), we get $\frac{E_{2k}}{(1-\varepsilon^2)^{2k}} > 2$, and consequently, $\frac{1}{2}\frac{E_{2k}}{(1-\varepsilon^2)^{2k}} - 1 > 0$. Therefore, both in the case $x_1 < x_2 < 0$ and $0 < x_1 < x_2$ we have

$$D_2 = \frac{1}{x_0^{2k}}\left[\frac{1}{2}\frac{E_{2k}}{(1-\varepsilon^2)^{2k}} - 1\right] > 0,$$

that is, the function is concave upward both on the interval $(-\infty, 0)$ and on $(0, +\infty)$. Hence, there is no inflection point.

7. Intersection and sign. There is no intersection with the y-axis, because the function is not defined at $x = 0$. There is no intersection with the x-axis either, because the function takes only positive values: $f(x) = \frac{1}{x^{2k}} > 0, \forall x \neq 0$.
8. Inverse. The function $f(x) = \frac{1}{x^{2k}} : X = \mathbb{R}\backslash\{0\} \to Y = (0, +\infty)$ is surjective, but not injective, since each $y > 0$ is associated with the two different points of the domain: $x_{\pm} = \pm\frac{1}{\sqrt[2k]{y}}$. Restricting the domain to the interval $(0, +\infty)$ or to $(-\infty, 0)$, we obtain an injective function. In the former case we have the bijective function $f(x) = \frac{1}{x^{2k}} : X = (0, +\infty) \to Y = (0, +\infty)$ with the inverse $f^{-1}(x) = \frac{1}{\sqrt[2k]{x}} : (0, +\infty) \to (0, +\infty)$. In the latter case we get the bijective function $f(x) = \frac{1}{x^{2k}} : X = (-\infty, 0) \to Y = (0, +\infty)$ whose inverse is $f^{-1}(x) = -\frac{1}{\sqrt[2k]{x}} : (0, +\infty) \to (-\infty, 0)$.
9. Based on the investigated properties we can sketch the graph of the function similar to that of $\frac{1}{x^2}$.

Complimentary Properties of the Function $y = \frac{1}{x^{2k}}$

Behavior at infinity and near the origin.

Like the function $\frac{1}{x^2}$, the function $\frac{1}{x^{2k}}$ converges to 0 when $|x|$ goes to infinity, that is, the values of $\frac{1}{|x|^{2k}}$ become smaller than any constant $m > 0$ chosen beforehand if we choose sufficiently large values of $|x|$. This implies that the line $y = 0$ is a horizontal asymptote of the function $y = \frac{1}{x^{2k}}$.

When x approaches 0 (both from the left and from the right), the values of the function $\frac{1}{x^{2k}}$ increase without restriction, that is, the function goes to $+\infty$. Therefore, the line $x = 0$ is a vertical asymptote of its graph.

Comparison Between Functions $y = \frac{1}{x^{2k}}$

It follows from the properties of powers that on the interval $(1, +\infty)$ we have $x^{2k} > x^{2m}$ when $k > m$, whence $y_k = \frac{1}{x^{2k}} < \frac{1}{x^{2m}} = y_m$. Besides, $\left(\frac{x_2}{x_1}\right)^{2k} > \left(\frac{x_2}{x_1}\right)^{2m}$ for any $x_2 > x_1 \geq 1$ and $k > m$, and consequently, $\frac{y_k(x_1)}{y_k(x_2)} = \left(\frac{x_2}{x_1}\right)^{2k} > \left(\frac{x_2}{x_1}\right)^{2m} = \frac{y_m(x_1)}{y_m(x_2)}$. This shows that on the interval $(1, +\infty)$ the function $y_k = \frac{1}{x^{2k}}$ has smaller values and decreases faster than $y_m = \frac{1}{x^{2m}}$ if $k > m$. On the interval $(0, 1)$ we have $x^{2k} < x^{2m}$ when $k > m$, that is, $y_k = \frac{1}{x^{2k}}$ has greater values than $y_m = \frac{1}{x^{2m}}$. Since the functions are even, similar relationships hold on the intervals $(-\infty, -1)$ and $(-1, 0)$. At the points $x = -1, 1$ the values of all the functions of this type coincide and are equal to $y = 1$. This comparison is illustrated in Fig. 3.19.

The functions $y = \frac{a}{x^n}$, $a > 0$, $n \in \mathbb{N}$ behave in a similar manner as $y = \frac{1}{x^n}$ and their graphs can be obtained using vertical extension by the factor a of the graph of $\frac{1}{x^n}$ (recall that this extension is understood as stretching a times from the x-axis if $a > 1$ and compressing $\frac{1}{a}$ times if $0 < a < 1$). If $a < 0$, we should add the reflection of the graph about the x-axis with the corresponding change of the properties.

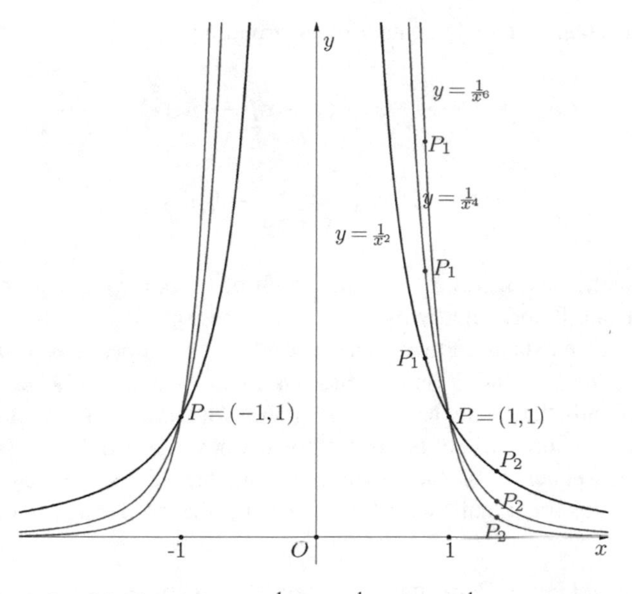

Fig. 3.19 The graphs of the functions $y=\frac{1}{x^2}$, $y=\frac{1}{x^4}$ and $y=\frac{1}{x^6}$

2.2 Fractional Linear Functions

Definition A *fractional linear function* is the ratio of two linear functions in the form $y=\frac{ax+b}{cx+d}$. It is usually assumed that $ad\neq bc$ (otherwise the function becomes a constant) and $c\neq 0$ (otherwise the function is linear). In the case $a=d=0$ and $b=c\neq 0$ we have the already studied function $y=\frac{1}{x}$.

Study of fractional linear functions $y=\frac{ax+b}{cx+d}$

To analyze different properties of this function it is convenient to use the following representation:

$$y=\frac{ax+b}{cx+d}=\frac{ax+b}{c\left(x+\frac{d}{c}\right)}=\frac{a\left(x+\frac{d}{c}\right)-\frac{ad}{c}+b}{c\left(x+\frac{d}{c}\right)}=\frac{a}{c}+\frac{bc-ad}{c^2\left(x+\frac{d}{c}\right)}=y_0+\frac{k}{x-x_0},$$

where $y_0=\frac{a}{c}$, $x_0=-\frac{d}{c}$ and $k=\frac{bc-ad}{c^2}$. This form indicates that this function has an intimate relationship with the hyperbola $y=\frac{1}{x}$. More specifically, the function $y=y_0+\frac{k}{x-x_0}$ can be obtained from the simpler function (hyperbola) $y=\frac{1}{x}$ using the following chain of the elementary transformations:

$$h(x) = \frac{1}{x} \to h_1(x) = \frac{1}{x - x_0} = h(x - x_0) \to h_2(x) = \frac{k}{x - x_0} = kh_1(x)$$

$$\to f(x) = y_0 + \frac{k}{x - x_0} = h_2(x) + y_0.$$

Hence, geometrically, starting from the graph of the known function $h(x) = \frac{1}{x}$, we first translate it horizontally by x_0 units (to the right if $x_0 > 0$ or to the left if $x_0 < 0$); then, we extend vertically the second graph k times if $k > 0$ or extend it vertically $|k|$ times and additionally reflect it about the x-axis if $k < 0$; finally, we translate vertically the graph of $h_2(x)$ by y_0 units (upward if $y_0 > 0$ or downward if $y_0 < 0$). See the illustration of these transformations in Fig. 3.20.

Using this sequence of the geometric transformations, we can deduce the properties of the fractional linear function from the properties of the hyperbola $y = \frac{1}{x}$.

1. Domain and range. Recalling the results of the geometric transformations (see Sect. 10 of Chap. 2), we can assert that the domain $X_h = \mathbb{R}\backslash\{0\}$ and the range $Y_h = \mathbb{R}\backslash\{0\}$ of the function $h(x) = \frac{1}{x}$ turn into the domain $X = \mathbb{R}\backslash\{x_0\} = \mathbb{R}\backslash\{-\frac{d}{c}\}$ and the range $Y = \mathbb{R}\backslash\{y_0\} = \mathbb{R}\backslash\{\frac{a}{c}\}$ of $f(x)$.
2. Boundedness. From the specification of the range it follows that the function is unbounded both below and above.

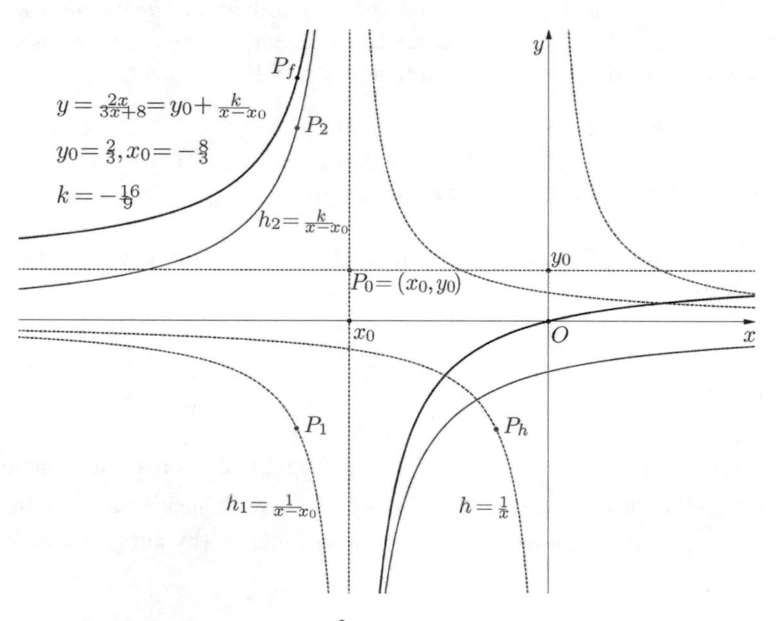

Fig. 3.20 The graph of the function $y = \frac{2x}{3x+8}$

3. Parity. Solving the relation of even functions $f(-x) = \frac{-ax+b}{-cx+d} = \frac{ax+b}{cx+d} = f(x), \forall x \in X$, we obtain the equality between the two polynomials: $(-ax + b)(cx + d) = (ax + b)(-cx + d)$. Recalling that two polynomials are equal if, and only if, the coefficients of the corresponding powers are the same, we obtain $bc - ad = -bc + ad$ (coefficient of x), whence $bc = ad$, that is, we arrive at the condition that was eliminated since the beginning (because it reduces the function to a constant). Therefore, the function is not even. Analogously, solving the relation of odd functions $f(-x) = \frac{-ax+b}{-cx+d} = -\frac{ax+b}{cx+d} = -f(x), \forall x \in X$, we obtain the following equality between the polynomials: $(-ax + b)(cx + d) = -(ax + b)(-cx + d)$. Again equating the coefficients of the corresponding powers, we obtain $ac = 0$ (coefficient of x^2) and $bd = 0$ (constant term). From the beginning it was posed the restriction $c \neq 0$, which implies that $ac = 0$ is reduced to $a = 0$. Consequently, it should be $b \neq 0$ (otherwise, the function $f(x)$ is simplified to a constant function eliminated from consideration), and therefore, $d = 0$. Hence, the function is odd only if it has the form $y = \frac{b}{cx}$. In all other cases the function is neither even nor odd.

   However, the function possesses the symmetry revealed by the representation $y = y_0 + \frac{k}{x-x_0}$, $y_0 = \frac{a}{c}$, $x_0 = -\frac{d}{c}$ and $k = \frac{bc-ad}{c^2}$. Indeed, we have $f(x_0 + x) = y_0 + \frac{k}{x_0+x-x_0} = y_0 + \frac{k}{x}$ and also $f(x_0 - x) = y_0 + \frac{k}{x_0-x-x_0} = y_0 - \frac{k}{x}$. This means that the points $P_1 = (x_0 + x, y_0 + \frac{k}{x})$ and $P_2 = (x_0 - x, y_0 - \frac{k}{x})$ belong to the graph of the function. Notice that these two points are symmetric with respect to the point $P_0 = (x_0, y_0)$, because their coordinates satisfy the conditions of the symmetry (see formulas (1.9) of Sect. 7.4 in Chap. 1 with $x_1 = x$ and $y_1 = \frac{k}{x}$). Since this symmetry takes place for each x of the domain, the graph is symmetric about the point $P_0 = (x_0, y_0)$.

4,5. Monotonicity and extrema, concavity and inflection. Again we use the representation $y = y_0 + \frac{k}{x-x_0}$, $y_0 = \frac{a}{c}$, $x_0 = -\frac{d}{c}$, $k = \frac{bc-ad}{c^2}$ and the sequence of geometric transformations starting from the hyperbola $h(x) = \frac{1}{x}$ and ending with $f(x) = y_0 + \frac{k}{x-x_0}$. Using the results of the transformation of graphs (see Sect. 10 of Chap. 2), we can assert that in the case when $k > 0$, the monotonicity and concavity of $h(x) = \frac{1}{x}$ are maintained on the corresponding intervals of $f(x)$, namely: since $h(x) = \frac{1}{x}$ is decreasing and concave downward on the interval $(-\infty, 0)$, the function $f(x)$ is decreasing and concave downward on the interval $(-\infty, x_0)$; since $h(x) = \frac{1}{x}$ is decreasing and concave upward on the interval $(0, +\infty)$, the function $f(x)$ is decreasing and concave upward on the interval $(x_0, +\infty)$. In the case when $k < 0$, the monotonicity and concavity of $h(x) = \frac{1}{x}$ are inverted under the transformation into $f(x)$, namely: since $h(x) = \frac{1}{x}$ is decreasing and concave downward on $(-\infty, 0)$, the function $f(x)$ is increasing and concave upward on $(-\infty, x_0)$; since $h(x) = \frac{1}{x}$ is decreasing and concave upward on $(0, +\infty)$, the function $f(x)$ is increasing and concave downward on $(x_0, +\infty)$. Finally, since $h(x) = \frac{1}{x}$ has no extrema and inflection points, the same is true for $f(x) = \frac{ax+b}{cx+d}$.

6. Periodicity. Monotonicity on the infinite interval $(-\frac{d}{c}, +\infty)$ indicates that the function cannot be periodic. (The same can be derived directly from the geometric transformations.)
7. Intersection and sign. If $a = 0$, then the graph does not cross the x-axis, but if $a \neq 0$, then there is the only point of intersection $(-\frac{b}{a}, 0)$ with the x-axis. If $d = 0$, the graph does not cross the y-axis (in this case the point $x = 0$ is out of the domain), but if $d \neq 0$, then intersection with the y-axis occurs at the only point $(0, \frac{b}{d})$.

   Employing the formula $y = y_0 + \frac{k}{x-x_0}$, we make the following conclusions regarding the values of the function:

   (1) if $k > 0$, then $y > y_0$ when $x > x_0$, and $y < y_0$ when $x < x_0$;
   (2) if $k < 0$, then $y < y_0$ when $x > x_0$, and $y > y_0$ when $x < x_0$.

8. Inverse. The function $f(x) = \frac{ax+b}{cx+d} : X = \mathbb{R}\backslash\{-\frac{d}{c}\} \to Y = \mathbb{R}\backslash\{\frac{a}{c}\}$ is surjective and injective. The latter property is the consequence of monotonicity on the intervals $(-\infty, -\frac{d}{c})$ and $(-\frac{d}{c}, +\infty)$ and of the fact that all the values of the function on one of these two intervals are greater than all the values on another interval. Therefore, the function $f(x)$ is bijective and has the inverse, which is also a fractional linear function $f^{-1}(x) = \frac{dx-b}{a-cx} : \mathbb{R}\backslash\{\frac{a}{c}\} \to \mathbb{R}\backslash\{-\frac{d}{c}\}$.
9. Based on the investigated properties we can plot the graph of the function. An example of such function $y = \frac{2x}{3x+8} = \frac{2}{3} - \frac{16/9}{x+8/3}$ is shown in Fig. 3.20. In this case, $x_0 = -\frac{8}{3}$, $y_0 = \frac{2}{3}$, $k = -\frac{16}{9}$.

Complimentary Properties of the Function $y = \frac{ax+b}{cx+d}$

Using the form $y = y_0 + \frac{k}{x-x_0}$, where $y_0 = \frac{a}{c}$, $x_0 = -\frac{d}{c}$ and $k = \frac{bc-ad}{c^2}$ and recalling the complimentary properties of the hyperbola $y = \frac{1}{x}$, we can deduce that the function $y = \frac{ax+b}{cx+d}$ approaches y_0 when x goes to $+\infty$ or to $-\infty$, which implies that $y = y_0$ is a horizontal asymptote of $f(x)$.

Notice also that $y = y_0 + \frac{k}{x-x_0}$ does not converges to infinity at any point of its domain. The only point at which the function goes to infinity is $x = x_0$, but this point is out of the domain. Making use of the properties of $y = \frac{1}{x}$, we conclude that if $k > 0$, then $y = y_0 + \frac{k}{x-x_0}$ goes to $+\infty$ when x approaches x_0 from the right (by the values greater than x_0) and to $-\infty$ when x approaches x_0 from the left (by the values smaller than x_0); in the case $k < 0$ the situation is opposite: the function goes to $-\infty$ when x approaches x_0 from the right and to $+\infty$ when x approaches x_0 from the left. The existence of at least one of the convergence to infinity ($+\infty$ or $-\infty$) under one-sided approximation of x_0 is already sufficient to assert that $x = x_0$ is a vertical asymptote. As we have seen, in the case of $y = y_0 + \frac{k}{x-x_0}$ the function converges to this asymptote from both sides.

2.3 Some General Properties of Rational Functions

The study of rational functions in a general form can be quite complicated and in many cases all the properties cannot be revealed even employing more advanced tools of Calculus and Analysis.

However, unexpectedly at first glance, the properties of qualitative behavior at infinity and convergence to infinity at finite points can be described in different general cases, albeit in an approximated form accessible for the level of this text.

1. Behavior at Infinity

It was discussed in Sect. 1.3 that the properties of a general polynomial $P_n(x) = a_n x^n + a_{n-1}x^{n-1} + \ldots + a_0$ can be difficult to determine on the finite intervals, but its behavior at infinity is defined only by the leading term $a_n x^n$, which converges to infinity as $|x|$ goes to $+\infty$. Recall that it was shown in Sect. 1.3 that $\lim\limits_{|x|\to\infty} \frac{1}{|x|^k} = 0$, $\forall k \in \mathbb{N}$, and consequently, $\lim\limits_{|x|\to\infty} \frac{a_k}{a_n}\frac{1}{|x|^k} = 0$ for any coefficients a_k and a_n. These results were used to deduce that among all the terms of a polynomial in the representation $P_n(x) = a_n x_n \left(1 + \frac{a_{n-1}}{a_n}\frac{1}{x} + \ldots + \frac{a_0}{a_n}\frac{1}{x^n}\right)$, the only summand inside parentheses which does not converge to 0 as $|x|$ goes to infinity is the constant 1. For this reason, the behavior of $P_n(x)$ at infinity is determined by the leading term $a_n x^n$, that is, $\lim\limits_{|x|\to\infty} P_n(x) = \lim\limits_{|x|\to\infty} a_n x^n$.

Taking this into account, we can guess that a behavior of a rational function $R(x) = \frac{P_n(x)}{Q_m(x)}$ at infinity is determined by the leading terms of the polynomials $P_n(x)$ and $Q_m(x)$, which indeed happens. To perform an approximate analysis, we represent both polynomials in the form

$$P_n(x) = a_n x^n + a_{n-1}x^{n-1} + \ldots + a_0 = a_n x^n \left(1 + \frac{a_{n-1}}{a_n}\frac{1}{x} + \ldots + \frac{a_0}{a_n}\frac{1}{x^n}\right)$$

and

$$Q_m(x) = b_m x^m + b_{m-1}x^{m-1} + \ldots + b_0 = b_m x^m \left(1 + \frac{b_{m-1}}{b_m}\frac{1}{x} + \ldots + \frac{b_0}{b_m}\frac{1}{x^m}\right),$$

where the leading terms $a_n x^n$ and $b_m x^m$ determine the behavior of $P_n(x)$ and $Q_m(x)$ at infinity (see details in Sect. 1.3). Then, for a rational function we obtain

$$R(x) = \frac{P_n(x)}{Q_m(x)} = \frac{a_n x^n \left(1 + \frac{a_{n-1}}{a_n}\frac{1}{x} + \ldots + \frac{a_0}{a_n}\frac{1}{x^n}\right)}{b_m x^m \left(1 + \frac{b_{m-1}}{b_m}\frac{1}{x} + \ldots + \frac{b_0}{b_m}\frac{1}{x^m}\right)}$$

$$= \frac{a_n}{b_m} x^{n-m} \frac{1 + \frac{a_{n-1}}{a_n}\frac{1}{x} + \ldots + \frac{a_0}{a_n}\frac{1}{x^n}}{1 + \frac{b_{m-1}}{b_m}\frac{1}{x} + \ldots + \frac{b_0}{b_m}\frac{1}{x^m}}.$$

Noting that each term $\frac{1}{|x|^k}$, $k \in \mathbb{N}$ becomes very small for sufficiently large values of $|x|$, that is, $\lim\limits_{|x|\to\infty} \frac{1}{|x|^k} = 0, \forall k \in \mathbb{N}$, we conclude that the main term in the numerator of the last fraction is 1 and all other terms are insignificant comparing with 1 when $|x|$ is sufficiently large. The same is true for the denominator. Therefore, the rational function $R(x)$ behaves at infinity in the same manner as the monomial $\frac{a_n}{b_m}x^{n-m}$, in other words, $\lim\limits_{|x|\to\infty} R(x) = \lim\limits_{|x|\to\infty} \frac{a_n}{b_m}x^{n-m}$. Taking advantage of the discussion about behavior of the terms $\frac{a_n}{b_m}x^{n-m}$ at infinity (see Sect. 1.3), we can specify the behavior of $R(x)$ at infinity as follows:

(1) if $n - m < 0$, then $\lim\limits_{|x|\to\infty} R(x) = \lim\limits_{|x|\to\infty} \frac{a_n}{b_m}x^{n-m} = 0$, which indicates that $y = 0$ is a horizontal asymptote of $R(x)$;
(2) if $n - m = 0$, then $\lim\limits_{|x|\to\infty} R(x) = \lim\limits_{|x|\to\infty} \frac{a_n}{b_m}x^{n-m} = \frac{a_n}{b_m}$, which implies that $y = \frac{a_n}{b_m}$ is a horizontal asymptote of $R(x)$;
(3) if $n - m > 0$, then $\lim\limits_{|x|\to\infty} R(x) = \lim\limits_{|x|\to\infty} \frac{a_n}{b_m}x^{n-m} = (\pm)\infty$, which means that there is no horizontal asymptote; in this last case, if $a_n b_m > 0$ and $n - m = k$ is even, then $\lim\limits_{x\to\pm\infty} R(x) = +\infty$; if $a_n b_m < 0$ and $n - m = k$ is even, then $\lim\limits_{x\to\pm\infty} R(x) = -\infty$; if $a_n b_m > 0$ and $n - m = k$ is odd, then $\lim\limits_{x\to+\infty} R(x) = +\infty$ and $\lim\limits_{x\to-\infty} R(x) = -\infty$; finally, if $a_n b_m < 0$ and $n - m = k$ is odd, then $\lim\limits_{x\to+\infty} R(x) = -\infty$ and $\lim\limits_{x\to-\infty} R(x) = +\infty$.

The three different cases of the relationship between powers of the numerator and denominator are shown in Figs. 3.21, 3.22, and 3.23.

It is worth noting that this analysis of a behavior of $R(x)$ is valid only when $|x|$ goes to infinity, that is, for sufficiently large values of $|x|$. It does not apply to values at finite points x or on finite intervals.

2. Convergence to Infinity

At each point of the domain of $R(x)$ the function does not tend to infinity. Therefore, a convergence to infinity can occur only at the points where the denominator vanishes and $R(x)$ is not defined. Let $x = c$ be a zero of degree $q \le m$ of the denominator $Q_m(x)$ and let us see what can happen with $R(x)$ near this point.

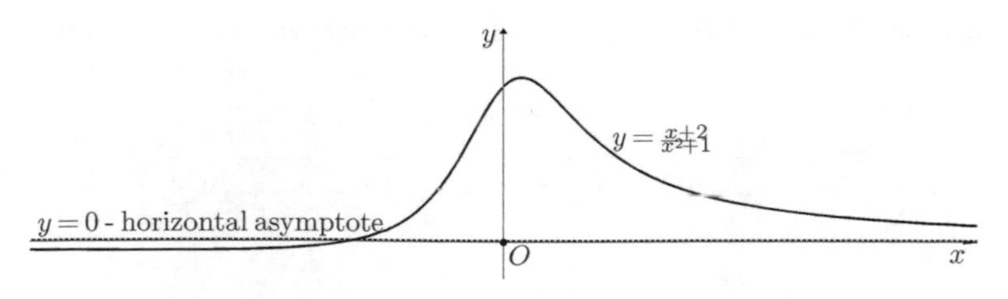

Fig. 3.21 The graph of the function $y = \frac{x+2}{x^2+1}$, case $n - m = 1 - 2 < 0$

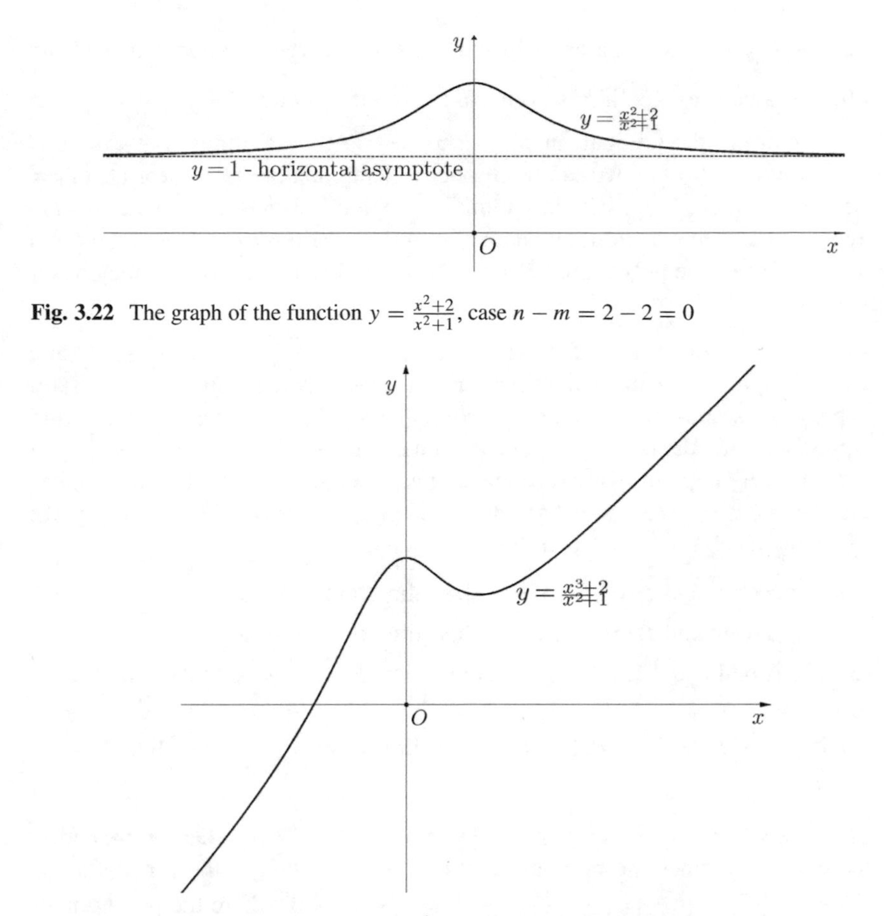

Fig. 3.22 The graph of the function $y = \frac{x^2+2}{x^2+1}$, case $n - m = 2 - 2 = 0$

Fig. 3.23 The graph of the function $y = \frac{x^3+2}{x^2+1}$, case $n - m = 3 - 2 > 0$

In the same way as the behavior of $R(x) = \frac{P_n(x)}{Q_m(x)}$ at infinity is determined by the term x^{n-m} when $|x|$ goes to ∞, the behavior of $R(x)$ near $x = c$ is defined by the term $\frac{1}{(x-c)^k}$, $k \in \mathbb{N}$ when x approaches c. Therefore, first we clarify what happens with this term near $x = c$. When x approaches c, the values of $x - c$ and also of $(x - c)^k$, $k \in \mathbb{N}$ become very small, and consequently, the fraction $\frac{1}{|x-c|^k}$ becomes very large exceeding any given constant, that is, it goes to infinity. To specify the considerations, assume, for the moment, that $x > c$ (in other words, x approaches c from the right). In this case, for any given beforehand constant $M > 0$, we can choose the values of x so close to c that $\frac{1}{(x-c)^k}$ becomes greater than M. In a bit more exact formulation we have: for any given $M > 0$, choosing $\delta = \frac{1}{M^{1/k}}$ we guarantee that for all x such that $c < x < c + \delta$ the inequality $\frac{1}{(x-c)^k} > M$ holds. This means that $\lim\limits_{x \to c^+} \frac{1}{(x-c)^k} = +\infty$. Analogously, we can see that when x approaches c from the left, $\frac{1}{(x-c)^k}$ goes to $+\infty$ if k is even and to $-\infty$ if k is odd.

Let us return to the rational function $R(x) = \frac{P_n(x)}{Q_m(x)}$. Assuming that $x = c$ is zero of degree q of $Q_m(x)$, we can represent the denominator in the form $Q_m(x) = (x-c)^q T_l(x)$, where $T_l(x)$ is a polynomial of degree $l = m - q$ whose zeros are different from c. The situation now depends on the relationship between the point $x = c$ and zeros of the polynomial $P_n(x)$. The three different situations may occur as shown below.

(1) If $x = c$ is not a zero of $P_n(x)$, then $R(x) = \frac{P_n(x)}{Q_m(x)} = \frac{P_n(x)}{(x-c)^q T_l(x)}$ and when x approaches c the denominator becomes very small, while the numerator approaches a non-zero number, and consequently, keeps its values away from zero. Therefore, the ratio $\frac{P_n(x)}{Q_m(x)}$ behaves near c in a similar manner as $\frac{1}{(x-c)^q}$: it becomes very large in absolute value, that is, $R(x)$ goes to infinity. This implies that the line $x = c$ is a vertical asymptote of $R(x)$. The form of this convergence to infinity can be specified as follows:

   (1) if q is even and $T_l(c)P_n(c) > 0$, then $\lim\limits_{x \to c} R(x) = +\infty$;
   (2) if q is even and $T_l(c)P_n(c) < 0$, then $\lim\limits_{x \to c} R(x) = -\infty$;
   (3) if q is odd and $T_l(c)P_n(c) > 0$, then $\lim\limits_{x \to c^+} R(x) = +\infty$ and $\lim\limits_{x \to c^-} R(x) = -\infty$;
   (4) if q is odd and $T_l(c)P_n(c) < 0$, then $\lim\limits_{x \to c^+} R(x) = -\infty$ and $\lim\limits_{x \to c^-} R(x) = +\infty$.

   The case when $x = c$ is a zero of degree $p < q$ of $P_n(x)$ can be treated in the same way, since the rational function can be written in the form $R(x) = \frac{P_n(x)}{Q_m(x)} = \frac{(x-c)^p S_e(x)}{(x-c)^q T_l(x)} = \frac{S_e(x)}{(x-c)^{q-p} T_l(x)}$, $k = q - p > 0$, where the polynomials $S_e(x)$ and $T_l(x)$ do not vanish at c, and consequently, keep their values away from 0 when x approaches c.

(2) The next case is when $x = c$ is zero of degree $p = q$ of $P_n(x)$. Then $R(x) = \frac{P_n(x)}{Q_m(x)} = \frac{(x-c)^q S_e(x)}{(x-c)^q T_l(x)} = \frac{S_e(x)}{T_l(x)}$, where the polynomials $S_e(x)$ and $T_l(x)$ do not vanish at c, and consequently, keep their values away from 0 when x is close to c. Therefore, $R(x)$ does not increase its absolute values without restriction, but simply approaches the ratio of the polynomials $S_e(x)$ and $T_l(x)$ at c: $\lim\limits_{x \to c} R(x) = \frac{S_e(c)}{T_l(c)}$. In this case, $R(x)$ does not have vertical asymptote at the point c.

(3) Finally, if $x = c$ is zero of degree $p > q$ $(p \le n)$ of $P_n(x)$, then $R(x) = \frac{P_n(x)}{Q_m(x)} = \frac{(x-c)^p S_e(x)}{(x-c)^q T_l(x)} = \frac{(x-c)^{p-q} S_e(x)}{T_l(x)}$, $p - q > 0$, where the polynomials $S_e(x)$ and $T_l(x)$ do not vanish at c, and consequently, their values stay away from 0 when x is close to c. Then, the ratio $\frac{S_e(x)}{T_l(x)}$ approaches $\frac{S_e(c)}{T_l(c)}$, while $(x-c)^{p-q}$ approaches 0. Therefore, $\lim\limits_{x \to c} R(x) = 0 \cdot \frac{S_e(c)}{T_l(c)} = 0$. In this case, $R(x)$ does not have vertical asymptote at the point c.

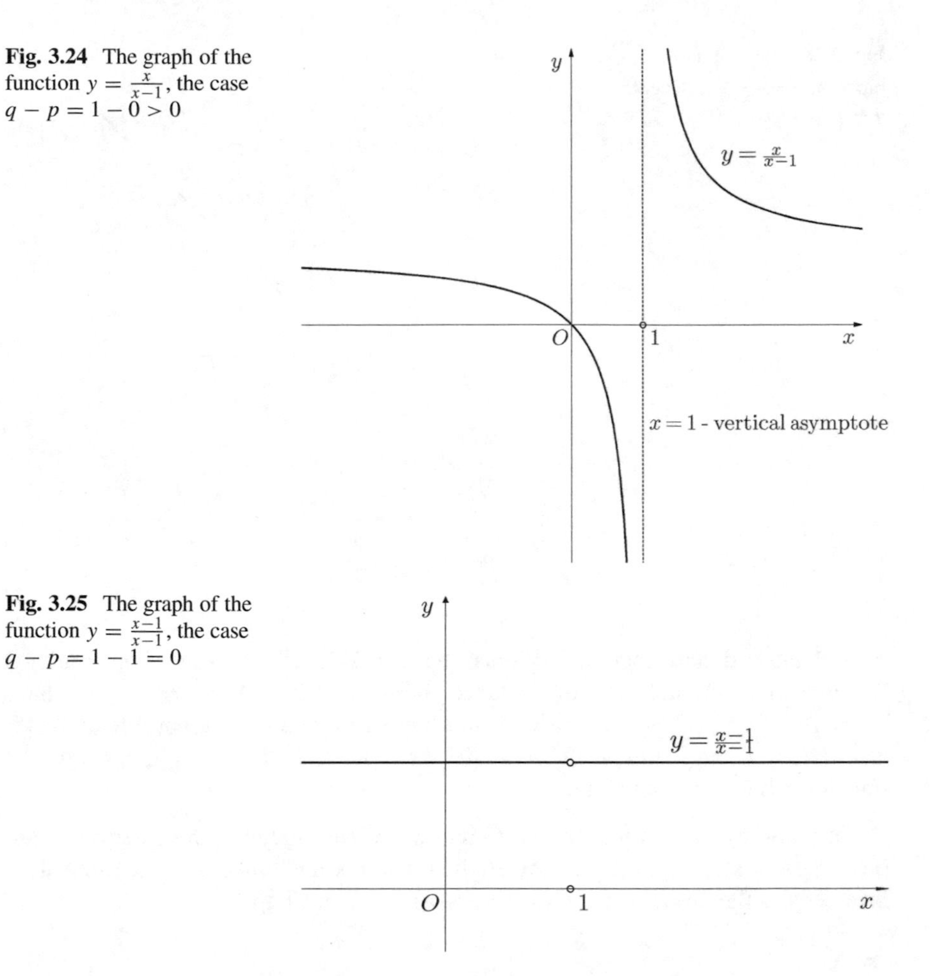

Fig. 3.24 The graph of the function $y = \frac{x}{x-1}$, the case $q - p = 1 - 0 > 0$

Fig. 3.25 The graph of the function $y = \frac{x-1}{x-1}$, the case $q - p = 1 - 1 = 0$

Notice that in the last two cases the rational function $R(x)$ is not defined at the point $x = c$, but it tends to a specific number as x approaches c.

Elementary illustrations of the three different cases of the behavior of $R(x)$ when x approaches one of zeros of the denominator are shown in Figs. 3.24, 3.25, and 3.26.

3 Irrational Functions

Definition of an Irrational Function In general form, an *irrational function* is defined by the formula $y = \sqrt[n]{R(x)}$, where $R(x)$ is a rational function and $n \in \mathbb{N}$ is degree of the root. Obviously, this definition is a generalization of rational functions, which are a particular case of this definition when $n = 1$. Notice that even when

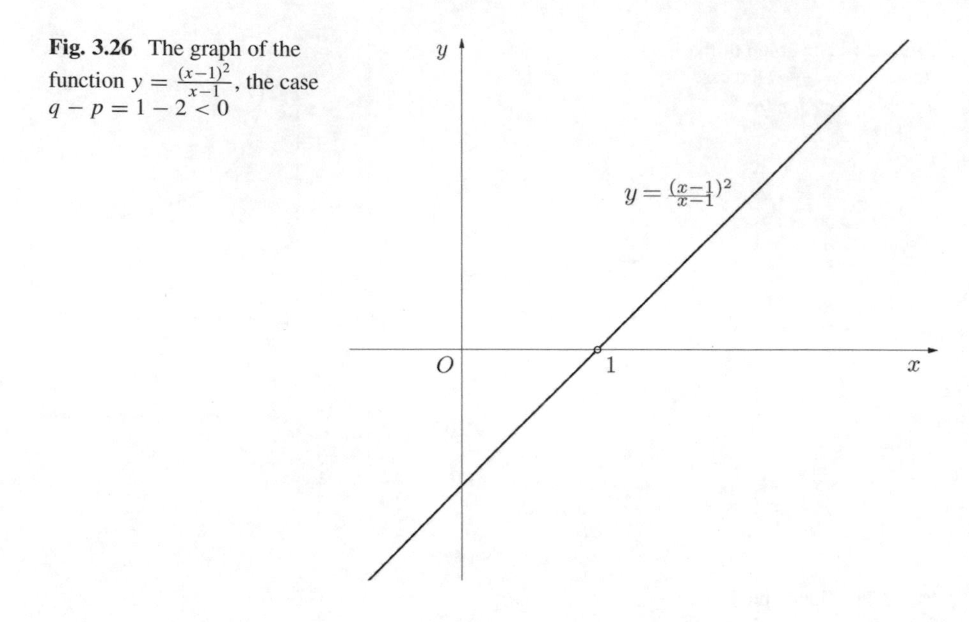

Fig. 3.26 The graph of the function $y = \frac{(x-1)^2}{x-1}$, the case $q - p = 1 - 2 < 0$

$n > 1$ the rational function $R(x)$ may have a form which reduces an irrational function to a rational one: for instance, if $n = 3$ and $R(x) = x^3$, we have $y = \sqrt[3]{x^3} = x$. Of course, this does not happen with all irrational functions: for example, if $n = 3$ and $R(x) = x$, the irrational function $y = \sqrt[3]{x}$ cannot be transformed into a rational one.

The simplest way to introduce different irrational functions and determine their properties is starting with already studied polynomial functions and using their inverses and the properties of these inverses proved in Chap. 2.

3.1 Roots $y = \sqrt[n]{x}$, $n \in \mathbb{N}$

Definition of the Function $y = \sqrt{x}$ The function $y = \sqrt{x}$ can be defined as the inverse of the function $f(x) = x^2 : [0, +\infty) \to [0, +\infty)$.

Study of $y = \sqrt{x}$

1. Domain and range. By the definition, the domain of $y = \sqrt{x}$ is $X = [0, +\infty)$ (the range of the original function $y = x^2$) and the range is $Y = [0, +\infty)$ (the domain of the original function). Notice that the same result we obtain applying the definition of the operation of a square root. Indeed, first, the square root is defined for any non-negative number and is not defined for a negative number, which means that the domain of the function is $X = [0, +\infty)$. Second, by the

definition, $\sqrt{x}$, $\forall x \geq 0$ is such a non-negative number y whose square is equal to x: $y^2 = x, y \geq 0$. This implies that the range of the function is a subset of $[0, +\infty)$, and the fact that the squares of non-negative numbers take all non-negative values (in other words, the function $x = y^2$ has the image $[0, +\infty)$) guarantees that the range of the function is $[0, +\infty)$.
2. Boundedness. From the specification of the range, it follows that the function is bounded below (for instance, by $m = 0$) and not bounded above.
3. Parity. The function is not even or odd, because its domain is not symmetric about the origin.
4. Periodicity. The function is not periodic, because its domain is bounded on the left. (Another reason why the function cannot be periodic is that it is the inverse of a function, and therefore, all its values are distinct.)
5. Monotonicity and extrema. Since the function $y = x^2$ increases on $[0, +\infty)$, by the Theorem of monotonicity of the inverse (Sect. 12.4 in Chap. 2), the function $y = \sqrt{x}$ increases over the entire domain $X = [0, +\infty)$.

   Since each point $x \neq 0$ is increasing, none of these points can be extremum of any type. The only remaining point $x = 0$ is a strict global minimum, because $f(x) = \sqrt{x} > 0 = f(0)$, $\forall x > 0$. Notice that $x = 0$ is not a local minimum, because there is no a neighborhood of this point contained in the domain.
6. Concavity and inflection. Since the function $y = x^2$ is increasing and has upward concavity on $[0, +\infty)$, by Theorem of concavity of the inverse (Sect. 12.4 in Chap. 2), the function $y = \sqrt{x}$ is concave downward on its domain $X = [0, +\infty)$. Hence, there is no inflection point.
7. Intersection and sign. Intersection with the y-axis occurs at the origin and the same is true for intersection with the x-axis. There are no other points of intersection with the x-axis, because $f(x) = \sqrt{x} > 0$, $\forall x > 0$.
8. Inverse. The function $f(x) = \sqrt{x} : X = [0, +\infty) \to Y = [0, +\infty)$ is bijective and its inverse is $f^{-1}(x) = x^2 : [0, +\infty) \to [0, +\infty)$.
9. Based on the investigated properties, we can sketch the graph of the function $y = \sqrt{x}$ as shown in Fig. 3.27. Alternatively, we can obtain the same graph by applying the Theorem about the graph of the inverse (Sect. 12.5 in Chap. 2): we need to reflect the known graph of the function $y = x^2 : [0, +\infty) \to [0, +\infty)$ with respect to the line $y = x$.

Complimentary Properties of $y = \sqrt{x}$

The graph of the function suggests that $y = \sqrt{x}$ increases without restriction when x goes to $+\infty$. Indeed, for any given constant $M > 0$, choosing $x > M^2$ we ensure that $\sqrt{x} > \sqrt{M^2} = M$, that is, the function also goes to $+\infty$. In this way, we confirm the behavior at infinity observed in Fig. 3.27. Since the function is not defined for negative values, there is no way to send x to $-\infty$ using the points of the domain. Hence, the function $y = \sqrt{x}$ has no horizontal asymptote.

Since the function $y = \sqrt{x}$ has no infinite limit at the points of the domain $X = [0, +\infty)$, and any point out of the domain cannot be approached by the points of the domain, the graph of $y = \sqrt{x}$ has no vertical asymptote.

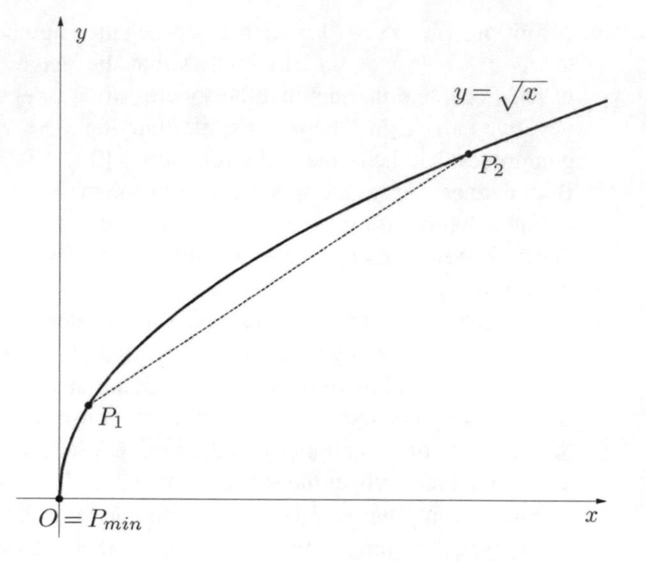

Fig. 3.27 The graph of the function $y = \sqrt{x}$

Definition of the Function $y = \sqrt[3]{x}$ The function $y = \sqrt[3]{x}$ can be defined as the inverse to the function $f(x) = x^3 : \mathbb{R} \to \mathbb{R}$.

Study of $y = \sqrt[3]{x}$

1. Domain and range. By the definition, the domain of the function $y = \sqrt[3]{x}$ is $X = \mathbb{R}$ (the range of the original function $y = x^3$) and the range is $Y = \mathbb{R}$ (the domain of the original function). Notice that the same conclusion can be obtained using the definition of the operation of a cubic root. Indeed, a cubic root is defined for any real number, which means that the domain is $X = \mathbb{R}$. The result of this operation is such a real number y whose cube is equal to x: $y^3 = x$. This shows that the range contains both positive and negative numbers, and the fact that cubes of real numbers can take any real value guarantees that the range is the entire real axis $\mathbb{R}$.
2. Boundedness. It follows from the specification of the range that the function is unbounded both below and above.
3. Parity. The function $y = \sqrt[3]{x}$ is odd because $y = x^3$ is an odd function. Another way to arrive at the same result is by applying the definition of the operation of a cubic root.
4. Periodicity. The function is not periodic, since it is defined as the inverse of a function, and consequently, all its values are distinct. (Another proof of non-periodicity follows from the property of the increase shown in the next point of the investigation.)

5. Monotonicity and extrema. Since the function $y = x^3$ is increasing in $\mathbb{R}$, by the Theorem of monotonicity of the inverse (Sect. 12.4 in Chap. 2) the function $y = \sqrt[3]{x}$ is increasing over the entire domain $X = \mathbb{R}$.
   Since each point x is increasing, there is no extremum of any type.
6. Concavity and inflection. Since the function $y = x^3$ is increasing and concave downward on $(-\infty, 0]$ (whose image under $y = x^3$ is $(-\infty, 0]$), by the Theorem of concavity of the inverse (Sect. 12.4 in Chap. 2), the function $y = \sqrt[3]{x}$ is concave upward on $(-\infty, 0]$. Analogously, since the function $y = x^3$ is increasing and concave upward on $[0, +\infty)$ (whose image under $y = x^3$ is $[0, +\infty)$), by the same Theorem of concavity of the inverse, the function $y = \sqrt[3]{x}$ is concave downward on $[0, +\infty)$.
   Since the function changes the type of concavity at 0 (and is defined at this point), this is an inflection point.
7. Intersection and sign. Intersection with the y-axis occurs at the origin and this point is also the only point of intersection with the x-axis. There are no other points of intersection with the x-axis, because $f(x) = \sqrt[3]{x} < 0, \forall x < 0$ and $f(x) = \sqrt[3]{x} > 0, \forall x > 0$.
8. Inverse. The function $f(x) = \sqrt[3]{x} : X = \mathbb{R} \to Y = \mathbb{R}$ is bijective and its inverse is $f^{-1}(x) = x^3 : \mathbb{R} \to \mathbb{R}$.
9. Based on these properties, we can plot the graph of the function $y = \sqrt[3]{x}$ as shown in Fig. 3.28. Alternatively, the same graph can be constructed using the known graph of the function $y = x^3 : \mathbb{R} \to \mathbb{R}$: by the Theorem about the graph of the inverse (Sect. 12.5 in Chap. 2) we need to reflect the graph of $y = x^3 : \mathbb{R} \to \mathbb{R}$ with respect to the line $y = x$.

Complimentary Properties of $y = \sqrt[3]{x}$
The shape of the graph of $y = \sqrt[3]{x}$ suggests that the function increases without restriction when x goes to $+\infty$. Indeed, for any given constant $M > 0$, choosing $x > M^3$ we ensure that $\sqrt[3]{x} > \sqrt[3]{M^3} = M$, that is, the values of the function become arbitrarily large. Since the function is odd, we can conclude that when x

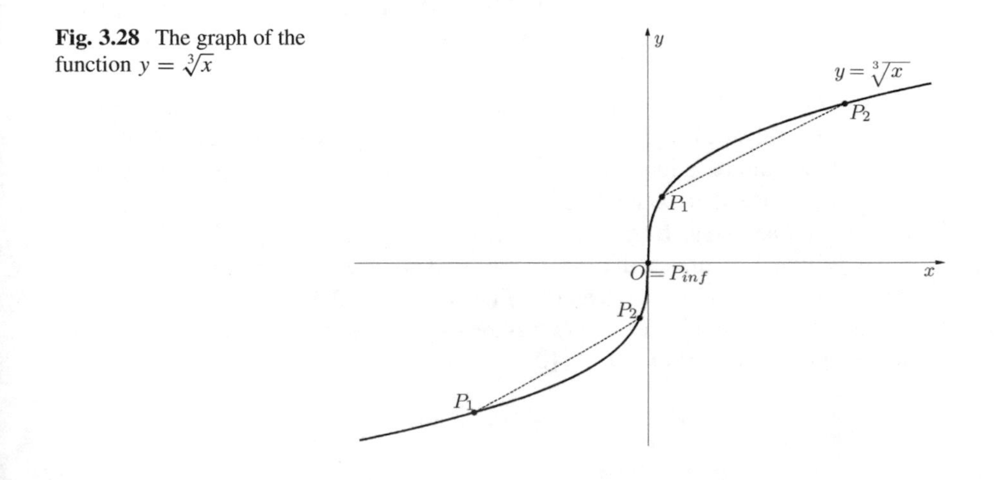

Fig. 3.28 The graph of the function $y = \sqrt[3]{x}$

goes to $-\infty$ the function converges to $-\infty$. Therefore, the function $y = \sqrt[3]{x}$ has no horizontal asymptote. In this way, we confirm the behavior at infinity observed in Fig. 3.28.

Since the function $y = \sqrt[3]{x}$ has no infinite limit at the points of the domain $X = \mathbb{R}$, its graph has no vertical asymptote.

In a similar manner the inversion of the functions $y = x^{2k}$, $k \in \mathbb{N}$ generates the root functions of even degree and of the functions $y = x^{2k+1}$, $k \in \mathbb{N}$ the root functions of odd degree. We present below just a brief discussion of the properties of $y = \sqrt[2k]{x}$ and $y = \sqrt[2k+1]{x}$, since the investigation of their properties follows the same reasoning as for the functions $y = \sqrt{x}$ and $y = \sqrt[3]{x}$.

Definition of the Function $y = \sqrt[2k]{x}$, $k \in \mathbb{N}$ The function $y = \sqrt[2k]{x}$ can be defined as the inverse of $f(x) = x^{2k} : [0, +\infty) \to [0, +\infty)$.

Study of $y = \sqrt[2k]{x}$, $k \in \mathbb{N}$

1. Domain and range. By the definition, the domain of the function $y = \sqrt[2k]{x}$ is $X = [0, +\infty)$ (the range of the original function $y = x^{2k}$) and the range is $Y = [0, +\infty)$ (the domain of the original function).
2. Boundedness. The specification of the range implies that the function is bounded below, but unbounded above.
3. Parity. The function is neither even nor odd, since its domain is not symmetric about the origin.
4. Periodicity. The function is not periodic, since its domain is bounded on the left.
5. Monotonicity and extrema. Since the function $y = x^{2k}$ increases on $[0, +\infty)$, by the Theorem of monotonicity of the inverse (Sect. 12.4 in Chap. 2), the function $y = \sqrt[2k]{x}$ increases over the entire domain $X = [0, +\infty)$.

   Since each point $x \neq 0$ is increasing, none of these points can be extremum of any type. The only remaining point is $x = 0$, which is a strict global minimum because $f(x) = \sqrt[2k]{x} > 0 = f(0)$, $\forall x > 0$. Notice that $x = 0$ is not a local minimum, because there is no a neighborhood of this point contained in the domain.
6. Concavity and inflection. Since the function $y = x^{2k}$ is increasing and concave upward on $[0, +\infty)$, by the Theorem of concavity of the inverse (Sect. 12.4 in Chap. 2), the function $y = \sqrt[2k]{x}$ is concave downward on its domain $X = [0, +\infty)$. Hence, there is no inflection point.
7. Intersection and sign. Intersection with the y-axis occurs at the origin and the same is true for intersection with the x-axis. There are no other points of intersection with the x-axis, because $f(x) = \sqrt[2k]{x} > 0$, $\forall x > 0$.
8. Inverse. The function $f(x) = \sqrt[2k]{x} : X = [0, +\infty) \to Y = [0, +\infty)$ is bijective and its inverse is $f^{-1}(x) = x^{2k} : [0, +\infty) \to [0, +\infty)$.

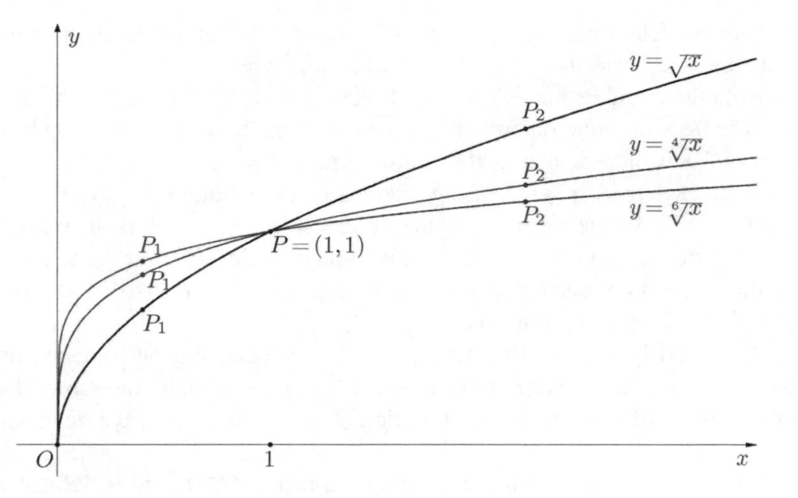

Fig. 3.29 The graphs of the functions $y = \sqrt{x}$, $y = \sqrt[4]{x}$ and $y = \sqrt[6]{x}$

9. Based on the investigated properties (or using the Theorem about the graph of the inverse), we can sketch the graph of the function $y = \sqrt[2k]{x}$, which has the same shape as that of $y = \sqrt{x}$.

Comparison Between the Functions $y = \sqrt[2k]{x}$
It follows from the properties of the roots that on the interval $(1, +\infty)$ we have $y_k = \sqrt[2k]{x} < \sqrt[2m]{x} = y_m$ when $k > m$. Besides, for any pair $x_2 > x_1 \geq 1$ and $k > m$, we get $\frac{y_k(x_2)}{y_k(x_1)} = \sqrt[2k]{\frac{x_2}{x_1}} < \sqrt[2m]{\frac{x_2}{x_1}} = \frac{y_m(x_2)}{y_m(x_1)}$. This shows that on the interval $(1, +\infty)$ the graph of the function $y_k = \sqrt[2k]{x}$ lies below and increases slower than the graph of $y_m = \sqrt[2m]{x}$ if $k > m$. On the interval $(0, 1)$ we have $y_k = \sqrt[2k]{x} > \sqrt[2m]{x} = y_m$ when $k > m$, that is, the graph of $y_k = \sqrt[2k]{x}$ lies above of the graph of $y_m = \sqrt[2m]{x}$. At the points $x = 0, 1$ the values of all the functions $y = \sqrt[2k]{x}$ coincide and are equal to $y = 0, 1$, respectively. This comparison is illustrated in Fig. 3.29.

Definition of the Function $y = \sqrt[2k+1]{x}$, $k \in \mathbb{N}$**.** The function $y = \sqrt[2k+1]{x}$ can be defined as the inverse of the function $f(x) = x^{2k+1} : \mathbb{R} \to \mathbb{R}$.

Study of $y = \sqrt[2k+1]{x}$, $k \in \mathbb{N}$

1. Domain and range. By the definition, the domain of $y = \sqrt[2k+1]{x}$ is $X = \mathbb{R}$ (the range of the original function $y = x^{2k+1}$) and the range is $Y = \mathbb{R}$ (the domain of the original function).
2. Boundedness. It follows from the specification of the range that the function is unbounded both below and above.
3. Parity. The function is odd since $y = x^{2k+1}$ is odd.

4. Periodicity. The function is not periodic, since it is defined as the inverse of a function, and consequently, all its values are distinct.
5. Monotonicity and extrema. Since the function $y = x^{2k+1}$ is increasing on $\mathbb{R}$, by the Theorem of monotonicity of the inverse (Sect. 12.4 in Chap. 2) the function $y = \sqrt[2k+1]{x}$ is increasing over the entire domain $X = \mathbb{R}$.

   Since each point x is increasing, there is no extremum of any type.
6. Concavity and inflection. Since the function $y = x^{2k+1}$ is increasing and concave downward on $(-\infty, 0]$ (whose image under $y = x^{2k+1}$ is $(-\infty, 0]$), by the Theorem of concavity of the inverse (Sect. 12.4 in Chap. 2), the function $y = \sqrt[2k+1]{x}$ is concave upward on $(-\infty, 0]$.

   Analogously, since the function $y = x^{2k+1}$ is increasing and concave upward on $[0, +\infty)$ (whose image under $y = x^{2k+1}$ is $[0, +\infty)$), by the same Theorem of concavity of the inverse, the function $y = \sqrt[2k+1]{x}$ is concave downward on $[0, +\infty)$.

   Since the function changes the type of concavity at 0 (and is defined at this point), this is an inflection point.
7. Intersection and sign. Intersection with the y-axis occurs at the origin and this point is also the only point of intersection with the x-axis. There are no other points of intersection with the x-axis, because $f(x) = \sqrt[2k+1]{x} < 0, \forall x < 0$ and $f(x) = \sqrt[2k+1]{x} > 0, \forall x > 0$.
8. Inverse. The function $f(x) = \sqrt[2k+1]{x} : X = \mathbb{R} \to Y = \mathbb{R}$ is bijective and its inverse is $f^{-1}(x) = x^{2k+1} : \mathbb{R} \to \mathbb{R}$.
9. Based on the investigated properties (or using the Theorem about the graph of the inverse), we can sketch the graph of the function $y = \sqrt[2k+1]{x}$, which has the same shape as that of $y = \sqrt[3]{x}$.

Comparison Between the Functions $y = \sqrt[2k+1]{x}$

It follows from the properties of the roots that on the interval $(1, +\infty)$ we have $y_k = \sqrt[2k+1]{x} < \sqrt[2m+1]{x} = y_m$ when $k > m$. Besides, for any pair $x_2 > x_1 \geq 1$ and $k > m$, we get $\frac{y_k(x_2)}{y_k(x_1)} = \sqrt[2k+1]{\frac{x_2}{x_1}} < \sqrt[2m+1]{\frac{x_2}{x_1}} = \frac{y_m(x_2)}{y_m(x_1)}$. This shows that on the interval $(1, +\infty)$, the graph of the function $y_k = \sqrt[2k+1]{x}$ lies below and increases slower than the graph of $y_m = \sqrt[2m+1]{x}$ if $k > m$. On the interval $(0, 1)$ we have $y_k = \sqrt[2k+1]{x} > \sqrt[2m+1]{x} = y_m$ when $k > m$, that is, the graph of $y_k = \sqrt[2k+1]{x}$ lies above of the graph of $y_m = \sqrt[2m+1]{x}$. Since the function is odd, similar properties are observed on the intervals $(-\infty, -1)$ and $(-1, 0)$. At the points $x = -1, 0, 1$ the values of all the functions $y = \sqrt[2k+1]{x}$ coincide and are equal to $y = -1, 0, 1$, respectively. This comparison is illustrated in Fig. 3.30.

3.2 *Absolute Value Function*

One more important irrational function which admits a complete study of its properties is the *absolute value function* $f(x) = |x|$. The fact that this function

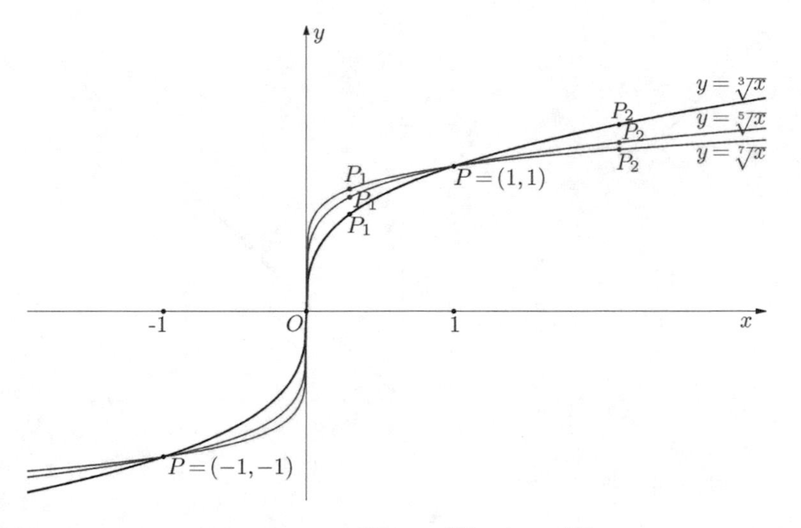

Fig. 3.30 The graphs of the functions $y = \sqrt[3]{x}$, $y = \sqrt[5]{x}$ and $y = \sqrt[7]{x}$

is irrational follows from the representation of the absolute value in the form $|x| = \sqrt{x^2}$, $\forall x \in \mathbb{R}$. Let us perform analysis of this function based on the definition of the operation of the absolute value.

Study of $y = |x|$

1. Domain and range. The absolute value is defined for any real x, and consequently, the domain of the function is $X = \mathbb{R}$. The absolute value of any number is non-negative, which means that the range is contained in $[0, +\infty)$. Noting that for each $y \geq 0$ there exists $x = y$ such that $f(x) = |x| = |y| = y$, we conclude that the range is the entire interval $Y = [0, +\infty)$.
2. Boundedness. Knowing the range we can state that the function is bounded below (for instance, by the constant $m = 0$), but is not bounded above.
3. Parity. The function is even, since $f(-x) = |-x| = |x| = f(x)$, $\forall x \in \mathbb{R}$.
4. Periodicity. The function is not periodic because on the infinite interval $[0, +\infty)$ it coincides with the increasing function $y = x$, which does not repeat its values.
5. Monotonicity and extrema. By the definition, on the interval $(-\infty, 0]$ the function has the form $f(x) = |x| = -x$ and represents decreasing function, while on the interval $[0, +\infty)$ it is the increasing function $f(x) = |x| = x$.

   Therefore, each point $x \neq 0$ is monotonic and cannot be an extremum of any type. The only remaining point is $x = 0$, which is strict global and local minimum since $f(x) = |x| > 0 = f(0)$, $\forall x \neq 0$.

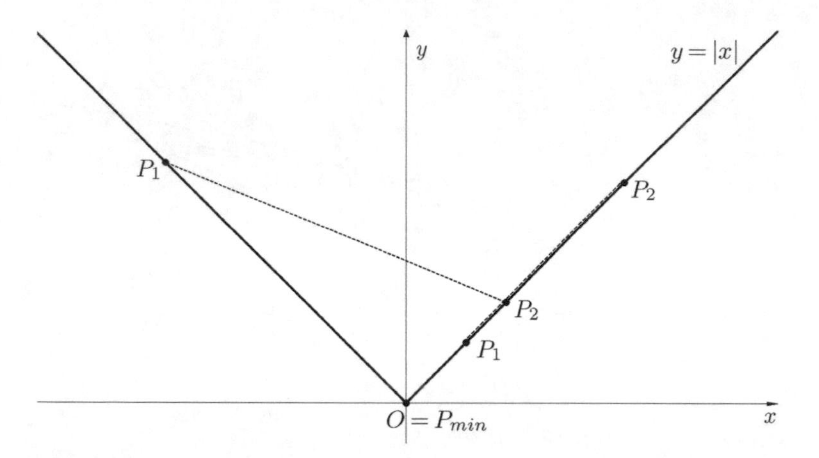

Fig. 3.31 The graph of the function $y = |x|$

6. Concavity and inflection. Taking two different points $x_1 < x_2$, we find

$$D_2 \equiv \frac{f(x_1) + f(x_2)}{2} - f\left(\frac{x_1 + x_2}{2}\right) = \frac{1}{2}(|x_1| + |x_2|) - \frac{1}{2}(|x_1 + x_2|) \geq 0,$$

which shows that the function is non-strictly concave upward on the entire domain. If we restrict the location of the two points to be in the interval $(-\infty, 0]$, then we have the formula of the line $y = |x| = -x$, and consequently, $D_2 = 0$. Similarly, on the interval $[0, +\infty)$ the function takes the form of another line $f(x) = |x| = x$ and again $D_2 = 0$. Finally, taking two arbitrary points in two different parts of the domain—$x_1 < 0 < x_2$—we find that $D_2 > 0$ (see Fig. 3.31). It follows from this result that there is no inflection point.
7. Intersection and sign. Intersection with the y-axis occurs at the origin and this is also the only point of intersection with the x-axis. There are no other points of intersection with the x-axis since $f(x) = |x| > 0, \forall x \neq 0$.
8. Inverse. The function $f(x) = |x| : X = \mathbb{R} \to Y = [0, +\infty)$ is surjective, but not injective, since for each $y_1 > 0$ there exist two different points $x_1 = y_1$ and $x_2 = -y_1$ such that $f(x_1) = |y_1| = y_1 = |-y_1| = f(x_2)$. If the domain is restricted to the interval $[0, +\infty)$, then the function becomes bijective and is simplified to the form of the line $f(x) = |x| = x : [0, +\infty) \to [0, +\infty)$ with the inverse $f^{-1}(x) = x : [0, +\infty) \to [0, +\infty)$. Analogously, restricting the domain to the interval $(-\infty, 0]$ we also obtain a bijective function in the form of another line $f(x) = |x| = -x : (-\infty, 0] \to [0, +\infty)$ whose inverse is $f^{-1}(x) = -x : [0, +\infty) \to (-\infty, 0]$.
9. Based on the investigated properties, we can plot the graph of the function $y = |x|$ as shown in Fig. 3.31.

Complimentary Properties of $y = |x|$
The graph of $y = |x|$ suggests that the function increases without restriction when $|x|$ goes to $+\infty$ (that is, x itself goes to $+\infty$ or to $-\infty$). Indeed, taking $x > M$, where $M > 0$ is any large constant, we guarantee that $|x| = x > M$, that is, the values of function become arbitrary large, that is, the function tends to $+\infty$. Since the function is even, the same is true when $|x|$ increases without restriction, but keeping negative values (x goes to $-\infty$). This implies that the function $y = |x|$ has no horizontal asymptote. Hence, we confirm the behavior of the graph at infinity observed in Fig. 3.31. Since the function $y = |x|$ has no infinite limit at the points of the domain, its graph has no vertical asymptote.

Study of irrational functions in a general form usually is very complicated and, in the most cases, all the properties cannot be evaluated even using advanced techniques of Calculus and Analysis.

4 Examples of the Study of Algebraic Functions

4.1 Preliminary Considerations

Before we start the study of functions, it is important to note that in various cases the analysis of monotonicity and concavity is technically complicated if made on an original interval, but this analysis becomes much simpler if performed on smaller intervals. For this reason, we first prove the three results which allow us to transfer the properties of monotonicity and concavity from parts of a given interval onto the entire interval.

Property 1 (Monotonicity on Subintervals) *If $f(x)$ is increasing on the intervals $[a, b]$ and $[b, c]$, then it is increasing on the entire interval $[a, c]$. The same result is true for decreasing: if $f(x)$ is decreasing on $[a, b]$ and $[b, c]$, then it is decreasing on $[a, c]$. Similar statements are also valid for non-strict monotonicity.*

Proof We provide the proof only in the case of increasing, since other three cases can be shown following the same reasoning. Take two arbitrary points $x_1 < x_2$ of $[a, c]$. It may occur three cases:

(1) if $x_2 \leq b$, then from the increase on $[a, b]$ it follows that $f(x_1) < f(x_2)$;
(2) if $x_1 \geq b$, then from the increase on $[b, c]$ it follows that $f(x_1) < f(x_2)$;
(3) if $x_1 < b < x_2$, then we can compare $f(x_1)$ and $f(x_2)$ by using the value of the function at the intermediate point b, taking into account increasing of $f(x)$ on both subintervals: $f(x_1) < f(b) < f(x_2)$.

Hence, for any pair $x_1 < x_2$ in $[a, c]$ we have $f(x_1) < f(x_2)$, that is, $f(x)$ is increasing on the entire interval $[a, c]$. □

Remark The condition of intersection of the two intervals at least at one point is important. Otherwise, the statement is not true: for instance, the function $f(x) =$

$x - H(x)$ increases on the intervals $(-\infty, 0)$ and $[0, +\infty)$, but it does not increase (even in non-strict sense) on the interval $(-\infty, +\infty)$ (or on any interval (a, b) containing the origin). (Recall that $H(x) = \begin{cases} 0, & x < 0 \\ 1, & x \geq 0 \end{cases}$ is the Heaviside function.)

Property 2 (Concavity on Intersecting Subintervals) *If $f(x)$ is concave upward on the intervals $[a, b]$ and $[c, d]$, where $a < c < b < d$, then it is concave upward on the entire interval $[a, d]$. The same result can be formulated for downward concavity: if $f(x)$ is concave downward on $[a, b]$ and $[c, d]$, where $a < c < b < d$, then it is concave downward on $[a, d]$. Similar statements can be made for non-strict concavity.*

Proof We prove only the result for upward concavity, since other situations can be shown in the same way. In what follows we will assume that $x_1 < x_2$ (without loss of generality).

If we consider $x_2 \leq b$, then both points are contained in $[a, b]$, where the concavity is upward. Analogously, if $x_1 \geq c$, then both points lie in $[c, d]$, where the function is also concave upward. Therefore, it remains to consider the situation when $x_1 < c < b < x_2$. Take an arbitrary point $x \in (x_1, x_2)$ and let us analyze different cases that may occur, by using the condition of upward concavity expressed through the slopes of secant lines (formula (2.9) of Sect. 7.3 in Chap. 2).

First, if $x \in (c, b]$, then we consider the concavity on the following intervals. On the interval $[c, x_2] \subset [c, d]$, with the point x inside it, we use the upward concavity in the form of the slopes of secant lines: $k_2 = \frac{f(x_2)-f(x)}{x_2-x} > \frac{f(x)-f(c)}{x-c} = k_c$. On the interval $[x_1, x] \subset [a, b]$, with the point c inside it, we again use the upward concavity in the form of the slopes of secant lines: $k_c > k_1 = \frac{f(x)-f(x_1)}{x-x_1}$. It follows from these two inequalities that $k_2 > k_1$, which corresponds to the upward concavity (see illustration in Fig. 3.32).

Second, if $x \in (b, x_2)$, then we use the conditions of the upward concavity on the following intervals. On the interval $[c, x_2] \subset [c, d]$, we have

$$k_0 = \frac{f(b) - f(c)}{b - c} < k_b = \frac{f(x) - f(b)}{x - b} < \frac{f(x_2) - f(x)}{x_2 - x} = k_2.$$

On the interval $[x_1, b] \subset [a, b]$, we get

$$k_{-1} = \frac{f(b) - f(x_1)}{b - x_1} < \frac{f(b) - f(c)}{b - c} = k_0.$$

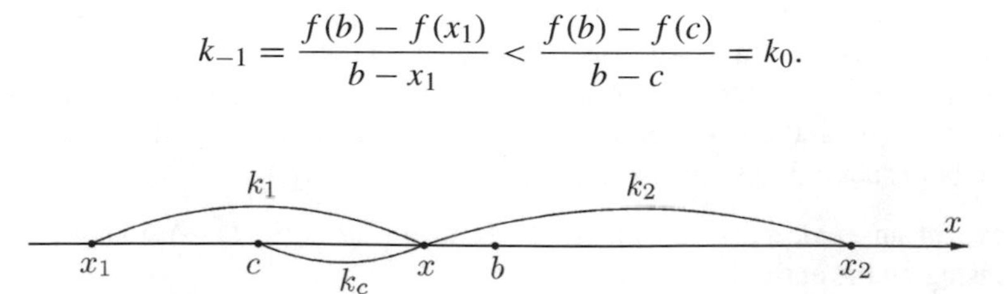

Fig. 3.32 Slopes in the case $x \in (c, b]$

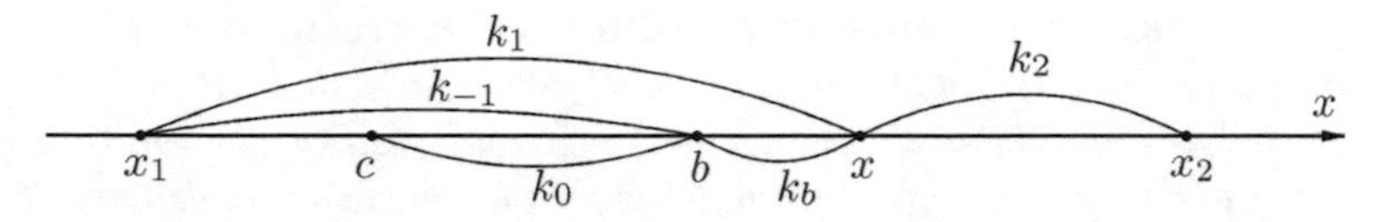

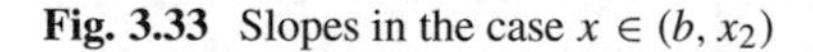
Fig. 3.33 Slopes in the case $x \in (b, x_2)$

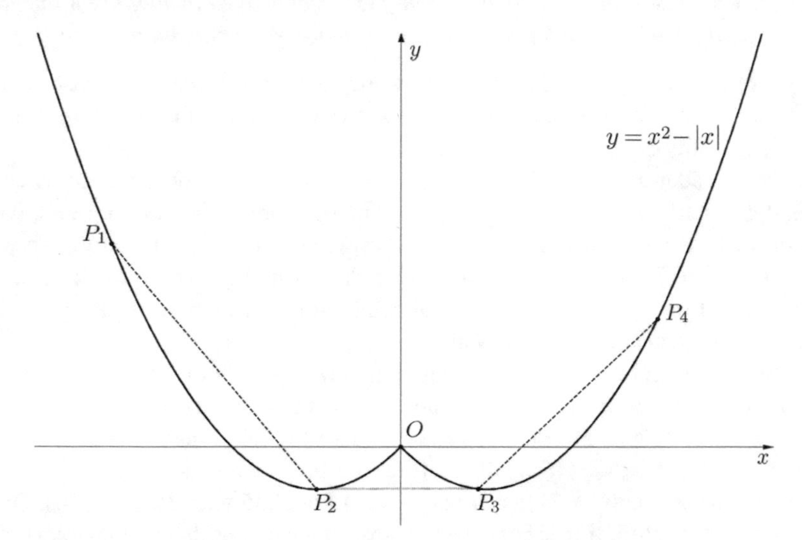

Fig. 3.34 Concavity of the function $f(x) = x^2 - |x|$

(See illustration in Fig. 3.33). Using these inequalities, we derive the following evaluation:

$$k_1 = \frac{f(x) - f(x_1)}{x - x_1} = \frac{f(x) - f(b)}{x - b} \cdot \frac{x - b}{x - x_1} + \frac{f(b) - f(x_1)}{b - x_1} \cdot \frac{b - x_1}{x - x_1}$$

$$= k_b \frac{x - b}{x - x_1} + k_{-1} \frac{b - x_1}{x - x_1}$$

$$< k_2 \frac{x - b}{x - x_1} + k_0 \frac{b - x_1}{x - x_1} < k_2 \frac{x - b}{x - x_1} + k_2 \frac{b - x_1}{x - x_1} = k_2.$$

This corresponds to the upward concavity.

Finally, in the case $x \in (x_1, c]$ the prove is similar. □

Remark The condition of intersection of the two intervals along a subinterval and not simply at a point (as in the case of monotonicity) is important. Otherwise, the statement is incorrect as shown in the example with the function $f(x) = x^2 - |x|$, which is concave upward on the intervals $(-\infty, 0]$ and $[0, +\infty)$, but does not concave upward (even in non-strict sense) on the interval $(-\infty, +\infty)$ (or any interval (a, b) containing the origin). See the illustration in Fig. 3.34.

Property 3 (Concavity on Subintervals with Local Extremum) *If $f(x)$ is concave upward on the intervals $[a, b]$ and $[b, d]$, where $a < b < d$, and the point $x = b$ is a local minimum, then $f(x)$ is concave upward on the entire interval $[a, d]$. The same statement can be formulated for the downward concavity: if $f(x)$ is concave downward on the intervals $[a, b]$ and $[b, d]$, where $a < b < d$, and the point $x = b$ is a local maximum, then $f(x)$ is concave downward on the entire interval $[a, d]$. Similar statements are true for non-strict concavity.*

Proof We prove only the case of the upward concavity, since other cases can be proved using similar arguments. In what follows we assume that $x_1 < x_2$ (without loss of generality).

If we consider $x_2 \le b$, then both points lie in $[a, b]$, where the concavity is upward. Similarly, if we consider $x_1 \ge b$, then both points belong to $[b, d]$, where the concavity is also upward. Therefore, it remains to consider the situation when $x_1 < b < x_2$. It is sufficient to consider this location of the points in such a neighborhood of b, where b is a global minimum, and then apply Property 2 of concavity on intersecting subintervals.

Choose a neighborhood of b such that $f(b) < f(x)$ for each x of this neighborhood, $x \neq b$. The coordinates of the three points $P_1 = (x_1, f(x_1))$, $P_2 = (x_2, f(x_2))$ and $P_b = (x_b, f(x_b))$ satisfy the relations $x_1 < b < x_2$ and $f(x_1) > f(b)$, $f(x_2) > f(b)$. Therefore, the line segments P_1P_b and P_bP_2 lie below the line segment P_1P_2 (see illustration in Fig. 3.35 with $P_b = (0, 0)$). On the interval $[x_1, b]$ the function is concave upward, which means that all the points of the graph lie below the corresponding points of the segment P_1P_b, and consequently, also below the corresponding points of P_1P_2. The same happens on the interval $[b, x_2]$. Hence, all the points of the graph on the interval $[x_1, x_2]$ lie below the corresponding points of the segment P_1P_2, or in other words, the part of the graph between the points P_1 and P_2 lies below the line segment P_1P_2 (see Fig. 3.35). Due to arbitrariness of the points x_1, x_2, this means that the function is concave upward on a chosen neighborhood of b. Therefore, by Property 2, $f(x)$ is concave upward

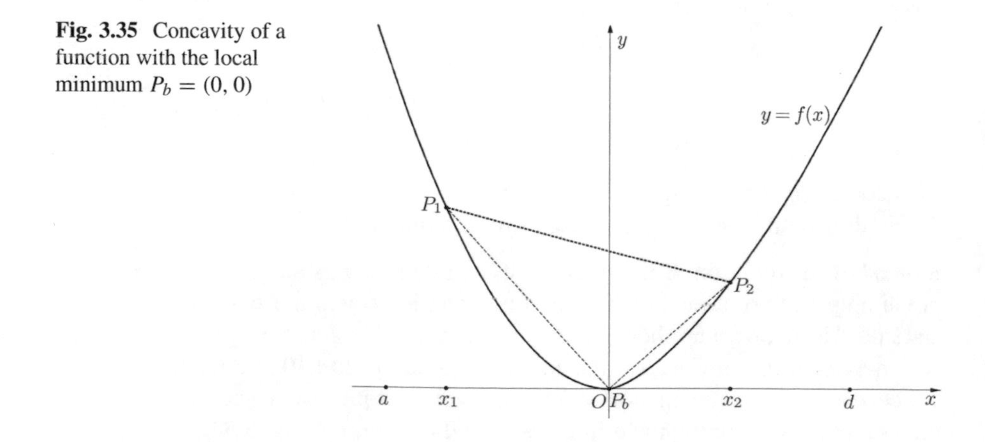

Fig. 3.35 Concavity of a function with the local minimum $P_b = (0, 0)$

on $[a, x_2]$ and on $[x_1, d]$, where $x_1 < x_2$. Hence, applying once more Property 2, we conclude that $f(x)$ is concave upward on $[a, d]$.

□

4.2 *Study of Polynomial Functions*

A. Study of $f(x) = x^3 - x^2 - x + 1$

1. Domain and range. Since the domain is not given explicitly, it consists of all the points for which the formula of the function is defined. For any polynomial function all the operations involved in the formula are defined for any real number. Therefore, the domain of any polynomial is $X = \mathbb{R}$. In particular, the domain of $f(x) = x^3 - x^2 - x + 1$ is $X = \mathbb{R}$.

   Recalling from the theory of polynomial equations that any cubic equation has at least one real root, we can conclude that any cubic polynomial has the range $Y = \mathbb{R}$. In particular, the equation $x^3 - x^2 - x + 1 = y$, where y is a given value, has at least one root whatever y is chosen. This means that the range of $f(x)$ is $Y = \mathbb{R}$.
2. Intersection and sign. The point of intersection with the y-axis is $P_0 = (0, 1)$. To find the points of intersection with the x-axis we should solve the equation $x^3 - x^2 - x + 1 = 0$. The solution of a cubic equation can be found through algebraic operations, although general formulas are not quite simple. However, in this specific case, the polynomial admits an elementary factorization $x^3 - x^2 - x + 1 = (x - 1)^2(x + 1)$ which means that the roots of the polynomial equation are $x = 1$ (double root) and $x = -1$. Therefore, the points of intersection with the x-axis are $P_1 = (1, 0)$ and $P_2 = (-1, 0)$.

   Using the factored form it is easy to determine the sign of the polynomial:

   (1) if $x < -1$, then $f(x) < 0$;
   (2) if $x > -1, x \neq 1$, then $f(x) > 0$.
3. Boundedness. Since the range is $Y = \mathbb{R}$, the function is unbounded both below and above.
4. Parity. Comparing the values of $f(-2) = -9$ and $f(2) = 3$, we conclude that the function is neither even nor odd. (Notice that accidentally $f(-1) = 0 = f(1)$, but this equality valid only for one pair of points, or some pairs of points, does not mean that the function is even or odd.)
5. Monotonicity and extrema.

Take a pair of the points $x_1 < x_2$ and evaluate the difference $D_1 = f(x_2) - f(x_1)$:

$$D_1 = f(x_2) - f(x_1) = x_2^3 - x_2^2 - x_2 + 1 - \left(x_1^3 - x_1^2 - x_1 + 1\right)$$
$$= (x_2 - x_1)(x_2^2 + x_2x_1 + x_1^2) - (x_2 - x_1)(x_2 + x_1) - (x_2 - x_1)$$
$$= (x_2 - x_1)\left(x_2^2 + x_2x_1 + x_1^2 - x_2 - x_1 - 1\right).$$

Using the central point $x_0 = \frac{x_1+x_2}{2}$ and the increment $h = x_2 - x_0 = x_0 - x_1 > 0$, write D_1 in the following form

$$D_1 = 2h\left[(x_0+h)^2 + (x_0+h)(x_0-h) + (x_0-h)^2 - 2x_0 - 1\right]$$
$$= 2h\left[3x_0^2 - 2x_0 - 1 + h^2\right] = 2h\left[(x_0-1)(3x_0+1) + h^2\right].$$

From the last expression it is clear that the sign of D_1 is well defined on the two sets: if $x_0 < -\frac{1}{3}$ ($x_1 < x_2 \le -\frac{1}{3}$) or $x_0 > 1$ ($1 \le x_1 < x_2$), then both terms in brackets are positive, and consequently $D_1 > 0$, which means that $f(x)$ increases on the intervals $(-\infty, -\frac{1}{3}]$ and $[1, +\infty)$. When $x_0 \in (-\frac{1}{3}, 1)$, the first term in brackets is negative, while the second is positive. To specify the sign of D_1 on this interval, it is convenient to divide it into the two parts—$(-\frac{1}{3}, 0]$ and $[0, 1)$. On the subinterval $(-\frac{1}{3}, 0]$, the condition $x_1 = x_0 - h > -\frac{1}{3}$ implies that $h < x_0 + \frac{1}{3}$, and consequently,

$$(x_0-1)(3x_0+1) + h^2 < (x_0-1)(3x_0+1) + (x_0+\frac{1}{3})^2 = (x_0+\frac{1}{3})(4x_0-\frac{8}{3}).$$

The first factor is positive, while the second is negative, which shows that $D_1 < 0$, that is, $f(x)$ decreases on $(-\frac{1}{3}, 0]$. On the second subinterval $[0, 1)$, using the inequality $x_2 = x_0+h < 1$ (or equivalently, $h < 1-x_0$), we obtain the following evaluation:

$$(x_0-1)(3x_0+1) + h^2 < (x_0-1)(3x_0+1) + (1-x_0)^2 = (x_0-1)\cdot 4x_0.$$

Here, the first factor is negative, while the second is positive, and again $D_1 < 0$, which means that $f(x)$ decreases on $[0, 1)$. Since $f(x)$ decreases on the two subintervals with the common point 0, by Property 1, $f(x)$ is decreasing on $[-\frac{1}{3}, 1]$.

Thus, we arrive at the following results regarding monotonicity:

(1) the function is increasing on the interval $(-\infty, -\frac{1}{3}]$;
(2) the function is decreasing on the interval $[-\frac{1}{3}, 1]$;
(3) the function is increasing on the interval $[1, +\infty)$.

Using these results, we can conclude that the point $x_3 = -\frac{1}{3}$ is a local maximum and $x_4 = 1$ is a local minimum. Notice that these points are not global extrema though, because the range of $f(x)$ contains all the real numbers. The fact that x_3 and x_4 are not global extrema can be corroborated by the evaluation of the leading term of $f(x)$. Indeed, for sufficiently large values of $|x|$, its main term is x^3, and the function satisfies the evaluation $f(x) > \frac{x^3}{2}$ for all $x > 3$ and $f(x) < \frac{x^3}{2}$ for all $x < -3$. Then, taking constant $M > 3^3$ we can find $x_M = \sqrt[3]{2M} > 3$ such that $f(x) > \frac{x^3}{2} > M, \forall x > x_M$. In the same manner, taking any constant $m < -3^3$ we can find $x_m = \sqrt[3]{2m} < -3$ such that $f(x) < \frac{x^3}{2} < m, \forall x < x_m$. This shows that $f(x)$ takes the values larger and smaller than any constant, and consequently, x_3 and x_4 cannot be global extrema.

6. Periodicity. The increase of the function on the unbounded interval $[1, +\infty)$ indicates that the function cannot be periodic.
7. Concavity and inflection.

   Take two different points $x_1 < x_2$, define the midpoint $x_0 = \frac{x_1+x_2}{2}$ and the increment $h = x_2 - x_0 = x_0 - x_1 > 0$, and use the quantity $D_2 \equiv f(x_1) + f(x_2) - 2f(x_0)$ to evaluate the difference between the mean value $\frac{f(x_1)+f(x_2)}{2}$ and the value at the midpoint $f(x_0)$. Employing the expression found for D_1, we obtain:

$$D_2 = f(x_1) + f(x_2) - 2f(x_0) = (f(x_2) - f(x_0)) - (f(x_0) - f(x_1))$$

$$= (x_2 - x_0)\left[x_2^2 + x_2x_0 + x_0^2 - x_2 - x_0 - 1\right]$$

$$- (x_0 - x_1)\left[x_0^2 + x_0x_1 + x_1^2 - x_0 - x_1 - 1\right]$$

$$= h\left[x_2^2 - x_1^2 + x_2x_0 - x_1x_0 - x_2 + x_1\right]$$

$$= 2h^2\left[x_2 + x_1 + x_0 - 1\right] = 2h^2(3x_0 - 1).$$

   Therefore, we get the following results about concavity:

   (1) the function is concave downward on the interval $(-\infty, \frac{1}{3}]$;
   (2) the function is concave upward on the interval $[\frac{1}{3}, +\infty)$.

   Consequently, the point $x = \frac{1}{3}$ is an inflection point.
8. Complimentary properties. When $|x|$ increases without restriction, that is, x goes to $\pm\infty$ $(x \to \pm\infty)$, the behavior of the polynomial is determined by its leading term, and for this reason we have (for sufficiently large $|x|$) $f(x) = x^3 - x^2 - x + 1 \approx x^3$ (see discussion of such a behavior in Sect. 1.3). From this approximation it follows that the function increases without restriction when $x \to +\infty$ and decreases without restriction when $x \to -\infty$, that is, $\lim_{x\to+\infty} f(x) = +\infty$ and $\lim_{x\to-\infty} f(x) = -\infty$. This implies that the function has no horizontal asymptote.

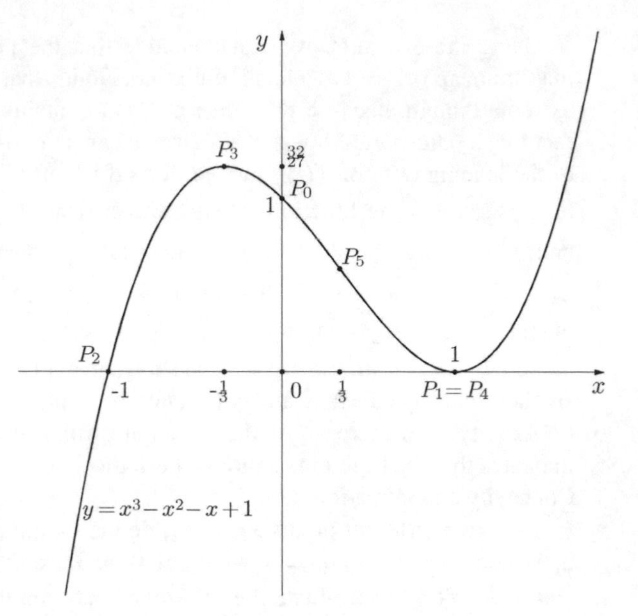

Fig. 3.36 The graph of the function $f(x) = x^3 - x^2 - x + 1$

At any point of the domain the function is continuous, that is, $f(x) \to f(x_0)$ when $x \to x_0$. Hence, there is no convergence of $f(x)$ to infinity at any point, which implies that the function has no vertical asymptote.

9. Using the investigated properties and marking some important points, such as $P_0 = (0, 1)$ (intersection with the y-axis), $P_1 = (1, 0)$ and $P_2 = (-1, 0)$ (intersection with the x-axis), $P_3 = (-\frac{1}{3}, \frac{32}{27})$ and $P_4 = P_1 = (1, 0)$ (local maximum and minimum), and $P_5 = (\frac{1}{3}, \frac{16}{27})$ (the inflection point), we can plot the graph shown in Fig. 3.36.

B. Study of $f(x) = \frac{x^4}{4} + x^3 - 4x - 1$

1. Domain and range. The formula of any polynomial function is defined for all real x. Therefore, the domain of the function $f(x)$ is $X = \mathbb{R}$.

   We postpone the examination of the range until after the investigation of monotonicity and extrema.
2. Intersection and sign. The point of intersection with the y-axis is $P_0 = (0, -1)$. To find the points of intersection with the x-axis we should solve the equation $\frac{x^4}{4} + x^3 - 4x - 1 = 0$. Although the solution of a quartic equation can be found through algebraic operations, the general formulas are quite complicated. For this reason we postpone the specification of the points of intersection with the x-axis and the sign of the function until after the study of monotonicity.
3. Boundedness. The study of boundedness is postponed until after the investigation of monotonicity and extrema.

4. Parity. Comparing the values $f(-1) = \frac{1}{4} - 1 + 4 - 1 = \frac{9}{4}$ and $f(1) = \frac{1}{4} + 1 - 4 - 1 = -\frac{15}{4}$, we conclude that the function is neither even nor odd.
5. Periodicity. We will study periodicity after the analysis of monotonicity.
6. Monotonicity and extrema.

   Take $x_1 < x_2$ and evaluate the difference $D_1 = f(x_2) - f(x_1)$:

$$D_1 = f(x_2) - f(x_1) = \frac{x_2^4}{4} + x_2^3 - 4x_2 - 1 - \left(\frac{x_1^4}{4} + x_1^3 - 4x_1 - 1\right)$$

$$= \frac{1}{4}(x_2^4 - x_1^4) + (x_2^3 - x_1^3) - 4(x_2 - x_1)$$

$$= \frac{1}{4}(x_2 - x_1)(x_2 + x_1)(x_2^2 + x_1^2) + (x_2 - x_1)(x_2^2 + x_2x_1 + x_1^2) - 4(x_2 - x_1)$$

$$= (x_2 - x_1)\left[\frac{1}{4}(x_2 + x_1)(x_2^2 + x_1^2) + (x_2^2 + x_2x_1 + x_1^2) - 4\right].$$

Using the midpoint $x_0 = \frac{x_1+x_2}{2}$ and the increment $h = x_2 - x_0 = x_0 - x_1 > 0$, write D_1 in the form

$$D_1 = 2h\left[\frac{1}{2}x_0((x_0 + h)^2 + (x_0 - h)^2)\right.$$

$$\left. + ((x_0 + h)^2 + (x_0 + h)(x_0 - h) + (x_0 - h)^2) - 4\right]$$

$$= 2h\left[x_0(x_0^2 + h^2) + (2x_0^2 + 2h^2 + x_0^2 - h^2) - 4\right]$$

$$= 2h[x_0^3 + 3x_0^2 - 4 + h^2(x_0 + 1)]$$

$$= 2h[(x_0 + 2)^2(x_0 - 1) + h^2(x_0 + 1)].$$

It is clear from the last expression that the sign of D_1 is well determined in the two parts of the domain:

(1) if $x_0 > 1$ ($1 \le x_1 < x_2$), then both terms in brackets are positive, and consequently $D_1 > 0$, which means that $f(x)$ increases on the interval $[1, +\infty)$;
(2) if $x_0 < -1$ ($x_1 < x_2 \le -1$), then both terms in brackets are negative, and consequently $D_1 < 0$, which indicates that $f(x)$ decreases on the interval $(-\infty, -1]$.

   In the case $-1 < x_0 < 1$, the first term in brackets is negative, while the second is positive. To specify the sign of D_1 on this interval, notice that the condition $x_2 = x_0 + h \le 1$ implies that $h \le 1 - x_0$, and using also the inequality $x_0 + 1 > 0$, we can evaluate the expression in brackets as follows:

$$(x_0+2)^2(x_0-1)+h^2(x_0+1) \le (x_0+2)^2(x_0-1)+(1-x_0)^2(x_0+1)$$
$$= (x_0-1)((x_0+2)^2+x_0^2-1) < 0.$$

In the last inequality we took into account that $x_0 - 1 < 0$ and $(x_0+2)^2 + x_0^2 - 1 > 1 + x_0^2 - 1 \ge 0$. Therefore, on the third part of the domain we have:

(3) if $-1 < x_0 < 1$ $(-1 \le x_1 < x_2 \le 1)$ then $D_1 < 0$, which means that $f(x)$ decreases on the interval $[-1, 1]$.

Since the function decreases on the intervals $(-\infty, -1]$ and $[-1, 1]$, which have the common point -1, we can conclude that, according to Property 1, $f(x)$ decreases on the entire interval $(-\infty, 1]$. Hence, we have the two intervals of the monotonicity:

(1) $f(x)$ decreases on $(-\infty, 1]$;
(2) $f(x)$ increases on $[1, +\infty)$.

It follows from the study of monotonicity that the only local and global (strict) extremum is the point $x_1 = 1$ where the function takes the minimum value: $P_1 = (1, -\frac{15}{4})$.

One of the implications of monotonicity and extremum is that the range Y of the function is contained in the interval $[-\frac{15}{4}, +\infty)$. To see that the values of the function become as large as it is desired when $|x|$ have the sufficiently large values, we recall that the behavior of $f(x)$ at infinity is determined by its leading term $\frac{x^4}{4}$ (see Sect. 1.3). In particular, taking an arbitrary constant $M > 3^4$ we can choose $x_M = \sqrt[4]{4M} > 3$ such that $f(x) > M, \forall x > x_M$. This shows that $Y \subset [-\frac{15}{4}, +\infty)$. Additionally, noting that $f(x)$ is continuous at each point of its domain, we conclude that $Y = [-\frac{15}{4}, +\infty)$. From this specification of the range it follows immediately that the function is bounded below (for instance, by the constant $-\frac{15}{4}$) and unbounded above.

The next consequence of monotonicity is the non-periodicity of $f(x)$ due to the increase on the infinite interval $[1, +\infty)$.

Finally, the properties of monotonicity help to specify the points of intersection with the x-axis. Indeed, noting that $f(x)$ takes positive values for sufficiently large values of $|x|$ and has negative value at the point $x_1 = 1$, we can deduce, based on the continuity of $f(x)$, that there are two points, one to the left and another to the right of x_1, where $f(x)$ is equal 0. The specification of exact locations of these two points requires to solve the quartic equation $\frac{x^4}{4} + x^3 - 4x - 1 = 0$. Although there are algebraic formulas to represent the roots of this equation (which can be both real and complex), but the involved algebraic expressions are quite complicated and of difficult application in practice. For this reasons, we find an approximate location of the intersection point, which is sufficient for our purposes. For the root x_2 located to the left of $x_1 = 1$ we find first that $f(0) = -1$, and consequently, $x_2 < 0$. At the point -1 the function has positive value $f(-1) = \frac{1}{4} - 1 + 4 - 1 = \frac{9}{4}$, which means that the change of sign happens on the interval $(-1, 0)$ and the zero value is located in between -1 and 0. If we need to specify the location of x_2 better, then we take the midpoint $x = -\frac{1}{2}$ of the interval $[-1, 0]$ and calculate the value of the function $f(-\frac{1}{2}) = \frac{57}{64}$, which means that x_2 is located in $(-\frac{1}{2}, 0)$. If we need to specify x_2 even better, then we bisect the interval $[-\frac{1}{2}, 0]$ at the midpoint $-\frac{1}{4}$, and so on. In a similar manner, we can specify the location of the root x_3, to the right of $x_1 = 1$. First we calculate $f(2) = 3 > 0$, which indicate that x_3 lies in $(1, 2)$. To specify better its location, we take the midpoint $\frac{3}{2}$ and calculate $f(\frac{3}{2}) = -\frac{151}{64} < 0$, which means that x_3 belongs to $(\frac{3}{2}, 2)$. Bisecting the last interval at the midpoint once more, we get the point $\frac{7}{4}$ and proceed in the same way until we specify the location of x_3 with required accuracy.

7. Concavity and inflection.

We take two arbitrary points $x_1 < x_2$, define the midpoint $x_0 = \frac{x_1+x_2}{2}$ and the increment $h = x_2 - x_0 = x_0 - x_1 > 0$, and use the quantity $D_2 \equiv f(x_1) + f(x_2) - 2f(x_0)$ to evaluate the difference between the mean value $\frac{f(x_1)+f(x_2)}{2}$ and the value at the midpoint $f(x_0)$. Taking advantage of the derived expression of D_1, we obtain:

$$D_2 = f(x_1) + f(x_2) - 2f(x_0) = (f(x_2) - f(x_0)) - (f(x_0) - f(x_1))$$

$$= (x_2 - x_0)\left[\frac{1}{4}(x_2 + x_0)(x_2^2 + x_0^2) + (x_2^2 + x_2x_0 + x_0^2) - 4\right]$$

$$- (x_0 - x_1)\left[\frac{1}{4}(x_0 + x_1)(x_0^2 + x_1^2) + (x_0^2 + x_0x_1 + x_1^2) - 4\right]$$

$$= h\Big[\frac{1}{4}(x_2 - x_1)(x_2^2 + x_2x_1 + x_1^2) + \frac{1}{4}(x_2 - x_1)(x_2 + x_1)x_0$$

$$+ \frac{1}{4}(x_2 - x_1)x_0^2 + (x_2 - x_1)(x_2 + x_1) + (x_2 - x_1)x_0\Big]$$

$$= 2h^2\left[\frac{1}{4}(x_2^2 + x_2x_1 + x_1^2) + (\frac{1}{4}x_0 + 1)(x_2 + x_1 + x_0)\right].$$

In the terms of x_0 and h this expression takes the form more convenient for analysis:

$$D_2 = 2h^2\left[\frac{1}{4}(3x_0^2 + h^2) + (\frac{1}{4}x_0 + 1) \cdot 3x_0\right] = h^2\left[3x_0^2 + 6x_0 + \frac{h^2}{2}\right]$$

$$= h^2\left[3x_0(x_0 + 2) + \frac{h^2}{2}\right].$$

It is evident that if $x_0 > 0$ ($0 \le x_1 < x_2$) or $x_0 < -2$ ($x_1 < x_2 \le -2$) then both terms in brackets are positive, and consequently $D_2 > 0$, which means that $f(x)$ has upward concavity. In the case $-2 < x_0 < 0$, the first term in brackets is negative, while the second is positive. To specify the sign of D_2 on $[-2, 0]$, we split this interval into two parts: $[-2, -1]$ and $[-1, 0]$. On the subinterval $[-2, -1]$, the condition $x_1 = x_0 - h \ge -2$ implies that $h \le x_0 + 2$, and consequently,

$$3x_0(x_0+2) + \frac{h^2}{2} = 3x_0(x_0+2) + h^2 - \frac{h^2}{2} \le 3x_0(x_0+2) + (x_0+2)^2 - \frac{h^2}{2}$$
$$= (x_0+2)(4x_0+2) - \frac{h^2}{2} < 0.$$

Then, $D_2 < 0$ and $f(x)$ has downward concavity on $[-2, -1]$. Notice that the same conclusion (following the very same arguments) is valid also on the interval $[-2, -0.9]$. On the subinterval $[-1, 0]$, the condition $x_2 = x_0 + h \le 0$ implies that $h \le -x_0$, and then

$$3x_0(x_0+2) + h^2 - \frac{h^2}{2} \le 3x_0(x_0+2) + x_0^2 - \frac{h^2}{2} = x_0(4x_0+6) - \frac{h^2}{2} < 0.$$

Again, $D_2 < 0$ and $f(x)$ is concave downward on $[-1, 0]$. Since $f(x)$ has downward concavity on the two intervals $[-2, -0.9]$ and $[-1, 0]$, whose union is $[-2, 0]$ and intersection is $[-1, -0.9]$, by Property 2, it follows that $f(x)$ has downward concavity on the entire interval $[-2, 0]$. Hence, we arrive at the following results regarding concavity:

(1) $f(x)$ is concave upward on $(-\infty, -2)$;
(2) $f(x)$ is concave downward on $(-2, 0)$;
(3) $f(x)$ is concave upward on $(0, +\infty)$.

Due to the change of the type of concavity at the points $x = -2$ and $x = 0$ (which belong to the domain), these two points are inflection points.

8. Complimentary properties. When $|x|$ goes to infinity ($x \to \pm\infty$), then the function can be approximated through its leading term $f(x) = \frac{x^4}{4} + x^3 - 4x - 1 \approx \frac{x^4}{4}$ (see Sect. 1.3 for details), which shows that $f(x)$ increases without restriction, that is, $\lim\limits_{x\to\pm\infty} f(x) = +\infty$. This implies that the function has no horizontal asymptote.

   At each real point the function is continuous, that is, $f(x) \to f(x_0)$ when $x \to x_0$. Therefore, there is no point at which the function converges to infinity. This implies that there is no vertical asymptote.

9. Using the investigated properties and marking some important points, such as $P_0 = (0, -1)$ (intersection with the y-axis), $P_1 = (1, -\frac{15}{4})$ (global and local minimum), $P_2 \approx (-\frac{1}{4}, 0)$, $P_3 \approx (\frac{7}{4}, 0)$ (intersection with the x-axis), $P_4 =$

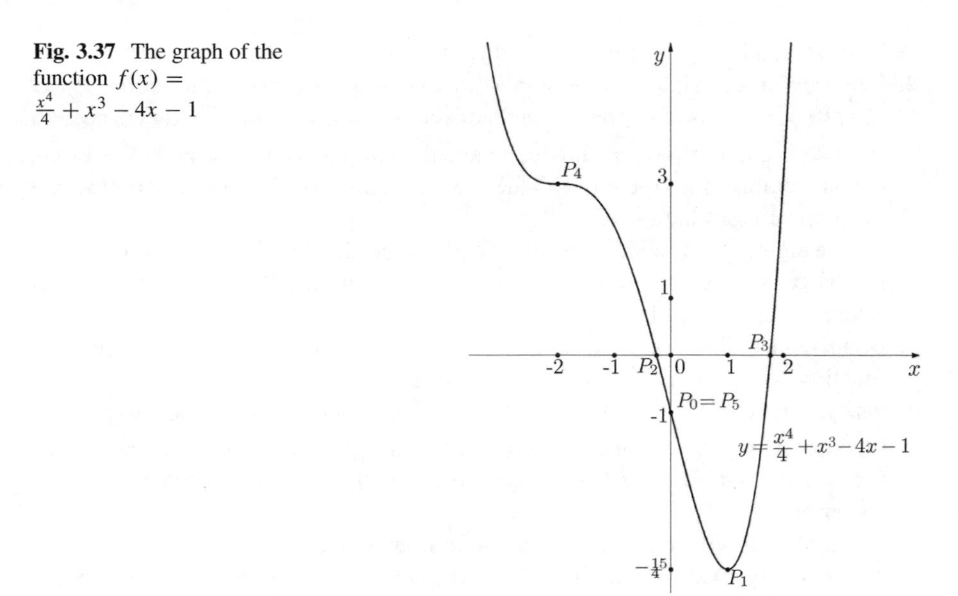

Fig. 3.37 The graph of the function $f(x) = \frac{x^4}{4} + x^3 - 4x - 1$

$(-2, 3)$ and $P_5 = P_0 = (0, -1)$ (inflection points), we can sketch the graph shown in Fig. 3.37.

4.3 *Study of Rational Functions*

A. Study of $f(x) = \frac{x^2}{x^2-1}$

1. Domain and range. The formula of the function is applicable for any real number except for the zeros of the denominator. Solving the trivial equation $x^2 = 1$, we specify the two zeros of the denominator $x = \pm 1$ and find that the domain is $X = \mathbb{R}\backslash\{-1, 1\}$.

   To determine the range Y, let us analyze for what values of y the equation $y = \frac{x^2}{x^2-1}$ with respect to unknown x has at least one solution in the domain X. Notice that the formula of $f(x)$ (or the relation $x^2 = (x^2 - 1)y$ following from this formula) shows that $y = 1$ should be excluded from Y, since there is no x such that $x^2 = x^2 - 1$. To verify other values of y, we transform the relation $y = \frac{x^2}{x^2-1}$ into the form $x^2 = \frac{y}{y-1}$ (we can divide by $y - 1$ because $y = 1$ was already excluded form the range). It is clear that this equation has solutions if, and only if, the right-hand side is non-negative. Therefore, the values admissible in Y satisfy the inequality $\frac{y}{y-1} \geq 0$, which has two parts of the solution:

   (1) $y > 1$ (which follows from the case $y \geq 0$ and $y - 1 > 0$);
   (2) $y \leq 0$ (which follows from the case $y \leq 0$ and $y - 1 < 0$).

Therefore, the range is $Y = (-\infty, 0] \cup (1, +\infty)$.

2. Intersection and sign. Intersection with the y-axis occurs at the point $P_0 = (0, f(0)) = (0, 0)$. To find the points of intersection with the x-axis we need to solve the equation $\frac{x^2}{x^2-1} = 0$, which have the unique solution $x = 0$. Therefore, the only point of intersection with the x-axis is $P_0 = (0, 0)$, the same point where the graph crosses the y-axis.

   The sign of the function was practically determined in the study of the range: if $0 < |x| < 1$, then $y < 0$; if $x = 0$, then $y = 0$; and if $|x| > 1$, then $y > 0$ (more precisely, $y > 1$).
3. Boundedness. The specified range $Y = (-\infty, 0] \cup (1, +\infty)$ indicates that the function is unbounded both below and above.
4. Parity. The relation $f(-x) = \frac{(-x)^2}{(-x)^2-1} = \frac{x^2}{x^2-1} = f(x), \forall x \in X$ shows that the function is even. Consequently, its graph is symmetric with respect to the y-axis.
5. Periodicity. Since only two points are missing in the domain, the function cannot be periodic.

   Another proof comes from the definition. Let us verify if there exists a constant $T \neq 0$ such that $f(x+T) = f(x)$. By the formula of the function, we have $\frac{(x+T)^2}{(x+T)^2-1} = \frac{x^2}{x^2-1}$ that can be written in the form $x^2[(x+T)^2 - 1] = (x^2-1)(x+T)^2$ or $x^2(x+T)^2 - x^2 = x^2(x+T)^2 - (x+T)^2$, or simplifying, $2xT + T^2 = 0$, whence $T(2x+T) = 0$. The last equation has two solutions—$T = 0$ and $T = -2x$, but both cannot be a period (the first is 0 and the second is not a constant). Therefore, $f(x)$ is non-periodic function.
6. Monotonicity and extrema.

   Since the function is even, it is sufficient to analyze the monotonicity only on the interval $[0, +\infty)$ (or, alternatively, on $(-\infty, 0]$). Taking $0 \le x_1 < x_2$, $x_1, x_2 \neq 1$ we have

$$D_1 = f(x_2) - f(x_1) = \frac{x_2^2}{x_2^2-1} - \frac{x_1^2}{x_1^2-1} = \frac{x_2^2(x_1^2-1) - x_1^2(x_2^2-1)}{(x_2^2-1)(x_1^2-1)}$$

$$= \frac{-x_2^2 + x_1^2}{(x_2^2-1)(x_1^2-1)} = \frac{(x_1-x_2)(x_1+x_2)}{(x_2^2-1)(x_1^2-1)}.$$

Under the choice $0 \le x_1 < x_2$, the numerator $(x_1 - x_2)(x_1 + x_2)$ is negative and the sign of the fraction depends only on denominator. Since the point $x = 1$ is excluded from the domain, it is natural to consider separately the behavior on $[0, 1)$ and on $(1, +\infty)$.

(1) On the interval $[0, 1)$ we have $x_2^2 - 1 < 0$ and $x_1^2 - 1 < 0$. Then, $D_1 < 0$, which shows that $f(x)$ is decreasing on $[0, 1)$.
(2) On the interval $(1, +\infty)$ we have $x_2^2 - 1 > 0$ and $x_1^2 - 1 > 0$. Then, $D_1 < 0$ and again $f(x)$ is decreasing on $(1, +\infty)$.

Using the property of the symmetry of even function, we conclude that $f(x)$ is increasing on $(-1, 0]$ and also on $(-\infty, -1)$.

It follows from the specified properties of monotonicity that each point in the intervals $(-\infty, -1)$, $(-1, 0)$, $(0, 1)$ and $(1, +\infty)$ is monotonic, and consequently, it cannot be extremum of any type. The only remaining point is $x = 0$, which is strict local maximum, since the function increases to the left of this point (in a left-side neighborhood of 0) and decreases to the right (in a right-side neighborhood of 0). This point is not a global maximum on the entire domain, since $f(0) = 0$ while the range includes positive values of the function (although $x = 0$ is a global maximum on the interval $(-1, 1)$).

7. Concavity and inflection.

We take two arbitrary points $x_1 < x_2$, both belonging to one of the intervals of the domain—$(-\infty, -1)$, $(-1, 1)$ or $(1, +\infty)$ –, define the midpoint $x_0 = \frac{x_1+x_2}{2}$ and increment $h = x_2 - x_0 = x_0 - x_1 > 0$, and use the quantity $D_2 \equiv f(x_1) + f(x_2) - 2f(x_0)$ to evaluate the difference between the mean value $\frac{f(x_1)+f(x_2)}{2}$ and the value at the midpoint $f(x_0)$. Taking advantage of the derived expression of D_1, we obtain:

$$
\begin{aligned}
D_2 &= (f(x_2) - f(x_0)) - (f(x_0) - f(x_1)) \\
&= \frac{(x_0 - x_2)(x_0 + x_2)}{(x_2^2 - 1)(x_0^2 - 1)} - \frac{(x_1 - x_0)(x_1 + x_0)}{(x_0^2 - 1)(x_1^2 - 1)} \\
&= -\frac{h}{x_0^2 - 1}\left[\frac{x_0 + x_2}{x_2^2 - 1} - \frac{x_1 + x_0}{x_1^2 - 1}\right] \\
&= -\frac{h}{x_0^2 - 1} \cdot \frac{(x_0 + x_2)(x_1^2 - 1) - (x_0 + x_1)(x_2^2 - 1)}{(x_2^2 - 1)(x_1^2 - 1)} \\
&= -\frac{h}{x_0^2 - 1} \cdot \frac{x_0x_1^2 - x_0x_2^2 + x_2x_1^2 - x_1x_2^2 - x_2 + x_1}{(x_2^2 - 1)(x_1^2 - 1)} \\
&= 2h^2 \frac{x_0x_1 + x_0x_2 + x_1x_2 + 1}{(x_0^2 - 1)(x_1^2 - 1)(x_2^2 - 1)}.
\end{aligned}
$$

Now we consider separately the three intervals of the domain.

(1) On $(-\infty, -1)$ we have $x_1 < x_0 < x_2 < -1$, and consequently, all the summands of the numerator are positive, as well as all the factors of the denominator. Therefore, $D_2 > 0$, which means that $f(x)$ is concave upward.

(2) On $(-1, 1)$ we have $-1 < x_1 < x_0 < x_2 < 1$, and writing the numerator in the form $x_0x_1 + x_0x_2 + x_1x_2 + 1 = \frac{x_1+x_2}{2}(x_1 + x_2) + x_1x_2 + 1 = \frac{(x_1+x_2)^2}{2} + (x_1x_2 + 1)$ we certify that it is positive. At the same time, the denominator is negative, since each of the three factors is negative. Therefore, $D_2 < 0$, which means that $f(x)$ is concave downward.

(3) On $(1, +\infty)$ we have $1 < x_1 < x_0 < x_2$, and consequently, all the summands of the numerator are positive, as well as all the factors of the denominator. Therefore, $D_2 > 0$ and $f(x)$ is concave upward.

Since the change of the type of concavity occurs only at the points $x = \pm 1$, which are not included in the domain, there is no point of inflection.

8. Complimentary properties.

   Let us see how the function behaves when $|x|$ goes to infinity ($x \to \pm\infty$). In this case, the function can be approximated through the leading terms of the numerator and denominator $f(x) = \frac{x^2}{x^2-1} \approx \frac{x^2}{x^2} = 1$ (see Sect. 2.3 for details). This shows that $f(x)$ approaches 1 when $x \to \pm\infty$, that is, $\lim\limits_{x \to \pm\infty} \frac{x^2}{x^2-1} = 1$. We can add that this convergence occurs by the values greater than 1, because $\frac{x^2}{x^2-1} > 1, \forall |x| > 1$. This implies that the function has the only horizontal asymptote $y = 1$.

   At each point of the domain, $x \neq \pm 1$, the function is continuous, since the numerator and denominator are defined and the denominator is different from 0 for any $x \neq \pm 1$. In other words, at each point $x = x_0 \neq \pm 1$ we have $f(x_0) = \frac{x_0^2}{x_0^2-1}$ and $f(x) \to f(x_0)$ when $x \to x_0$. Hence, no convergence to infinity is observed at the points of the domain. This implies that there is no vertical asymptote at any point of the domain X.

   The situation is different at the points $x = \pm 1$, where the denominator vanishes. At these points the function is not defined, but the behavior of the function when $x \to \pm 1$ can be analyzed. If $x \to 1^+$, that is, x approaches 1 from the right (by the values of x greater than 1), then $x^2 - 1 \to 0$ and $x^2 - 1 > 0$. This means that the ratio $\frac{x^2}{x^2-1}$ keeps positive values which increase without restriction, since the numerator is close to the constant 1 and the denominator converges to 0. In the symbols: $\lim\limits_{x \to 1^+} \frac{x^2}{x^2-1} = +\infty$. When $x \to 1^-$, that is, x approaches 1 from the left (by the values of x smaller than 1), then $x^2 - 1 \to 0$ and $x^2 - 1 < 0$. This shows that the ratio $\frac{x^2}{x^2-1}$ keeps negative values whose absolute value increase without restriction, since the numerator is close to the constant 1 and the denominator converges to 0. In the symbols: $\lim\limits_{x \to 1^-} \frac{x^2}{x^2-1} = -\infty$. Hence, $x = 1$ is a vertical asymptote of the graph of the function.

   Due to the symmetry of the graph with respect to the y-axis, we can complete that $\lim\limits_{x \to -1^+} \frac{x^2}{x^2-1} = -\infty$ and $\lim\limits_{x \to -1^-} \frac{x^2}{x^2-1} = +\infty$. Therefore, $x = -1$ is the second vertical asymptote of the graph of the function.

9. Using the investigated properties and marking some important points, such as $P_0 = (0, 0)$ (intersection with the coordinate axes and also local maximum), $P_1 = (\frac{1}{2}, -\frac{1}{3})$ (the value of x on the interval $(0, 1)$), $P_2 = (2, \frac{4}{3})$ (the value of x on the interval $(1, +\infty)$), we can plot the graph shown in Fig. 3.38.

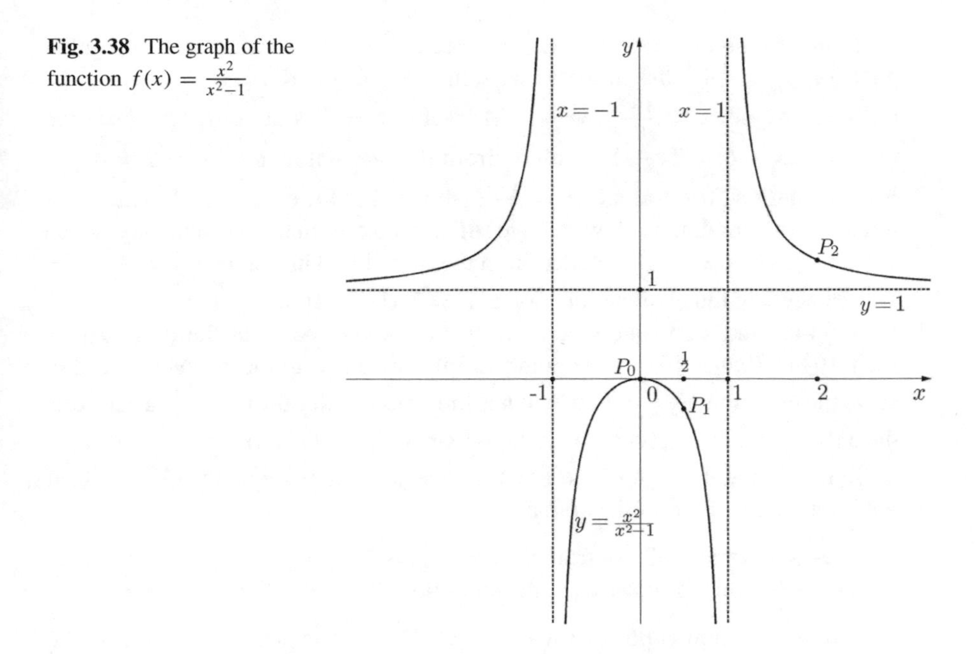

Fig. 3.38 The graph of the function $f(x) = \frac{x^2}{x^2-1}$

B. Study of $f(x) = \frac{2x^2+3x+1}{x^2+4x+3}$

1. Domain and range. The formula of the function is applicable for any real number except for the zeros of the denominator. Solving the elementary quadratic equation $x^2 + 4x + 3 = 0$, we specify the two zeros of the denominator $x = -3$ and $x = -1$ and determine that the domain is $X = \mathbb{R}\backslash\{-3, -1\}$.

   Using the found roots of the denominator, we can represent it in the form $x^2 + 4x + 3 = (x + 3)(x + 1)$. Before to proceed to specification of the range, let us find the roots of the numerator: $2x^2 + 3x + 1 = 0$, whence $x_3 = -1$ and $x_4 = -\frac{1}{2}$, which generates the following factorization of the numerator: $2x^2+3x+1 = (2x+1)(x+1)$. Noting that the numerator and denominator have the same root $x = -1$, we can simplify the formula of $f(x)$ as follows: $f(x) = \frac{2x^2+3x+1}{x^2+4x+3} = \frac{(x+1)(2x+1)}{(x+1)(x+3)} = \frac{2x+1}{x+3}$, $x \neq -1$ (-1 stay out of the domain, as well as -3). In what follows we will use this simplified form of the definition of $f(x)$, which is equivalent to the original formula on the domain $X = \mathbb{R}\backslash\{-3, -1\}$.

   As the result of the simplification, we obtain a fractional linear function $y = \frac{2x+1}{x+3}$ with the additional condition that $x = -1$ is excluded from the domain. The properties of fractional linear functions have been investigated in a general form in Sect. 2.2, and we have now the two options: specify general results in the case of the given function or perform an independent analysis of this function. We choose here to carry out an independent investigation and leave the use of general properties to the reader.

To determine the range of $f(x)$, we need to check for what values of y the equation $y = \frac{2x+1}{x+3}$ has at least one solution x in the domain X. Solving the equation, we get $x = \frac{1-3y}{y-2}$, $y \neq 2$. The value $y = 2$ is out of range, since the equation $2x + 6 = 2x + 1$, resulting from the original relation $y = 2 = \frac{2x+1}{x+3}$, has no solution. The formula $x = \frac{1-3y}{y-2}$ shows that for each $y \neq 2$ there is the unique corresponding x. However, we still should eliminate the value of y which corresponds to $x = -1$ (excluded from the domain). This value is $y = \frac{-2+1}{-1+3} = -\frac{1}{2}$. Hence, the range of the function is $Y = \mathbb{R}\backslash\{-\frac{1}{2}, 2\}$.

2. Intersection and sign. Intersection with the y-axis occurs at the point $P_0 = (0, f(0)) = (0, \frac{1}{3})$. To find the points of intersection with the x-axis we need to solve the equation $\frac{2x+1}{x+3} = 0$, which has the unique solution $x = -\frac{1}{2}$. Therefore, the only point of intersection with the x-axis is $P_1 = (-\frac{1}{2}, 0)$.

   To determine the sign of the function, we solve the inequality $\frac{2x+1}{x+3} > 0$. It holds on the two parts of the domain:

   (1) $x > -\frac{1}{2}$ (which follows from the conditions $2x + 1 > 0$ and $x + 3 > 0$);
   (2) $x < -3$ (which follows from the conditions $2x + 1 < 0$ and $x + 3 < 0$).

   Hence, the function is positive when $x < -3$ and also when $x > -\frac{1}{2}$. At the point $x = -\frac{1}{2}$ the function vanishes, and on the remaining points, that is, on the intervals $(-3, -1)$ and $(-1, -\frac{1}{2})$ the function is negative.

3. Boundedness. Knowing the range $Y = \mathbb{R}\backslash\{-\frac{1}{2}, 2\}$, we can immediately conclude that $f(x)$ is not bounded either below or above.

4. Parity. The calculation of the two values $f(-\frac{1}{2}) = 0$ and $f(\frac{1}{2}) = \frac{4}{7}$ shows that $f(-\frac{1}{2}) \neq \pm f(\frac{1}{2})$, and consequently, the function is neither even nor odd. (Another reason is that the domain is not symmetric about the origin.)

   Nevertheless, we can reveal a symmetry of the graph of the function using the following representation: $y = \frac{2x+1}{x+3} = \frac{2x+6-5}{x+3} = 2 - \frac{5}{x+3}$ (comparing with the general representation $y = y_0 + \frac{k}{x-x_0}$ in Sect. 2.2, we have here $y_0 = 2$, $x_0 = -3$ and $k = -5$). For each $x \in X$ we have $f(x-3) = 2 - \frac{5}{x}$ and $f(-x-3) = 2 + \frac{5}{x}$, which means that the points $\tilde{P}_1 = (x-3, 2-\frac{5}{x})$ and $\tilde{P}_2 = (-x-3, 2+\frac{5}{x})$ belong to the graph of $f(x)$. Notice that these points are symmetric about the point $\tilde{P}_0 = (-3, 2)$ (recall formulas (1.9) in Sect. 7.4, Chap. 1). Therefore, the graph of the function $y = \frac{2x+1}{x+3}$ is symmetric about the point $\tilde{P}_0 = (-3, 2)$. The only additional detail related to the original function is that the point $(-1, -\frac{1}{2})$ and its pair $(-5, \frac{9}{2})$ (symmetric about $\tilde{P}_0$) should be excluded from this symmetry, because $x = -1$ does not belong to the original domain.

5. Periodicity. Since only two points are missing in the domain, the function cannot be periodic.

   Another proof comes from the definition. Indeed, using the simplified formula of $f(x)$ and the condition of periodicity, we get $\frac{2(x+T)+1}{(x+T)+3} = \frac{2x+1}{x+3}$, which can be rewritten in the form $(2x + 2T + 1)(x + 3) = (x + T + 3)(2x + 1)$, whence $T = 0$, which is not admissible for a periodic function.

6. Monotonicity and extrema.

   Differently from the geometric approach used in study of general fractional linear function, let us apply here the analytic method.

   Taking $x_1 < x_2$, $x_1, x_2 \neq -3, -1$ we have

$$D_1 = f(x_2) - f(x_1) = \frac{2x_2 + 1}{x_2 + 3} - \frac{2x_1 + 1}{x_1 + 3}$$
$$= \frac{(2x_2 + 1)(x_1 + 3) - (2x_1 + 1)(x_2 + 3)}{(x_2 + 3)(x_1 + 3)} = \frac{5(x_2 - x_1)}{(x_2 + 3)(x_1 + 3)}.$$

The numerator $5(x_2 - x_1)$ is positive, and the sign of the expression depends only on the denominator. Since the points $x = -3$ and $x = -1$ are out of the domain, it is natural to consider the monotonicity separately on the intervals $(-\infty, -3)$, $(-3, -1)$ and $(-1, +\infty)$.

(1) On $(-\infty, -3)$ we have $x_1 + 3 < 0$, $x_2 + 3 < 0$. Then $D_1 > 0$, which means that $f(x)$ increases on $(-\infty, -3)$.
(2) On $(-3, -1)$ we have $x_1 + 3 > 0$, $x_2 + 3 > 0$. Consequently, $D_1 > 0$ and $f(x)$ increases on $(-3, -1)$.
(3) On $(-1, +\infty)$ we get $x_1 + 3 > 0$, $x_2 + 3 > 0$. Then $D_1 > 0$, which means that $f(x)$ increases on $(-1, +\infty)$.

Based on the properties of the monotonicity, we can conclude that $f(x)$ has no extremum of any type. Indeed, each point of the intervals $(-\infty, -3)$, $(-3, -1)$ and $(-1, +\infty)$ is increasing and cannot be an extremum. There is no other point of the domain. Therefore, the function has no extremum.

7. Concavity and inflection.

   For two arbitrary points $x_1 < x_2$, both belonging to one of the intervals of the domain—$(-\infty, -3)$, $(-3, -1)$ or $(-1, +\infty)$—we use the midpoint $x_0 = \frac{x_1+x_2}{2}$ and increment $h = x_2 - x_0 = x_0 - x_1 > 0$ to evaluate the difference between the mean value $\frac{f(x_1)+f(x_2)}{2}$ and the value at the midpoint $f(x_0)$ through the quantity $D_2 \equiv f(x_1) + f(x_2) - 2f(x_0)$:

$$D_2 = (f(x_2) - f(x_0)) - (f(x_0) - f(x_1))$$
$$= \frac{5(x_2 - x_0)}{(x_2+3)(x_0+3)} - \frac{5(x_0 - x_1)}{(x_0+3)(x_1+3)} = \frac{5h}{x_0 + 3}\left[\frac{1}{x_2+3} - \frac{1}{x_1+3}\right]$$
$$= 5h\frac{x_1 - x_2}{(x_0 + 3)(x_1 + 3)(x_2 + 3)} = \frac{-10h^2}{(x_0 + 3)(x_1 + 3)(x_2 + 3)}.$$

The numerator is negative, and the sign of the denominator is easy to determine separately on each of the three intervals which compose the domain.

(1) On $(-\infty, -3)$ we have $x_0 + 3 < 0$, $x_1 + 3 < 0$ and $x_2 + 3 < 0$. Therefore, $D_2 > 0$, which means that $f(x)$ is concave upward on $(-\infty, -3)$.
(2) On $(-3, -1)$ we get $x_0 + 3 > 0$, $x_1 + 3 > 0$ and $x_2 + 3 > 0$. Consequently, $D_2 < 0$ and $f(x)$ has downward concavity on $(-3, -1)$.
(3) On $(-1, +\infty)$ we have $x_0 + 3 > 0$, $x_1 + 3 > 0$ and $x_2 + 3 > 0$. This results in $D_2 < 0$, and consequently, $f(x)$ is concave downward on $(-1, +\infty)$.

Since the change of the concavity occurs only at the point $x = -3$, which is out of the domain, there is no point of inflection.

8. Complimentary properties.

Let us analyze what happens when $|x|$ goes to infinity ($x \to \pm\infty$). In this case, the function can be approximated through the leading terms of the numerator and denominator: $f(x) = \frac{2x+1}{x+3} \approx \frac{2x}{x} = 2$ (see Sect. 2.3 for details). This shows that $f(x)$ approaches 2 when $x \to \pm\infty$, in symbols, $\lim\limits_{x\to\pm\infty} \frac{2x+1}{x+3} = 2$. This implies that the function has the only horizontal asymptote $y = 2$. We can add that when $x \to +\infty$ the convergence to 2 occurs by the values smaller than 2, because $\frac{2x+1}{x+3} < \frac{2x+6}{x+3} = 2$, $\forall x > 0$. While when $x \to -\infty$ this convergence happens by the values greater than 2, because $\frac{2x+1}{x+3} = \frac{-2x-1}{-x-3} > \frac{-2x-6}{-x-3} = 2$, $\forall x < -3$.

At each point of the domain, $x \neq -3$, $x \neq -1$, the function is continuous, since the numerator and denominator are defined and the denominator is different from 0. In other words, at each point $x = x_0 \neq -3, \neq -1$ we have $f(x_0) = \frac{2x_0+1}{x_0+3}$ and $f(x) \to f(x_0)$ when $x \to x_0$. Hence, no convergence to infinity is observed at the points of the domain. This implies that there is no vertical asymptote at any point of the domain X.

The situation is different at the points $x = -3$ and $x = -1$, where the original denominator vanishes. At these points the function is not defined, but the behavior of the function when x approaches one of these points can be analyzed. At the point $x = -1$ the function is not defined, but $\frac{2x+1}{x+3}$ approaches $-\frac{1}{2}$ as $x \to -1$ (it is sufficient to calculate some values of the function at the points near $x = -1$ to note this behavior). Then, $f(x)$ has no vertical asymptote at the point $x = -1$.

At the point $x = -3$ both the denominator of the original and simplified formulas vanish. Notice that both numerators (original and simplified) are different from 0. Analyzing the behavior of the function when $x \to -3^+$, that is, x approaches -3 from the right, we have that $2x + 1 \to -5$ and $x + 3 \to 0$ by positive values. Therefore, the ratio $\frac{2x+1}{x+3}$ keeps negative values whose absolute value increases without restriction. In other words, $f(x) \to -\infty$ when $x \to -3^+$: $\lim\limits_{x\to-3^+} \frac{2x+1}{x+3} = -\infty$. When $x \to -3^-$, that is, x approaches -3 from the left, then $2x + 1 \to -5$ and $x + 3 \to 0$ by negative values. This implies that the ratio $\frac{2x+1}{x+3}$ keeps positive values which increase without restriction. In other words, $\lim\limits_{x\to-3^-} \frac{2x+1}{x+3} = +\infty$. Consequently, $x = -3$ is a vertical asymptote (the unique) of the graph of $f(x)$.

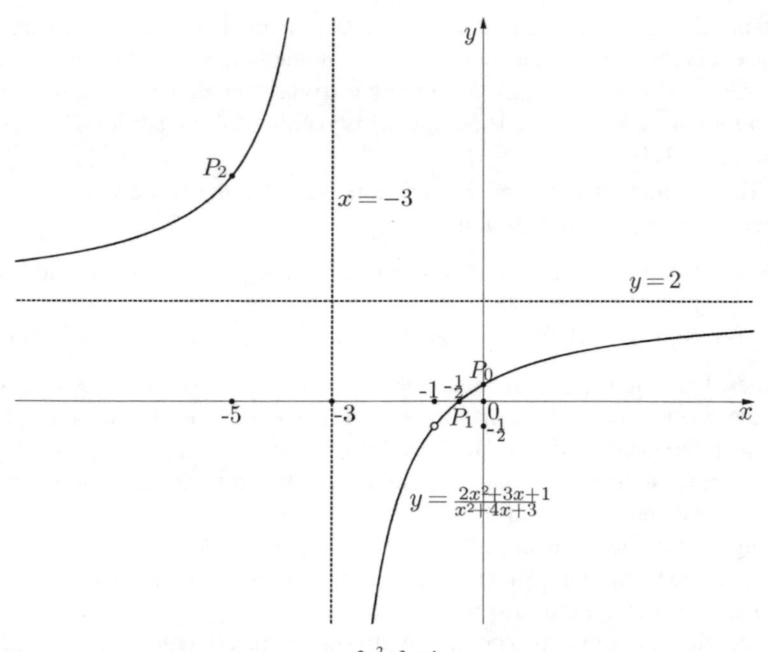

Fig. 3.39 The graph of the function $f(x) = \frac{2x^2+3x+1}{x^2+4x+3}$

9. Using the investigated properties and marking some important points, such as $P_0 = (0, \frac{1}{3})$ (intersection with the y-axis), $P_1 = (-\frac{1}{2}, 0)$ (intersection with the x-axis), $P_2 = (-5, \frac{9}{2})$ (the point to the left of the vertical asymptote), we can sketch the graph shown in Fig. 3.39.

C. Study of $f(x) = \frac{x^2-1}{x^2-3x}$

1. Domain and image. The formula of the function is applicable for any real number except for the zeros of the denominator. Solving the trivial equation $x^2 - 3x = 0$, we specify the two zeros $x = 0$ and $x = 3$ of the denominator and determine that the domain is $X = \mathbb{R}\backslash\{0, 3\}$.

   To determine the range Y, let us analyze for what values of y the equation $y = \frac{x^2-1}{x^2-3x}$ with respect to unknown x has at least one solution in the domain X. Rewriting this relation in the form of the quadratic equation $(1-y)x^2+3yx-1 = 0$ and noting that its discriminant is positive for any real y: $9y^2 + 4(1 - y) = (y - 2)^2 + 8y^2 > 0$, we see that the equation for x has real solutions for any real y. Since the transfer of $x^2 - 3x$ to the numerator does not generate any additional value of y (for $x = 0$ we have the impossible relation $-1 = 0$ and for $x = 3$ another impossible relation $9 - 1 = 0$), we conclude that the range contains all the real numbers: $Y = \mathbb{R}$.

2. Intersection and sign. The point $x = 0$ is excluded from the domain, and consequently, there is no intersection with the y-axis. To find the points of intersection with the x-axis we solve the trivial equation $x^2 - 1 = 0$, whose solutions are $x = \pm 1$, and determine the two points of intersection $P_1 = (-1, 0)$ and $P_2 = (1, 0)$.

   To determine the sign of the function we solve the inequality $\frac{x^2-1}{x^2-3x} > 0$. There are two parts of the solution:

   (1) $x \in (-\infty, -1) \cup (3, +\infty)$ (which is the consequence of the conditions $x^2 - 1 > 0$ and $x^2 - 3x > 0$);
   (2) $x \in (0, 1)$ (which follows from the conditions $x^2 - 1 < 0$ and $x^2 - 3x < 0$).

   Hence, the function is positive on the intervals $(-\infty, -1)$, $(3, +\infty)$ and $(0, 1)$. At the points $x = \pm 1$ the function vanishes, and at the remaining points, that is, on the intervals $(-1, 0)$ and $(1, 3)$ the function is negative.
3. Boundedness. Since the range includes all the real points, the function is unbounded both below and above.
4. Parity. It is sufficient to calculate the two values $f(-2) = \frac{3}{10}$ and $f(2) = -\frac{3}{2}$ to conclude that the function is neither even nor odd. (Another reason is that the domain is not symmetric about the origin.)
5. Periodicity. Since only two points are missing in the domain, the function cannot be periodic. (Another argument comes from monotonicity.)
6. Monotonicity and extrema.

   Take $x_1 < x_2$, $x_1, x_2 \neq 0, 3$ and obtain

$$D_1 \equiv f(x_2) - f(x_1) = \frac{x_2^2 - 1}{x_2^2 - 3x_2} - \frac{x_1^2 - 1}{x_1^2 - 3x_1}$$
$$= \frac{(x_2^2 - 1)(x_1^2 - 3x_1) - (x_1^2 - 1)(x_2^2 - 3x_2)}{(x_2^2 - 3x_2)(x_1^2 - 3x_1)}$$
$$= (x_1 - x_2)\frac{3x_1x_2 - x_1 - x_2 + 3}{x_1x_2(x_1 - 3)(x_2 - 3)}.$$

   The factor $(x_1 - x_2)$ is negative. Besides, it is easy to see that the products x_1x_2 and $(x_1 - 3)(x_2 - 3)$ are positive in each of the natural intervals of the domain: $(-\infty, 0)$, $(0, 3)$ and $(3, +\infty)$, which means that the denominator is also positive in each of these three intervals. It remains to analyze the sign of the numerator $A_1 = 3x_1x_2 - x_1 - x_2 + 3$, that we make separately in each of the three intervals $(-\infty, 0)$, $(0, 3)$ and $(3, +\infty)$.

   (1) On the interval $(-\infty, 0)$ we have $x_1x_2 > 0$ and $3 - x_1 - x_2 > 0$. Then $A_1 > 0$, whence $D_1 < 0$.
   (2) On the interval $(0, 3)$ we regroup the terms of A_1 and make the evaluation in the form $A_1 = (x_1 - 1)(x_2 - 1) + 2(x_1x_2 + 1) > (0 - 1)(3 - 1) + 2(0 + 1) = 0$. Again, $A_1 > 0$, whence $D_1 < 0$.

(3) On the interval $(3, +\infty)$ we have $A_1 = (x_1 - 1)(x_2 - 1) + 2(x_1x_2 + 1) > 0$, whence $D_1 < 0$.

Hence, $f(x)$ decreases on each of the three intervals $(-\infty, 0)$, $(0, 3)$ and $(3, +\infty)$.

Using the established properties of monotonicity, we conclude that $f(x)$ has no extremum of any type. Indeed, each point in any of the three intervals $(-\infty, 0)$, $(0, 3)$ and $(3, +\infty)$ is decreasing and cannot be an extremum. There is no other point of the domain, which means that $f(x)$ has no extremum.

From the decrease on the infinite interval $(3, +\infty)$ it follows that the function is not periodic.

7. Concavity and inflection.

To evaluate the difference between the mean value $\frac{f(x_1)+f(x_2)}{2}$ and the value at the midpoint $f(x_0)$, $x_0 = \frac{x_1+x_2}{2}$ we use the quantity $D_2 = f(x_2) + f(x_1) - 2f(x_0)$, which we express employing the result derived for $D_1 = f(x_2) - f(x_1)$:

$$\begin{aligned}D_2 &= (f(x_2) - f(x_0)) - (f(x_0) - f(x_1))\\ &= (x_0 - x_2)\frac{3x_2x_0 - x_2 - x_0 + 3}{x_2x_0(x_2 - 3)(x_0 - 3)} - (x_1 - x_0)\frac{3x_1x_0 - x_1 - x_0 + 3}{x_1x_0(x_1 - 3)(x_0 - 3)}\\ &= -\frac{h}{x_0(x_0 - 3)}\left(\frac{3x_2x_0 - x_2 - x_0 + 3}{x_2(x_2 - 3)} - \frac{3x_1x_0 - x_1 - x_0 + 3}{x_1(x_1 - 3)}\right)\\ &= -h\frac{(3x_2x_0 - x_2 - x_0 + 3)x_1(x_1 - 3) - (3x_1x_0 - x_1 - x_0 + 3)x_2(x_2 - 3)}{x_0x_1x_2(x_0 - 3)(x_1 - 3)(x_2 - 3)}.\end{aligned}$$

(here $h = x_2 - x_0 = x_0 - x_1 > 0$). The main problem in determining the sign of D_2 is the analysis of the numerator. Let us regroup its terms in the following form:

$$\begin{aligned}\tilde{A} &\equiv (3x_2x_0 - x_2 - x_0 + 3)x_1(x_1 - 3) - (3x_1x_0 - x_1 - x_0 + 3)x_2(x_2 - 3)\\ &= [x_2(3x_0 - 1) + (3 - x_0)]x_1(x_1 - 3) - [x_1(3x_0 - 1) + (3 - x_0)]x_2(x_2 - 3)\\ &= (3x_0 - 1)[x_1x_2(x_1 - 3) - x_1x_2(x_2 - 3)] + (3 - x_0)[x_1(x_1 - 3) - x_2(x_2 - 3)]\\ &= (3x_0 - 1)x_1x_2(x_1 - x_2) + (3 - x_0)[x_1^2 - x_2^2 + 3(x_2 - x_1)]\\ &= (x_1 - x_2)[(3x_0 - 1)x_1x_2 + (3 - x_0)(x_1 + x_2 - 3)].\end{aligned}$$

Since $x_1 - x_2 = -2h$, we still write D_2 as follows:

$$D_2 = 2h^2\frac{(3x_0 - 1)x_1x_2 + (3 - x_0)(x_1 + x_2 - 3)}{x_0x_1x_2(x_0 - 3)(x_1 - 3)(x_2 - 3)} \equiv 2h^2\frac{A_2}{B_2}.$$

Using a preliminary numerical calculations of A_2 at some points, we can see that the sign is different on the intervals $(0, 1)$ and $(1, 3)$ (it is sufficient to take x_1, x_2 near $\frac{1}{2}$ and near 2), which suggests the additional division of the interval $(0, 3)$ into to subintervals. Therefore, let us evaluate the sign of D_2 separately on the intervals $(-\infty, 0)$, $(0, 1)$, $(1, 3)$ and $(3, +\infty)$.

(1) On $(-\infty, 0)$ we have $x_0x_1x_2 < 0$, $(x_0-3)(x_1-3)(x_2-3) < 0$, which shows that $B_2 > 0$. At the same time, $(3x_0-1)x_1x_2 < 0$ and $(3-x_0)(x_1+x_2-3) < 0$, which implies that $A_2 < 0$. Then, $D_2 < 0$, which means that $f(x)$ is concave downward on $(-\infty, 0)$.

(2) On $(0, 1)$ we have $x_0x_1x_2 > 0$, $(x_0 - 3)(x_1 - 3)(x_2 - 3) < 0$, and consequently, $B_2 < 0$. To evaluate A_2 we represent it in the form

$$A_2 = (3x_0 - 1)x_1x_2 + (3 - x_0)(x_1 + x_2 - 3)$$
$$= (3x_0 - 3 + 2)x_1x_2 + (3 - x_0)(x_1 + x_2) - 3(3 - 3x_0 + 2x_0)$$
$$= 3(x_0 - 1)x_1x_2 + 2x_1x_2 + (3 - x_0)(x_1 + x_2) - 9(1 - x_0) - 3(x_1 + x_2)$$
$$= 3(x_0 - 1)x_1x_2 + (2x_1x_2 - x_1 - x_2) + (1 - x_0)(x_1 + x_2) - 9(1 - x_0)$$
$$= 3(x_0 - 1)x_1x_2 + [x_1(x_2 - 1) + x_2(x_1 - 1)] + (1 - x_0)(x_1 + x_2 - 9).$$

Now it is seen that all the three summands are negative, and consequently, $A_2 < 0$. Therefore, $D_2 > 0$ and $f(x)$ has upward concavity on $(0, 1)$.

(3) On $(1, 3)$ we have $x_0x_1x_2 > 0$, $(x_0 - 3)(x_1 - 3)(x_2 - 3) < 0$, which gives $B_2 < 0$. For A_2 we use another representation:

$$A_2 = (3x_0 - 1)x_1x_2 + (3 - x_0)(x_1 + x_2 - 3)$$
$$= (3x_0 - 1)x_1x_2 + (3 - x_0)(x_1 + x_2 - 2) - (3 - x_0)$$
$$= [(3x_0 - 1)x_1x_2 - 2] + (3 - x_0)(x_1 + x_2 - 2) + (x_0 - 1).$$

Now it is clear that all the three summands are positive, and consequently, $A_2 > 0$. Therefore, $D_2 < 0$ and $f(x)$ has downward concavity on $(1, 3)$.

(4) On $(3, +\infty)$ we have $x_0x_1x_2 > 0$, $(x_0 - 3)(x_1 - 3)(x_2 - 3) > 0$, which shows that $B_2 > 0$. For A_2 we use one more representation

$$A_2 = (3x_0 - 1)x_1x_2 + (3 - x_0)(x_1 + x_2 - 3)$$
$$= (x_0 - 1)x_1x_2 + 2x_0x_1x_2 + 3(x_1 + x_2 - 3) - x_0(x_1 + x_2 - 3)$$
$$= (x_0 - 1)x_1x_2 + 3(x_1 + x_2 - 3) + x_0(2x_1x_2 - x_1 - x_2 + 3)$$
$$= (x_0 - 1)x_1x_2 + 3(x_1 + x_2 - 3) + x_0[x_1(x_2 - 1) + x_2(x_1 - 1) + 3).$$

Since all the three summands are positive, it follows that $A_2 > 0$. Therefore, $D_2 > 0$ and $f(x)$ is concave upward on $(3, +\infty)$.

The change of the type of concavity occurs at the three points $x = 0$, $x = 1$ and $x = 3$, but the first and the third are out of the domain. Hence, the unique inflection point is $x = 1$.

8. Complimentary properties.

Let us see the behavior of function when $|x|$ goes to infinity ($x \to \pm\infty$). In this case, the function can be approximated through the leading terms of the numerator and denominator $f(x) = \frac{x^2-1}{x^2-3x} \approx \frac{x^2}{x^2} = 1$ (see Sect. 2.3 for details). This shows that $f(x)$ approaches 1 when $x \to \pm\infty$, that is, $\lim\limits_{x\to\pm\infty} \frac{x^2-1}{x^2-3x} = 1$. This implies that the function has the only horizontal asymptote $y = 1$. We can add that the convergence to 1 as x goes to $+\infty$ occurs by the values greater than 1, because $\frac{x^2-1}{x^2-3x} > \frac{x^2-1}{x^2-1} = 1, \forall x > 3$. When x goes to $-\infty$ the convergence to 1 occurs by the values smaller than 1, because $\frac{x^2-1}{x^2-3x} < \frac{x^2-1}{x^2} < 1, \forall x < 0$.

At each point of the domain $x \neq 0, 3$, the function is continuous, since the numerator and denominator are defined and the denominator is different from 0 for any $x \neq 0, 3$. In other words, at each point $x = x_0 \neq 0, 3$ we have $f(x_0) = \frac{x_0^2}{x_0^2-1}$ and $f(x) \to f(x_0)$ when $x \to x_0$. Hence, no convergence to infinity is observed at the points of the domain, which implies that there is no vertical asymptote at any point of the domain X.

The situation is different at the points $x = 0$ and $x = 3$, where the denominator vanishes. At these points the function is not defined, but the behavior of the function when x approaches one of these points can be analyzed. When $x \to 3^+$, that is, x approaches 3 from the right, we have that $x^2 - 1 \to 8$ and $x^2 - 3x \to 0$ by positive values. This shows that the ratio $\frac{x^2-1}{x^2-3x}$ keeps positive values which increase without restriction. In other words, $f(x) \to +\infty$ when $x \to 3^+$: $\lim\limits_{x\to3^+} \frac{x^2-1}{x^2-3x} = +\infty$. When $x \to 3^-$, that is, x approaches 3 from the left, then $x^2 - 1 \to 8$ and $x^2 - 3x \to 0$ by negative values. This implies that the ratio $\frac{x^2-1}{x^2-3x}$ keeps negative values whose absolute value increases without restriction. In other words, $\lim\limits_{x\to3^-} \frac{x^2-1}{x^2-3x} = -\infty$. Consequently, $x = 3$ is a vertical asymptote of the graph of $f(x)$.

In a similar manner we investigate the behavior of $f(x)$ when x approaches 0. If $x \to 0^+$, that is, x approaches 0 from the right, then $x^2 - 1 \to -1$ and $x^2 - 3x \to 0$ by negative values. This indicates that the ratio $\frac{x^2-1}{x^2-3x}$ keeps positive values which increase without restriction, that is, $\lim\limits_{x\to0^+} \frac{x^2-1}{x^2-3x} = +\infty$. If $x \to 0^-$, that is, x approaches 0 from the left, then $x^2-1 \to -1$ and $x^2-3x \to 0$ by positive values. This implies that the ratio $\frac{x^2-1}{x^2-3x}$ keeps negative values whose absolute value increases without restriction, that is, $\lim\limits_{x\to0^-} \frac{x^2-1}{x^2-3x} = -\infty$. Consequently, $x = 0$ is the second vertical asymptote of the graph of $f(x)$.

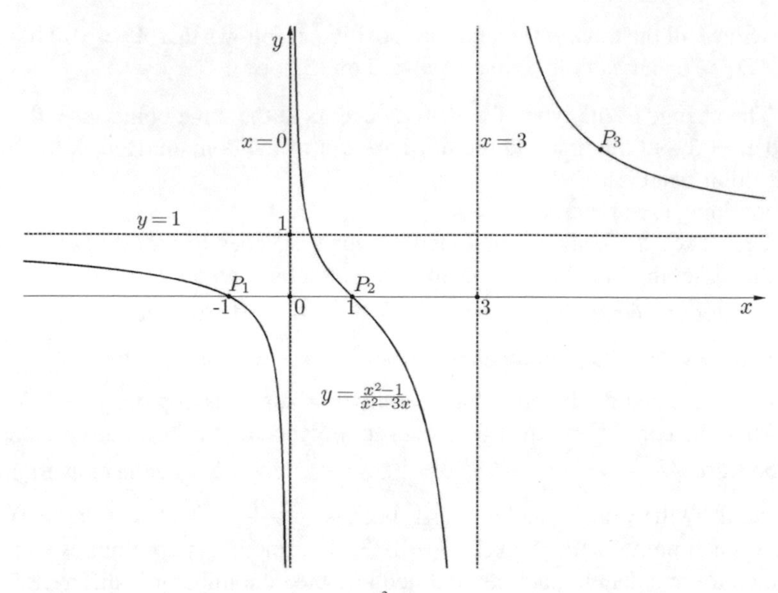

Fig. 3.40 The graph of the function $f(x) = \frac{x^2-1}{x^2-3x}$

9. Using the investigated properties and marking some important points, such as $P_1 = (-1, 0)$ (intersection with the x-axis), $P_2 = (1, 0)$ (intersection with the x-axis and inflection) and $P_3 = (5, \frac{12}{5})$ (a point on the right part of the graph), we sketch the graph shown in Fig. 3.40.

4.4 Study of Irrational Functions

A. Study of $f(x) = \sqrt{2x+6}$

The properties of this function can be derived through geometric transformations starting from the simpler function $\sqrt{x}$ which was studied in Sect. 3.1. However, here we choose to apply the analytic method of investigation.

1. Domain and range. The square root is defined only for non-negative values, which leads to the restriction $x \geq -3$ and to the domain $X = [-3, +\infty)$.

   On the other hand, by the definition, the value of a square root is a non-negative number, which means that the range Y is contained in $[0, +\infty)$. To show that each y of this interval belongs to the range, we need to solve the equation $\sqrt{2x+6} = y$, $\forall y \geq 0$ for unknown x. For any $y \geq 0$, the solution is trivial $x = \frac{y^2-6}{2}$ and it belongs to the domain X. Therefore, the range is $Y = [0, +\infty)$.
2. Intersection and sign. Intersection with the y-axis occurs at the point $P_0 = (0, \sqrt{6})$. According to the definition of square root, the function is positive over

the entire domain, except for $x = -3$, where it vanishes. Therefore, the unique point of intersection with the x-axis is $P_1 = (-3, 0)$.

3. Boundedness. Knowing that the range is $Y = [0, +\infty)$, we immediately conclude that $f(x)$ is bounded below and unbounded above.
4. Parity. Since the domain $X = [-3, +\infty)$ is not symmetric with respect to the origin, the function is neither even nor odd.
5. Periodicity. The form of the domain (the fact that it is bounded on the left) implies that the function is not periodic.
6. Monotonicity and extrema. Taking $-3 \le x_1 < x_2$ and using the properties of square root, we can see that $2x_1 + 6 < 2x_2 + 6$ ensures that $f(x_1) = \sqrt{2x_1+6} < \sqrt{2x_2+6} = f(x_2)$, that is, the function is increasing over the entire domain. Another option to obtain the same result is by applying the property of composition of monotonic functions (Sect. 9.3 of Chap. 2): the function $\tilde{f}(x) = 2x + 6$ is increasing on $\mathbb{R}$ and the function $\tilde{g}(y) = \sqrt{y}$ is increasing on $[0, +\infty)$, which implies that the composite function $f(x) = \tilde{g}(\tilde{f}(x)) = \sqrt{2x+6}$ is increasing on its domain.

   The immediate consequence of the increase on the interval $(-3, +\infty)$ is that no point of this interval can be an extremum of any type. It remains the only point of the domain, $x = -3$, which is a strict global minimum (although it is not a local minimum, because it does not belong to the domain together with some neighborhood).
7. Concavity and inflection.

   To evaluate the difference between the mean value $\frac{f(x_1)+f(x_2)}{2}$ and the value at the midpoint $f(x_0)$, $x_0 = \frac{x_1+x_2}{2}$, we use the quantity $D_2 = f(x_2) + f(x_1) - 2f(x_0)$:

$$D_2 = f(x_2) + f(x_1) - 2f(x_0) = \sqrt{2x_2+6} + \sqrt{2x_1+6} - 2\sqrt{2x_0+6}$$

$$= \frac{(\sqrt{2x_2+6} - \sqrt{2x_0+6})(\sqrt{2x_2+6} + \sqrt{2x_0+6})}{\sqrt{2x_2+6} + \sqrt{2x_0+6}}$$

$$+ \frac{(\sqrt{2x_1+6} - \sqrt{2x_0+6})(\sqrt{2x_1+6} + \sqrt{2x_0+6})}{\sqrt{2x_1+6} + \sqrt{2x_0+6}}$$

$$= \frac{2(x_2 - x_0)}{\sqrt{2x_2+6} + \sqrt{2x_0+6}} + \frac{2(x_1 - x_0)}{\sqrt{2x_1+6} + \sqrt{2x_0+6}}$$

$$= 2h \frac{\sqrt{2x_1+6} - \sqrt{2x_2+6}}{(\sqrt{2x_2+6} + \sqrt{2x_0+6})(\sqrt{2x_1+6} + \sqrt{2x_0+6})}$$

$$= 2h \frac{(\sqrt{2x_1+6} - \sqrt{2x_2+6})(\sqrt{2x_1+6} + \sqrt{2x_2+6})}{(\sqrt{2x_2+6} + \sqrt{2x_0+6})(\sqrt{2x_1+6} + \sqrt{2x_0+6})(\sqrt{2x_1+6} + \sqrt{2x_2+6})}$$

$$= \frac{-8h^2}{(\sqrt{2x_2+6} + \sqrt{2x_0+6})(\sqrt{2x_1+6} + \sqrt{2x_0+6})(\sqrt{2x_1+6} + \sqrt{2x_2+6})}$$

(here, $h = x_2 - x_0 = x_0 - x_1 > 0$). This representation shows that $D_2 < 0$ over the entire domain, and consequently, $f(x)$ has downward concavity over the entire domain. It follows that there is no inflection point.

Another way to investigate concavity is by using the property of composition of concave functions (Sect. 9.3 of Chap. 2). Indeed, the function $\tilde{f}(x) = 2x+6$ is concave downward (non-strictly) on $\mathbb{R}$ and $\tilde{g}(y) = \sqrt{y}$ is concave downward on $[0, +\infty)$ and is increasing. Therefore, the composite function $f(x) = \tilde{g}(\tilde{f}(x)) = \sqrt{2x+6}$ is concave downward on its domain.

8. Complimentary properties. If x goes to $+\infty$, the values of the function increase without restriction, that is, $\lim_{x\to+\infty} \sqrt{2x+6} = +\infty$. On the other side, x cannot goes to $-\infty$, since the domain contains only the points $x \geq -3$. Therefore, the function has no horizontal asymptote.

   At each point $x \geq -3$ the function is continuous, that is, $f(x) \to f(x_0)$ when $x \to x_0 \geq -3$. Therefore, there is no convergence to infinity at the points of the domain X. On the other hand, x cannot goes to any number $a < -3$, since there is a neighborhood of any such number which is not contained in the domain. Therefore, the function has no vertical asymptote.
9. Using the investigated properties and marking some important points, such as $P_0 = (0, \sqrt{6})$ (intersection with the y-axis), $P_1 = (-3, 0)$ (intersection with the x-axis) and $P_2 = (1, \sqrt{8})$ (a point to the right of the y-axis), we plot the graph shown in Fig. 3.41.

We leave to the reader the geometric approach, transforming $f_0(x) = \sqrt{x}$ into $f(x) = \sqrt{2x+6}$ through the chain $f_0(x) = \sqrt{x} \to f_1(x) = f_0(2x) = \sqrt{2x} \to f(x) = f_1(x+3) = \sqrt{2(x+3)}$, which leads to the same results.

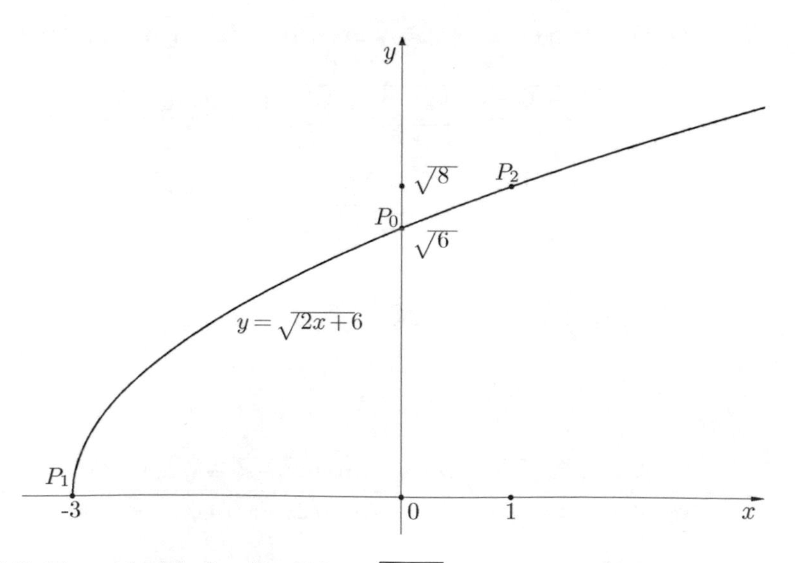

Fig. 3.41 The graph of the function $f(x) = \sqrt{2x+6}$

B. Study of $f(x) = \frac{1}{\sqrt[3]{4-2x}}$

1. Domain and range. Since the cubic root is defined for any real number, the only restriction for the domain goes from the denominator, which vanishes at the point $x = 2$. Hence, the domain of $f(x)$ is $X = \mathbb{R}\backslash\{2\}$.

   To determine the range Y, notice first that the value 0 is out of the range, and then consider the equation $\frac{1}{\sqrt[3]{4-2x}} = y$, $\forall y \neq 0$. Solving for x, we get $x = 2 - \frac{1}{2y^3}$, that is, for any real $y \neq 0$ there exists corresponding x in the domain. This means that $Y = \mathbb{R}\backslash\{0\}$.
2. Intersection and sign. Intersection with the y-axis occurs at the point $P_0 = (0, \frac{1}{\sqrt[3]{4}})$. Since $y = 0$ is out of the range, there is no intersection with the x-axis.

   According to the definition of the cubic root, the function is positive on the interval $(-\infty, 2)$ and negative on $(2, +\infty)$.
3. Boundedness. Since the range is determined, we conclude that $f(x)$ is unbounded both below and above.
4. Parity. Since the domain is not symmetric about the origin, the function is neither even nor odd.

   However, writing the function in the form $f(x) = \frac{1}{\sqrt[3]{2(2-x)}}$, we can see that its points are symmetric with respect to the point $(2, 0)$. Indeed, for any $x \in X$ we have $f(2-x) = \frac{1}{\sqrt[3]{2x}} = -\frac{1}{\sqrt[3]{-2x}} = -f(2+x)$. This means that the points $P_1 = (2-x, \frac{1}{\sqrt[3]{2x}})$ and $P_2 = (2+x, -\frac{1}{\sqrt[3]{2x}})$, symmetric with respect to the point $(2, 0)$, belong to the graph of the function (see the formulas of symmetry (1.9) in Sect. 7.4, Chap. 1).
5. Periodicity. The fact that the domain has the only real point excluded implies that the function cannot be periodic.
6. Monotonicity and extrema. Let us verify first the monotonicity of the simpler function $g(x) = \frac{1}{\sqrt[3]{x}}$. For any pair $x_1 < x_2$, $x_1, x_2 \neq 0$ we get $g(x_2) - g(x_1) = \frac{1}{\sqrt[3]{x_2}} - \frac{1}{\sqrt[3]{x_1}} = \frac{\sqrt[3]{x_1} - \sqrt[3]{x_2}}{\sqrt[3]{x_1}\sqrt[3]{x_2}}$. The numerator is negative and the denominator is positive when $x_1 < x_2 < 0$ and when $0 < x_1 < x_2$. Therefore, $g(x)$ is decreasing separately on each of the intervals $(-\infty, 0)$ and $(0, +\infty)$ of its domain $\mathbb{R}\backslash\{0\}$.

   Now we can apply the property of composition of monotonic functions (see Sect. 9.3 of Chap. 2): the function $\tilde{f}(x) = 4 - 2x$ is decreasing on $\mathbb{R}$ and the function $\tilde{g}(y) = \frac{1}{\sqrt[3]{y}}$ is decreasing on the intervals $(-\infty, 0)$ and $(0, +\infty)$, which results in the increase of the composite function $f(x) = \tilde{g}(\tilde{f}(x)) = \frac{1}{\sqrt[3]{4-2x}}$ on the intervals $(-\infty, 2)$ and $(2, +\infty)$.

   According to the monotonicity of $f(x)$, each point of its domain is increasing, and consequently, cannot be an extremum of any type. Therefore, the function has no extremum.
7. Concavity and inflection.

We start again with the study of concavity of the simpler function $g(x) = \frac{1}{\sqrt[3]{x}}$. For any points $x_1 < x_2$, $x_1, x_2 \neq 0$ and $x_0 = \frac{x_1+x_2}{2}$, $x_0 \neq 0$, $h = x_2 - x_0 = x_0 - x_1 > 0$ we obtain:

$$D_g = g(x_2) + g(x_1) - 2g(x_0) = \frac{1}{\sqrt[3]{x_2}} - \frac{1}{\sqrt[3]{x_0}} + \frac{1}{\sqrt[3]{x_1}} - \frac{1}{\sqrt[3]{x_0}}$$

$$= \frac{\sqrt[3]{x_0} - \sqrt[3]{x_2}}{\sqrt[3]{x_0 x_2}} + \frac{\sqrt[3]{x_0} - \sqrt[3]{x_1}}{\sqrt[3]{x_0 x_1}}$$

$$= \frac{x_0 - x_2}{\sqrt[3]{x_0 x_2}\left(\sqrt[3]{x_0^2} + \sqrt[3]{x_0 x_2} + \sqrt[3]{x_2^2}\right)} + \frac{x_0 - x_1}{\sqrt[3]{x_0 x_1}\left(\sqrt[3]{x_0^2} + \sqrt[3]{x_0 x_1} + \sqrt[3]{x_1^2}\right)}$$

$$= \frac{h}{\sqrt[3]{x_0}} \cdot \frac{\sqrt[3]{x_2}\left(\sqrt[3]{x_0^2} + \sqrt[3]{x_0 x_2} + \sqrt[3]{x_2^2}\right) - \sqrt[3]{x_1}\left(\sqrt[3]{x_0^2} + \sqrt[3]{x_0 x_1} + \sqrt[3]{x_1^2}\right)}{\sqrt[3]{x_1 x_2}\left(\sqrt[3]{x_0^2} + \sqrt[3]{x_0 x_2} + \sqrt[3]{x_2^2}\right)\left(\sqrt[3]{x_0^2} + \sqrt[3]{x_0 x_1} + \sqrt[3]{x_1^2}\right)}$$

$$= \frac{h}{\sqrt[3]{x_0}} \cdot \frac{\sqrt[3]{x_0^2}\left(\sqrt[3]{x_2} - \sqrt[3]{x_1}\right) + \sqrt[3]{x_0}\left(\sqrt[3]{x_2^2} - \sqrt[3]{x_1^2}\right) + (x_2 - x_1)}{\sqrt[3]{x_1 x_2}\left(\sqrt[3]{x_0^2} + \sqrt[3]{x_0 x_2} + \sqrt[3]{x_2^2}\right)\left(\sqrt[3]{x_0^2} + \sqrt[3]{x_0 x_1} + \sqrt[3]{x_1^2}\right)}$$

$$= \frac{h}{\sqrt[3]{x_0}} \cdot \frac{\left(\sqrt[3]{x_2} - \sqrt[3]{x_1}\right)\left(\sqrt[3]{x_0^2} + \sqrt[3]{x_0 x_2} + \sqrt[3]{x_0 x_1}\right) + (x_2 - x_1)}{\sqrt[3]{x_1 x_2}\left(\sqrt[3]{x_0^2} + \sqrt[3]{x_0 x_2} + \sqrt[3]{x_2^2}\right)\left(\sqrt[3]{x_0^2} + \sqrt[3]{x_0 x_1} + \sqrt[3]{x_1^2}\right)}.$$

In the second fraction, the numerator is positive for any pair $x_1 < x_2$ and the denominator is positive when x_1 and x_2 have the same sign, that is, on the intervals $(-\infty, 0)$ and $(0, +\infty)$ separately. The only term that changes sign is the denominator $\sqrt[3]{x_0}$ of the first fraction, which is negative on the interval $(-\infty, 0)$ and positive on $(0, +\infty)$. Therefore, $g(x)$ is concave downward on the interval $(-\infty, 0)$ and upward on $(0, +\infty)$.

Next, we use the property of composition of concave functions (see Sect. 9.3 of Chap. 2). The function $\tilde{f}(x) = 4 - 2x$ is concave downward (non-strictly) on the interval $(-\infty, 2)$ and transforms this interval into $(0, +\infty)$. The function $\tilde{g}(y) = \frac{1}{\sqrt[3]{y}}$, in turn, is concave upward and decreasing on $(0, +\infty)$. Therefore, the composite function $f(x) = \tilde{g}(\tilde{f}(x)) = \frac{1}{\sqrt[3]{4-2x}}$ has upward concavity on $(-\infty, 2)$. On the second part of the domain, on the interval $(2, +\infty)$, the situation is inverted. The function $\tilde{f}(x) = 4 - 2x$ is concave upward (non-strictly) on $(2, +\infty)$ and maps this interval into $(-\infty, 0)$, where the second function

$\tilde{g}(y) = \frac{1}{\sqrt[3]{y}}$ is concave downward and decreasing. Therefore, the composite function $f(x) = \tilde{g}(\tilde{f}(x)) = \frac{1}{\sqrt[3]{4-2x}}$ has downward concavity on $(2, +\infty)$.

Since the function changes the type of concavity only at the point $x = 2$, which is out of the domain, there is no inflection point.

8. Complimentary properties.

Let us analyze the behavior of function when $|x|$ goes to infinity ($x \to \pm\infty$). Using the properties of the cubic root we obtain $\lim\limits_{x\to+\infty} \sqrt[3]{4-2x} = -\infty$ and $\lim\limits_{x\to-\infty} \sqrt[3]{4-2x} = +\infty$. Therefore, $\lim\limits_{x\to+\infty} \frac{1}{\sqrt[3]{4-2x}} = 0$ and $\lim\limits_{x\to-\infty} \frac{1}{\sqrt[3]{4-2x}} = 0$. The first convergence to 0 occurs by negative values and the second—by positive values. Consequently, the function has the only horizontal asymptote $y = 0$.

At each point of the domain, $x \neq 2$, the function is continuous, that is, $f(x) \to f(x_0)$ when $x \to x_0 \neq 2$. Therefore, no convergence to infinity occurs at the points of the domain, which implies that there is no vertical asymptote at any point of the domain.

The situation is different at the only point $x = 2$ excluded from the domain. At this point the denominator vanishes and the function is not defined, but still the behavior of the function when $x \to 2$ can be investigated. If $x \to 2^+$, that is, x approaches 2 from the right, then $4 - 2x < 0$ and $4 - 2x \to 0$. This implies that $\sqrt[3]{4-2x} < 0$ and $\sqrt[3]{4-2x} \to 0$, and consequently, the quotient $\frac{1}{\sqrt[3]{4-2x}}$ has negative values whose absolute value increases without restriction. Using the symbols, we write: $\lim\limits_{x\to2+} \frac{1}{\sqrt[3]{4-2x}} = -\infty$. When $x \to 2^-$, that is, x approaches 2 from the left, then $4 - 2x > 0$ and $4 - 2x \to 0$. This implies that $\sqrt[3]{4-2x} > 0$ and $\sqrt[3]{4-2x} \to 0$, and consequently, the quotient $\frac{1}{\sqrt[3]{4-2x}}$ has positive values which increase without restriction, or in symbols: $\lim\limits_{x\to2^-} \frac{1}{\sqrt[3]{4-2x}} = +\infty$. Hence, the line $x = 2$ is a vertical asymptote of the graph of $f(x)$.

9. Using the investigated properties and marking some important points, such as $P_0 = (0, \frac{1}{\sqrt[3]{4}})$ (intersection with the y-axis) and $P_1 = (3, -\frac{1}{\sqrt[3]{2}})$ (a point on the part of the graph below the x-axis), we sketch the graph shown in Fig. 3.42.

The properties of this function can also be obtained by investigating first the simpler function $g(x) = \frac{1}{\sqrt[3]{x}}$ (which was partially made in the study of monotonicity and concavity) and then applying the following sequence of geometric transformations:

$$g(x) = \frac{1}{\sqrt[3]{x}} \to f_1(x) = g(-x) = \frac{1}{\sqrt[3]{-x}} \to f_2(x)$$
$$= f_1(2x) = \frac{1}{\sqrt[3]{-2x}} \to f(x) = f_2(x-2) = \frac{1}{\sqrt[3]{4-2x}}.$$

We leave to the reader the task to apply this approach and compare the results with those obtained through analytic method.

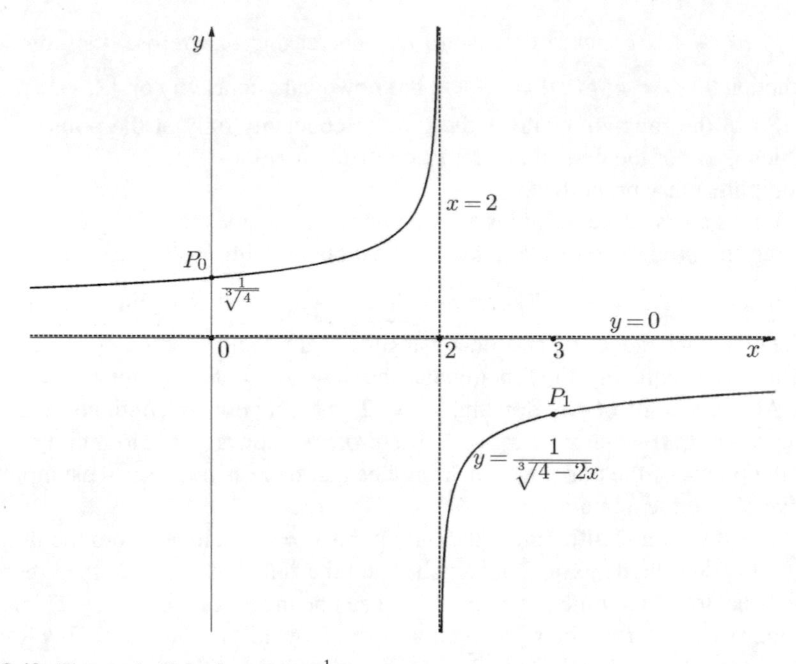

Fig. 3.42 The graph of the function $\frac{1}{\sqrt[3]{4-2x}}$

C. Study of $f(x) = |x^2 - 4x|$

There are two approaches to study this function: geometric, based on the use of the elementary transformations applied to the basic function x^2, and analytic, based on the investigation of the properties of the given function. We perform here the analytic study.

1. Domain and range. The formula of the function is defined for any real number, which means that the domain is $X = \mathbb{R}$.

   By the definition of the absolute value, the range is contained in the interval $[0, +\infty)$. To show that any y of this interval belongs to the range, let us solve the equation $|x^2 - 4x| = y$, $\forall y \geq 0$ for unknown x. Recalling solution of the equations with absolute value, we have two options $x^2 - 4x = \pm y$. Since it is sufficient to find a solution of one of them, we choose $x^2 - 4x = y$. The last equation is quadratic in x with the positive discriminant $2 + y > 0$, which means that there are real roots $x = 2 \pm \sqrt{2 + y}$ for any non-negative y. Therefore, the range is $Y = [0, +\infty)$.
2. Intersection and sign. Intersection with the y-axis occurs at the point $P_0 = (0, 0)$. To find the points of intersection with the x-axis, we need to solve the equation $|x^2 - 4x| = 0$. Writing this equation in an equivalent form $x(x - 4) = 0$ we get the two roots $x = 0$ and $x = 4$. Therefore, the graph crosses the x-axis at the points $P_0 = (0, 0)$ and $P_1 = (4, 0)$.

The function is positive at all the points of the domain, except for $x = 0$ and $x = 4$ where it vanishes.

3. Boundedness. Knowing the range of $f(x)$ we immediately conclude that the function is bounded below and unbounded above.
4. Parity. Calculating the values of the function at the points -2 and 2 we see that the function is neither even nor odd: $f(-2) = 12 \neq \pm 4 = \pm f(2)$.

   However, the graph of the function has a symmetry about $x = 2$ since $f(2-x) = |(2-x)^2 - 4(2-x)| = |(2-x)(-2-x)| = |(x-2)(x+2)| = |(2+x)^2 - 4(2+x)| = f(2+x)$ (recall formulas (1.11) in Sect. 7.5, Chap. 1).
5. Periodicity. We leave the study of periodicity until after the investigation of monotonicity.
6. Monotonicity and extrema.

   To investigate monotonicity, it is convenient to open the absolute value operation by the definition: $f(x) = |x^2 - 4x| = \begin{cases} x^2 - 4x, \ x \leq 0 \\ -x^2 + 4x, \ 0 < x < 4 \\ x^2 - 4x, \ x \geq 4 \end{cases}$.

Naturally, we analyze monotonicity separately on each of the three intervals.

(1) For $x_1 < x_2 \leq 0$ we have $f(x_2) - f(x_1) = x_2^2 - 4x_2 - (x_1^2 - 4x_1) = (x_2 - x_1)(x_2 + x_1 - 4)$. The first factor is positive and the second is negative, which shows that $f(x)$ is decreasing.

(2) When $4 \leq x_1 < x_2$ we have the same representation $f(x_2) - f(x_1) = (x_2 - x_1)(x_2 + x_1 - 4)$, but now both factors are positive and the function is increasing.

(3) If $0 \leq x_1 < x_2 \leq 4$, then $f(x_2) - f(x_1) = -x_2^2 + 4x_2 - (-x_1^2 + 4x_1) = (x_2 - x_1)(4 - x_2 - x_1)$. The sign of this expression depends on whether the points x_1 and x_2 are chosen to the left or to the right of 2. In the case $0 \leq x_1 < x_2 \leq 2$ we have the two positive factors and the function is increasing, while in the case $2 \leq x_1 < x_2 \leq 4$ the second factor becomes negative and the function is decreasing.

Another way to investigate monotonicity is by using the property of composition of monotonic functions (see Sect. 9.3 of Chap. 2) on the four intervals of the domain.

(1) On the interval $X_1 = (-\infty, 0]$ the function $\tilde{f}(x) = x^2 - 4x$ decreases and has the image $[0, +\infty)$, which is the set of increase of the function $\tilde{g}(y) = |y|$. Therefore, $f(x) = \tilde{g}(\tilde{f}(x)) = |x^2 - 4x|$ decreases on $X_1 = (-\infty, 0]$.

(2) On the next interval, $X_2 = [0, 2]$, the function $\tilde{f}(x) = x^2 - 4x$ continue to be decreasing, but its image is the interval $[-4, 0]$, where the function $\tilde{g}(y) = |y|$ is decreasing. Consequently, $f(x) = \tilde{g}(\tilde{f}(x)) = |x^2 - 4x|$ increases on $X_2 = [0, 2]$.

(3) On the third interval, $X_3 = [2, 4]$, the function $\tilde{f}(x) = x^2 - 4x$ increases and has the image $[-4, 0]$, on which the function $\tilde{g}(y) = |y|$ decreases. Therefore, $f(x) = \tilde{g}(\tilde{f}(x)) = |x^2 - 4x|$ decreases on $X_3 = [2, 4]$.

(4) Finally, on the interval $X_4 = [4, +\infty)$ the function $\tilde{f}(x) = x^2 - 4x$ increases and has the image $[0, +\infty)$, which is the set of increase of the function $\tilde{g}(y) = |y|$. Consequently, $f(x) = \tilde{g}(\tilde{f}(x)) = |x^2 - 4x|$ increases on $X_4 = [4, +\infty)$.

It follows from the results of monotonicity that each point $x \neq 0, 2, 4$ is monotonic, and consequently, it cannot be an extremum of any type. The points $x = 0$ and $x = 4$ are strict local minima (because $f(x)$ decreases to the left and increases to the right of these points) and also global minima (because at the remaining points $f(x) > 0 = f(0) = f(4)$). The point $x = 2$ is a strict local maximum ($f(x)$ increases to the left and decreases to the right of this point), but it is not a global maximum (since the range is $[0, +\infty)$).

One more consequence of the properties of monotonicity is that the function cannot be periodic, since it increases on the infinite interval $[4, +\infty)$.

7. Concavity and inflection.

To evaluate the difference between the mean value $\frac{f(x_1)+f(x_2)}{2}$ and the value at the midpoint $f(x_0)$, where $x_1 < x_2$, $x_0 = \frac{x_1+x_2}{2}$ and $h = x_2 - x_0 = x_0 - x_1 > 0$, we use the quantity $D_2 \equiv f(x_1) + f(x_2) - 2f(x_0)$ and consider separately the three intervals of the domain.

(1) When $x_1 < x_2 \le 0$, we obtain

$$D_2 = f(x_2) + f(x_1) - 2f(x_0) = (x_2^2 - 4x_2) + (x_1^2 - 4x_1) - 2(x_0^2 - 4x_0)$$
$$= (x_2^2 + x_1^2 - 2x_0^2) - 4(x_2 + x_1 - 2x_0)$$
$$= (x_0 + h)^2 + (x_0 - h)^2 - 2x_0^2 = 2h^2 > 0.$$

Therefore, $f(x)$ is concave upward on $(-\infty, 0]$.

(2) In the same way, when $4 \le x_1 < x_2$, we get

$$D_2 = f(x_2) + f(x_1) - 2f(x_0) = (x_2^2 - 4x_2) + (x_1^2 - 4x_1) - 2(x_0^2 - 4x_0) = 2h^2 > 0.$$

This means that $f(x)$ is concave upward on $[4, +\infty)$.

(3) Similarly, when $0 \le x_1 < x_2 \le 4$, we have

$$D_2 = f(x_2) + f(x_1) - 2f(x_0)$$
$$= (-x_2^2 + 4x_2) + (-x_1^2 + 4x_1) - 2(-x_0^2 + 4x_0) = -2h^2 < 0,$$

that is, $f(x)$ is concave downward on $[0, 4]$.

Since $f(x)$ changes the type of concavity at the points $x = 0$ and $x = 4$, both belonging to the domain, these two points are inflection points.

The same result about concavity can be obtained applying the property of composition of concave functions on the three intervals of the domain (see

Sect. 9.3 of Chap. 2). Notice first that the function $\tilde{f}(x) = x^2 - 4x$ has upward concavity over the entire domain $\mathbb{R}$, and the function $\tilde{g}(y) = |y|$ has concavity (non-strict) of both types on the intervals $(-\infty, 0]$ and $[0, +\infty)$. Therefore, the concavity of the composite function $f(x) = \tilde{g}(\tilde{f}(x)) = |x^2 - 4x|$ depends on the type of monotonicity of $\tilde{g}(y) = |y|$. On the interval $X_1 = (-\infty, 0]$, the function $\tilde{f}(x) = x^2 - 4x$ has the image $[0, +\infty)$, which is the set of increase of $\tilde{g}(y) = |y|$, and consequently, the composite function has upward concavity. On the interval $X_2 \cup X_3 = [0, 4]$, the image of the function $\tilde{f}(x) = x^2 - 4x$ is the interval $[-4, 0]$, where $\tilde{g}(y) = |y|$ decreases, which ensures downward concavity of the composite function. Finally, on the interval $X_4 = [4, +\infty)$, the image of the function $\tilde{f}(x) = x^2 - 4x$ is the interval $[0, +\infty)$, on which $\tilde{g}(y) = |y|$ increases, which guarantees upward concavity of the composite function.

8. Complimentary properties. For sufficiently large values of $|x|$ the main term of the function is x^2. Therefore, $\lim\limits_{|x|\to+\infty} f(x) = \lim\limits_{|x|\to+\infty} x^2 = +\infty$. Consequently, the function has no horizontal asymptote.

   The function is continuous at each real number, that is, $f(x) \to f(x_0)$ when $x \to x_0$. Therefore, no convergence to infinity can occur at the points of the domain X, which implies that there is no vertical asymptote.
9. Using the investigated properties and marking some important points, such as $P_0 = (0, 0)$ (intersection with the coordinate axes), $P_1 = (4, 0)$ (intersection with the x-axis), $P_2 = P_0$ and $P_3 = P_1$ (local and global minima), $P_4 = (2, 4)$ (local maximum), we sketch the graph shown in Fig. 3.43.

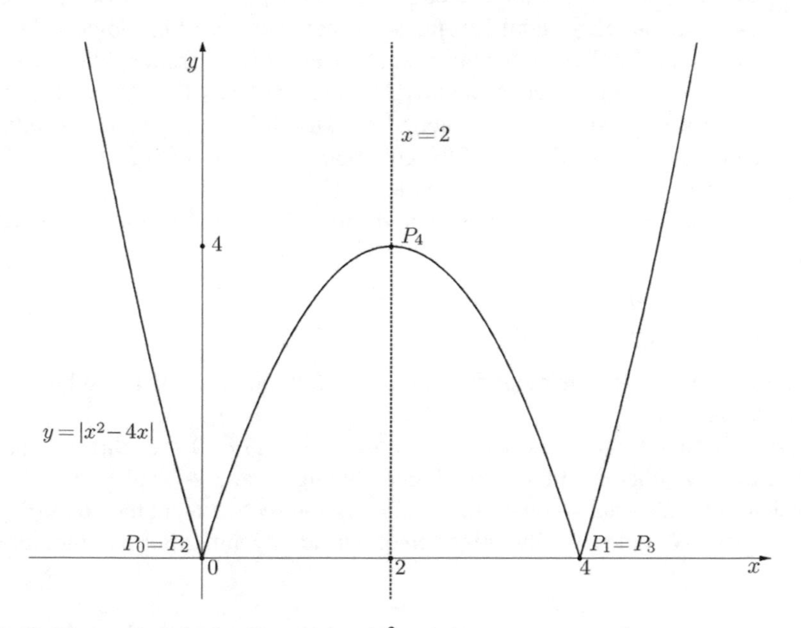

Fig. 3.43 The graph of the function $f(x) = |x^2 - 4x|$

The properties of the given function can also be analyzed using the following geometric algorithm:

(1) transform $f_0(x) = x^2$ into $f_2(x) = x^2 - 4x$ through the chain $f_0(x) = x^2 \to f_1(x) = f_0(x-2) = (x-2)^2 \to f_2(x) = f_1(x) - 4 = (x-2)^2 - 4 = x^2 - 4x$;
(2) on the part $[0, 4]$ of the domain $\mathbb{R}$ of $f_2(x)$, transform $f_2(x)$ into $-f_2(x)$.

We leave to the reader the task to apply this approach and compare the results with those obtained through analytic method.

5 Solved Exercises

5.1 *Study of Polynomial Functions*

A. Study of $f(x) = x^3 + 2x - 3$

1. Domain and range. Like any polynomial function, this function has the domain $X = \mathbb{R}$.

   Recalling from the theory of polynomial equations that any cubic equation has at least one real root, we can conclude that any cubic polynomial, in particular the given one, has the range $Y = \mathbb{R}$.
2. Intersection and sign. The point of intersection with the y-axis is $P_0 = (0, -3)$. To find the points of intersection with the x-axis we should solve the equation $x^3 + 2x - 3 = 0$. The solution of a cubic equation can be found through algebraic operations, although general formulas are not quite simple. However, in this specific case, the polynomial admits an elementary factorization $x^3 + 2x - 3 = (x-1)(x^2 + x + 3)$, and consequently, its roots are easily found: $x = 1$ from the first factor and no root from the second, because it has a negative discriminant. Therefore, $x = 1$ the only root of the equation $x^3 + 2x - 3 = 0$, and $P_1 = (1, 0)$ is the only point of intersection with the x-axis.

   Using the factored form, it is easy to decide where the function is positive and negative:

   (1) if $x < 1$, then $f(x) < 0$;
   (2) if $x > 1$, then $f(x) > 0$.

3. Boundedness. Since the range is $Y = \mathbb{R}$, the function is unbounded both below and above.
4. Parity. Comparing the values $f(-1) = -6$ and $f(1) = 0$, we conclude that the function is neither even nor odd. Notice although that $3 + f(x) = 3 - f(-x)$, which shows that the graph of the function is symmetric about the point $(0, -3)$.
5. Periodicity. We leave the investigation of periodicity until after monotonicity.

6. Monotonicity and extrema. Take a pair of the points $x_1 < x_2$ and evaluate the difference $D_1 = f(x_2) - f(x_1)$:

$$D_1 = f(x_2) - f(x_1) = x_2^3 + 2x_2 - 3 - \left(x_1^3 + 2x_1 - 3\right)$$

$$= (x_2 - x_1)(x_2^2 + x_2x_1 + x_1^2 + 2) = (x_2 - x_1)\left(\left(x_2 + \frac{1}{2}x_1\right)^2 + \frac{3}{4}x_1^2 + 2\right) > 0.$$

This means that $f(x)$ increases over the entire domain $X = \mathbb{R}$. The immediate consequence of this is that there is no extremum of any type.

Another implication of the monotonicity on $(-\infty, +\infty)$ is that the function is not periodic.

7. Concavity and inflection. Take two arbitrary points $x_1 < x_2$, define the midpoint $x_0 = \frac{x_1+x_2}{2}$ and the increment $h = x_2 - x_0 = x_0 - x_1 > 0$, and use the quantity $D_2 \equiv f(x_1) + f(x_2) - 2f(x_0)$ to evaluate the difference between the mean value $\frac{f(x_1)+f(x_2)}{2}$ and the value at the midpoint $f(x_0)$. Employing the expression found for D_1, we obtain:

$$D_2 = f(x_1) + f(x_2) - 2f(x_0) = (f(x_2) - f(x_0)) - (f(x_0) - f(x_1))$$

$$= (x_2 - x_0)\left(x_2^2 + x_2x_0 + x_0^2 + 2\right) - (x_0 - x_1)\left(x_0^2 + x_1x_0 + x_1^2 + 2\right)$$

$$= h\left(x_2^2 - x_1^2 + x_2x_0 - x_1x_0\right) = 2h^2(x_2 + x_1 + x_0) = 6h^2x_0.$$

Therefore, we get the following results about concavity:

(1) the function is concave downward on the interval $(-\infty, 0]$;
(2) the function is concave upward on the interval $[0, +\infty)$.

Consequently, the point $x = 0$ is an inflection point.

8. Complimentary properties. When $|x|$ increases without restriction, that is, $x \to \pm\infty$, the behavior of the polynomial is determined by its leading term, and for this reason we have (for sufficiently large $|x|$) $f(x) = x^3 + 2x - 3 \approx x^3$ (see discussion of such a behavior in Sect. 1.3). From this approximation it follows that the function increases without restriction when $x \to +\infty$ and decreases without restriction when $x \to -\infty$, that is, $\lim_{x\to+\infty} f(x) = +\infty$ and $\lim_{x\to-\infty} f(x) = -\infty$. This implies that the function has no horizontal asymptote.

At any point of the domain the function is continuous, that is, $f(x) \to f(x_0)$ when $x \to x_0$. Hence, there is no convergence of $f(x)$ to infinity at any point, which implies that the function has no vertical asymptote.

9. Using the investigated properties and marking some important points, such as $P_0 = (0, -3)$ (intersection with the y-axis and inflection), $P_1 = (1, 0)$ (intersection with the x-axis) and $P_2 = (-\frac{1}{2}, -\frac{33}{8})$ (a point with a negative x-coordinate), we can plot the graph shown in Fig. 3.44.

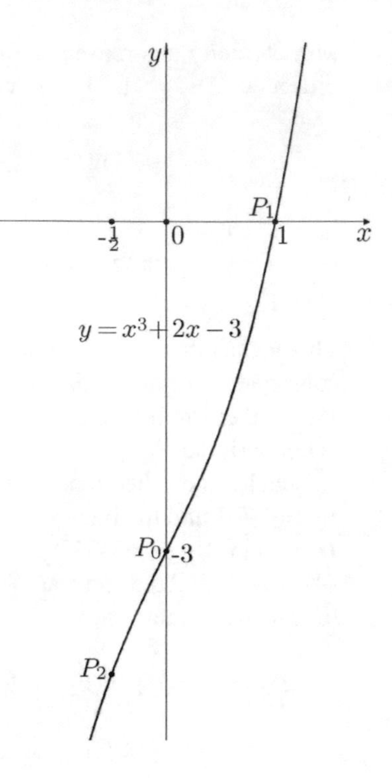

Fig. 3.44 The graph of the function $f(x) = x^3 + 2x - 3$

B. Study of $f(x) = x^3 - 3x + 2$

1. Domain and range. Like any cubic polynomial, the given one has the domain $X = \mathbb{R}$ and the range $Y = \mathbb{R}$.
2. Intersection and sign. The point of intersection with the y-axis is $P_0 = (0, 2)$. To find the points of intersection with the x-axis we should solve the equation $x^3 - 3x + 2 = 0$. This polynomial admits an elementary factorization $x^3 - 3x + 2 = (x - 1)^2(x + 2)$, and consequently, its roots are easily found: $x = 1$ from the first factor (double root) and $x = -2$ from the second. Therefore, there are two points of intersection with the x-axis: $P_1 = (1, 0)$ and $P_2 = (-2, 0)$.

   Using the factored form it is easy to decide where the function is positive and negative:

   (1) if $x < -2$, then $f(x) < 0$;
   (2) if $x > -2, x \neq 1$, then $f(x) > 0$.

3. Boundedness. Since the range is $Y = \mathbb{R}$, the function is unbounded both below and above.
4. Parity. Comparing the values $f(-1) = 4$ and $f(1) = 0$, we conclude that the function is neither even nor odd. Notice although that $f(x) - 2 = -f(-x) - 2$, which means that the graph of the function is symmetric about the point $(0, 2)$.
5. Periodicity. We leave the investigation of periodicity until after monotonicity.

6. Monotonicity and extrema. Take a pair of the points $x_1 < x_2$ and evaluate the difference $D_1 = f(x_2) - f(x_1)$:

$$D_1 = f(x_2) - f(x_1) = x_2^3 - 3x_2 + 2 - \left(x_1^3 - 3x_1 + 2\right)$$
$$= (x_2 - x_1)(x_2^2 + x_2x_1 + x_1^2 - 3).$$

The first factor is positive, and the form of the second suggests the partition of the domain into the three parts:

(1) if $x_1 < x_2 \le -1$, then $x_2^2 + x_2x_1 + x_1^2 - 3 > 0$, and consequently, $D_1 > 0$, which means that $f(x)$ increases on $(-\infty, -1]$;
(2) if $-1 \le x_1 < x_2 \le 1$, then $x_2^2 + x_2x_1 + x_1^2 - 3 < 0$, and consequently, $D_1 < 0$, that is, $f(x)$ decreases on $[-1, 1]$;
(3) if $1 \le x_1 < x_2$, then $x_2^2 + x_2x_1 + x_1^2 - 3 > 0$, and consequently, $D_1 > 0$, that is, $f(x)$ increases on $[1, +\infty)$.

It follows from this that $x = -1$ is a strict local maximum and $x = 1$ is a strict local minimum (both points are not global extrema since the range is not bounded below and above).

Another implication of the monotonicity is that the increase on the infinite interval $[1, +\infty)$ guarantees that $f(x)$ is not periodic.

7. Concavity and inflection. Take two arbitrary points $x_1 < x_2$, define the midpoint $x_0 = \frac{x_1+x_2}{2}$ and the increment $h = x_2 - x_0 = x_0 - x_1 > 0$, and use the quantity $D_2 \equiv f(x_1)+f(x_2)-2f(x_0)$ to evaluate the difference between the mean value $\frac{f(x_1)+f(x_2)}{2}$ and the value at the midpoint $f(x_0)$. Employing the expression found for D_1, we obtain:

$$D_2 = f(x_1) + f(x_2) - 2f(x_0) = (f(x_2) - f(x_0)) - (f(x_0) - f(x_1))$$
$$= (x_2 - x_0)(x_2^2 + x_2x_0 + x_0^2 - 3) - (x_0 - x_1)(x_0^2 + x_0x_1 + x_1^2 - 3)$$
$$= h(x_2^2 - x_1^2 + x_2x_0 - x_1x_0) = 2h^2(x_2 + x_1 + x_0) = 6h^2x_0.$$

Therefore, we get the following results:

(1) the function is concave downward on the interval $(-\infty, 0]$;
(2) the function is concave upward on the interval $[0, +\infty)$.

Consequently, the point $x = 0$ is an inflection point.

8. Complimentary properties. For sufficiently large $|x|$ the behavior of the polynomial is determined by its leading term: $f(x) = x^3 - 3x + 2 \approx x^3$, which implies that the function increases without restriction when $x \to +\infty$ and decreases without restriction when $x \to -\infty$, that is, $\lim_{x\to+\infty} f(x) = +\infty$ and $\lim_{x\to-\infty} f(x) = -\infty$. Therefore, there is no horizontal asymptote.

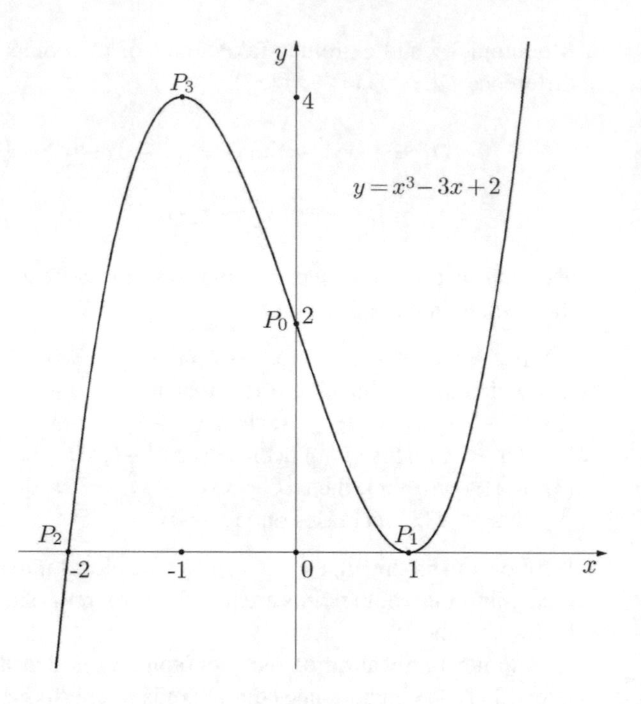

Fig. 3.45 The graph of the function $f(x) = x^3 - 3x + 2$

At any point of the domain the function is continuous, that is, $f(x) \to f(x_0)$ when $x \to x_0$. Hence, there is no convergence of $f(x)$ to infinity at any point, which implies that the function has no vertical asymptote.

9. Using the investigated properties and marking some important points, such as $P_0 = (0, 2)$ (intersection with the y-axis and inflection), $P_1 = (1, 0)$ and $P_2 = (-2, 0)$ (intersection with the x-axis), $P_3 = (-1, 4)$ and $P_1 = (1, 0)$ (local extrema), we can sketch the graph of the function shown in Fig. 3.45.

C. Study of $f(x) = 4 - 3x^2 - x^3$

1. Domain and range. Like any cubic polynomial, the given function has the domain $X = \mathbb{R}$ and the range $Y = \mathbb{R}$.
2. Intersection and sign. The point of intersection with the y-axis is $P_0 = (0, 4)$. To find the points of intersection with the x-axis we should solve the equation $4 - 3x^2 - x^3 = 0$. This polynomial admits an elementary factorization $4 - 3x^2 - x^3 = (1 - x)(x + 2)^2$, and consequently, its roots are easily found: $x = 1$ from the first factor and $x = -2$ from the second (double root). Therefore, there are two points of intersection with the x-axis: $P_1 = (1, 0)$ and $P_2 = (-2, 0)$.

Using the factored form it is easy to decide where the function is positive and negative:

(1) if $x < 1, x \neq -2$, then $f(x) > 0$;
(2) if $x > 1$, then $f(x) < 0$.

3. Boundedness. Since the range is $Y = \mathbb{R}$, the function is unbounded both below and above.
4. Parity. Comparing the values $f(-1) = 2$ and $f(1) = 0$, we conclude that the function is neither even nor odd.

   However, if we represent the function in the form $f(x) = 4 - 3x^2 - x^3 = -(x+1)^3 + 3(x+1) + 2$, we can see that $f(x-1) - 2 = -f(-x-1) - 2$, and the last relation reveals the symmetry of the graph with respect to the point $(-1, 2)$.
5. Periodicity. We leave the investigation of periodicity until after monotonicity.
6. Monotonicity and extrema. Take a pair of the points $x_1 < x_2$ and evaluate the difference $D_1 = f(x_2) - f(x_1)$:

$$D_1 = f(x_2) - f(x_1) = 4 - 3x_2^2 - x_2^3 - \left(4 - 3x_1^2 - x_1^3\right)$$
$$= (x_1 - x_2)(x_2^2 + x_2x_1 + x_1^2 + 3(x_2 + x_1)).$$

The first factor is negative, and the form of the second suggests the partition of the domain into the three parts.

(1) If $x_1 < x_2 \leq -2$, then $x_1x_2 > -2x_2$ and $x_1x_2 \geq -2x_1$, whence $2x_1x_2 > -2(x_1 + x_2)$ or $x_1x_2 + (x_1 + x_2) > 0$. Consequently, the second factor can be evaluated as follows: $x_2^2 + x_2x_1 + x_1^2 + 3(x_2 + x_1) = (x_2 - x_1)^2 + 3x_2x_1 + 3(x_2 + x_1) > 0$. Then $D_1 < 0$, which means that $f(x)$ decreases on $(-\infty, -2]$.
(2) If $-2 \leq x_1 < x_2 \leq 0$, then $x_1x_2 \leq -2x_2$ and $x_1x_2 < -2x_1$, whence $x_1x_2 < -(x_1 + x_2)$. Therefore, we get the following evaluation of the second factor: $x_2^2 + x_2x_1 + x_1^2 + 3(x_2 + x_1) = x_2^2 + 2x_2 + x_1^2 + 2x_1 + x_2x_1 + (x_2 + x_1) = x_2(x_2 + 2) + x_1(x_1 + 2) + (x_2x_1 + x_2 + x_1) < 0$ (the first two summands are non-positive and the third is negative). Consequently, $D_1 > 0$ and the function increases on $[-2, 0]$.
(3) If $0 \leq x_1 < x_2$, then $x_2^2 + x_2x_1 + x_1^2 + 3(x_2 + x_1) > 0$ implying that $D_1 < 0$ and the function is decreasing on $[0, +\infty)$.

   It follows from this that $x = -2$ is a strict local minimum and $x = 0$ is a strict local maximum (both points are not global extrema since the range is not bounded below and above).

   Another implication of the monotonicity is that the decrease on the infinite interval $[0, +\infty)$ guarantees that $f(x)$ is not periodic.
7. Concavity and inflection. Take two arbitrary points $x_1 < x_2$, define the midpoint $x_0 = \frac{x_1 + x_2}{2}$ and the increment $h = x_2 - x_0 = x_0 - x_1 > 0$, and use the quantity $D_2 \equiv f(x_1) + f(x_2) - 2f(x_0)$ to evaluate the difference between the mean value

$\frac{f(x_1)+f(x_2)}{2}$ and the value at the midpoint $f(x_0)$. Employing the expression found for D_1, we obtain:

$$D_2 = f(x_1) + f(x_2) - 2f(x_0) = (f(x_2) - f(x_0)) - (f(x_0) - f(x_1))$$

$$= (x_0-x_2)(x_2^2+x_2x_0+x_0^2+3(x_2+x_0))-(x_1-x_0)(x_0^2+x_0x_1+x_1^2+3(x_0+x_1))$$

$$= -h\left[x_2^2 - x_1^2 + x_0(x_2 - x_1) + 3(x_2 - x_1)\right] = -2h^2(x_2 + x_1 + x_0 + 3)$$

$$= -6h^2(x_0 + 1).$$

Therefore, we get the following results:

(1) the function is concave upward on the interval $(-\infty, -1]$;
(2) the function is concave downward on the interval $[-1, +\infty)$.

Consequently, the point $x = -1$ is an inflection point.

8. Complimentary properties. For sufficiently large $|x|$ the behavior of the polynomial is determined by its leading term: $f(x) = 4 - 3x^2 - x^3 \approx -x^3$, which implies that the function decreases without restriction when $x \to +\infty$ and increases without restriction when $x \to -\infty$, that is, $\lim_{x\to+\infty} f(x) = -\infty$ and $\lim_{x\to-\infty} f(x) = +\infty$. Therefore, there is no horizontal asymptote.

   At any point of the domain the function is continuous, that is, $f(x) \to f(x_0)$ when $x \to x_0$. Hence, there is no convergence of $f(x)$ to infinity at any point, which implies that the function has no vertical asymptote.
9. Using the investigated properties and marking some important points, such as $P_0 = (0, 4)$ (intersection with the y-axis), $P_1 = (1, 0)$ and $P_2 = (-2, 0)$ (intersection with the x-axis), $P_0 = (0, 4)$ and $P_2 = (-2, 0)$ (local extrema) and $P_3 = (-1, 2)$ (inflection), we sketch the graph shown in Fig. 3.46.

D. Study of $f(x) = x^4 + x^2 - 2$

1. Domain and range. Any polynomial function, including this one, has the domain $X = \mathbb{R}$.

   To determine the range Y, notice first that $x^4 + x^2 - 2 \geq -2$, that is, $Y \subset [-2, +\infty)$. To show that any point of the interval $[-2, +\infty)$ belongs to the range, let us solve the bi-quadratic equation $y = x^4 + x^2 - 2$ for unknown x when $y \geq -2$ is a given number. We rewrite this equation in the form $(x^2+\frac{1}{2})^2-\frac{9}{4} = y$ or $(x^2 + \frac{1}{2})^2 = y + \frac{9}{4}$ and notice that the last right-hand side is positive for $y \geq -2$, and consequently, $x^2 = \sqrt{y + \frac{9}{4}} - \frac{1}{2}$. Noticing again that the new right-hand side is non-negative for any $y \geq -2$, we find the following solution:

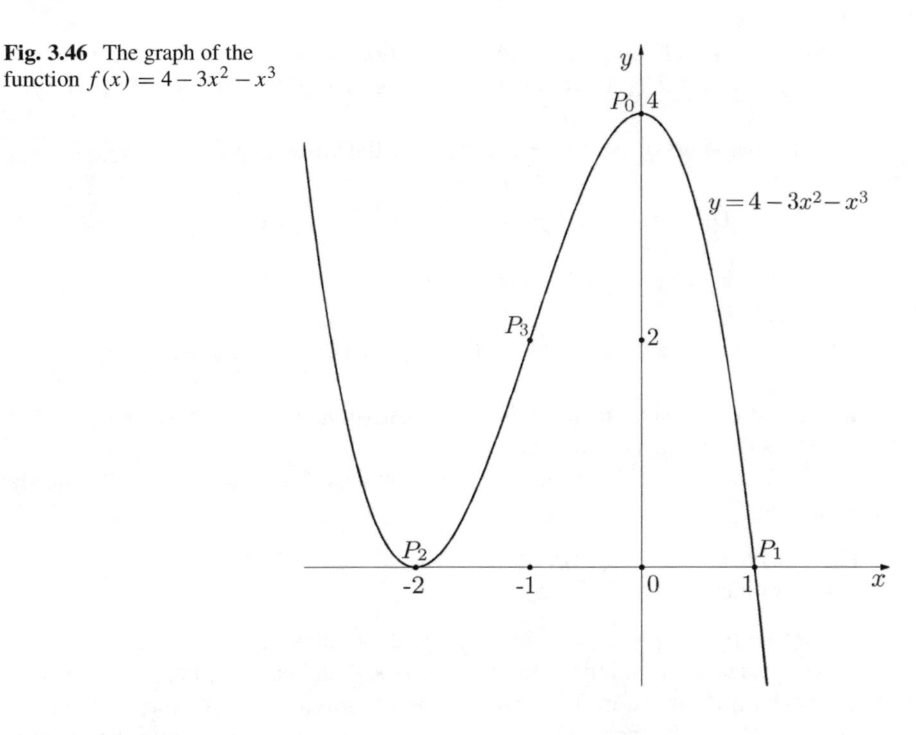

Fig. 3.46 The graph of the function $f(x) = 4 - 3x^2 - x^3$

$x = \pm\sqrt{\sqrt{y + \frac{9}{4}} - \frac{1}{2}}$. Hence, any $y \geq -2$ is associated with at least one x of the domain under the function rule. This means that $Y = [-2, +\infty)$.

2. Intersection and sign. The point of intersection with the y-axis is $P_0 = (0, -2)$. To find the points of intersection with the x-axis we should solve the bi-quadratic equation $(x^2 + \frac{1}{2})^2 - \frac{9}{4} = 0$. The algorithm of the solution is the same as for finding the range and it gives $x = \pm\sqrt{\sqrt{\frac{9}{4}} - \frac{1}{2}} = \pm 1$. Therefore, there are two points of intersection: $P_1 = (-1, 0)$ and $P_2 = (1, 0)$.

   Using the factored form $f(x) = (x^2 - 1)(x^2 + 2)$ it is easy to see that:

   (1) if $x < -1$, then $f(x) > 0$;
   (2) if $-1 < x < 1$, then $f(x) < 0$;
   (3) if $x > 1$, then $f(x) > 0$.

3. Boundedness. Since the range is $Y = [-2, +\infty)$, the function is bounded below and unbounded above.
4. Parity. From the formula $f(-x) = (-x)^4 + (-x)^2 - 2 = x^4 + x^2 - 2 = f(x)$, $\forall x \in \mathbb{R}$, it follows that the function is even and its graph is symmetric about the y-axis.
5. Periodicity. We leave the investigation of periodicity until after monotonicity.

6. Monotonicity and extrema. Since the function is even, we can evaluate monotonicity only for $x \geq 0$, completing the negative part later, by the property of symmetry.

   Take the points $0 \leq x_1 < x_2$ and evaluate the difference $D_1 = f(x_2) - f(x_1)$:

$$D_1 = f(x_2) - f(x_1) = (x_2^4 + x_2^2 - 2) - \left(x_1^4 + x_1^2 - 2\right)$$
$$= (x_2^4 - x_1^4) + (x_2^2 - x_1^2)$$
$$= (x_2^2 - x_1^2)(x_2^2 + x_1^2) + (x_2^2 - x_1^2) = (x_2^2 - x_1^2)(x_2^2 + x_1^2 + 1).$$

   In the last expression, both factors are positive, and consequently, the function increases on the interval $[0, +\infty)$.

   The symmetry of the even function allows us to extend this result onto the entire domain:

   (1) $f(x)$ decreases on $(-\infty, 0]$;
   (2) $f(x)$ increases on $[0, +\infty)$.

   Consequently, the point $x = 0$ is a local and global minimum of the function.

   The increase on the infinite interval $[0, +\infty)$ implies that $f(x)$ is not periodic.

7. Concavity and inflection. Take two arbitrary points $x_1 < x_2$, define the midpoint $x_0 = \frac{x_1+x_2}{2}$ and the increment $h = x_2 - x_0 = x_0 - x_1 > 0$, and use the quantity $D_2 \equiv f(x_1) + f(x_2) - 2f(x_0)$ to evaluate the difference between the mean value $\frac{f(x_1)+f(x_2)}{2}$ and the value at the midpoint $f(x_0)$. Employing the expression found for D_1, we obtain:

$$D_2 = f(x_1) + f(x_2) - 2f(x_0) = (f(x_2) - f(x_0)) - (f(x_0) - f(x_1))$$
$$= (x_2^2 - x_0^2)(x_2^2 + x_0^2 + 1) - (x_0^2 - x_1^2)(x_0^2 + x_1^2 + 1)$$
$$= h(x_2 + x_0)\left(x_2^2 + x_0^2 + 1\right) - h(x_0 + x_1)\left(x_0^2 + x_1^2 + 1\right)$$
$$= h\left[(2x_0 + h)\left((x_0 + h)^2 + x_0^2 + 1\right) - (2x_0 - h)\left(x_0^2 + (x_0 - h)^2 + 1\right)\right]$$
$$= h\Big[2x_0\left((x_0 + h)^2 + x_0^2 + 1 - x_0^2 - (x_0 - h)^2 - 1\right)$$
$$+ h\left((x_0 + h)^2 + x_0^2 + 1 + x_0^2 + (x_0 - h)^2 + 1\right)\Big]$$
$$= h\left[2x_0 \cdot 4x_0 h + h\left(4x_0^2 + 2h^2 + 2\right)\right] = 2h^2(6x_0^2 + h^2 + 1).$$

   Both factors are positive, and consequently, the function is concave upward over the entire domain $X = \mathbb{R}$ and does not possess inflection points.

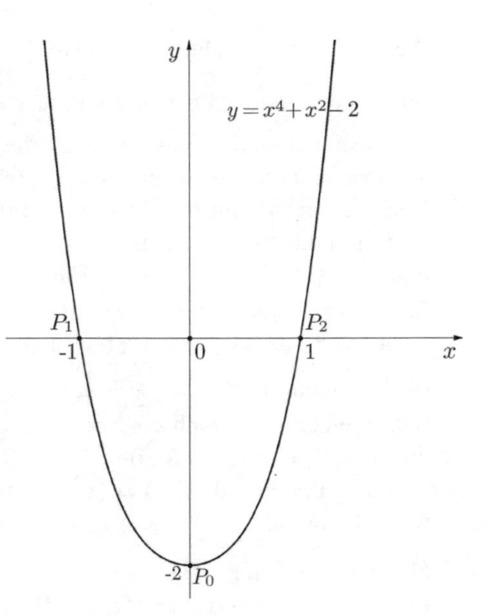

Fig. 3.47 The graph of the function $f(x) = x^4 + x^2 - 2$

8. Complimentary properties. For sufficiently large $|x|$ the behavior of the polynomial is determined by its leading term: $f(x) = x^4 + x^2 - 2 \approx x^4$, which implies that the function increases without restriction when $x \to \pm\infty$. Consequently, the function has no horizontal asymptote.

   At any point of the domain the function is continuous, that is, $f(x) \to f(x_0)$ when $x \to x_0$. Hence, there is no convergence of $f(x)$ to infinity at any point, which implies that the function has no vertical asymptote.
9. Using the investigated properties and marking some important points, such as $P_0 = (0, -2)$ (intersection with the y-axis and global minimum), $P_1 = (-1, 0)$, $P_2 = (1, 0)$ (intersection with the x-axis), we plot the graph shown in Fig. 3.47.

E. Study of $f(x) = x^4 - 5x^2 + 4$

1. Domain and range. Any polynomial function, including this one, has the domain $X = \mathbb{R}$.

   To determine the range, we represent the function in the form $f(x) = (x^2 - \frac{5}{2})^2 - \frac{9}{4}$. This makes evident that $f(x) \geq -\frac{9}{4}$ and the minimum value the function $f(x)$ takes at the points $x = \pm\sqrt{\frac{5}{2}}$: $f(\pm\sqrt{\frac{5}{2}}) = -\frac{9}{4}$. Then, the range Y is contained in the interval $[-\frac{9}{4}, +\infty)$. To show that any point of this interval belongs to the range, we need to solve the bi-quadratic equation $y = (x^2 - \frac{5}{2})^2 - \frac{9}{4}$ for x when an arbitrary $y \geq -\frac{9}{4}$ is given. Write this equation in the form $(x^2 - \frac{5}{2})^2 = y + \frac{9}{4}$ and then $x^2 - \frac{5}{2} = \pm\sqrt{y + \frac{9}{4}}$ or $x^2 = \frac{5}{2} \pm \sqrt{y + \frac{9}{4}}$. To avoid the cases without solutions when the right-hand side is negative, we choose only the positive root and get the solution $x = \pm\sqrt{\frac{5}{2} + \sqrt{y + \frac{9}{4}}}$. Hence, any $y \geq -\frac{9}{4}$ is associated with at least one x of the domain (actually, there are at least two such x) under the given function. Therefore, $Y = [-\frac{9}{4}, +\infty)$.

2. Intersection and sign. The point of intersection with the y-axis is $P_0 = (0, 4)$. To find the points of intersection with the x-axis we should solve the bi-quadratic equation $(x^2 - \frac{5}{2})^2 - \frac{9}{4} = 0$. The algorithm of the solution is the same as for finding the range and it reduces the equation to the form $x^2 = \frac{5}{2} \pm \frac{3}{2}$. Here, the right-hand side is positive for both roots, and consequently, we find the four zeros of the function: $x = \pm\sqrt{\frac{5}{2} \pm \frac{3}{2}}$, that is, $x = -2, -1, 1, 2$. The corresponding points where the graph crosses the x-axis are $P_1 = (-2, 0)$, $P_2 = (-1, 0)$, $P_3 = (1, 0)$ and $P_4 = (2, 0)$.

   Using the formula $f(x) = (x^2 - 1)(x^2 - 4)$ it is easy to determine the sign of the polynomial:

   (1) if $x < -2$, then $f(x) > 0$;
   (2) if $-2 < x < -1$, then $f(x) < 0$;
   (3) if $-1 < x < 1$, then $f(x) > 0$;
   (4) if $1 < x < 2$, then $f(x) < 0$;
   (5) if $x > 2$, then $f(x) > 0$.

3. Boundedness. Since the range is $Y = [-\frac{9}{4}, +\infty)$, the function is bounded below and unbounded above.
4. Parity. Since $f(-x) = (-x)^4 - 5(-x)^2 + 4 = x^4 - 5x^2 + 4 = f(x), \forall x \in \mathbb{R}$, the function is even and its graph is symmetric with respect to the y-axis.
5. Periodicity. We leave the investigation of periodicity until after monotonicity.
6. Monotonicity and extrema.

   Since the function is even, we can evaluate monotonicity only for $x \geq 0$, completing the negative part later, by the property of symmetry.

   Take the points $0 \leq x_1 < x_2$ and evaluate the difference $D_1 = f(x_2) - f(x_1)$:

$$D_1 = f(x_2) - f(x_1) = (x_2^4 - 5x_2^2 + 4) - \left(x_1^4 - 5x_1^2 + 4\right) = (x_2^4 - x_1^4) - 5(x_2^2 - x_1^2)$$

$$= (x_2^2 - x_1^2)(x_2^2 + x_1^2) - 5(x_2^2 - x_1^2) = (x_2^2 - x_1^2)(x_2^2 + x_1^2 - 5).$$

   In the last expression, the first factor is positive, and consequently, the sign of D_1 depends only on the second factor. From the form of this factor it follows that its sign is well determined on the two intervals: on $[0, \sqrt{\frac{5}{2}}]$ and on $[\sqrt{\frac{5}{2}}, +\infty)$. On the former, $x_2 \leq \sqrt{\frac{5}{2}}$, whence $x_2^2 + x_1^2 - 5 < 0$, while on the second $x_1 \geq \sqrt{\frac{5}{2}}$, and then $x_2^2 + x_1^2 - 5 > 0$. Therefore, on the interval $[0, \sqrt{\frac{5}{2}}]$ we have $D_1 < 0$

and $f(x)$ is decreasing, while on the interval $[\sqrt{\frac{5}{2}}, +\infty)$ we have $D_1 > 0$ and $f(x)$ is increasing.

The symmetry of the even function allows us to extend this result onto the entire domain:

(1) $f(x)$ decreases on $(-\infty, -\sqrt{\frac{5}{2}}]$;

(2) $f(x)$ increases on $[-\sqrt{\frac{5}{2}}, 0]$;

(3) $f(x)$ decreases on $[0, \sqrt{\frac{5}{2}}]$;

(4) $f(x)$ increases on $[\sqrt{\frac{5}{2}}, +\infty)$.

Consequently, the points $x = \pm\sqrt{\frac{5}{2}}$ are strict local and global minima, while the point $x = 0$ is a strict local (but not global) maximum.

The monotonicity on the infinite interval $[\sqrt{\frac{5}{2}}, +\infty)$ implies that $f(x)$ is not periodic.

7. Concavity and inflection.

We will start with the investigation of concavity on the interval $[0, +\infty)$.

Take two arbitrary points $0 \le x_1 < x_2$, define the midpoint $x_0 = \frac{x_1+x_2}{2}$ and the increment $h = x_2 - x_0 = x_0 - x_1 > 0$, and use the quantity $D_2 \equiv f(x_1) + f(x_2) - 2f(x_0)$ to evaluate the difference between the mean value $\frac{f(x_1)+f(x_2)}{2}$ and the value at the midpoint $f(x_0)$. Using the expression found for D_1, we obtain:

$$D_2 = f(x_1) + f(x_2) - 2f(x_0) = (f(x_2) - f(x_0)) - (f(x_0) - f(x_1))$$

$$= (x_2^2 - x_0^2)\left(x_2^2 + x_0^2 - 5\right) - (x_0^2 - x_1^2)\left(x_0^2 + x_1^2 - 5\right)$$

$$= h(x_2 + x_0)\left(x_2^2 + x_0^2 - 5\right) - h(x_0 + x_1)\left(x_0^2 + x_1^2 - 5\right)$$

$$= h\left[(2x_0 + h)\left((x_0 + h)^2 + x_0^2 - 5\right) - (2x_0 - h)\left(x_0^2 + (x_0 - h)^2 - 5\right)\right]$$

$$= h\left[2x_0\left((x_0 + h)^2 + x_0^2 - 5 - x_0^2 - (x_0 - h)^2 + 5\right)\right.$$

$$\left. + h\left((x_0 + h)^2 + x_0^2 - 5 + x_0^2 + (x_0 - h)^2 - 5\right)\right]$$

$$= h\left[2x_0 \cdot 4x_0 h + h\left(4x_0^2 + 2h^2 - 10\right)\right] = 2h^2(6x_0^2 + h^2 - 5).$$

The choice of x_0 defines the sign of the second factor and the division occurs at the points $6x_0^2 = 5$, that is, $x_0 = \pm\sqrt{\frac{5}{6}}$. Since we choose to consider non-negative

x, we get the only point $x_0 = \sqrt{\frac{5}{6}}$. Then, on the interval $[0, \sqrt{\frac{5}{6}}]$ we have $x_2 \le \sqrt{\frac{5}{6}}$ or $x_0 + h \le \sqrt{\frac{5}{6}}$, which implies that $(x_0 + h)^2 \le \frac{5}{6}$ or $x_0^2 + h^2 \le \frac{5}{6} - 2x_0h$, and consequently, $6x_0^2 + h^2 < 5$. In this case, the second factor is negative, which ensures that $D_2 < 0$ and $f(x)$ is concave downward on $[0, \sqrt{\frac{5}{6}}]$. On the interval $[\sqrt{\frac{5}{6}}, +\infty)$ we have $x_1 \ge \sqrt{\frac{5}{6}}$ and then $x_0 > \sqrt{\frac{5}{6}}$ and $6x_0^2 > 5$. Therefore, the second factor is positive, and consequently, $D_2 > 0$ and $f(x)$ is concave upward on $[\sqrt{\frac{5}{6}}, +\infty)$.

Extending this results onto the entire domain, we obtain:

(1) $f(x)$ has upward concavity on $(-\infty, -\sqrt{\frac{5}{6}}]$;

(2) $f(x)$ has downward concavity on $[-\sqrt{\frac{5}{6}}, \sqrt{\frac{5}{6}}]$;

(3) $f(x)$ has upward concavity on $[\sqrt{\frac{5}{6}}, +\infty)$.

Notice that the upward concavity on $(-\infty, -\sqrt{\frac{5}{6}}]$ is the consequence of the upward concavity on $[\sqrt{\frac{5}{6}}, +\infty)$ and the evenness of the function. On the other hand, the downward concavity on $[-\sqrt{\frac{5}{6}}, \sqrt{\frac{5}{6}}]$ follows from the downward concavity on $[0, \sqrt{\frac{5}{6}}]$, the fact that $f(x)$ is even and that $x = 0$ is a local maximum (the last is used in Property 3 of Sect. 4.1 to extend concavity from the two subintervals onto the entire interval).

Since the function changes the type of concavity at the points $x = \pm\sqrt{\frac{5}{6}}$, these are inflection points.

8. Complimentary properties. For sufficiently large $|x|$ the behavior of the polynomial is determined by its leading term: $f(x) = x^4 - 5x^2 + 4 \approx x^4$, which implies that the function increases without restriction when $x \to \pm\infty$. Consequently, the function has no horizontal asymptote.

   At any point of the domain the function is continuous, that is, $f(x) \to f(x_0)$ when $x \to x_0$. Hence, there is no convergence of $f(x)$ to infinity at any point, which implies that the function has no vertical asymptote.
9. Using the investigated properties and marking some important points, such as $P_0 = (0, 4)$ (intersection with the y-axis and local maximum), $P_1 = (-2, 0)$, $P_2 = (-1, 0)$, $P_3 = (1, 0)$ and $P_4 = (2, 0)$ (intersection with the x-axis), $P_5 = (-\sqrt{\frac{5}{2}}, -\frac{9}{4})$ and $P_6 = (\sqrt{\frac{5}{2}}, -\frac{9}{4})$ (local and global minima), $P_7 = (-\sqrt{\frac{5}{6}}, \frac{19}{36})$ and $P_8 = (\sqrt{\frac{5}{6}}, \frac{19}{36})$ (inflection points), we can plot the graph shown in Fig. 3.48.

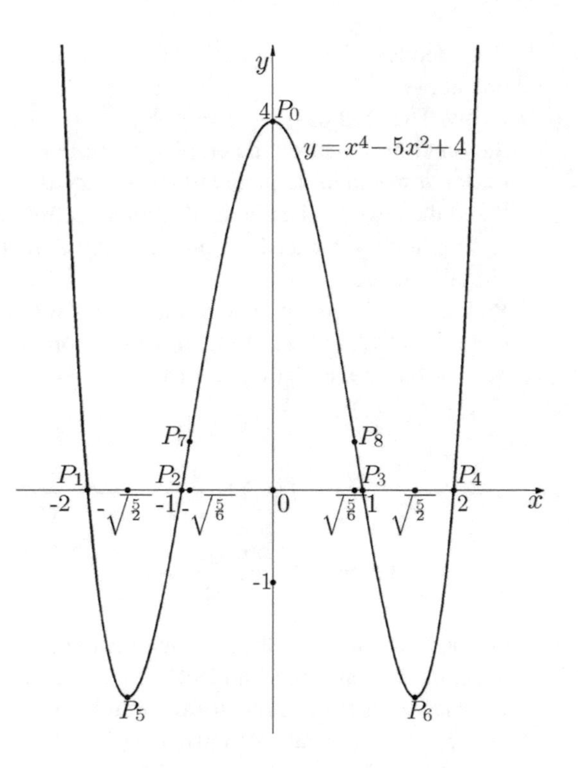

Fig. 3.48 The graph of the function $f(x) = x^4 - 5x^2 + 4$

5.2 *Study of Rational Functions*

A. Study of $f(x) = \frac{1-x^2}{x}$

1. Domain and range. Any rational function is defined at each point where its denominator does not vanish. For the given function it means that $X = \mathbb{R}\backslash\{0\}$.

   To determine the range Y, let us analyze for what values of y the equation $\frac{1-x^2}{x} = y$ for unknown x has at least one solution in the domain X. Rewriting this equation in the canonical form of a quadratic equation $x^2 + yx - 1 = 0$, we find the solutions by the standard formula: $x = \frac{1}{2}(-y \pm \sqrt{y^2 + 4})$. Therefore, the real roots x different from 0 (that is, belonging to the domain X) exist for any real y. Hence, the range is $Y = \mathbb{R}$.
2. Intersection and sign. There is no intersection with the y-axis because $x = 0$ is out of domain. The points of intersection with the x-axis are $P_1 = (-1, 0)$ and $P_2 = (1, 0)$.

   The sign of the function is easily determined:

   (1) if $x \in (-\infty, -1)$, then $f(x) > 0$;
   (2) if $x \in (-1, 0)$, then $f(x) < 0$;
   (3) if $x \in (0, 1)$, then $f(x) > 0$;
   (4) if $x \in (1, +\infty)$, then $f(x) < 0$.

3. Boundedness. Since the range is $Y = \mathbb{R}$, the function is unbounded both below and above.
4. Parity. The relation $f(-x) = \frac{1-(-x)^2}{-x} = -\frac{1-x^2}{x} = -f(x)$, $\forall x \in X$, shows that the function is odd and its graph is symmetric with respect to the origin. For this reason it would be sufficient to investigate the remaining properties only for $x > 0$ and then extend the obtained results according to the symmetry. However, due to simplicity of the analysis on the entire domain we do not apply this additional simplification.
5. Periodicity. Since only one point is missing in the domain, the function cannot be periodic. (Another argument comes from monotonicity.)
6. Monotonicity and extrema. Take $x_1 < x_2$ and evaluate the difference $D_1 = f(x_2) - f(x_1)$:

$$D_1 = f(x_2) - f(x_1) = \frac{1-x_2^2}{x_2} - \frac{1-x_1^2}{x_1} = \frac{x_1 - x_1x_2^2 - (x_2 - x_2x_1^2)}{x_2x_1}$$

$$= (x_1 - x_2) \cdot \frac{1 + x_1x_2}{x_1x_2}.$$

On the right-hand side the first factor is negative, and consequently, the sign of D_1 depends only on the second factor. Since the domain consists of the two intervals, we analyze the monotonicity separately on $(-\infty, 0)$ and on $(0, +\infty)$. In each of these two intervals we have $x_1x_2 > 0$, which means that the second factor is positive. Therefore, in each of the two intervals $D_1 < 0$, and consequently, $f(x)$ decreases on $(-\infty, 0)$ and on $(0, +\infty)$. Notice that the function does not decrease on the entire domain (for example, $f(-1) = 0 < \frac{3}{2} = f(\frac{1}{2})$).

It follows from the monotonicity that there is no extremum of any type, since each point of the domain is decreasing.

Another implication of the monotonicity is that the function is not periodic since it decreases on the infinite interval $(0, +\infty)$.
7. Concavity and inflection. Take two arbitrary points $x_1 < x_2$ both belonging to one of the intervals $(-\infty, 0)$ or $(0, +\infty)$, define the midpoint $x_0 = \frac{x_1+x_2}{2}$ and the increment $h = x_2 - x_0 = x_0 - x_1 > 0$, and use the quantity $D_2 \equiv f(x_1) + f(x_2) - 2f(x_0)$ to evaluate the difference between the mean value $\frac{f(x_1)+f(x_2)}{2}$ and the value at the midpoint $f(x_0)$. Employing the expression found for D_1, we obtain:

$$D_2 = f(x_1) + f(x_2) - 2f(x_0) = (f(x_2) - f(x_0)) - (f(x_0) - f(x_1))$$

$$= (x_0 - x_2)\frac{1 + x_0x_2}{x_0x_2} - (x_1 - x_0)\frac{1 + x_1x_0}{x_1x_0} = -h\frac{(1 + x_0x_2)x_1 - (1 + x_1x_0)x_2}{x_0x_2x_1}$$

$$= -h\frac{x_1 - x_2}{x_0x_2x_1} = \frac{2h^2}{x_1x_0x_2}.$$

The numerator is positive and the denominator has different sign in $(-\infty, 0)$ and in $(0, +\infty)$: if $x_2 < 0$, then $x_1x_0x_2 < 0$ and $D_2 < 0$, while if $x_1 > 0$, then $x_1x_0x_2 > 0$ and $D_2 > 0$. Therefore, $f(x)$ is concave downward on $(-\infty, 0)$ and upward on $(0, +\infty)$.

Although the function has different types of concavity to the left and to the right of 0, but this point is not inflection because it does not belong to the domain.

8. Complimentary properties. For sufficiently large $|x|$ the behavior of the rational function is determined by the leading terms of the polynomials in the numerator and denominator: $f(x) = \frac{1-x^2}{x} \approx \frac{-x^2}{x} = -x$. This implies that the function increases without restriction when $x \to -\infty$ and decreases without restriction when $x \to +\infty$. Consequently, the function has no horizontal asymptote.

   At any point of the domain the function is continuous, that is, $f(x) \to f(x_0)$ when $x \to x_0$. Hence, there is no convergence of $f(x)$ to infinity at any point of the domain, and consequently, there is no vertical asymptote among the points of the domain. The only remaining point is the origin, which is out of the domain, but can be approximated by the points of the domain. Evaluating the behavior of $f(x)$ when x approaches 0, we notice that $f(x) = \frac{1-x^2}{x} \approx \frac{1}{x}$. Therefore, the function goes to $-\infty$ as x approaches 0 from the left and goes to $+\infty$ as x approaches 0 from the right. Therefore, $x = 0$ is a vertical asymptote.

9. Using the investigated properties, marking some important points, such as $P_1 = (-1, 0)$ and $P_2 = (1, 0)$ (intersection with the x-axis), and taking advantage of the symmetry about the origin, we plot the graph shown in Fig. 3.49.

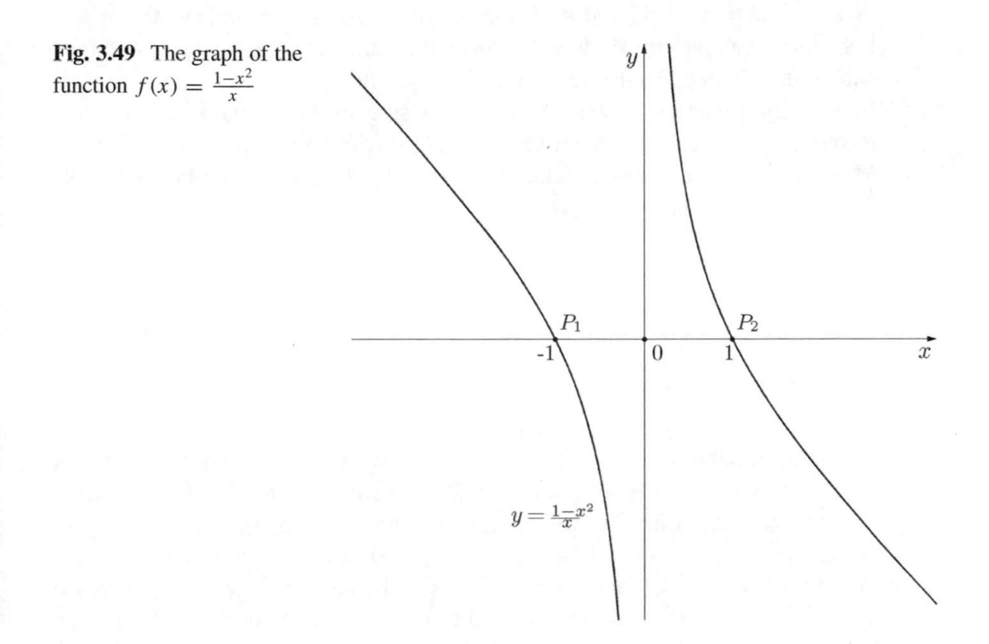

Fig. 3.49 The graph of the function $f(x) = \frac{1-x^2}{x}$

B. Study of $f(x) = \frac{x}{1-x^2}$

1. Domain and range. Any rational function is defined at each point where its denominator does not vanish. For the given function it means that $X = \mathbb{R}\backslash\{-1, 1\}$.
   To determine the range Y, let us analyze for what values of y the equation $\frac{x}{1-x^2} = y$ for unknown x has at least one solution in the domain X. Rewriting this equation in the canonical form of a quadratic equation $yx^2 + x - y = 0$ and assuming for the moment that $y \neq 0$, we find the solutions by the standard formula: $x = \frac{1}{2y}(-1 \pm \sqrt{1+4y^2})$. Notice that $x = \pm 1$ are not solutions, since the quadratic equation becomes a false relation $y+1-y = 0$ in the case $x = 1$ and another false formula $y - 1 - y = 0$ in the case $x = -1$. Therefore, for any $y \neq 0$ there exists at least one x in the domain X ($x \neq \pm 1$) such that $f(x) = y$. For $y = 0$ we return to the original equation $\frac{x}{1-x^2} = y$ and obtain the solution $x = 0$ (also in the domain X). Hence, the range is $Y = \mathbb{R}$.
2. Intersection and sign. The point of intersection with the y-axis is $P_0 = (0, 0)$ and the very same point is also intersection with the x-axis. There are no other points of intersection with the x-axis, since the equation $\frac{x}{1-x^2} = 0$ has the only root $x = 0$.
   The sign of the function is easier to define after the study of monotonicity.
3. Boundedness. Since the range is $Y = \mathbb{R}$, the function is unbounded both below and above.
4. Parity. The relation $f(-x) = \frac{-x}{1-(-x)^2} = -\frac{x}{1-x^2} = -f(x)$, $\forall x \in X$ shows that the function is odd and its graph is symmetric with respect to the origin. For this reason we can study some properties first for $x \geq 0$, $x \neq 1$ and then extend the obtained results according to the symmetry.
5. Periodicity. Since only two points are missing in the domain, the function cannot be periodic. (Another argument comes from monotonicity.)
6. Monotonicity and extrema. Take $0 \leq x_1 < x_2$, $x_1, x_2 \neq 1$ and evaluate the difference $D_1 = f(x_2) - f(x_1)$:

$$D_1 = f(x_2) - f(x_1) = \frac{x_2}{1-x_2^2} - \frac{x_1}{1-x_1^2} = \frac{x_2 - x_1^2 x_2 - (x_1 - x_1 x_2^2)}{(1-x_1^2)(1-x_2^2)}$$
$$= (x_2 - x_1) \cdot \frac{1 + x_1 x_2}{(1-x_1^2)(1-x_2^2)}.$$

   On the right-hand side the first factor is positive, and consequently, the sign of D_1 depends only on the second factor. Its numerator is positive, but the terms of the denominator have different signs on the intervals $[0, 1)$ and $(1, +\infty)$. On the former, $x_1 < x_2 < 1$, whence $1 - x_1^2 > 0$ and $1 - x_2^2 > 0$, which results in the positive sign of the second factor. On the later, $1 < x_1 < x_2$, whence $1 - x_1^2 < 0$, $1 - x_2^2 < 0$, and again the second factor is positive. Therefore, on both intervals $[0, 1)$ and $(1, +\infty)$ we get $D_1 > 0$, which means that $f(x)$

increases on $[0, 1)$ and on $(1, +\infty)$. Notice that the function does not increase on $[0, 1) \cup (1, +\infty)$ (for example, $f(\frac{1}{2}) = \frac{2}{3} > -\frac{2}{3} = f(2)$).

It follows from the increase on the infinite interval $(1, +\infty)$ that $f(x)$ is not periodic.

7. Concavity and inflection. Take two arbitrary points $x_1 < x_2$ both belonging to one of the intervals $[0, 1)$ or $(1, +\infty)$, define the midpoint $x_0 = \frac{x_1+x_2}{2}$ and the increment $h = x_2 - x_0 = x_0 - x_1 > 0$, and use the quantity $D_2 \equiv f(x_1) + f(x_2) - 2f(x_0)$ to evaluate the difference between the mean value $\frac{f(x_1)+f(x_2)}{2}$ and the value at the midpoint $f(x_0)$. Employing the expression found for D_1, we obtain:

$$D_2 = f(x_1) + f(x_2) - 2f(x_0) = (f(x_2) - f(x_0)) - (f(x_0) - f(x_1))$$

$$= (x_2 - x_0) \cdot \frac{1 + x_0 x_2}{(1 - x_0^2)(1 - x_2^2)} - (x_0 - x_1) \cdot \frac{1 + x_1 x_0}{(1 - x_1^2)(1 - x_0^2)}$$

$$= h \frac{(1 + x_0 x_2)(1 - x_1^2) - (1 + x_1 x_0)(1 - x_2^2)}{(1 - x_0^2)(1 - x_2^2)(1 - x_1^2)}$$

$$= h \frac{x_0 x_2 - x_0 x_1 + x_2^2 - x_1^2 + x_1 x_0 x_2^2 - x_0 x_2 x_1^2}{(1 - x_0^2)(1 - x_2^2)(1 - x_1^2)}$$

$$= 2h^2 \frac{x_0 + x_2 + x_1 + x_1 x_0 x_2}{(1 - x_0^2)(1 - x_2^2)(1 - x_1^2)}.$$

The numerator is positive and the denominator has different sign in $[0, 1)$ and in $(1, +\infty)$: if $x_2 < 1$, then $(1 - x_0^2)(1 - x_2^2)(1 - x_1^2) > 0$ and $D_2 > 0$, while if $x_1 > 1$, then $(1 - x_0^2)(1 - x_2^2)(1 - x_1^2) < 0$ and $D_2 < 0$. Therefore, $f(x)$ is concave upward on $[0, 1)$ and downward on $(1, +\infty)$.

Although the function changes the type of concavity passing the point 1, this is not an inflection point, because it does not belong to the domain.

6–7. Applying the symmetry of the function about the origin we can extend the obtained results of monotonicity as follows:

(1) $f(x)$ increases on $(-\infty, -1)$;
(2) $f(x)$ increases on $(-1, 1)$;
(3) $f(x)$ increases on $(1, +\infty)$.

(Notice that $f(x)$ does not increase on the entire domain.) Therefore, there are no extrema of any type (each point of the domain is increasing and the points $x = \pm 1$ are out of the domain).

From the properties of monotonicity it also follows that $f(x) > 0$ on the intervals $(-\infty, -1)$ and $(0, 1)$ and that $f(x) < 0$ on the intervals $(-1, 0)$ and $(1, +\infty)$.

Similarly, we can extend the result of the concavity in the following way:

(1) $f(x)$ is concave upward on $(-\infty, -1)$;
(2) $f(x)$ is concave downward on $(-1, 0]$;
(3) $f(x)$ is concave upward on $[0, 1)$;
(4) $f(x)$ is concave downward on $(1, +\infty)$.

Therefore, there is only one inflection point $x = 0$ (the points $x = \pm 1$ are out of the domain).

8. Complimentary properties. For sufficiently large $|x|$ the behavior of a rational function is determined by the leading terms of the polynomials in the numerator and denominator: $f(x) = \frac{x}{1-x^2} \approx \frac{x}{-x^2} = -\frac{1}{x}$. This implies that the function converges to 0 when $x \to \pm\infty$, and consequently, the line $y = 0$ is the only horizontal asymptote.

   At each point of the domain the function is continuous, that is, $f(x) \to f(x_0)$ when $x \to x_0$. Hence, there is no convergence of $f(x)$ to infinity at any point of the domain, and consequently, there is no vertical asymptote among the points of the domain. The only remaining points are $x = \pm 1$. Near the point 1 the function can be evaluate in the form $f(x) = \frac{x}{1-x^2} \approx \frac{1}{2(1-x)}$. This implies that $f(x)$ goes to $+\infty$ as x approaches 1 from the left and goes to $-\infty$ as x approaches 1 from the right. A similar behavior is observed near the point -1. Therefore, the lines $x = 1$ and $x = -1$ are vertical asymptotes.
9. Using the investigated properties, marking some important points, such as $P_0 = (0, 0)$ (intersection with the coordinate axes), $P_1 = (\frac{1}{2}, \frac{2}{3})$ (a point with the x-coordinate in the interval $(0, 1)$) and $P_2 = (2, -\frac{2}{3})$ (a point with the x-coordinate in the interval $(1, +\infty)$), and taking advantage of the symmetry about the origin, we plot the graph shown in Fig. 3.50.

C. Study of $f(x) = \frac{1}{x^2+1}$

1. Domain and range. The denominator of this rational function is positive for any x which means that the domain is $X = \mathbb{R}$.

   Since the denominator satisfies the inequality $x^2 + 1 \geq 1$, the range Y contains only positive values smaller than or equal to 1. In other words, the range is contained in the interval $(0, 1]$. Besides, for each $y \in (0, 1]$ the equation $\frac{1}{x^2+1} = y$ has solutions $x = \pm\sqrt{\frac{1}{y} - 1}$ which show that $Y = (0, 1]$.
2. Intersection and sign. The point of intersection with the y-axis is $P_0 = (0, 1)$. There is no point of intersection with the x-axis since the range does not contain the point $y = 0$.

   The function takes only positive values.
3. Boundedness. Since the range is $Y = (0, 1]$, the function is bounded both below and above.

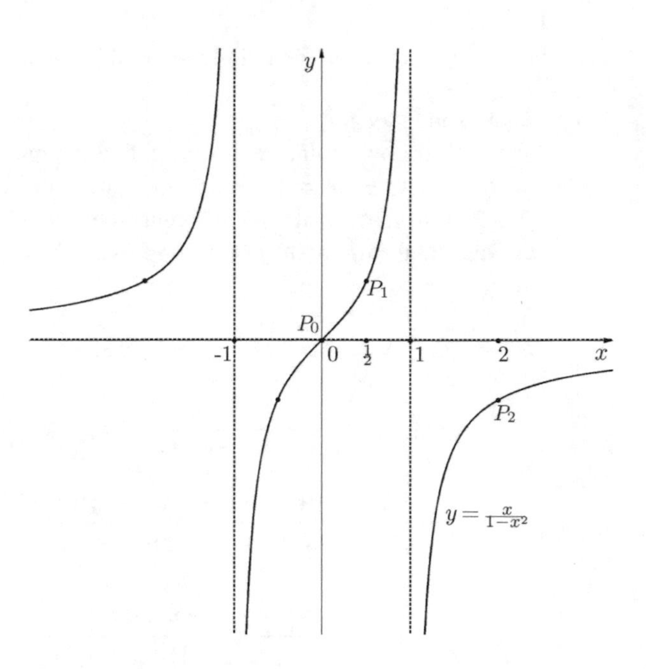

Fig. 3.50 The graph of the function $f(x) = \frac{x}{1-x^2}$

4. Parity. The relation $f(-x) = \frac{1}{1+(-x)^2} = \frac{1}{1+x^2} = f(x)$, $\forall x \in X$ shows that the function is even and its graph is symmetric with respect to the y-axis. For this reason it is sufficient to investigate monotonicity for the values $x \geq 0$, completing the negative part by the symmetry property.
5. Periodicity. The investigation of periodicity is deferred until after monotonicity.
6. Monotonicity and extrema.

   Take $0 \leq x_1 < x_2$ and evaluate the difference $D_1 = f(x_2) - f(x_1)$:

$$D_1 = f(x_2) - f(x_1) = \frac{1}{x_2^2+1} - \frac{1}{x_1^2+1} = \frac{x_1^2 - x_2^2}{(x_2^2+1)(x_1^2+1)}$$

$$= \frac{x_1 - x_2}{(x_2^2+1)(x_1^2+1)} \cdot (x_1 + x_2).$$

On the right-hand side the first factor is negative and the second is positive for $x_1 \geq 0$. Therefore, $D_1 < 0$ and $f(x)$ decreases on $[0, +\infty)$.

Extending this result by the symmetry about the y-axis, we obtain the following properties:

(1) $f(x)$ increases on $(-\infty, 0]$;
(2) $f(x)$ decreases on $[0, +\infty)$.

Consequently, $x = 0$ is a strict local and global maximum. There are no other extrema.

From the decrease on the infinite interval $[0, +\infty)$ it follows that $f(x)$ is not periodic.

7. Concavity and inflection.

Take two arbitrary points $x_1 < x_2$, define the midpoint $x_0 = \frac{x_1+x_2}{2}$ and the increment $h = x_2 - x_0 = x_0 - x_1 > 0$, and use the quantity $D_2 \equiv f(x_1) + f(x_2) - 2f(x_0)$ to evaluate the difference between the mean value $\frac{f(x_1)+f(x_2)}{2}$ and the value at the midpoint $f(x_0)$. Employing the expression found for D_1, we obtain:

$$D_2 = f(x_1) + f(x_2) - 2f(x_0) = (f(x_2) - f(x_0)) - (f(x_0) - f(x_1))$$

$$= (x_0 - x_2)\frac{x_0 + x_2}{(x_2^2+1)(x_0^2+1)} - (x_1 - x_0)\frac{x_1 + x_0}{(x_0^2+1)(x_1^2+1)}$$

$$= -h\frac{(x_0 + x_2)(x_1^2 + 1) - (x_1 + x_0)(x_2^2 + 1)}{(x_2^2+1)(x_0^2+1)(x_1^2+1)}$$

$$= -h\frac{x_0(x_1^2 - x_2^2) + x_1x_2(x_1 - x_2) + (x_2 - x_1)}{(x_2^2+1)(x_0^2+1)(x_1^2+1)}$$

$$= 2h^2\frac{x_0(x_1 + x_2) + x_1x_2 - 1}{(x_2^2+1)(x_0^2+1)(x_1^2+1)}$$

$$= 2h^2\frac{2x_0^2 + (x_0 - h)(x_0 + h) - 1}{(x_2^2+1)(x_0^2+1)(x_1^2+1)} = 2h^2\frac{3x_0^2 - h^2 - 1}{(x_2^2+1)(x_0^2+1)(x_1^2+1)}.$$

The denominator is positive, while the expression of the numerator reveals that it has different sign at the points where $3x_0^2 < 1$ and where $3x_0^2 > 1$ as shown below.

(1) On the interval $(-\infty, -\frac{1}{\sqrt{3}}]$ we have $x_1 < x_0 < x_2 \le -\frac{1}{\sqrt{3}}$, whence $x_0(x_1 + x_2) > \frac{2}{3}$ and $x_1x_2 > \frac{1}{3}$. Therefore, $x_0(x_1 + x_2) + x_1x_2 - 1 > \frac{2}{3} + \frac{1}{3} - 1 = 0$, and consequently, $D_2 > 0$ which means that $f(x)$ is concave upward.

(2) On the interval $[-\frac{1}{\sqrt{3}}, \frac{1}{\sqrt{3}}]$ we have $-\frac{1}{\sqrt{3}} \le x_1 < x_0 < x_2 \le \frac{1}{\sqrt{3}}$, whence $|x_0(x_1 + x_2)| < \frac{2}{3}$ and $|x_1x_2| < \frac{1}{3}$. Therefore, $x_0(x_1 + x_2) + x_1x_2 - 1 < \frac{2}{3} + \frac{1}{3} - 1 = 0$, and consequently, $D_2 < 0$, which means that $f(x)$ is concave downward.

(3) On the interval $[\frac{1}{\sqrt{3}}, +\infty)$ we have $\frac{1}{\sqrt{3}} \le x_1 < x_0 < x_2$, whence $x_0(x_1 + x_2) > \frac{2}{3}$ and $x_1x_2 > \frac{1}{3}$. Therefore, $x_0(x_1 + x_2) + x_1x_2 - 1 > \frac{2}{3} + \frac{1}{3} - 1 = 0$, and consequently, $D_2 > 0$ which means that $f(x)$ is concave upward.

According to the concavity properties the function has the two inflection points $x = \pm\frac{1}{\sqrt{3}}$.

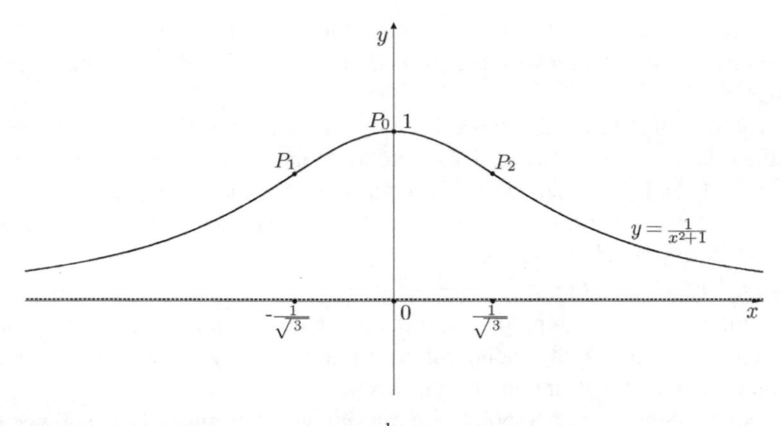

Fig. 3.51 The graph of the function $f(x) = \frac{1}{x^2+1}$

8. Complimentary properties. For sufficiently large $|x|$ the behavior of the rational function is determined by the leading terms of the polynomials in the numerator and denominator: $f(x) = \frac{1}{x^2+1} \approx \frac{1}{x^2}$. This implies that the function converges to 0 when $x \to \pm\infty$. Consequently, the function has the only horizontal asymptote $y = 0$.

   At any point of the domain (that is, at any real point) the function is continuous, that is, $f(x) \to f(x_0)$ when $x \to x_0$. Hence, there is no convergence of $f(x)$ to infinity at any real point, and consequently, there is no vertical asymptote.
9. Using the investigated properties, marking some important points, such as $P_0 = (0, 1)$ (intersection with the y-axis and global maximum), $P_1 = (-\frac{1}{\sqrt{3}}, \frac{3}{4})$ and $P_2 = (\frac{1}{\sqrt{3}}, \frac{3}{4})$ (inflection points), and taking advantage of the symmetry about the y-axis, we plot the graph shown in Fig. 3.51.

D. Study of $f(x) = \frac{1}{x^2-1}$

1. Domain and range. Any rational function is defined at each point where its denominator does not vanish. For the given function it means that $X = \mathbb{R}\backslash\{-1, 1\}$.

   To determine the range Y, let us check for what values of y the equation $\frac{1}{x^2-1} = y$ for unknown x has at least one solution in the domain X. Notice first that there is no x corresponding to $y = 0$, in other words, $y = 0$ is out of the range. For any $y \neq 0$ the equation can be written in the equivalent form $x^2 - 1 = \frac{1}{y}$ or $x^2 = \frac{1}{y} + 1$. There are solutions only when $\frac{1}{y} + 1 \geq 0$, that is, for $y > 0$ and for $y \leq -1$. Hence, the range of the function is $Y = (-\infty, -1] \cup (0, +\infty)$.

2. Intersection and sign. The point of intersection with the y-axis is $P_0 = (0, -1)$. There is no intersection with the x-axis, because $y = 0$ does not belong to the range.

   The sign of the function depends on the denominator: if $|x| > 1$ it is positive, while if $|x| < 1$ it is negative. Therefore, the function is positive on the intervals $(-\infty, -1)$ and $(1, +\infty)$, while it is negative on the interval $(-1, 1)$.
3. Boundedness. Since the range is $Y = (-\infty, -1] \cup (0, +\infty)$, the function is unbounded both below and above.
4. Parity. The relation $f(-x) = \frac{1}{(-x)^2-1} = \frac{1}{x^2-1} = f(x), \forall x \in X$ shows that the function is even and its graph is symmetric with respect to the y-axis. For this reason we can first study monotonicity for $x \geq 0$, $x \neq 1$, and then extend the obtained results according to the symmetry.
5. Periodicity. Since only two points are missing in the domain, the function cannot be periodic. (Another argument comes from monotonicity.)
6. Monotonicity and extrema.

   Take $0 \leq x_1 < x_2$, $x_1, x_2 \neq 1$ and evaluate the difference $D_1 = f(x_2) - f(x_1)$:

$$D_1 = f(x_2) - f(x_1) = \frac{1}{x_2^2 - 1} - \frac{1}{x_1^2 - 1} = \frac{(x_1 - x_2)(x_1 + x_2)}{(x_2^2 - 1)(x_1^2 - 1)}.$$

   The numerator of the last quotient is negative, and consequently, the sign of D_1 depends on the denominator. If $x_1 < x_2 < 1$, then $x_1^2 - 1 < 0$ and $x_2^2 - 1 < 0$, which gives the positive denominator. If $1 < x_1 < x_2$, then $x_1^2 - 1 > 0$ and $x_2^2 - 1 > 0$, which leads again to the positive denominator. Hence, on both intervals $[0, 1)$ and $(1, +\infty)$ we have $D_1 < 0$, which means that $f(x)$ decreases on $[0, 1)$ and $(1, +\infty)$. Notice that the function does not decrease on $[0, 1) \cup (1, +\infty)$ (for instance, $f(\frac{1}{2}) = -\frac{4}{3} < \frac{1}{3} = f(2)$).

   Applying the symmetry of the function about the y-axis, we extend the obtained results of monotonicity as follows:

   (1) $f(x)$ increases on $(-\infty, -1)$;
   (2) $f(x)$ increases on $(-1, 0]$;
   (3) $f(x)$ decreases on $[0, 1)$;
   (4) $f(x)$ decreases on $(1, +\infty)$.

   (Notice that $f(x)$ does not increase on $(-\infty, -1) \cup (-1, 0]$ or decrease on $[0, 1) \cup (1, +\infty)$.) Consequently, there exists the only extremum (local maximum) at the point $x = 0$ (the points $x = \pm 1$ are out of the domain).

   The decrease on the infinite interval $(1, +\infty)$ ensures that the function is not periodic.
7. Concavity and inflection.

   Take two arbitrary points $x_1 < x_2$ both belonging to one of the intervals of the domain $(-\infty, -1)$, $(-1, 1)$ or $(1, +\infty)$, define the midpoint $x_0 = \frac{x_1+x_2}{2}$ and the increment $h = x_2 - x_0 = x_0 - x_1 > 0$, and use the quantity $D_2 \equiv f(x_1) +$

$f(x_2) - 2f(x_0)$ to evaluate the difference between the mean value $\frac{f(x_1)+f(x_2)}{2}$ and the value at the midpoint $f(x_0)$. Employing the expression found for D_1, we obtain:

$$\begin{aligned}
D_2 = f(x_1) + f(x_2) - 2f(x_0) &= (f(x_2) - f(x_0)) - (f(x_0) - f(x_1)) \\
&= \frac{(x_0 - x_2)(x_0 + x_2)}{(x_2^2 - 1)(x_0^2 - 1)} - \frac{(x_1 - x_0)(x_1 + x_0)}{(x_0^2 - 1)(x_1^2 - 1)} \\
&= -h \frac{(x_0 + x_2)(x_1^2 - 1) - (x_1 + x_0)(x_2^2 - 1)}{(x_2^2 - 1)(x_0^2 - 1)(x_1^2 - 1)} \\
&= -h \frac{x_0(x_1^2 - x_2^2) + x_2 x_1 (x_1 - x_2) + (x_1 - x_2)}{(x_2^2 - 1)(x_0^2 - 1)(x_1^2 - 1)} \\
&= h^2 \frac{(x_1 + x_2)^2 + 2x_2 x_1 + 2}{(x_2^2 - 1)(x_0^2 - 1)(x_1^2 - 1)}.
\end{aligned}$$

Let us consider separately three intervals of the domain.

(1) On $(-\infty, -1)$ we have $x_1 < x_0 < x_2 < -1$, and consequently, all the summands of the numerator are positive as well as all the factors of the denominator. Therefore, $D_2 > 0$, which means that $f(x)$ is concave upward.
(2) On $(-1, 1)$ we have $-1 < x_1 < x_0 < x_2 < 1$, whence $2(x_1 x_2 + 1) > 0$, which shows that the numerator is positive. At the same time the denominator is negative, because each of the three factors is negative. Therefore, $D_2 < 0$ and $f(x)$ is concave downward.
(3) On $(1, +\infty)$ we have $1 < x_1 < x_0 < x_2$, and consequently, all the summands of the numerator are positive as well as all the factors of the denominator. Therefore, $D_2 > 0$, which means that $f(x)$ is concave upward.

Although the type of concavity changes at the points $x = \pm 1$, but these points are out of the domain and the function has no inflection points.

8. Complimentary properties. For sufficiently large $|x|$ the behavior of the rational function is determined by the leading terms of the polynomials in the numerator and denominator: $f(x) = \frac{1}{x^2-1} \approx \frac{1}{x^2}$. This implies that the function converges to 0 when $x \to \pm\infty$. Consequently, the function has the only horizontal asymptote $y = 0$.

   At each point of the domain the function is continuous, that is, $f(x) \to f(x_0)$ when $x \to x_0$. Hence, there is no convergence of $f(x)$ to infinity at any point of the domain, and consequently, there is no vertical asymptote among the points of the domain. The only remaining points are $x = \pm 1$. Near the point 1 the function can be evaluate in the form $f(x) = \frac{1}{x^2-1} \approx \frac{1}{2(x-1)}$. This implies that $f(x)$ goes to $-\infty$ as x approaches 1 from the left and goes to $+\infty$ as x approaches 1 from

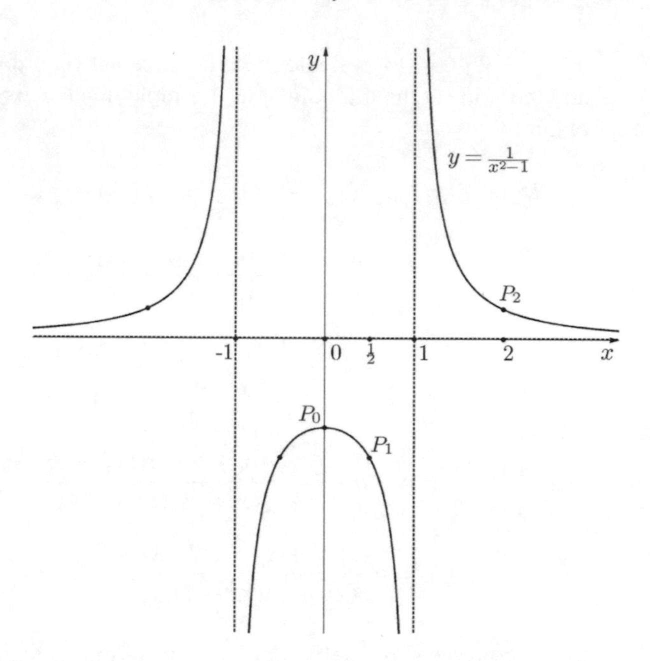

Fig. 3.52 The graph of the function $f(x) = \frac{1}{x^2-1}$

the right. A similar behavior is observed near the point -1. Therefore, the lines $x = 1$ and $x = -1$ are vertical asymptotes.

9. Using the investigated properties, marking some important points, such as $P_0 = (0, -1)$ (intersection with the y-axis and local maximum), $P_1 = (\frac{1}{2}, -\frac{4}{3})$ (a point with the x-coordinate in $(0, 1)$) and $P_2 = (2, \frac{1}{3})$ (a point with the x-coordinate in $(1, +\infty)$), and taking advantage of the symmetry about the y-axis, we plot the graph shown in Fig. 3.52.

E. Study of $f(x) = \frac{1}{x^3+2}$

1. Domain and range. Any rational function is defined at each point where its denominator does not vanish. For the given function it means that $X = \mathbb{R}\backslash\{-\sqrt[3]{2}\}$.

   To find the range Y, we need to determine for what values of y the equation $\frac{1}{x^3+2} = y$ for unknown x has at least one solution in the domain X. It is easy to see that this equation has the solution $x = \sqrt[3]{\frac{1}{y} - 2}$ for any $y \neq 0$ and has no solution if $y = 0$. Therefore, the range is $Y = \mathbb{R}\backslash\{0\}$.
2. Intersection and sign. Intersection with the y-axis occurs at the point $P_0 = (0, \frac{1}{2})$. There is no point of intersection with the x-axis, because $y = 0$ does not belong to the range.

The sign of the function depends on the denominator: if $x < -\sqrt[3]{2}$ it is negative, while if $x > -\sqrt[3]{2}$ it is positive. Then, on the interval $(-\infty, -\sqrt[3]{2})$ the function is negative, while on the interval $(-\sqrt[3]{2}, +\infty)$ it is positive.

3. Boundedness. Since the range is $Y = \mathbb{R}\backslash\{0\}$, the function is unbounded both below and above.
4. Parity. Since $f(-1) = 1$ and $f(1) = \frac{1}{3}$, the function is neither even nor odd. (Another elementary proof of this fact is that the domain is not symmetric about the origin.)
5. Periodicity. Since only one point is missing in the domain, the function cannot be periodic. (Another argument comes from monotonicity.)
6. Monotonicity and extrema.

Take $x_1 < x_2$ such that $x_1, x_2 \neq -\sqrt[3]{2}$ and evaluate the difference $D_1 = f(x_2) - f(x_1)$:

$$D_1 = f(x_2) - f(x_1) = \frac{1}{x_2^3+2} - \frac{1}{x_1^3+2} = \frac{(x_1-x_2)(x_1^2+x_1x_2+x_2^2)}{(x_2^3+2)(x_1^3+2)}.$$

The numerator of the last quotient is negative, and consequently, the sign depends on the denominator. If $x_1 < x_2 < -\sqrt[3]{2}$, then $x_1^3+2 < 0$ and $x_2^3+2 < 0$, which results in the positive denominator. If $-\sqrt[3]{2} < x_1 < x_2$, then $x_1^3+2 > 0$ and $x_2^3+2 > 0$, which again gives the positive denominator. Therefore, on both intervals $(-\infty, -\sqrt[3]{2})$ and $(-\sqrt[3]{2}, +\infty)$ we have $D_1 < 0$, which means that $f(x)$ decreases on each of these two intervals. Notice that the function does not decrease on the entire domain (for instance, $f(-2) = -\frac{1}{6} < \frac{1}{3} = f(1)$).

It follows from the monotonicity that there is no extremum of any type, since each point of the domain is decreasing.

Another implication of the monotonicity is that the function is not periodic, since it decreases on the infinite interval $(-\sqrt[3]{2}, +\infty)$.

7. Concavity and inflection.

Take two arbitrary points $x_1 < x_2$ both contained in one of the intervals $(-\infty, -\sqrt[3]{2})$ or $(-\sqrt[3]{2}, +\infty)$, define the midpoint $x_0 = \frac{x_1+x_2}{2}$ and the increment $h = x_2 - x_0 = x_0 - x_1 > 0$, and use the quantity $D_2 \equiv f(x_1) + f(x_2) - 2f(x_0)$ to evaluate the difference between the mean value $\frac{f(x_1)+f(x_2)}{2}$ and the value at the midpoint $f(x_0)$. Employing the expression found for D_1, we obtain:

$$D_2 = f(x_1) + f(x_2) - 2f(x_0) = (f(x_2) - f(x_0)) - (f(x_0) - f(x_1))$$

$$= \frac{(x_0-x_2)(x_0^2+x_0x_2+x_2^2)}{(x_2^3+2)(x_0^3+2)} - \frac{(x_1-x_0)(x_1^2+x_1x_0+x_0^2)}{(x_0^3+2)(x_1^3+2)}$$

$$= -h\frac{(x_0^2+x_0x_2+x_2^2)(x_1^3+2) - (x_1^2+x_1x_0+x_0^2)(x_2^3+2)}{(x_2^3+2)(x_0^3+2)(x_1^3+2)}$$

$$= -h\frac{x_0^2(x_1^3-x_2^3)+x_0x_1x_2(x_1^2-x_2^2)+x_1^2x_2^2(x_1-x_2)+2(x_2^2-x_1^2)+2x_0(x_2-x_1)}{(x_2^3+2)(x_0^3+2)(x_1^3+2)}$$

$$= 2h^2\frac{x_0^2(x_1^2+x_1x_2+x_2^2)+x_0x_1x_2(x_1+x_2)+x_1^2x_2^2-2(x_2+x_1)-2x_0}{(x_2^3+2)(x_0^3+2)(x_1^3+2)}$$

$$= 2h^2\frac{x_0^2(x_1^2+3x_1x_2+x_2^2)+x_1^2x_2^2-6x_0}{(x_2^3+2)(x_0^3+2)(x_1^3+2)}.$$

The denominator is negative on $(-\infty, -\sqrt[3]{2})$ and positive on $(-\sqrt[3]{2}, +\infty)$. Let us analyze the numerator. First, if $x_2 < -\sqrt[3]{2}$, then all the three summands are positive, and consequently, the numerator is positive. The same happens when $-\sqrt[3]{2} < x_1 < x_2 < 0$. On the interval $[0, 1]$ the numerator can be evaluated as follows: $x_0^2(x_1^2+3x_1x_2+x_2^2)+x_1^2x_2^2-6x_0 = x_0(x_0(x_1^2+3x_1x_2+x_2^2)-5)+x_1^2x_2^2-x_0 < x_1^2x_2^2-x_0 = (x_0^2-h^2)^2-x_0 < x_0^4-x_0 < 0$. Finally, if $1 \le x_1$, then regrouping the terms of the numerator we get $x_0^2(x_1^2+3x_1x_2+x_2^2)+x_1^2x_2^2-6x_0 = x_0(x_0(x_1^2+3x_1x_2+x_2^2)-5)+x_2(x_1^2x_2-1)+(x_2-x_0) > 0$. Hence, we arrive at the following results:

(1) on the interval $(-\infty, -\sqrt[3]{2})$ the numerator is positive and the denominator is negative, which means that $D_2 < 0$ and $f(x)$ is concave downward;
(2) on the interval $(-\sqrt[3]{2}, 0)$ both numerator and denominator are positive, and then $D_2 > 0$ and $f(x)$ is concave upward;
(3) on the interval $(0, 1)$ the numerator is negative and the denominator is positive, which means that $D_2 < 0$ and $f(x)$ is concave downward;
(4) on the interval $(1, +\infty)$ both numerator and denominator are positive, and then $D_2 > 0$ and $f(x)$ is concave upward.

Consequently, the points $x = 0$ and $x = 1$ are inflection points (the point $x = -\sqrt[3]{2}$ is not inflection since it is out of the domain).

8. Complimentary properties. For sufficiently large $|x|$ the behavior of the rational function is determined by the leading terms of the polynomials in the numerator and denominator: $f(x) = \frac{1}{x^3+2} \approx \frac{1}{x^3}$. This implies that the function converges to 0 when x goes to $\pm\infty$, and consequently, it has the only horizontal asymptote $y = 0$.

At any point of the domain the function is continuous, that is, $f(x) \to f(x_0)$ when $x \to x_0$. Hence, there is no convergence of $f(x)$ to infinity at any point of the domain, and consequently, there is no vertical asymptote among the points of the domain. The only remaining point is $x = -\sqrt[3]{2}$, which is out of the domain, but can be approximated by the points of the domain. Evaluating the behavior of $f(x)$ when x approaches $-\sqrt[3]{2}$, we notice that the function goes to $-\infty$ as x approaches $-\sqrt[3]{2}$ from the left and goes to $+\infty$ when x approaches $-\sqrt[3]{2}$ from the right. Therefore, the line $x = -\sqrt[3]{2}$ is a vertical asymptote.

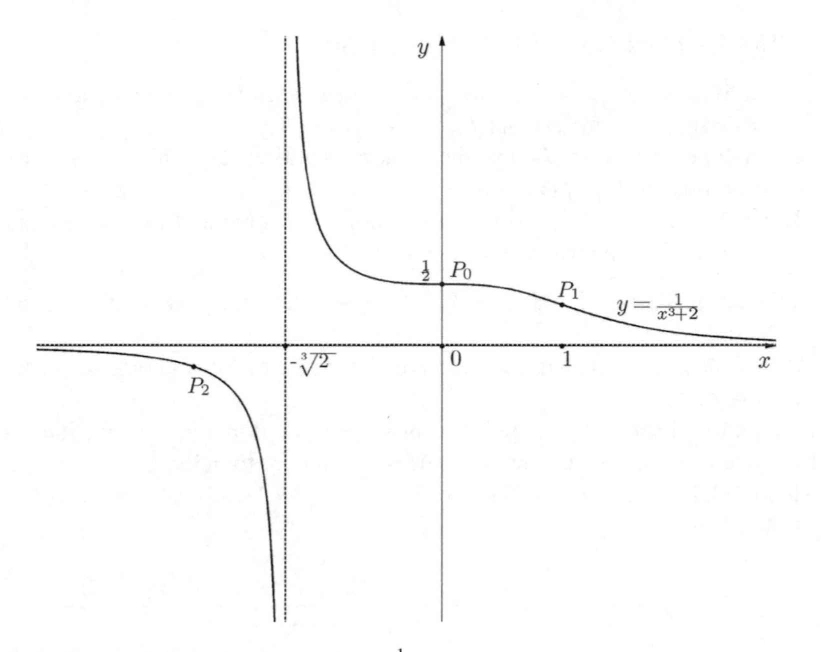

Fig. 3.53 The graph of the function $f(x) = \frac{1}{x^3+2}$

9. Using the investigated properties, marking some important points, such as $P_0 = (0, \frac{1}{2})$ (intersection with the y-axis and inflection point), $P_1 = (1, \frac{1}{3})$ (inflection point) and $P_2 = (-2, -\frac{1}{6})$ (a point with the x-coordinate in $(-\infty, -\sqrt[3]{2})$), we sketch the graph shown in Fig. 3.53.

F. Study of $f(x) = \frac{2x+3}{1-x}$

This is a fractional linear function and its study could be reduced to specification of the properties of fractional linear functions investigated in a general form in Sect. 2.2. However, we choose to carry out an independent investigation and the reader can compare the scheme of study with a general case.

1. Domain and range. The denominator vanishes at the point $x = 1$, which means that the domain is $X = \mathbb{R}\backslash\{1\}$.

   To find the range, we need to determine for what values of y the equation $\frac{2x+3}{1-x} = y$ for unknown x has a solution in the domain X. Notice first that if $y = -2$, then the equation $\frac{2x+3}{1-x} = -2$ is equivalent to the false relation $2x+3 = 2x-2$, which means that $y = -2$ is out of the range. If $y \neq -2$, then the equation has the solution $x = \frac{y-3}{y+2}$. Therefore, the range is $Y = \mathbb{R}\backslash\{-2\}$.
2. Intersection and sign. The graph crosses the y-axis at the point $P_0 = (0, 3)$ and the x-axis at the only point $P_1 = (-\frac{3}{2}, 0)$.

The sign of the function is easily determined:

(1) on the interval $(-\infty, -\frac{3}{2})$ the numerator is negative and the denominator is positive, which means that $f(x) < 0$;
(2) on the interval $(-\frac{3}{2}, 1)$ both the numerator and the denominator are positive, which means that $f(x) > 0$;
(3) on the interval $(1, +\infty)$ the numerator is positive and the denominator is negative, which means that $f(x) < 0$.

3. Boundedness. Since the range is $Y = \mathbb{R}\backslash\{-2\}$, the function is unbounded both below and above.
4. Parity. Since the domain is not symmetric about the origin, the function is neither even nor odd.
5. Periodicity. Since only one point is missing in the domain, the function cannot be periodic. (Another justification comes from monotonicity.)
6. Monotonicity and extrema. Take $x_1 < x_2, x_1, x_2 \neq 1$ and evaluate the difference $D_1 = f(x_2) - f(x_1)$:

$$D_1 = f(x_2) - f(x_1) = \frac{2x_2 + 3}{1 - x_2} - \frac{2x_1 + 3}{1 - x_1} = \frac{5(x_2 - x_1)}{(1 - x_2)(1 - x_1)}.$$

The numerator of the last quotient is positive and the denominator is positive on the two parts of the domain $(-\infty, 1)$ and $(1, +\infty)$ separately. Therefore, $f(x)$ increases on $(-\infty, 1)$ and on $(1, +\infty)$. Notice that the function is not increasing on the entire domain (for example, $f(0) = 3 > -7 = f(2)$).

The property of monotonicity implies that each point of the domain is increasing, and consequently, the function has no extremum of any type.

From increase on the infinite interval $(1, +\infty)$ it follows that $f(x)$ is not periodic.

7. Concavity and inflection. Take two arbitrary points $x_1 < x_2$ both contained in one of the intervals $(-\infty, 1)$ or $(1, +\infty)$, define the midpoint $x_0 = \frac{x_1+x_2}{2}$ and the increment $h = x_2 - x_0 = x_0 - x_1 > 0$, and use the quantity $D_2 \equiv f(x_1) + f(x_2) - 2f(x_0)$ to evaluate the difference between the mean value $\frac{f(x_1)+f(x_2)}{2}$ and the value at the midpoint $f(x_0)$. Employing the expression found for D_1, we obtain:

$$\begin{aligned} D_2 = f(x_1) + f(x_2) - 2f(x_0) &= (f(x_2) - f(x_0)) - (f(x_0) - f(x_1)) \\ &= \frac{5(x_2 - x_0)}{(1 - x_2)(1 - x_0)} - \frac{5(x_0 - x_1)}{(1 - x_0)(1 - x_1)} \\ &= 5h\frac{x_2 - x_1}{(1 - x_2)(1 - x_0)(1 - x_1)} = \frac{10h^2}{(1 - x_2)(1 - x_0)(1 - x_1)}. \end{aligned}$$

The numerator is positive and the denominator has different sign to the left and to the right of the point 1: it is positive on $(-\infty, 1)$ and negative on $(1, +\infty)$.

Therefore, $f(x)$ has upward concavity on $(-\infty, 1)$ and downward concavity on $(1, +\infty)$. Notice that $x = 1$ is not an inflection point, because it does not belong to the domain.

8. Complimentary properties. For sufficiently large $|x|$ the behavior of the rational function is determined by the leading terms of the polynomials in the numerator and denominator: $f(x) = \frac{2x+3}{1-x} \approx \frac{2x}{-x} = -2$. Therefore, the function approaches -2 as x goes to $\pm\infty$. This means that the line $y = -2$ is the only horizontal asymptote.

   At any point of the domain the function is continuous, that is, $f(x) \to f(x_0)$ when $x \to x_0$. Hence, there is no convergence of $f(x)$ to infinity at any point of the domain, and consequently, there is no vertical asymptote among the points of the domain. The only remaining point is $x = 1$, which is out of the domain, but can be approximated by the points of the domain. Evaluating the behavior of $f(x)$ when x approaches 1, we notice that the function goes to $+\infty$ when x approaches 1 from the left and goes to $-\infty$ when x approaches 1 from the right. Therefore, the line $x = 1$ is the only vertical asymptote.

8. Using the investigated properties, marking some important points, such as $P_0 = (0, 3)$ (intersection with the y-axis), $P_1 = (-\frac{3}{2}, 0)$ (intersection with the x-axis) and $P_2 = (2, -7)$ (a point with the x-coordinate in $(1, +\infty)$), we sketch the graph shown in Fig. 3.54.

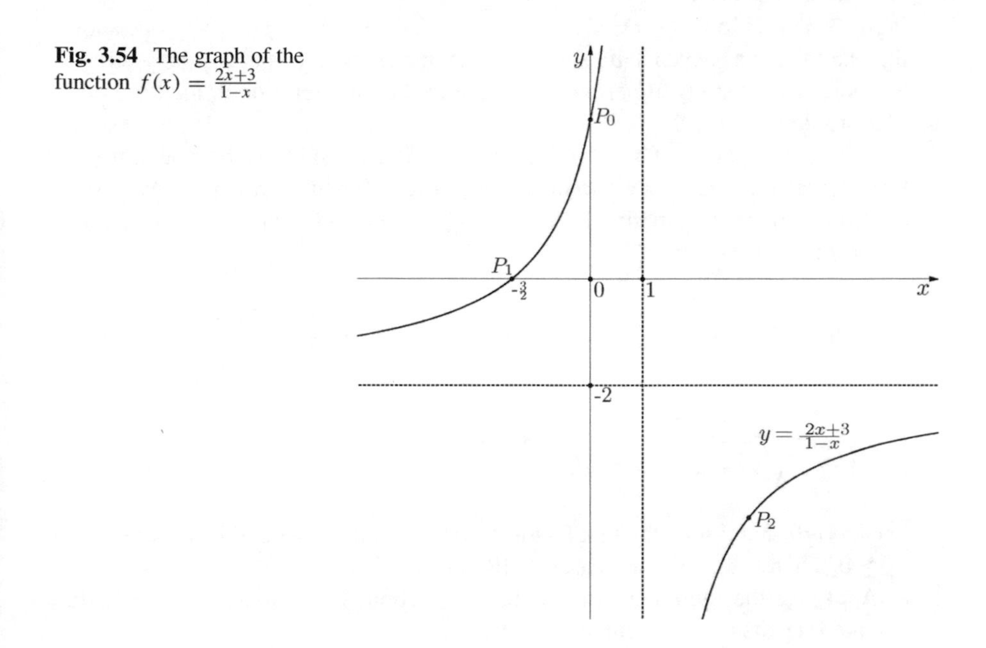

Fig. 3.54 The graph of the function $f(x) = \frac{2x+3}{1-x}$

5.3 *Study of Irrational Functions*

A. Study of $f(x) = \sqrt{x^2 + 1}$

1. Domain and range. The argument $x^2 + 1$ of the square root is positive, and consequently, all the operations can be performed for any real x, that is, the domain is $X = \mathbb{R}$.

   By the definition, the values of the square root are non-negative. Besides, its argument satisfies the inequality $x^2 + 1 \geq 1$, whence $\sqrt{x^2 + 1} \geq 1$. Therefore, the image Y is contained in the interval $[1, +\infty)$. To show that each point of $[1, +\infty)$ belongs to the image, we need to solve the equation $\sqrt{x^2 + 1} = y$ for any given $y \geq 1$. Since the right-hand side is positive, we can square both sides and get $x^2 + 1 = y^2$ or $x^2 = y^2 - 1$. For $y \geq 1$, we have $y^2 - 1 \geq 0$, and consequently, there are solutions $x = \pm\sqrt{y^2 - 1}$. Hence, for any $y \geq 1$ there are x such that $f(x) = \sqrt{x^2 + 1} = y$, which means that $Y = [1, +\infty)$.
2. Intersection and sign. The point of intersection with the y-axis is $P_0 = (0, 1)$. There is no intersection with the x-axis, because $y = 0$ does not belong to the range.

   Since the range is $Y = [1, +\infty)$, the function takes only positive values.
3. Boundedness. Since the range is $Y = [1, +\infty)$, the function is bounded below and unbounded above.
4. Parity. The relation $f(-x) = \sqrt{(-x)^2 + 1} = \sqrt{x^2 + 1} = f(x), \forall x \in X$ shows that the function is even and its graph is symmetric with respect to the y-axis.
5. Periodicity. The study of periodicity is deferred until after monotonicity.
6. Monotonicity and extrema.

   Due to the parity of the function it is sufficient to study monotonicity on the interval $[0, +\infty)$ (or alternatively on $(-\infty, 0]$) and then extend the results according to the symmetry. Take $0 \leq x_1 < x_2$ and evaluate the difference $D_1 = f(x_2) - f(x_1)$:

$$D_1 = f(x_2) - f(x_1) = \sqrt{x_2^2 + 1} - \sqrt{x_1^2 + 1} = \frac{(x_2^2 + 1) - (x_1^2 + 1)}{\sqrt{x_2^2 + 1} + \sqrt{x_1^2 + 1}}$$

$$= \frac{x_2 - x_1}{\sqrt{x_2^2 + 1} + \sqrt{x_1^2 + 1}} \cdot (x_2 + x_1).$$

On the right-hand side, the first factor is positive and the second is positive when $x_1 \geq 0$. Therefore, $f(x)$ increases on $[0, +\infty)$.

Applying the symmetry of the function about the y-axis, we extend the obtained results of monotonicity as follows:

(1) $f(x)$ decreases on $(-\infty, 0]$;
(2) $f(x)$ increases on $[0, +\infty)$.

Consequently, $x = 0$ is the local and global minimum, and there is no other local or global extremum.

The increase of $f(x)$ on the infinite interval $[0, +\infty)$ implies that the function is not periodic.

7. Concavity and inflection.

In the investigation of concavity we use the parity of the function and the fact that $x = 0$ is a local minimum. Take two arbitrary points such that $0 \le x_1 < x_2$, define the midpoint $x_0 = \frac{x_1+x_2}{2}$ and the increment $h = x_2 - x_0 = x_0 - x_1 > 0$, and use the quantity $D_2 \equiv f(x_1)+f(x_2)-2f(x_0)$ to evaluate the difference between the mean value $\frac{f(x_1)+f(x_2)}{2}$ and the value at the midpoint $f(x_0)$. Employing the expression found for D_1, we obtain:

$$D_2 = f(x_1) + f(x_2) - 2f(x_0) = (f(x_2) - f(x_0)) - (f(x_0) - f(x_1))$$

$$= \sqrt{x_1^2+1} + \sqrt{x_2^2+1} - 2\sqrt{x_0^2+1}$$

$$= \frac{(x_2-x_0)(x_2+x_0)}{\sqrt{x_2^2+1}+\sqrt{x_0^2+1}} - \frac{(x_0-x_1)(x_0+x_1)}{\sqrt{x_0^2+1}+\sqrt{x_1^2+1}}$$

$$= h\frac{(x_2+x_0)(\sqrt{x_0^2+1}+\sqrt{x_1^2+1}) - (x_0+x_1)(\sqrt{x_2^2+1}+\sqrt{x_0^2+1})}{(\sqrt{x_2^2+1}+\sqrt{x_0^2+1})(\sqrt{x_0^2+1}+\sqrt{x_1^2+1})}$$

$$= h\frac{(x_2-x_1)\sqrt{x_0^2+1} + x_2\sqrt{x_1^2+1} - x_1\sqrt{x_2^2+1} + x_0(\sqrt{x_1^2+1} - \sqrt{x_2^2+1})}{(\sqrt{x_2^2+1}+\sqrt{x_0^2+1})(\sqrt{x_0^2+1}+\sqrt{x_1^2+1})}$$

$$= \frac{h}{(\sqrt{x_2^2+1}+\sqrt{x_0^2+1})(\sqrt{x_0^2+1}+\sqrt{x_1^2+1})}$$

$$\times \left[(x_2-x_1)\sqrt{x_0^2+1} + \frac{x_2^2(x_1^2+1) - x_1^2(x_2^2+1)}{x_2\sqrt{x_1^2+1}+x_1\sqrt{x_2^2+1}} + \frac{x_0(x_1^2-x_2^2)}{\sqrt{x_1^2+1}+\sqrt{x_2^2+1}}\right]$$

$$= \frac{2h^2}{(\sqrt{x_2^2+1}+\sqrt{x_0^2+1})(\sqrt{x_0^2+1}+\sqrt{x_1^2+1})}$$

$$\times \left[\sqrt{x_0^2+1} + \frac{x_2+x_1}{x_2\sqrt{x_1^2+1}+x_1\sqrt{x_2^2+1}} - \frac{x_0(x_1+x_2)}{\sqrt{x_1^2+1}+\sqrt{x_2^2+1}}\right].$$

In the last expression, the first factor is positive. To evaluate the second factor (inside brackets), notice that $\frac{x_0(x_1+x_2)}{\sqrt{x_1^2+1}+\sqrt{x_2^2+1}} < \frac{x_0(x_1+x_2)}{x_1+x_2} = x_0 < \sqrt{x_0^2+1}$. Therefore, the second factor is also positive. Consequently, $f(x)$ is concave upward on $[0, +\infty)$.

Employing the parity, we deduce that $f(x)$ is also concave upward on $(-\infty, 0]$.

Hence, $f(x)$ is concave upward on $(-\infty, 0]$ and on $[0, +\infty)$, and besides $x = 0$ is its local minimum. Therefore, by Property 3 of concavity (Sect. 4.1), we conclude that $f(x)$ is concave upward on the entire domain $X = \mathbb{R}$. Consequently, there is no inflection point.

8. Complimentary properties. For sufficiently large $|x|$ the function can be approximated in the form $f(x) = \sqrt{x^2+1} \approx |x|$, from which it follows that the function goes to $+\infty$ when $x \to \pm\infty$, and consequently, there is no horizontal asymptote.

   At each point of the domain the function is continuous, that is, $f(x) \to f(x_0)$ when $x \to x_0$. Hence, there is no convergence of $f(x)$ to infinity at any point of the domain $X = \mathbb{R}$, and consequently, there is no vertical asymptote.
9. Using the investigated properties, marking some important points, such as $P_0 = (0, 1)$ (intersection with the y-axis and global minimum) and $P_1 = (1, \sqrt{2})$ (a point with a positive x-coordinate), and taking advantage of the symmetry about the y-axis, we plot the graph shown in Fig. 3.55.

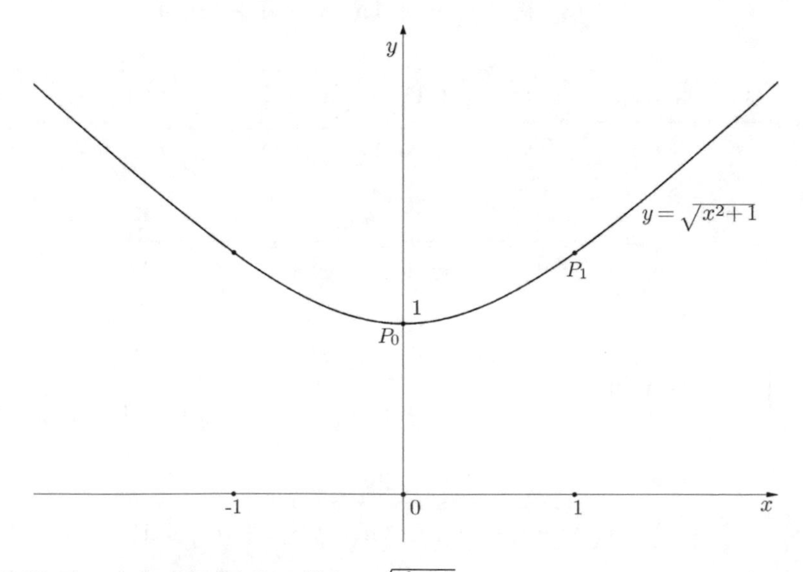

Fig. 3.55 The graph of the function $f(x) = \sqrt{x^2+1}$

B. Study of $f(x) = \sqrt{x^2+1} - |x|$

1. Domain and range. The argument $x^2 + 1$ of the square root is positive, and consequently, all the operations can be performed for any real x, that is, the domain is $X = \mathbb{R}$.

   Writing the function in the form $f(x) = \sqrt{x^2+1} - |x| = \frac{1}{\sqrt{x^2+1}+|x|}$ and noting that $\sqrt{x^2+1} \geq 1$, $\forall x$, we deduce that the range Y is contained in the interval $(0, 1]$ (the function does not take the value 0, but takes the value 1 at the point $x = 0$). To verify that the entire interval $(0, 1]$ is the range of the function, we should solve the equation $\sqrt{x^2+1} - |x| = y$ for any $y \in (0, 1]$. Writing this equation in the form $\sqrt{x^2+1} = y + |x|$ and noting that the right-hand side is positive, we can square both sides $x^2 + 1 = (y + |x|)^2$ and simplify $1 = y^2 + 2y|x|$. Then, we have the following equation for $|x|$: $|x| = \frac{1-y^2}{2y}$. Since the right-hand side is non-negative, there are solutions $x = \pm\frac{1-y^2}{2y}$, which shows that $Y = (0, 1]$.
2. Intersection and sign. The point of intersection with the y-axis is $P_0 = (0, 1)$. There is no intersection with the x-axis, because the range does not contain $y = 0$.

   Since the range is $Y = (0, 1]$, the function takes only positive values.
3. Boundedness. Since the range is $Y = (0, 1]$, the function is bounded below and above.
4. Parity. The relation $f(-x) = \sqrt{(-x)^2+1} - |-x| = \sqrt{x^2+1} - |x| = f(x)$, $\forall x \in X$, shows that the function is even and its graph is symmetric with respect to the y-axis.
5. Periodicity. The study of periodicity is postponed until after monotonicity.
6. Monotonicity and extrema.

   Due to the parity of the function it is sufficient to study monotonicity on the interval $[0, +\infty)$ (or alternatively on $(-\infty, 0]$) and then extend the results according to the symmetry. Take $0 \leq x_1 < x_2$ and evaluate the difference $D_1 = f(x_2) - f(x_1)$:

$$D_1 = f(x_2) - f(x_1) = \sqrt{x_2^2+1} - |x_2| - (\sqrt{x_1^2+1} - |x_1|)$$

$$= \frac{(x_2^2+1) - (x_1^2+1)}{\sqrt{x_2^2+1} + \sqrt{x_1^2+1}} - (x_2 - x_1)$$

$$= (x_2 - x_1)\left(\frac{x_2 + x_1}{\sqrt{x_2^2+1} + \sqrt{x_1^2+1}} - 1\right).$$

   On the right-hand side, the first factor is positive and the second is negative when $x_1 \geq 0$. Therefore, $f(x)$ decreases on $[0, +\infty)$.

Applying the symmetry of the function about the y-axis, we extend the obtained results of monotonicity as follows:

(1) $f(x)$ increases on $(-\infty, 0]$;
(2) $f(x)$ decreases on $[0, +\infty)$.

Consequently, $x = 0$ is the local and global maximum, and there is no other local or global extremum.

The decrease of $f(x)$ on the infinite interval $[0, +\infty)$ implies that the function is not periodic.

7. Concavity and inflection.

Take two arbitrary points such that $0 \le x_1 < x_2$, define the midpoint $x_0 = \frac{x_1+x_2}{2}$ and the increment $h = x_2 - x_0 = x_0 - x_1 > 0$, and use the quantity $D_2 \equiv f(x_1) + f(x_2) - 2f(x_0)$ to evaluate the difference between the mean value $\frac{f(x_1)+f(x_2)}{2}$ and the value at the midpoint $f(x_0)$:

$$D_2 = f(x_1) + f(x_2) - 2f(x_0) = (f(x_2) - f(x_0)) - (f(x_0) - f(x_1))$$

$$= (\sqrt{x_1^2+1} - x_1) + (\sqrt{x_2^2+1} - x_2) - 2(\sqrt{x_0^2+1} - x_0)$$
$$= \sqrt{x_1^2+1} + \sqrt{x_2^2+1} - 2\sqrt{x_0^2+1}.$$

In this way, we obtain the same expression as in the previous exercise (for the function $f(x) = \sqrt{x^2+1}$). Therefore, we can conclude that $f(x)$ is concave upward on $[0, +\infty)$.

Employing the parity, we deduce that $f(x)$ is also concave upward on $(-\infty, 0]$. Consequently, there is no inflection point.

Notice that the expression of D_2 in this exercise and the previous one is the same on $[0, +\infty)$ (and also on $(-\infty, 0]$), but the property of concavity on the entire domain $X = \mathbb{R}$ is different in these two cases. In the previous exercise it was possible to apply Property 3 of concavity (Sect. 4.1) and conclude that $f(x) = \sqrt{x^2+1}$ is concave upward on $X = \mathbb{R}$. In the current example, the point $x = 0$ is not a local minimum (it is a local maximum), and consequently, Property 3 is not applicable. Moreover, it is easy to show that $f(x) = \sqrt{x^2+1} - |x|$ is not concave upward on $X = \mathbb{R}$. For example, taking the points $x_1 = -1$, $x_2 = 1$ and $x_0 = 0$, we get $D_2 = f(x_1) + f(x_2) - 2f(x_0) = \sqrt{2} - 1 + \sqrt{2} - 1 - 2 \cdot 1 = 2\sqrt{2} - 4 < 0$, that is, the segment of the secant line that passes through the points $(-1, f(-1))$ and $(1, f(1))$ lies below the graph of the function on the interval $(-1, 1)$.

8. Complimentary properties. For sufficiently large $|x|$ the function can be approximated in the form $f(x) = \sqrt{x^2+1} - |x| \approx |x| - |x| = 0$, from which it follows that the function approaches 0 when $x \to \pm\infty$. Consequently, the line $y = 0$ is the only horizontal asymptote.

At each point of the domain the function is continuous, that is, $f(x) \to f(x_0)$ when $x \to x_0$. Hence, there is no convergence of $f(x)$ to infinity at any point of the domain $X = \mathbb{R}$, and consequently, there is no vertical asymptote.

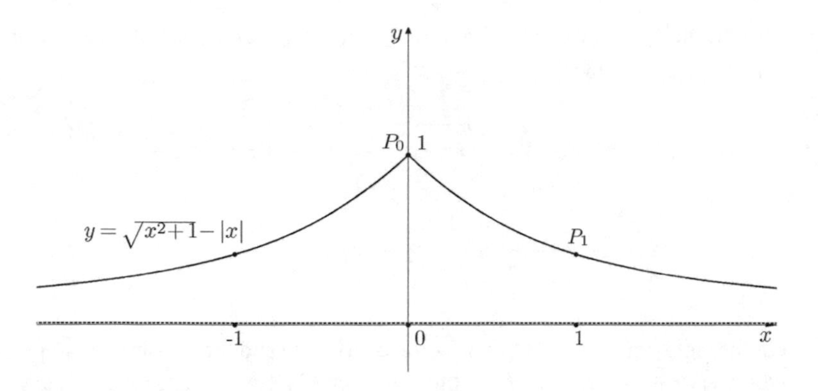

Fig. 3.56 The graph of the function $f(x) = \sqrt{x^2+1} - |x|$

9. Using the investigated properties, marking some important points, such as $P_0 = (0, 1)$ (intersection with the y-axis and global maximum) and $P_1 = (1, \sqrt{2} - 1)$ (a point with a positive x-coordinate), and taking advantage of the symmetry about the y-axis, we plot the graph shown in Fig. 3.56.

C. Study of $f(x) = \sqrt{x^2 - 1}$

1. Domain and range. The square root imposes restrictions both on the domain and the range. On the one hand, its argument $x^2 - 1$ should be non-negative, which results in the condition $|x| \geq 1$, that is, $X = (-\infty, -1] \cup [1, +\infty)$. On the other hand, the values of the square root are non-negative, and consequently, the range Y is contained in $[0, +\infty)$. Let us show that $Y = [0, +\infty)$. Indeed, for any $y \geq 0$, the equation $\sqrt{x^2 - 1} = y$ has the solutions $x = \pm\sqrt{y^2 + 1}$ which belong to the domain X.
2. Intersection and sign. There is no intersection with the y-axis since $x = 0$ is out of the domain. The points of intersection with the x-axis are found from the solution of the equation $\sqrt{x^2 - 1} = 0$, whose roots are $x = \pm 1$. Therefore, these points are $P_1 = (-1, 0)$ and $P_2 = (1, 0)$.

   In study of the range it was already determined that $f(x) \geq 0$, $\forall x \in X$. Notice additionally that $f(x) > 0$, $\forall x \in (-\infty, -1) \cup (1, +\infty)$.
3. Boundedness. Since the range is $Y = [0, +\infty)$, the function is bounded below and unbounded above.
4. Parity. The relation $f(-x) = \sqrt{(-x)^2 - 1} = \sqrt{x^2 - 1} = f(x)$, $\forall x \in X$ shows that the function is even and its graph is symmetric with respect to the y-axis. Since the function is even and the domain does not included a neighborhood of the origin (the interval $(-1, 1)$), we can study the remaining properties only for positive values of x and then extend the results onto the negative values.
5. Periodicity. The study of periodicity is postponed until after monotonicity.

6. Monotonicity and extrema. Take $1 \le x_1 < x_2$ and evaluate the difference $D_1 = f(x_2) - f(x_1)$:

$$D_1 = f(x_2) - f(x_1) = \sqrt{x_2^2-1} - \sqrt{x_1^2-1} = \frac{(x_2^2-1)-(x_1^2-1)}{\sqrt{x_2^2-1}+\sqrt{x_1^2-1}}$$

$$= \frac{x_2 - x_1}{\sqrt{x_2^2-1}+\sqrt{x_1^2-1}} \cdot (x_2 + x_1).$$

On the right-hand side, the first factor is positive and the second is also positive when $x_1 \ge 1$. Therefore, $f(x)$ increases on $[1, +\infty)$. Consequently, $x = 1$ is a global (but not local) minimum of $f(x)$.

The increase of $f(x)$ on the infinite interval $[1, +\infty)$ implies that the function is not periodic.

7. Concavity and inflection. Take two arbitrary points such that $1 \le x_1 < x_2$, define the midpoint $x_0 = \frac{x_1+x_2}{2}$ and the increment $h = x_2 - x_0 = x_0 - x_1 > 0$, and use the quantity $D_2 \equiv f(x_1) + f(x_2) - 2f(x_0)$ to evaluate the difference between the mean value $\frac{f(x_1)+f(x_2)}{2}$ and the value at the midpoint $f(x_0)$. Employing the expression found for D_1, we obtain:

$$D_2 = f(x_1) + f(x_2) - 2f(x_0) = (f(x_2) - f(x_0)) - (f(x_0) - f(x_1))$$

$$= \sqrt{x_1^2-1} + \sqrt{x_2^2-1} - 2\sqrt{x_0^2-1}$$

$$= \frac{(x_2-x_0)(x_2+x_0)}{\sqrt{x_2^2-1}+\sqrt{x_0^2-1}} - \frac{(x_0-x_1)(x_0+x_1)}{\sqrt{x_0^2-1}+\sqrt{x_1^2-1}}$$

$$= h\frac{(x_2+x_0)(\sqrt{x_0^2-1}+\sqrt{x_1^2-1}) - (x_0+x_1)(\sqrt{x_2^2-1}+\sqrt{x_0^2-1})}{(\sqrt{x_2^2-1}+\sqrt{x_0^2-1})(\sqrt{x_0^2-1}+\sqrt{x_1^2-1})}$$

$$= h\frac{(x_2-x_1)\sqrt{x_0^2-1} + x_2\sqrt{x_1^2-1} - x_1\sqrt{x_2^2-1} + x_0(\sqrt{x_1^2-1} - \sqrt{x_2^2-1})}{(\sqrt{x_2^2-1}+\sqrt{x_0^2-1})(\sqrt{x_0^2-1}+\sqrt{x_1^2-1})}$$

$$= \frac{h}{(\sqrt{x_2^2-1}+\sqrt{x_0^2-1})(\sqrt{x_0^2-1}+\sqrt{x_1^2-1})}$$

$$\times\left[(x_2-x_1)\sqrt{x_0^2-1} + \frac{x_2^2(x_1^2-1) - x_1^2(x_2^2-1)}{x_2\sqrt{x_1^2-1}+x_1\sqrt{x_2^2-1}} + \frac{x_0(x_1^2-x_2^2)}{\sqrt{x_1^2-1}+\sqrt{x_2^2-1}}\right]$$

$$= \frac{2h^2}{(\sqrt{x_2^2-1}+\sqrt{x_0^2-1})(\sqrt{x_0^2-1}+\sqrt{x_1^2-1})}$$

$$\times \left[\sqrt{x_0^2-1} - \frac{x_2+x_1}{x_2\sqrt{x_1^2-1}+x_1\sqrt{x_2^2-1}} - \frac{x_0(x_1+x_2)}{\sqrt{x_1^2-1}+\sqrt{x_2^2-1}}\right].$$

In the last expression, the first factor is positive. To evaluate the second factor (inside brackets), notice that $x_1 > \sqrt{x_1^2-1}$, $x_2 > \sqrt{x_2^2-1}$, whence $\frac{x_1+x_2}{\sqrt{x_1^2-1}+\sqrt{x_2^2-1}} > 1$, and consequently, $\sqrt{x_0^2-1} - x_0\frac{x_1+x_2}{\sqrt{x_1^2-1}+\sqrt{x_2^2-1}} < 0$, which shows that the second factor is negative. Therefore, $D_2 < 0$ and $f(x)$ is concave downward on $[1, +\infty)$.

6–7. Extending the obtained results according to the symmetry of the function, we obtain the following properties of monotonicity:

(1) $f(x)$ decreases on $(-\infty, -1]$;
(2) $f(x)$ increases on $[1, +\infty)$.

The points $x = \pm 1$ are global (but not local) minima of the function, and there is no other extremum of any type.

In a similar way, we obtain the following results about concavity: $f(x)$ is concave downward on $(-\infty, -1]$ and on $[1, +\infty)$. Consequently, there are no inflection points.

8. Complimentary properties. For sufficiently large $|x|$ the function can be approximated in the form $f(x) = \sqrt{x^2-1} \approx |x|$, from which it follows that the function goes to $+\infty$ when $x \to \pm\infty$, and consequently, there is no horizontal asymptote.

At each point $|x_0| > 1$ (an interior point of the domain) the function is continuous, that is, $f(x) \to f(x_0)$ when $x \to x_0$. The two remaining points of the domain $x_0 = \pm 1$ are the boundary points and can be approached only from one side of the domain: $x_0 = -1$ from the left and $x_0 = 1$ from the right. Still the function is continuous at these two points, because $f(x) \to f(-1) = 0$ when $x \to -1^-$ and $f(x) \to f(1) = 0$ when $x \to 1^+$. All other points, that is, the points $|x_0| < 1$, are exterior points of the domain $X = (-\infty, -1] \cup [1, +\infty)$ and cannot be approached by the points of X. Hence, there is no convergence of $f(x)$ to infinity at any point, and consequently, there is no vertical asymptote.

9. Using the investigated properties, marking some important points, such as $P_1 = (-1, 0)$ and $P_2 = (1, 0)$ (intersections with the x-axis and global minima) and $P_3 = (2, \sqrt{3})$ (a point with a positive x-coordinate), and taking advantage of the symmetry about the y-axis, we plot the graph shown in Fig. 3.57.

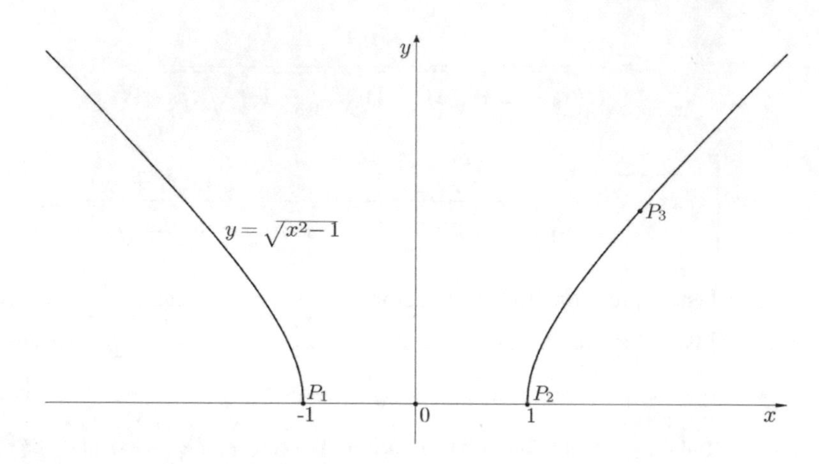

Fig. 3.57 The graph of the function $f(x) = \sqrt{x^2 - 1}$

Notice that another manner to investigate the properties of the function $f(x) = \sqrt{x^2 - 1}$ is to link it to the function $f(x) = \sqrt{x^2 + 1}$ of the exercise A. Restricting the domain of the last function by the interval $[0, +\infty)$, we obtain the bijective function $y = f(x) = \sqrt{x^2 + 1} : [0, +\infty) \to [1, +\infty)$. Therefore, this function has the inverse $x = f^{(-1)}(y) : [1, +\infty) \to [0, +\infty)$, whose formula is obtained by solving the equation $y = \sqrt{x^2 + 1}$ with respect to $x \geq 0$. The solution is found in the form $x = \sqrt{y^2 - 1}$ which is the function of the current exercise with the independent variable y and the domain restricted to $[1, +\infty)$. Therefore, the properties of the function $y = f^{(-1)}(x) = \sqrt{x^2 - 1} : [1, +\infty) \to [0, +\infty)$ can be deduced from the relationship between the direct and inverse functions, and then the properties of the function $y = \sqrt{x^2 - 1}$ on the extended domain $X = (-\infty, -1] \cup [1, +\infty)$ can be completed in accordance with the parity of the last function. The details of this algorithm of the investigation are left to the reader.

D. Study of $f(x) = x + \sqrt{x^2 - 1}$

1. Domain and range. The argument $x^2 - 1$ of the square root is non-negative when $|x| \geq 1$, that is, $X = (-\infty, -1] \cup [1, +\infty)$.

   Let us investigate separately the images of the intervals $(-\infty, -1]$ and $[1, +\infty)$. On the latter, the values of the function satisfy the inequality $y \geq 1$. We can show that the entire interval $[1, +\infty)$ is the image of $[1, +\infty)$, that is, $f([1, +\infty)) = [1, +\infty)$. Indeed, solving the equation $x + \sqrt{x^2 - 1} = y$ for a given $y \geq 1$, we notice first that the equation itself implies that $y \geq x$. Then, $\sqrt{x^2 - 1} = y - x$, where $y - x \geq 0$, and consequently, squaring both sides we get $x^2 - 1 = y^2 - 2xy + x^2$ or $2xy = y^2 + 1$. The solution x on the interval $[1, +\infty)$ is found in the form: $x = \frac{1}{2}(y + \frac{1}{y}) \geq 1$. Therefore, the interval $[1, +\infty)$ is a part of the range Y.

On the interval $(-\infty,-1]$, the function can be written in the form $y = x + \sqrt{x^2-1} = -|x| + \sqrt{x^2-1} = \frac{-1}{|x|+\sqrt{x^2-1}}$. The denominator is greater than or equal to 1, and for this reason the values of the function satisfy the inequality $-1 \le y < 0$. Now we take an arbitrary $y \in [-1, 0)$ and show that it belongs to the image of $(-\infty,-1]$. Solving the equation $x + \sqrt{x^2-1} = y$ for a given $y \in [-1, 0)$, we find $\sqrt{x^2-1} = y - x \ge 0$, whence $x^2 - 1 = y^2 - 2xy + x^2$ and finally $x = \frac{1}{2}(y + \frac{1}{y}) \le -1$. Therefore, the interval $[-1, 0)$ is the image of $(-\infty,-1]$: $f((-\infty,-1]) = [-1, 0)$. Hence, $Y = [-1, 0) \cup [1,+\infty)$.

2. Intersection and sign. There is no intersection point either with the y-axis or with the x-axis, since $x = 0$ is out of the domain and $y = 0$ is out of the range.

   On the interval $[1,+\infty)$ the function is positive and on $(-\infty,-1]$ it is negative.
3. Boundedness. The determined above range $Y = [-1, 0)\cup[1,+\infty)$ indicates that the function is bounded below and unbounded above.
4. Parity. The calculation $f(-2) = -2 + \sqrt{3} \ne \pm(2 + \sqrt{3}) = \pm f(2)$ shows that the function is neither even nor odd.
5. Periodicity. The study of periodicity is deferred until after monotonicity.
6. Monotonicity and extrema. Take $x_1 < x_2, x_1, x_2 \in X$ and evaluate the difference $D_1 = f(x_2) - f(x_1)$:

$$D_1 = f(x_2) - f(x_1) = x_2 + \sqrt{x_2^2 - 1} - (x_1 + \sqrt{x_1^2 - 1})$$
$$= (x_2 - x_1) + \frac{(x_2 - x_1)(x_2 + x_1)}{\sqrt{x_2^2 - 1} + \sqrt{x_1^2 - 1}}$$
$$= (x_2 - x_1)\left(1 + \frac{x_2 + x_1}{\sqrt{x_2^2 - 1} + \sqrt{x_1^2 - 1}}\right).$$

   If $x_2 \le -1$, then $\frac{x_2+x_1}{\sqrt{x_2^2-1}+\sqrt{x_1^2-1}} = -\frac{|x_2|+|x_1|}{\sqrt{x_2^2-1}+\sqrt{x_1^2-1}} < -1$, whence $D_1 < 0$ and $f(x)$ decreases on $(-\infty,-1]$. If $1 \le x_1$, then all the terms inside parentheses are positive and $D_1 > 0$, which means that $f(x)$ increases on $[1,+\infty)$. Consequently, $x = -1$ is a global (but not local) minimum of $f(x)$ and there is no other extrema of any type. Notice that $x = 1$ is a global (but not local) minimum on the interval $[1,+\infty)$.

   The increase on the infinite interval $[1,+\infty)$ implies that $f(x)$ is not periodic.
7. Concavity and inflection. Take two arbitrary points $x_1 < x_2$, both belonging to one of the two intervals $(-\infty,-1]$ or $[1,+\infty)$, define the midpoint $x_0 = \frac{x_1+x_2}{2}$ and the increment $h = x_2 - x_0 = x_0 - x_1 > 0$, and use the quantity $D_2 \equiv f(x_1) + f(x_2) - 2f(x_0)$ to evaluate the difference between the mean

value $\frac{f(x_1)+f(x_2)}{2}$ and the value at the midpoint $f(x_0)$:

$$D_2 = f(x_1) + f(x_2) - 2f(x_0) = (f(x_2) - f(x_0)) - (f(x_0) - f(x_1))$$
$$= x_1 + \sqrt{x_1^2 - 1} + x_2 + \sqrt{x_2^2 - 1} - 2x_0 - 2\sqrt{x_0^2 - 1}$$
$$= \sqrt{x_1^2 - 1} + \sqrt{x_2^2 - 1} - 2\sqrt{x_0^2 - 1}.$$

We arrive at the same expression obtained in the study of concavity in Exercise C (for the function $f(x) = \sqrt{x^2 - 1}$), and taking advantage of the results obtained there, we conclude that $D_2 < 0$ both on the interval $(-\infty, -1]$ and on $[1, +\infty)$. Therefore, $f(x)$ is concave downward both on $(-\infty, -1]$ and on $[1, +\infty)$. Consequently, there is no inflection point.

8. Complimentary properties. For sufficiently large x the function can be approximated in the form $f(x) = x + \sqrt{x^2 - 1} \approx 2x$, from which it follows that the function goes to $+\infty$ when $x \to +\infty$. For negative values of x, which have sufficiently large values $|x|$, the function has approximation $f(x) = x + \sqrt{x^2 - 1} \approx x + |x| = 0$, which shows that the function approaches 0 as $x \to -\infty$. Therefore, the function has a horizontal asymptote $y = 0$ when x goes to $-\infty$.

   At each point $|x_0| > 1$ (an interior point of the domain) the function is continuous, that is, $f(x) \to f(x_0)$ when $x \to x_0$. The two remaining points of the domain $x_0 = \pm 1$ are the boundary points and can be approached only from one side of the domain: $x_0 = -1$ from the left and $x_0 = 1$ from the right. Still the function is continuous at these two points, because $f(x) \to f(-1) = -1$ when $x \to -1^-$ and $f(x) \to f(1) = 1$ when $x \to 1^+$. All other points, that is, the points $|x_0| < 1$, are exterior points of the domain $X = (-\infty, -1] \cup [1, +\infty)$ and cannot be approached by the points of X. Hence, there is no convergence of $f(x)$ to infinity at any point, and consequently, there is no vertical asymptote.

9. Using the investigated properties and marking some important points, such as $P_1 = (-1, -1)$ (global minimum) and $P_2 = (1, 1)$ (global minimum on the interval $[1, +\infty)$), we sketch the graph shown in Fig. 3.58.

E. Study of $f(x) = \sqrt{x + 4} - \sqrt{x}$

1. Domain and range. The arguments $x + 4$ and x of the square roots should be non-negative, which requires the inequality $x \geq 0$. Therefore, the domain is $X = [0, +\infty)$.

   To determine the range, we represent the function in the form $f(x) = \sqrt{x + 4} - \sqrt{x} = \frac{4}{\sqrt{x+4}+\sqrt{x}}$, which shows that the points of the range Y should satisfy the restriction $0 < y \leq 2$. Let us prove that the entire interval $(0, 2]$ is the range of the function. Take an arbitrary y in this interval and solve the equation

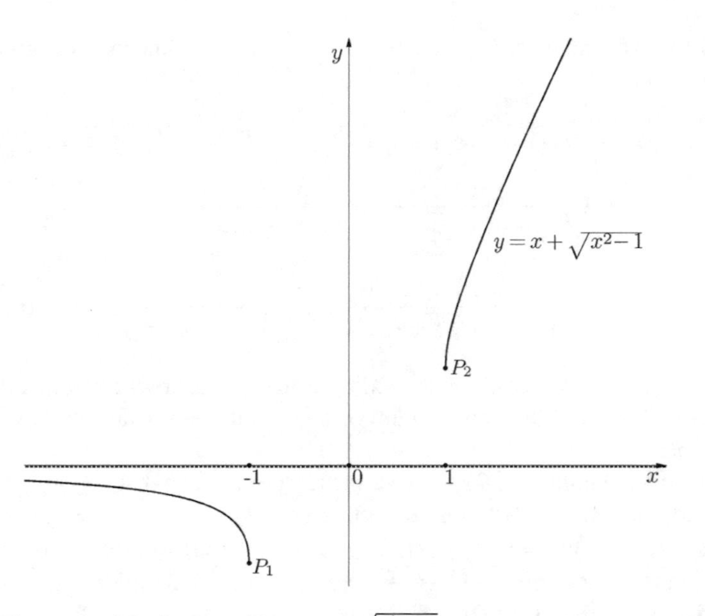

Fig. 3.58 The graph of the function $f(x) = x + \sqrt{x^2 - 1}$

$\sqrt{x+4} - \sqrt{x} = y$ for unknown x. To this end, transfer one of the square roots to another side of the equation: $\sqrt{x+4} = y + \sqrt{x}$, and then square both sides of this relation: $x + 4 = y^2 + 2y\sqrt{x} + x$. Now isolate the remaining square root: $\sqrt{x} = \frac{1}{2y}(4 - y^2)$, and square the relation once more to obtain the solution $x = \left(\frac{1}{2y}(4 - y^2)\right)^2$, which belongs to the domain $X = [0, +\infty)$. Therefore, $Y = (0, 2]$.

2. Intersection and sign. The point of intersection with the y-axis is $P_0 = (0, 2)$. There is no intersection with the x-axis, because $y = 0$ does not belong to the range.

   Since the range is $Y = (0, 2]$, the function takes only positive values.
3. Boundedness. Since the range is $Y = (0, 2]$, the function is bounded both below and above.
4. Parity. Since the domain is not symmetric about the origin, the function cannot be even or odd.
5. Periodicity. Since the domain is bounded from the left, the function cannot be periodic.

6. Monotonicity and extrema. Take $0 \leq x_1 < x_2$ and evaluate the difference $D_1 = f(x_2) - f(x_1)$:

$$D_1 = f(x_2) - f(x_1) = \sqrt{x_2+4} - \sqrt{x_2} - (\sqrt{x_1+4} - \sqrt{x_1})$$
$$= \frac{x_2 - x_1}{\sqrt{x_2+4} + \sqrt{x_1+4}} - \frac{x_2 - x_1}{\sqrt{x_2} + \sqrt{x_1}}$$
$$= (x_2 - x_1)\left(\frac{1}{\sqrt{x_2+4} + \sqrt{x_1+4}} - \frac{1}{\sqrt{x_2} + \sqrt{x_1}}\right) < 0.$$

Therefore, $f(x)$ decreases on the entire domain. It follows from this that $x = 0$ is a global (but not local) maximum of $f(x)$ and there is no other extremum of any type.

7. Concavity and inflection. Take two arbitrary points such that $0 \leq x_1 < x_2$, define the midpoint $x_0 = \frac{x_1+x_2}{2}$ and the increment $h = x_2 - x_0 = x_0 - x_1 > 0$, and use the quantity $D_2 \equiv f(x_1) + f(x_2) - 2f(x_0)$ to evaluate the difference between the mean value $\frac{f(x_1)+f(x_2)}{2}$ and the value at the midpoint $f(x_0)$. Using the expression found for D_1, we obtain:

$$D_2 = f(x_1) + f(x_2) - 2f(x_0) = (f(x_2) - f(x_0)) - (f(x_0) - f(x_1))$$
$$= \left(\sqrt{x_2+4} - \sqrt{x_2} - (\sqrt{x_0+4} - \sqrt{x_0})\right)$$
$$- \left(\sqrt{x_0+4} - \sqrt{x_0} - (\sqrt{x_1+4} - \sqrt{x_1})\right)$$
$$= \left(\frac{x_2 - x_0}{\sqrt{x_2+4} + \sqrt{x_0+4}} - \frac{x_2 - x_0}{\sqrt{x_2} + \sqrt{x_0}}\right)$$
$$- \left(\frac{x_0 - x_1}{\sqrt{x_0+4} + \sqrt{x_1+4}} - \frac{x_0 - x_1}{\sqrt{x_0} + \sqrt{x_1}}\right)$$
$$= h\left(\frac{\sqrt{x_1+4} - \sqrt{x_2+4}}{(\sqrt{x_2+4} + \sqrt{x_0+4})(\sqrt{x_0+4} + \sqrt{x_1+4})}\right.$$
$$\left. + \frac{\sqrt{x_2} - \sqrt{x_1}}{(\sqrt{x_2} + \sqrt{x_0})(\sqrt{x_0} + \sqrt{x_1})}\right)$$
$$= 2h^2\left(\frac{1}{(\sqrt{x_2} + \sqrt{x_0})(\sqrt{x_0} + \sqrt{x_1})(\sqrt{x_2} + \sqrt{x_1})}\right.$$
$$\left. - \frac{1}{(\sqrt{x_2+4} + \sqrt{x_0+4})(\sqrt{x_0+4} + \sqrt{x_1+4})(\sqrt{x_2+4} + \sqrt{x_1+4})}\right) > 0.$$

This means that $f(x)$ has upward concavity on the entire domain. Consequently, there is no inflection point.

8. Complimentary properties. For sufficiently large x the following approximation can be used: $f(x) = \sqrt{x+4} - \sqrt{x} \approx \sqrt{x} - \sqrt{x} = 0$, from which it follows that the function converges to 0 when $x \to +\infty$. Therefore, the line $y = 0$ is a horizontal asymptote.

   Each point $x_0 > 0$ is an interior point of the domain $X = [0, +\infty)$ and the function is continuous at any such point since $f(x) \to f(x_0)$ when $x \to x_0$. The remaining point of the domain $x_0 = 0$ can be approached from the right and again $f(x) \to f(0) = 2$ when $x \to 0^+$. All other points, that is, the points $x_0 < 0$ are exterior points of the domain and cannot be approached by the points of the set $X = [0, +\infty)$. Hence, there is no convergence of $f(x)$ to infinity at any point, and consequently, there is no vertical asymptote.
9. Using the investigated properties, marking some important points, such as $P_0 = (0, 2)$ (intersection with the y-axis and global maximum) and $P_1 = (1, \sqrt{5} - 1)$ (a point with a positive x-coordinate), we sketch the graph shown in Fig. 3.59.

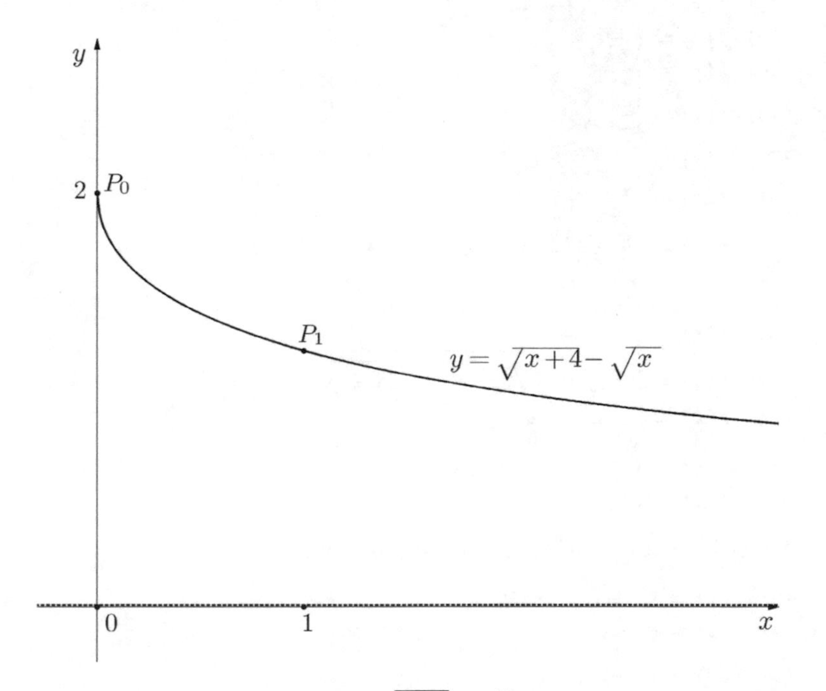

Fig. 3.59 The graph of the function $f(x) = \sqrt{x+4} - \sqrt{x}$

Problems

1. Find the domain of the function:

   (a) $f(x) = \sqrt{3x - x^3}$;
   (b) $f(x) = \frac{x-1}{x^3-3x^2+2x}$;
   (c) $f(x) = \sqrt{x^2 - 3x + 2}$;
   (d) $f(x) = \frac{x}{\sqrt{x^2-4x+3}}$;
   (e) $f(x) = \frac{1}{\sqrt{|x|-2|x-1|}}$;
   (f) $f(x) = \frac{1}{x^3-x|x|+4|x|-4}$;
   (g) $f(x) = \sqrt{\frac{5-x}{x+1}} + \sqrt{\frac{x}{x-3}}$;
   (h) $f(x) = \frac{\sqrt{(x^2+3x-10)\cdot|x+1|}}{\sqrt{2-x^2-x}}$.

2. Determine the range and the global extrema (if they exist) of the function:

   (a) $f(x) = x^4 + (1-x)^4$;
   (b) $f(x) = \frac{x+2}{x-3}$;
   (c) $f(x) = \frac{x}{1+x^2}$;
   (d) $f(x) = \frac{1-x^2}{1+x^2}$;
   (*e) $f(x) = \frac{x^2+x+2}{x^2-x+2}$;
   (f) $f(x) = \frac{x(x-1)}{x(x-1)+2}$;
   (g) $f(x) = \frac{2x^2-4x+9}{x^2-2x+4}$;
   (*h) $f(x) = \frac{x^4+x^2+5}{x^4+2x^2+1}$;
   (i) $f(x) = \sqrt{2 + x - x^2}$;
   (j) $f(x) = 2 - \sqrt{1 - \sqrt{3x^2 + 4\sqrt{3}x + 4}}$;
   (k) $f(x) = \frac{1}{\sqrt{x^2-6x+10}}$;
   (*l) $f(x) = 3x + 4\sqrt{1 - x^2}$;
   (m) $f(x) = x^2\sqrt{4 - x^2}$.

3. Verify if the function is even, odd or none of these:

   (a) $f(x) = x^4 + (1-x)^4$;
   (b) $f(x) = \frac{x}{1+x^2}$;
   (c) $f(x) = \frac{1-x^2}{1+x^2}$;
   (d) $f(x) = \frac{x(x-1)}{x(x-1)+2}$;
   (e) $f(x) = \frac{x^4+x^2+5}{x^4+2x^2+1}$;
   (f) $f(x) = \sqrt{2 + x - x^2}$;
   (g) $f(x) = \sqrt{2 + |x| - x^2}$;
   (h) $f(x) = x^3\sqrt{4 - x^2}$.

4. Answer whether two points $P_1 = (x_1, y_1)$ and $P_2 = (x_2, y_2)$ define uniquely a parabola? What if one of these points is its vertex? Does three points always determine a unique parabola? Provide a proof if a parabola is unique or give examples if it is not.
5. Show that a cubic parabola $y = ax^3 + bx^2 + cx + d, a \neq 0$ has different types of concavity to the left and to the right of $x_0 = -\frac{b}{3a}$, and consequently, x_0 is its inflection point.
6. Answer whether three points $P_1 = (x_1, y_1)$, $P_2 = (x_2, y_2)$ and $P_3 = (x_3, y_3)$ define uniquely a cubic parabola? What if one of this points is its inflection point? Does four points always determine a unique cubic parabola? Provide a proof if a cubic parabola is unique or give examples if it is not.

*7. Show that $n + 1$ points $P_i = (x_i, y_i)$, $i = 0, \ldots, n$ define uniquely a polynomial of degree at most n.

(Hint. One approach is of a "brute force": plug the given points into a polynomial $L(x) = a_n x^n + \ldots + a_1 x + a_0$ and obtain the linear system of $n+1$ equations for $n+1$ unknowns $a_n, \ldots, a_1, a_0$; the matrix of this system is the Vandermonde matrix whose determinant is not zero, and consequently, the system has a unique solution. Another approach is less straightforward and more elegant: first, construct a set of polynomials $L_i(x)$, $i = 0, \ldots, n + 1$ such that $L_i(x_j) = 0, \forall j \neq i$ and $L_i(x_i) = 1$ (these polynomials are called the Lagrangian elementary polynomials); then the required polynomial $L(x)$ is found as the linear combination of the Lagrangian elementary polynomials— $L(x) = y_0 L_0(x) + y_1 L_1(x) + \ldots + y_n L_n(x)$ (it is called the Lagrange interpolating polynomial). Verify if $L(x)$ is determined uniquely and what is a degree of this polynomial.

8. Show that a composition of fractional linear functions is again a fractional linear function.
9. Find fractional linear functions whose inverses are themselves.
10. Derive the properties of $f(x) = -\sqrt{x}$ as the inverse to $g(x) = x^2 : (-\infty, 0] \to [0, +\infty)$. Do the same for the function $f(x) = \sqrt{-x}$ considering an appropriate branch of $g(x) = -x^2$.
11. Find the axis of symmetry of the graph of the function $f(x) = |ax+b|, a \neq 0$. Is this graph symmetric about some point?
12. Derive the equation of the curve (called ellipse) with the following property: the sum of the distances from any point of this curve to the two points $F_1 = (-c, 0)$ and $F_2 = (c, 0)$ (called foci) is a constant. Make an analysis of the properties of this curve.

(Hint. Consider an arbitrary point $P = (x, y)$ of the curve and write the definition of the curve in the coordinate form $\sqrt{(x+c)^2 + y^2} + \sqrt{(x-c)^2 + y^2} = 2a$, where a is a positive constant. Regrouping terms and squaring twice, eliminate both roots and simplify this equation to the form $\frac{x^2}{a^2} + \frac{y^2}{b^2} = 1$, $b^2 = a^2 - c^2 > 0$, which is a canonical form of ellipse. To reveal the properties of the curve, divide it into the upper and lower parts, which satisfy the relations $y = \pm\frac{b}{a}\sqrt{a^2 - x^2}$. Analyze one of the functions, say $y = \frac{b}{a}\sqrt{a^2 - x^2}$, using the studied algorithm of investigation of a function, and then employ the symmetry between the two functions to get the properties of the second function. Join the obtained results to make conclusions about the properties of the ellipse.)

13. Show that a non-constant polynomial function cannot be bounded. Can such a function be periodic?

    (Hint: consider principal term of a polynomial for large values of $|x|$.)

*14. Show that a non-constant rational function cannot be periodic. Can such a function be bounded?

    (Hint: consider separately the case when numerator or denominator has zeros, and the case when none of them vanishes.)

15. The counterclockwise rotation of the Cartesian plane about the origin through the angle $\frac{\pi}{4}$ is given by the formulas $\begin{cases} \bar{x} = \frac{1}{\sqrt{2}}(y + x) \\ \bar{y} = \frac{1}{\sqrt{2}}(y - x) \end{cases}$, where $\bar{x}, \bar{y}$ are the new Cartesian coordinates. Show that in the new Cartesian coordinates $\bar{x}, \bar{y}$ the formula of the hyperbola $y = \frac{1}{x}$ takes the form $\bar{x}^2 - \bar{y}^2 = 2$, more common in analytic geometry. Identify the functions given by the last formula and derive its geometric characterization.

    (Hint: notice that a rotation does not change the distance between any two points.)

16. Determine the properties of a function $f(x)$ and sketch its graph by using a chain of elementary transformations starting from a known function $g(x)$:

    (a) $f(x) = \frac{7x-5}{2-x}$ $(g(x) = \frac{1}{x})$;
    (b) $f(x) = \frac{x+4}{3x-2}$ $(g(x) = \frac{1}{x})$;
    (c) $f(x) = \sqrt[3]{7 - 3x}$ $(g(x) = \sqrt[3]{x})$;
    (d) $f(x) = \frac{1}{\sqrt[3]{7-3x}}$ $(g(x) = \frac{1}{\sqrt[3]{x}})$.

17. Perform analytic investigation of a given function and sketch its graph:

    (a) $f(x) = \frac{7x-5}{2-x}$;
    (b) $f(x) = \frac{x+4}{3x-2}$;
    (c) $f(x) = -|x^2 - 5x + 4|$;
    (d) $f(x) = |x^2 - 5|x| + 4|$;
    (e) $f(x) = \sqrt[3]{7 - 3x}$;
    (f) $f(x) = \frac{1}{\sqrt[3]{7-3x}}$;
    (g) $f(x) = x^3 - 4x + 3$;
    (h) $f(x) = 5 - 4x - x^3$;
    (i) $f(x) = 2x^3 - 3x^2 + 5$;
    (j) $f(x) = x^4 - 3x^2 + 2$;
    (k) $f(x) = |x^4 - 3x^2 + 2|$;
    (l) $f(x) = x^4 - 2x^3 + 2x - 4$;
    (m) $f(x) = \sqrt{x^2 + 5}$;
    (n) $f(x) = \sqrt{x^2 - 4}$;

(o) $f(x) = \sqrt{|x| + 9} - \sqrt{|x|}$;
(*p) $f(x) = \frac{1}{x^3-2}$;
(*q) $f(x) = x - \sqrt{x^2 - 1}$;
(r) $f(x) = \frac{1}{x^2+3}$;
(s) $f(x) = \frac{1}{x^2-4}$;
(t) $f(x) = \frac{x^2}{x^2+3}$;
(u) $f(x) = \frac{x}{x^2+3}$;
(*v) $f(x) = \frac{x-1}{x^3}$;
(*w) $f(x) = \frac{1}{x^3+2}$;
(x) $f(x) = \frac{x^3}{x^3+2}$;
(y) $f(x) = \frac{1}{x^4-1}$;
(z) $f(x) = \frac{x^4}{x^4-1}$.

Chapter 4
Transcendental Functions: Exponential, Logarithmic, Trigonometric

In this chapter we analyze the transcendental functions divided into two groups: first, exponential and logarithmic functions, and second, trigonometric functions.

The algorithm of the study follows the pattern of Chap. 3: given an analytic form (formula) of a function, first we identify its domain and range, then we determine if the function is bounded and possesses the symmetry properties (parity and periodicity), next we investigate monotonicity and extrema, followed by the study of concavity and inflection, after this we find the points of intersection with the coordinate axes and the intervals where the function maintains the same sign, proceed with verification of bijectivity and consider additional properties (continuity and asymptotes) and finalize by sketching the graph of the function, which reflects geometrically the properties revealed analytically.

Recall that not all of these steps are obligatory and executable for any elementary function. Besides, the order of the steps can be interchanged, in the cases when some more advanced properties (like monotonicity and extrema) can immediately provide information about preceding properties, dismissing their separate investigation.

1 Exponential and Logarithmic Functions

1.1 Preliminary Notions: Power with Real Exponent and its Properties

Natural Exponent

Definition If $a \in \mathbb{R}$ and $n \in \mathbb{N}$, the *n-th power of a* a^n (also called the *n-th exponent of* a) is defined by the formula $a^n = a \cdot \ldots \cdot a$ (a is multiplied by itself n times). In this operation, a is called the *base* and n the *exponent*.

A. Bourchtein, L. Bourchtein, *Elementary Functions*,
https://doi.org/10.1007/978-3-031-29075-6_4

Integer Exponent

Definition If $a \neq 0$ and $n \in \mathbb{N}$, the *power* a^{-n} is defined by the formula $a^{-n} = \frac{1}{a^n}$; for any $a \neq 0$, $a^0 = 1$ (the expression 0^0 is undetermined).

There are two types of the properties of powers: those of equality and those of comparison.

Properties of Equality The main *properties of equality* of integer powers are as follows: for any two real numbers $a \neq 0$ and $b \neq 0$, and for any integers n and m we have

(1) $(ab)^n = a^n b^n$;
(2) $(\frac{a}{b})^n = \frac{a^n}{b^n}$;
(3) $a^n a^m = a^{n+m}$;
(4) $\frac{a^n}{a^m} = a^{n-m}$;
5) $(a^n)^m = a^{nm}$.

The proofs follow directly from the definition of the integer power and are elementary.

Properties of Comparison The main *properties of comparison* are as follows: for any two integers n and m we have

(1) in the case $a > 1$, if $n > m$ then $a^n > a^m$; in particular, $a^n > 1$ for $n > 0$, and $a^n < 1$ for $n < 0$;
(2) in the case $0 < a < 1$, if $n > m$ then $a^n < a^m$; in particular, $a^n < 1$ for $n > 0$, and $a^m > 1$ for $m < 0$;
(3) in the case $n > 0$, if $b > a > 0$, then $b^n > a^n$, and vice-verse;
(4) in the case $n < 0$, if $b > a > 0$, then $b^n < a^n$, and vice-verse.

The proofs are very simple and follow directly from the definition of the integer power.

Radicals: n-th Root $\sqrt[n]{a}$—Exponent $\frac{1}{n}$, $n \in \mathbb{N}$

Definition For any $a \in \mathbb{R}$ and $n = 2k-1$, $k \in \mathbb{N}$, the *n-th root* (or the *root of degree n*) $\sqrt[n]{a}$ is such a number b that $b^n = a$. For any $a \geq 0$ and $n = 2k$, $k \in \mathbb{N}$, the *n-th root* $\sqrt[n]{a}$ is such a number $b \geq 0$ that $b^n = a$ (notice that in this case the negative number $-b$ also has the property that $(-b)^n = a$). For $a < 0$ and $n = 2k$, $k \in \mathbb{N}$ the operation is not defined (there is no real number whose even power gives a negative number).

Hence, for the roots of even degree there are two restrictions: they are defined only for non-negative numbers and the result of the operation is also a non-negative number. The quadratic root has simplified notation $\sqrt{a}$. If there is no ambiguity, a n-th root can be called simply root.

According to the definition, $(\sqrt[n]{a})^n = a$ in all the cases when the root is defined. Therefore, the same operation is naturally denoted by the power with the exponent $\frac{1}{n}$, $n \in \mathbb{N}$: $\sqrt[n]{a} \equiv a^{\frac{1}{n}}$. Then, the property of the definition takes the form $(a^{\frac{1}{n}})^n = a$

in the cases accepted in the definition, namely: if n is odd, then $a^{\frac{1}{n}}$ is such a number that $(a^{\frac{1}{n}})^n = a$ for any real a; if n is even, then $a^{\frac{1}{n}}$ is such a non-negative number that $(a^{\frac{1}{n}})^n = a$ for any $a \geq 0$. The power $a^{\frac{1}{n}}$ is not defined in the case of even n and $a < 0$.

Notice that in the case of the reversed order of the exponents $\frac{1}{n}$ and n, an analogous property holds in the same cases. Indeed, if n is odd, then $(a^n)^{\frac{1}{n}} = a$ for any $a \in \mathbb{R}$, because a^n is the base of the n-th root. If n is even and $a \geq 0$, then we have the equality $(a^n)^{\frac{1}{n}} = a$ due to the same reasons. However, if n is even and $a < 0$, then, differently from $(a^{\frac{1}{n}})^n$, which has no sense in this case, the operation $(a^n)^{\frac{1}{n}}$ is defined (because $a^n > 0$), although the result will not be a but $-a$. Indeed, the root of an even degree should be a non-negative number, and consequently, it cannot be a. At the same time, $-a$ is the correct result, since $-a > 0$ and $(-a)^n = a^n$ for even n. Joining the results for even n, we have $(a^n)^{\frac{1}{n}} = |a|$ for any $a \in \mathbb{R}$. In particular, $\sqrt{a^2} = |a|$ for any $a \in \mathbb{R}$.

Due to restrictions on the argument of a root of even degree (there is no such root of a negative number) and on its result (it should be a non-negative number), the properties of integer exponents are not transferred directly to the fractional exponents $\frac{1}{n}$, $n \in \mathbb{N}$. Indeed, different results can be found for $(a^{\frac{1}{n}})^n$ and $(a^n)^{\frac{1}{n}}$ when n is even (as was just seen), and this suggests that the situation for roots is more complicated. Exemplifying this difference, we can take $a = -1$, $n = 2$ and obtain non-existent expression $((-1)^{\frac{1}{2}})^2 = (\sqrt{-1})^2$, while $((-1)^2)^{\frac{1}{2}} = 1$. This makes incorrect straightforward application of the properties of integer exponents in the case of roots. For example, in the case of the first property, we can arrive at the following absurd result: $1 = \sqrt{1} = \sqrt{(-1)\cdot(-1)} = \sqrt{-1}\cdot\sqrt{-1}$, where the expression on the right-hand side does not exist. Using the third property, we can fall in a similar absurd: $-1 = (-1)^1 = (-1)^{\frac{1}{2}+\frac{1}{2}} = (-1)^{\frac{1}{2}}\cdot(-1)^{\frac{1}{2}}$, where the expression on the right-hand side does not exist. An application of the fifth property to the exponent $\frac{1}{n}$ also can give erroneous results: $-1 = (-1)^1 = (-1)^{\frac{1}{2}\cdot 2} = ((-1)^{\frac{1}{2}})^2$ with non-existent expression on the right-hand side, or $-1 = (-1)^1 = (-1)^{2\cdot\frac{1}{2}} = ((-1)^2)^{\frac{1}{2}} = 1$ with the false statement.

Nevertheless, the properties of integer exponents can be extended to roots if we allow only positive bases $a > 0$. In this case, the following three properties are maintained:

(1) $\sqrt[n]{ab} = \sqrt[n]{a}\sqrt[n]{b}$;

(2) $\sqrt[n]{\frac{a}{b}} = \frac{\sqrt[n]{a}}{\sqrt[n]{b}}$;

(3) $\sqrt[m]{\sqrt[n]{a}} = \sqrt[mn]{a}$.

The remaining two properties involve the operations which are not yet introduced for roots and they will be added when we will consider rational exponents.

Rational Exponent a^p, $p \in \mathbb{Q}$

Definition For any $a \in \mathbb{R}$ and $p = \frac{m}{n}$, $m \in \mathbb{Z}$, $n \in \mathbb{N}$, the *power* a^p is defined by the formula $a^p = \sqrt[n]{a^m} = (\sqrt[n]{a})^m$ in all the cases when both involved operations can be performed and the two expressions have the same result.

The conditions of the definition are satisfied, in particular, when a is an arbitrary real and n is odd, and also when $a > 0$ and p is an arbitrary rational.

For a validity of the five properties of integer powers in the case of rational exponents, we have to use the restriction on the base inherited from the roots.

Properties of Equality If $a > 0, b > 0$, then for any $p, q \in \mathbb{Q}$ the following *properties of equality* are valid:

(1) $(ab)^p = a^p b^p$;
(2) $(\frac{a}{b})^p = \frac{a^p}{b^p}$;
(3) $a^p a^q = a^{p+q}$;
(4) $\frac{a^p}{a^q} = a^{p-q}$;
(5) $(a^p)^q = a^{pq}$.

Properties of Comparison For rational exponents, the following *properties of comparison* are true:

(1) in the case $a > 1$, if $p > q$ then $a^p > a^q$, and vice-verse;
(2) in the case $0 < a < 1$, if $p > q$ then $a^p < a^q$, and vice-verse;
(3) in the case $p > 0$, if $b > a > 0$, then $b^p > a^p$, and vice-verse;
(4) in the case $p < 0$, if $b > a > 0$, then $b^p < a^p$, and vice-verse.

Real Exponents

Turning to consideration of real exponents, from the very beginning we suppose that the base a is positive. Since the powers with rational exponents were already defined, we focus on the case when the exponent x is an irrational number. It is sufficient to define a^x when x is positive, because, by the definition, $a^{-x} = \frac{1}{a^x}$.

Definition Consider, first, the situation when $a > 1$. Take an arbitrary positive irrational number $x = x_0.x_1 \dots x_n \dots$ and consider different rational approximations of x, both superior and inferior. We can start with $x_0 < x < x_0 + 1$, then we move to the first decimal digit $x_0 \leq x_0.x_1 < x < x_0.x_1 + 0.1 \leq x_0 + 1$, then to the second one $x_0 \leq x_0.x_1 \leq x_0.x_1x_2 < x < x_0.x_1x_2 + 0.01 \leq x_0.x_1 + 0.1 \leq x_0 + 1$, and so on. Each subsequent approximation is closer to x then preceding, both for superior and inferior approximations. Under this process, the n-th inferior approximation $p_n = x_0.x_1x_2 \dots x_n$ satisfies the inequality $p_{n-1} \leq p_n < x$, while the n-th superior approximation $q_n = x_0.x_1x_2 \dots x_n + 10^{-n}$ satisfies the inequality $x < q_n \leq q_{n-1}$. These rational approximations of x generates the corresponding approximations of a^x, which satisfy the inequalities $a^{p_0} \leq \dots \leq a^{p_{n-1}} \leq a^{p_n} \leq \dots \leq a^{q_n} \leq a^{q_{n-1}} \leq \dots \leq a^{q_0}$, according to the first property of comparison of rational powers. Since the difference between p_n and q_n is equal to 10^{-n} and it decreases as n increases, it is natural to expect that these rational approximations of powers stay each time closer to the unique number which is called a^x. Although intuitively the existence and uniqueness of this number can appear to be evident, an exact demonstration of this fact is based on the concepts and results, which are well beyond the scope of this text.

When $0 < a < 1$, we follow the same algorithm of approximations, with the only difference that the inequalities of the powers have inverted sign (according to the second property of comparison between rational powers).

The *properties of equality* of real powers are the same as those of rational powers and can be formulated as follows.

Properties of Equality If $a > 0, b > 0$, then for any $x, y \in \mathbb{R}$ the following properties are valid:

(1) $(ab)^x = a^x b^x$;
(2) $(\frac{a}{b})^x = \frac{a^x}{b^x}$;
(3) $a^x a^y = a^{x+y}$;
(4) $\frac{a^x}{a^y} = a^{x-y}$;
(5) $(a^x)^y = a^{xy}$.

The *properties of comparison* are also maintained.

Properties of Comparison For any real x and y, we have:

(1) in the case $a > 1$, if $x > y$ then $a^x > a^y$, and vice-verse;
(2) in the case $0 < a < 1$, if $x > y$ then $a^x < a^y$, and vice-verse;
(3) in the case $x > 0$, if $b > a > 0$, then $b^x > a^x$, and vice-verse;
(4) in the case $x < 0$, if $b > a > 0$, then $b^x < a^x$, and vice-verse.

A proof of these properties requires a knowledge of advanced concepts and results, which are well beyond the scope of this text.

1.2 Exponential Function

Definition of Exponential Function The *exponential function* $y = f(x) = a^x$, $a > 0$ is a function with the domain $\mathbb{R}$ which assigns to each real x the only number $y = a^x$. The constant $a > 0$ is called a *base* and variable x an *exponent*.

Before we start a general investigation, let us discard the case when $a = 1$, since the corresponding function is a singular and trivial case of the constant function $y = 1^x = 1, \forall x \in \mathbb{R}$, already considered in Chap. 3. From now on we assume that $a \neq 1$.

Study of $y = a^x$

1. Domain and range. By the definition, the domain is $X = \mathbb{R}$ (since the power of a positive base is defined for any real exponent). Since any number a^x, $a > 0$ is

positive (by the definition of power), the range is contained in $(0, +\infty)$. Besides, we assume without a proof that a^x can take any positive value, which means that $Y = (0, +\infty)$.

2. Boundedness. The range of the function determines that the function is bounded below (for example, by the constant $m = 0$) and is not bounded above.
3. Monotonicity and extrema. If $a > 1$, then, by the properties of powers, $a^{x_2} > a^{x_1}$ for any pair $x_2 > x_1$, and consequently, $f(x)$ increases over the entire domain $X = \mathbb{R}$. If $0 < a < 1$, then, by the properties of powers, $a^{x_2} < a^{x_1}$ for any pair $x_2 > x_1$, and consequently, $f(x)$ decreases over the entire domain $X = \mathbb{R}$.

   Alternatively, we can consider the difference of values of the function. For $x_1 < x_2$ we have $f(x_2) - f(x_1) = a^{x_2} - a^{x_1} = a^{x_1}(a^{x_2-x_1} - 1)$, where the first factor is positive, which means that the sign of $f(x_2) - f(x_1)$ depends only on the second factor. If $a > 1$, then, by the properties of powers, $a^t > 1, \forall t > 0$, that is, the second factor is positive, and therefore, the function increases over its domain $X = \mathbb{R}$. If $0 < a < 1$, then, by the properties of powers, $a^t < 1, \forall t > 0$, that is, the second factor is negative, and therefore, the function decreases over its domain $X = \mathbb{R}$.

   Hence, any point of the domain is monotonic (increasing if $a > 1$ and decreasing if $a < 1$), and consequently, there is no extrema of any type, whether global or local, strict or not.
4. Parity. Monotonicity of the exponential function over the entire domain guarantees that the function is not even. On the other hand, positivity of the range guarantees that the function is not odd.
5. Periodicity. Monotonicity of the exponential function over the entire domain guarantees that the function is not periodic, since it does not repeat any of its values.
6. Concavity and inflection. For any two points $x_1, x_2 \in \mathbb{R}$, the following inequality holds: $f(x_1) + f(x_2) - 2f\left(\frac{x_1+x_2}{2}\right) = a^{x_1} + a^{x_2} - 2a^{\frac{x_1+x_2}{2}} = \left(a^{\frac{x_1}{2}} - a^{\frac{x_2}{2}}\right)^2 > 0$, which shows upward concavity on the entire domain. This implies that there is no inflection point.
7. Intersection and sign. The graph of the function crosses the y-axis at the point $(0, 1)$, but it does not cross the x-axis (because the range contains only positive values). As was noted, the function has only positive values.
8. The function $f(x) = a^x : X = \mathbb{R} \to Y = (0, +\infty)$ is bijective. Indeed, it is surjective because the range is $Y = (0, +\infty)$, and monotonicity ensures that it is injective. The inverse function is called logarithmic and is considered in the next section.
9. Using the investigated properties and marking some important points such as $P = (0, 1)$ (intersection with the y-axis), we can plot the graphs of the exponential functions in the two distinct cases $a = e > 1$ and $0 < a = e^{-1} < 1$ as shown in Fig. 4.1.

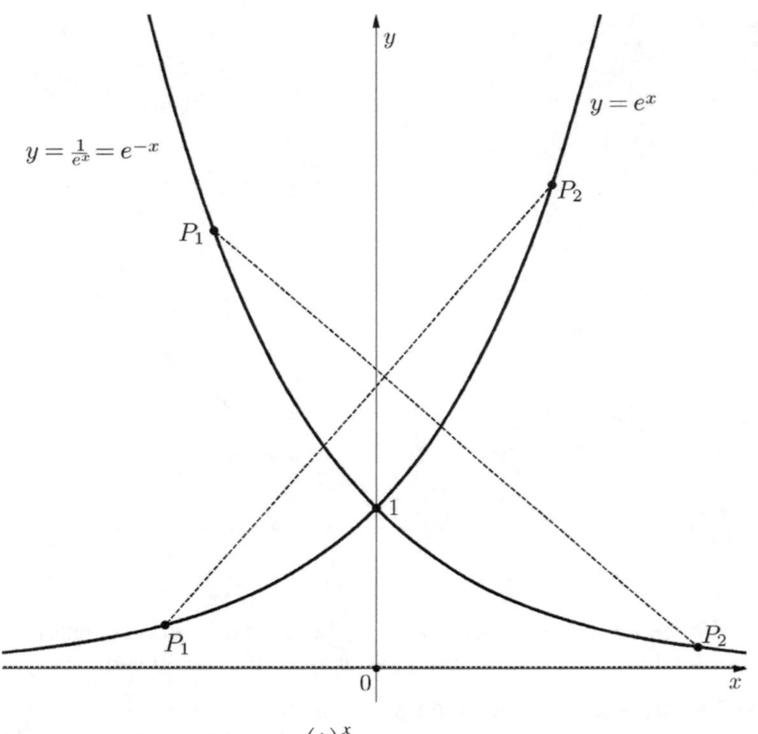

Fig. 4.1 The graphs of $y = e^x$ and $y = \left(\frac{1}{e}\right)^x$

Complimentary Properties

The graph of the function $f(x) = a^x$, $a > 1$ suggests that its values growth without restriction when x goes to $+\infty$ and approaches 0 keeping positive values when x goes to $-\infty$. Indeed, for any given $M > 0$, we can solve the equation $a^x = M$ and find the solution $x_0 = \log_a M$. Then, using increase of a^x, $a > 1$, we conclude that $a^x > M$ for all $x > x_0$ and $a^x < M$ for all $x < x_0$. Hence, for sufficiently large values of the variable x, the values of the function become greater than any constant given in advance, which means that $f(x)$ goes to $+\infty$ when x goes to $+\infty$. On the other hand, for sufficiently small values of x (large in absolute value but negative), the values of the function becomes smaller than any given positive constant (keeping the positive sign), which means that $f(x)$ approaches 0 when x goes to $-\infty$. Geometrically, these two results mean that $f(x)$ has no horizontal asymptote at $+\infty$, but it has the horizontal asymptote $y = 0$ at $-\infty$.

Additionally, the function is continuous at every point of its domain $X = \mathbb{R}$, that is, $f(x) \to f(x_0)$ when $x \to x_0$. Therefore, no infinite behavior at the points of the domain is observed, which means that there is no vertical asymptote.

In a similar way, the function $f(x) = a^x$, $a < 1$ goes to $+\infty$ when x goes to $-\infty$, and approaches 0 when x goes to $+\infty$. Besides, $f(x)$ approaches $f(x_0)$ as

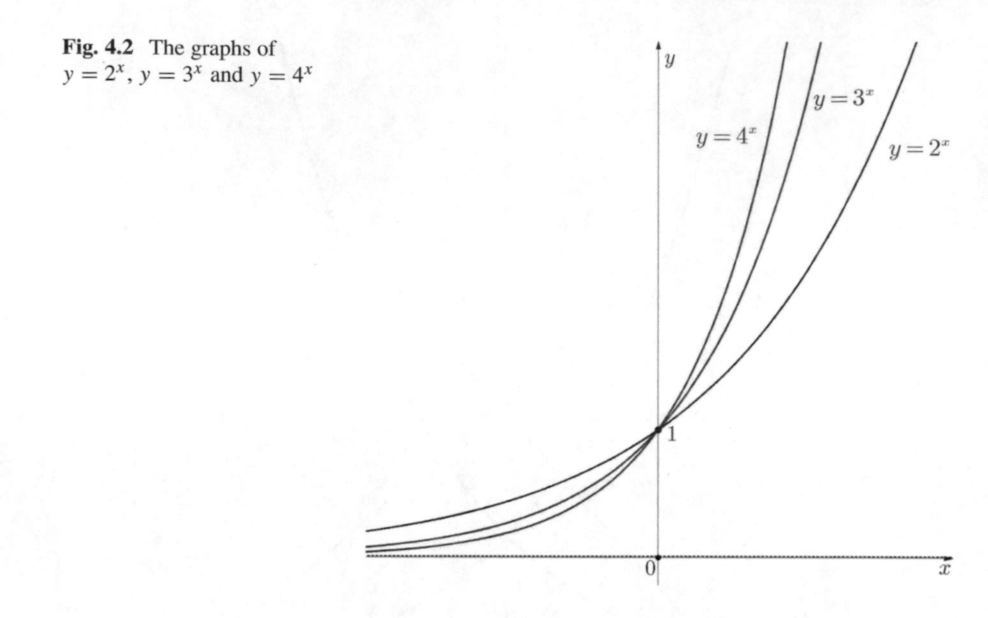

Fig. 4.2 The graphs of $y = 2^x$, $y = 3^x$ and $y = 4^x$

$x \to x_0$. Geometrically this means that $f(x)$ has the horizontal asymptote $y = 0$ at $+\infty$, does not have a horizontal asymptote at $-\infty$, and has no vertical asymptote.

Comparison Between Exponential Functions

Consider first the situation when $1 < a < b$. In this case, it follows from the properties of powers that $y_a = a^x < b^x = y_b$ on the interval $(0, +\infty)$. Besides, for any $0 \le x_1 < x_2$, we have $\frac{y_a(x_2)}{y_a(x_1)} = a^{x_2-x_1} < b^{x_2-x_1} = \frac{y_b(x_2)}{y_b(x_1)}$. This shows that on the interval $(0, +\infty)$ the graph of $y_b = b^x$ is located above and grows faster than the graph of $y_a = a^x$. On the interval $(-\infty, 0)$ we have $y_a = a^x > b^x = y_b$, that is, the graph of $y_b = b^x$ lies below the graph of $y_a = a^x$. This relationship is illustrated in Fig. 4.2.

When $0 < a < b < 1$, the relationship between the functions $y_a = a^x$ and $y_b = b^x$ is as follows: on the interval $(0, +\infty)$ the function a^x is smaller than b^x, while on the interval $(-\infty, 0)$ the function a^x is greater and decreases faster than b^x. This follows from the properties of powers or simply noting that $\left(\frac{1}{a}\right)^x = a^{-x}$, that is, the graph of $\left(\frac{1}{a}\right)^x$ can be obtained by reflecting the graph of a^x about the y-axis. The relationship between exponential functions in the case $0 < a < b < 1$ is shown in Fig. 4.3.

Notice that at the point $x = 0$ the value of each exponential function a^x is the same and equal to $y = 1$.

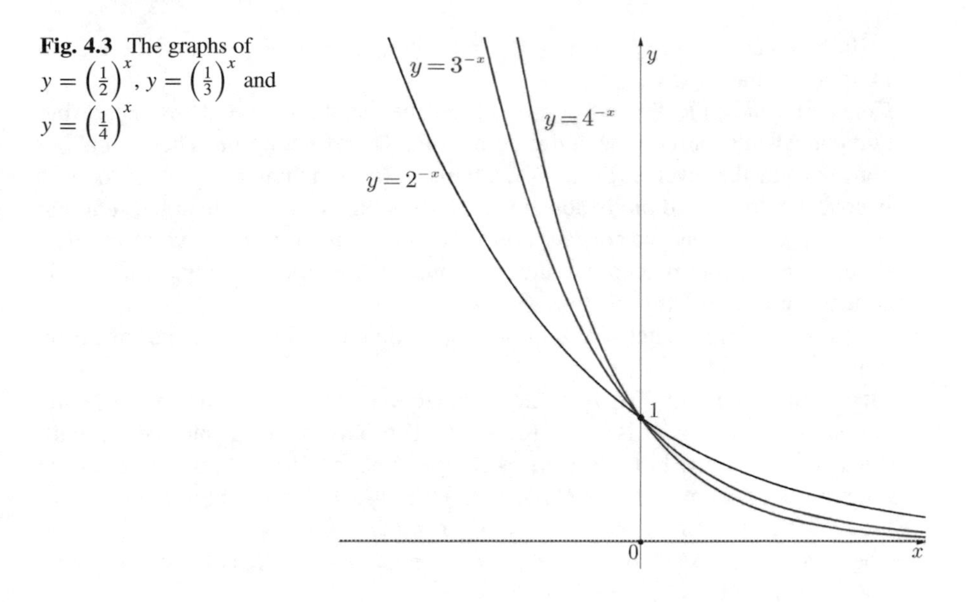

Fig. 4.3 The graphs of $y = \left(\frac{1}{2}\right)^x$, $y = \left(\frac{1}{3}\right)^x$ and $y = \left(\frac{1}{4}\right)^x$

1.3 Logarithmic Function

Definition of Logarithmic Function A *logarithmic function* $y = \log_a x$ can be defined as the inverse to the exponential function $f(x) = a^x : \mathbb{R} \to (0, +\infty)$, where the number $a > 0$, $a \neq 1$ is called a *base*.

Study of $y = \log_a x$

1. Domain and range. By the definition, the domain of $y = \log_a x$ is $X = (0, +\infty)$ (the range of the original function $y = a^x$) and the range is $Y = \mathbb{R}$ (the domain of the original function).
2. Boundedness. The specified range implies that the function is unbounded both below and above.
3. Parity. The function is neither even nor odd, since its domain is not symmetric about the origin.
4. Periodicity. The function is not periodic because its domain is bounded from the left. (Another reason is that $y = \log_a x$ is defined as the inverse function, and consequently, it has no repeated values.)
5. Monotonicity and extrema. For $a > 1$, the function $y = a^x$ is increasing over the entire domain $\mathbb{R}$. Therefore, by the Theorem about monotonicity of the inverse (Sect. 12.4, Chap. 2), the function $y = \log_a x$, $a > 1$ is increasing on its domain $X = (0, +\infty)$. In the case $a < 1$, the function $y = a^x$ is decreasing on $\mathbb{R}$, and consequently, by the same theorem about the inverse, the function $y = \log_a x$, $a < 1$ is decreasing on its domain $X = (0, +\infty)$.

In both cases, every point $x > 0$ is monotonic, and consequently, there is no extremum of any type.

6. Concavity and inflection. If $a > 1$, then the function $y = a^x$ is increasing and concave upward on the entire domain $\mathbb{R}$. Therefore, by the Theorem about concavity of the inverse (Sect. 12.4, Chap. 2), the function $y = \log_a x$, $a > 1$ is concave downward on its domain $X = (0, +\infty)$. If $a < 1$, then the function $y = a^x$ is decreasing and concave upward on the entire domain $\mathbb{R}$, which implies, by the same theorem about the inverse, that the function $y = \log_a x$, $a < 1$ is concave upward on its domain $X = (0, +\infty)$.

   It follows immediately from these results that $y = \log_a x$ has no inflection point.
7. Intersection and sign. There is no intersection with the y-axis, because the domain of $y = \log_a x$ is $X = (0, +\infty)$. The only point of intersection with the x-axis is $(1, 0)$. In the case $a > 1$, the function $\log_a x$ is negative on the interval $(0, 1)$ and positive on $(1, +\infty)$. In the case $a < 1$, the signs are inverted: $\log_a x > 0, \forall x \in (0, 1)$ and $\log_a x < 0, \forall x \in (1, +\infty)$.
8. The function $f(x) = \log_a x : X = (0, +\infty) \to Y = \mathbb{R}$ is bijective and its inverse is $f^{-1}(x) = a^x : \mathbb{R} \to (0, +\infty)$.
9. Using the investigated properties and marking some important points such as $P = (1, 0)$ (intersection with the x-axis), we can draw the graph of the function $y = \log_a x$ for any $a > 0, a \neq 1$. The graph for $a = e$, called the natural logarithm and denoted by $y = \ln x$, and the graph for $a = e^{-1}$ are shown in Fig. 4.4.

Another option to plot the same graphs is to apply the Theorem about the graph of the inverse (Sect. 12.5, Chap. 2): reflecting the graph of $y = e^x : \mathbb{R} \to (0, +\infty)$ about the line $y = x$ we obtain the graph of $y = \ln x$, and reflecting $y = \left(\frac{1}{e}\right)^x :$ $\mathbb{R} \to (0, +\infty)$ about the same symmetry axis $y = x$ we get the graph of $y = \log_{1/e} x$. For $y = \ln x$ the illustration of such reflection is given in Fig. 4.5.

Notice that the properties of $y = \log_a x$, $a < 1$ can also be derived from the properties of $y = \log_a x$, $a > 1$ using the formula $y = \log_{a^{-1}} x = -\log_a x$, $\forall x > 0$.

Complimentary Properties

The graph of the function $f(x) = \log_a x$, $a > 1$ suggests that its values grow without restriction when x goes to $+\infty$ and decrease without restriction (increase in absolute value without restriction keeping the negative values) when x approaches 0 from the right (by positive values contained in the domain). Indeed, for any given constant M, we can solve the equation $\log_a x = M$ and find the solution $x_0 = a^M$. Then, using increase of $\log_a x$, $a > 1$, we conclude that $\log_a x > M$ for all $x > x_0$ and $\log_a x < M$ for all $x < x_0$. Hence, for sufficiently large values of the variable x, the values of the function exceed any constant chosen in advance, which means that $f(x)$ goes to $+\infty$ when x goes to $+\infty$. On the other hand, for sufficiently small (and positive) values of x, the values of $f(x)$ are smaller than any chosen constant, which means that $f(x)$ goes to $-\infty$ as x approaches 0. These two results

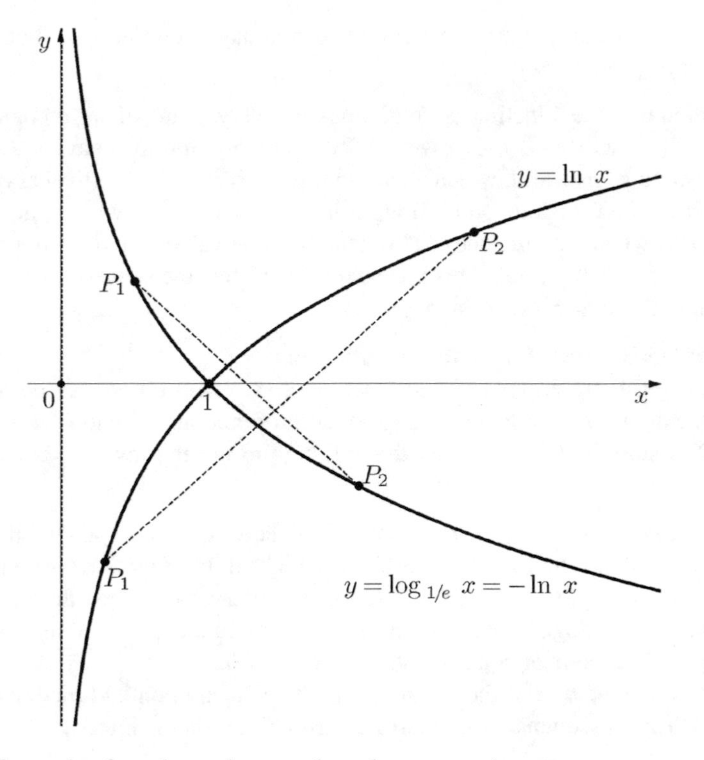

Fig. 4.4 The graphs of $y = \ln x = \log_e x$ and $y = \log_{1/e} x = -\ln x$

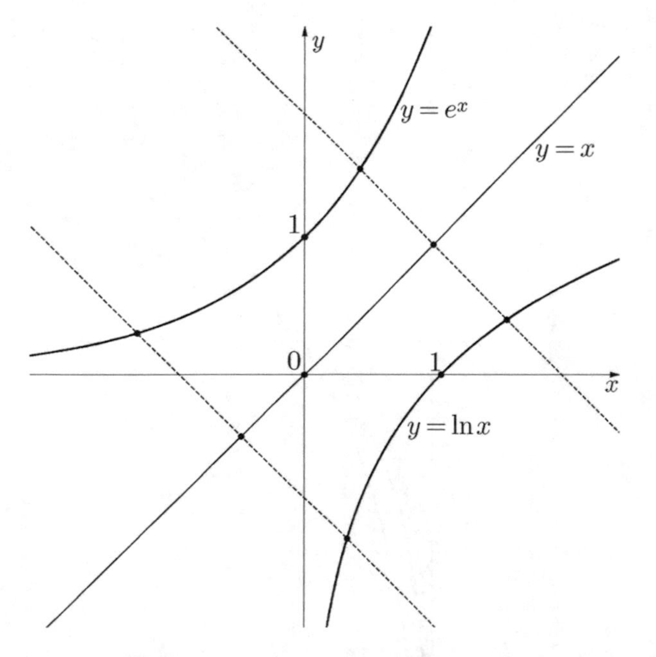

Fig. 4.5 Construction of the graph of $y = \ln x$ by reflecting the graph of $y = e^x$ about $y = x$

geometrically mean that $f(x)$ has no horizontal asymptote at $+\infty$, but it has the vertical asymptote $x = 0$.

Additionally, the function is continuous at every point of its domain, that is, $f(x) \to f(x_0)$ as $x \to x_0 \in (0, +\infty)$. Therefore, no infinite behavior at the points of the domain is observed, which means that there is no other vertical asymptote.

Similarly, the function $f(x) = \log_a x$, $a < 1$ goes to $-\infty$ when x goes to $+\infty$, goes to $+\infty$ when x approaches 0 (from the positive values), and approaches $f(x_0)$ when $x \to x_0 \in (0, +\infty)$. This means that $f(x)$ has the only vertical asymptote $x = 0$ and has no horizontal asymptote.

Comparison Between Logarithmic Functions

A comparison between logarithmic functions $\log_a x$ and $\log_b x$ can be reduced to the results about the corresponding exponential functions a^x and b^x (presented in Sect. 1.2), using the fact that logarithmic functions are the inverses of exponential ones.

In the case $1 < a < b$, the comparison between the exponential functions a^x and b^x leads to the following results for logarithmic functions: on the interval $(0, 1)$ we have $\log_a x < \log_b x$, while on the interval $(1, +\infty)$ the inequality is inverted $\log_a x > \log_b x$; besides, on $(1, +\infty)$ the function $\log_a x$ increases faster than $\log_b x$. This comparison is illustrated in Fig. 4.6.

When $0 < a < b < 1$, the comparison between exponential functions provides the following consequences for logarithmic functions: on the interval $(0, 1)$ we have

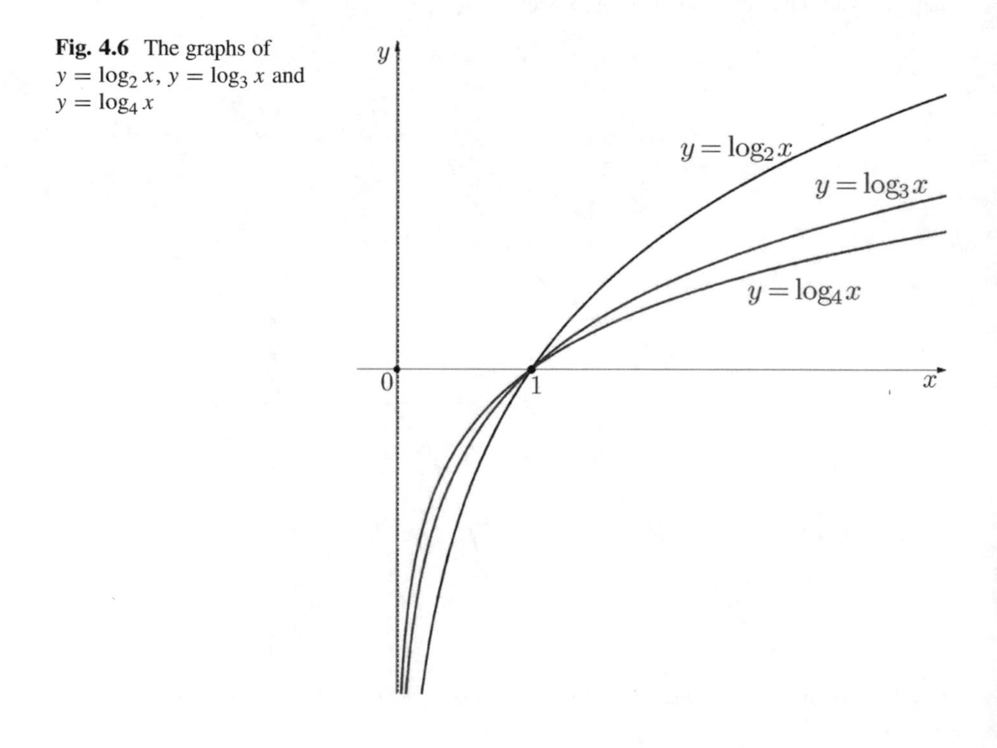

Fig. 4.6 The graphs of $y = \log_2 x$, $y = \log_3 x$ and $y = \log_4 x$

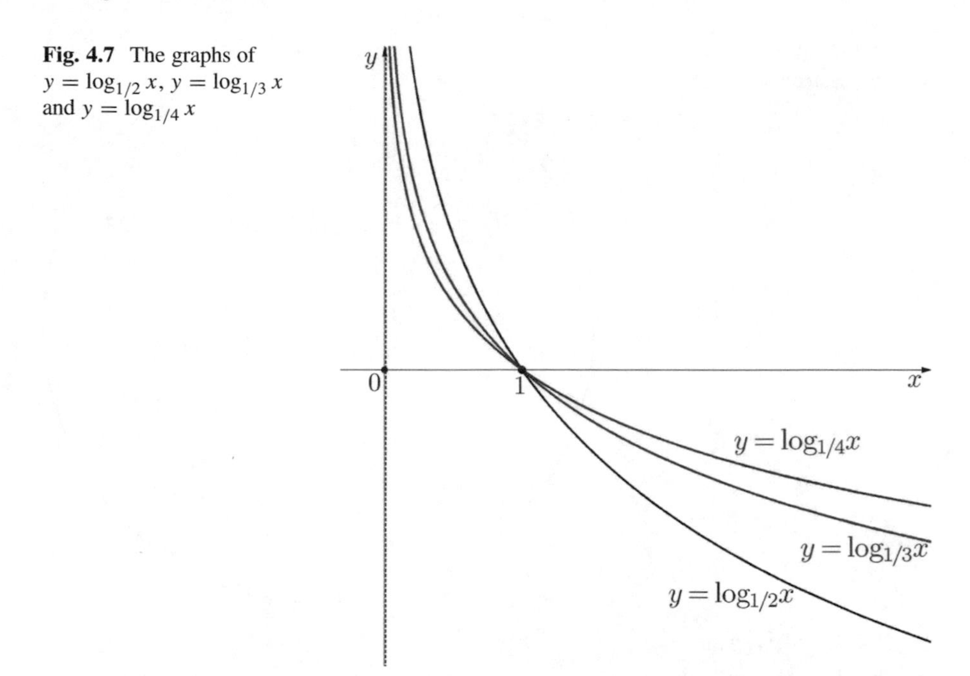

Fig. 4.7 The graphs of $y = \log_{1/2} x$, $y = \log_{1/3} x$ and $y = \log_{1/4} x$

$\log_a x < \log_b x$, while on the interval $(1, +\infty)$ the inequality is opposite $\log_a x > \log_b x$; besides, on $(1, +\infty)$ the function $\log_a x$ decreases slower than $\log_b x$. The relationship in the case $0 < a < b < 1$ follows also from the formula $\log_{1/a} x = -\log_a x$, that is, the graph of $\log_{1/a} x$ can be obtained by reflecting the graph of $\log_a x$ about the x-axis. This comparison is illustrated in Fig. 4.7.

Notice that all the graphs of logarithmic functions pass through the point $(1, 0)$.

2 Trigonometric Functions

2.1 Preliminary Notions and Results of Trigonometry

Definition of $\cos a$ and $\sin a$

Circle The set of all points of the Cartesian plane equidistant from a fixed point O is called a *circle*. The point O is the *center of a circle* and the distance r from O to the points of a circle is the *radius*. Recalling the formula of the distance between two points of the Cartesian plane, we arrive at the following *equation of a circle* centered at $O = (x_0, y_0)$ with the radius r: $(x - x_0)^2 + (y - y_0)^2 = r^2$ (see Fig. 4.8). In particular, the set of all points of the Cartesian plane whose distance to the origin is equal 1 is called the *unit circle centered at the origin* (in this specific case, $O = (0, 0)$ and $r = 1$). Obviously, the equation of this unit circle is $x^2 + y^2 = 1$.

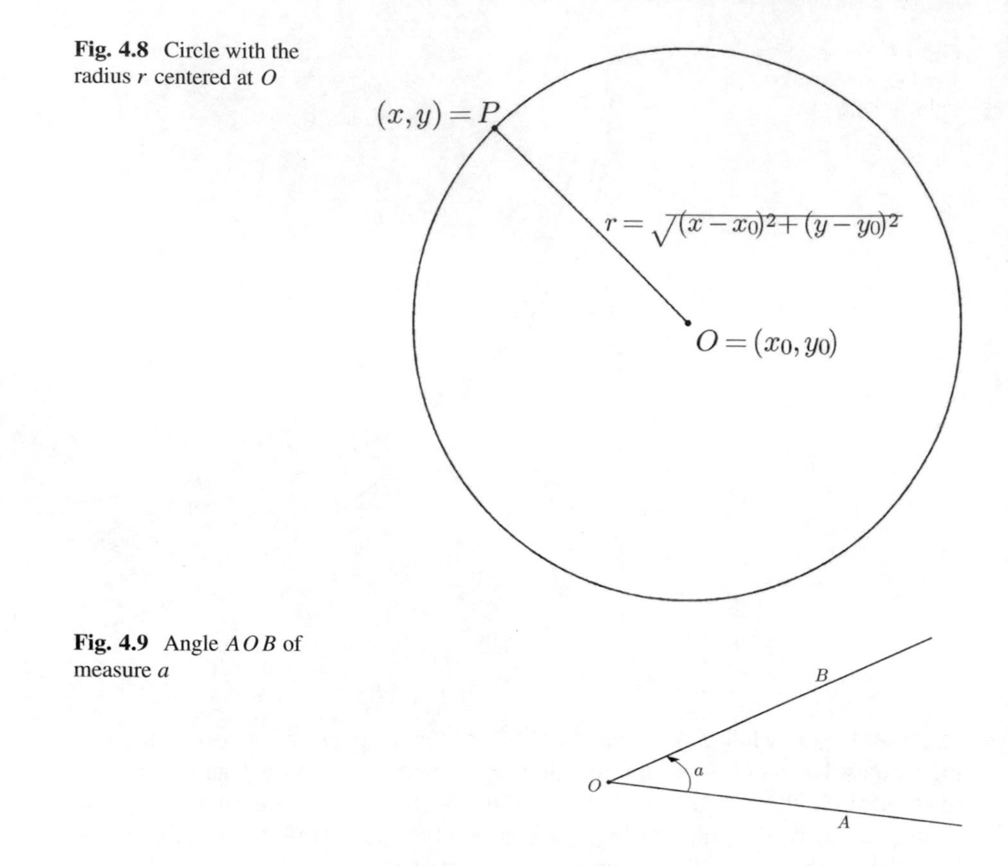

Fig. 4.8 Circle with the radius r centered at O

Fig. 4.9 Angle AOB of measure a

Angles and Their Measures An angle AOB consists of two rays OA and OB with the common vertex O. The measure of an angle AOB is the number of rotations about O needed to rotate OA until OB (see Fig. 4.9).

One of common units of measure of angles is the degree which represents $\frac{1}{360}$ fraction of the complete rotation about the vertex O. If an angle measures a degrees, then this is denoted as a^0. Another unit of measure, more important and natural in mathematics, is the radian. One radian is a non-dimensional measure of an angle AOB which corresponds to the arc (part) of the circle of the radius r, contained between the rays OA and OB, whose length is equal to r. In other words, the radian measure of AOB represents the ratio between the length of the arc AB and the radius of the circle. If an angle measures a radians it is denoted by $a\ rad$ or simply a. In the case of the unit circle, an angle between the rays OA and OB has a radians if the length of the arc AB is equal to a. Although the radian is a non-dimensional measure, frequently the notation is $a\ rad$ to distinguish from the measure in degrees.

Since one complete rotation around the point O corresponds to 360^0 and $2\pi\ rad$, it is easy to see that the relationship between the two measures is given by the formula $a^0 = \frac{\pi}{180} a\ rad$ or $b\ rad = \frac{180}{\pi} b^0$.

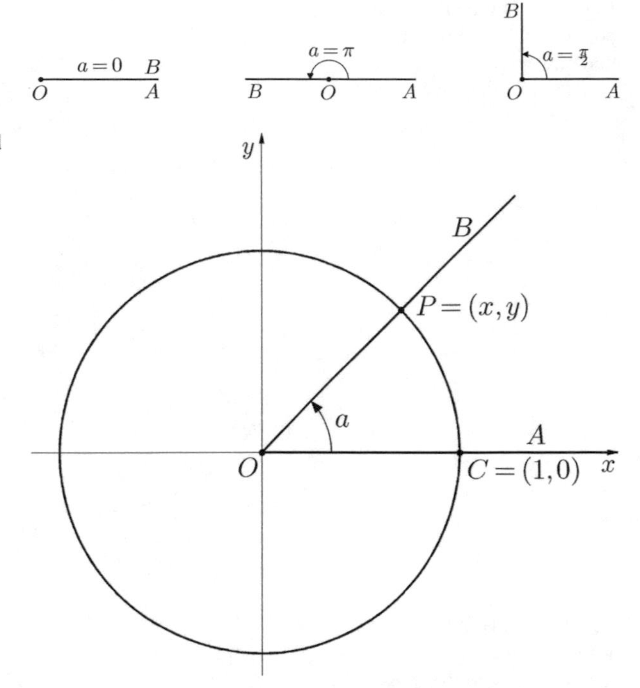

Fig. 4.10 Angles AOB of the three specific measures a

Fig. 4.11 The unit circle and the corresponding angle AOB

According to the presented definitions, the angle AOA measures 0^0 and $0\ rad$. The angle AOB that forms a line has the measure of 180^0 and $\pi\ rad$. A right angle is such that it measures 90^0 and $\frac{\pi}{2}\ rad$. (These three situations are shown in Fig. 4.10.) Any angle with the measure $a \in (0, 90^0)$, or equivalently, $a \in (0, \frac{\pi}{2})$, is called acute. Any angle with the measure $a \in (90^0, 180^0)$, or equivalently, $a \in (\frac{\pi}{2}, \pi)$, is called obtuse.

Unit Circle and Angles in Trigonometry

From now on we will consider only the unit circle centered at the origin, which we will call simply the unit circle. Besides, we will fix the position of the ray OA (called the initial ray or initial side) to be a positive part of the x-axis (with the initial point of OA to be the origin). Then, the angle is determined by the second ray OB (see Fig. 4.11). Notice that any ray OB has the only point P of intersection with the unit circle, and vice-verse, every point P of the unit circle determines the unique ray OB which passes through this point. Therefore, we can use the points P of the unit circle to characterize the rays OB, which we will denote by OP. Since the only not fixed element that determines the angle is the ray OP, the corresponding angle will also be denoted by OP. We will systematically use the measure of angles in radians and will not specify this anymore. We will also identify an angle with its measure.

One more specificity of the angles in trigonometry is that they are directed, which means that it is important to know if the second ray OP is obtained by

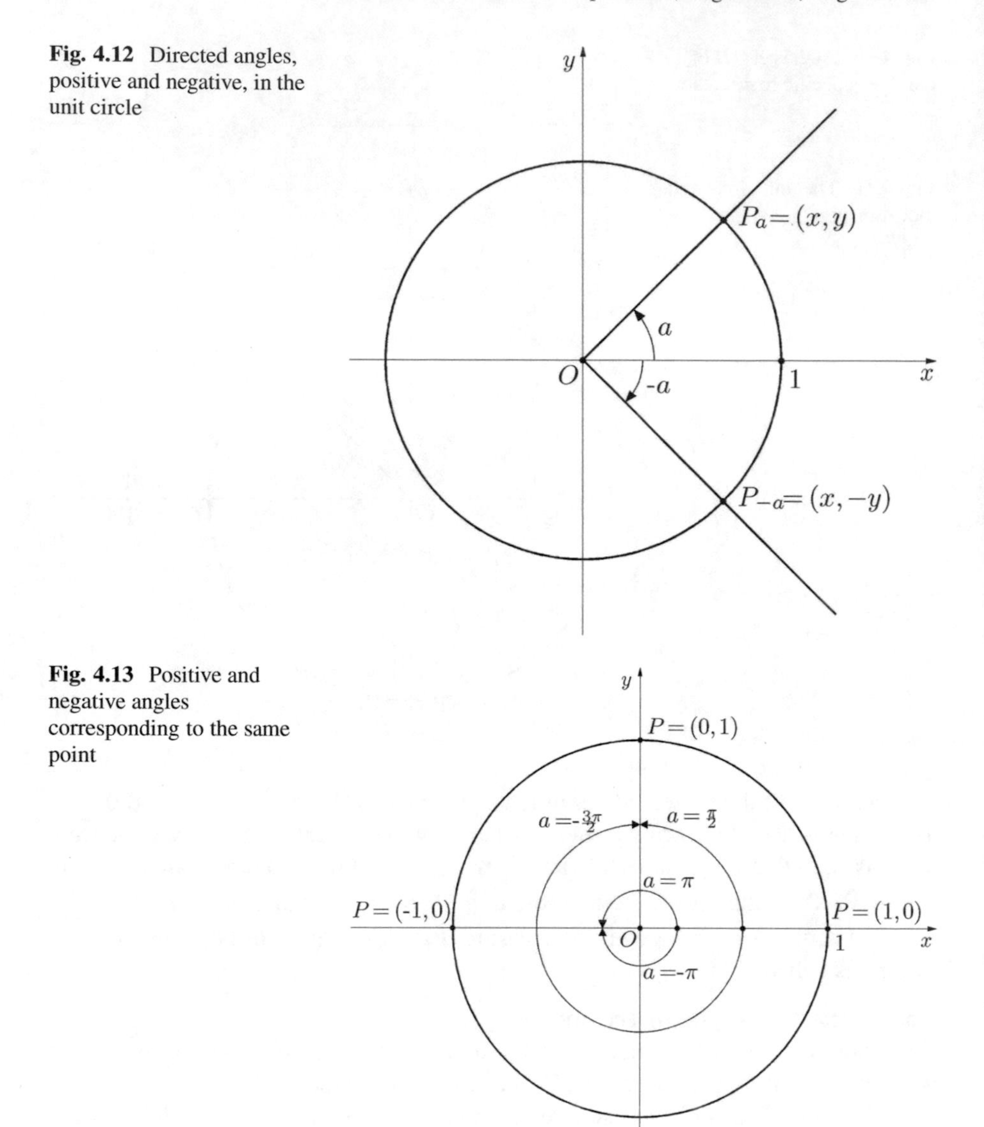

Fig. 4.12 Directed angles, positive and negative, in the unit circle

Fig. 4.13 Positive and negative angles corresponding to the same point

a counterclockwise or clockwise rotation of the initial ray Ox. According to the direction of rotation of the ray Ox, the measures of angles have positive or negative values: if a directed angle is obtained by a counterclockwise rotation of Ox, then it is a positive angle (having a positive value of its measure), otherwise it is a negative angle (see Fig. 4.12).

For example, the angle OP, $P = (-1, 0)$, which forms the x-axis, can correspond to the angle π or to the angle $-\pi$. The right angle OP, $P = (0, 1)$, whose second side coincides with the positive part of the y-axis, can be related to the angle $\frac{\pi}{2}$ or to the angle $-\frac{3\pi}{2}$ (see Fig. 4.13).

According to the specification of angles in trigonometry, an angle $a = OP$ determines uniquely the point P of the intersection of the ray OP with the unit circle, but the converse is not true. Indeed, adding any number of rotations (in any direction, positive or negative) about the origin to the given angle corresponding to the point P, we obtain another angle which corresponds to the same point P. For example, the point $P = (1, 0)$, that generates the ray $OP = Ox$, corresponds to the angle 0 and also to the angles 2π and -2π, etc. The point $P = (-1, 0)$, that generates the negative part of the x-axis, corresponds to the angle π and also to the angles $\pi + 2\pi$ and $\pi - 2\pi$, etc. The point $P = (0, 1)$ of the right angle $a = \frac{\pi}{2}$ (which generates the positive part of the y-axis), can be obtained also using the angles $\frac{\pi}{2} + 2\pi$ or $\frac{\pi}{2} - 2\pi$, etc.

Definition and Basic Properties of $\cos a$ and $\sin a$
Using the preliminary concepts introduced above, we can define now $\cos a$ and $\sin a$ for an arbitrary angle a.

Definition of $\cos a$ and $\sin a$ Given an angle a, which determines uniquely the point $P_a = (x_a, y_a)$ of the unit circle, the *cosine* of the angle a ($\cos a$) is the first coordinate of P_a and the *sine* of a ($\sin a$) is the second coordinate of P_a: $\cos a = x_a$, $\sin a = y_a$. (See illustration in Fig. 4.14.)

According to this definition, $\cos a$ and $\sin a$ are determined for any angle a and always have the values between -1 and 1, since the coordinates of the unit circle satisfy the relation $x_a^2 + y_a^2 = 1$. Moreover, from the last formula it follows the fundamental relation between $\cos a$ and $\sin a$ called the trigonometric identity: $\cos^2 a + \sin^2 a = 1$.

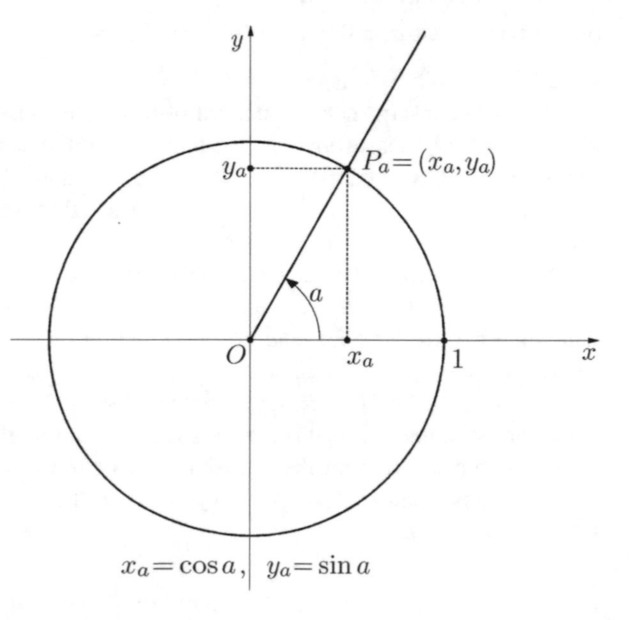

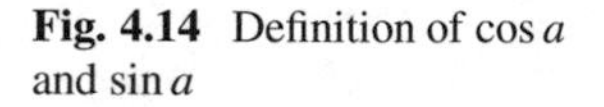
Fig. 4.14 Definition of $\cos a$ and $\sin a$

It follows also from the definition that $\cos(a+2k\pi) = \cos a$ and $\sin(a+2k\pi) = \sin a$, $\forall k \in \mathbb{Z}$. Indeed, these formulas are merely an analytic representation of the fact that the angles obtained from a given angle a by adding k complete rotations (in positive or negative direction) determine the same point P_a of the unit circle. This property is called 2π-periodicity.

The next direct consequence of the definition is a specification of the sign of $\cos a$ and $\sin a$. For the angle $a = 0$ we have the point $P_a = (1, 0)$, and consequently, $\cos a = 1$, $\sin a = 0$. For $a = \frac{\pi}{2}$ we get the point $P_a = (0, 1)$, and consequently, $\cos a = 0$, $\sin a = 1$. The angle $a = \pi$ determines the point $P_a = (-1, 0)$, which means that $\cos a = -1$, $\sin a = 0$. Finally, the angle $a = \frac{3\pi}{2}$ determines the point $P_a = (0, -1)$, which means that $\cos a = 0$, $\sin a = -1$. For the angles $a \in (0, \frac{\pi}{2})$, the corresponding points P_a lie in the first quadrant, and consequently, $\cos a > 0$, $\sin a > 0$. For the angles $a \in (\frac{\pi}{2}, \pi)$, the points P_a are located in the second quadrant where $\cos a < 0$, $\sin a > 0$. The angles $a \in (\pi, \frac{3\pi}{2})$ determine the points in the third quadrant where $\cos a < 0$, $\sin a < 0$. Finally, the angles $a \in (\frac{3\pi}{2}, 2\pi)$ determine the points in the fourth quadrant, and consequently, $\cos a > 0$, $\sin a < 0$.

To finalize the study of the sign, we use the 2π-periodicity and obtain the following results:

(1) $\cos a = 1$, $\sin a = 0$ if $a = 2k\pi$;
(2) $\cos a = 0$, $\sin a = 1$ if $a = \frac{\pi}{2} + 2k\pi$;
(3) $\cos a = -1$, $\sin a = 0$ if $a = \pi + 2k\pi$;
(4) $\cos a = 0$, $\sin a = -1$ if $a = \frac{3\pi}{2} + 2k\pi$;
(5) $\cos a > 0$, $\sin a > 0$ if $a \in (0, \frac{\pi}{2}) + 2k\pi$;
(6) $\cos a < 0$, $\sin a > 0$ if $a \in (\frac{\pi}{2}, \pi) + 2k\pi$;
(7) $\cos a < 0$, $\sin a < 0$ if $a \in (\pi, \frac{3\pi}{2}) + 2k\pi$;
(8) $\cos a > 0$, $\sin a < 0$ if $a \in (\frac{3\pi}{2}, 2\pi) + 2k\pi$.

Here k is an arbitrary integer.

One more consequence of the definition is the evenness of $\cos a$ and the oddness of $\sin a$. Indeed, measuring angles a and $-a$ (of the same absolute value, but with opposite signs), we find the points $P_a = (x_a, y_a)$ and $P_{-a} = (x_{-a}, y_{-a})$, which are symmetric with respect to the x-axis. This means that $\cos(-a) = x_{-a} = x_a = \cos a$ and $\sin(-a) = y_{-a} = -y_a = -\sin a$.

Other simple formulas can be deduced from the definition. For example, measuring angles a and $b = a + \pi$ we find the points $P_a = (x_a, y_a)$ and $P_b = (x_b, y_b)$, which are symmetric about the origin. Then, $\cos b = x_b = -x_a = -\cos a$ and $\sin b = y_b = -y_a = -\sin a$, that is, $\cos(a+\pi) = -\cos a$ and $\sin(a+\pi) = -\sin a$. If we use angles a and $b = \pi - a$, we get the points $P_a = (x_a, y_a)$ and $P_b = (x_b, y_b)$, which are symmetric about the y-axis (the second angle can be considered as the angle $-a$ measured from the negative part of the x-axis). Then, $\cos b = x_b = -x_a = -\cos a$ and $\sin b = y_b = y_a = \sin a$, that is, $\cos(\pi - a) = -\cos a$ and $\sin(\pi - a) = \sin a$.

Fig. 4.15 Geometric characterization of $\cos a$ and $\sin a$, *I* quadrant

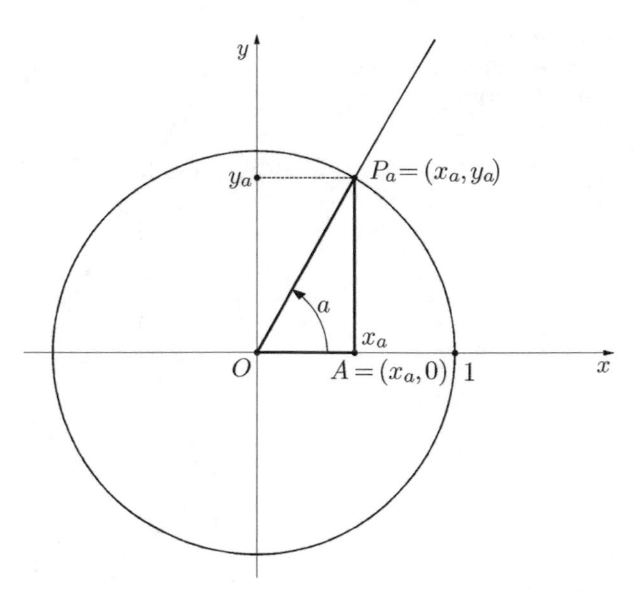

Geometric Characterization of $\cos a$ **and** $\sin a$

Let us consider the *geometric interpretation* of $\cos a$ and $\sin a$ separately in each quadrant of the Cartesian plane. For the angles $0 < a < \frac{\pi}{2}$, within the first quadrant, consider the right triangle OP_aA, where $A = (x_a, 0)$ is the projection of P_a on the x-axis. The side OA is called adjacent and AP_a opposite to the angle a. Notice that the length of the side OA coincides with the x-coordinate of P_a (since A is the projection of P_a on the positive part of the x-axis) and the length of the side AP_a is equal to the y-coordinate of P_a (since $d(A, P_a) = d(O, (0, y_a))$ and y_a is positive). (See illustration in Fig. 4.15.) Recalling the definition of $\cos a$ and $\sin a$, we conclude that $\cos a = x_a = d(O, A)$ and $\sin a = y_a = d(A, P_a)$. Noting additionally that the length of the hypotenuse OP_a is 1, we can write these formulas in the form $\cos a = \frac{d(O,A)}{d(O,P_a)}$ and $\sin a = \frac{d(A,P_a)}{d(O,P_a)}$, or in the words, $\cos a = \frac{adjacent\ side}{hypotenuse}$ and $\sin a = \frac{opposite\ side}{hypotenuse}$.

In what follows we will see that the formulas in other quadrants are not quite the same. Consider, for example, the next range of the angles $\frac{\pi}{2} < a < \pi$, located in the second quadrant. Construct the right triangle OP_aA, where $A = (x_a, 0)$ is the projection of P_a on the x-axis, and introduce the internal angle b in this triangle, supplementary to a: $b = \pi - a$, measured from the negative part of the x-axis clockwise. Naturally, the side OA is called adjacent and AP_a opposite to the angle b. Then, we have the following relationships between the length of the sides of OP_aA and coordinates of P_a: $d(O, A) = -x_a$ (since A is the projection of P_a on the negative part of the x-axis) and $d(A, P_a) = y_a$ (since the y-coordinate of the points in the second quadrant is positive). (See the illustration in Fig. 4.16.) Recalling the definition of $\cos a$ and $\sin a$, we obtain that $\cos a = x_a = -d(O, A)$ and $\sin a = y_a = d(A, P_a)$. Taking into account that $d(O, P_a) = 1$, we have $\cos a =$

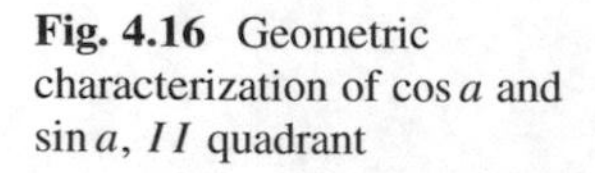

Fig. 4.16 Geometric characterization of $\cos a$ and $\sin a$, II quadrant

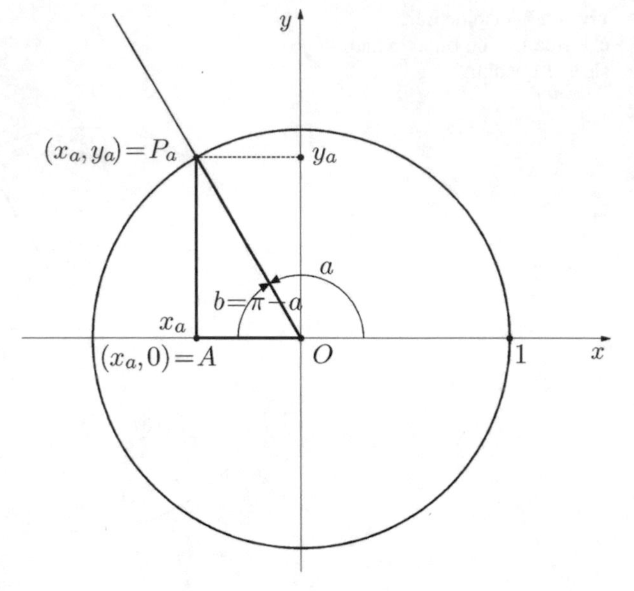

$\frac{-d(O,A)}{d(O,P_a)}$ and $\sin a = \frac{d(A,P_a)}{d(O,P_a)}$, or expressing in the words, $\cos a = \frac{-adjacent\ side}{hypotenuse}$ and $\sin a = \frac{opposite\ side}{hypotenuse}$. Comparing with the relations in the first quadrant, we see that the last two formulas involve the sides of the triangle based on the internal angle b (supplementary to the original angle a) and the formula for cosine has the opposite sine.

In a similar mode, it can be shown that for the angles $\pi < a < \frac{3\pi}{2}$ of the third quadrant the following formulas hold: $\cos a = \frac{-d(O,A)}{d(O,P_a)}$ and $\sin a = \frac{-d(A,P_a)}{d(O,P_a)}$, where the right triangle OP_aA, with $A = (x_a, 0)$, is related to the internal angle $b = a - \pi$, measured from the negative part of the x-axis counterclockwise. In the words, $\cos a = \frac{-adjacent\ side}{hypotenuse}$ and $\sin a = \frac{-opposite\ side}{hypotenuse}$. Analogously, for the angles $\frac{3\pi}{2} < a < 2\pi$ of the fourth quadrant we can derive the following relations: $\cos a = \frac{d(O,A)}{d(O,P_a)}$ and $\sin a = \frac{-d(A,P_a)}{d(O,P_a)}$, where the right triangle OP_aA, with $A = (x_a, 0)$, has the internal angle $b = 2\pi - a$ measured from the positive part of the x-axis clockwise. In the words, $\cos a = \frac{adjacent\ side}{hypotenuse}$ and $\sin a = \frac{-opposite\ side}{hypotenuse}$.

We leave the task to derive these last formulas to the reader.

Formula $\cos(a - b) = \cos a \cos b + \sin a \sin b$ **and the Consequent Results**

Let us demonstrate the formula for the cosine of the difference $\cos(a - b) = \cos a \cos b + \sin a \sin b$. Consider an angle a and the corresponding point P, and also an angle b and the corresponding point Q. Suppose, without loss of generality, that $a > b$ (otherwise we can simply interchange the notation of the angles and points). Consider also the angle $a - b$ with the corresponding point A. Since the angles COA ($C = (1, 0)$) and QOP have the same measure $a - b$, the corresponding

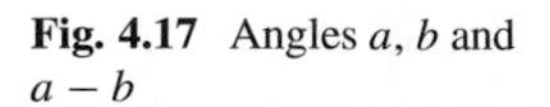

Fig. 4.17 Angles a, b and $a - b$

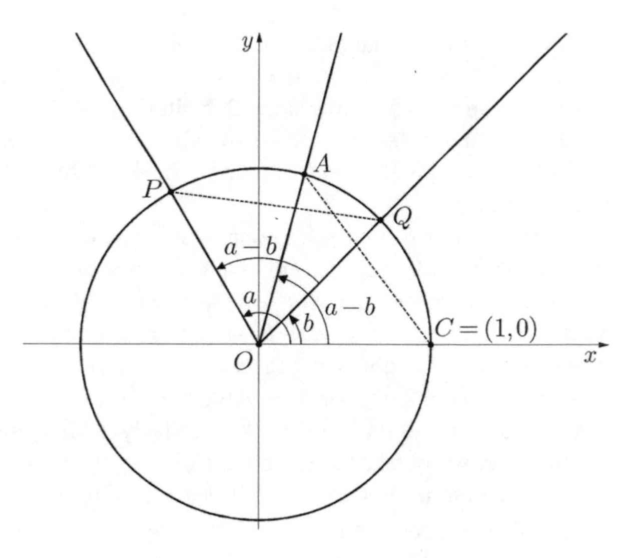

arcs have the same length, and consequently, the corresponding distances are equal: $d(C, A) = d(Q, P)$. (See illustration in Fig. 4.17.)

The first distance is found by the formula

$$d^2(C, A) = (x_A - x_C)^2 + (y_A - y_C)^2 = (\cos(a - b) - 1)^2 + (\sin(a - b) - 0)^2$$
$$= \cos^2(a - b) - 2\cos(a - b) + 1 + \sin^2(a - b) = 2 - 2\cos(a - b).$$

Similarly, the second distance is

$$d^2(Q, P) = (x_P - x_Q)^2 + (y_P - y_Q)^2 = (\cos a - \cos b)^2 + (\sin a - \sin b)^2$$
$$= \cos^2 a - 2\cos a \cos b + \cos^2 b + \sin^2 a - 2\sin a \sin b + \sin^2 b$$
$$= 2 - 2\cos a \cos b - 2\sin a \sin b.$$

Then, the equality between these two distances results in the formula $2 - 2\cos(a - b) = 2 - 2\cos a \cos b - 2\sin a \sin b$, or simplifying, $\cos(a - b) = \cos a \cos b + \sin a \sin b$.

Using the angle $-b$ instead of b in the last formula, we find the formula for the cosine of the sum: $\cos(a + b) = \cos(a - (-b)) = \cos a \cos(-b) + \sin a \sin(-b) = \cos a \cos b - \sin a \sin b$.

Setting $a = b$ in the last formula, we get $\cos 2a = \cos^2 a - \sin^2 a$ (the cosine of the double angle).

Specifying $a = \frac{\pi}{2}$ in the formula $\cos(a - b)$, we have $\cos(\frac{\pi}{2} - b) = \cos\frac{\pi}{2}\cos b + \sin\frac{\pi}{2}\sin b = \sin b$. Using in this formula $b = \frac{\pi}{2} - a$, we get $\cos a = \sin(\frac{\pi}{2} - a)$.

Applying these relations, we find the formula for the sine of the difference: $\sin(a-b) = \cos(\frac{\pi}{2} - (a-b)) = \cos((\frac{\pi}{2} - a) + b)) = \cos(\frac{\pi}{2} - a)\cos b - \sin(\frac{\pi}{2} - a)\sin b = \sin a \cos b - \cos a \sin b$. Substituting $-b$ for b in the last formula, we find the expression for the sine of the sum: $\sin(a+b) = \sin a \cos b + \cos a \sin b$.

Setting $a = b$ in the last formula, we have $\sin 2a = 2\sin a \cos a$ (the sine of the double angle).

From the four formulas for cosine and sine of the sum and difference, we can deduce the formulas for the sum and difference of cosine and sine. Adding the first two formulas, we obtain $\cos(a-b) + \cos(a+b) = 2\cos a \cos b$. Introducing the angles $c = a - b$ e $d = a + b$, we can rewrite this formula as $\cos c + \cos d = 2\cos\frac{c+d}{2}\cos\frac{c-d}{2}$. Subtracting the second formula from the first one, we get $\cos(a-b) - \cos(a+b) = 2\sin a \sin b$, or using the angles $c = a-b$ and $d = a+b$, we have $\cos c - \cos d = -2\sin\frac{c+d}{2}\sin\frac{c-d}{2}$. Similarly, adding and subtracting the formulas for the sine, we obtain the relations $\sin(a-b) + \sin(a+b) = 2\sin a \cos b$ and $\sin(a-b) - \sin(a+b) = -2\sin b \cos a$ or $\sin c + \sin d = 2\sin\frac{c+d}{2}\cos\frac{c-d}{2}$ and $\sin c - \sin d = 2\sin\frac{c-d}{2}\cos\frac{c+d}{2}$.

Definition and Basic Properties of $\tan a$ and $\cot a$

Definition of $\tan a$ and $\cot a$ Given an angle a, which determines uniquely the point $P_a = (x_a, y_a)$ of the unit circle, the *tangent* of the angle a ($\tan a$) is the quotient $\frac{y_a}{x_a}$ and the *cotangent* ($\cot a$) is the inverted quotient $\frac{x_a}{y_a}$, that is: $\tan a = \frac{y_a}{x_a} = \frac{\sin a}{\cos a}$ and $\cot a = \frac{x_a}{y_a} = \frac{\cos a}{\sin a}$.

According to this definition, $\tan a$ is defined for all angles $a \neq \frac{\pi}{2} + k\pi, \forall k \in \mathbb{Z}$ for which $\cos a \neq 0$, and $\cot a$ is defined for all angles $a \neq k\pi, \forall k \in \mathbb{Z}$ for which $\sin a \neq 0$. Besides, it follows directly from the definition that $\tan a \cot a = 1$ for any angle that can be used both for $\tan a$ and $\cot a$, that is, for $a \neq \frac{k\pi}{2}, \forall k \in \mathbb{Z}$. One more formula of the relationship between $\tan a$ and $\cot a$ can be derived using the properties of $\cos a$ and $\sin a$: $\tan(\frac{\pi}{2} - a) = \frac{\sin(\frac{\pi}{2}-a)}{\cos(\frac{\pi}{2}-a)} = \frac{\cos a}{\sin a} = \cot a, a \neq k\pi, \forall k \in \mathbb{Z}$, and analogously $\cot(\frac{\pi}{2} - a) = \tan a, a \neq \frac{\pi}{2} + k\pi, \forall k \in \mathbb{Z}$.

Recalling the properties of $\cos a$ and $\sin a$, we can show that $\tan a$ and $\cot a$ are π-periodic. Indeed, $\tan(a+\pi) = \frac{\sin(a+\pi)}{\cos(a+\pi)} = \frac{-\sin a}{-\cos a} = \tan a$ and $\cot(a+\pi) = \frac{\cos(a+\pi)}{\sin(a+\pi)} = \frac{-\cos a}{-\sin a} = \cot a$.

By the definition, $\tan a$ vanishes at the same angles where $\sin a$ is zero, that is, at $a = k\pi, \forall k \in \mathbb{Z}$, and $\cot a$ vanishes at the same angles where $\cos a$ is zero, that is, at $a = \frac{\pi}{2} + k\pi, \forall k \in \mathbb{Z}$. The values of $\tan a$ and $\cot a$ are positive when $\cos a$ and $\sin a$ have the same sign, which happens for the angles $a \in (0, \frac{\pi}{2}) + k\pi, \forall k \in \mathbb{Z}$. If $\cos a$ and $\sin a$ have opposite signs, which happens for $a \in (-\frac{\pi}{2}, 0) + k\pi, \forall k \in \mathbb{Z}$, then $\tan a$ and $\cot a$ are negative.

The oddness of $\tan a$ and $\cot a$ also follows directly from the definition: $\tan(-a) = \frac{\sin(-a)}{\cos(-a)} = \frac{-\sin a}{\cos a} = -\tan a$ and $\cot(-a) = \frac{\cos(-a)}{\sin(-a)} = \frac{\cos a}{-\sin a} = -\cot a$.

Two more simple formulas follows from the definition: $\tan(\pi - a) = \frac{\sin(\pi-a)}{\cos(\pi-a)} = \frac{\sin a}{-\cos a} = -\tan a$ and $\cot(\pi - a) = \frac{\cos(\pi-a)}{\sin(\pi-a)} = \frac{-\cos a}{\sin a} = -\cot a$.

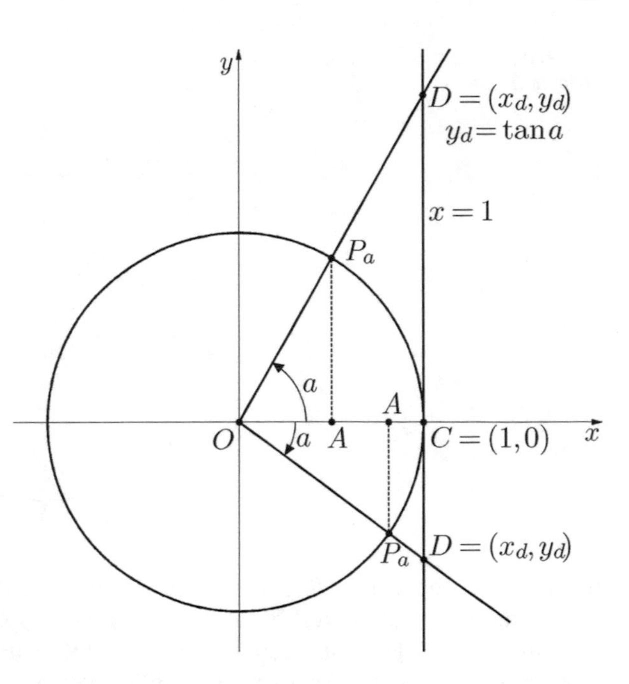

Fig. 4.18 Geometric characterization of $\tan a$

Geometric Characterization of $\tan a$ **and** $\cot a$

The *geometric interpretation* of the definition of $\tan a$ can be given as follows. First, taking into account the π-periodicity of $\tan a$, we can restrict the variation of angle a to the interval $(-\frac{\pi}{2}, \frac{\pi}{2})$ (called the principal interval of tangent). Now we construct an auxiliary line (called the tangent line), which passes through the point $C = (1, 0)$ and is parallel to the y-axis, and associate with each point P_a the only point $D = (x_d, y_d)$ obtained at the intersection of the ray OP_a with the auxiliary line (see the illustration in Fig. 4.18). It follows from this construction that the triangles OP_aA and ODC are similar, because they have the same angles. Considering the subinterval $(0, \frac{\pi}{2})$, we have $\tan a = \frac{y_a}{x_a} = \frac{d(A,P_a)}{d(O,A)} = \frac{d(C,D)}{d(O,C)} = d(C, D) = y_d$. On the second subinterval $(-\frac{\pi}{2}, 0)$ we have $\tan a = \frac{y_a}{x_a} = \frac{-d(A,P_a)}{d(O,A)} = \frac{-d(C,D)}{d(O,C)} = -d(C, D) = y_d$. Finally, for the angle $a = 0$ we get $\tan 0 = 0 = d(C, D) = y_d$. Joining these results, we arrive to the formula $\tan a = y_d$ for any angle $a \in (-\frac{\pi}{2}, \frac{\pi}{2})$. (See the illustration in Fig. 4.18.)

This geometric interpretation allows us to establish the relationship between $\tan a$ and the distance from $C = (1, 0)$ to the point D of the tangent line and it also reveals clearly the range of the values of the tangent. Indeed, since each point P_a is associated with the only point D and vice-verse, to any point D of the auxiliary line corresponds the unique angle $a \in (-\frac{\pi}{2}, \frac{\pi}{2})$ such that $\tan a = y_d$. Taking into account that the coordinate y_d can take any real value, we conclude that $\tan a$ takes all the real values when $a \in (-\frac{\pi}{2}, \frac{\pi}{2})$.

In a similar way, we can provide the *geometric interpretation* of $\cot a$ by introducing the cotangent line, which passes through the point $B = (0, 1)$ and

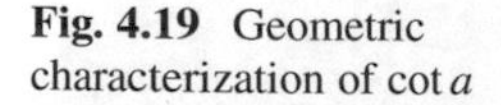

Fig. 4.19 Geometric characterization of $\cot a$

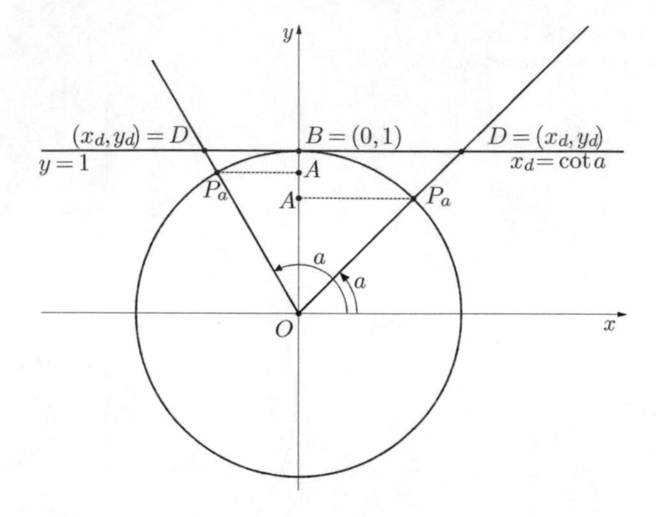

is parallel to the x-axis. Restricting the variation of the angle a to the interval $(0, \pi)$ (the principal interval of cotangent), we can associate with each point P_a the only point D of the intersection of the ray OP_a with the cotangent line (see the illustration in Fig. 4.19). Then, due to similarity of the triangles OP_aA and ODB, we have $\cot a = d(B, D) = x_d$ when $a \in (0, \frac{\pi}{2})$ and $\cot a = -d(B, D) = x_d$ when $a \in (\frac{\pi}{2}, \pi)$. This shows that $\cot a$ takes all the real values when $a \in (0, \pi)$. (See the illustration in Fig. 4.19.)

2.2 *Function* sin *x*

Definition of $\sin x$ The function $f(x) = \sin x$ is a rule that associates with each $x \in \mathbb{R}$ the only number $f(x) = \sin x$.

Study of $y = \sin x$

1. Domain and range. By the definition of the function, the domain is $X = \mathbb{R}$, since the operation $\sin x$ is defined for any real x. By the definition of the sine, the values of $\sin x$ belong to the interval $[-1, 1]$, and when x varies on the interval $[-\frac{\pi}{2}, \frac{\pi}{2}]$ the values of $\sin x$ take all the numbers in $[-1, 1]$. Therefore, the range is $Y = [-1, 1]$.
2. Boundedness. It follows from the specification of the range that the function is bounded. The lower bound can be $m = -1$ and the upper $M = 1$, which are the minimum and maximum values of the function, respectively.
3. Parity. It follows from the definition of sine that $\sin(-x) = -\sin x, \forall x \in \mathbb{R}$, that is, $y = \sin x$ is an odd function.

4. Periodicity. It follows from the definition of sine that $\sin(x + 2\pi) = \sin x$, $\forall x \in \mathbb{R}$, that is, $y = \sin x$ is a 2π-periodic function. Moreover, 2π is its fundamental (minimum) period. To see this, it is sufficient to note that the function takes its maximum value 1 only at the points $x_k = \frac{\pi}{2} + 2k\pi$, $\forall k \in \mathbb{Z}$, and the minimum distance between two points of this set is 2π.
5. Intersection and sign. Intersection with the y-axis occurs at the point $(0, 0)$, which is also the point of intersection with the x-axis. Besides, all the points $x_k = k\pi$, $\forall k \in \mathbb{Z}$ are the points of intersection with the x-axis. On the intervals $(0, \pi) + 2k\pi$, $\forall k \in \mathbb{Z}$ the function has positive values, while on the intervals $(\pi, 2\pi) + 2k\pi$, $\forall k \in \mathbb{Z}$ its values are negative.
6. Monotonicity and extrema.

Recalling that $\sin\alpha$ was defined as the y-coordinate of the point P of intersection of the second ray of the angle α with the unit circle, we see that increase of α from $-\frac{\pi}{2}$ (when this ray coincides with the negative part of the y-axis) to $\frac{\pi}{2}$ (when this ray attains the positive part of the y-axis) causes a "climb" of the corresponding point P from the position $P_1 = (0, -1)$ until $P_2 = (0, 1)$, that is, the y-coordinate of this point increases from -1 to 1. In terms of the properties of the function, this means that $\sin x$ is increasing on the interval $[-\frac{\pi}{2}, \frac{\pi}{2}]$ with the values growing from -1 until 1. On the next interval of the variation of the angle α, from $\frac{\pi}{2}$ to $\frac{3\pi}{2}$ (for the latter, the corresponding ray coincides again with the negative part of the y-axis), increase of the angle leads to a "descent" of the point P from the position $P_2 = (0, 1)$ to $P_1 = (0, -1)$, that is, the y-coordinate of this point decreases from 1 to -1. For the properties of the function, this means that $\sin x$ is decreasing on the interval $[\frac{\pi}{2}, \frac{3\pi}{2}]$, diminishing its values from 1 to -1. The illustration can be found in Fig. 4.20.

Alternatively, we can derive the same result considering the difference between the values of the function. For any two points we have $\sin x_2 - \sin x_1 =$

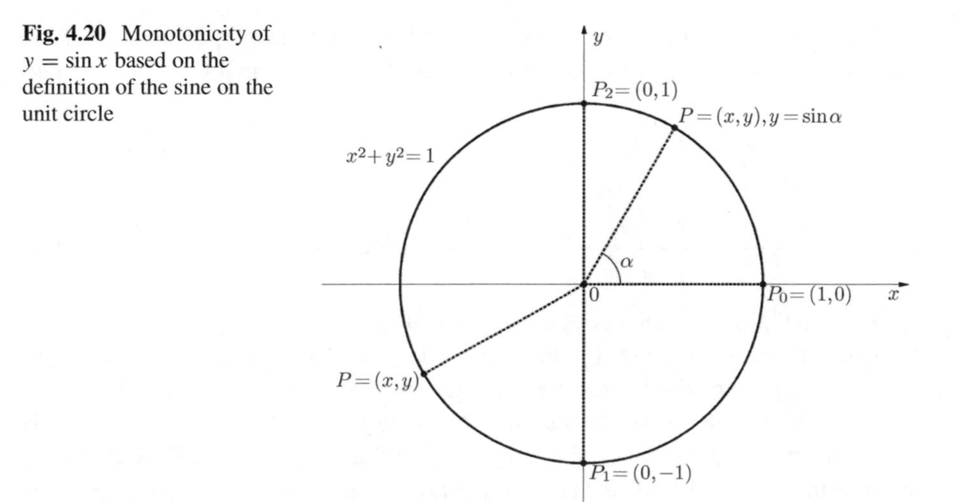

Fig. 4.20 Monotonicity of $y = \sin x$ based on the definition of the sine on the unit circle

$2\sin\frac{x_2-x_1}{2}\cos\frac{x_2+x_1}{2}$. Using this trigonometric formula, let us show first that the function $\sin x$ increases on the interval $[-\frac{\pi}{2},\frac{\pi}{2}]$. To verify this, take two points satisfying the inequality $-\frac{\pi}{2}\le x_1<x_2\le\frac{\pi}{2}$ and notice that for these points we have, on the one side, $0<\frac{x_2-x_1}{2}\le\frac{\pi}{2}$, which implies that $\sin\frac{x_2-x_1}{2}>0$, and on the other side, $-\frac{\pi}{2}<\frac{x_2+x_1}{2}<\frac{\pi}{2}$, which ensures that $\cos\frac{x_2+x_1}{2}>0$. Therefore, $\sin x_2-\sin x_1>0$, which means that the function $\sin x$ increases on $[-\frac{\pi}{2},\frac{\pi}{2}]$. Using the same formula we can also show that $\sin x$ decreases on the interval $[\frac{\pi}{2},\frac{3\pi}{2}]$. Indeed, taking two points such that $\frac{\pi}{2}\le x_1<x_2\le\frac{3\pi}{2}$ we get $0<\frac{x_2-x_1}{2}\le\frac{\pi}{2}$, which implies that $\sin\frac{x_2-x_1}{2}>0$, and at the same time, $\frac{\pi}{2}<\frac{x_2+x_1}{2}<\frac{3\pi}{2}$, which ensures that $\cos\frac{x_2+x_1}{2}<0$. Then, $\sin x_2-\sin x_1<0$, which means that the function $\sin x$ decreases on $[\frac{\pi}{2},\frac{3\pi}{2}]$.

From these results and from the 2π-periodicity it follows that the function $\sin x$ is not monotonic on its domain, but it is increasing on each interval $[-\frac{\pi}{2},\frac{\pi}{2}]+2k\pi$ and decreasing on each interval $[\frac{\pi}{2},\frac{3\pi}{2}]+2k\pi$, $\forall k\in\mathbb{Z}$.

From the study of monotonicity, it follows that every point of the intervals $(-\frac{\pi}{2},\frac{\pi}{2})+2k\pi$ and $(\frac{\pi}{2},\frac{3\pi}{2})+2k\pi$, $\forall k\in\mathbb{Z}$ is monotonic, and consequently, none of these points can be an extremum, local or global, strict or non-strict. The remaining (non-monotonic) points are $x_k=\frac{\pi}{2}+2k\pi$, $\forall k\in\mathbb{Z}$ and $x_n=\frac{3\pi}{2}+2n\pi$, $\forall n\in\mathbb{Z}$. Since $\sin x_k=1$, $\forall k\in\mathbb{Z}$, which is the maximum value that the sine can take, all these points are (non-strict) global maxima over the entire domain. Additionally, on the intervals $(-\frac{\pi}{2},\frac{\pi}{2})$ and $(\frac{\pi}{2},\frac{3\pi}{2})$ we have $\sin x<1$ which shows that $\frac{\pi}{2}$ is a strict global maximum on the interval $(-\frac{\pi}{2},\frac{3\pi}{2})$ (and also on $(-\pi,2\pi)$ and on $(-\frac{3\pi}{2},\frac{5\pi}{2})$). Therefore, the point $\frac{\pi}{2}$ is a strict local maximum, and, due to the 2π-periodicity, the same is true for all the points x_k. Similarly, each point x_n is a (non-strict) global minimum over the entire domain, a strict global minimum on the interval $(-\frac{\pi}{2},\frac{7\pi}{2})+2n\pi$ and a strict local minimum.

7. Concavity and inflection.

For two arbitrary points $x_1<x_2\in\mathbb{R}$ we have the following evaluation of the difference between the double values of the secant and the function at the midpoint:

$$2D_2=f(x_1)+f(x_2)-2f\left(\frac{x_1+x_2}{2}\right)=\sin x_1+\sin x_2-2\sin\frac{x_1+x_2}{2}$$

$$=2\sin\frac{x_1+x_2}{2}\cos\frac{x_1-x_2}{2}-2\sin\frac{x_1+x_2}{2}=2\sin\frac{x_1+x_2}{2}\left(\cos\frac{x_1-x_2}{2}-1\right).$$

The second factor is always non-positive, which means that the sign of this expression depends only on the first factor. Let us verify for what angles x_1 and x_2 this factor is positive. Recalling that the sine has positive values on $(0,\pi)$, we have to find the interval of the variation of x_1,x_2 such that $0<\frac{x_1+x_2}{2}<\pi$. It is clear that this happens if $0\le x_1<x_2\le\pi$. In this case, $0<\frac{x_1+x_2}{2}<\pi$, which ensures that $\sin\frac{x_1+x_2}{2}>0$, and at the same time, $0<\frac{x_2-x_1}{2}\le\frac{\pi}{2}$, which guarantees that the second factor is negative. Therefore, $D_2 < 0$ and concavity is downward on the interval $[0, \pi]$.

In a similar way, we justify the choice of the next interval of the variation of the angles $\pi \leq x_1 < x_2 \leq 2\pi$, which implies that $\sin \frac{x_1+x_2}{2} < 0$ since $\pi < \frac{x_1+x_2}{2} < 2\pi$. At the same time, $0 < \frac{x_2-x_1}{2} \leq \frac{\pi}{2}$, which implies that the second factor is negative. Therefore, $D_2 > 0$, which means that the function is concave upward on the interval $[\pi, 2\pi]$.

Applying the 2π-periodicity, we conclude that $\sin x$ is concave downward on each interval $[0, \pi] + 2k\pi, \forall k \in \mathbb{Z}$ and concave upward on each interval $[\pi, 2\pi] + 2n\pi, \forall n \in \mathbb{Z}$.

Since the function change the type of concavity at the points $x_k = k\pi, \forall k \in \mathbb{Z}$, all these points are inflection ones.

8. The function $f(x) = \sin x : X = \mathbb{R} \to Y = [-1, 1]$ is surjective, but not injective. Notice that the function has repeated values on each interval where it is not monotonic. Therefore, to obtain an injective function, we have to choose an interval of monotonicity of $\sin x$. The traditional choice is the interval $[-\frac{\pi}{2}, \frac{\pi}{2}]$ of increase. With this restriction of the domain, the function $f(x) = \sin x : X = [-\frac{\pi}{2}, \frac{\pi}{2}] \to Y = [-1, 1]$ is bijective, and consequently, invertible. Its inverse is called arcsine: $f^{-1}(x) = \arcsin x : [-1, 1] \to [-\frac{\pi}{2}, \frac{\pi}{2}]$. This inverse will be analyzed in one of the next sections.
9. Based on the investigated properties and marking some important points, we can represent the graph of $f(x) = \sin x$ as shown in Fig. 4.21.

Complimentary Properties

The function $f(x) = \sin x$ keeps the values between -1 and 1, which implies that it does not have vertical asymptotes. When x goes to $+\infty$, the function does not approach any specific value, which means that there is no horizontal asymptote. Indeed, if x is sent to $+\infty$ by the path $x_k = k\pi, k \in \mathbb{N}$, the function keeps all its values equal to 0, while along the path $x_n = \frac{\pi}{2} + 2n\pi, n \in \mathbb{N}$, all its values are equal to 1. The same happens when x goes to $-\infty$. Hence, the function $f(x) = \sin x$ has no asymptote, either vertical or horizontal.

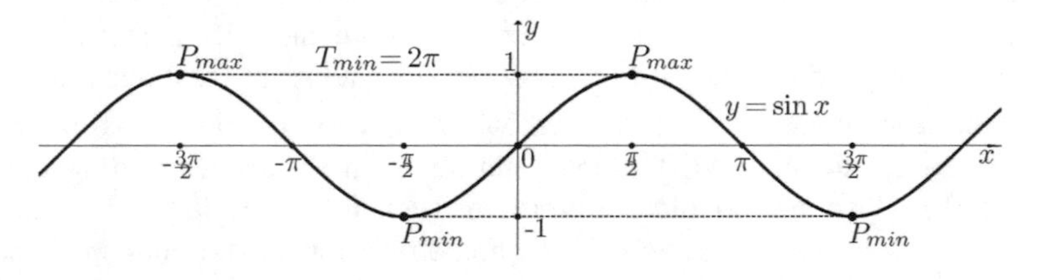

Fig. 4.21 The graph of $y = \sin x$

2.3 *Function* cos *x*

Definition of $\cos x$ The function $f(x) = \cos x$ is a rule that associates with each $x \in \mathbb{R}$ the only number $f(x) = \cos x$.

Study of $y = \cos x$

The properties of $f(x) = \cos x$ can be derived in the same manner as the properties of $\sin x$ in the previous study. Alternatively, all the properties of the cosine function can be deduced from the corresponding properties of the sine function using the trigonometric formula $\cos x = \sin(x + \frac{\pi}{2})$, which means geometrically that the graph of $\cos x$ can be obtained by horizontal translation of the graph of $\sin x$ by $\frac{\pi}{2}$ units to the left. To diversify the methods of investigation, we apply below this relationship between cosine and sine.

1. Domain and range. The domain of $\cos x$ is $X = \mathbb{R}$ and the range is $Y = [-1, 1]$.
2. Boundedness. The function is bounded with the lower bound $m = -1$ and the upper bound $M = 1$, which are the minimum and maximum values of the function, respectively.
3. Parity. The function is even since $\cos(-x) = \sin(\frac{\pi}{2} - x) = \sin\frac{\pi}{2}\cos x - \cos\frac{\pi}{2}\sin x = 1 \cdot \cos x + 0 = \cos x$, $\forall x \in \mathbb{R}$.
4. Periodicity. The function $\cos x$ is periodic with the fundamental period 2π: $\cos(x + 2\pi) = \sin(x + \frac{\pi}{2} + 2\pi) = \sin(x + \frac{\pi}{2}) = \cos x$, $\forall x \in \mathbb{R}$.
5. Intersection and sign. Intersection with the y-axis occurs at the point $(0, 1)$. The points of intersection with the x-axis are $x_k = \frac{\pi}{2} + k\pi$, $\forall k \in \mathbb{Z}$. On the intervals $(-\frac{\pi}{2}, \frac{\pi}{2}) + 2k\pi$, $\forall k \in \mathbb{Z}$ the function has positive values, while on the intervals $(\frac{\pi}{2}, \frac{3\pi}{2}) + 2k\pi$, $\forall k \in \mathbb{Z}$ it takes negative values.
6. Monotonicity and extrema. Recalling that $\sin x$ is increasing on each interval $[-\frac{\pi}{2}, \frac{\pi}{2}] + 2k\pi$ and decreasing on each interval $[\frac{\pi}{2}, \frac{3\pi}{2}] + 2k\pi$, $\forall k \in \mathbb{Z}$, we conclude that $\cos x$ is increasing on each interval $[-\pi, 0] + 2k\pi$ and decreasing on each interval $[0, \pi] + 2k\pi$, $\forall k \in \mathbb{Z}$.

   Recall also that for the function $\sin x$ each of the points $x_k = \frac{\pi}{2} + 2k\pi$, $\forall k \in \mathbb{Z}$ is a non-strict global maximum on the entire domain, strict global maximum on the corresponding interval $(-\frac{3\pi}{2}, \frac{5\pi}{2}) + 2k\pi$ and strict local maximum; and also that each point $x_n = \frac{3\pi}{2} + 2n\pi$, $\forall n \in \mathbb{Z}$ is a non-strict global minimum on the entire domain, strict global minimum on the corresponding interval $(-\frac{\pi}{2}, \frac{7\pi}{2}) + 2n\pi$ and strict local minimum. Then, it follows that for the function $\cos x$ each point $x_k = 2k\pi$, $\forall k \in \mathbb{Z}$ is a non-strict global maximum on the entire domain, strict global maximum on the corresponding interval $(-2\pi, 2\pi) + 2k\pi$ and strict local maximum; and also that each point $x_n = \pi + 2n\pi$, $\forall n \in \mathbb{Z}$ is a non-strict global minimum on the entire domain, strict global minimum on the corresponding interval $(-\pi, 3\pi) + 2n\pi$ and strict local minimum.

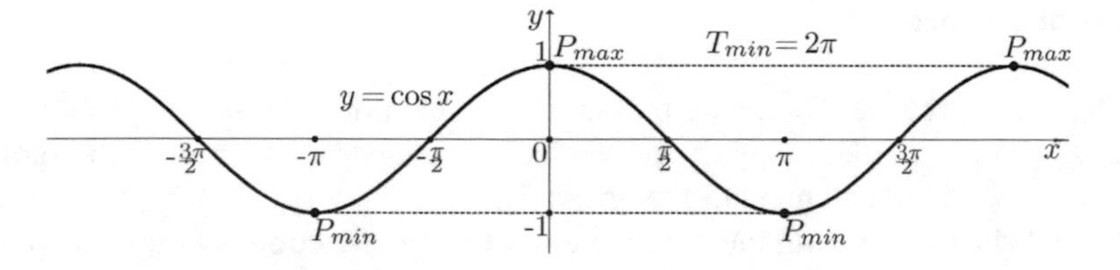

Fig. 4.22 The graph of $y = \cos x$

7. Concavity and inflection. Recalling that $\sin x$ is concave downward on each interval $[0, \pi]+2k\pi, \forall k \in \mathbb{Z}$ and upward on each interval $[\pi, 2\pi]+2n\pi, \forall n \in \mathbb{Z}$, we conclude that $\cos x$ is concave downward on each interval $[-\frac{\pi}{2}, \frac{\pi}{2}] + 2k\pi$, $\forall k \in \mathbb{Z}$ and upward on each interval $[\frac{\pi}{2}, \frac{3\pi}{2}] + 2n\pi$, $\forall n \in \mathbb{Z}$. Correspondingly, the points $x_k = \frac{\pi}{2} + k\pi$, $\forall k \in \mathbb{Z}$ are inflection points of $\cos x$.
8. The function $f(x) = \cos x : X = \mathbb{R} \to Y = [-1, 1]$ is surjective, but not injective. Reducing its domain to the interval $[0, \pi]$, we obtain bijective function $f(x) = \cos x : X = [0, \pi] \to Y = [-1, 1]$, whose inverse is called arccosine: $f^{-1}(x) = \arccos x : [-1, 1] \to [0, \pi]$. This inverse will be analyzed in one of the next sections.
9. Shifting the graph of $\sin x$ by $\frac{\pi}{2}$ units to the left, we obtain the graph of $f(x) = \cos x$ as shown in Fig. 4.22.

Notice once more that all these properties of cosine can be derived independently from the properties of sine, using the same method of study as for the sine function. This approach is left to the reader.

Complimentary Properties

The function $f(x) = \cos x$ keeps all its values between -1 and 1, which shows that it has no vertical asymptote. When x goes to $+\infty$, the function does not approach any specific value, which means that it does not have a horizontal asymptote. It can be proved using similar paths as for the sine function: along the path $x_n = \frac{\pi}{2} + n\pi$, $n \in \mathbb{N}$ the function has all the values equal to 0, while along the path $x_k = 2k\pi$, $k \in \mathbb{N}$ the function takes only the value 1. The same happens when x goes to $-\infty$. Hence, the function $f(x) = \cos x$ has no asymptote, either vertical or horizontal.

2.4 *Function* arcsin x

Definition of the Function $\arcsin x$ The function $y = \arcsin x$ is usually defined as the inverse to the function $f(x) = \sin x : [-\frac{\pi}{2}, \frac{\pi}{2}] \to [-1, 1]$.

Study of $y = \arcsin x$

1. Domain and range. By the definition, the domain of the function $y = \arcsin x$ is $X = [-1, 1]$ (the range of the original function $y = \sin x$) and the range is $Y = [-\frac{\pi}{2}, \frac{\pi}{2}]$ (the domain of the original).
2. Boundedness. Knowing the range, we see that the function is bounded with the lower bound $m = -\frac{\pi}{2}$ and the upper bound $M = \frac{\pi}{2}$, which are the minimum and maximum values of the function, respectively.
3. Parity. The function $\arcsin x$ is odd, since the original function $\sin x$ is odd (recall the Theorem on the oddness of the inverse in Sect. 7.4, Chap. 2).
4. Periodicity. The function is not periodic since its domain is bounded. (Another reason why the function is not periodic is that it is the inverse function, and consequently, it cannot have repeated values.)
5. Intersection and sign. The graph crosses the y- and x-axes at the only point $(0, 0)$. Since $y = \sin x$ transforms $[-\frac{\pi}{2}, 0)$ into $[-1, 0)$ and $(0, \frac{\pi}{2}]$ into $(0, 1]$, the inverse $y = \arcsin x$ has negative values on $[-1, 0)$ and positive on $(0, 1]$.
6. Monotonicity and extrema. Since $y = \sin x$ increases on its domain $[-\frac{\pi}{2}, \frac{\pi}{2}]$, its inverse $y = \arcsin x$ increases on the domain $[-1, 1]$ (recall the Theorem on monotonicity of the inverse in Sect. 12.4, Chap. 2). Therefore, the point $x_1 = -1$ is a strict global minimum of $\arcsin x$ where $\arcsin(-1) = -\frac{\pi}{2}$, and the point $x_2 = 1$ is a strict global maximum where $\arcsin 1 = \frac{\pi}{2}$. Notice that x_1 and x_2 are not local extrema, because there is no neighborhood of these points contained in the domain of the function.

   Any point different from x_1 and x_2 is increasing and interior in the interval $[-1, 1]$. For this reason, it cannot be an extremum of any type.
7. Concavity and inflection. Applying the Theorem about concavity of the inverse (Sect. 12.4 in Chap. 2), we obtain the following results. On the subinterval $[-\frac{\pi}{2}, 0]$ the function $y = \sin x$ is increasing and concave upward, which implies that $y = \arcsin x$ is concave downward on the subinterval $[-1, 0]$. On the subinterval $[0, \frac{\pi}{2}]$, the function $y = \sin x$ is also increasing, but concave downward, which results in upward concavity of the function $y = \arcsin x$ on the subinterval $[0, 1]$. Consequently, $x = 0$ is an inflection point of $\arcsin x$.
8. The function $f(x) = \arcsin x : X = [-1, 1] \to Y = [-\frac{\pi}{2}, \frac{\pi}{2}]$ is bijective (like any inverse function) and its inverse is $f(x) = \sin x : [-\frac{\pi}{2}, \frac{\pi}{2}] \to [-1, 1]$.
9. The graph of $y = \arcsin x$ can be obtained applying the Theorem about the graph of the inverse (see Sect. 12.5 in Chap. 2) that is illustrated in Fig. 4.23.

Alternatively, the same graph can be constructed using the investigated properties of the function $y = \arcsin x$.

Complimentary Properties

The values of the function $f(x) = \arcsin x$ lie between $-\frac{\pi}{2}$ and $\frac{\pi}{2}$, which ensures that the function does not have a vertical asymptote. At the same time, the domain of the function is the bounded interval $X = [-1, 1]$, and consequently, x cannot approach $+\infty$ or $-\infty$, which means that $f(x) = \arcsin x$ does not have a horizontal asymptote either.

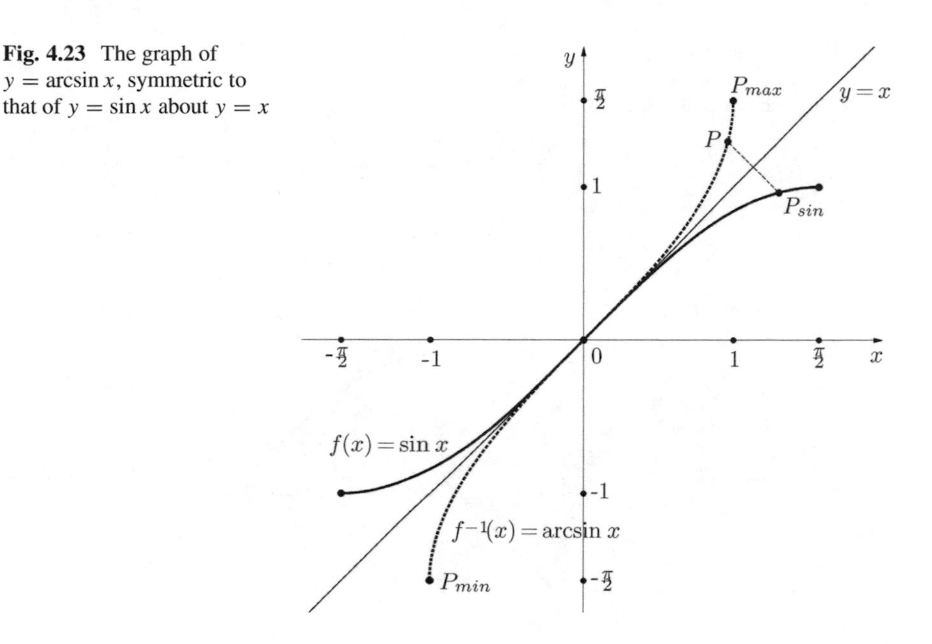

Fig. 4.23 The graph of $y = \arcsin x$, symmetric to that of $y = \sin x$ about $y = x$

2.5 *Function* arccos *x*

Definition of the Function $\arccos x$ The function $y = \arccos x$ is usually defined as the inverse to the function $f(x) = \cos x : [0, \pi] \to [-1, 1]$.

Study of $y = \arccos x$

The study of this function can be performed in the same way as for $y = \arcsin x$, using as the basis the properties of the original function $\cos x : [0, \pi] \to [-1, 1]$ and applying the theorems about the properties of the inverses (Sect. 12.4 of Chap. 2). Hence, we leave this analytic study to the reader and represent below only the list of the properties of the function $y = \arccos x$ and its graph shown in Fig. 4.24.

1. The domain is $X = [-1, 1]$ and the range is $Y = [0, \pi]$.
2. The function is bounded with the lower bound $m = 0$ and the upper bound $M = \pi$.

3–4. The function is not even or odd and it is not periodic.

5. The graph crosses the y-axis at the point $(0, \frac{\pi}{2})$ and the x-axis at the point $(1, 0)$. All the values of the function are positive, except at the point 1 where it vanishes.
6. The function is decreasing over the entire domain. It has the strict global maximum at $x = -1$ and the strict global minimum at $x = 1$. There is no local extrema.

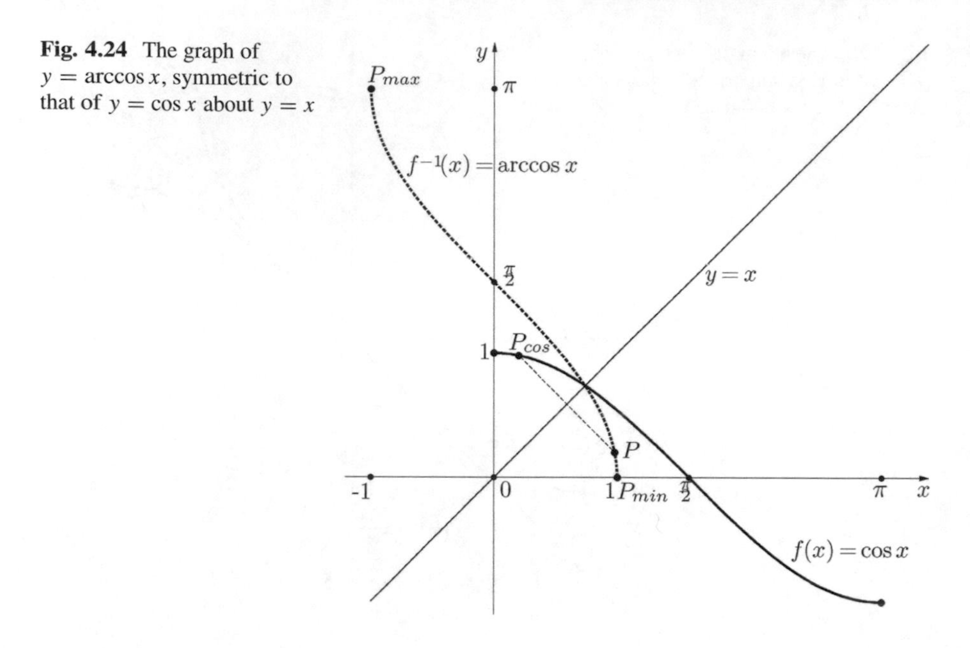

Fig. 4.24 The graph of $y = \arccos x$, symmetric to that of $y = \cos x$ about $y = x$

7. The function is concave upward on $[-1, 0]$ and downward on $[0, 1]$, with the point $x = 0$ being an inflection.
8. The function $f(x) = \arccos x : X = [-1, 1] \to Y = [0, \pi]$ is bijective and its inverse is $f(x) = \cos x : [0, \pi] \to [-1, 1]$.
9. The graph of $f(x) = \arccos x$ has no asymptote.

2.6 *Function* tan *x*

Definition of $\tan x$ The function $f(x) = \tan x$ is defined by the formula $\tan x = \frac{\sin x}{\cos x}$.

Study of $y = \tan x$

1. Domain and range. The formula $\tan x = \frac{\sin x}{\cos x}$ is valid for any x for which $\cos x \neq 0$, that is, the domain is $X = \mathbb{R} \backslash \{x_k = \frac{\pi}{2} + k\pi, \forall k \in \mathbb{Z}\}$. When x approaches $\frac{\pi}{2}$ from the left, the numerator approaches 1 while the denominator converges to 0 keeping positive values. This implies that the values of the function goes to $+\infty$ and become greater than any constant given in advance. Similarly, when x approaches $\frac{\pi}{2}$ from the right, the numerator approaches 1 while the denominator converges to 0 keeping negative values. This shows that the function goes to $-\infty$ and its values are negative and become smaller than any given constant. It can

also be shown that the range of $\tan x$ is $Y = \mathbb{R}$: the proof of this is provided by the geometric characterization of tangent represented in Fig. 4.18 of Sect. 2.1.

2. Boundedness. Knowing the range, we immediately conclude that the function is unbounded both below and above.
3. Parity. It follows from the properties of sine and cosine that $\tan(-x) = \frac{\sin(-x)}{\cos(-x)} = \frac{-\sin x}{\cos x} = -\tan x, \forall x \in X$, that is, $y = \tan x$ is an odd function.
4. Periodicity. It follows from the properties of sine and cosine that $\tan(x + \pi) = \frac{\sin(x+\pi)}{\cos(x+\pi)} = \frac{-\sin x}{-\cos x} = \tan x, \forall x \in X$, that is, $y = \tan x$ is a periodic function with the period π. Moreover, π is the minimum period, since the function vanishes only at the points $x_k = k\pi, \forall k \in \mathbb{Z}$, and the minimum distance between two points of this set is π. (The fact that π is the minimum period can also be proved by applying the properties of monotonicity studied shortly after.)
5. Intersection and sign. Intersection with the y-axis occurs at the point $(0, 0)$, which is also the point of intersection with the x-axis. Besides, all the points $x_k = k\pi, \forall k \in \mathbb{Z}$ are the points of intersection with the x-axis. The function is negative on each interval $(-\frac{\pi}{2}, 0) + k\pi, \forall k \in \mathbb{Z}$ and positive on each interval $(0, \frac{\pi}{2}) + k\pi, \forall k \in \mathbb{Z}$.
6. Monotonicity and extrema.

   Choose a "natural" interval of the domain (without "holes", with the length equal to the minimum period) $(-\frac{\pi}{2}, \frac{\pi}{2})$, take any two points such that $-\frac{\pi}{2} < x_1 < x_2 < \frac{\pi}{2}$, and obtain, using the trigonometric formulas, the following evaluation $D_1 = \tan x_2 - \tan x_1 = \frac{\sin x_2}{\cos x_2} - \frac{\sin x_1}{\cos x_1} = \frac{\sin x_2 \cos x_1 - \sin x_1 \cos x_2}{\cos x_2 \cos x_1} = \frac{\sin(x_2 - x_1)}{\cos x_2 \cos x_1}$. For $x_1, x_2 \in (-\frac{\pi}{2}, \frac{\pi}{2})$ we have $\cos x_2 > 0, \cos x_1 > 0$, which ensures that the denominator is positive. Besides, $0 < x_2 - x_1 < \pi$, and consequently, $\sin(x_2 - x_1) > 0$. Therefore, $D_1 > 0$ and the function is increasing on the interval $(-\frac{\pi}{2}, \frac{\pi}{2})$.

   Extending this result with the use of the π-periodicity, we conclude that $\tan x$ is increasing on each interval $(-\frac{\pi}{2}, \frac{\pi}{2}) + k\pi, \forall k \in \mathbb{Z}$. Although the union of all these intervals gives the entire domain, but this does not mean that the function increases on its domain or even on a union of some of these intervals. For example, taking two subsequent intervals $(-\frac{\pi}{2}, \frac{\pi}{2})$ and $(\frac{\pi}{2}, \frac{3\pi}{2})$, we can choose the point $x_1 = \frac{\pi}{4}$ in the former and $x_2 = \frac{3\pi}{4}$ in the latter to obtain $\tan \frac{\pi}{4} = 1 > -1 = \tan \frac{3\pi}{4}$, which clearly demonstrates that the function is not increasing on the union of these two intervals.

   It follows from the shown monotonicity that any point of the intervals $(-\frac{\pi}{2}, \frac{\pi}{2}) + k\pi, \forall k \in \mathbb{Z}$ is increasing, and for this reason it cannot be an extremum of any kind. All the remaining points are out of the domain, which means that the function has no extremum, global or local, strict or not.

   (Notice also that increasing of $\tan x$ on an interval of the length π eliminate a possibility that the function has a period less than π.)
7. Concavity and inflection.

   Choose again the natural interval $(-\frac{\pi}{2}, \frac{\pi}{2})$ of the domain and consider two points $-\frac{\pi}{2} < x_1 < x_2 < \frac{\pi}{2}$ with the midpoint $x_0 = \frac{x_1 + x_2}{2}$ and increment

$h = x_2 - x_0 = x_0 - x_1 > 0$. Taking advantage of the derived formula for D_1, we obtain

$$2D_2 = f(x_1) + f(x_2) - 2f\left(\frac{x_1 + x_2}{2}\right) = \tan x_2 + \tan x_1 - 2\tan x_0$$

$$= (\tan x_2 - \tan x_0) - (\tan x_0 - \tan x_1)$$

$$= \frac{\sin(x_2 - x_0)}{\cos x_2 \cos x_0} - \frac{\sin(x_0 - x_1)}{\cos x_0 \cos x_1} = \sin h \frac{\cos x_1 - \cos x_2}{\cos x_2 \cos x_1 \cos x_0}.$$

Since $x_1, x_0, x_2 \in (-\frac{\pi}{2}, \frac{\pi}{2})$ and $h \in (0, \frac{\pi}{2})$, the denominator and the factor $\sin h$ are positive. So, the sign of the expression depends only on the difference $\cos x_1 - \cos x_2$ that does not keep the same sign on the entire interval $(-\frac{\pi}{2}, \frac{\pi}{2})$. For this reason, we divide the interval $(-\frac{\pi}{2}, \frac{\pi}{2})$ into two subintervals $(-\frac{\pi}{2}, 0]$ and $[0, \frac{\pi}{2})$. On the former, we have $\cos x_1 < \cos x_2$, and consequently, $D_2 < 0$, which means downward concavity. On the latter, $\cos x_1 > \cos x_2$, which ensures that $D_2 > 0$ and shows upward concavity. Hence, $\tan x$ is concave downward on the interval $(-\frac{\pi}{2}, 0]$ and upward on $[0, \frac{\pi}{2})$. Consequently, 0 is the inflection point.

Using the π-periodicity, we can extend these results onto the entire domain and obtain that $\tan x$ is concave downward on each interval $(-\frac{\pi}{2}, 0]+k\pi, \forall k \in \mathbb{Z}$ and upward on each interval $[0, \frac{\pi}{2}) + k\pi, \forall k \in \mathbb{Z}$. The points $k\pi, \forall k \in \mathbb{Z}$ are inflection points. (The points $\frac{\pi}{2} + k\pi, \forall k \in \mathbb{Z}$ are out of the domain, and consequently, are not inflection points, although the function has different types of concavity to the left and to the right of these points.)

8. The function $f(x) = \tan x : X = \mathbb{R}\backslash\{x_k = \frac{\pi}{2} + k\pi, \forall k \in \mathbb{Z}\} \to Y = \mathbb{R}$ is surjective, but not injective. Notice that on each interval of the length greater than π the function repeats its values. Therefore, to obtain an injection, we need to chose a domain contained in the interval of the maximum length π. The traditional choice is the interval $(-\frac{\pi}{2}, \frac{\pi}{2})$ where $\tan x$ is increasing, that guarantees the property of injection. With this choice, the function $f(x) = \tan x : X = (-\frac{\pi}{2}, \frac{\pi}{2}) \to Y = \mathbb{R}$ is bijective and admits the inverse called arctangent: $f^{-1}(x) = \arctan x : \mathbb{R} \to (-\frac{\pi}{2}, \frac{\pi}{2})$. The inverse function will be analyzed in one of the next sections.
9. Based on the investigated properties, we can represent the graph of $f(x) = \tan x$ as shown in Fig. 4.25.

Complimentary Properties

The function $f(x) = \tan x$ goes to $+\infty$ as x approaches $\frac{\pi}{2} + k\pi, \forall k \in \mathbb{Z}$ from the left, and goes to $-\infty$ as x approaches any of these points from the right. Therefore, all the lines $x = \frac{\pi}{2} + k\pi, \forall k \in \mathbb{Z}$ are vertical asymptotes. At the remaining points, that is, at any point of the domain, the function is continuous, meaning that $f(x)$ approaches $f(x_0)$ as x approaches x_0. Consequently, there are no other vertical asymptotes.

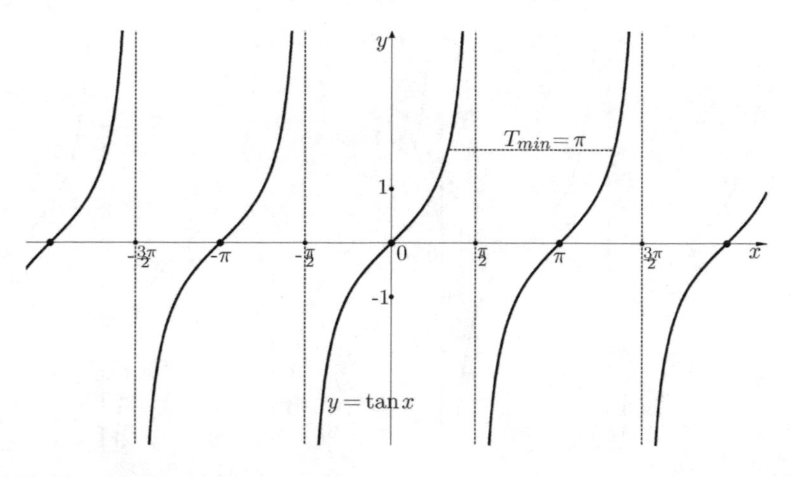

Fig. 4.25 The graph of $y = \tan x$

When x goes to $+\infty$, the function does not approach any specific value: for example, using the path $x_k = k\pi$, $k \in \mathbb{N}$, the function has all the values 0, while at all the points of the path $x_n = \frac{\pi}{4} + n\pi$, $n \in \mathbb{N}$ the function equals 1. This means that there is no horizontal asymptote when x goes to $+\infty$. The same happens when x goes to $-\infty$.

2.7 *Function* cot *x*

Definition of $\cot x$ The function $f(x) = \cot x$ is defined by the formula $\cot x = \frac{\cos x}{\sin x}$.

Study of $y = \cot x$

The study of this function can be performed in the same way as for $y = \tan x$. Another approach, geometric, is based on the relation $\cot x = -\tan(x + \frac{\pi}{2})$, which involves elementary transformations of the graphs and the corresponding properties of functions. This approach was already illustrated in different cases. So, we leave an application of the analytic or geometric method to the reader and represent below only the list of the properties of the function $y = \cot x$.

1. Domain and range. The domain is $X = \mathbb{R} \backslash \{x_k = k\pi, \forall k \in \mathbb{Z}\}$ and the range is $Y = \mathbb{R}$.
2. Boundedness. The function is unbounded both below and above.
3. Parity. The function is odd.
4. Periodicity. The function is π-periodic and π is the minimum period.

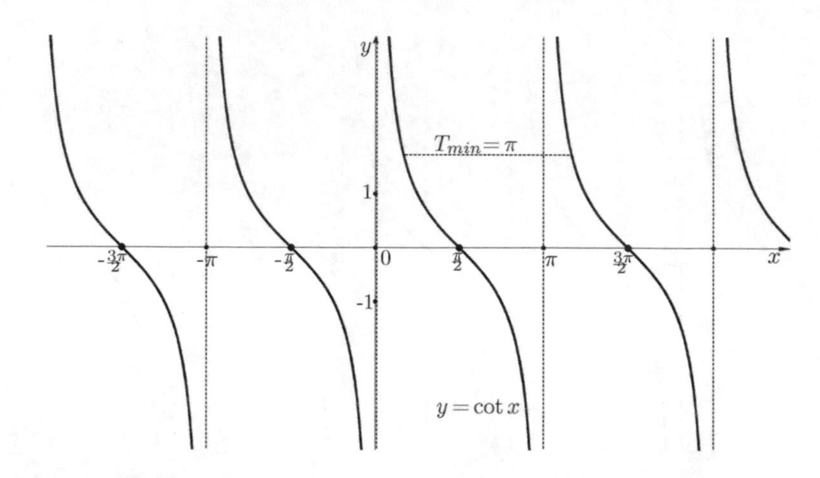

Fig. 4.26 The graph of $y = \cot x$

5. Intersection and sign. There is no intersection with the y-axis. Intersection with the x-axis occurs at the points $x_k = \frac{\pi}{2} + k\pi$, $\forall k \in \mathbb{Z}$. The function is positive on each interval $(0, \frac{\pi}{2}) + k\pi$, $\forall k \in \mathbb{Z}$ and negative on each interval $(\frac{\pi}{2}, \pi) + k\pi$, $\forall k \in \mathbb{Z}$.
6. Monotonicity and extrema. The function $\cot x$ is decreasing on each interval $(0, \pi) + k\pi$, $\forall k \in \mathbb{Z}$ and it does not have extrema of any type.
7. Concavity and inflection. The function is concave upward on each interval $(0, \frac{\pi}{2}] + k\pi$, $\forall k \in \mathbb{Z}$ and downward on each interval $[\frac{\pi}{2}, \pi) + k\pi$, $\forall k \in \mathbb{Z}$. The points $\frac{\pi}{2} + k\pi$, $\forall k \in \mathbb{Z}$ are inflection points.
8. The function $f(x) = \cot x : X = \mathbb{R}\backslash\{x_k = k\pi, \forall k \in \mathbb{Z}\} \to Y = \mathbb{R}$ is surjective, but not injective. Restricting the domain to the interval $(0, \pi)$ we obtain the bijective function $f(x) = \cot x : X = (0, \pi) \to Y = \mathbb{R}$ that admits the inverse called arccotangent: $f^{-1}(x) = \operatorname{arccot} x : \mathbb{R} \to (0, \pi)$. The inverse function will be considered in one of the next sections.
9. The graph of $f(x) = \cot x$ is shown in Fig. 4.26.

Complimentary Properties

The function $f(x) = \cot x$ goes to $+\infty$ as x approaches $k\pi$, $\forall k \in \mathbb{Z}$ from the right, and goes to $-\infty$ as x approaches any of these points from the left. Therefore, all the lines $x = k\pi$, $\forall k \in \mathbb{Z}$ are vertical asymptotes. At the remaining points, that is, at any point of the domain, the function is continuous, meaning that $f(x)$ approaches $f(x_0)$ as x approaches x_0. Consequently, there are no other vertical asymptotes.

When x goes to $+\infty$, the function does not approach any specific value: for example, at all the points of the path $x_k = \frac{\pi}{2} + k\pi$, $k \in \mathbb{N}$, the function is equal to 0, while at the points of the path $x_n = \frac{\pi}{4} + n\pi$, $n \in \mathbb{N}$ the function keeps the value 1. This means that there is no horizontal asymptote when x goes to $+\infty$. The same happens when x goes to $-\infty$.

2.8 *Function* arctan x

Definition of the Function arctan x The function $y = \arctan x$ is usually defined as the inverse of the function $f(x) = \tan x : (-\frac{\pi}{2}, \frac{\pi}{2}) \to \mathbb{R}$.

Study of $y = \arctan x$

1. Domain and range. By the definition, the domain of $y = \arctan x$ is $X = \mathbb{R}$ (the range of the original function $y = \tan x$) and the range is $Y = (-\frac{\pi}{2}, \frac{\pi}{2})$ (the domain of the original).
2. Boundedness. Knowing the range, we immediately conclude that the function is bounded both below (with the lower bound $m = -\frac{\pi}{2}$) and above (with the upper bound $M = \frac{\pi}{2}$). Notice that both indicated bounds are not the values of the function: $-\frac{\pi}{2} < f(x) < \frac{\pi}{2}, \forall x \in \mathbb{R}$.
3. Parity. The function arctan x is odd, because the original function is odd (recall the Theorem on the oddness of the inverse in Sect. 12.4, Chap. 2).
4. Periodicity. As any inverse function, this function cannot be periodic.
5. Intersection and sign. The graph crosses both axes at the only point $(0, 0)$. Since $y = \tan x$ maps $(-\frac{\pi}{2}, 0)$ onto $(-\infty, 0)$ and $(0, \frac{\pi}{2})$ onto $(0, \infty)$, the inverse $y = \arctan x$ is negative on $(-\infty, 0)$ and positive on $(0, \infty)$.
6. Monotonicity and extrema. Since $y = \tan x$ increases on its domain $(-\frac{\pi}{2}, \frac{\pi}{2})$, the inverse $y = \arctan x$ also increases on its domain $\mathbb{R}$ (recall the Theorem on monotonicity of the inverse in Sect. 12.4, Chap. 2). Consequently, all the points of the domain of the inverse are increasing, which implies that there is no extremum of any type.
7. Concavity and inflection. Applying the Theorem about concavity of the inverse (Sect. 12.4 in Chap. 2), we obtain the following results. On the subinterval $(-\frac{\pi}{2}, 0]$ the function $y = \tan x$ is increasing and concave downward, which implies that the inverse $y = \arctan x$ has upward concavity on the subinterval $(-\infty, 0]$. On the subinterval $[0, \frac{\pi}{2})$, the original $y = \tan x$ is also increasing, but concave upward, which results in downward concavity of the inverse $y = \arctan x$ on the subinterval $[0, +\infty)$. Consequently, $x = 0$ is an inflection point of arctan x.
8. The function $f(x) = \arctan x : X = \mathbb{R} \to Y = (-\frac{\pi}{2}, \frac{\pi}{2})$ is bijective and its inverse is $f(x) = \tan x : (-\frac{\pi}{2}, \frac{\pi}{2}) \to \mathbb{R}$.
9. The graph of $y = \arctan x$ can be obtained by applying the Theorem about the graph of the inverse (see Sect. 12.5 in Chap. 2) that is illustrated in Fig. 4.27.

Alternatively, the same graph can be constructed using the investigated properties of $y = \arctan x$.

Complimentary Properties
Since $f(x) = \arctan x$ is the inverse to $f(x) = \tan x$, it approximates $\frac{\pi}{2}$ (from the below) when x goes to $+\infty$, and approximates $-\frac{\pi}{2}$ (from the above) when x goes to

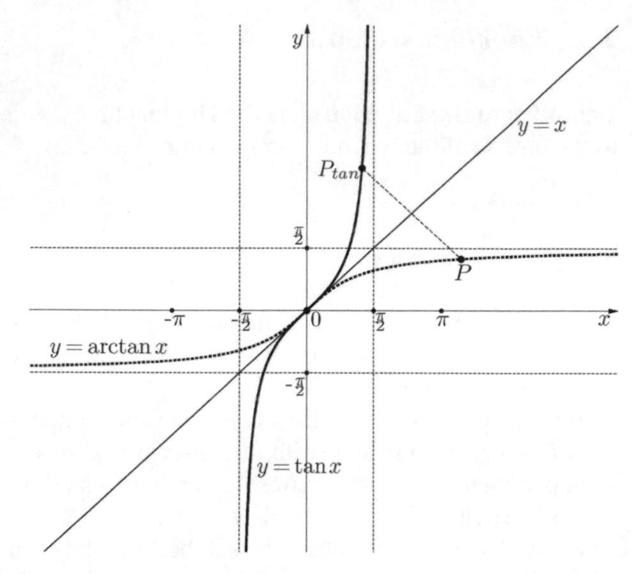

Fig. 4.27 The graph of $y = \arctan x$, symmetric to that of $y = \tan x$ about $y = x$

$-\infty$. Then, $y = \frac{\pi}{2}$ and $y = -\frac{\pi}{2}$ are horizontal asymptotes. Since $f(x) = \arctan x$ is continuous over the entire domain, meaning that $f(x) \to f(x_0)$ when $x \to x_0 \in \mathbb{R}$, there is no infinite increase/decrease at any point of the domain, and consequently, there is no vertical asymptote.

2.9 *Function* arccot x

Definition of the Function arccot x The function $y = \operatorname{arccot} x$ is usually defined as the inverse to the function $f(x) = \cot x : (0, \pi) \to \mathbb{R}$.

Study of $y = \operatorname{arccot} x$

The study of this function can be made in the same manner as for $y = \arctan x$, using the properties of the original function $\cot x$ and applying the theorems about the properties of the inverse (Sect. 12.4 of Chap. 2). We leave this task to the reader and show only the graph of the function $y = \operatorname{arccot} x$ in Fig. 4.28.

3 Examples of Study of Transcendental Functions

In some cases the analysis of a function can be made using both analytic and geometric approach (the latter by applying elementary transformations of graphs). Here we focus on the analytic approach, which can be applied both to simpler and more complex functions.

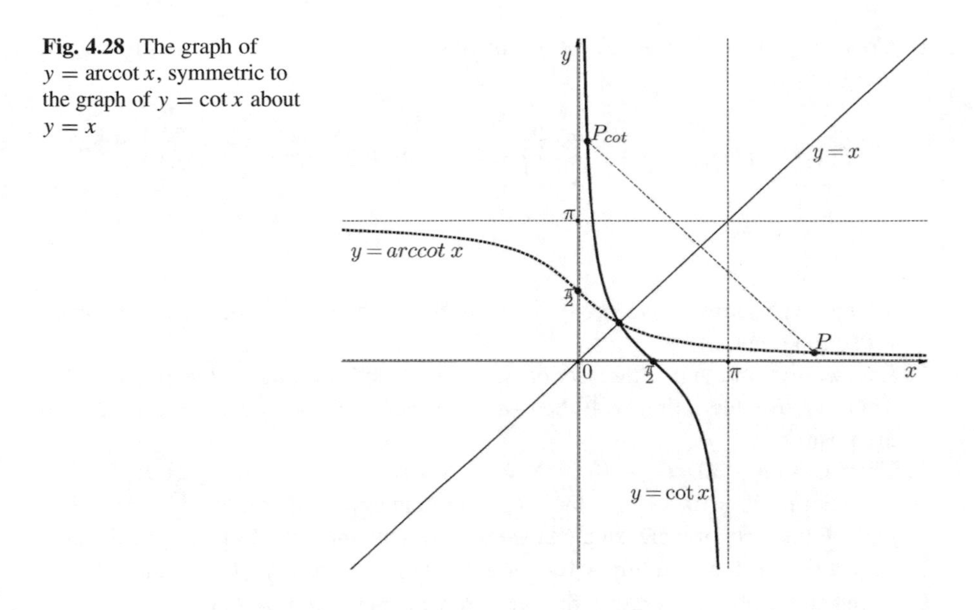

Fig. 4.28 The graph of $y = \operatorname{arccot} x$, symmetric to the graph of $y = \cot x$ about $y = x$

3.1 *Study of Exponential Functions*

A. Study of $f(x) = 2 \cdot 10^{3x}$

1. Domain and range. By the definition of powers, the domain is $X = \mathbb{R}$, since the operation 10^t is defined for any real exponent. The range is contained in $(0, +\infty)$, because $2 \cdot 10^{3x} > 0$ for any $x \in \mathbb{R}$, by the definition of powers. Moreover, it can be shown that $f(x) = 2 \cdot 10^{3x}$ takes any positive value. Indeed, pick any $y_0 > 0$ and consider the equation $f(x) = 2 \cdot 10^{3x} = y_0$. It can be written as $10^{3x} = \frac{y_0}{2}$, and using the inverse we find $3x = \log_{10} \frac{y_0}{2}$, whence $x_0 = \frac{1}{3} \log_{10} \frac{y_0}{2}$. Therefore, for any $y_0 > 0$ there exists exactly one point x_0 of the domain such that $f(x_0) = y_0$, which means that the range is $Y = (0, +\infty)$.
2. Boundedness. It follows from a specification of the range that the function is bounded below (for example, by $m = 0$) and is not bounded above.
3. Monotonicity and extrema. By the properties of powers, we have $2 \cdot 10^{3x_2} > 2 \cdot 10^{3x_1}$ for any pair $x_2 > x_1$, which means that the function increases on its domain $X = \mathbb{R}$. Consequently, any point of the domain is increasing, which implies that the function has no extremum of any type.
4. Parity. Increase of $2 \cdot 10^{3x}$ on the entire domain eliminates a possibility of the function being even. On the other hand, positivity of the range discards an option of the function being odd.
5. Periodicity. Increase of the function on the entire domain ensures that it is not periodic, since it cannot repeat any of its values.

6. Concavity and inflection. For any two points $x_1, x_2 \in \mathbb{R}$, we get the following inequality:

$$f(x_1) + f(x_2) - 2f\left(\frac{x_1+x_2}{2}\right) = 2 \cdot 10^{3x_1} + 2 \cdot 10^{3x_2} - 2 \cdot 2 \cdot 10^{3\frac{x_1+x_2}{2}}$$
$$= 2\left(10^{\frac{3x_1}{2}} - 10^{\frac{3x_2}{2}}\right)^2 > 0,$$

which reveals upward concavity on the entire domain. Consequently, there is no inflection point.

7. Intersection and sign. Intersection with the y-axis occurs at the point $(0, 2)$. There is no intersection with the x-axis, because all the values of the function are positive.
8. The function $f(x) = 2 \cdot 10^{3x} : X = \mathbb{R} \to Y = (0, +\infty)$ is bijective. Indeed, the fact that $Y = (0, +\infty)$ is the range guarantees that the function is surjective, and the property of increasing on the entire domain guarantees that it is injective. The inverse is the logarithmic function $f^{-1}(x) = \frac{1}{3}\log_{10}\frac{x}{2} : (0, +\infty) \to \mathbb{R}$.
9. Using the investigated properties we can draw the graph of $f(x) = 2 \cdot 10^{3x}$ as shown in Fig. 4.29.

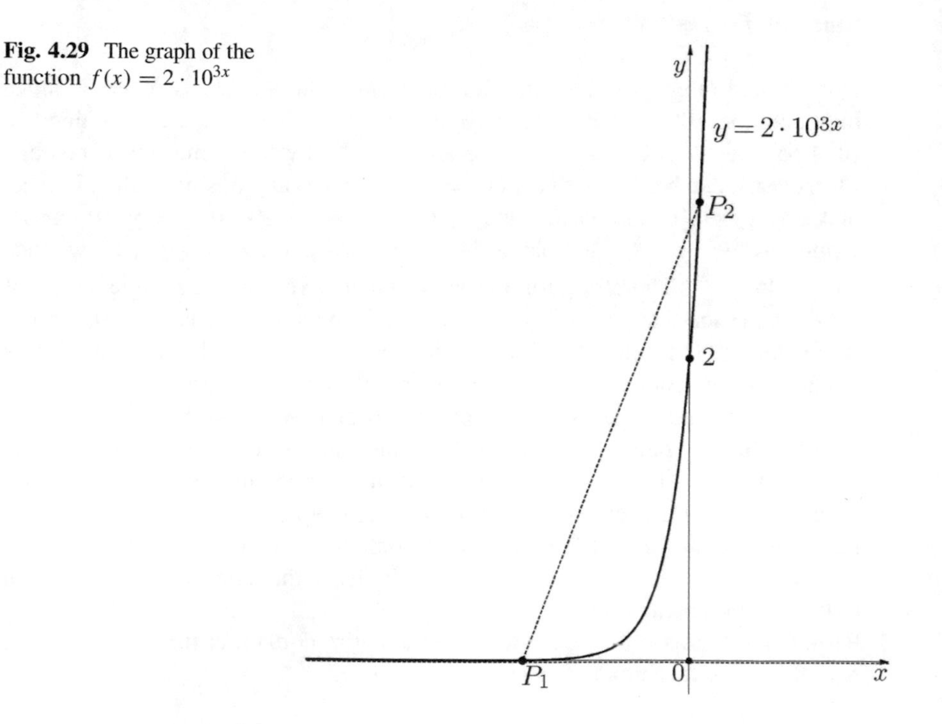

Fig. 4.29 The graph of the function $f(x) = 2 \cdot 10^{3x}$

Complimentary Properties

The graph of the function $f(x) = 2 \cdot 10^{3x}$ suggests that its values increase without restriction when x goes to $+\infty$ and converge to 0 from the above when x goes to $-\infty$. Indeed, for any given $M > 0$, we can solve the equation $f(x) = 2 \cdot 10^{3x} = M$ and find the solution $x_0 = \frac{1}{3}\log_{10}\frac{M}{2}$. Then, using increase of $f(x) = 2 \cdot 10^{3x}$, we conclude that $2 \cdot 10^{3x} > M$ for all $x > x_0$ and $2 \cdot 10^{3x} < M$ for all $x < x_0$. Hence, for all sufficiently large values of x the values of the function become greater than any constant given in advance, which means that $2 \cdot 10^{3x}$ goes to $+\infty$ when x goes to $+\infty$. On the other hand, for sufficiently small values of x (large in absolute value but negative), the values of the function become smaller than any given positive constant, keeping the positive sign, which means that $2 \cdot 10^{3x}$ approaches 0 when x goes to $-\infty$. These two results show that $f(x)$ has no horizontal asymptote at $+\infty$, but has the horizontal asymptote $y = 0$ at $-\infty$.

Additionally, the function is continuous at each point of its domain, that is, $f(x) \to f(x_0)$ as $x \to x_0$. Therefore, there is no infinite increase/decrease at any point of the domain $X = \mathbb{R}$, and consequently, there is no vertical asymptote.

Another way to determine the properties of the function $f(x) = 2 \cdot 10^{3x}$ is by finding elementary transformations from $g(x) = 10^x$ to $f(x) = 2 \cdot 10^{3x}$. The following chain of the transformation can be employed:

$$g(x) = 10^x \to h(x) = 10^{3x} = g(3x) \to f(x) = 2 \cdot 10^{3x} = 2h(x).$$

Geometrically, the first transformation is a horizontal contraction by a factor of 3 of the graph of $g(x) = 10^x$. Under this, the properties 10^x are maintained: like $g(x) = 10^x$, the function $h(x) = 10^{3x}$ has the domain $X = \mathbb{R}$ and the range $Y = (0, +\infty)$, it is not periodic, neither even nor odd, it is increasing and concave upward on the entire domain, it does not have extrema and inflection points. However, the increase of $h(x)$ happens 3 times as fast as that of $g(x)$. The second transformation is a vertical stretching by a factor of 2 of the graph of $h(x) = 10^{3x}$. Again, all the properties of 10^{3x} are maintained, but the increase of $f(x)$ is 2 times as fast as that of $h(x)$.

See the illustration of the graph transformations in Fig. 4.30.

B. Study of $f(x) = 2 - 5e^{-x/3} = 2 - 5\left(\frac{1}{e}\right)^{x/3}$

1. Domain and range. By the definition of powers, the domain includes all real numbers, $X = \mathbb{R}$, since a power with a positive base is defined for any real exponent. The range of $e^{-x/3}$ is the interval $(0, +\infty)$, whence the range of $-5e^{-x/3}$ is the interval $(-\infty, 0)$, and finally, the range of $f(x)$ is $Y = (-\infty, 2)$.
2. Boundedness. It follows from the specification of the range that the function is bounded above (for example, by the constant $M = 2$) and is not bounded below.

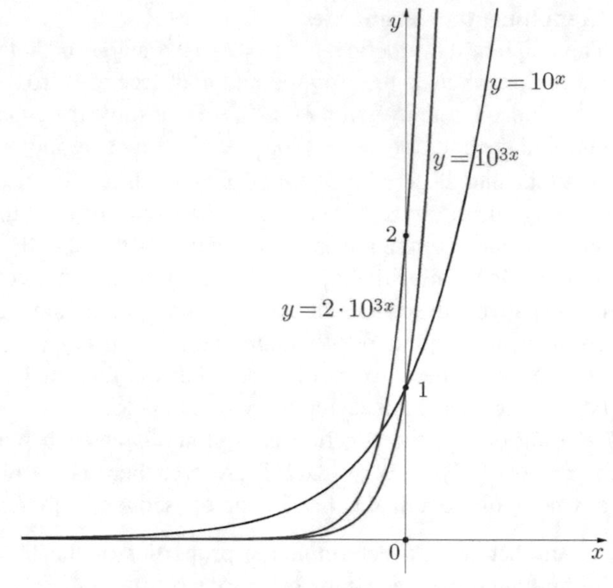

Fig. 4.30 Transformation of the graphs from $g(x) = 10^x$ to $f(x) = 2 \cdot 10^{3x}$

3. Monotonicity and extrema. By the properties of powers, we have $f(x_2) - f(x_1) = (2 - 5e^{-x_2/3}) - (2 - 5e^{-x_1/3}) = -5e^{-x_2/3}(1 - e^{(x_2-x_1)/3})$. For any pair $x_2 > x_1$ we have $\frac{x_2-x_1}{3} > 0$ and then, by the properties of powers, $e^{(x_2-x_1)/3} > 1$. Hence, the first factor -5 is negative, the second is positive, and the third is negative, which results in $f(x_2) - f(x_1) > 0$. Therefore, the function increases on its domain $X = \mathbb{R}$.

   Consequently, each point of the domain is increasing, and for this reason, the function $f(x)$ has no extremum of any type.
4. Parity. Increase of $f(x)$ over the entire domain guarantees that the function is not even. On the other hand, the fact that $f(x)$ is bounded above and unbounded below implies that the function cannot be odd.
5. Periodicity. Increase of $f(x)$ over the entire domain implies that the function does not repeat any of its values, which discards a possibility of periodicity.
6. Concavity and inflection. For two arbitrary points $x_1, x_2 \in \mathbb{R}$, we have

$$f(x_1)+f(x_2)-2f\left(\frac{x_1+x_2}{2}\right)=(2-5e^{-x_1/3})+(2-5e^{-x_2/3})-2(2--5e^{-(x_1+x_2)/6})$$

$$= -5\left(e^{-x_1/6} - e^{-x_2/6}\right)^2 < 0,$$

   which reveals downward concavity on the entire domain. Consequently, there is no inflection point.

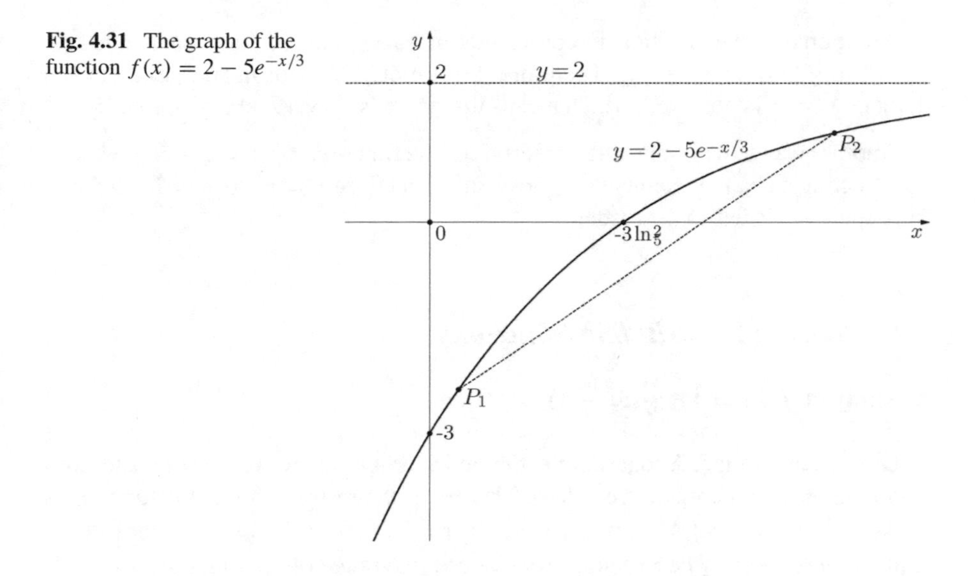

Fig. 4.31 The graph of the function $f(x) = 2 - 5e^{-x/3}$

7. Intersection and sign. Intersection with the y-axis occurs at the point $(0, -3)$, and intersection with the x-axis at the point $\left(-3\ln\frac{2}{5}, 0\right)$. Due to increase, the function is negative when $x < -3\ln\frac{2}{5}$ and positive when $x > -3\ln\frac{2}{5}$.
8. The function $f(x) = 2 - 5e^{-x/3} : X = \mathbb{R} \to Y = (-\infty, 2)$ is bijective. Indeed, since $Y = (-\infty, 2)$ is its range, the function is surjective, and increase on $X = \mathbb{R}$ guarantees that the function is injective. The inverse function is logarithmic $f^{-1}(x) = -3\ln\frac{2-x}{5} : (-\infty, 2) \to \mathbb{R}$.
9. Based on the investigated properties, we can construct the graph of $f(x) = 2 - 5e^{-x/3}$ as shown in Fig. 4.31.

Complimentary Properties

The graph of the function $f(x) = 2 - 5e^{-x/3}$ suggests that the values of the function decrease without restriction when x goes to $-\infty$, and approach 2 from the below when x goes to $+\infty$. Indeed, for any given constant $M \in Y = (-\infty, 2)$, we can solve the equation $f(x) = 2 - 5e^{-x/3} = M$ and find the solution $x_0 = -3\ln\frac{2-M}{5}$. Then, using increase of $f(x) = 2 - 5e^{-x/3}$, we conclude that $2 - 5e^{-x/3} > M$ for all $x > x_0$ and $2 - 5e^{-x/3} < M$ for all $x < x_0$. Hence, for negative and large in the absolute value x, the values of the function become smaller than any constant $M \in (-\infty, 2)$, including any negative constant, which means that $f(x)$ goes to $-\infty$ when x goes to $-\infty$. On the other hand, for sufficiently large values of x, the function surpasses any constant $M < 2$. Together with increase, this implies that $f(x)$ approaches 2 from the below as x goes to $+\infty$. Consequently, the line $y = 2$ is a horizontal asymptote of $f(x)$ when $x \to +\infty$.

Additionally, the function is continuous at each point of its domain, that is, $f(x) \to f(x_0)$ as $x \to x_0$. Therefore, no infinite behavior at the points of the domain $X = \mathbb{R}$ is observed, which means that there is no vertical asymptote.

Another way to determine the properties of the function $f(x) = 2 - 5e^{-x/3}$ is by constructing a chain of elementary transformations from $y = e^x$ to $y = 2 - 5e^{-x/3}$. This approach is left to the reader.

3.2 *Study of Logarithmic Functions*

A. Study of $f(x) = \frac{1}{3}\log_2(5x - 1)$

1. Domain and range. A logarithm is defined for any positive real value, and does not defined for non-positive values. Therefore, the restriction for this function is $5x - 1 > 0$, whence $X = (\frac{1}{5}, +\infty)$. Since $5x - 1$ takes all the values in the interval $(0, +\infty)$ when x varies in $(\frac{1}{5}, +\infty)$, the range of the function is $Y = \mathbb{R}$.
2. Boundedness. Knowing the range, we conclude that the function is unbounded both below and above.
3. Parity. Since the domain is not symmetric about the origin, the function cannot be even or odd.
4. Periodicity. The function is not periodic because its domain is bounded from the left.
5. Monotonicity and extrema. For any pair of the points such as $\frac{1}{5} < x_1 < x_2$, we can represent the difference of the values of the function in the following form: $f(x_2) - f(x_1) = \frac{1}{3}\left[\log_2(5x_2 - 1) - \log_2(5x_1 - 1)\right] = \frac{1}{3}\log_2\frac{5x_2-1}{5x_1-1}$. Since $\frac{1}{5} < x_1 < x_2$, we have $0 < 5x_1 - 1 < 5x_2 - 1$, whence $\frac{5x_2-1}{5x_1-1} > 1$. Consequently, $\log_2\frac{5x_2-1}{5x_1-1} > 0$ and the function increases over the entire domain. Then, each point of the domain is increasing, which implies that the function has no extremum of any type, global or local, strict or non-strict.

   (Increase of the function on the entire domain also guarantees that the function cannot be periodic.)
6. Concavity and inflection. For any two points $x_1, x_2 \in (\frac{1}{5}, +\infty)$, the following representation can be used: $f(x_1) + f(x_2) - 2f\left(\frac{x_1+x_2}{2}\right) = \frac{1}{3}\left[\log_2(5x_1-1) + \log_2(5x_2-1) - 2\log_2\left(5\frac{x_1+x_2}{2} - 1\right)\right] = \frac{1}{3}\log_2\frac{(5x_1-1)(5x_2-1)}{\left(5\frac{x_1+x_2}{2}-1\right)^2}$.

   Let us evaluate the expression $t \equiv \frac{(5x_1-1)(5x_2-1)}{\left(5\frac{x_1+x_2}{2}-1\right)^2}$ in the last logarithm. We can show that $(5x_1 - 1)(5x_2 - 1) < \left(5\frac{x_1+x_2}{2} - 1\right)^2$. Indeed, this inequality is equivalent to $4(25x_1x_2 - 5x_1 - 5x_2 + 1) < (5(x_1 + x_2) - 2)^2$, which can be simplified to $4x_1x_2 < (x_1 + x_2)^2$, which in turn can be written as the inequality $0 < (x_1 - x_2)^2$ satisfied for any $x_1 \neq x_2$. Therefore, $0 < t < 1$, and consequently, $f(x_1) + f(x_2) - 2f\left(\frac{x_1+x_2}{2}\right) = \frac{1}{3}\log_2 t < 0$, which means that the function is concave downward over the entire domain. Then, there is no inflection point.

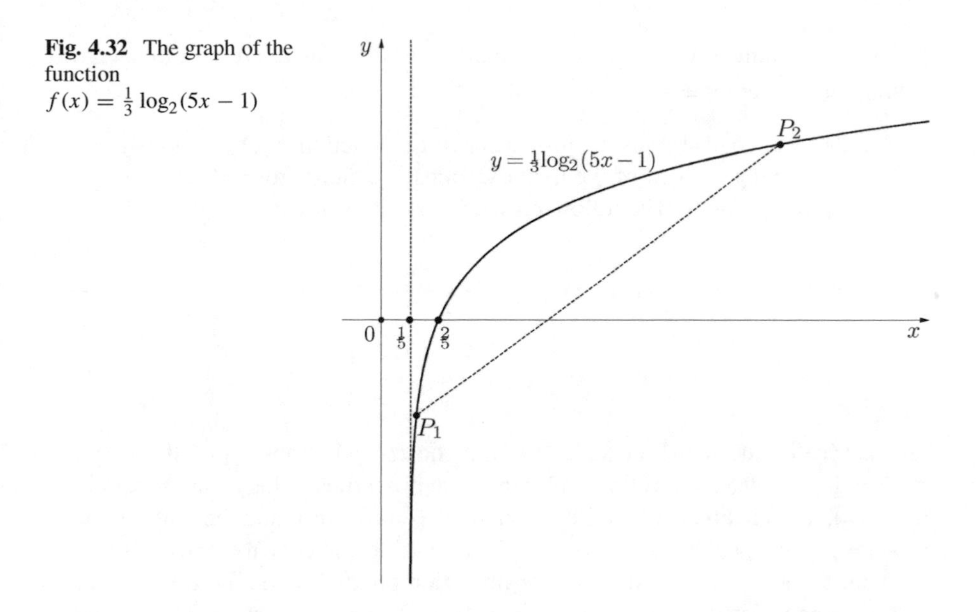

Fig. 4.32 The graph of the function $f(x) = \frac{1}{3}\log_2(5x-1)$

7. Intersection and sign. There is no intersection with the y-axis, since the domain does not contain the origin. Intersection with the x-axis occurs at the point $(\frac{2}{5}, 0)$. Due to increase, the function is negative for $x < \frac{2}{5}$ and positive for $x > \frac{2}{5}$.
8. The function $f(x) = \frac{1}{3}\log_2(5x-1) : X = (\frac{1}{5}, +\infty) \to Y = \mathbb{R}$ is bijective, because $Y = \mathbb{R}$ is its range and injection follows from increase on the entire domain. The inverse function is exponential: $f^{-1}(x) = \frac{1}{5}\left(2^{3x}+1\right) : \mathbb{R} \to (\frac{1}{5}, +\infty)$.
9. Using the revealed properties, we can construct the graph of $f(x) = \frac{1}{3}\log_2(5x-1)$ in the form shown in Fig. 4.32.

Complimentary Properties

The graph of the function $f(x) = \frac{1}{3}\log_2(5x-1)$ suggests that the function increases without restriction when x goes to $+\infty$ and decreases without restriction when x approaches $\frac{1}{5}$ from the right. Indeed, for any given M, we can solve the equation $f(x) = \frac{1}{3}\log_2(5x-1) = M$ and find the solution $x_0 = \frac{1}{5}\left(2^{3M}+1\right) > \frac{1}{5}$. Then, using increase of $f(x)$, we conclude that $\frac{1}{3}\log_2(5x-1) > M$ for all $x > x_0$ and $\frac{1}{3}\log_2(5x-1) < M$ for all $x < x_0$. Hence, for sufficiently great values of x, the values of the function surpass any constant given in advance, which means that $f(x)$ goes to $+\infty$ when x goes to $+\infty$. On the other hand, when x approaches $\frac{1}{5}^+$ the values of the function become smaller than any given constant, that is, $f(x)$ goes to $-\infty$. These results shows that $f(x)$ has no horizontal asymptote, but is has a vertical asymptote $x = \frac{1}{5}$.

Additionally, the function is continuous at each point of its domain, that is, $f(x) \to f(x_0)$ when $x \to x_0 \in X$. Therefore, no infinite behavior exists at any point of the domain $X = (\frac{1}{5}, +\infty)$, which means that there are no other vertical asymptotes, except $x = \frac{1}{5}$.

Another way to determine the properties of the function $f(x) = \frac{1}{3}\log_2(5x - 1)$ is by constructing a chain of elementary transformations from $g(x) = \log_2 x$ to $f(x) = \frac{1}{3}\log_2(5x - 1)$. The following sequence can be used:

$$g(x) = \log_2 x \to g_1(x) = \log_2(5x) = g(5x) \to g_2(x) = \log_2(5x-1) = g_1(x-\frac{1}{5})$$

$$\to f(x) = \frac{1}{3}\log_2(5x - 1) = \frac{1}{3}g_2(x).$$

Geometrically, the first transformation is a horizontal shrinking of the graph of $g(x) = \log_2 x$ by a factor of 5. Under this, the properties of $\log_2 x$ are maintained: like $g(x)$, the function $g_1(x)$ has the domain $X_1 = (0, +\infty)$ and the range $Y_1 = \mathbb{R}$, it is not periodic, neither even nor odd, is increasing and concave downward on its domain, it does not have extremum or inflection point. The second transformation is a horizontal shifting of the graph of $g_1(x)$ by $\frac{1}{5}$ units to the right. The properties of $g_2(x)$ coincide with those of $g_1(x)$, except for the domain which is now $X_2 = (\frac{1}{5}, +\infty)$. The third transformation is a vertical shrinking of the graph of $g_2(x)$ by a factor of 3. Again, the properties of $g_2(x)$ are maintained, but $f(x)$ increases $\frac{1}{3}$-th of the rate of $g_2(x)$.

B. Study of $f(x) = 1 - \log_{1/10}(4 - 2x)$

1. Domain and range. A logarithm is only defined for positive real numbers, that leads to the restriction $4 - 2x > 0$, whence $X = (-\infty, 2)$. Since $4 - 2x$ takes all the values in $(0, +\infty)$ when x varies in $(-\infty, 2)$, the range contains all real numbers: $Y = \mathbb{R}$.
2. Boundedness. Knowing the range, we conclude that the function is unbounded both below and above.
3. Parity. Since the domain is not symmetric about the origin, the function cannot be even or odd.
4. Periodicity. The function is not periodic because its domain is bounded from the right.
5. Monotonicity and extrema. For any pair of the points such as $x_1 < x_2 < 2$, we can represent the difference of the values of the function in the following form: $f(x_2) - f(x_1) = (1 - \log_{1/10}(4 - 2x_2)) - (1 - \log_{1/10}(4 - 2x_1)) = \log_{1/10}\frac{4-2x_1}{4-2x_2}$. Since $x_1 < x_2 < 2$, we have $\frac{4-2x_1}{4-2x_2} > 1$. Then, $\log_{1/10}\frac{4-2x_1}{4-2x_2} < 0$ and the function decreases on the entire domain. Consequently, each point of the domain is decreasing, which implies that the function has no extremum of any type, global or local, strict or non-strict.

(Decrease of the function on the entire domain also guarantees that the function cannot be periodic.)

6. Concavity and inflection. For any two points $x_1, x_2 \in (-\infty, 2)$, the following representation can be used: $f(x_1)+f(x_2)-2f\left(\frac{x_1+x_2}{2}\right) = [1-\log_{1/10}(4-2x_1)] + [1-\log_{1/10}(4-2x_2)] - 2[1-\log_{1/10}(4-(x_1+x_2))] = \log_{1/10}\frac{(4-(x_1+x_2))^2}{(4-2x_1)(4-2x_2)}$. Let us evaluate the expression $t \equiv \frac{(4-(x_1+x_2))^2}{(4-2x_1)(4-2x_2)}$ in the last logarithm. We can show that $(4-2x_1)(4-2x_2) < (4-(x_1+x_2))^2$. Indeed, this inequality is equivalent to $16 - 8(x_1+x_2) + 4x_1x_2 < 16 - 8(x_1+x_2) + (x_1+x_2)^2$, which can be simplified to $4x_1x_2 < (x_1+x_2)^2$, which in turn can be written as the inequality $0 < (x_1-x_2)^2$ that is true for any $x_1 \neq x_2$. Therefore, $t > 1$, and consequently, $f(x_1) + f(x_2) - 2f\left(\frac{x_1+x_2}{2}\right) = \log_{1/10} t < 0$, which means that the function is concave downward over the entire domain. Then, there is no inflection point.
7. Intersection and sign. Intersection with the y-axis occurs at the point $(0, y_e)$, where $y_e = 1 - \log_{1/10} 4$, and intersection with the x-axis occurs at the point $(x_e, 0)$, where $x_e = \frac{4-\frac{1}{10}}{2}$ is found from the equation $4 - 2x = \frac{1}{10}$. Due to decrease, the function is positive when $x < x_e$ and negative when $x > x_e$.
8. The function $f(x) = 1 - \log_{1/10}(4-2x) : X = (-\infty, 2) \to Y = \mathbb{R}$ is bijective, because its range is $Y = \mathbb{R}$ and injection is guaranteed by decrease on the entire domain. The inverse function is exponential: $f^{-1}(x) = 2 - \frac{1}{2}\left(\frac{1}{10}\right)^{1-x} : \mathbb{R} \to (-\infty, 2)$.
9. Using the revealed properties, we can construct the graph of $f(x) = 1 - \log_{1/10}(4-2x)$ in the form shown in Fig. 4.33.

Complimentary Properties

Looking at the graph of the function $f(x) = 1 - \log_{1/10}(4-2x)$ we can suppose that its values increase without restriction when x goes to $-\infty$ and decrease without restriction when x approaches 2 from the left. Indeed, for any given M, we can solve the equation $f(x) = 1 - \log_{1/10}(4-2x) = M$ and find the solution $x_0 = 2 - \frac{1}{2}\left(\frac{1}{10}\right)^{1-M} < 2$. Then, using decrease of $f(x)$, we conclude that $f(x) > M$ for all $x < x_0$ and $f(x) < M$ for all $x > x_0$. Hence, for negative x, which are sufficiently large in the absolute value, the function becomes greater than any constant given in advance, which means that $f(x)$ goes to $+\infty$ when x goes to $-\infty$. On the other hand, when x approaches 2^- the values of the function become smaller than any given constant, that is, $f(x)$ goes to $-\infty$. These results show that $f(x)$ has no horizontal asymptote, but is has a vertical asymptote $x = 2$.

Additionally, the function is continuous at each point of its domain, that is, $f(x) \to f(x_0)$ when $x \to x_0 \in X$. Therefore, no infinite behavior exists at any point of the domain $X = (-\infty, 2)$, which means that there are no other vertical asymptotes, except $x = 2$.

Another way to determine the properties of the function $f(x) = 1 - \log_{1/10}(4-2x)$ is by constructing a chain of elementary transformations from $g(x) = \log_{1/10} x$

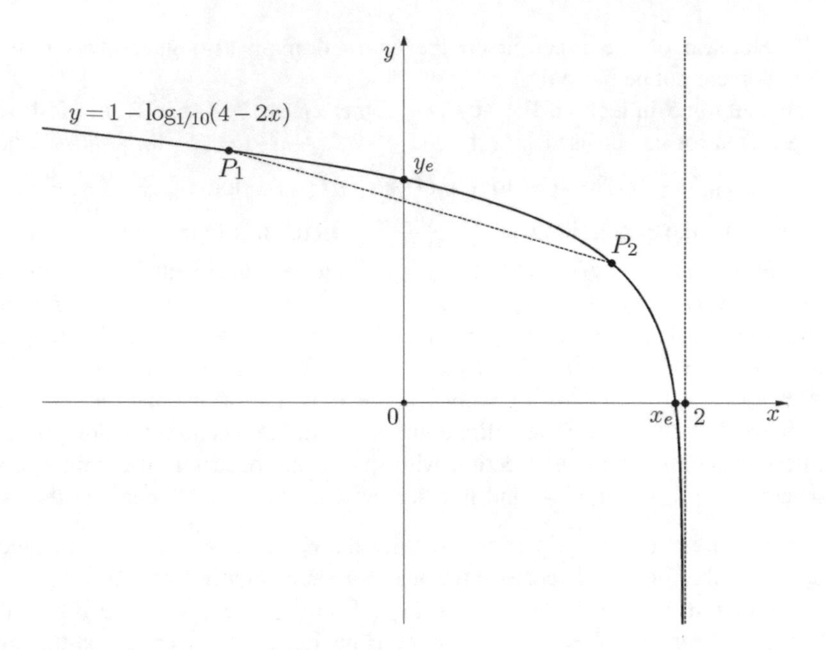

Fig. 4.33 The graph of the function $f(x) = 1 - \log_{1/10}(4 - 2x)$

to $f(x) = 1 - \log_{1/10}(4 - 2x)$. The following sequence can be used:

$$g(x) = \log_{1/10} x \rightarrow g_1(x) = \log_{1/10}(-2x)$$
$$= g(-2x) \rightarrow g_2(x) = \log_{1/10}(4 - 2x) = g_1(x - 2)$$

$$\rightarrow g_3(x) = -\log_{1/10}(4-2x) = -g_2(x) \rightarrow f(x)=1-\log_{1/10}(4-2x) = 1+g_3(x).$$

Geometrically, the first transformation is a horizontal shrinking of the graph of $g(x) = \log_{1/10} x$ by a factor of 2 with the subsequent reflection about the y-axis. Under this, the properties of $g(x)$ are changed as follows: the domain $X_g = (0, +\infty)$ of $g(x)$ turns into $X_1 = (-\infty, 0)$ of $g_1(x)$; the range continues to be the same $Y_g = Y_1 = \mathbb{R}$; decrease on the interval $X_g = (0, +\infty)$ turns into increase of $g_1(x)$ on $X_1 = (-\infty, 0)$; $g_1(x)$ continues to be non-periodic, neither even nor odd; the upward concavity of $g(x)$ on the entire domain is maintained by $g_1(x)$; like $g(x)$, the function $g_1(x)$ has no extremum or inflection point.

The second transformation is a horizontal translation 2 units to the right. This results in the following properties of $g_2(x)$: the domain $X_1 = (-\infty, 0)$ of $g_1(x)$ turns into $X_2 = (-\infty, 2)$ of $g_2(x)$; the range continues the same $Y_1 = Y_2 = \mathbb{R}$; increase of $g_1(x)$ on the interval $X_1 = (-\infty, 0)$ is converted in increase of $g_2(x)$ on $X_2 = (-\infty, 2)$; $g_2(x)$ is still non-periodic, neither even nor odd; upward concavity of $g_1(x)$ is maintained by $g_2(x)$; like $g_1(x)$, the function $g_2(x)$ has no extremum or inflection point.

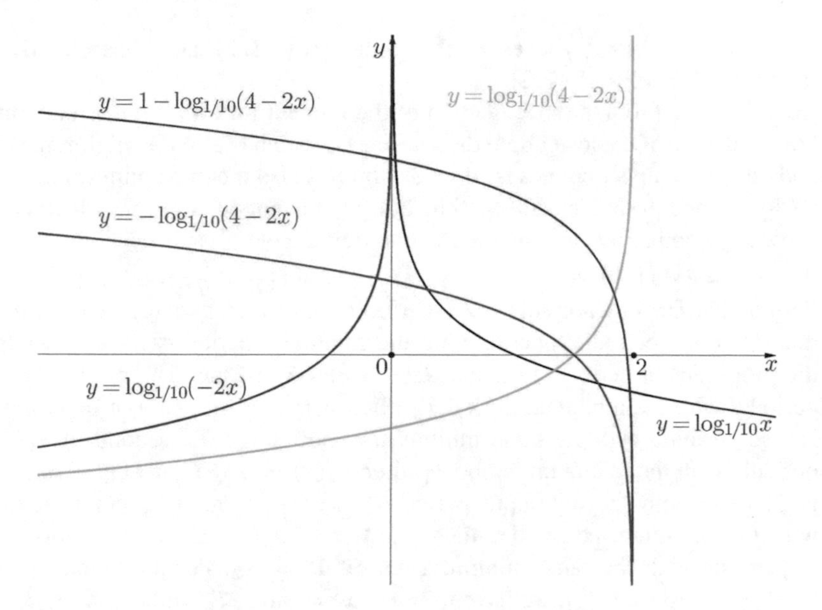

Fig. 4.34 The elementary transformations from $g(x) = \log_{1/10} x$ to $f(x) = 1 - \log_{1/10}(4 - 2x)$

The third transformation is a reflection about the x-axis, which leads to the following properties of $g_3(x)$: the domain and range of $g_2(x)$ and $g_3(x)$ are the same, $X_2 = X_3 = (-\infty, 2)$ and $Y_2 = Y_3 = \mathbb{R}$; increase of $g_2(x)$ is converted in decrease of $g_3(x)$; the function continues to be non-periodic, neither even nor odd; upward concavity of $g_2(x)$ is transformed into downward concavity of $g_3(x)$; both functions do not have extrema and inflection points.

The last transformation is a vertical translation 1 unit upward. Consequently, $f(x)$ has the following properties: the domain and range of $g_3(x)$ and $f(x)$ are the same, $X_3 = X_f = (-\infty, 2)$ and $Y_3 = Y_f = \mathbb{R}$; decrease of $g_3(x)$ is maintained by $f(x)$, as well as downward concavity; $f(x)$ is still non-periodic, neither even nor odd; $f(x)$ continues without extrema and inflection points.

See illustration of the transformations of graphs in Fig. 4.34.

3.3 Study of Trigonometric Functions

A. Study of $y = 2\sin\left(3x + \frac{\pi}{6}\right)$

1. Domain and range. By the definition of sine, the domain is $X = \mathbb{R}$ (the operation $\sin t$ is defined for any real number t) and the values of $\sin t$ belong to the interval $[-1, 1]$. Notice also that when x ranges over $[-\frac{2\pi}{9}, \frac{\pi}{9}]$ the values of $t = 3x + \frac{\pi}{6}$ varies on $[-\frac{\pi}{2}, \frac{\pi}{2}]$, and $\sin t$ takes all the numbers on $[-1, 1]$, which implies that

$f(x) = 2\sin\left(3x + \frac{\pi}{6}\right)$ takes all the values on $[-2, 2]$, and consequently, the range of $f(x)$ is $Y = [-2, 2]$.

2. Boundedness. From the specification of the range it follows that the functions is bounded. One of the lower bounds is $m = -2$, which is also the minimum value, and one of the upper bounds is $M = 2$, which is also the maximum value.
3. Parity. Although $\sin t$ is odd function, but $f(x)$ is not odd or even, which can be shown by comparing the values at the two points $x_1 = \frac{\pi}{18}$ and $x_2 = -x_1 = -\frac{\pi}{18}$: $f(-x_1) = 2\sin\left(-\frac{\pi}{6} + \frac{\pi}{6}\right) = 0 \neq \pm\sqrt{3} = \pm 2\sin\left(\frac{\pi}{6} + \frac{\pi}{6}\right) = \pm f(x_1)$.
4. Periodicity. The function $\sin t$ is 2π-periodic: $\sin(t + 2\pi) = \sin t$, which implies that $f(x)$ is a periodic function with the minimum period $\frac{2\pi}{3}$. It follows from the properties of periodic functions (see Sect. 4.5 in Chap. 2): first, if $g(x)$ is a periodic with the minimum period T_g, then $g_1(x) = g(x + c)$, $c = constant$ is also periodic with the same minimum period $T_1 = T_g$; second, if $g_1(x)$ is periodic with the minimum period T_1, then $g_2(x) = g_1(ax)$, $a = constant \neq 0$ is periodic with the minimum period $T_2 = \frac{T_1}{|a|}$; third, if $g_2(x)$ is periodic with the minimum period T_2, then $f(x) = Cg_2(x)$, $C = constant \neq 0$ is periodic with the same minimum period $T = T_2$. For the given function $f(x) = 2\sin 3\left(x + \frac{\pi}{18}\right)$, we have to use $g(x) = \sin x$, $T_g = 2\pi$, $a = 3$, $c = \frac{\pi}{18}$ and $C = 2$.

   Of course, the same result can be obtained directly for the specific function using the definition of periodicity and properties of $\sin x$.
5. Intersection and sign. Intersection with the y-axis occurs at the point $(0, 1)$. The points of intersection with the x-axis are found from the equation $2\sin\left(3x + \frac{\pi}{6}\right) = 0$. Recalling that solutions of $\sin t = 0$ are $t_k = k\pi$, $\forall k \in \mathbb{Z}$, we conclude that $3x + \frac{\pi}{6} = k\pi$, $\forall k \in \mathbb{Z}$, and consequently, $x_k = -\frac{\pi}{18} + k\frac{\pi}{3}$, $\forall k \in \mathbb{Z}$ are zeros of $f(x)$, that is, the points of intersection with the x-axis are $P_k = (x_k, 0)$, $k \in \mathbb{Z}$. On the intervals $(-\frac{\pi}{18}, -\frac{\pi}{18} + \frac{\pi}{3}) + 2k\frac{\pi}{3}$, $\forall k \in \mathbb{Z}$ the function is positive, while on $(-\frac{\pi}{18} + \frac{\pi}{3}, -\frac{\pi}{18} + \frac{2\pi}{3}) + 2k\frac{\pi}{3}$, $\forall k \in \mathbb{Z}$ it is negative.
6. Monotonicity and extrema.

   For any pair of points $x_1 < x_2$ we have $f(x_2) - f(x_1) = 2\sin\left(3x_2 + \frac{\pi}{6}\right) - 2\sin\left(3x_1 + \frac{\pi}{6}\right) = 2\sin t_2 - 2\sin t_1 = 4\sin\frac{t_2 - t_1}{2}\cos\frac{t_2 + t_1}{2}$, where $t_1 = 3x_1 + \frac{\pi}{6}$, $t_2 = 3x_2 + \frac{\pi}{6}$. Choosing $-\frac{\pi}{2} \leq t_1 < t_2 \leq \frac{\pi}{2}$, we obtain $0 < \frac{t_2 - t_1}{2} \leq \frac{\pi}{2}$ and $-\frac{\pi}{2} < \frac{t_2 + t_1}{2} < \frac{\pi}{2}$. The first double inequality guarantees that $\sin\frac{t_2 - t_1}{2} > 0$ and the second that $\cos\frac{t_2 + t_1}{2} > 0$. Then, $f(x_2) - f(x_1) > 0$ and the function is increasing. It remains to transform the restriction on t into the conditions on x: $-\frac{\pi}{2} \leq t_1 = 3x_1 + \frac{\pi}{6}$ is equivalent to $-\frac{\pi}{18} - \frac{\pi}{6} \leq x_1$, and $t_2 = 3x_2 + \frac{\pi}{6} \leq \frac{\pi}{2}$ is equivalent to $x_2 \leq -\frac{\pi}{18} + \frac{\pi}{6}$. Joining these conditions, we get $-\frac{\pi}{18} - \frac{\pi}{6} \leq x_1 < x_2 \leq -\frac{\pi}{18} + \frac{\pi}{6}$. Therefore, on the interval $\left[-\frac{\pi}{18} - \frac{\pi}{6}, -\frac{\pi}{18} + \frac{\pi}{6}\right]$ the function $f(x)$ is increasing. Analogously, it can be shown that on the interval $\left[-\frac{\pi}{18} + \frac{\pi}{6}, -\frac{\pi}{18} + \frac{\pi}{2},\right]$ the function is decreasing. Applying the $\frac{2\pi}{3}$-periodicity we can extend these results on the entire domain and conclude that $f(x)$ increases on the intervals $\left[-\frac{\pi}{18} - \frac{\pi}{6}, -\frac{\pi}{18} + \frac{\pi}{6}\right] + k\frac{2\pi}{3}$, $\forall k \in \mathbb{Z}$ and decreases on the intervals $\left[-\frac{\pi}{18} + \frac{\pi}{6}, -\frac{\pi}{18} + \frac{\pi}{2},\right] + k\frac{2\pi}{3}$, $\forall k \in \mathbb{Z}$.

Therefore, any point of the intervals $\left(-\frac{\pi}{18}-\frac{\pi}{6}, -\frac{\pi}{18}+\frac{\pi}{6}\right)+k\frac{2\pi}{3}$ and $\left(-\frac{\pi}{18}+\frac{\pi}{6}, -\frac{\pi}{18}+\frac{\pi}{2}\right)+k\frac{2\pi}{3}$, $\forall k \in \mathbb{Z}$ is monotonic and cannot be an extremum of any type. The remaining points are $x_k = -\frac{\pi}{18}+\frac{\pi}{6}+k\frac{2\pi}{3}$, $\forall k \in \mathbb{Z}$ and $x_n = -\frac{\pi}{18}+\frac{\pi}{2}+n\frac{2\pi}{3}$, $\forall n \in \mathbb{Z}$. The points of the first group are strict local and non-strict global maxima: each point x_k is a strict local maximum because $f(x)$ increases in its left neighborhood and decreases in right neighborhood; each x_k is a non-strict global maximum because $f(x)$ attains the maximum value of the range at each of these points: $f(x_k)=2$, $\forall k \in \mathbb{Z}$. Similarly, the points x_n are strict local and non-strict global minima.

7. Concavity and inflection.

   Using again the parameter $t = 3x+\frac{\pi}{6}$, for any pair of the points $x_1, x_2 \in \mathbb{R}$ we obtain

$$2D_2 = f(x_1)+f(x_2)-2f\left(\frac{x_1+x_2}{2}\right) = 2\sin t_1 + 2\sin t_2 - 4\sin\frac{t_1+t_2}{2}$$

$$= 4\sin\frac{t_1+t_2}{2}\cos\frac{t_1-t_2}{2} - 4\sin\frac{t_1+t_2}{2} = 4\sin\frac{t_1+t_2}{2}\left(\cos\frac{t_1-t_2}{2}-1\right).$$

The second factor is always non-positive, which implies that the sign of the expression depends only on the first factor. If we choose $0 \le t_1 < t_2 \le \pi$, then $0 < \frac{t_1+t_2}{2} < \pi$, which guarantees that $\sin\frac{t_1+t_2}{2} > 0$, and at the same time $0 < \frac{t_2-t_1}{2} \le \frac{\pi}{2}$, which ensures that the second factor is negative. Therefore, $D_2 < 0$ and $f(x)$ is concave downward. Translating the inequality of t_1, t_2 into the terms of x_1, x_2, we have $0 \le 3x_1+\frac{\pi}{6} < 3x_2+\frac{\pi}{6} \le \pi$ or $-\frac{\pi}{18} \le x_1 < x_2 \le \frac{\pi}{3}-\frac{\pi}{18}$, that is, $f(x)$ is concave downward on the interval $\left[-\frac{\pi}{18}, \frac{\pi}{3}-\frac{\pi}{18}\right]$. In a similar manner, choosing $\pi \le t_1 < t_2 \le 2\pi$ we have $\sin\frac{t_1+t_2}{2} < 0$ because $\pi < \frac{t_1+t_2}{2} < 2\pi$, and at the same time the second factor is negative because $0 < \frac{t_2-t_1}{2} \le \frac{\pi}{2}$. Then, $D_2 > 0$ and $f(x)$ is concave upward. In terms of x, the condition for t means that $\frac{\pi}{3}-\frac{\pi}{18} \le x_1 < x_2 \le \frac{2\pi}{3}-\frac{\pi}{18}$.

   Using the $\frac{2\pi}{3}$-periodicity, we conclude that $f(x)$ is concave downward on the intervals $\left[-\frac{\pi}{18}, \frac{\pi}{3}-\frac{\pi}{18}\right]+k\frac{2\pi}{3}$, $\forall k \in \mathbb{Z}$ and concave upward on the intervals $\left[\frac{\pi}{3}-\frac{\pi}{18}, \frac{2\pi}{3}-\frac{\pi}{18}\right]+n\frac{2\pi}{3}$, $\forall n \in \mathbb{Z}$. Since the function changes the type of concavity at the points $x_k = -\frac{\pi}{18}+k\frac{\pi}{3}$, $\forall k \in \mathbb{Z}$, each of these points is an inflection point.

8. The function $f(x) = 2\sin\left(3x+\frac{\pi}{6}\right) : X = \mathbb{R} \to Y = [-2,2]$ is surjective (since $Y = [-2,2]$ is its range), but not injective. Notice that in each interval where the function is not monotonic it repeats its values. Therefore, to obtain an injection we need to choose an interval of monotonicity of $f(x)$. One of the choices is the interval of increase $X_f = \left[-\frac{\pi}{18}-\frac{\pi}{6}, -\frac{\pi}{18}+\frac{\pi}{6}\right]$. Under this restriction of the domain, the function $f(x) = 2\sin\left(3x+\frac{\pi}{6}\right) : X_f \to Y = [-2,2]$ is bijective, and consequently, it admits the inverse: $f^{-1}(x) : [-2,2] \to$

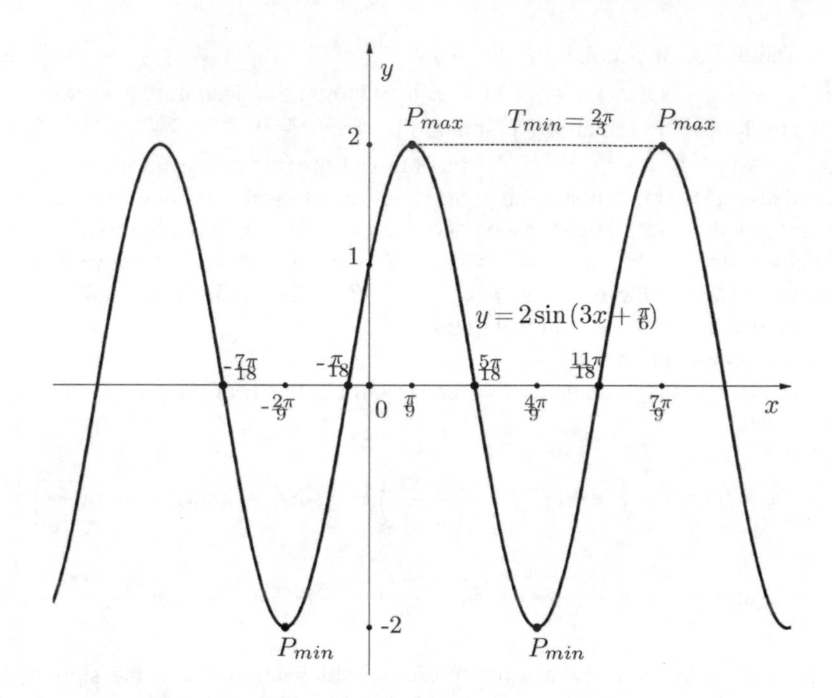

Fig. 4.35 The graph of $y = 2\sin\left(3x + \frac{\pi}{6}\right)$

X_f. The interval X_f corresponds to the interval $T_f = \left[-\frac{\pi}{2}, \frac{\pi}{2}\right]$ of the variable $t = 3x + \frac{\pi}{6}$, which means that the inverse can be represented in the form $f^{-1}(x) = -\frac{\pi}{18} + \frac{1}{3}\arcsin\frac{x}{2}$.

9. Using the revealed properties, we can plot the graph of $f(x) = 2\sin\left(3x + \frac{\pi}{6}\right)$ as shown in Fig. 4.35.

Complimentary Properties

All the values of $f(x) = 2\sin\left(3x + \frac{\pi}{6}\right)$ are contained in the interval $[-2, 2]$, which implies that the function has no vertical asymptote. When x goes to $+\infty$, the function does not approach any specific value: sending x to $+\infty$ by the path $x_k = 2k\pi$, $k \in \mathbb{N}$, we keep the values of the function equal to 1, but using another path $x_n = -\frac{\pi}{18} + n\frac{\pi}{3}$, $\forall n \in \mathbb{N}$, we have all the values equal to 0. The same happens when x goes to $-\infty$. This means that $f(x)$ has no horizontal asymptote.

Another way to determine the properties of the function $f(x) = 2\sin\left(3x + \frac{\pi}{6}\right)$, is by using elementary transformations from $g(x) = \sin x$ to $f(x) = 2\sin\left(3x + \frac{\pi}{6}\right)$. The following chain of transformations can be employed:

$$g(x) = \sin x \to g_1(x) = \sin 3x = g(3x) \to g_2(x) = \sin\left(3x + \frac{\pi}{6}\right) = g_1(x + \frac{\pi}{18})$$

$$\to f(x) = 2\sin\left(3x + \frac{\pi}{6}\right) = 2g_2(x).$$

Geometrically, the first transformation is horizontal shrinking of the graph of $g(x)$ by a factor of 3. Under this, the properties of $g(x)$ are changed as follows. The domain and the range are the same: $X_1 = X_g = \mathbb{R}$ and $Y_1 = Y_g = [-1, 1]$. The periodicity with the minimum period 2π turns into the periodicity with the minimum period $\frac{2\pi}{3}$. The function $g_1(x)$ continues to be odd. Increase on the intervals $C_g = \left[-\frac{\pi}{2}, \frac{\pi}{2}\right] + 2k\pi$, $\forall k \in \mathbb{Z}$ turns into increase on the intervals $C_1 = \left[-\frac{\pi}{6}, \frac{\pi}{6}\right] + k\frac{2\pi}{3}$, $\forall k \in \mathbb{Z}$, while decrease on the intervals $D_g = \left[\frac{\pi}{2}, \frac{3\pi}{2}\right] + 2k\pi$, $\forall k \in \mathbb{Z}$ becomes decrease on the intervals $D_1 = \left[\frac{\pi}{6}, \frac{\pi}{2}\right] + k\frac{2\pi}{3}$, $\forall k \in \mathbb{Z}$. Notice that the rate of increase/decrease on each of intervals of monotonicity is greater comparing with $g(x)$. The maxima $a_g = \frac{\pi}{2} + 2k\pi$ and minima $b_g = -\frac{\pi}{2} + 2k\pi$, $\forall k \in \mathbb{Z}$ of $g(x)$ are converted into the maxima $a_1 = \frac{\pi}{6} + k\frac{2\pi}{3}$ and minima $b_1 = -\frac{\pi}{6} + k\frac{2\pi}{3}$, $\forall k \in \mathbb{Z}$ of $g_1(x)$. The intervals of upward concavity $A_g = [-\pi, 0] + 2k\pi$ and downward concavity $B_g = [0, \pi] + 2k\pi$, $\forall k \in \mathbb{Z}$ of $g(x)$ are transformed into $A_1 = \left[-\frac{\pi}{3}, 0\right] + k\frac{2\pi}{3}$ and $B_1 = \left[0, \frac{\pi}{3}\right] + k\frac{2\pi}{3}$, $\forall k \in \mathbb{Z}$ of $g_1(x)$. The inflection points $c_g = k\pi$, $\forall k \in \mathbb{Z}$ of $g(x)$ become the inflection points $c_1 = k\frac{\pi}{3}$, $\forall k \in \mathbb{Z}$ of $g_1(x)$.

The next transformation is a horizontal translation $\frac{\pi}{18}$ units to the left. This results in the following properties of the function $g_2(x)$. The domain and range are the same, $X_2 = \mathbb{R}$ and $Y_2 = [-1, 1]$, as well as the $\frac{2\pi}{3}$-periodicity. However, the oddness is lost, although the graph of $g_2(x)$ is still symmetric about the point $(-\frac{\pi}{18}, 0)$. The function $g_2(x)$ is increasing on the intervals $C_2 = \left[-\frac{\pi}{6} - \frac{\pi}{18}, \frac{\pi}{6} - \frac{\pi}{18}\right] + k\frac{2\pi}{3}$ and decreasing on the intervals $D_2 = \left[\frac{\pi}{6} - \frac{\pi}{18}, \frac{\pi}{2} - \frac{\pi}{18}\right] + k\frac{2\pi}{3}$, $\forall k \in \mathbb{Z}$. The maxima and minima of $g_2(x)$ are $a_2 = \frac{\pi}{6} - \frac{\pi}{18} + k\frac{2\pi}{3}$ and $b_2 = -\frac{\pi}{6} - \frac{\pi}{18} + k\frac{2\pi}{3}$, $\forall k \in \mathbb{Z}$, respectively. The function $g_2(x)$ is concave upward on the intervals $A_2 = \left[-\frac{\pi}{3} - \frac{\pi}{18}, -\frac{\pi}{18}\right] + k\frac{2\pi}{3}$ and concave downward on the intervals $B_2 = \left[-\frac{\pi}{18}, \frac{\pi}{3} - \frac{\pi}{18}\right] + k\frac{2\pi}{3}$, $\forall k \in \mathbb{Z}$. The points of inflection of $g_2(x)$ are $c_2 = -\frac{\pi}{18} + k\frac{\pi}{3}$, $\forall k \in \mathbb{Z}$.

The last transformation is a vertical stretching by a factor of 2, which leads to the following properties of the function $f(x)$. The domain remains the same $X_f = \mathbb{R}$, but the range is extended to $Y_f = [-2, 2]$. The function $f(x)$ continues to be periodic with the minimum period $\frac{2\pi}{3}$ and non-odd and non-even (although its graph keeps the symmetry of $g_2(x)$ about the point $(-\frac{\pi}{18}, 0)$). Increase and decrease occur on the same intervals as for the function $g_2(x)$: $C_f = C_2 = \left[-\frac{\pi}{6} - \frac{\pi}{18}, \frac{\pi}{6} - \frac{\pi}{18}\right] + k\frac{2\pi}{3}$ and $D_f = D_2 = \left[\frac{\pi}{6} - \frac{\pi}{18}, \frac{\pi}{2} - \frac{\pi}{18}\right] + k\frac{2\pi}{3}$, $\forall k \in \mathbb{Z}$. However, both increase and decrease of $f(x)$ are faster than those of $g_2(x)$. The maxima and minima of $f(x)$ are attained at the same points $a_f = a_2 = \frac{\pi}{6} - \frac{\pi}{18} + k\frac{2\pi}{3}$ and $b_f = b_2 = -\frac{\pi}{6} - \frac{\pi}{18} + k\frac{2\pi}{3}$, $\forall k \in \mathbb{Z}$, but the values of the function at these points change from 1 to 2 for maxima and from -1 to -2 for minima. The intervals of concavity of $f(x)$ are the same as those of $g_2(x)$: on $A_f = A_2 = \left[-\frac{\pi}{3} - \frac{\pi}{18}, -\frac{\pi}{18}\right] + k\frac{2\pi}{3}$ both functions have upward concavity and on $B_f = B_2 = \left[-\frac{\pi}{18}, \frac{\pi}{3} - \frac{\pi}{18}\right] + k\frac{2\pi}{3}$ downward concavity, $\forall k \in \mathbb{Z}$. The points of inflection of the functions $f(x)$ and $g_2(x)$ also coincide: $c_f = c_2 = -\frac{\pi}{18} + k\frac{\pi}{3}$, $\forall k \in \mathbb{Z}$.

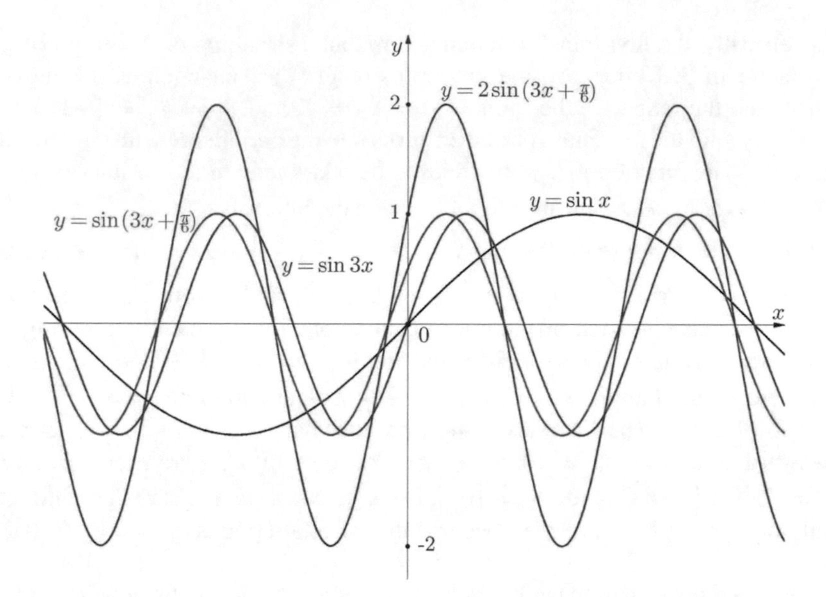

Fig. 4.36 Elementary transformations from $g(x) = \sin x$ to $f(x) = 2\sin\left(3x + \frac{\pi}{6}\right)$

See transformations of the graphs in Fig. 4.36.

B. Study of $y = \frac{1}{2}\arccos(2 - 4x) - \frac{\pi}{4}$

1. Domain and range. Since the domain of $\arccos t$ is the interval $[-1, 1]$, the values of x of the function $f(x) = \frac{1}{2}\arccos(2 - 4x) - \frac{\pi}{4}$ have to satisfy the double inequality $-1 \le t = 2 - 4x \le 1$, whence $\frac{1}{4} \le x \le \frac{3}{4}$, that is, the domain of $f(x)$ is $X = \left[\frac{1}{4}, \frac{3}{4}\right]$. Similarly, recalling that the range of $\arccos t$ is the interval $[0, \pi]$, we deduce that the range of $f(x) = \frac{1}{2}\arccos t - \frac{\pi}{4}$ is $Y = \left[-\frac{\pi}{4}, \frac{\pi}{4}\right]$.
2. Boundedness. Knowing the range, we conclude that $f(x)$ is bounded with the lower bound $m = -\frac{\pi}{4}$ and the upper bound $M = \frac{\pi}{4}$, the former being the minimum value of the function and the latter the maximum value.
3. Parity. The function is not even or odd, since its domain is not symmetric about the origin.
4. Periodicity. The function is not periodic, since its domain is bounded.
5. Intersection and sign. There is no intersection with the y-axis, because the point 0 is out of the domain. To find intersections with the x-axis, we need to solve the equation $y = \frac{1}{2}\arccos(2 - 4x) - \frac{\pi}{4} = 0$, that is, $\arccos(2 - 4x) = \frac{\pi}{2}$. Recall that $\arccos t$ equals to $\frac{\pi}{2}$ at the only point $t = 0$, which gives the relation $2 - 4x = 0$, whence $x = \frac{1}{2}$. Hence, the only point of intersection with the x-axis is $P = \left(\frac{1}{2}, 0\right)$.

The sign of the function is easier to determine after investigation of monotonicity.

6. Monotonicity and extrema.

Take two arbitrary points $x_1 < x_2$ of the domain and use the decrease of $\arccos t$ (with respect to t) to establish the following chain of implications: from $x_1 < x_2$ it follows that $2-4x_1 > 2-4x_2$, whence $\arccos(2-4x_1) < \arccos(2-4x_2)$. Therefore, $f(x_1) < f(x_2)$, which means that $f(x)$ is increasing over the entire domain.

This result of monotonicity has some immediate consequences. First, the strict global minimum of the function is the point $P_{min} = \left(\frac{1}{4}, -\frac{\pi}{4}\right)$ and the strict global maximum is $P_{max} = \left(\frac{3}{4}, \frac{\pi}{4}\right)$ (the minimum and maximum values of the function were already identified in the course of the study of the range, but it was not specified at which points of the domain these extrema are attained and whether they are strict or non-strict). Second, the function has no local extremum, since any point of the interval $\left(\frac{1}{4}, \frac{3}{4}\right)$ is increasing, and the two remaining points, $\frac{1}{4}$ and $\frac{3}{4}$ are the endpoints of the domain. Third, it is easy now to determine the sign of the function: since $f(x)$ vanishes at the point $\frac{1}{2}$ and increase on its domains, it follows that $f(x)$ is negative on the interval $\left[\frac{1}{4}, \frac{1}{2}\right)$ and positive on the interval $\left(\frac{1}{2}, \frac{3}{4}\right]$.

7. Concavity and inflection.

For any pair of the points x_1, x_2 of the domain $X = \left[\frac{1}{4}, \frac{3}{4}\right]$ we have

$$2D_2 = f(x_1) + f(x_2) - 2f\left(\frac{x_1+x_2}{2}\right)$$

$$= \left[\frac{1}{2}\arccos(2-4x_1) - \frac{\pi}{4}\right] + \left[\frac{1}{2}\arccos(2-4x_2) - \frac{\pi}{4}\right]$$

$$-2\cdot\left[\frac{1}{2}\arccos\left(2-4\frac{x_1+x_2}{2}\right) - \frac{\pi}{4}\right]$$

$$= \frac{1}{2}\left[\arccos(2-4x_1) + \arccos(2-4x_2) - 2\arccos\left(2-4\frac{x_1+x_2}{2}\right)\right].$$

Using the variable $t = (2-4x) \in [-1, 1]$ to rewrite the expression in the brackets in the form $4D_2 = \arccos t_1 + \arccos t_2 - 2\arccos\left(\frac{t_1+t_2}{2}\right)$, we obtain the same expression that determines the type of concavity of the function $\arccos t$. Recalling that $\arccos t$ is concave upward on $[-1, 0]$, we conclude that $D_2 > 0$ for any $t_1 \neq t_2$ in this interval. The inequality $t \geq -1$ is equivalent to $2-4x \geq -1$ or $x \leq \frac{3}{4}$, and inequality $t \leq 0$ is equivalent to $2-4x \leq 0$ or $x \geq \frac{1}{2}$. Then, for any pair $x_1 \neq x_2$ in the interval $\left[\frac{1}{2}, \frac{3}{4}\right]$ we have $D_2 > 0$, which means that

Fig. 4.37 The graph of $f(x) = \frac{1}{2}\arccos(2-4x) - \frac{\pi}{4}$

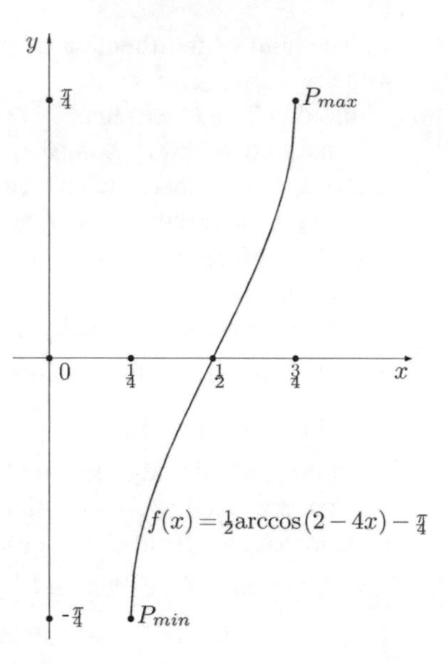

$f(x)$ is concave upward on the interval $\left[\frac{1}{2}, \frac{3}{4}\right]$. Similarly, using the information that $\arccos t$ is concave downward on the interval $[0, 1]$, we obtain that $D_2 < 0$ for any $t_1 \neq t_2$ in this interval. The double inequality $0 \leq t \leq 1$ is equivalent to $\frac{1}{4} \leq x \leq \frac{1}{2}$, which means that $D_2 < 0$ for any pair $x_1 \neq x_2$ in the interval $\left[\frac{1}{4}, \frac{1}{2}\right]$, that is, $f(x)$ is concave downward on the interval $\left[\frac{1}{4}, \frac{1}{2}\right]$.

Since the function changes the type of concavity at the point $\frac{1}{2}$, this is an inflection point.

8. The function $f(x) = \frac{1}{2}\arccos(2-4x) - \frac{\pi}{4} : X = \left[\frac{1}{4}, \frac{3}{4}\right] \to Y = \left[-\frac{\pi}{4}, \frac{\pi}{4}\right]$ is bijective and has the inverse $f^{-1}(x) = \frac{1}{4}\left[2 - \cos\left(2x + \frac{\pi}{2}\right)\right] : \left[-\frac{\pi}{4}, \frac{\pi}{4}\right] \to \left[\frac{1}{4}, \frac{3}{4}\right]$.
9. Based on the investigated properties we can draw the graph of the function shown in Fig. 4.37.

Another way to determine the properties of the function $f(x) = \frac{1}{2}\arccos(2 - 4x) - \frac{\pi}{4}$ is by applying elementary transformations which convert $g(x) = \arccos x$ into $f(x) = \frac{1}{2}\arccos(2-4x) - \frac{\pi}{4}$. The following chain of the transformations can be used:

$$g(x) = \arccos x \to g_1(x) = \arccos(-4x) = g(-4x) \to g_2(x)$$
$$= \arccos(2-4x) = g_1(x - \frac{1}{2})$$

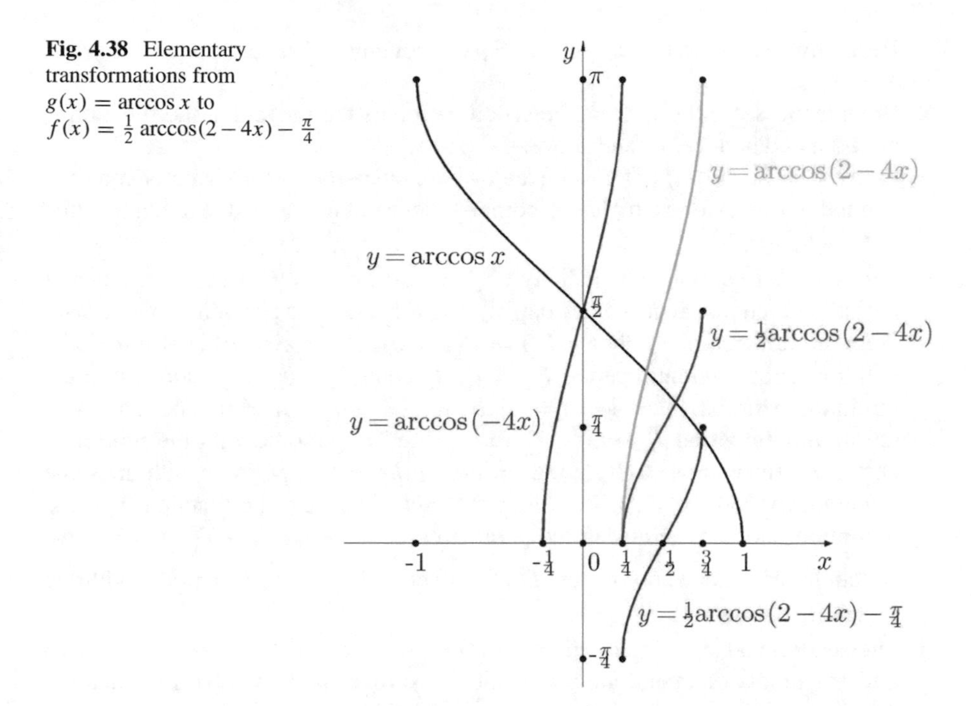

Fig. 4.38 Elementary transformations from $g(x) = \arccos x$ to $f(x) = \frac{1}{2}\arccos(2-4x) - \frac{\pi}{4}$

$$\rightarrow g_3(x) = \frac{1}{2}\arccos\,(2-4x) = \frac{1}{2}g_2(x) \rightarrow f(x)$$

$$= \frac{1}{2}\arccos(2-4x) - \frac{\pi}{4} = g_3(x) - \frac{\pi}{4}.$$

The corresponding transformations of the graphs are shown in Fig. 4.38. We leave to the reader the task to specify the modifications of the properties of the functions which correspond to the proposed transformations.

C. Study of $y = 2\cot\left(\frac{\pi}{4} - \frac{x}{3}\right)$

1. Domain and range. The formula $f(x) = 2\cot\left(\frac{\pi}{4} - \frac{x}{3}\right)$ is valid for any x that does not vanish $\sin\left(\frac{\pi}{4} - \frac{x}{3}\right)$, that is, we should exclude the points $\frac{\pi}{4} - \frac{x}{3} = k\pi$, $\forall k \in \mathbb{Z}$ or $x = \frac{3\pi}{4} + 3k\pi$, $\forall k \in \mathbb{Z}$. Therefore, the domain is $X = \mathbb{R}\backslash\{x_k = \frac{3\pi}{4} + 3k\pi, \forall k \in \mathbb{Z}\}$. When x approaches $\frac{3\pi}{4}$ from the left, the argument of cotangent $t = \frac{\pi}{4} - \frac{x}{3}$ approaches 0 from above (keeping positive values), and consequently, $y = 2\cot t$ goes to $+\infty$. Analogously, when x approaches $\frac{3\pi}{4}$ from the right, the argument $t = \frac{\pi}{4} - \frac{x}{3}$ approaches 0 from below (keeping negative values), and therefore, $y = 2\cot t$ goes to $-\infty$. Besides, when x ranges over the interval $\left(-\frac{9\pi}{4}, \frac{3\pi}{4}\right)$, the variable $t = \frac{\pi}{4} - \frac{x}{3}$ takes all the values on the interval $(0, \pi)$.

Recalling that the range of $\cot t$ is $\mathbb{R}$, we conclude that the range of $f(x)$ is $Y = \mathbb{R}$.

2. Boundedness. It follows from the specification of the range that the function is not bounded both below and above.
3. Parity. The function $f(x)$ is not even or odd, since its domain is not symmetric about the origin (for example, the point $\frac{3\pi}{4}$ does not belong to the domain, while $-\frac{3\pi}{4}$ belongs).
4. Periodicity. To establish periodicity of $f(x)$ we can apply general properties of periodic functions (Sect. 4.5 in Chap. 2). First, if $g(x)$ is a periodic function with the minimum period T_g, then $g_1(x) = g(x+c)$, $c = constant$ is also periodic with the same minimum period $T_1 = T_g$. Second, if $g_1(x)$ is periodic with the minimum period T_1, then $g_2(x) = g_1(ax)$, $a = constant \neq 0$ is periodic with the minimum period $T_2 = \frac{T_1}{|a|}$. Third, if $g_2(x)$ is periodic with the minimum period T_2, then $f(x) = Cg_2(x)$, $C = constant \neq 0$ is periodic with the same minimum period $T = T_2$. Recalling that $\cot t$ is a periodic function with the minimum period π and using the parameters $a = -\frac{1}{3}$, $c = -\frac{3\pi}{4}$, $C = 2$, we conclude that the given function $f(x) = 2\cot\frac{1}{3}\left(\frac{3\pi}{4} - x\right)$ is periodic with the minimum period 3π.
5. Intersection and sign. Intersection with the y-axis occurs at the point $(0, 2)$. To find the points of intersection with the x-axis we have to solve the equation $2\cot\left(\frac{\pi}{4} - \frac{x}{3}\right) = 0$. Using the auxiliary variable $t = \frac{\pi}{4} - \frac{x}{3}$ we rewrite the equation in the form $\cot t = 0$ and recall that $\cot t$ vanishes at the points $t_k = \frac{\pi}{2} + k\pi$, $\forall k \in \mathbb{Z}$. Therefore, we have $\frac{\pi}{4} - \frac{x_k}{3} = \frac{\pi}{2} + k\pi$ or $x_k = -\frac{3\pi}{4} + 3k\pi$, $\forall k \in \mathbb{Z}$. Hence, the points of intersection with the x-axis are $P_k = (x_k, 0)$, $\forall k \in \mathbb{Z}$.

   A specification of the sign of the function is postponed until after the study of monotonicity.
6. Monotonicity and extrema.

   We use again the auxiliary variable $t = \frac{\pi}{4} - \frac{x}{3}$ and employ the properties of monotonicity of the function $\cot t$. Choose the principal interval $(-\frac{9\pi}{4}, \frac{3\pi}{4})$ (a part of the domain without "holes" of the length 3π equal to the minimum period of the function) and take two arbitrary points $x_1 < x_2$ in this interval. The condition $-\frac{9\pi}{4} < x_1 < x_2 < \frac{3\pi}{4}$ is equivalent to $\pi > t_1 = \frac{\pi}{4} - \frac{x_1}{3} > t_2 = \frac{\pi}{4} - \frac{x_2}{3} > 0$. Since $\cot t$ is decreasing on $(0, \pi)$, we have $\cot t_1 < \cot t_2$, and consequently, $f(x_1) < f(x_2)$, that is, $f(x)$ increases on the interval $(-\frac{9\pi}{4}, \frac{3\pi}{4})$.

   Using the 3π-periodicity, we conclude that $f(x)$ is increasing on each interval $(-\frac{9\pi}{4}, \frac{3\pi}{4}) + 3k\pi$, $\forall k \in \mathbb{Z}$. Notice that this does not mean that the function increases over the entire domain (although the union of all these intervals represents the domain) or that it increases on the union of some of these intervals. Since any point of the intervals $(-\frac{9\pi}{4}, \frac{3\pi}{4}) + 3k\pi$, $\forall k \in \mathbb{Z}$ is increasing and the union of all these intervals gives the domain, there is no extremum of any type.

   Recalling that on the principal interval $(-\frac{9\pi}{4}, \frac{3\pi}{4})$ there exists the only zero $x = -\frac{3\pi}{4}$ of $f(x)$ and taking into account increase of $f(x)$ on this interval, we can see that $f(x)$ is negative on $(-\frac{9\pi}{4}, -\frac{3\pi}{4})$ and positive on $(-\frac{3\pi}{4}, \frac{3\pi}{4})$. Then, according to the 3π-periodicity, $f(x)$ is negative on each interval $(-\frac{9\pi}{4}, -\frac{3\pi}{4}) + 3k\pi$ and positive on each interval $(-\frac{3\pi}{4}, \frac{3\pi}{4}) + 3k\pi, \forall k \in \mathbb{Z}$.

Notice also that increase of $f(x)$ on the interval of the length 3π eliminates a possibility of the existence of a period less than 3π (this fact was already proved by using the properties of periodic functions).

7. Concavity and inflection.

   We follow the same line of reasoning as for the study of monotonicity. Choose the left half of the principal interval $(-\frac{9\pi}{4}, -\frac{3\pi}{4}]$ and notice that the inequality $-\frac{9\pi}{4} < x_1 < x_2 < -\frac{3\pi}{4}$ is equivalent to $\pi > t_1 = \frac{\pi}{4} - \frac{x_1}{3} > t_2 = \frac{\pi}{4} - \frac{x_2}{3} > \frac{\pi}{2}$. Since $\cot t$ is concave downward on the interval $[\frac{\pi}{2}, \pi)$, the function $f(x)$ has the same type of concavity on the interval $(-\frac{9\pi}{4}, -\frac{3\pi}{4}]$. In a similar way, it can be shown that $f(x)$ is concave upward on the right half of the principal interval $[-\frac{3\pi}{4}, \frac{3\pi}{4})$. Consequently, $-\frac{3\pi}{4}$ is inflection point.

   Applying the 3π-periodicity, we conclude that $f(x)$ is concave downward on each interval $(-\frac{9\pi}{4}, -\frac{3\pi}{4}] + 3k\pi$, concave upward on each interval $[-\frac{3\pi}{4}, \frac{3\pi}{4}) + 3k\pi$, and has the inflection points $-\frac{3\pi}{4} + 3k\pi, \forall k \in \mathbb{Z}$.

8. The function $f(x) : X = \mathbb{R}\backslash\{x_k = \frac{3\pi}{4} + 3k\pi, \forall k \in \mathbb{Z}\} \to Y = \mathbb{R}$ is surjective, but not injective. Notice that on any interval of the length greater than 3π the function repeats some of its values. Therefore, to obtain an injection we need to choose an interval of the length at most 3π. Since $f(x)$ is increasing on the principal interval $(-\frac{9\pi}{4}, \frac{3\pi}{4})$, it is injective on this interval, and consequently, invertible. Recalling that the principal interval corresponds to the interval $(0, \pi)$ of the variation of the variable $t = \frac{\pi}{4} - \frac{x}{3}$ and that the inverse of $\cot t$, $t \in (0, \pi)$ is arccotangent, we can express the inverse of $f(x) : X = (-\frac{9\pi}{4}, \frac{3\pi}{4}) \to Y = \mathbb{R}$ in the form $f^{-1}(x) = \frac{3\pi}{4} - 3 \operatorname{arccot} \frac{x}{2} : \mathbb{R} \to (-\frac{9\pi}{4}, \frac{3\pi}{4})$.

9. Based on the investigated properties, we can construct the graph of $f(x) = 2\cot\left(\frac{\pi}{4} - \frac{x}{3}\right)$ in the form shown in Fig. 4.39.

Complimentary Properties

The function $f(x) = 2\cot\left(\frac{\pi}{4} - \frac{x}{3}\right)$ goes to $+\infty$ as x approaches $\frac{3\pi}{4} + 3k\pi, \forall k \in \mathbb{Z}$ from the left, and it goes to $-\infty$ as x approaches any of these points from the right. Therefore, each line $x = \frac{3\pi}{4} + 3k\pi, \forall k \in \mathbb{Z}$ is a vertical asymptote. At the remaining points (that is, at any point of the domain) the function is continuous, and consequently, it does not have any other vertical asymptote.

When x goes to $+\infty$, the function does not approach any specific value: for example, at all the points of the path $x_k = -\frac{3\pi}{4} + 3k\pi, k \in \mathbb{N}$ the function has the value 0, while along the path $x_n = 3n\pi, n \in \mathbb{N}$ the function keeps the value 2. This means that there is no horizontal asymptote when x goes to $+\infty$. The same happens when x goes to $-\infty$.

Another way to determine the properties of the function $f(x) = 2\cot\left(\frac{\pi}{4} - \frac{x}{3}\right)$ is by constructing the sequence of elementary transformations which goes from $g(x) = \cot x$ to $f(x) = 2\cot\left(\frac{\pi}{4} - \frac{x}{3}\right)$. This problem is left to the reader.

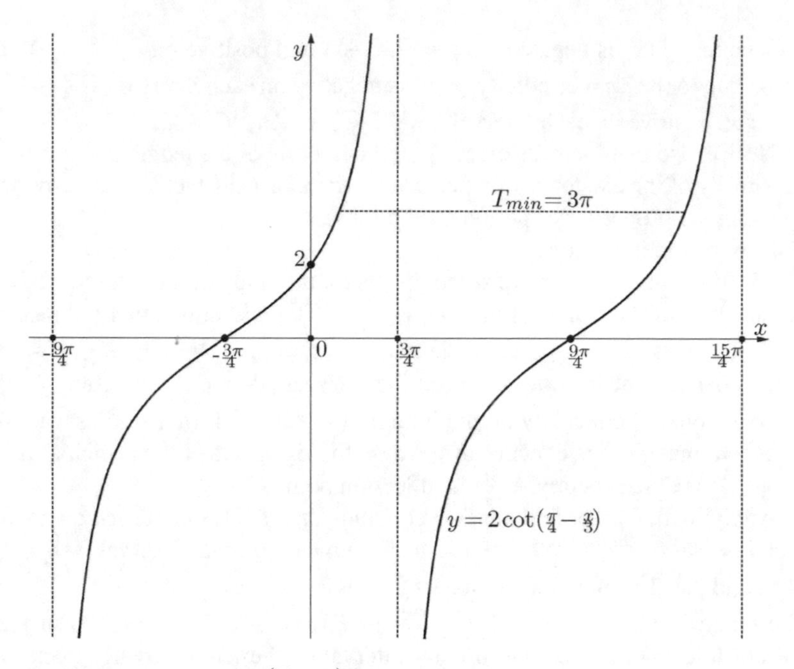

Fig. 4.39 The graph of $y = 2\cot\left(\frac{\pi}{4} - \frac{x}{3}\right)$

4 Solved Exercises

4.1 Study of Exponential Functions

A. Study of $f(x) = 1 - 2^{3x}$

1. Domain and range. All the operations can be made for any real x, which means that the domain is $X = \mathbb{R}$.

   From the fact that $2^{3x} > 0$, $\forall x$ it follows that $y = 1 - 2^{3x} < 1$. Let us show that every number of the interval $(-\infty, 1)$ belongs to the range. Solving the equation $1 - 2^{3x} = y$, $y < 1$, we have $2^{3x} = 1 - y$, whence $3x = \log_2(1-y)$ or $x = \frac{1}{3}\log_2(1-y)$. Therefore, for any $y < 1$ there exists x such that $f(x) = y$, which means that the range is $Y = (-\infty, 1)$.
2. Boundedness. Since the range is $Y = (-\infty, 1)$, the function is unbounded below and bounded above (by the upper bound $M = 1$).
3. Intersection and sign. We leave a specification of these properties until after the study of monotonicity.
4. Parity. Comparing the values $f(-1) = 1 - 2^{-3}$ and $f(1) = 1 - 2^3$, we scc that the function is neither even nor odd.
5. Periodicity. We leave verification of periodicity until after the study of monotonicity.

6. Monotonicity and extrema. Using the properties of powers, for any pair $x_1 < x_2$ we have $2^{3x_1} < 2^{3x_2}$, and consequently, $1 - 2^{3x_1} > 1 - 2^{3x_2}$, which shows that the function is decreasing on $\mathbb{R}$. It follows immediately from this result that there is no extremum of any type.

   Another direct implication of monotonicity on the infinite interval $(-\infty, +\infty)$ is that the function cannot be periodic.

   Besides, it is easy to specify now the sign of the function. Intersection with both coordinate axes occurs at the same point $P_0 = (0, 0)$. Since the function decreases on $X = \mathbb{R}$, it follows that $f(x)$ is positive on the interval $(-\infty, 0)$ and is negative on the interval $(0, +\infty)$.
7. Concavity and inflection. For two arbitrary points $x_1 < x_2$, we can evaluate the quantity $2D_2 \equiv f(x_1) + f(x_2) - 2f(\frac{x_1+x_2}{2})$ as follows:

$$2D_2 = (1 - 2^{3x_1}) + (1 - 2^{3x_2}) - 2(1 - 2^{3(x_1+x_2)/2})$$

$$= 2 \cdot 2^{3(x_1+x_2)/2} - 2^{3x_1} - 2^{3x_2} = -\left(2^{3x_1/2} - 2^{3x_2/2}\right)^2 < 0.$$

   Therefore, the function is concave downward over the entire domain $(-\infty, +\infty)$.
8. Complimentary properties

   When $x \to -\infty$, the function approaches 1, which implies that $y = 1$ is a horizontal asymptote. When $x \to +\infty$, the function goes to $-\infty$.

   At each point of the domain the function is continuous, that is, $f(x) \to f(x_0)$ when $x \to x_0$. Therefore, no infinite behavior is detected at the points of the domain $X = \mathbb{R}$, which means that there is no vertical asymptote.
9. Using the analyzed properties and marking some important points such as $P_0 = (0, 0)$ (intersection with the coordinate axes), $P_1 = (-\frac{1}{3}, \frac{1}{2})$ (a point with a negative x-coordinate) and $P_2 = (\frac{1}{3}, -1)$ (a point with a positive x-coordinate), we can plot the graph shown in Fig. 4.40.

B. Study of $f(x) = 1 - 2^{-3|x|}$

1. Domain. All the operations are valid for any real x, and consequently, the domain is $X = \mathbb{R}$.

1–8. The remaining properties can be derived from the relationship between this function and the previous one $g(x) = 1 - 2^{3x}$. Indeed, the function $f(x) = 1 - 2^{-3|x|}$ is even, because $f(-x) = 1 - 2^{-3|-x|} = 1 - 2^{-3|x|} = f(x)$, $\forall x \in \mathbb{R}$, and on the part $(-\infty, 0]$ of the domain it coincides with $g(x)$. So, extending the known properties of $g(x)$ from the interval $(-\infty, 0]$ onto the interval $(0, +\infty)$, according to the evennes of $f(x)$, we can obtain all the properties of $f(x)$ listed below:

(1) the range is $Y = [0, 1)$;
(2) the function is bounded with the lower bound $m = 0$ and the upper bound $M = 1$;

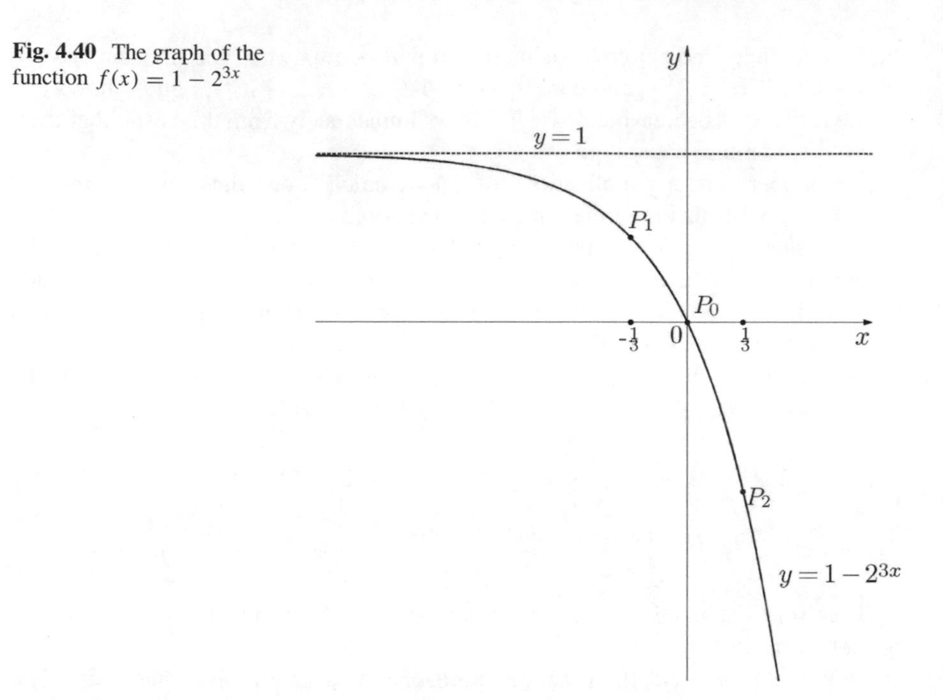

Fig. 4.40 The graph of the function $f(x) = 1 - 2^{3x}$

(3) the only point of intersection with the coordinate axes is the origin; for any $x \neq 0$ the function is positive;
(4) the function is even, as was already shown;
(5) the function is not periodic;
(6) $f(x)$ decreases on $(-\infty, 0]$ and increases on $[0, +\infty)$; consequently, $P_0 = (0, 0)$ is a (strict) local and global minimum;
(7) $f(x)$ is concave downward on $(-\infty, 0]$ and on $[0, +\infty)$; however, it does not concave downward on $(-\infty, +\infty)$, which is shown by choosing the points $x_1 = -1$ and $x_2 = 1$ for which $2D_2 = f(x_1) + f(x_2) - 2f(\frac{x_1+x_2}{2}) = f(-1) + f(1) - 2f(0) = 2(1 - 2^{-3}) > 0$; there is no inflection point;
(8) when $x \to \pm\infty$, the function approaches 1, which means that $y = 1$ is a horizontal asymptote; at each point of the domain the function is continuous, and consequently, there is no vertical asymptote.

9. Using the investigated properties, and marking some important points, such as $P_0 = (0, 0)$ (intersection with the coordinate axes, and also the local and global minimum) and $P_1 = (-1, \frac{7}{8})$ (a point with a negative x-coordinate), and applying the property of the evenness, we can construct the graph of the function as shown in Fig. 4.41.

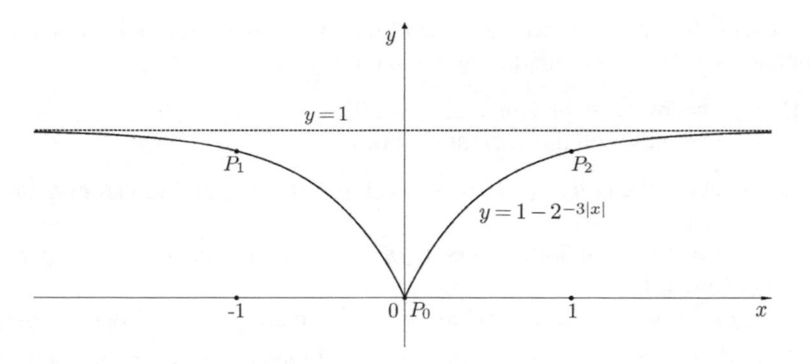

Fig. 4.41 The graph of the function $f(x) = 1 - 2^{-3|x|}$

C. Study of $f(x) = 2^{3x^2} - 4$

1. Domain and range. All the operations can be performed for any real x, which means that the domain is $X = \mathbb{R}$.

   Since $3x^2 \geq 0$, $\forall x \in \mathbb{R}$, we get $2^{3x^2} \geq 1$ and $2^{3x^2} - 4 \geq -3$, $\forall x$. Taking any $y \geq -3$, we transform the equation $2^{3x^2} - 4 = y$ into $3x^2 = \log_2(y+4)$ or $x^2 = \frac{1}{3}\log_2(y+4)$, where the right-hand side is non-negative. Consequently, the last equation has the solution $x = \pm\sqrt{\frac{1}{3}\log_2(y+4)}$, which means that $Y = [-3, +\infty)$ is the range.
2. Boundedness. Since the range is $Y = [-3, +\infty)$, the function is bounded below (by the lower bound $m = -3$, which is the global minimum) and unbounded above.
3. Intersection and sign. The point of intersection with the y-axis is $P_0 = (0, -3)$. To find the x-coordinates of the points of intersection with the x-axis we substitute $y = 0$ in the obtained above relation for x^2 and get the two solutions: $x = \pm\sqrt{\frac{1}{3}\log_2(0+4)} = \pm\sqrt{\frac{2}{3}}$. We leave a specification of the sign until after the study of monotonicity.
4. Parity. The function is even: $f(-x) = 2^{3(-x)^2} - 4 = 2^{3x^2} - 4 = f(x)$, $\forall x \in \mathbb{R}$. Therefore, the investigation of monotonicity and extrema can be made first on the interval $[0, +\infty)$, and then extended on the entire domain due to the symmetry of the function about the y-axis.
5. Periodicity. The verification of periodicity is left until after the study of monotonicity.
6. Monotonicity and extrema. For any pair of points $0 \leq x_1 < x_2$ we have $3x_1^2 < 3x_2^2$, whence, by the properties of powers, $2^{3x_1^2} - 4 < 2^{3x_2^2} - 4$, that is, the function increases on $[0, +\infty)$.

Using the evenness, we can extend this result onto the entire domain and obtain the following properties of monotonicity and extrema:

(1) $f(x)$ decreases on the interval $(-\infty, 0]$;
(2) $f(x)$ increases on the interval $[0, +\infty)$.

Consequently, the point $P_0 = (0, -3)$ is a (strict) local and global minimum of the function.

Increase on the infinite interval $[0, +\infty)$ implies that $f(x)$ cannot be a periodic function.

Besides, it is easy to specify the sign using monotonicity. The two points of intersection with the x-axis are $P_1 = (-\sqrt{\frac{2}{3}}, 0)$ and $P_2 = (\sqrt{\frac{2}{3}}, 0)$. Since $f(x)$ decreases on $(-\infty, 0]$ and increases on $[0, +\infty)$, the function is positive when $x < -\sqrt{\frac{2}{3}}$ and $x > \sqrt{\frac{2}{3}}$, and negative when $-\sqrt{\frac{2}{3}} < x < \sqrt{\frac{2}{3}}$.

7. Concavity and inflection. Take two arbitrary points $x_1 < x_2$ and evaluate the quantity $2D_2 \equiv f(x_1) + f(x_2) - 2f(\frac{x_1+x_2}{2})$:

$$2D_2 = 2^{3x_1^2} + 2^{3x_2^2} - 2 \cdot 2^{3(\frac{x_1+x_2}{2})^2} \geq 2^{3x_1^2} + 2^{3x_2^2} - 2 \cdot 2^{3\frac{x_1^2+x_2^2}{2}}$$
$$= \left(2^{3\frac{x_1^2}{2}} - 2^{3\frac{x_2^2}{2}}\right)^2 > 0.$$

(The first inequality follows from the evaluation $(x_1 + x_2)^2 \leq 2(x_1^2 + x_2^2)$.) Therefore, the function is concave upward over the entire domain $\mathbb{R}$, and consequently, has no inflection point.

8. Complimentary properties

When $x \to \pm\infty$, the function goes to $+\infty$, which implies that there is no horizontal asymptote.

At each point of the domain the function is continuous, that is, $f(x) \to f(x_0)$ when $x \to x_0$. Hence, it does not exist an infinite behavior at the points of the domain $X = \mathbb{R}$, which implies that there is no vertical asymptote.

9. Using the revealed properties, and marking some important points, such as $P_0 = (0, -3)$ (intersection with the y-axis and local and global minimum), $P_1 = (-\sqrt{\frac{2}{3}}, 0)$ and $P_2 = (\sqrt{\frac{2}{3}}, 0)$ (intersection with the x-axis), and employing the evenness of the function, we construct the graph shown in Fig. 4.42.

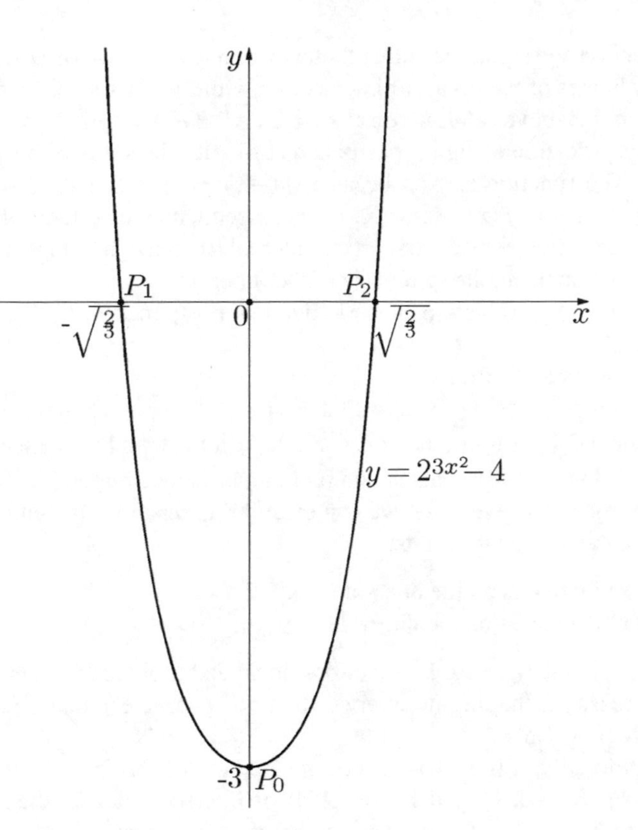

Fig. 4.42 The graph of the function $f(x) = 2^{3x^2} - 4$

4.2 Study of Logarithmic Functions

A. Study of $f(x) = 1 - \log_5(x^2 + 1)$

1. Domain and range. The argument $t = x^2 + 1$ is positive for any real x, and consequently, all the operations can be made for any x, that is, the domain is $X = \mathbb{R}$.

   Since $x^2+1 \geq 1, \forall x \in \mathbb{R}$, we have $\log_5(x^2+1) \geq 0$ and $1-\log_5(x^2+1) \leq 1$, $\forall x$. This shows that the range Y is contained in the interval $(-\infty, 1]$. For any $y \leq 1$, the equation $1 - \log_5(x^2 + 1) = y$ can be written as $\log_5(x^2 + 1) = 1 - y$ and, exponentiating each side, transformed into $x^2 + 1 = 5^{1-y}$ or $x^2 = 5^{1-y} - 1$. When $y \leq 1$ the right-hand side is non-negative and the last equation has solutions $x = \pm\sqrt{5^{1-y} - 1}$, which means that $Y = (-\infty, 1]$ is the range of the function.
2. Boundedness. Since the range is $Y = (-\infty, 1]$, the function is unbounded below and bounded above.

3. Intersection and sign. The point of intersection with the y-axis is $P_0 = (0, 1)$. The x-coordinates of the points of intersection with the x-axis are found substituting $y = 0$ in the above solution for x: $x = \pm\sqrt{5^{1-0} - 1} = \pm 2$.
   A specification of sign is postponed until after the study of monotonicity.
4. Parity. The function is even because $f(-x) = 1 - \log_5((-x)^2 + 1) = 1 - \log_5(x^2 + 1) = f(x)$, $\forall x \in \mathbb{R}$. As a consequence, the study of monotonicity and extrema can be made first on the interval $[0, +\infty)$, and then extended on the entire domain using the symmetry of the function.
5. Periodicity. A verification of periodicity is postponed until after the study of monotonicity.
6. Monotonicity and extrema.
   For any pair $0 \le x_1 < x_2$ we have $x_1^2 + 1 < x_2^2 + 1$, which implies, by the properties of logarithms, that $\log_5(x_1^2+1) < \log_5(x_2^2+1)$, whence $1-\log_5(x_1^2+1) > 1 - \log_5(x_2^2 + 1)$, that is, the function is decreasing on $[0, +\infty)$.
   Employing the evenness, we can extend this result on the entire domain and arrive at the following results:

   (1) $f(x)$ increases on the interval $(-\infty, 0]$;
   (2) $f(x)$ decreases on the interval $[0, +\infty)$.

   Then, the point $P_0 = (0, 1)$ is a (strict) local and global maximum.
   Decrease on the infinite interval $[0, +\infty)$ guarantees that $f(x)$ cannot be a periodic function.
   Additionally, it is easy to determine the sign of the function using the found points $P_1 = (-2, 0)$ and $P_2 = (2, 0)$ of intersection with the x-axis and the specified properties of monotonicity. Since $f(x)$ increases on $(-\infty, 0]$, it takes negative values on $(-\infty, -2)$ and positive on $(-2, 0]$. Since $f(x)$ decreases on $[0, +\infty)$, it is positive on $[0, 2)$ and negative on $(2, +\infty)$.
7. Concavity and inflection.
   Take two arbitrary points $x_1 < x_2$ and evaluate the quantity $2D_2 \equiv f(x_1) + f(x_2) - 2f(x_0)$, where $x_0 = \frac{x_1+x_2}{2}$:

$$2D_2 = (1 - \log_5(x_1^2 + 1)) + (1 - \log_5(x_2^2 + 1)) - 2 \cdot \left[1 - \log_5(x_0^2 + 1)\right]$$
$$= \log_5 \frac{(x_0^2 + 1)^2}{(x_1^2 + 1)(x_2^2 + 1)}.$$

To assess the expression inside the last logarithm, let us evaluate the auxiliary parameter A:

$$A = (x_0^2 + 1)^2 - (x_1^2 + 1)(x_2^2 + 1) = x_0^4 + 2x_0^2 - x_1^2 x_2^2 - x_1^2 - x_2^2$$
$$= \left(\frac{x_1 + x_2}{2}\right)^4 + 2\left(\frac{x_1 + x_2}{2}\right)^2 - x_1^2 x_2^2 - x_1^2 - x_2^2$$

$$= \left(\left(\frac{x_1+x_2}{2}\right)^2 - x_1x_2\right)\left(\left(\frac{x_1+x_2}{2}\right)^2 + x_1x_2\right)$$

$$+ \frac{1}{2}\left(x_1^2 + 2x_1x_2 + x_2^2 - 2x_1^2 - 2x_2^2\right)$$

$$= \frac{(x_1-x_2)^2}{4}\left(\left(\frac{x_1+x_2}{2}\right)^2 + x_1x_2\right) - \frac{(x_1-x_2)^2}{2}$$

$$= \frac{(x_1-x_2)^2}{4}\left(\left(\frac{x_1+x_2}{2}\right)^2 + x_1x_2 - 2\right).$$

There are two different situations.

(1) If $x_1, x_2 \in [-1, 1]$, then $\left(\frac{x_1+x_2}{2}\right)^2 < 1$ and $x_1x_2 < 1$, which implies that $A < 0$. Then $\frac{(x_0^2+1)^2}{(x_1^2+1)(x_2^2+1)} < 1$, and consequently, $D_2 < 0$, which means that $f(x)$ is concave downward on $[-1, 1]$.

(2) If $x_1, x_2 \in (-\infty, -1]$ or $x_1, x_2 \in [1, +\infty)$, then $\left(\frac{x_1+x_2}{2}\right)^2 > 1$ and $x_1x_2 > 1$, which implies that $A > 0$. Then $\frac{(x_0^2+1)^2}{(x_1^2+1)(x_2^2+1)} > 1$, and consequently, $D_2 > 0$, which means that $f(x)$ is concave upward on $(-\infty, -1]$ and also on $[1, +\infty)$.

The immediate implication of these results is that $P_3 = (-1, 1 - \log_5 2)$ and $P_4 = (1, 1 - \log_5 2)$ are inflection points.

8. Complimentary properties

   When $x \to \pm\infty$, the function goes to $-\infty$, and consequently, there is no horizontal asymptote.

   At every point of the domain $X = \mathbb{R}$ the function is continuous, that is, $f(x) \to f(x_0)$ as $x \to x_0$. Therefore, there is no infinite behavior at the points of the domain, which implies that there is no vertical asymptote.

9. Using the investigated properties, and marking some important points, like $P_0 = (0, 1)$ (intersection with the y-axis and local/global maximum), $P_1 = (-2, 0)$ and $P_2 = (2, 0)$ (intersection with the x-axis), $P_3 = (-1, 1 - \log_5 2)$ and $P_4 = (1, 1 - \log_5 2)$ (inflection points), and employing the symmetry about the y-axis, we plot the graph shown in Fig. 4.43.

B. Study of $f(x) = \log_3(|x| - 1) - 1$

1. Domain and range. The argument of a logarithm must be positive, which leads to the inequality $|x| > 1$. There is no other restriction, and therefore, the domain is $X = (-\infty, -1) \cup (1, +\infty)$.

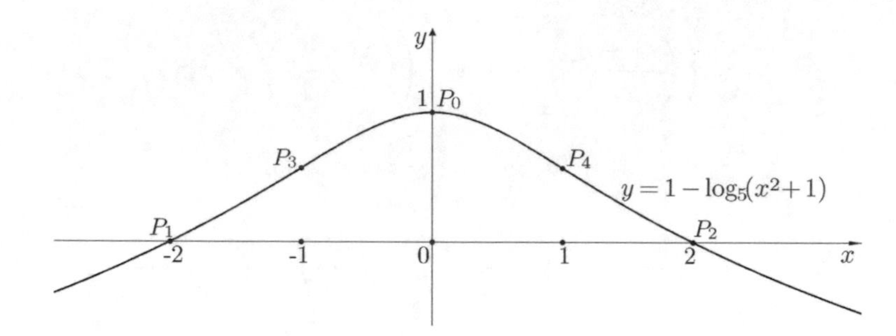

Fig. 4.43 The graph of the function $f(x) = 1 - \log_5(x^2 + 1)$

Solving the equation $\log_3(|x| - 1) - 1 = y, \forall y \in \mathbb{R}$, we find $|x| = 3^{y+1} + 1 > 1$. Since the right-hand side is positive, we have the two solutions $x = \pm(3^{y+1} + 1) \in X$ for any real y, which means that $Y = \mathbb{R}$ is the range of the function.

2. Boundedness. Since the range is $Y = \mathbb{R}$, the function is unbounded both below and above.
3. Intersection and sign. There is no intersection with the y-axis, because the point $x = 0$ does not belong to the domain. The x-coordinates of the points of intersection with the x-axis are found by substituting $y = 0$ in the above solution for x: $x = \pm(3^{0+1} + 1) = \pm 4$.

   A specification of sign is postponed until after the study of monotonicity.
4. Parity. The function is even because $f(-x) = \log_3(|-x|-1)-1 = \log_3(|x|-1) - 1 = f(x), \forall x \in X$. As a consequence, the study of monotonicity and extrema, as well as concavity and inflection can be made first on the interval $(1, +\infty)$, and then extended on the entire domain using the symmetry of the function.
5. Periodicity. A verification of periodicity is postponed until after the study of monotonicity.
6. Monotonicity. For any pair $1 < x_1 < x_2$ we have $f(x_2) - f(x_1) = \log_3(|x_2| - 1) - \log_3(|x_1| - 1) = \log_3 \frac{|x_2|-1}{|x_1|-1} > 0$, because $\frac{|x_2|-1}{|x_1|-1} > 1$. Therefore, the function is increasing on $(1, +\infty)$.

   Increase on the infinite interval $(1, +\infty)$ guarantees that $f(x)$ cannot be a periodic function.
7. Concavity.

   Take two points such that $1 < x_1 < x_2$ and evaluate the quantity $2D_2 \equiv f(x_1) + f(x_2) - 2f(x_0)$, where $x_0 = \frac{x_1+x_2}{2}$:

$$2D_2 = \log_3(|x_2| - 1) + \log_3(|x_1| - 1) - 2\log_3(|x_0| - 1)$$
$$= \log_3 \frac{(|x_1| - 1)(|x_2| - 1)}{(|x_0| - 1)^2}.$$

To assess the expression inside the last logarithm, let us evaluate the auxiliary parameter $A = (|x_1| - 1)(|x_2| - 1) - (|x_0| - 1)^2$ (notice that $|x_0| = x_0$, $|x_1| = x_1$, $|x_2| = x_2$ when $1 < x_1 < x_2$):

$$\begin{aligned} A &= (x_1 - 1)(x_2 - 1) - (x_0 - 1)^2 \\ &= x_1 x_2 - x_1 - x_2 - x_0^2 + 2x_0 \\ &= x_1 x_2 - x_1 - x_2 - \left(\frac{x_1 + x_2}{2}\right)^2 + 2\frac{x_1 + x_2}{2} \\ &= -\frac{(x_1 - x_2)^2}{4} < 0. \end{aligned}$$

Therefore, $\frac{(|x_1|-1)(|x_2|-1)}{(|x_0|-1)^2} < 1$, and consequently, $D_2 < 0$, which implies that $f(x)$ is concave downward on $(1, +\infty)$.

6–7. Monotonicity and extrema, concavity and inflection.

According to the evenness of the function, we can extend the obtained results of monotonicity and concavity on the entire domain. First, we obtain the following results about monotonicity and extrema:

(1) $f(x)$ decreases on $(-\infty, -1)$;
(2) $f(x)$ increases on $(1, +\infty)$.

Since every point of the domain is an interior point of one of the two intervals of monotonicity, there is no extremum of any type.

Next, we get the following results about concavity and inflection:

(1) $f(x)$ is concave downward on $(-\infty, -1)$;
(2) $f(x)$ is concave downward on $(1, +\infty)$.

Consequently, there is no inflection point.

Moreover, it is easy now to specify the sign of the function. Recall that the two points of intersection with the x-axis are $P_1 = (-4, 0)$ and $P_2 = (4, 0)$. Since $f(x)$ decreases on $(-\infty, -1)$, it is positive on $(-\infty, -4)$ and negative on $(-4, -1)$. Since $f(x)$ increases on $(1, +\infty)$, it is negative on $(1, 4)$ and positive on $(4, +\infty)$.

8. Complimentary properties.

When $x \to \pm\infty$, the function goes to $+\infty$, and consequently, there is no horizontal asymptote.

At each point of the domain the function is continuous, that is, $f(x) \to f(x_0)$ as $x \to x_0$. Therefore, there is no infinite behavior at the points of the domain X. However, when $x \to -1^-$, then $f(x) \to -\infty$, which shows that $x = -1$ is a vertical asymptote. In the same way, when $x \to 1^+$, then $f(x) \to -\infty$, which indicates that $x = 1$ is another vertical asymptote.

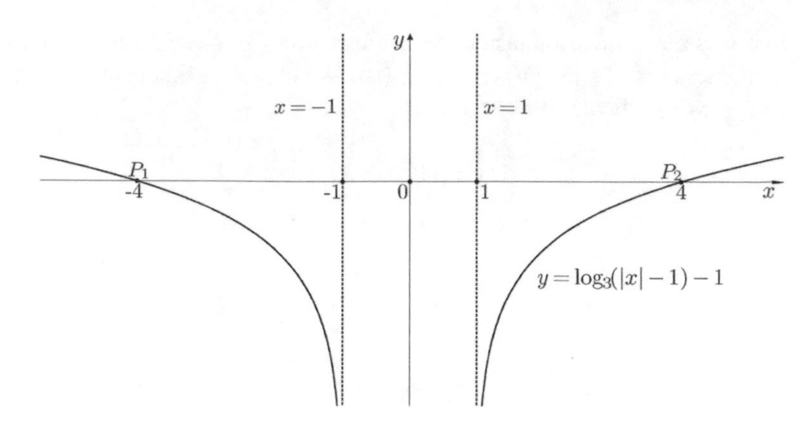

Fig. 4.44 The graph of the function $f(x) = \log_3(|x| - 1) - 1$

9. Using the investigated properties, and marking some important points, like $P_1 = (-4, 0)$ and $P_2 = (4, 0)$ (intersection with the x-axis), and applying the symmetry about the $y-$axis, we construct the graph shown in Fig. 4.44.

C. Study of $f(x) = \log_2(4 - |x|) - 1$

1. Domain and range. The argument $t = 4 - |x|$ of the logarithm must be positive, which leads to the inequality $|x| < 4$. There is no other restriction on the involved operations, which means that $X = (-4, 4)$ is the domain.

   Solving the equation $\log_2(4 - |x|) - 1 = y$ with respect to $|x|$, we find $|x| = 4 - 2^{y+1}$. This equation has solutions only if the right-hand side is non-negative, whence we get the restriction on the values of y: $2^{y+1} \leq 4$ or $y + 1 \leq 2$. Therefore, $Y = (-\infty, 1]$ is the range.
2. Boundedness. Since the range is $Y = (-\infty, 1]$, the function is unbounded below and bounded above.
3. Intersection and sign. Intersection with the y-axis occurs at the point $P_0 = (0, 1)$. The x-coordinates of the points of intersection with the x-axis are found by substituting $y = 0$ into the above solution for $|x|$: $|x| = 4 - 2^{0+1} = 2$, whence $x = \pm 2$. A specification of sign is deferred until after the study of monotonicity.
4. Parity. The function is even because $f(-x) = \log_2(4 - |-x|) - 1 = \log_2(4 - |x|) - 1 = f(x)$, $\forall x \in X$. As a consequence, the study of monotonicity and extrema can be made first on the interval $[0, +\infty)$, and then extended on the entire domain using the symmetry of the function.
5. Periodicity. Since the domain is bounded, the function cannot be periodic.
6. Monotonicity and extrema. For any two points such that $0 \leq x_1 < x_2 < 4$ we have $f(x_2) - f(x_1) = \log_2(4 - |x_2|) - \log_2(4 - |x_1|) = \log_2 \frac{4-|x_2|}{4-|x_1|} < 0$, because $\frac{4-|x_2|}{4-|x_1|} < 1$. Therefore, the function decreases on $[0, 4)$.

Employing the evenness, we can extend this result on the entire domain and arrive at the following results:

(1) $f(x)$ increases on the interval $(-4, 0]$;
(2) $f(x)$ decreases on the interval $[0, 4)$.

Then, the point $P_0 = (0, 1)$ is a (strict) local and global maximum.

Besides, using monotonicity it is easy to specify the sign of the function. Recall that the found two points of intersection with the x-axis are $P_1 = (-2, 0)$ and $P_2 = (2, 0)$. Since $f(x)$ increases on $(-4, 0]$, the function is negative on $(-4, -2)$ and positive on $(-2, 0]$. Since $f(x)$ decreases on $[0, 4)$, the function is positive on $[0, 2)$ and negative on $(2, 4)$.

7. Concavity and inflection.

Take two points such that $-4 < x_1 < x_2 < 4$ and evaluate the quantity $2D_2 \equiv f(x_1) + f(x_2) - 2f(x_0)$, where $x_0 = \frac{x_1+x_2}{2}$:

$$2D_2 = \log_2(4-|x_1|)+\log_2(4-|x_2|)-2\log_2(4-|x_0|) = \log_2 \frac{(4-|x_1|)(4-|x_2|)}{(4-|x_0|)^2}.$$

Taking into account that $|x_0| = \frac{|x_1+x_2|}{2} \le \frac{|x_1|+|x_2|}{2} < 4$, whence $4 - |x_0| \ge 4 - \frac{|x_1|+|x_2|}{2} > 0$, we can proceed with the evaluation of D_2 as follows:

$$2D_2 = \log_2 \frac{(4-|x_1|)(4-|x_2|)}{(4-|x_0|)^2} \le \log_2 \frac{(4-|x_1|)(4-|x_2|)}{\left(4-\frac{|x_1|+|x_2|}{2}\right)^2}.$$

To asses the argument of the last logarithm, let us evaluate the auxiliary parameter A:

$$A = (4-|x_1|)(4-|x_2|) - \left(4-\frac{|x_1|+|x_2|}{2}\right)^2$$

$$= 16 - 4|x_1| - 4|x_2| + |x_1||x_2| - [16 - 4|x_1| - 4|x_2|] - \left(\frac{|x_1|+|x_2|}{2}\right)^2$$

$$= |x_1||x_2| - \left(\frac{|x_1|+|x_2|}{2}\right)^2 = -\left(\frac{|x_1|-|x_2|}{2}\right)^2 < 0.$$

Therefore, $\frac{(4-|x_1|)(4-|x_2|)}{\left(4-\frac{|x_1|+|x_2|}{2}\right)^2} < 1$, which implies that $D_2 < 0$, and consequently, $f(x)$ is concave downward on $(-4, 4)$. Since the function keeps the same type of concavity over the entire domain, there is no inflection point.

8. Complimentary properties

The independent variable cannot go to $\pm\infty$ since the domain is bounded. This implies that there is no horizontal asymptote.

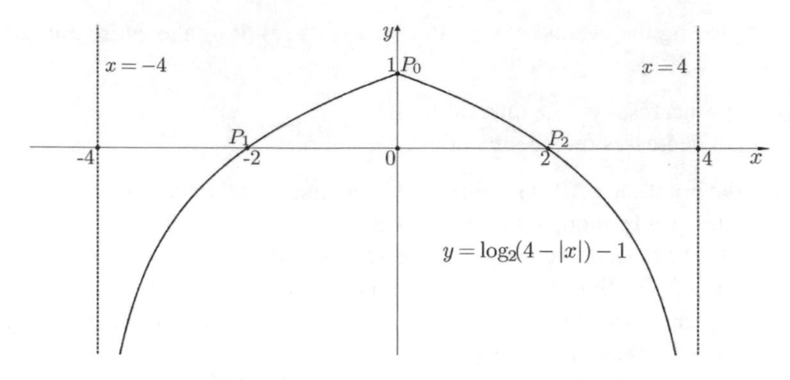

Fig. 4.45 The graph of the function $f(x) = \log_2(4 - |x|) - 1$

At each point of the domain the function is continuous, that is, $f(x) \to f(x_0)$ as $x \to x_0$. However, when $x \to -4^+$ and $x \to 4^-$ the function goes to $-\infty$, which implies that $x = -4$ and $x = 4$ are vertical asymptotes.

9. Using the investigated properties, and marking some important points, such as $P_0 = (0, 1)$ (intersection with the y-axis and also local and global maximum), $P_1 = (-2, 0)$ and $P_2 = (2, 0)$ (intersection with the x-axis), and employing the symmetry about the y-axis, we plot the graph shown in Fig. 4.45.

4.3 Study of Trigonometric Functions

A. Study of $f(x) = \tan x + \cot x$

1. Domain and range. The domain consists of all the real numbers at which both $\tan x$ and $\cot x$ are defined. Consequently, we have $X = \mathbb{R}\backslash\{x_k = \frac{k\pi}{2}, \forall k \in \mathbb{Z}\}$.

   To determine the range, we represent the function in the following equivalent form: $f(x) = \tan x + \cot x = \frac{\sin x}{\cos x} + \frac{\cos x}{\sin x} = \frac{1}{\sin x \cos x} = \frac{2}{\sin 2x}$. Since the range of $\sin 2x$ is $[-1, 1]$, the ratio $\frac{2}{\sin 2x}$ takes all the values whose absolute value is greater than or equal to 2, which means that $Y = (-\infty, -2] \cup [2, +\infty)$ is the range of the function.
2. Boundedness. Since the range is $Y = (-\infty, -2] \cup [2, +\infty)$, the function is unbounded both below and above.
3. Intersection and sign. The point $x = 0$ is out of the domain, and consequently, there is no intersection with the y-axis. On the other hand, the equation $\frac{2}{\sin 2x} = 0$ has no solutions, which means that there is no intersection with the x-axis also. We specify the sign after investigation of symmetries.
4. Parity. The function is odd as the sum of the two odd functions. (It can also be proved directly by the definition.)

5. Periodicity. The functions $\tan x$ and $\cot x$ have the same fundamental period π, which implies that the given function is π-periodic. The fact that π is the fundamental period of $f(x)$ follows from the representation $f(x) = \frac{2}{\sin 2x}$, where $\sin 2x$ has the fundamental period π.

   Now let us specify the sign of the function. Consider the set $(-\frac{\pi}{2}, 0) \cup (0, \frac{\pi}{2})$ of the length π (the points $-\frac{\pi}{2}, 0, \frac{\pi}{2}$ are excluded because they do not belong to the domain). From the formula $f(x) = \frac{2}{\sin 2x}$ it follows immediately that $f(x) < 0$ when $x \in (-\frac{\pi}{2}, 0)$ and $f(x) > 0$ when $x \in (0, \frac{\pi}{2})$. Using the π-periodicity of $f(x)$, we conclude that $f(x) < 0$ when $x \in (-\frac{\pi}{2}, 0) + k\pi, \forall k \in \mathbb{Z}$ and $f(x) > 0$ when $x \in (0, \frac{\pi}{2}) + k\pi, \forall k \in \mathbb{Z}$.
6. Monotonicity and extrema.

   Due to the oddness and π-periodicity, the investigation of monotonicity and extrema can be initially performed on the interval $(0, \frac{\pi}{2})$ and the obtained results subsequently extended on the entire domain.

   For the two points $0 < x_1 < x_2 < \frac{\pi}{2}$ we obtain $D_1 = f(x_2) - f(x_1) = \frac{2}{\sin 2x_2} - \frac{2}{\sin 2x_1} = 2\frac{\sin 2x_1 - \sin 2x_2}{\sin 2x_1 \sin 2x_2} = \frac{4\sin(x_1 - x_2)}{\sin 2x_1 \sin 2x_2}\cos(x_1 + x_2)$. The first factor (the quotient) is negative, since $\sin(x_1 - x_2) < 0$ for $-\frac{\pi}{2} < x_1 - x_2 < 0$ and $\sin t > 0$ for any t in $(0, \pi)$. Therefore, the sign of D_1 depends on the second factor. If $0 < x_1 < x_2 \le \frac{\pi}{4}$, then $0 < x_1 + x_2 < \frac{\pi}{2}$ and $\cos(x_1 + x_2) > 0$, which guarantees the negative sign of D_1 and decrease of $f(x)$. If $\frac{\pi}{4} \le x_1 < x_2 < \frac{\pi}{2}$, then $\frac{\pi}{2} < x_1 + x_2 < \pi$ and $\cos(x_1 + x_2) < 0$, which gives the positive sign of D_1 and increase of $f(x)$. Consequently, the point $x = \frac{\pi}{4}$ is a (strict) local minimum of the function.

   Applying the oddness, we conclude that the function increases on $(-\frac{\pi}{2}, -\frac{\pi}{4}]$ and decreases on $[-\frac{\pi}{4}, 0)$, having a (strict) local maximum at $x = -\frac{\pi}{4}$.

   Finally, using the π-periodicity, we arrive at the following results:

   (1) $f(x)$ increases on each of the intervals $(-\frac{\pi}{2}, -\frac{\pi}{4}] + k\pi, \forall k \in \mathbb{Z}$;
   (2) $f(x)$ decreases on each of the intervals $[-\frac{\pi}{4}, 0) + k\pi, \forall k \in \mathbb{Z}$;
   (3) $f(x)$ decreases on each of the intervals $(0, \frac{\pi}{4}] + k\pi, \forall k \in \mathbb{Z}$;
   (4) $f(x)$ increases on each of the intervals $[\frac{\pi}{4}, \frac{\pi}{2}) + k\pi, \forall k \in \mathbb{Z}$.

   Consequently, the points $x_k = \frac{\pi}{4} + k\pi, \forall k \in \mathbb{Z}$ are (strict) local minima and $x_n = -\frac{\pi}{4} + n\pi, \forall n \in \mathbb{Z}$ are (strict) local maxima.
7. Concavity and inflection.

   Again, we analyze first the concavity on the interval $(0, \frac{\pi}{2})$ and then extend the results using the oddness and periodicity of $f(x)$.

   Take two arbitrary points $0 < x_1 < x_2 < \frac{\pi}{2}$, define the midpoint $x_0 = \frac{x_1 + x_2}{2}$ and the increment $h = x_2 - x_0 = x_0 - x_1$, and evaluate the quantity $2D_2 \equiv f(x_1) + f(x_2) - 2f(x_0)$ applying the formula for D_1:

$$
\begin{aligned}
2D_2 &= f(x_2) - f(x_0) - (f(x_0) - f(x_1)) \\
&= \frac{2}{\sin 2x_2} - \frac{2}{\sin 2x_0} - \left(\frac{2}{\sin 2x_0} - \frac{2}{\sin 2x_1}\right)
\end{aligned}
$$

$$= \frac{4\sin(x_0 - x_2)}{\sin 2x_0 \sin 2x_2}\cos(x_0 + x_2) - \frac{4\sin(x_1 - x_0)}{\sin 2x_1 \sin 2x_0}\cos(x_1 + x_0)$$

$$= \frac{-4\sin h}{\sin 2x_0 \sin 2x_2 \sin 2x_1}\left[\sin 2x_1 \cos(x_0 + x_2) - \sin 2x_2 \cos(x_1 + x_0)\right].$$

The expression in the brackets can be represented as follows:

$$\begin{aligned} A &\equiv \sin 2x_1 \cos(x_0 + x_2) - \sin 2x_2 \cos(x_1 + x_0) \\ &= \frac{1}{2}\left[\sin(2x_1 + x_0 + x_2) + \sin(2x_1 - x_0 - x_2)\right. \\ &\quad \left. - \sin(2x_2 + x_0 + x_1) - \sin(2x_2 - x_0 - x_1)\right] \end{aligned}$$

$$= \sin\frac{x_1 - x_2}{2}\cos\frac{3x_1 + 3x_2 + 2x_0}{2} + \sin\frac{3x_1 - 3x_2}{2}\cos\frac{x_1 + x_2 - 2x_0}{2}$$

$$= -\sin h \cos 4x_0 - \sin 3h \cos 0 = -\sin h \cos 4x_0 - \sin 3h.$$

The last term can still be expressed in the form

$$\begin{aligned} \sin 3h &= \sin(h + 2h) = \sin h \cos 2h + \cos h \sin 2h \\ &= \sin h(\cos 2h + 2\cos^2 h) = \sin h(1 + 2\cos 2h), \end{aligned}$$

and substituting this in A, we obtain

$$A = -\sin h(1 + 2\cos 2h + \cos 4x_0).$$

Therefore,

$$2D_2 = \frac{4\sin^2 h}{\sin 2x_0 \sin 2x_2 \sin 2x_1}\left[2\cos 2h + 1 + \cos 4x_0\right].$$

The denominator of the quotient is positive, and consequently, the entire quotient is positive. Since $0 < 2h < \frac{\pi}{2}$, it follows that $\cos 2h > 0$ and together with $1 + \cos 4x_0 \geq 0$ this guarantees that the expression in the brackets is positive. Then, $D_2 > 0$ and $f(x)$ is concave upward on $(0, \frac{\pi}{2})$.

Using the oddness and π-periodicity, we arrive at the following results:

(1) $f(x)$ is concave upward on each interval $(0, \frac{\pi}{2}) + k\pi$, $\forall k \in \mathbb{Z}$;
(2) $f(x)$ is concave downward on each interval $(-\frac{\pi}{2}, 0) + k\pi$, $\forall k \in \mathbb{Z}$.

There are no inflection points (the points $x_k = \frac{k\pi}{2}$, $\forall k \in \mathbb{Z}$ are out of the domain).

8. Complimentary properties.

The function has no horizontal asymptote, because sending x to infinity by different paths, we have distinct results: for the path $x_k = \frac{\pi}{4} + k\pi, \forall k \in \mathbb{Z}$, we get $f(x_k) = 2 \underset{k\to\pm\infty}{\to} 2$, while for the path $x_n = -\frac{\pi}{4} + n\pi, \forall n \in \mathbb{Z}$, we obtain $f(x_n) = -2 \underset{n\to\pm\infty}{\to} -2$.

At each point of the domain the function $f(x) = \frac{2}{\sin 2x}$ is continuous, that is, $f(x) \to f(x_0)$ as $x \to x_0$. Hence, an infinite increase/decrease is not observed at any point of the domain, which means that there are no vertical asymptotes at these points.

It remains to consider the points $x_k = \frac{k\pi}{2}, \forall k \in \mathbb{Z}$. Let us choose one of these points, for instance, $x = 0$. If x approaches 0 from the right, then $\sin 2x$ approaches 0 keeping positive values, and consequently, $f(x) = \frac{2}{\sin 2x} \underset{x\to 0+}{\to} +\infty$. If x approaches 0 from the left, then $\sin 2x$ approximates 0 from the below, and consequently, $f(x) = \frac{2}{\sin 2x} \underset{x\to 0^-}{\to} -\infty$. Therefore, $x = 0$ is a vertical asymptote. The same is true for any point $x_k = \frac{k\pi}{2}, \forall k \in \mathbb{Z}$, which means that each line $x_k = \frac{k\pi}{2}, \forall k \in \mathbb{Z}$ is a vertical asymptote.

9. Using the analyzed properties, marking some important points, such as $P_k = (\frac{\pi}{4} + k\pi, 2), \forall k \in \mathbb{Z}$ (local minima) and $Q_n = (-\frac{\pi}{4} + n\pi, -2), \forall n \in \mathbb{Z}$ (local maxima), and employing the established properties of the symmetry, we can plot the graph shown in Fig. 4.46.

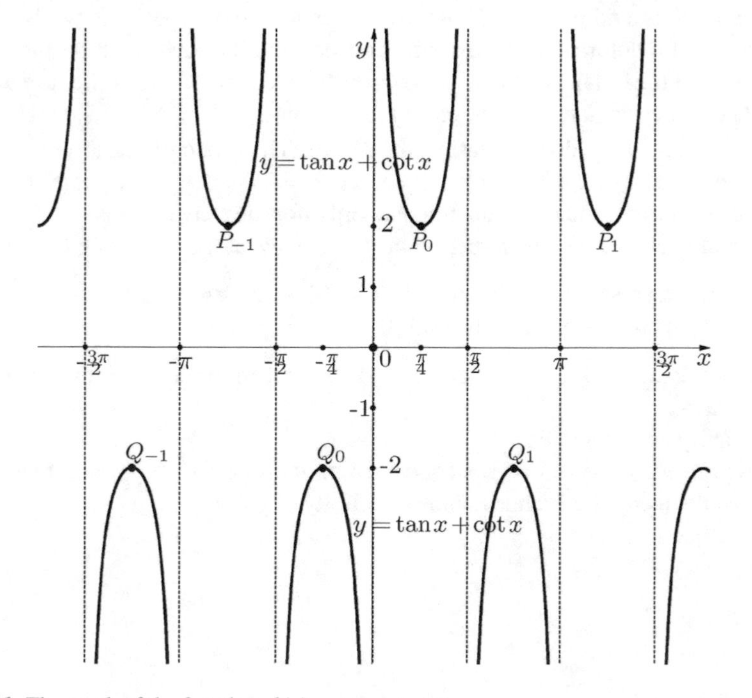

Fig. 4.46 The graph of the function $f(x) = \tan x + \cot x$

B. Study of $f(x) = \tan^2 x$

1. Domain and range. The domain X consists of all the points at which $\tan x$ is defined, that is, $X = \mathbb{R}\backslash\{x_k = \frac{\pi}{2} + k\pi, \forall k \in \mathbb{Z}\}$.
   The square of any real number is non-negative, which means that the range does not contain negative values. For any $y \geq 0$ we can solve the equation $\tan^2 x = y$ by taking the square root $\tan x = \pm\sqrt{y}$, and then restricting x to the interval $[0, \frac{\pi}{2})$ in order to apply the inverse of tangent: $x = \arctan\sqrt{y}$. Hence, the equation $\tan^2 x = y$ has solution for any $y \geq 0$, which means that the range is $Y = [0, +\infty)$.
2. Boundedness. Knowing the range, we conclude that $f(x)$ is bounded below and unbounded above.
3. Intersection and sign. Intersection with the y-axis occurs at the point $(0, 0)$, which is also the point of intersection with the x-axis. Solving the equation $\tan^2 x = 0$, we find the x-coordinates of all the points of intersection with the x-axis: $x_k = k\pi, \forall k \in \mathbb{Z}$. Except for these points, the function is positive over the entire domain.
4. Parity. The function is even: $\tan^2(-x) = \tan^2 x, \forall x \in X$.
5. Periodicity. The π-periodicity of $\tan x$ implies that $f(x)$ is π-periodic. To show that π is the minimum period, notice that on the interval $(-\frac{\pi}{2}, \frac{\pi}{2})$ of the length π (the points $-\frac{\pi}{2}$ and $\frac{\pi}{2}$ are excluded because they do not belong to the domain), the function takes the value 0 only at the point $x = 0$.
6. Monotonicity and extrema. Due to the evenness and π-periodicity, the investigation of monotonicity and extrema can be initially performed on the interval $[0, \frac{\pi}{2})$ and the obtained results subsequently extended on the entire domain.
   The function $\tan x$ is increasing and non-negative on $[0, \frac{\pi}{2})$. Therefore, $f(x) = \tan^2 x$ is also increasing on $[0, \frac{\pi}{2})$. Then, according to the evenness, $f(x)$ decreases on $(-\frac{\pi}{2}, 0]$. Consequently, $x = 0$ is a local minimum (it is also a global minimum since the function has only non-negative values).
   Finally, employing the π-periodicity, we arrive at the following results:

   (1) $f(x)$ decreases on each interval $(-\frac{\pi}{2}, 0] + k\pi, \forall k \in \mathbb{Z}$;
   (2) $f(x)$ increases on each interval $[0, \frac{\pi}{2}) + k\pi, \forall k \in \mathbb{Z}$.

   The points $x_k = k\pi, \forall k \in \mathbb{Z}$ are (strict) local minima (and non-strict global minima).
7. Concavity and inflection.
   Again, we analyze first the concavity on the interval $[0, \frac{\pi}{2})$ and then extend the results using the evenness and periodicity of $f(x)$.

Take two arbitrary points $0 \le x_1 < x_2 < \frac{\pi}{2}$, define the midpoint $x_0 = \frac{x_1+x_2}{2}$, and evaluate the quantity $2D_2 \equiv f(x_1) + f(x_2) - 2f(x_0)$:

$$2D_2 = f(x_2) - f(x_0) - (f(x_0) - f(x_1))$$
$$= \tan^2 x_2 - \tan^2 x_0 - \left(\tan^2 x_0 - \tan^2 x_1\right)$$
$$= (\tan x_2 - \tan x_0)(\tan x_2 + \tan x_0) - (\tan x_0 - \tan x_1)(\tan x_0 + \tan x_1).$$

On the interval $[0, \frac{\pi}{2})$ the function $\tan x$ is increasing, non-negative and concave upward. The upward concavity guarantees that $\tan x_2 + \tan x_1 - 2\tan x_0 > 0$, whence $\tan x_2 - \tan x_0 > \tan x_0 - \tan x_1$. The increase ensures that $\tan x_2 > \tan x_1$, and consequently, $\tan x_2 + \tan x_0 > \tan x_0 + \tan x_1$. Therefore, the factors of the first term in D_2 are greater than the corresponding factors of the second term. Besides, all the factors are positive, and consequently, $D_2 > 0$, that is, $f(x)$ is concave upward on $[0, \frac{\pi}{2})$.

It follows from the evenness that $f(x)$ is also concave upward on $(-\frac{\pi}{2}, 0]$. Taking into account that $x = 0$ is a local minimum, we conclude that $f(x)$ is concave upward on the entire interval $(-\frac{\pi}{2}, \frac{\pi}{2})$ (see Property 3 in Sect. 4.1 of Chap. 3).

Finally, employing the π-periodicity, we can show that $f(x)$ is concave upward on each interval $(-\frac{\pi}{2}, \frac{\pi}{2}) + k\pi, \forall k \in \mathbb{Z}$. Therefore, there is no inflection point (the points $x_k = \frac{\pi}{2} + k\pi, \forall k \in \mathbb{Z}$ are out of the domain).

8. Complimentary properties.

The function has no horizontal asymptote, because sending x to infinity by different paths, we have distinct results: for the path $x_k = k\pi, \forall k \in \mathbb{Z}$, we get $f(x_k) = 0 \underset{k\to\pm\infty}{\to} 0$, while for the path $x_n = \frac{\pi}{4} + n\pi, \forall n \in \mathbb{Z}$, we obtain $f(x_n) = 1 \underset{n\to\pm\infty}{\to} 1$.

At each point of the domain the function $f(x) = \tan^2 x$ is continuous, that is, $f(x) \to f(x_0)$ as $x \to x_0$. Hence, an infinite increase/decrease is not observed at any point of the domain, which means that there are no vertical asymptotes at these points.

It remains to consider the points $x_k = \frac{\pi}{2} + k\pi, \forall k \in \mathbb{Z}$. Let us choose one of these points, for instance, $x = \frac{\pi}{2}$. If x approaches $\frac{\pi}{2}$ from the right, then $\tan^2 x$ goes to $+\infty$. The same happens when x approaches $\frac{\pi}{2}$ from the left. Therefore, $x = \frac{\pi}{2}$ is a vertical asymptote. The same is true for any point $x_k = \frac{\pi}{2} + k\pi, \forall k \in \mathbb{Z}$, which means that each line $x_k = \frac{\pi}{2} + k\pi, \forall k \in \mathbb{Z}$ is a vertical asymptote.

9. Using the analyzed properties, marking some important points, such as $P_k = (k\pi, 0), \forall k \in \mathbb{Z}$ (strict local minima and non-strict global minima) and $Q_n = (\frac{\pi}{4}+n\pi, 1), \forall n \in \mathbb{Z}$ (additional points), and employing the established properties of the symmetry, we can plot the graph shown in Fig. 4.47.

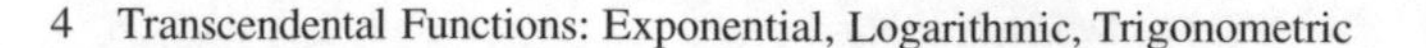

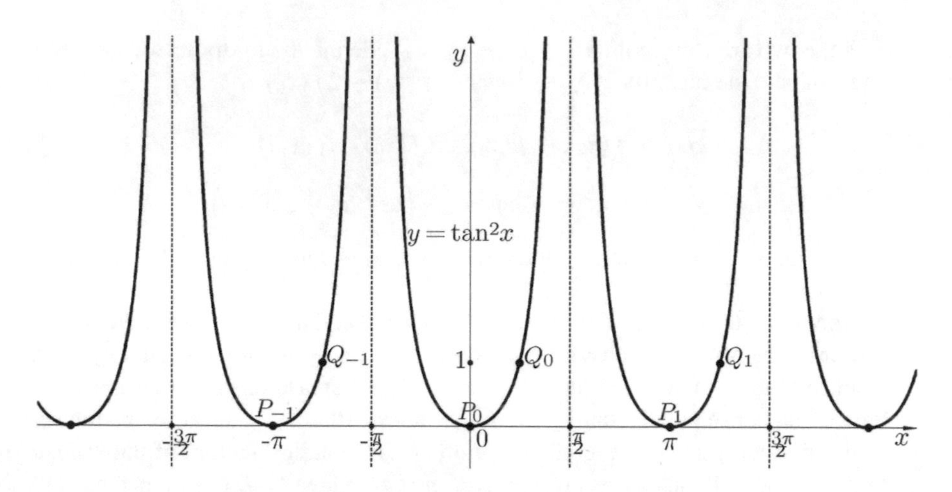

Fig. 4.47 The graph of the function $f(x) = \tan^2 x$

C. Study of $f(x) = \arctan x^2$

1. Domain and range. The domain of $\arctan x$ is the set of all real numbers. Consequently, the domain of $\arctan x^2$ is also $X = \mathbb{R}$.

   Recall that $\arctan t \in [0, \frac{\pi}{2})$, $\forall t \geq 0$, and for this reason $f(x)$ takes only the values between 0 and $\frac{\pi}{2}$. Let us take an arbitrary $y \in [0, \frac{\pi}{2})$ and solve the equation $\arctan x^2 = y$. Applying the inverse function we get $x^2 = \tan y$, whence $x = \pm\sqrt{\tan y}$ (notice that $\tan y \geq 0$ because $y \in [0, \frac{\pi}{2})$). Since the equation has solutions for any $y \in [0, \frac{\pi}{2})$, this means that the range of the function is $Y = [0, \frac{\pi}{2})$.
2. Boundedness. Since the range is $Y = [0, \frac{\pi}{2})$, the function is bounded below and above.
3. Intersection and sign. Intersection with the y-axis occurs at the point $(0, 0)$, which is also the point of intersection with the x-axis. Since the equation $\arctan x^2 = 0$ has the only solution $x = 0$, the origin is the only point of intersection with the x-axis. Except for the point $x = 0$, all the values of the function are positive.
4. Parity. The function is even: $f(-x) = \arctan(-x)^2 = \arctan x^2 = f(x)$, $\forall x \in \mathbb{R}$.
5. Periodicity. The study of periodicity is left until after the investigation of monotonicity.
6. Monotonicity and extrema. Due to the evenness, the investigation of monotonicity and extrema can be initially performed on the interval $[0, +\infty)$ with subsequent extension of the results to the entire domain.

   For two arbitrary points $0 \leq x_1 < x_2 < +\infty$ it follows that $x_1^2 < x_2^2$, and consequently, $\arctan x_1^2 < \arctan x_2^2$ due to increase of arctangent. Therefore, $f(x)$ is increasing on $[0, +\infty)$.

Then, it follows from the evenness that $f(x)$ is decreasing on $(-\infty, 0]$. Consequently, $x = 0$ is a strict local and global minimum.

Another consequence of increase on the infinite interval $[0, +\infty)$ is impossibility of the function to be periodic.

7. Concavity and inflection.

Again, we analyze first the concavity on the interval $[0, +\infty)$, and then extend the results according to the evenness.

Take two arbitrary points $0 \leq x_1 < x_2$, denote the midpoint $x_0 = \frac{x_1+x_2}{2}$ and the increment $h = x_2 - x_0 = x_0 - x_1$, and evaluate the quantity $2D_2 \equiv f(x_1) + f(x_2) - 2f(x_0)$ using the trigonometric formula $\arctan a - \arctan b = \arctan \frac{a-b}{1+ab}$:

$$2D_2 = f(x_2) - f(x_0) - (f(x_0) - f(x_1))$$

$$= \arctan x_2^2 - \arctan x_0^2 - \left(\arctan x_0^2 - \arctan x_1^2\right)$$

$$= \arctan \frac{x_2^2 - x_0^2}{1 + x_2^2 x_0^2} - \arctan \frac{x_0^2 - x_1^2}{1 + x_1^2 x_0^2}$$

$$= \arctan \frac{\frac{x_2^2-x_0^2}{1+x_2^2x_0^2} - \frac{x_0^2-x_1^2}{1+x_1^2x_0^2}}{1 + \frac{x_2^2-x_0^2}{1+x_2^2x_0^2} \cdot \frac{x_0^2-x_1^2}{1+x_1^2x_0^2}} \equiv \arctan \frac{A}{B}.$$

Since $0 \leq x_1 < x_0 < x_2$, it follows that $B > 0$. To assess A we use the following representation:

$$A = \frac{x_2^2 - x_0^2}{1 + x_2^2 x_0^2} - \frac{x_0^2 - x_1^2}{1 + x_1^2 x_0^2}$$

$$= \frac{h}{(1 + x_2^2 x_0^2)(1 + x_1^2 x_0^2)}[(x_2 + x_0)(1 + x_1^2 x_0^2) - (x_1 + x_0)(1 + x_2^2 x_0^2)]$$

$$\equiv \frac{h}{(1 + x_2^2 x_0^2)(1 + x_1^2 x_0^2)} C.$$

The expression C can be still simplified to the form:

$$C = x_2 + x_0 + x_2 x_1^2 x_0^2 + x_1^2 x_0^3 - x_1 - x_0 - x_1 x_2^2 x_0^2 - x_2^2 x_0^3$$

$$= (x_2 - x_1) + x_1 x_2 x_0^2 (x_1 - x_2) + x_0^3 (x_1 - x_2)(x_1 + x_2) = 2h(1 - x_1 x_2 x_0^2 - 2x_0^4).$$

If $x_2 \leq \frac{1}{\sqrt[4]{3}}$, then $1 - x_1 x_2 x_0^2 - 2x_0^4 > 0$, whence $C > 0$, and consequently, $A > 0$ and $D_2 > 0$, that is, $f(x)$ is concave upward. If $x_1 \geq \frac{1}{\sqrt[4]{3}}$, then $1 - x_1 x_2 x_0^2 -$

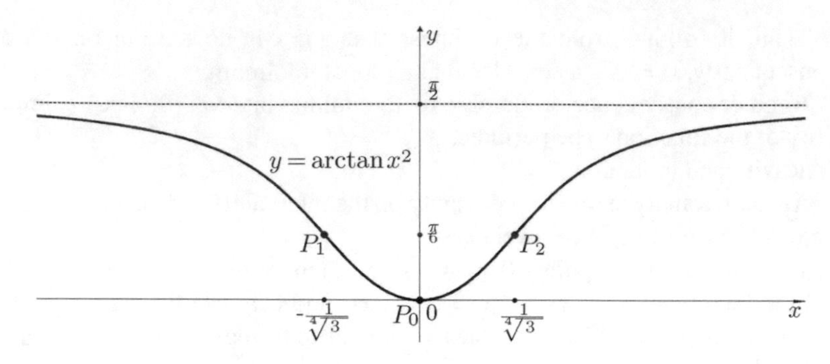

Fig. 4.48 The graph of the function $f(x) = \arctan x^2$

$2x_0^4 < 0$, whence $C < 0$, and consequently, $A < 0$ and $D_2 < 0$, that is, $f(x)$ is concave downward.

From the evenness and the fact that $x = 0$ is a local minimum it follows that (see Property 3 in Sect. 4.1 of Chap. 3):

(1) $f(x)$ is concave downward on $(-\infty, -\frac{1}{\sqrt[4]{3}}]$;
(2) $f(x)$ is concave upward on $[-\frac{1}{\sqrt[4]{3}}, \frac{1}{\sqrt[4]{3}}]$;
(3) $f(x)$ is concave downward on $[\frac{1}{\sqrt[4]{3}}, +\infty)$.

Consequently, the points $x = \pm\frac{1}{\sqrt[4]{3}}$ are inflection points.

8. Complimentary properties.

   When $x \to \pm\infty$, we have $x^2 \to +\infty$, and consequently, $\arctan x^2$ approaches $\frac{\pi}{2}$ from the below. This implies that $y = \frac{\pi}{2}$ is a horizontal asymptote.

   There is no vertical asymptote, because the range of the function is bounded.
9. Using the analyzed properties, marking some important points, such as $P_0 = (0, 0)$ (intersection with the coordinate axes and also local and global minimum) and $P_1 = (-\frac{1}{\sqrt[4]{3}}, \frac{\pi}{6})$ and $P_2 = (\frac{1}{\sqrt[4]{3}}, \frac{\pi}{6})$, (inflection points), and applying the symmetry about the y-axis, we construct the graph shown in Fig. 4.48.

D. Study of $f(x) = \sin 2x - 2\sin x$

1. Domain and range. The domain of the function is $X = \mathbb{R}$.

   Since both functions $\sin 2x$ and $\sin x$ have the values in the interval $[-1, 1]$, the range of $f(x)$ is contained in the interval $[-3, 3]$. A better specification of the range will be given after the study of monotonicity and extrema.
2. Boundedness. Since the range is contained in $[-3, 3]$, the function is bounded.
3. Intersection and sign. Intersection with the y-axis occurs at the point $(0, 0)$, which is also the point of intersection with the x-axis.

To find the x-coordinates of the points of intersection with the x-axis we need to solve the equation $\sin 2x - 2\sin x = 2\sin x(\cos x - 1) = 0$. There are two options: the first, $\sin x = 0$, whose solutions are $x_n = n\pi$, $\forall n \in \mathbb{Z}$, and the second, $\cos x = 1$, whose solutions $x_k = 2k\pi$, $\forall k \in \mathbb{Z}$ are included in the first set. Hence, intersection with the x-axis occurs at the points $(n\pi, 0)$, $\forall n \in \mathbb{Z}$.

Representing the function in the form $f(x) = 2\sin x(\cos x - 1)$, we notice that the factor $\cos x - 1$ is negative at all the points, except for $x_k = 2k\pi$, $\forall k \in \mathbb{Z}$ where it vanishes. Therefore, the sign of $f(x)$ depends on the first factor. If $x \in (0, \pi) + 2n\pi$, $\forall n \in \mathbb{Z}$, then $\sin x > 0$ and $f(x) < 0$. If $x \in (\pi, 2\pi) + 2n\pi$, $\forall n \in \mathbb{Z}$, then $\sin x < 0$ and $f(x) > 0$.

4. Parity. The function is odd as the difference of the two odd functions.
5. Periodicity. The functions $\sin 2x$ and $2\sin x$ are 2π-periodic, and consequently, $f(x)$ is also 2π-periodic. We will prove that 2π is the minimum period after the investigation of monotonicity and extrema.
6. Monotonicity and extrema.

Due to the oddness and 2π-periodicity, the investigation of monotonicity and extrema can be initially performed on the interval $[0, \pi]$ and the obtained results subsequently extended on the entire domain.

Take two arbitrary points $0 \leq x_1 < x_2 \leq \pi$, define $x_0 = \frac{x_1+x_2}{2}$ and $h = x_2 - x_0 = x_0 - x_1$, and obtain the following evaluation:

$$\begin{aligned} D_1 &= f(x_2) - f(x_1) = \sin 2x_2 - 2\sin x_2 - (\sin 2x_1 - 2\sin x_1) \\ &= 2\sin(x_2 - x_1)\cos(x_2 + x_1) - 4\sin\frac{x_2 - x_1}{2}\cos\frac{x_2 + x_1}{2} \\ &= 4\sin h(\cos h\cos 2x_0 - \cos x_0) \end{aligned}$$

$$= 4\sin h(\cos h(2\cos^2 x_0 - 1) - \cos x_0) = 4\sin h(2\cos h\cos^2 x_0 - \cos x_0 - \cos h).$$

Let us solve the equation $2\cos^2 t - \cos t - 1 = 0$. First, solving the quadratic equation, we find $\cos t = \frac{1\pm 3}{4}$, that is, $\cos t = 1$ and $\cos t = -\frac{1}{2}$. The solutions of the former are $t_k = 2k\pi$, $\forall k \in \mathbb{Z}$, and of the latter $t_n = \pm\frac{2\pi}{3} + 2n\pi$, $\forall n \in \mathbb{Z}$. Notice that only two of these solutions belong to the interval $[0, \pi]$: $t = 0$ and $t = \frac{2\pi}{3}$.

According to the find solutions, we divide $[0, \pi]$ into the two subintervals $[0, \frac{2\pi}{3}]$ and $[\frac{2\pi}{3}, \pi]$. In the first subinterval we have

$$\begin{aligned} A &= \cos h\cos 2x_0 - \cos x_0 = \frac{1}{2}\cos(h + 2x_0) + \frac{1}{2}\cos(h - 2x_0) - \cos x_0 \\ &= \frac{1}{2}\cos(x_2 - x_0 + 2x_0) + \frac{1}{2}\cos(2x_0 - x_0 + x_1) - \cos x_0 \\ &= \frac{1}{2}\cos(x_2 + x_0) + \frac{1}{2}\cos(x_1 + x_0) - \cos x_0 \end{aligned}$$

$$= \frac{1}{2}(\cos(x_2 + x_0) - \cos x_0) + \frac{1}{2}(\cos(x_1 + x_0) - \cos x_0)$$
$$= -\sin(\frac{x_2}{2} + x_0)\sin\frac{x_2}{2} - \sin(\frac{x_1}{2} + x_0)\sin\frac{x_1}{2}.$$

Since $0 \le \frac{x_1}{2} < \frac{x_2}{2} \le \frac{\pi}{3}$ and $0 < \frac{x_1}{2} + x_0 < \frac{x_2}{2} + x_0 < \pi$, all the four sines are positive, and consequently, $A < 0$, whence $D_1 < 0$, that is, $f(x)$ decreases on $[0, \frac{2\pi}{3}]$.

In the second subinterval we get

$$A = \cos h \cos 2x_0 - \cos x_0 = -\sin(\frac{x_2}{2} + x_0)\sin\frac{x_2}{2} - \sin(\frac{x_1}{2} + x_0)\sin\frac{x_1}{2},$$

where the arguments of the sines satisfy the inequalities $\frac{\pi}{3} \le \frac{x_1}{2} < \frac{x_2}{2} \le \frac{\pi}{2}$, which implies that $\sin\frac{x_1}{2} > 0$, $\sin\frac{x_2}{2} > 0$, and $\pi < \frac{x_1}{2} + x_0 < \frac{x_2}{2} + x_0 < \frac{3\pi}{2}$, and consequently, $\sin(\frac{x_1}{2} + x_0) < 0$, $\sin(\frac{x_2}{2} + x_0) < 0$. Then, $A > 0$, whence $D_1 > 0$, which shows that $f(x)$ increases on $[\frac{2\pi}{3}, \pi]$.

The oddness of the function insures that $f(x)$ increases on $[-\pi, -\frac{2\pi}{3}]$ and decreases on $[-\frac{2\pi}{3}, 0]$.

Finally, applying the 2π-periodicity, we arrive at the following results:

(1) $f(x)$ decreases on each interval $[-\frac{2\pi}{3}, \frac{2\pi}{3}] + 2k\pi, \forall k \in \mathbb{Z}$;
(2) $f(x)$ increases on each interval $[-\frac{4\pi}{3}, -\frac{2\pi}{3}] + 2k\pi, \forall k \in \mathbb{Z}$.

Therefore, the points $x_k = \frac{2\pi}{3} + 2k\pi, \forall k \in \mathbb{Z}$ are strict local and non-strict global minima, and the points $x_n = -\frac{2\pi}{3} + 2n\pi, \forall n \in \mathbb{Z}$ are strict local and non-strict global maxima. The corresponding values of the function are $f_{min} = f(x_k) = -\frac{\sqrt{3}}{2} - 2\frac{\sqrt{3}}{2} = -\frac{3\sqrt{3}}{2}$ and $f_{max} = f(x_n) = \frac{\sqrt{3}}{2} + 2\frac{\sqrt{3}}{2} = \frac{3\sqrt{3}}{2}$.

After determining the global extrema, we can specify the range: $Y = [-\frac{3\sqrt{3}}{2}, \frac{3\sqrt{3}}{2}]$.

Since the distance between the two nearest minima (and the two nearest maxima) is 2π, this shows that 2π is the minimum period of the function.

7. Concavity and inflection.

Again, we analyze first the concavity on the interval $[0, \pi]$ and then extend the results using the oddness and periodicity of $f(x)$.

Take two arbitrary points $0 \le x_1 < x_2 \le \frac{\pi}{2}$, denote the midpoint $x_0 = \frac{x_1+x_2}{2}$ and the increment $h = x_2 - x_0 = x_0 - x_1$, and evaluate the quantity $2D_2 \equiv f(x_1) + f(x_2) - 2f(x_0)$ using the formula for D_1:

$$2D_2 = f(x_2) - f(x_0) - (f(x_0) - f(x_1))$$
$$= 2\sin(x_2 - x_0)\cos(x_2 + x_0) - 4\sin\frac{x_2 - x_0}{2}\cos\frac{x_2 + x_0}{2}$$

$$
\begin{aligned}
&- \left(2\sin(x_0-x_1)\cos(x_0+x_1)-4\sin\frac{x_0-x_1}{2}\cos\frac{x_0+x_1}{2}\right)\\
&\quad= 4\sin\frac{h}{2}\left(\cos\frac{h}{2}\cos(x_2+x_0)-\cos\frac{x_2+x_0}{2}\right.\\
&\qquad\left.-\cos\frac{h}{2}\cos(x_0+x_1)+\cos\frac{x_0+x_1}{2}\right)\\
&\quad= 4\sin\frac{h}{2}\left(\cos\frac{h}{2}\cdot 2\sin\frac{x_2+x_1+2x_0}{2}\sin\frac{x_1-x_2}{2}\right.\\
&\qquad\left.+2\sin\frac{x_1+x_2+2x_0}{4}\sin\frac{x_2-x_1}{4}\right)\\
&= 8\sin\frac{h}{2}\left(-\cos\frac{h}{2}\sin 2x_0\cdot 2\sin\frac{h}{2}\cos\frac{h}{2}+\sin x_0\cdot\sin\frac{h}{2}\right)\\
&= 8\sin^2\frac{h}{2}\sin x_0\left(1-4\cos^2\frac{h}{2}\cos x_0\right).
\end{aligned}
$$

The first two factors are positive, which means that the sign depends on the last factor. The equation $1-4\cos x = 0$ has the only root $x = \arccos\frac{1}{4}$ which belongs to the interval $[0, \pi]$. Consider now separately the subintervals $[0, \arccos\frac{1}{4}]$ and $[\arccos\frac{1}{4}, \pi]$.

On the subinterval $[0, \arccos\frac{1}{4}]$, the last factor (the expression inside the parentheses) can be written in the form

$$
\begin{aligned}
&1-4\cos^2\frac{h}{2}\cos x_0 = 1-2(1+\cos h)\cos x_0 = 1-2\cos x_0-2\cos h\cos x_0\\
&= 1-2\cos x_0-\cos(h+x_0)-\cos(h-x_0) = 1-2\cos x_0-\cos x_2-\cos x_1.
\end{aligned}
$$

Since $\cos x_0 > \frac{1}{4}$, $\cos x_1 > \frac{1}{4}$ and $\cos x_2 \geq \frac{1}{4}$, the last expression is negative, and consequently, $D_2 < 0$, which indicates the downward concavity on $[0, \arccos\frac{1}{4}]$.

On the subinterval $[\arccos\frac{1}{4}, \pi]$, we have $\cos x_0 < \frac{1}{4}$, and consequently, $1-4\cos^2\frac{h}{2}\cos x_0 > 0$, which results in $D_2 > 0$, that is, the function has upward concavity on $[\arccos\frac{1}{4}, \pi]$.

Using the oddness, we conclude that $f(x)$ is concave downward on $[-\pi, -\arccos\frac{1}{4}]$ and upward on $[-\arccos\frac{1}{4}, 0]$.

Finally, employing the 2π-periodicity, we arrive at the following results:

(1) $f(x)$ is concave downward on each interval $[-\pi, -\arccos\frac{1}{4}]+2k\pi$, $\forall k \in \mathbb{Z}$;
(2) $f(x)$ is concave upward on each interval $[-\arccos\frac{1}{4}, 0]+2k\pi$, $\forall k \in \mathbb{Z}$;
(3) $f(x)$ is concave downward on each interval $[0, \arccos\frac{1}{4}]+2k\pi$, $\forall k \in \mathbb{Z}$;
(4) $f(x)$ is concave upward on each interval $[\arccos\frac{1}{4}, \pi]+2k\pi$, $\forall k \in \mathbb{Z}$.

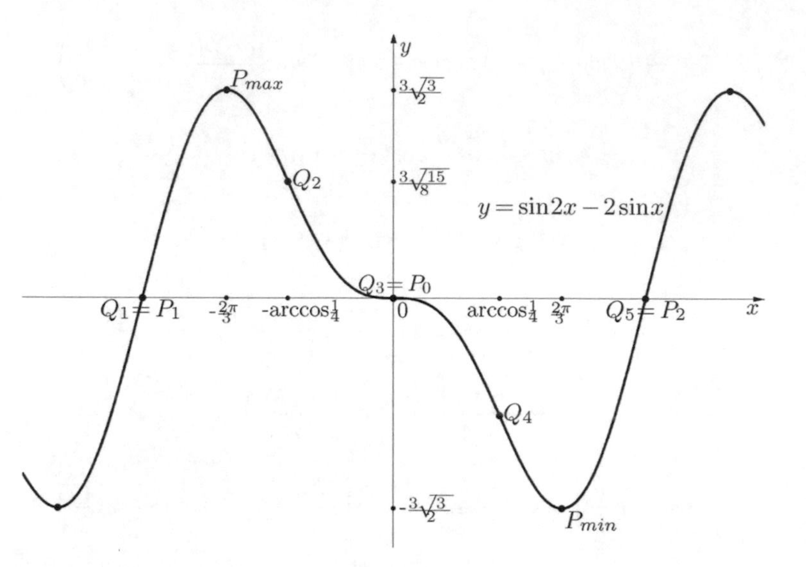

Fig. 4.49 The graph of the function $f(x) = \sin 2x - 2\sin x$

Consequently, the points $x_k = k\pi, \forall k \in \mathbb{Z}$ and $x_n = \pm \arccos \frac{1}{4} + 2n\pi, \forall n \in \mathbb{Z}$ are inflection points.

8. Complimentary properties.

   The function has no horizontal asymptote, because sending x to infinity by different paths, we have distinct results: for the path $x_k = k\pi, \forall k \in \mathbb{Z}$, we get $f(x_k) = 0 \underset{k\to\pm\infty}{\to} 0$, while for the path $x_n = -\frac{2\pi}{3} + 2n\pi, \forall n \in \mathbb{Z}$, we obtain $f(x_n) = \frac{3\sqrt{3}}{2} \underset{n\to\pm\infty}{\to} \frac{3\sqrt{3}}{2}$.

   There is no vertical asymptote, because the range is bounded.

9. Using the analyzed properties, marking some important points on the interval $[-\pi, \pi]$, such as $P_0 = (0, 0)$ (intersection with the coordinate axes), $P_1 = (-\pi, 0)$ and $P_2 = (\pi, 0)$ (intersection with the x-axis), $P_{max} = (-\frac{2\pi}{3}, \frac{3\sqrt{3}}{2})$ and $P_{min} = (\frac{2\pi}{3}, -\frac{3\sqrt{3}}{2})$ (local and global extrema), $Q_1 = P_1$, $Q_2 = (-\arccos \frac{1}{4}, \frac{3\sqrt{15}}{8})$, $Q_3 = P_0$, $Q_4 = (\arccos \frac{1}{4}, -\frac{3\sqrt{15}}{8})$ and $Q_5 = P_2$ (inflection points), and employing the established properties of the symmetry, we construct the graph shown in Fig. 4.49.

Problems

1. Find the domain of the function:

   (a) $f(x) = \log_2(\log_3(\log_4 x))$;

(b) $f(x) = \log_2(\sin x - \cos x)$;
(c) $f(x) = \arcsin \log_{10} \frac{x}{10}$;
(d) $f(x) = \cot \frac{1}{x^2}$;
(e) $f(x) = \sqrt{\sin(\cos x)}$;
(f) $f(x) = \arccos \frac{2}{1+x^2}$;
(g) $f(x) = \arccos \frac{x}{1+x^2}$;
(h) $f(x) = \sqrt{\arcsin(\ln x)}$;
(i) $f(x) = \sqrt{\ln(\arcsin x)}$;
(j) $f(x) = \arccos(\sin x)$;
(k) $f(x) = \sin(\arccos x)$;
(l) $f(x) = \arccos(\tan x)$.

2. Determine the range and the global extrema (if they exist) of the function:

(a) $f(x) = \log_3(1 - 2\cos x)$;
(b) $f(x) = \ln(1 - x^2)$;
(c) $f(x) = \frac{e^{x^2}-1}{e^{x^2}}$;
(d) $f(x) = 2^{\frac{x}{1+x^2}}$;
(e) $f(x) = 2^{\cos x}$;
(*f) $f(x) = \frac{2^x-2^{-x}}{2^x+2^{-x}}$;
(*g) $f(x) = \ln(e^x + e^{-x})$;
(h) $f(x) = \sin x + \cos x$;
(i) $f(x) = 2\cos 2x + \sin^2 x$;
(j) $f(x) = \cos^2 x + \cos x + 3$;
(k) $f(x) = \sin x - \cos^2 x - 1$;
(l) $f(x) = \frac{1}{\sin^2 x} + \frac{1}{\cos^2 x}$;
(m) $f(x) = \sin^4 x + \cos^4 x$;
(*n) $f(x) = \sin^4 x + \cos^6 x$;
(o) $f(x) = \sin^6 x + \cos^6 x$;
(p) $f(x) = 2\cos^2 x - 3\sqrt{3}\cos x - \sin^2 x + 5$;
(q) $f(x) = \sin 2x \cdot \sin(2x - \frac{\pi}{6})$;
(r) $f(x) = \sin x \cos^3 x - \sin^3 x \cos x$;
(s) $f(x) = \cos(\sin x)$;
(*t) $f(x) = \arcsin x \cdot \arccos x$.

3. Verify whether the function is even, odd or none of these:

(a) $f(x) = \frac{e^{-x}-1}{e^x+1}$;
(b) $f(x) = \frac{e^x-1}{e^x+1}$;
(c) $f(x) = \log_2(x + \sqrt{x^2+1})$;
(d) $f(x) = \frac{\sqrt[3]{(x+5)^2}-\sqrt[3]{(x-5)^2}}{x \cos x}$;
(e) $f(x) = 2\cos 2x + \sin^2 x$;
(f) $f(x) = \frac{1}{\sin^2 x} + \frac{1}{\cos^2 x}$;

(g) $f(x) = \arccos \frac{x}{1+x^2}$;
(h) $f(x) = \arcsin \frac{x}{1+x^2}$;
(i) $f(x) = \arcsin \frac{x^2}{1+x^2}$.

4. Verify whether the function is periodic; if so, find a period and, if possible, the minimum period:

(a) $f(x) = \sin^4 x + \cos^4 x$;
(b) $f(x) = \tan 2x + \tan \frac{x}{3}$;
(c) $f(x) = \cos \sqrt{2}x + \cos \frac{x}{\sqrt{8}}$;
(d) $f(x) = \tan x \cdot \cot x$;
(e) $f(x) = \sin \frac{x}{2} \cdot \cos^3 \frac{x}{2}$;
(*f) $f(x) = \cos x^2$;
(g) $f(x) = |\sin x|$;
(*h) $f(x) = \sin |x|$;
(*i) $f(x) = \sin x + \sin \pi x$;
(j) $f(x) = \arccos(\sin x)$;
(k) $f(x) = \sin(\arccos x)$;
(l) $f(x) = \arccos(\tan x)$;
(m) $f(x) = \frac{1}{\sin^2 x} + \frac{1}{\cos^2 x}$;
(n) $f(x) = 2^{\sin x}$;
(o) $f(x) = \sqrt{|\sin 2x|}$;
(p) $f(x) = \ln(\tan x)$;
(*q) $f(x) = \sin 2^x$.

5. Answer if the set $\mathbb{R}\backslash\mathbb{Z}$ can be the domain of a non-constant periodic function.
6. Show that the set $\mathbb{R}\backslash I$, where I is an interval, cannot be the domain of a periodic function.
7. Find the centerpoints of symmetry and the axes of symmetry of the function $f(x) = 2\sin(3x + \frac{\pi}{6})$.
8. Answer whether a non-periodic function can have an infinite number of vertical asymptotes.
9. Prove the formula $\arcsin x + \arccos x = \frac{\pi}{2}$, $\forall x \in [-1, 1]$ and use it to set up the properties of $\arccos x$ from the known properties of $\arcsin x$. Compare the results with those obtained in Sect. 2.5.
10. Prove the formula $\arctan x + \operatorname{arccot} x = \frac{\pi}{2}$, $\forall x \in \mathbb{R}$ and use it to set up the properties of $\operatorname{arccot} x$ from the known properties of $\arctan x$. Compare the results with those obtained using the properties of the inverse function as suggested in Sect. 2.9.
11. Compare the rate of growth of $f(x) = \ln x$ and $f(x) = \ln(3x^2)$.
12. Can the graph of a function cross its horizontal asymptote? If so, how many times can it happen?

*13. Consider the function defined in the parametric form $\begin{cases} x = a\cos t \\ y = b\sin t \end{cases}$, $t \in \mathbb{R}$, $a, b \neq 0$. Make an analysis of the properties of the curve determined by this function.
(Hint. Consider first a particular case when $a = b$ and reduce the parametric form to the implicit one, which represents the formula of the well-known figure. This procedure can be generalized to the case of arbitrary non-zero a and b.)

*14. Consider the function defined in the parametric form $\begin{cases} x = a\frac{e^t+e^{-t}}{2} \\ y = b\frac{e^t-e^{-t}}{2} \end{cases}$, $t \in \mathbb{R}$, $a, b \neq 0$. Make an analysis of the properties of the curve determined by this function.
(Hint. Consider first a particular case when $a = b$ and reduce the parametric form to the implicit one, which represents the formula of the well-known figure. This procedure can be generalized to the case of arbitrary non-zero a and b.)

15. Determine the properties of a function $f(x)$ and sketch its graph by using a chain of elementary transformations starting from a known function $g(x)$:

(a) $f(x) = 3^{2x} - 2$ $(g(x) = 3^x)$;
(b) $f(x) = 3\log_{1/2}\frac{4}{(x+1)^2}$ $(g(x) = \log_{1/2} x)$;
(c) $f(x) = \sqrt{2}\cos(2x - \frac{\pi}{8})$ $(g(x) = \cos x)$;
(d) $f(x) = 2\sin(\frac{x}{2} + \frac{\pi}{6})$ $(g(x) = \sin x)$.

16. Perform analytic investigation of a given function and sketch its graph:

(a) $f(x) = 3^{2x} - 2$;
(b) $f(x) = |3^{2x} - 2|$;
(c) $f(x) = |3^{2|x|} - 2|$;
(d) $f(x) = 5^x + 5^{-x}$;
(e) $f(x) = 5^{-x} - 5^x$;
(f) $f(x) = |5^{-x} - 5^x|$;
(g) $f(x) = 3\log_{1/2}\frac{4}{(x+1)^2}$;
(h) $f(x) = \log_2(x^2 + 4) - 3$;
(i) $f(x) = |\log_2(x^2 + 4) - 3|$;
(j) $f(x) = \log_4(x^2 - 2) + 1$;
(k) $f(x) = |\log_4(x^2 - 2) + 1|$;
(l) $f(x) = \log_4|x^2 - 2| + 1$;
(m) $f(x) = \log_2\frac{x-1}{x+1}$;
(*n) $f(x) = \log_2\left|\frac{x-1}{x+1}\right|$;
(o) $f(x) = \sin(2x + \frac{\pi}{8}) - \sin(2x - \frac{3\pi}{8})$;
(p) $f(x) = \cos(\frac{x}{2} - \frac{\pi}{3}) - \cos(\frac{x}{2} + \frac{2\pi}{3})$;
(q) $f(x) = \cos(\frac{x}{3} + \frac{\pi}{6}) + \sin(\frac{5\pi}{6} - \frac{x}{3})$;
(r) $f(x) = x + \sin x$ (hint: use the inequality $|\sin x| < |x|, \forall x \neq 0$);

(s) $f(x) = \tan x - \cot x$;

(*t) $f(x) = \cot^2 x$ (hint: use the theorem of concavity of a composite function);

(u) $f(x) = \arctan \sqrt{|x|}$;

(v) $f(x) = \arctan \frac{1}{x}$;

(w) $f(x) = \operatorname{arccot} |x|$;

(x) $f(x) = \arcsin^2 x$ (hint: use the theorem of concavity of a composite function);

(*y) $f(x) = \sin 2x + 2 \sin x$;

(*z) $f(x) = \sin 2x - 2 \cos x$.

Chapter 5
Epilogue: A Bridge to Calculus

1 Monotonicity and Extrema: First Derivative

1.1 Preliminary Considerations

In analytic study of monotonicity and extrema we have regularly used the definitions of these concepts and some elementary properties of monotonic functions (recall the techniques employed in Sect. 5.1 of Chap. 2 and in studies of algebraic and transcendental functions in Chaps. 3 and 4). For instance, to show that $f(x) = x^2$ is decreasing on the interval $(-\infty, 0]$ and increasing on the interval $[0, +\infty)$, we have evaluated the difference $f(x_2) - f(x_1) = x_2^2 - x_1^2 = (x_2 - x_1)(x_2 + x_1)$ at the points $x_1 < x_2$. For chosen points, the first factor is always positive, but the second factor is negative when $x_1 < x_2 \le 0$ and positive when $0 \le x_1 < x_2$, from which we have concluded that $f(x)$ is decreasing on $(-\infty, 0]$ and increasing on $[0, +\infty)$ according to the definition. As a consequence of these properties of monotonicity, we have deduced that $x_0 = 0$ is a local (and global) minimum. This is a pattern of the study of monotonicity and extrema we have used in this text.

Recall that its application is trivial for simple functions like $f(x) = x^2$, but it can be quite technically intricate if the formula of a function is slightly more sophisticated. So it is natural to ask if there is a simpler way to study monotonicity and extrema. To give some hints in this direction, let us make a few remarks on the used scheme. Notice first that in all studied cases we have represented the difference $D_1 = f(x_2) - f(x_1)$ in the form $f(x_2) - f(x_1) = (x_2 - x_1) \cdot g(x_1, x_2)$, where $g(x_1, x_2)$ is a function of the variables x_1, x_2, whose sign can usually be determined in a simpler way than that of D_1. However, although the original function $f(x)$ depends only on a single variable, both D_1 and g are functions of two variables, which is one of the major problem of their evaluation. Naturally, we can be curious whether there exists another characteristic, expressed as a function of one variable, whose sign can define monotonicity of the original function. The answer in many cases (for various classes of functions) is "yes". Although, restricted by elementary

A. Bourchtein, L. Bourchtein, *Elementary Functions*,
https://doi.org/10.1007/978-3-031-29075-6_5

concepts and techniques, we cannot justify completely this answer or determine exactly the class of functions which admits such a characterization, we can give some indications, based on reasonable suppositions, of how such characteristic can be constructed.

To get rid of the second variable, let us fix one of them, say x_1, and leave the second to be arbitrary. In this context, it will be more natural to use the notation x_0 for a fixed point and x for a variable one. From now on we will follow this notation. If we choose x as close to x_0 as we wish (in a small neighborhood of x_0), we can frequently separate the principal (primary) and secondary (negligible) terms in the expression of $g(x_1, x_2) = g(x_0, x)$. For this purpose, it is convenient to write x in the form $x = x_0 + \Delta x$, where Δx is a small increment of the independent variable, and also to use the standard notation $\Delta f = f(x) - f(x_0)$ for the increment of the function. Then, we can consider all the terms in $g(x_0, x_0 + \Delta x) = h(x_0, \Delta x)$ with a positive power of Δx as secondary ones. Let us denote the principal terms (not depending on Δx) by A and the secondary ones by α. For example, for $f(x) = x^2$, we have $\Delta f = f(x) - f(x_0) = \Delta x \cdot (2x_0 + \Delta x)$, where $A = 2x_0$ and $\alpha = \Delta x$. Whatever small value the primary term $2x_0$ takes, if $x_0 \neq 0$ we can always find such small Δx that the secondary term will be much less than the primary one. It means, in particular, that around the point $x_0 \neq 0$ the sign of Δf is determined by the sign of $A = 2x_0$.

Recalling other examples, we can see that the same happens for some other functions. For instance, for $f(x) = x^3$ at any point x_0 we have $f(x) - f(x_0) = \Delta x \cdot (3x_0^2 + 3\Delta x \cdot x_0 + (\Delta x)^2)$ with $A = 3x_0^2$ and $\alpha = 3\Delta x \cdot x_0 + (\Delta x)^2$; for $f(x) = \frac{1}{x}$ and $x_0 \neq 0$ we have $f(x) - f(x_0) = \Delta x \cdot \frac{-1}{x_0(x_0+\Delta x)} = \Delta x \cdot \frac{-1}{x_0^2}(1 - \frac{\Delta x}{x_0} + (\frac{\Delta x}{x_0})^2 + (-1)^n (\frac{\Delta x}{x_0})^n + \ldots)$, with $A = -\frac{1}{x_0^2}$ and $\alpha = \frac{-1}{x_0^2}(-\frac{\Delta x}{x_0} + (\frac{\Delta x}{x_0})^2 + (-1)^n (\frac{\Delta x}{x_0})^n + \ldots)$ (recall the formula of the sum of the infinite geometric progression $\sum_{k=0}^{\infty} q^k = \frac{1}{1-q}$ under assumption that $|q| = |\frac{\Delta x}{x_0}| < 1$).

1.2 Definition of Differentiability and Derivative

We can generalize this construction by allowing the secondary term α to be not only proportional to Δx, like in the above examples, but be evaluated in the form, which shows that α is negligible comparing to A around x_0 (unless $A = 0$). In this case, we arrive at the following important concept. A function whose increment in a neighborhood of x_0 can be represented in the form $\Delta f = f(x) - f(x_0) = \Delta x \cdot (A + \alpha)$, where $A = constant$ does not depend on $\Delta x = x - x_0$ (the principal term) and α can be evaluated in the form $|\alpha| \leq C|\Delta x|^k$, $k > 0$, $C = constant$ (the secondary term), is called differentiable at x_0. The constant A is called the derivative of $f(x)$ at x_0 and is denoted by $f'(x_0)$. Although A does depend on x_0, we will frequently omit this in notation when it will be clear from the context. Notice that the definition itself implies that $f(x)$ should be defined in a neighborhood of x_0 (in order to Δf be defined in such a neighborhood), but the mere existence of $f(x)$ in a neighborhood of x_0 does not guarantee differentiability.

For the readers familiar with this notion, usually studied in Calculus and Analysis, we would like to mention that the above definition is slightly more restrictive (in the form of evaluation of α) than the classical definition. It means that there exist functions differentiable in the classical sense, which do not satisfy the above definition. However, the given simplified definition still involves many classes of elementary functions and, what is the most important, it allows us to avoid the use of such fine and difficult concept like limit and describe differentiation at elementary pre-calculus level. Of course, the classical definition of differentiation and derivative used in Calculus and Analysis is based on the notion of the limit.

1.3 Relationship Between Derivative and Monotonicity

According to the introduced definition, $f'(x_0)$ depends only on x_0 and in a small neighborhood of x_0 the following approximation holds: $\Delta f \approx f'(x_0)\Delta x$. Using the difference (incremental) quotient, we can represent the same approximation in the form $\frac{\Delta f}{\Delta x} \approx f'(x_0)$. This implies that near the point x_0 the sign of $\frac{\Delta f}{\Delta x}$ is determined by the sign of $f'(x_0)$ (unless $f'(x_0) = 0$). Of course, this is true only in a sufficiently small neighborhood of x_0. However, if we need to establish the monotonicity on an interval I, we can examine the sign of $f'(x_0)$ at each point $x_0 \in I$ and, if it is the same, we can expect that Δf will keep the same sign on I, just like the monotonicity on smaller adjacent (or overlapping) subintervals was extended to greater intervals in previous studies (see Property 1 in Sect. 4.1 Chap. 3 and subsequent examples). If this algorithm works, we can simplify the investigation of monotonicity by evaluating the sign of a function $f'(x_0)$ of a single variable x_0, instead of working with a function Δf or $g(x_1, x_2)$ or $h(x_1, \Delta x)$ of two variables.

Notice also that at the points where $f'(x_0) = 0$ the derivative can change sign, like in the case of $f(x) = x^2$ with $f'(x) = 2x$, or can keep the same sign on both sides of x_0, like in the case of $f(x) = x^3$ with $f'(x) = 3x^2$. In the first case, the function changes the type of monotonicity, and consequently, x_0 is a local extremum. For example, the derivative $f'(x) = 2x$ changes sign from negative to positive at the origin, that is, $f(x) = x^2$ is decreasing to the left of 0 and increasing to the right of 0, which implies that $x_0 = 0$ is a local minimum. In the second case, the function does not change the type of monotonicity, and x_0 is in a monotonicity interval. For example, the derivative $f'(x) = 3x^2$ does not change sign at the origin, and $f(x)$ is increasing both to the left and to the right of 0, which implies that $x_0 = 0$ is a point of the increasing interval $(-\infty, +\infty)$. As a consequence, $x_0 = 0$ is not a local extremum in this case.

It is necessary to emphasize that the described relationship between monotonicity and local extrema on the one side and the properties of the derivatives on the other side is observed only in the case when a function is differentiable. There are many examples of elementary functions, which are not differentiable, at least at some points of its domain. For instance, the functions $|x|$, $\sqrt{|x|}$, $\sqrt[3]{x}$ are defined on $\mathbb{R}$, but are not differentiable at the origin (although differentiable at any point $x_0 \neq 0$); the function $\arccos x$ is defined on $[0, \pi]$, but it is not differentiable at the end points of this interval. Naturally, it may happen that the derivative changes sign at a point of non-differentiability or does not change it. Using the same arguments as in the case of vanishing derivative, we can conclude that if the derivative changes sign at x_0, this point is a local extremum, otherwise x_0 is not a local extremum. The first case occurs, for instance, with the function $\sqrt{|x|}$ and the second with $\sqrt[3]{x}$.

Let us specify considerations for the functions $\sqrt{|x|}$ and $\sqrt[3]{x}$. Consider first $f(x) = \sqrt[3]{x}$ and show that it is not differentiable at the origin. At $x_0 = 0$ we have $\Delta x = x - x_0 = x$ and the increment of $f(x)$ is $\Delta f = f(x) - f(0) = \sqrt[3]{x} - \sqrt[3]{0} = \sqrt[3]{x} = \sqrt[3]{\Delta x} = \Delta x \cdot (0 + \frac{1}{\sqrt[3]{(\Delta x)^2}})$. In the last expression the term independent of Δx is $A = 0$ and $\alpha = \frac{1}{\sqrt[3]{(\Delta x)^2}}$. However, this α does not satisfy the evaluation $|\alpha| \leq C|\Delta x|^k$, $k > 0$, $C = \mathit{constant}$ required by the definition. Indeed, $(\Delta x)^{-2/3}$ takes arbitrary large values for sufficiently small Δx, more precisely, for any constant C and $k > 0$, we can always find such small Δx that $|(\Delta x)^{-2/3}| > C|\Delta x|^k$: simply solving the last inequality for $|\Delta x|$ we find $|\Delta x| < \left(\frac{1}{C}\right)^{\frac{1}{2/3+k}}$. (This function is also non-differentiable at the origin in the classical sense, but we cannot prove this within the scope of our text.) Notice that the function $f(x) = \sqrt[3]{x}$ is strictly increasing on $\mathbb{R}$ and has no extremum at the origin (see Sect. 3.1 in Chap. 3 for detailed study of this function).

Now consider in a similar way the function $f(x) = \sqrt{|x|}$. At $x_0 = 0$ we have $\Delta x = x - x_0 = x$ and the increment of $f(x)$ is $\Delta f = f(x) - f(0) = \sqrt{|x|} - \sqrt{0} = \sqrt{|\Delta x|} = \Delta x \cdot (0 + \frac{\operatorname{sgn}(\Delta x)}{\sqrt{|\Delta x|}})$. In the last expression the term independent of Δx is $A = 0$ and $\alpha = \frac{\operatorname{sgn}(\Delta x)}{\sqrt{|\Delta x|}}$. Solving the inequality $|\alpha| = \frac{1}{\sqrt{|\Delta x|}} > C|\Delta x|^k$ for any given constant $C > 0$ and $k > 0$, we find the solution $|\Delta x| < \left(\frac{1}{C}\right)^{\frac{1}{1/2+k}}$. This means that in a small neighborhood of $x_0 = 0$ the inequality in the definition of differentiability does not hold, that is, $f(x) = \sqrt{|x|}$ is not differentiable at the origin. (The same is true if we consider the classical definition of differentiability.) Notice that, differently from $f(x) = \sqrt[3]{x}$, the function $f(x) = \sqrt{|x|}$ is decreasing on $(-\infty, 0]$ and increasing on $[0, +\infty)$, and consequently, it has a local (and global) minimum at the origin (see Sect. 7.1 of Chap. 2 for details).

Summarizing the above considerations, we can see that for a function $f(x)$ differentiable on an interval I the condition $f'(x) > 0, \forall x \in I$ guarantees that $f(x)$ is increasing on I, while the condition $f'(x) < 0, \forall x \in I$ guarantees that $f(x)$ is decreasing on I. Notice that these conditions of monotonicity are sufficient: first, a function can be monotonic without been differentiable, and second, a differentiable monotonic function may have zero derivative at some points. The first case can be illustrated with the function $f(x) = \sqrt[3]{x}$ and the second one with the function $f(x) = x^3$: both functions are increasing on the entire domain $\mathbb{R}$, but at the origin the first function is not differentiable and the second has vanishing derivative. It can also be shown that for a differentiable function $f(x)$ the condition $f'(x) \geq 0$ is necessary and sufficient for non-strict increase on I; and similarly, the condition $f'(x) \leq 0$ is necessary and sufficient for non-strict decrease on I. This result is less used in the study of functions and we do not comment on it further.

Another conclusion we can derive concerns local extrema. We have seen that the fact that $f'(x_0) \neq 0$ implies that Δf has different signs on the two sides of x_0, which means that on one side of x_0 we have $f(x) > f(x_0)$, while on the other side $f(x) < f(x_0)$ (in a small neighborhood of x_0). This shows that any point x_0, at which $f'(x_0) \neq 0$, cannot be a local extremum. Therefore, a necessary condition of an extremum is $f'(x_0) = 0$ or $\nexists f'(x_0)$. The points where the derivative vanishes or does not exist are called critical points of a function.

1.4 Rules of Differentiation and Derivatives of Rational Functions

Certainly, the usefulness of the concept of the derivative for study of monotonicity and other properties of a function depends crucially on possibility to calculate (in rather simple way) derivatives of different functions and on complexity of the form of these derivatives. The development of this concept and rules of its application took many centuries. In the form sufficiently close to modern treatment is was considered still in the works of Fermat, Barrow, Gregory and other mathematicians of the seventeenth century. Newton and Leibniz, the inventors of Calculus, have developed and used different rules of differentiation in almost modern form, but still without an exact definition of the derivative or differentiation. Johann Bernoulli, Euler and Lagrange pushed even more forward understanding and application of differential techniques. All the rules we learn now in the university course of Calculus were already known for these famous mathematicians and many other mathematicians of eighteenth century. However, only Bolzano and Cauchy in the first part of nineteenth century gave an exact definitions of differentiability and derivative based on the clearly formulated concept of the limit.

The study of any concepts and results of differential Calculus is out of scope of this text. However, we introduce (without exact justification) some derivatives of the functions and some rules of differentiation in order to show how the tools of differential Calculus, developed by generations of great mathematicians and taken for granted by us, dramatically simplify the study of many important properties of functions.

Arithmetic Rules of Differentiation

First, according our simplified definition, the differentiation formulas for $f(x) = C$, $C = constant$ and $f(x) = x$ are trivial: $\Delta(f(x) = C) = C - C = 0 = \Delta x \cdot (0+0)$ with $A = 0$ and $\alpha = 0$; $\Delta(f(x) = x) = \Delta x = \Delta x \cdot (1 + 0)$ with $A = 1$ and $\alpha = 0$. Consequently, if $f(x) = C$, then $f'(x) = 0$, $\forall x \in \mathbb{R}$; if $f(x) = x$, then $f'(x) = 1$, $\forall x \in \mathbb{R}$.

Next, let us see that the arithmetic rules of differentiation are quite simple. If $f(x)$ and $g(x)$ are differentiable at x_0, then $f(x) + g(x)$, $f(x) - g(x)$, $f(x) \cdot g(x)$ and $\frac{f(x)}{g(x)}$ are also differentiable at x_0 (the last one under the additional condition that $g(x_0) \neq 0$) and the following formulas for the derivatives are true:

(1) $(f(x) + g(x))'(x_0) = f'(x_0) + g'(x_0)$;
(2) $(f(x) - g(x))'(x_0) = f'(x_0) - g'(x_0)$;
(3) $(f(x) \cdot g(x))'(x_0) = f'(x_0)g(x_0) + g'(x_0)f(x_0)$;
(4) $\left(\frac{f(x)}{g(x)}\right)'(x_0) = \frac{f'(x_0)g(x_0) - g'(x_0)f(x_0)}{g^2(x_0)}$.

The reader familiar with differential Calculus will promptly recognize these rules as an analogue of arithmetic rules of classical derivatives.

We can even prove these statements using our simplified version of differentiation. Let us do it for the sum and product. According the hypothesis, $\Delta f = \Delta x \cdot (A + \alpha)$, where $A = constant$ and α satisfies the evaluation $|\alpha| \le C|\Delta x|^k$, $k > 0$, $C = constant$ and $\Delta g = \Delta x \cdot (B + \beta)$, where $B = constant$ and β is evaluated as $|\beta| \le D|\Delta x|^n$, $n > 0$, $D = constant$. Then, the increment of $h(x) = f(x) + g(x)$ can be represented as follows:

$$\Delta h = \Delta f + \Delta g = (A + \alpha)\Delta x + (B + \beta)\Delta x = [(A + B) + (\alpha + \beta)]\Delta x,$$

where $A + B = constant$ and $\gamma = \alpha + \beta$ has the evaluation

$$|\gamma| \le |\alpha| + |\beta| \le C|\Delta x|^k + D|\Delta x|^n \le E|\Delta x|^m,$$

where $m = \min\{k, n\}$ and $E = 2\max\{C, D\}$ (assuming that Δx is sufficiently small, $|\Delta x| < 1$). The rule for the subtraction can be proved in the same way.

For the product $h(x) = f(x)g(x)$ the proof is a bit more difficult, but still rather simple. We start with the following representation of the increment of the function $h(x)$:

$$\Delta h = f(x)g(x) - f(x_0)g(x_0) = f(x)g(x) - f(x_0)g(x) + f(x_0)g(x) - f(x_0)g(x_0)$$
$$= g(x) \cdot (f(x) - f(x_0)) + f(x_0) \cdot (g(x) - g(x_0))$$
$$= [g(x_0) + (B + \beta)\Delta x](A + \alpha)\Delta x + f(x_0)(B + \beta)\Delta x$$
$$= ([g(x_0)A + f(x_0)B] + [(B + \beta)(A + \alpha)\Delta x + g(x_0)\alpha + f(x_0)\beta])\,\Delta x.$$

Let us evaluate the second term $\gamma = (B + \beta)(A + \alpha)\Delta x + g(x_0)\alpha + f(x_0)\beta$ on the right-hand side, assuming that $|\Delta x| < 1$. First,

$$|(B + \beta)(A + \alpha)\Delta x| \le (|B| + D)(|A| + C)|\Delta x|$$
$$= C_1|\Delta x|,\ C_1 = (|B| + D)(|A| + C) = constant.$$

Second,

$$|g(x_0)\alpha| \le |g(x_0)| \cdot C|\Delta x|^k = C_2|\Delta x|^k,\ C_2 = |g(x_0)| \cdot C = constant,$$

and similarly

$$|f(x_0)\beta| \le |f(x_0)| \cdot D|\Delta x|^n = C_3|\Delta x|^n,\ C_3 = |f(x_0)| \cdot D = constant.$$

Joining these three inequalities, we have $|\gamma| \le C_1|\Delta x| + C_2|\Delta x|^k + C_3|\Delta x|^n$. Finally, choosing $m = \min\{k, n, 1\}$ and $E = 3\max\{C_1, C_2, C_3\}$ we arrive at the evaluation $|\gamma| \le E|\Delta x|^m$, which shows that $h(x)$ is differentiable at x_0 and $h'(x_0) = f'(x_0)g(x_0) + g'(x_0)f(x_0)$. The rule of division can be demonstrated in a similar manner.

To simplify notations, hereinafter we will omit the indication of the point of differentiation x_0 or its index 0, when it is clear from the context.

Applying these rules (together with the derivatives of constant and identity functions) we can easily find a derivative of any rational function. Indeed, for a polynomial function, we have

$$(P_n(x))' = (a_n x^n + \ldots + a_1 x + a_0)' = (a_n x^n)' + \ldots + (a_1 x)' + (a_0)'$$

$$= (a_n)' x^n + a_n (x^n)' \ldots + (a_1)' x + a_1 x' + (a_0)' = a_n n x^{n-1} + \ldots + a_1,\ \forall x \in \mathbb{R}.$$

For an arbitrary rational function $R(x) = \frac{P_n(x)}{Q_m(x)}$ at any point of its domain (all the points excluding the zeros of the polynomial $Q_m(x)$) we have

$$(R(x))' = \left(\frac{P_n(x)}{Q_m(x)}\right)' = \frac{(P_n(x))' Q_m(x) - (Q_m(x))' P_n(x)}{(Q_m(x))^2}.$$

Application of the Derivative: Rational Functions

Let us illustrate how the study of functions is simplified by employing the derivatives. It happens even in the study of polynomial and rational functions. Let us apply derivatives to some functions considered in the previous chapters.

Problem A, Sect. 14.9, Chap. 2: $f(x) = x^4 - 4x^2 + 3$

The derivative is $f'(x) = 4x^3 - 8x = 4x(x^2 - 2)$. Obviously, $f'(x) < 0$ when $x \in (-\infty, -\sqrt{2}) \cup (0, \sqrt{2})$ and $f'(x) > 0$ when $x \in (-\sqrt{2}, 0) \cup (\sqrt{2}, +\infty)$. This implies that $f(x)$ is decreasing on $(-\infty, -\sqrt{2})$ and also on $(0, \sqrt{2})$ and increasing on $(-\sqrt{2}, 0)$ and $(\sqrt{2}, +\infty)$. Consequently, the points $\pm\sqrt{2}$ are local (and global) minimum and the point 0 is a local (but not global) maximum.

Problem B, Sect. 14.9, Chap. 2: $f(x) = \frac{x}{1+x^2}$

The derivative is $f'(x) = \frac{x' \cdot (1+x^2) - (1+x^2)' \cdot x}{(1+x^2)^2} = \frac{1 \cdot (1+x^2) - (0+2x) \cdot x}{(1+x^2)^2} = \frac{1-x^2}{(1+x^2)^2}$. Obviously, $f'(x) < 0$ when $x \in (-\infty, -1) \cup (1, +\infty)$ and $f'(x) > 0$ when $x \in (-1, 1)$. Therefore, $f(x)$ is decreasing on $(-\infty, -1)$ and also on $(1, +\infty)$ and increasing on $(-1, 1)$. Consequently, the point -1 is a local minimum and the point 1 is a local maximum. It can be shown that both extrema are also global.

Problem A, Sect. 4.2, Chap. 3: $f(x) = x^3 - x^2 - x + 1$
The derivative is $f'(x) = 3x^2 - 2x - 1$. Solving the equation $f'(x) = 0$ we find the critical points $x_1 = -\frac{1}{3}$ and $x_2 = 1$, which allows us to factorize the derivative as $f'(x) = (3x + 1)(x - 1)$. Then, it is easy to see that $f'(x) > 0$ when $x \in (-\infty, -\frac{1}{3}) \cup (1, +\infty)$ and $f'(x) < 0$ when $x \in (-\frac{1}{3}, 1)$. This implies that $f(x)$ is increasing on $(-\infty, -\frac{1}{3})$ and also on $(1, +\infty)$ and decreasing on $(-\frac{1}{3}, 1)$. Consequently, the point $-\frac{1}{3}$ is a local maximum and $x = 1$ is a local minimum. Comparing with the method used in Sect. 4.2, Chap. 3, we can see that this approach simplify essentially the investigation of monotonicity and extrema.

Problem B, Sect. 4.2, Chap. 3: $f(x) = \frac{x^4}{4} + x^3 - 4x - 1$
The derivative is $f'(x) = x^3 + 3x^2 - 4$. Noting that $f'(x)$ vanishes at $x_1 = 1$ (which is a critical point), we can factorize the derivative in the form $f'(x) = (x - 1)(x^2 + 4x + 4)$. Solving the equation $x^2 + 4x + 4 = 0$, we find the remaining critical point $x_2 = -2$, which is a double root of the equation. Therefore, the derivative takes the form $f'(x) = (x - 1)(x + 2)^2$. The second factor is always non-negative, which means that $f'(x) < 0$ when $x \in (-\infty, 1)$, $x \neq -2$, and $f'(x) > 0$ when $x \in (1, +\infty)$. This implies that $f(x)$ is decreasing on $(-\infty, 1)$ and increasing on $(1, +\infty)$. Consequently, the point 1 is local (and global) minimum, while the point -2 is not a local extremum (it lies in the interval of decrease). Compare this simple procedure with the technicalities used in Sect. 4.2, Chap. 3 to arrive at the same results about monotonicity and extremum.

Problem A, Sect. 4.3, Chap. 3: $f(x) = \frac{x^2}{x^2-1}$
Notice first that the points $x = \pm 1$ are excluded from the domain: $X = \mathbb{R}\backslash\{\pm 1\}$. At any point of the domain the derivative can be found by the quotient rule: $f'(x) = \frac{(x^2)'(x^2-1)-(x^2-1)'x^2}{(x^2-1)^2} = \frac{-2x}{(x^2-1)^2}$. Since the denominator is always non-negative, the sign of $f'(x)$ is determined by the numerator. Then, $f'(x) > 0$ when $x \in (-\infty, -1) \cup (-1, 0)$ and $f'(x) < 0$ when $x \in (0, 1) \cup (1, +\infty)$ (the points $x = \pm 1$ are out of the domain of $f(x)$). Therefore, $f(x)$ is increasing on $(-\infty, -1)$ and also on $(-1, 0)$, and decreasing on $(0, 1)$ and on $(1, +\infty)$. Consequently, the point 0 is local maximum. One can see that this approach is simpler than that used in Sect. 4.3 of Chap. 3.

Problem B, Sect. 4.3, Chap. 3: $f(x) = \frac{2x^2+3x+1}{x^2+4x+3}$
Notice first that the points $x = -3$ and $x = -1$ are excluded from the domain: $X = \mathbb{R}\backslash\{-3, -1\}$. Next, we can simplify the formula of $f(x)$ by factoring numerator and denominator and canceling the equal factors: $f(x) = \frac{2x^2+3x+1}{x^2+4x+3} = \frac{(x+1)(2x+1)}{(x+1)(x+3)} = \frac{2x+1}{x+3}$, $\forall x \in X$. Then, at any point of the domain the derivative can be found by the quotient rule: $f'(x) = \frac{(2x+1)'(x+3)-(x+3)'(2x+1)}{(x+3)^2} = \frac{5}{(x+3)^2}$. Since the derivative is positive at each point of the domain, the function is increasing on

the intervals $(-\infty, -3)$, $(-3, -1)$ and $(-1, +\infty)$ (the points $x = -3$ and $x = -1$ should be excluded since both do not belong to the domain of $f(x)$). There is no critical point, which means that there is no extremum. Again this approach is simpler than that used in Sect. 4.3 of Chap. 3.

1.5 Differentiability of Irrational Functions

Differentiability of Root Functions

Consider first a differentiability of $f(x) = \sqrt{x}$ at $x_0 > 0$. Naturally, we have to consider only the points of the domain of $\sqrt{x}$, that is, $x \geq 0$. We have already shown that $\sqrt{|x|}$ is not differentiable (albeit defined) at $x_0 = 0$, and the same is true for $\sqrt{x}$ (with practically the same proof). Therefore, let us consider any point $x_0 > 0$ and corresponding $x = x_0 + \Delta x > 0$ (we can always choose such a small neighborhood of x_0 that the last inequality holds). Under these conditions, we have the following representation:

$$\Delta f = \sqrt{x} - \sqrt{x_0} = \frac{\Delta x}{\sqrt{x_0 + \Delta x} + \sqrt{x_0}}$$
$$= \Delta x \cdot \left[\frac{1}{2\sqrt{x_0}} + \left(\frac{1}{\sqrt{x_0 + \Delta x} + \sqrt{x_0}} - \frac{1}{2\sqrt{x_0}} \right) \right].$$

Now we show that $\alpha = \frac{1}{\sqrt{x_0+\Delta x}+\sqrt{x_0}} - \frac{1}{2\sqrt{x_0}}$ satisfies the definition of differentiability. For this, we write α in the form

$$\alpha = \frac{2\sqrt{x_0} - (\sqrt{x_0 + \Delta x} + \sqrt{x_0})}{(\sqrt{x_0 + \Delta x} + \sqrt{x_0}) \cdot 2\sqrt{x_0}} = \frac{\sqrt{x_0} - \sqrt{x_0 + \Delta x}}{(\sqrt{x_0 + \Delta x} + \sqrt{x_0}) \cdot 2\sqrt{x_0}}$$
$$= \frac{-\Delta x}{(\sqrt{x_0 + \Delta x} + \sqrt{x_0}) \cdot 2\sqrt{x_0} \cdot (\sqrt{x_0} + \sqrt{x_0 + \Delta x})}$$
$$= \frac{-\Delta x}{(\sqrt{x_0 + \Delta x} + \sqrt{x_0})^2 \cdot 2\sqrt{x_0}}.$$

Since $x = x_0 + \Delta x > 0$, the denominator can be evaluated as follows: $(\sqrt{x_0 + \Delta x} + \sqrt{x_0})^2 \cdot 2\sqrt{x_0} > (\sqrt{x_0})^2 \cdot 2\sqrt{x_0} = 2x_0^{3/2}$. Therefore, $|\alpha| = \frac{|\Delta x|}{(\sqrt{x_0+\Delta x}+\sqrt{x_0})^2 \cdot 2\sqrt{x_0}} < \frac{|\Delta x|}{2x_0^{3/2}}$. This shows that α satisfies the inequality of the definition with $k = 1$ and $C = \frac{1}{2x_0^{3/2}}$. Hence, the first term in the expression for Δf is the derivative $f'(x_0) = \frac{1}{2\sqrt{x_0}}$.

Similarly, it can be shown that the function $f(x) = \sqrt[3]{x}$ is differentiable at any $x_0 \neq 0$ and its derivative is $f'(x_0) = \frac{1}{3\sqrt[3]{x_0^2}}$. These two results can be extended to any root: the function $\sqrt[n]{x}$ with n even is differentiable at each $x > 0$ and the function $\sqrt[n]{x}$ with n odd is differentiable at each $x \neq 0$. In both cases the formula for the derivative is $(\sqrt[n]{x})' \equiv (x^{\frac{1}{n}})' = \frac{1}{n}x^{\frac{1}{n}-1} = \frac{1}{n\sqrt[n]{x^{n-1}}}$. However, the proof is much more technically involved for an arbitrary n and, for this reason, we do not provide it. Certainly, the reader familiar with differential Calculus will recognize here the formulas of classical derivatives of the root functions.

Differentiation of a Composite Function
Let us consider functions $y = f(x)$ and $g(y)$ such that the composite function $h(x) = g(f(x))$ is defined on a chosen set X. If $y = f(x)$ is differentiable at $x_0 \in X$ and $g(y)$ is differentiable at the corresponding point $y_0 = f(x_0)$, then the composite function $h(x) = g(f(x))$ is differentiable at x_0 and its derivative can be found by the formula $h'(x_0) = g'(y_0) \cdot f'(x_0)$ (here the derivative of each function is calculated with respect to its own independent variable).

Let us prove this statement. First, by the definition of differentiability, $\Delta y = \Delta f = \Delta x \cdot (A + \alpha)$, where $A = f'(x_0)$ and α satisfies the evaluation $|\alpha| \leq C|\Delta x|^k$, $k > 0$, $C = constant$ and $\Delta g = \Delta y \cdot (B + \beta)$, where $B = g'(y_0)$ and β is evaluated as $|\beta| \leq D|\Delta y|^n$, $n > 0$, $D = constant$. Then, the increment of $h(x) = g(f(x))$ can be represented as follows:

$$\Delta h = g(f(x)) - g(f(x_0)) = g(y) - g(y_0) = \Delta g = (B + \beta)\Delta y = (B + \beta)\Delta f$$

$$= (B + \beta)(A + \alpha)\Delta x = (AB + A\beta + B\alpha + \beta\alpha)\Delta x,$$

where $AB = constant$ and $\gamma = A\beta + B\alpha + \beta\alpha$ is a function of Δx. Separate terms in γ have the following evaluations: $|\gamma_1| = |A\beta| \leq |A| \cdot D|\Delta y|^n$, $|\gamma_2| = |B\alpha| \leq |B| \cdot C|\Delta x|^k$, $|\gamma_3| = |\beta\alpha| \leq D|\Delta y|^n \cdot C|\Delta x|^k$. It remains to evaluate $|\Delta y|$ through $|\Delta x|$. Employing once more differentiability of $y = f(x)$ and assuming that $|\Delta x| < 1$, we obtain

$$|\Delta y| = |(A + \alpha)\Delta x| \leq (|A| + |\alpha|)|\Delta x| \leq (|A| + C|\Delta x|^k)|\Delta x| < (|A| + C)|\Delta x|.$$

Using this inequality in the evaluations of γ_1 and γ_3 we have

$$|\gamma_1| \leq |A| \cdot D|\Delta y|^n \leq |A| \cdot D \cdot (|A| + C)^n |\Delta x|^n = C_1|\Delta x|^n,$$

$$|\gamma_3| - |\beta\alpha| \leq D \cdot (|A| + C)^n |\Delta x|^n \cdot C|\Delta x|^k = C_3|\Delta x|^{k+n},$$

where $C_1 = |A| \cdot D \cdot (|A| + C)^n$ and $C_3 = D \cdot C \cdot (|A| + C)^n$. Denoting additionally $C_2 = |B| \cdot C$, we obtain the following evaluation of γ:

$$|\gamma| \leq |\gamma_1| + |\gamma_2| + |\gamma_3| \leq C_1|\Delta x|^n + C_2|\Delta x|^k + C_3|\Delta x|^{k+n} \leq 3C_h|\Delta x|^m,$$

where $C_h = \max\{C_1, C_2, C_3\}$ and $m = \min\{k, n\}$. Thus, $h(x)$ is differentiable at x_0 and the constant AB is its derivative: $h'(x_0) = AB = g'(y_0) \cdot f'(x_0)$.

The reader familiar with differential Calculus can see that this result (and the used technique of its proof) resembles the chain rule of the classical differentiation.

Differentiability of Irrational Functions

Using the chain rule, we can find derivative of an irrational function at any interior point of its domain. Indeed, considering an irrational function $h(x) = \sqrt[n]{R(x)}$ as a composite function $h(x) = g(f(x))$, where $y = f(x) = R(x)$ is a rational function and $g(y) = \sqrt[n]{y}$ is a root function, each of which was already differentiated, we arrive at the following result:

$$h'(x) = g'(y) \cdot f'(x) = \frac{1}{n\sqrt[n]{y^{n-1}}} \cdot R'(x) = \frac{1}{n\sqrt[n]{(R(x))^{n-1}}} \cdot R'(x).$$

Application of the Derivative: Irrational Functions

Let us apply this result to the study of monotonicity and extrema of some investigated irrational functions.

Problem B, Sect. 4.4, Chap. 3: $h(x) = \frac{1}{\sqrt[3]{4-2x}}$

The function is defined on $X = \mathbb{R}\backslash\{2\}$ and we can consider its differentiability on the entire domain. Using the formula of the derivative of a quotient and applying the chain rule to $\sqrt[3]{4-2x}$ with $f(x) = 4 - 2x$ and $g(y) = \sqrt[3]{y}$, we get: $h'(x) = \frac{1' \cdot \sqrt[3]{4-2x} - (\sqrt[3]{4-2x})' \cdot 1}{(\sqrt[3]{4-2x})^2} = \frac{-\frac{1}{3\sqrt[3]{y^2}} \cdot (-2)}{\sqrt[3]{(4-2x)^2}} = \frac{\frac{1}{3\sqrt[3]{(4-2x)^2}} \cdot 2}{\sqrt[3]{(4-2x)^2}} = \frac{2}{3\sqrt[3]{(4-2x)^4}}$, $\forall x \in X$. Both the numerator and denominator are positive, which shows that the function increases on $(-\infty, 2)$ and also on $(2, +\infty)$. Notice that the function does not increase on the interval $(-\infty, +\infty)$ since $x = 2 \notin X$, or on the set $(-\infty, 2) \cup (2, +\infty)$ since $f(0) = \frac{1}{\sqrt[3]{4}} > -\frac{1}{\sqrt[3]{4}} = f(4)$.

Problem A, Sect. 5.3, Chap. 3: $h(x) = \sqrt{x^2 + 1}$

The derivative of $h(x)$ is found by applying the chain rule with $f(x) = x^2 + 1$ and $g(y) = \sqrt{y}$: $h'(x) = g'(y) \cdot f'(x) = \frac{1}{2\sqrt{y}} \cdot 2x = \frac{x}{\sqrt{x^2+1}}$, $\forall x \in \mathbb{R}$. The derivative is negative on $(-\infty, 0)$ and positive on $(0, +\infty)$, and the only point where it vanishes is $x = 0$. Therefore, $h(x)$ is decreasing on $(-\infty, 0)$, increasing on $(0, +\infty)$, and has local (and global) minimum at $x = 0$.

Problem D, Sect. 5.3, Chap. 3: $h(x) = x + \sqrt{x^2 - 1}$

The function is defined on $X = (-\infty, -1] \cup [1, +\infty)$ and we can consider its differentiability on the intervals $(-\infty, -1)$ and $(1, +\infty)$. The derivative of $h(x)$ can be found by using the formula of the derivative of a sum and applying the chain rule to the second term with $f(x) = x^2 - 1$ and $g(y) = \sqrt{y}$: $h'(x) = (x)' + (\sqrt{x^2 - 1})' = 1 + \frac{1}{2\sqrt{y}} \cdot 2x = 1 + \frac{x}{\sqrt{x^2-1}}$, $\forall x \in (-\infty, -1) \cup (1, +\infty)$. On the interval $(-\infty, -1)$ the derivative is negative and $h(x)$ decreases. On the interval $(1, +\infty)$ the derivative is positive and the function increases. The points $x = \pm 1$ are not local extrema, because these points are not interior. However, one of them, $x = -1$, is a global minimum (it follows from the monotonicity of $h(x)$ and the comparison between $h(-1)$ and $h(1)$).

1.6 Differentiability of Transcendental Functions

Derivatives of Transcendental Functions

It is very much harder (if possible) to derive formulas of differentiation of transcendental functions using rudimentary (pre-limit) techniques. For this reason we just give below some basic results of the derivatives of some transcendental functions: $(a^x)' = a^x \ln a$, $\forall x \in \mathbb{R}$; $(\log_a x)' = \frac{1}{x \ln a}$, $\forall x \in (0, +\infty)$; $(\sin x)' = \cos x$, $\forall x \in \mathbb{R}$; $(\cos x)' = -\sin x$, $\forall x \in \mathbb{R}$. Using these formulas together with the arithmetic rules and the chain rule, we can investigate the monotonicity and extrema of some previously seen functions in a simpler way.

Application of the Derivative: Exponential and Logarithmic Functions

Problem A, Sect. 3.1, Chap. 4: $f(x) = 2 \cdot 10^{3x}$

Using the arithmetic rules, the chain rule and the derivative of the exponential function, we obtain: $f'(x) = 2 \cdot 10^{3x} \ln 10 \cdot 3 = 6 \ln 10 \cdot 10^{3x}$, $\forall x \in \mathbb{R}$. Since all the factors are positive, $f'(x) > 0$, which means that the function increases on the entire domain $\mathbb{R}$. Consequently, it has no extremum, local or global.

Problem B, Sect. 3.1, Chap. 4: $f(x) = 2 - 5e^{-x/3}$

Using the arithmetic rules, the chain rule and the derivative of the exponential function, we obtain: $f'(x) = -5e^{-x/3} \ln e \cdot \frac{-1}{3} = \frac{5}{3} e^{-x/3}$, $\forall x \in \mathbb{R}$. Since all the factors are positive, $f'(x) > 0$, which means that the function increases on the entire domain $\mathbb{R}$. Consequently, it has no extremum, local or global.

Problem B, Sect. 3.2, Chap. 4: $f(x) = 1 - \log_{1/10}(4 - 2x)$

The domain of this function is $X = (-\infty, 2)$ and it is differentiable on X. Using the arithmetic rules, the chain rule and the derivative of the logarithmic function, we get: $f'(x) = -\frac{1}{(4-2x) \ln \frac{1}{10}} \cdot (-2) = \frac{-2}{(4-2x) \ln 10}$, $\forall x \in X$. Since the derivative is negative on the entire domain, the function is decreasing on $X = (-\infty, 2)$, and consequently, there is no extremum, local or global.

Problem A, Sect. 4.2, Chap. 4: $f(x) = 1 - \log_5(x^2 + 1)$

The domain of this function is $X = \mathbb{R}$ and it is differentiable on X. Using arithmetic rules, the chain rule and the derivative of the logarithmic function, we obtain: $f'(x) = -\frac{1}{(x^2+1) \ln 5} \cdot 2x = \frac{-2}{\ln 5} \frac{x}{x^2+1}$, $\forall x \in X$. Clearly, the derivative is positive when $x < 0$ and negative when $x > 0$. Therefore, the function is increasing on $(-\infty, 0)$ and decreasing on $(0, +\infty)$, and $x = 0$ is local (and global) maximum.

Application of the Derivative: Trigonometric Functions

Problem A, Sect. 3.3, Chap. 4: $f(x) = 2\sin(3x + \frac{\pi}{6})$

Using the arithmetic rules, the chain rule and the derivative of the sine function, we obtain: $f'(x) = 2\cos(3x + \frac{\pi}{6}) \cdot 3 = 6\cos(3x + \frac{\pi}{6})$, $\forall x \in \mathbb{R}$. Recalling that $\cos t$ has zeros $t_k = \frac{\pi}{2} + k\pi$, $k \in \mathbb{Z}$, and that it is positive when $t \in (-\frac{\pi}{2}, \frac{\pi}{2}) + 2k\pi$ and negative when $t \in (\frac{\pi}{2}, \frac{3\pi}{2}) + 2k\pi$, $k \in \mathbb{Z}$, we conclude that $f'(x) = 0$ at the points $x_k = \frac{\pi}{9} + k\frac{\pi}{3}$, $k \in \mathbb{Z}$, and that $f'(x) > 0$ when $x \in (-\frac{2\pi}{9}, \frac{\pi}{9}) + 2k\frac{\pi}{3}$ and $f'(x) < 0$ when $x \in (\frac{\pi}{9}, \frac{4\pi}{9}) + 2k\frac{\pi}{3}$, $k \in \mathbb{Z}$. Therefore, the function is increasing on the intervals $(-\frac{2\pi}{9}, \frac{\pi}{9}) + 2k\frac{\pi}{3}$ and decreasing on the intervals $(\frac{\pi}{9}, \frac{4\pi}{9}) + 2k\frac{\pi}{3}$, $k \in \mathbb{Z}$; the points $x_k = \frac{\pi}{9} + 2k\frac{\pi}{3}$, $k \in \mathbb{Z}$ are local maxima and $x_n = \frac{-2\pi}{9} + 2k\frac{\pi}{3}$, $k \in \mathbb{Z}$ are local minima.

Problem A, Sect. 4.3, Chap. 4: $f(x) = \tan x + \cot x$

First, recall that $\tan x = \frac{\sin x}{\cos x}$ and $\cot x = \frac{\cos x}{\sin x}$. Since $\tan x$ is defined on $X_t = \mathbb{R}\backslash\{\frac{\pi}{2} + k\pi, k \in \mathbb{Z}\}$ and $\cot x$ is defined on $X_c = \mathbb{R}\backslash\{n\pi, n \in \mathbb{Z}\}$, the function $f(x)$ is defined on $X = X_t \cap X_c = \mathbb{R}\backslash\{\frac{k\pi}{2}, k \in \mathbb{Z}\}$. Let us calculate separately the derivatives of $\tan x$ and $\cot x$. At any point of the domain, using the arithmetic rules and the derivatives of the sine and cosine functions, we obtain:

$$
\begin{aligned}
(\tan x)' &= \frac{(\sin x)' \cdot \cos x - (\cos x)' \cdot \sin x}{\cos^2 x} \\
&= \frac{\cos x \cdot \cos x - (-\sin x) \cdot \sin x}{\cos^2 x} = \frac{1}{\cos^2 x}, \ \forall x \in X; \\
(\cot x)' &= \frac{(\cos x)' \cdot \sin x - (\sin x)' \cdot \cos x}{\sin^2 x} \\
&= \frac{(-\sin x) \cdot \sin x - (\cos x) \cdot \cos x}{\sin^2 x} = -\frac{1}{\sin^2 x}, \ \forall x \in X.
\end{aligned}
$$

Therefore,

$$
f'(x) = (\tan x)' + (\cot x)' = \frac{1}{\cos^2 x} - \frac{1}{\sin^2 x}, \ \forall x \in X.
$$

Both the original function and its derivative are π-periodic functions, which allows us to consider first the properties on an interval of the length π, say $(-\frac{\pi}{2}, \frac{\pi}{2})$ (the interval should be open, since the endpoints are out of the domain). More precisely, we consider the set $(-\frac{\pi}{2}, 0) \cup (0, \frac{\pi}{2})$, since 0 does not belong to the domain. Let us see where $f'(x) = 0$ on this set. Solving the equation $\cos^2 x = \sin^2 x$, or equivalently, $\tan^2 x = 1$ we find the two points $x = \pm\frac{\pi}{4}$. The derivative is positive on the intervals $(-\frac{\pi}{2}, -\frac{\pi}{4})$ and $(\frac{\pi}{4}, \frac{\pi}{2})$, and is negative on $(-\frac{\pi}{4}, 0)$ and $(0, \frac{\pi}{4})$. Therefore, the function increases on $(-\frac{\pi}{2}, -\frac{\pi}{4})$ and $(\frac{\pi}{4}, \frac{\pi}{2})$, and decreases on $(-\frac{\pi}{4}, 0)$ and $(0, \frac{\pi}{4})$. This implies that $x = -\frac{\pi}{4}$ is a local maximum and $x = \frac{\pi}{4}$ is a local minimum. The inequality $f(-\frac{\pi}{4}) = -2 < 2 = f(\frac{\pi}{4})$ shows that the former is not a global maximum and the latter is not a global minimum. Extending the obtained results according to π-periodicity, we obtain that $f(x)$ increases on $(-\frac{\pi}{2}, -\frac{\pi}{4}) + k\pi$ and on $(\frac{\pi}{4}, \frac{\pi}{2}) + k\pi$, $k \in \mathbb{Z}$ and decreases on $(-\frac{\pi}{4}, 0) + n\pi$ and on $(0, \frac{\pi}{4}) + n\pi$ $n \in \mathbb{Z}$. The points $x_k = -\frac{\pi}{4} + k\pi$, $k \in \mathbb{Z}$ are local maxima and $x_n = \frac{\pi}{4} + n\pi$, $n \in \mathbb{Z}$ are local minima.

Problem D, Sect. 4.3, Chap. 4: $f(x) = \sin 2x - 2\sin x$
Using the arithmetic rules, the chain rule and the derivative of the sine function, we obtain: $f'(x) = \cos 2x \cdot 2 - 2\cos x = 2(2\cos^2 x - \cos x - 1)$, $\forall x \in \mathbb{R}$. Denoting $t = \cos x$ and solving the quadratic equation $2t^2 - t - 1 = 0$, we find the roots $t = -\frac{1}{2}$ and $t = 1$. Therefore, the polynomial $2t^2 - t - 1$ is positive when $t \in (-\infty, -\frac{1}{2}) \cup (1, +\infty)$ and negative when $t \in (-\frac{1}{2}, 1)$. Returning to x, this means that $f'(x) = 0$ when $\cos x = -\frac{1}{2}$ and $\cos x = 1$, $f'(x) > 0$ when $\cos x \in (-\infty, -\frac{1}{2}) \cup (1, +\infty)$, and $f'(x) < 0$ when $\cos x \in (-\frac{1}{2}, 1)$. Since all the values of cosine lie between -1 and 1, the last conditions should be rewritten in the form: $f'(x) > 0$ when $\cos x \in [-1, -\frac{1}{2})$ and $f'(x) < 0$ when $\cos x \in (-\frac{1}{2}, 1)$. Using 2π-periodicity of $\cos x$, we can solve these restrictions on x on the interval $[-\pi, \pi]$ and then extend the obtained results according to periodicity. On the chosen interval, $\cos x = -\frac{1}{2}$ at the points $x_1 = -\frac{2\pi}{3}$ and $x_2 = \frac{2\pi}{3}$, while $\cos x = 1$ at the only point $x_3 = 0$. Therefore, $\cos x \in [-1, -\frac{1}{2})$ if $x \in [-\pi, -\frac{2\pi}{3}) \cup (\frac{2\pi}{3}, \pi]$, and $\cos x \in (-\frac{1}{2}, 1)$ if $x \in (-\frac{2\pi}{3}, 0) \cup (0, \frac{2\pi}{3})$. For the derivative, this means that $f'(x) > 0$ if $x \in [-\pi, -\frac{2\pi}{3}) \cup (\frac{2\pi}{3}, \pi]$, and $f'(x) < 0$ if $x \in (-\frac{2\pi}{3}, 0) \cup (0, \frac{2\pi}{3})$. Therefore, $f(x)$ increases on the intervals $[-\pi, -\frac{2\pi}{3})$ and $(\frac{2\pi}{3}, \pi]$ and decreases on the interval $(-\frac{2\pi}{3}, \frac{2\pi}{3})$. Consequently, $x_1 = -\frac{2\pi}{3}$ is a local (and global) maximum, and $x_2 = \frac{2\pi}{3}$ is a local (and global) minimum, while $x_3 = 0$ is not extremum of any type (it lies in the interval of decrease). The properties on the entire domain are obtained by applying 2π-periodicity: $f(x)$ increases on the intervals $(\frac{2\pi}{3}, \frac{4\pi}{3}) + 2k\pi$, $k \in \mathbb{Z}$ and decreases on the intervals $(-\frac{2\pi}{3}, \frac{2\pi}{3}) + 2n\pi$, $n \in \mathbb{Z}$; the points $x_k = -\frac{2\pi}{3} + 2k\pi$, $k \in \mathbb{Z}$ are local (and non-strict global) maxima, and $x_n = \frac{2\pi}{3} + 2n\pi$, $n \in \mathbb{Z}$ are local (and non-strict global) minima.

2 Concavity and Inflection: Second Derivative

2.1 Definition of Second Derivative

The next concept of Calculus, commonly used for study of concavity and inflection, is called the second order derivative (or just the second derivative, for brevity) and is rather simple. Looking at the introduced earlier concept of derivative (or the first derivative, to distinguish from the second one) we can see that in many cases the first derivative can be considered on the domain of the original function (as for polynomial or rational functions or exponential and logarithmic functions or for sine and cosine) or on the major part of this domain (as for some irrational functions or some inverse trigonometric functions like arcsine and arccosine). In all these cases, starting from the original function $f(x)$ and calculating its derivative, we obtain a new function $g(x) = f'(x)$ defined either on the original domain or on a principal part of this domain. So, we can naturally put the question of differentiability of $g(x)$, and if it is the case, the first derivative of $g(x)$ is called the second derivative of $f(x)$ and the following notation is used: $g'(x) = f''(x)$. In this case, the function $f(x)$ is called twice differentiable. This is true for both our simplified definition of differentiability and derivative and the classical counterparts. Then, it follows immediately from this definition that the considered rules (arithmetic and chain) of calculation of derivatives are valid also for the second derivatives when we apply these rules to the function $g(x) = f'(x)$.

2.2 Relationship Between Second Derivative and Concavity

In spite of simplicity of the concept of the second derivative, its relationship with concavity and inflection is not so transparent, and we give here only an intuitive idea how the second derivative can be connected with these properties of a graph. First recall the definition of the midpoint concavity, to specify consideration of upward concavity: $f(x)$ is (midpoint) concave upward on an interval I of its domain, if for any two different points $x_1, x_2 \in I$, the following inequality is satisfied: $f\left(\frac{x_1+x_2}{2}\right) < \frac{f(x_1)+f(x_2)}{2}$. Denoting the central point $x_0 = \frac{x_1+x_2}{2}$ and the increment $\Delta x = x_2 - x_0 = x_0 - x_1$, we can also write this inequality as $f(x_0 + \Delta x) + f(x_0 - \Delta x) - 2f(x_0) > 0$, or, dividing by $(\Delta x)^2$, we write it as $\frac{f(x_0+\Delta x)+f(x_0-\Delta x)-2f(x_0)}{(\Delta x)^2} > 0$ or still $\frac{\frac{f(x_0+\Delta x)-f(x_0)}{\Delta x} - \frac{f(x_0)-f(x_0-\Delta x)}{\Delta x}}{\Delta x} > 0$. Now recall that for a differentiable function $f(x)$ the derivative represents an approximation to the difference quotient: $f' \approx \frac{\Delta f}{\Delta x}$. Then, introducing the midpoints $x_0 + \frac{\Delta x}{2}$ and $x_0 - \frac{\Delta x}{2}$, we can evaluate the difference quotients in the inequality of concavity as $\frac{f(x_0+\Delta x)-f(x_0)}{\Delta x} \approx f'(x_0+\frac{\Delta x}{2}) = g(x_0+\frac{\Delta x}{2})$ and $\frac{f(x_0)-f(x_0-\Delta x)}{\Delta x} \approx f'(x_0-\frac{\Delta x}{2}) = g(x_0 - \frac{\Delta x}{2})$. Consequently, the inequality of concavity takes (approximately) the form $\frac{g(x_0+\frac{\Delta x}{2})-g(x_0-\frac{\Delta x}{2})}{\Delta x} > 0$. Again using the relationship between the difference quotient of $g(x)$ and its derivative $g'(x)$ we can expect that the inequality of concavity is represented by $g'(x_0) > 0$, that is, $f''(x_0) > 0$. In this way, assuming that $f(x)$ is twice differentiable on an interval I of its domain, we apparently arrive at the following condition of upward concavity: if the inequality $f''(x) > 0$ holds for any $x \in I$, then $f(x)$ is concave upward on an interval I. Similarly, the condition of downward concavity goes as follows: if the inequality $f''(x) < 0$ holds for any $x \in I$, then $f(x)$ is concave downward on an interval I. Consequently, x_0 is an inflection point if $f''(x)$ changes the sign at this point, which may happen only if $f''(x_0) = 0$ or $\nexists f''(x_0)$.

The readers familiar with differential Calculus will immediately connect these results with the classical concavity test and inflection point test.

2.3 *Applications of the Second Derivative*

Let us illustrate application of the second derivative to the study of concavity and inflection.

Rational Functions

Problem A, Sect. 14.9, Chap. 2: $f(x) = x^4 - 4x^2 + 3$
The first derivative $f'(x) = 4x^3 - 8x$ was already found, and calculating the derivative of $f'(x)$, we find the second derivative $f''(x) = 12x^2 - 8$. Obviously, $f''(x) > 0$ when $x \in (-\infty, -\sqrt{\frac{2}{3}}) \cup (\sqrt{\frac{2}{3}}, +\infty)$ and $f''(x) < 0$ when $x \in (-\sqrt{\frac{2}{3}}, \sqrt{\frac{2}{3}})$. This implies that $f(x)$ is concave upward on $(-\infty, -\sqrt{\frac{2}{3}})$ and also on $(\sqrt{\frac{2}{3}}, +\infty)$ and concave downward on $(-\sqrt{\frac{2}{3}}, \sqrt{\frac{2}{3}})$. Since $f''(x)$ changes the sign at $\pm\sqrt{\frac{2}{3}}$, these points are inflection ones. Comparing with the rudimentary techniques used in Sect. 14.9, Chap. 2, we can see that this approach simplify essentially the investigation of concavity and inflection.

Problem B, Sect. 14.9, Chap. 2: $f(x) = \frac{x}{1+x^2}$
The first derivative was found in the form $f'(x) = \frac{1-x^2}{(1+x^2)^2}$. Calculating the derivative of the last formula, we find the second derivative $f''(x) = \frac{-2x(1+x^2)^2 - 2(1+x^2)\cdot 2x(1-x^2)}{(1+x^2)^4} = \frac{-2x(1+x^2) - 4x(1-x^2)}{(1+x^2)^3} = 2\frac{x^3-3x}{(1+x^2)^3}$. The denominator is always positive and the numerator vanishes at $x = 0$ and $x = \pm\sqrt{3}$. Then, $f''(x) < 0$ when $x \in (-\infty, -\sqrt{3}) \cup (0, \sqrt{3})$ and $f''(x) > 0$ when $x \in (-\sqrt{3}, 0) \cup (\sqrt{3}, +\infty)$. Therefore, $f(x)$ is concave downward on the intervals $(-\infty, -\sqrt{3})$ and $(0, \sqrt{3})$, and concave upward on $(-\sqrt{3}, 0)$ and $(\sqrt{3}, +\infty)$. Consequently, $x = 0$ and $x = \pm\sqrt{3}$ are inflection points. Again this approach is much simpler than the elementary techniques used before in Sect. 14.9, Chap. 2.

Problem A, Sect. 4.2, Chap. 3: $f(x) = x^3 - x^2 - x + 1$
The first derivative $f'(x) = 3x^2 - 2x - 1$ was already calculated. The second derivative has very simple expression $f''(x) = 6x - 2$. Obviously, $f''(x) < 0$ when $x < \frac{1}{3}$ and $f''(x) > 0$ when $x > \frac{1}{3}$. This implies that $f(x)$ is concave downward on $(-\infty, \frac{1}{3})$, and concave upward on $(\frac{1}{3}, +\infty)$. Consequently, the point $\frac{1}{3}$ is an inflection point.

Problem B, Sect. 4.2, Chap. 3: $f(x) = \frac{x^4}{4} + x^3 - 4x - 1$
The first derivative was found in the form $f'(x) = x^3 + 3x^2 - 4$. Then, the second derivative is $f''(x) = 3x^2 + 6x = 3x(x + 2)$. Obviously, $f''(x)$ vanishes at the

points $x = -2$ and $x = 0$, which makes them the only candidates for inflection points. The second derivative is positive, and consequently, the function is concave upward on $(-\infty, -2)$ and also on $(0, +\infty)$; it is negative, and consequently, the function is concave downward on $(-2, 0)$. Since $f''(x)$ changes the sign at $x = -2$ and $x = 0$, these two points are inflection ones. Notice that this straightforward (in application but not in justification) method is visibly simpler than techniques used in Sect. 4.2, Chap. 3.

Problem A, Sect. 4.3, Chap. 3: $f(x) = \frac{x^2}{x^2-1}$
The first derivative was already found in the form $f'(x) = \frac{-2x}{(x^2-1)^2}$ for any x of the domain $X = \mathbb{R}\backslash\{\pm 1\}$. The second derivative can be found on the same set in the form $f''(x) = \frac{-2(x^2-1)^2-2(x^2-1)\cdot 2x\cdot(-2x)}{(x^2-1)^4} = \frac{-2(x^2-1)+8x^2}{(x^2-1)^3} = \frac{6x^2+2}{(x^2-1)^3}$. Since the numerator is always positive, the sign of $f''(x)$ is determined by the denominator. Then, $f''(x) > 0$ when $x \in (-\infty, -1) \cup (1, +\infty)$ and $f''(x) < 0$ when $x \in (-1, 1)$. Therefore, $f(x)$ is concave upward on $(-\infty, -1)$ and also on $(1, +\infty)$, and concave downward on $(-1, 1)$. Although $f(x)$ changes concavity at $x = \pm 1$ but these two points are out of the domain and cannot be inflection points. One can see that this approach is simpler (in application but not in justification) than that used in Sect. 4.3 of Chap. 3.

Problem B, Sect. 4.3, Chap. 3: $f(x) = \frac{2x^2+3x+1}{x^2+4x+3}$
This function admits a simplification in the form $f(x) = \frac{2x+1}{x+3}$, $\forall x \in X = \mathbb{R}\backslash\{-3, -1\}$. Its first derivative was found on the domain X in the form $f'(x) = \frac{5}{(x+3)^2}$. Calculating the derivative of the last quotient, we obtain the second derivative $f''(x) = \frac{-10}{(x+3)^3}$, $\forall x \in X$. Obviously, $f''(x) > 0$ when $x < -3$ and $f''(x) < 0$ when $x > -3$, $x \neq -1$. Therefore, $f(x)$ is concave upward on $(-\infty, -3)$, and concave downward on $(-3, -1)$ and $(-1, +\infty)$ (the points $x = -3$ and $x = -1$ do not belong to the domain of $f(x)$). There is no inflection point. Again this approach is simpler than that used in Sect. 4.3 of Chap. 3.

Irrational Functions

Problem B, Sect. 4.4, Chap. 3: $h(x) = \frac{1}{\sqrt[3]{4-2x}}$
The function is defined on $X = \mathbb{R}\backslash\{2\}$ and its first derivative exists on the entire domain and is represented by the formula $h'(x) = \frac{2}{3\sqrt[3]{(4-2x)^4}}$. Calculating the derivative of $h'(x)$ we obtain $h''(x) = \frac{2}{3}\frac{-\frac{1}{3\sqrt[3]{(4-2x)^8}}\cdot 4(4-2x)^3\cdot(-2)}{(\sqrt[3]{(4-2x)^4})^2} = \frac{16}{9}\frac{1}{\sqrt[3]{(4-2x)^7}}$. Obviously, $h''(x) > 0$ when $x < 2$ and $h''(x) < 0$ when $x > 2$. Consequently, the function is concave upward on $(-\infty, 2)$ and concave downward on $(2, +\infty)$. Although the second derivative has different signs to the left and to the right of $x = 2$, but this point is not an inflection one, because it does not belong to the domain.

Problem A, Sect. 5.3, Chap. 3: $h(x) = \sqrt{x^2+1}$
The derivative of $h(x)$ was found in the form $h'(x) = \frac{x}{\sqrt{x^2+1}}$, $\forall x \in \mathbb{R}$. The second derivative is $h''(x) = \frac{\sqrt{x^2+1} - \frac{x}{\sqrt{x^2+1}} \cdot x}{x^2+1} = \frac{1}{\sqrt{(x^2+1)^3}}$, $\forall x \in \mathbb{R}$. The second derivative is always positive, which means that the function has upward concavity on the entire domain $X = \mathbb{R}$. Consequently, there is no inflection point.

Problem D, Sect. 5.3, Chap. 3: $h(x) = x + \sqrt{x^2-1}$
The function is defined on $X = (-\infty, -1] \cup [1, +\infty)$ and its first derivative exists on the intervals $(-\infty, -1)$ and $(1, +\infty)$ and is defined by the formula $h'(x) = 1 + \frac{x}{\sqrt{x^2-1}}$. The second derivative is $h''(x) = \frac{\sqrt{x^2-1} - \frac{x}{\sqrt{x^2-1}} \cdot x}{x^2-1} = \frac{-1}{\sqrt{(x^2-1)^3}}$. Therefore, $h''(x)$ is negative on both intervals $(-\infty, -1)$ and $(1, +\infty)$, and the function has downward concavity on each of these intervals. Consequently, there is no inflection point.

Exponential and Logarithmic Functions

Problem A, Sect. 3.1, Chap. 4: $f(x) = 2 \cdot 10^{3x}$
The first derivative was found in the form $f'(x) = 6 \ln 10 \cdot 10^{3x}$, $\forall x \in \mathbb{R}$. The second derivative is calculated analogously: $f''(x) = 6 \ln 10 \cdot 10^{3x} \cdot 3 \ln 10 = 18 \ln^2 10 \cdot 10^{3x}$, $\forall x \in \mathbb{R}$. Since $f''(x) > 0$ over the entire domain, the function is concave upward on $X = \mathbb{R}$. Consequently, there is no inflection point.

Problem B, Sect. 3.1, Chap. 4: $f(x) = 2 - 5e^{-x/3}$
The first derivative was found in the form $f'(x) = \frac{5}{3}e^{-x/3}$, $\forall x \in \mathbb{R}$. The second derivative is calculated in the same way: $f''(x) = -\frac{5}{9}e^{-x/3}$. Since $f''(x) < 0$ over the entire domain, the function is concave downward on $X = \mathbb{R}$. Consequently, there is no inflection point.

Problem B, Sect. 3.2, Chap. 4: $f(x) = 1 - \log_{1/10}(4-2x)$
The function is differentiable on the entire domain $X = (-\infty, 2)$ and its derivative was found in the form $f'(x) = \frac{-2}{(4-2x)\ln 10}$, $\forall x \in X$. Calculating the derivative of $f'(x)$ we find the second derivative $f''(x) = \frac{-2}{\ln 10} \frac{2}{(4-2x)^2}$, $\forall x \in X$. Therefore, $f''(x) < 0$, $\forall x \in X$, which means that the function is concave downward over the entire domain. Consequently, there is no inflection point.

Problem A, Sect. 4.2, Chap. 4: $f(x) = 1 - \log_5(x^2+1)$
The function is differentiable on the entire domain $X = \mathbb{R}$ and its derivative was found in the form $f'(x) = \frac{-2}{\ln 5} \frac{x}{x^2+1}$, $\forall x \in X$. Calculating the derivative of $f'(x)$ we find the second derivative $f''(x) = \frac{-2}{\ln 5} \frac{x^2+1-x \cdot 2x}{(x^2+1)^2} = \frac{-2}{\ln 5} \frac{1-x^2}{(x^2+1)^2}$, $\forall x \in X$. Therefore, $f''(x) > 0$ when $|x| > 1$ and $f''(x) < 0$ when $|x| < 1$. This means that $f(x)$ is concave upward on $(-\infty, -1)$ and on $(1, +\infty)$, and is concave downward on $(-1, 1)$. Consequently, there are two inflection points $x = -1$ and $x = 1$.

Trigonometric Functions

Problem A, Sect. 3.3, Chap. 4: $f(x) = 2\sin(3x + \frac{\pi}{6})$

The first derivative was found in the form $f'(x) = 6\cos(3x + \frac{\pi}{6})$, $\forall x \in \mathbb{R}$. Then, the second derivative is $f''(x) = -18\sin(3x + \frac{\pi}{6})$, $\forall x \in \mathbb{R}$. Recalling that $\sin t$ has zeros $t_n = n\pi$, $n \in \mathbb{Z}$, and that it is positive when $t \in (0, \pi) + 2n\pi$ and negative when $t \in (\pi, 2\pi) + 2n\pi$, $n \in \mathbb{Z}$, we conclude that $f''(x) = 0$ at the points $x_n = -\frac{\pi}{18} + n\frac{\pi}{3}$, $n \in \mathbb{Z}$, and that $f''(x) > 0$ when $x \in (-\frac{\pi}{18}, \frac{5\pi}{18}) + 2n\frac{\pi}{3}$ and $f''(x) < 0$ when $x \in (\frac{5\pi}{18}, \frac{11\pi}{18}) + 2n\frac{\pi}{3}$, $n \in \mathbb{Z}$. Therefore, the function is concave upward on the intervals $(-\frac{\pi}{18}, \frac{5\pi}{18}) + 2n\frac{\pi}{3}$ and concave downward on the intervals $(\frac{5\pi}{18}, \frac{11\pi}{18}) + 2n\frac{\pi}{3}$, $n \in \mathbb{Z}$; consequently, all the points $x_n = -\frac{\pi}{18} + n\frac{\pi}{3}$, $n \in \mathbb{Z}$ are inflection points.

Problem A, Sect. 4.3, Chap. 4: $f(x) = \tan x + \cot x$
The first derivative exists on the domain $X = \mathbb{R}\backslash\{\frac{k\pi}{2}, k \in \mathbb{Z}\}$ and has the form $f'(x) = \frac{1}{\cos^2 x} - \frac{1}{\sin^2 x}$, $\forall x \in X$. The second derivative can be calculated at each point of X as follows: $f''(x) = \frac{2\cos x \cdot \sin x}{\cos^4 x} + \frac{2\sin x \cdot \cos x}{\sin^4 x} = \sin 2x(\frac{1}{\cos^4 x} + \frac{1}{\sin^4 x})$. Using π-periodicity of the original function and its derivatives, we first determine the zeros of $f''(x)$ and its sign on the part of the domain whose length is π, say on the set $(-\frac{\pi}{2}, 0) \cup (0, \frac{\pi}{2})$ (the point 0 does not belong to the domain). Let us see where the second derivative vanishes on this set. The expression in the parentheses $\frac{1}{\cos^4 x} + \frac{1}{\sin^4 x}$ is always positive, so the zeros of $f''(x)$ are found from the equation $\sin 2x = 0$. On the chosen set there is no solution to this equation. The sign of $f''(x)$ is also defined by $\sin 2x$ and we see that $f''(x) < 0$ on $(-\frac{\pi}{2}, 0)$ and $f''(x) > 0$ on $(0, \frac{\pi}{2})$. Therefore, $f(x)$ has downward concavity on $(-\frac{\pi}{2}, 0)$ and upward concavity on $(0, \frac{\pi}{2})$. The points $x = 0$ and $x = \pm\frac{\pi}{2}$ are not inflection points, since they are out of the domain. Extending these results on the entire domain, we obtain the following properties: $f(x)$ has downward concavity on $(-\frac{\pi}{2}, 0) + k\pi$, $k \in \mathbb{Z}$ and upward concavity on $(0, \frac{\pi}{2}) + n\pi$, $n \in \mathbb{Z}$. There is no inflection point. The obtained results coincide with those of Problem A, Sect. 4.3, Chap. 4, but are derived much faster.

Problem D, Sect. 4.3, Chap. 4: $f(x) = \sin 2x - 2\sin x$
The first derivative was found in the form $f'(x) = 2\cos 2x - 2\cos x$, $\forall x \in \mathbb{R}$. Then, the second derivative is $f''(x) = -4\sin 2x + 2\sin x = 2\sin x(1 - 4\cos x)$. Using 2π-periodicity of the original function and its derivatives, we solve initially the equation $f''(x) = 0$ and determine the sign of $f''(x)$ on the interval $[-\pi, \pi]$. For solution of $f''(x) = 0$ we have the two options: $\sin x = 0$ and $\cos x = \frac{1}{4}$. The first has the solutions $x = -\pi, 0, \pi$ and the second $x = \pm\arccos\frac{1}{4}$. Using these solutions, we can specify the sign of the second derivative: $f''(x) < 0$ when $x \in (-\pi, -\arccos\frac{1}{4}) \cup (0, \arccos\frac{1}{4})$, and $f''(x) > 0$ when $x \in (-\arccos\frac{1}{4}, 0) \cup (\arccos\frac{1}{4}, \pi)$. Therefore, $f(x)$ is concave downward on $(-\pi, -\arccos\frac{1}{4})$ and $(0, \arccos\frac{1}{4})$, and concave upward on $(-\arccos\frac{1}{4}, 0)$ and $(\arccos\frac{1}{4}, \pi)$. Consequently, all the solutions of the equation $f''(x) = 0$, that is,

$x = -\pi, 0, \pi$ and $x = \pm \arccos \frac{1}{4}$ are inflection points. Extending these results on the entire domain, we obtain the following properties: $f(x)$ is concave downward on $(-\pi, -\arccos \frac{1}{4}) + 2k\pi$ and $(0, \arccos \frac{1}{4}) + 2k\pi$, $k \in \mathbb{Z}$ and concave upward on $(-\arccos \frac{1}{4}, 0) + 2n\pi$ and $(\arccos \frac{1}{4}, \pi) + 2n\pi$, $n \in \mathbb{Z}$. The inflection points are $x_k = k\pi$ and also $x_m = -\arccos \frac{1}{4} + 2m\pi$ and $x_n = \arccos \frac{1}{4} + 2n\pi$, $k, m, n \in \mathbb{Z}$. The obtained results coincide with those of Problem D, Sect. 4.3, Chap. 4, but are derived much faster.

3 Epilogue to Epilogue

Differential Calculus and Analysis study different analytic properties of functions based on the concept of the limit, such as continuity, differentiability, analyticity. Each of these concepts, in turn, has important applications both in development of Calculus and Analysis and in the solution of various science and engineering problems. For instance, a continuous on an interval function, which has the values of different signs at two points of the interval, certainly has at least one zero in the same interval. This allows us to verify the existence of roots of complicated equations (which cannot be solved by standard methods) and also gives rise to a number of methods of approximated (numerical) solution of such equations, frequently used in practice (for instance, the bisection method, the fixed point method, the Newton method). Another important property of continuous functions is the extremal property: if a function is continuous on a closed interval, then it attains its global minimum and maximum on this interval. Further, in Integral Calculus, the continuity is used to establish the fundamental formula for calculation of definite integrals, called the Fundamental Theorem of Calculus. This result allows us, in particular, to find the areas of many plane figures with curved boundaries.

The concept of differentiability is of major importance both in development of theory of functions and in applications. The first and second derivatives are the principal tools in a study of monotonicity and concavity, which was exemplified (in a simplified form) in this chapter. A natural consequence of this is the use of derivatives in formulation and solution of problems of optimization, where finding the maximum and minimum values is a main task. If a function has derivatives of a higher order, it can be approximated by a specific type of polynomial (called the Taylor polynomial) with a guaranteed accuracy within a chosen interval. The derivatives of functions are systematically involved in formulations of scientific problems, which lead to so-called ordinary differential equations. This can be found in a great variety of problems, such as dynamic of a population and cell growth, radioactive disintegration and chemical kinetics, trajectory of a missile and of a ball, the rates of change of a particular bank deposit and of a gross domestic product, just to mention a few.

These topics are related to functions of a single variable. Naturally, all these lines of mathematical research and areas of application are multiplied when we proceed to functions of several variables.

However, all this study of functions, in historical and logical sequence, starts with learning and understanding elementary properties of functions by using simple (pre-calculus) techniques. This creates a solid basis for further analysis of functions (elementary and non-elementary) and develops the important elements of logical reasoning relevant not only for courses of Calculus and Analysis, but also for any mathematical discipline.

Remarks on Bibliography

Although our approach to study of functions and their elementary properties is quite different from the standard one, and our text contains all the material required to learn the covered topics without additional sources, we feel that it may be useful for some readers to have a few references to the subject considered in the text. The goal of this is manifold. First, the reader can find additional explanations for some pre-requisite topics, like in the case of the basic notions of trigonometry and exponents and logarithms. Second, the reader can compare the two approaches (ours and standard) and determine for him/herself what are the principal differences between them and where the logic reasoning enters in scene in our approach, substituting unnecessary additional suppositions and/or "evident" believes. Third, in the provided references the reader can find many exercises solved in the standard style and we recommend to "re-solve" some of these examples using an appropriate mathematical reasoning explored and applied in this text. Finally, a few Calculus and Analysis books included in the references are recommended for study of more advanced techniques of an investigation of functions, touched in the Epilogue.

We start with a few Pre-Calculus books, which are quite representative for an extensive bibliography on this subject. Usually, the Pre-Calculus books contain all the information necessary for development of the theory. In particular, this is the case of the following popular books:

1. Axler S., *Precalculus. A Prelude to Calculus*. Wiley, Hoboken NJ, 2016.
2. Stewart J., Redlin L., Watson S. *Precalculus: Mathematics for Calculus*. Cengage Learning, Boston, 2015.
3. Sullivan M., *Precalculus*. Pearson, Hoboken NJ, 2020.
4. Zill D.G., Dewar J.M., *Precalculus with Calculus Preview*. Jones Barlett Learning, Burlington MA, 2017.

All these books have the sections related to general characteristics of functions and to specific types of elementary functions, algebraic and transcendental. They also contain an extensive parts of trigonometry and exponents and logarithms,

A. Bourchtein, L. Bourchtein, *Elementary Functions*,
https://doi.org/10.1007/978-3-031-29075-6

including definitions, basic properties, relevant formulas, solution of equations and inequalities. There is a lot of examples of elementary functions and various applications of these functions in geometry, physics, chemistry, biology, engineering, and even in economics and social sciences. However, the approach to study of these functions is usually standard, which frequently leads to logical gaps when some properties are imposed or assumed instead of being derived and others are evaluated approximately by numerical computations and graphing, even when the form of functions is quite simple for analytic investigation. In these cases we recommend to "re-study" the functions by applying the analytic methods studied in this text.

Do not be disappointed if you were not succeeded in solving some problems, since there are many elementary functions whose exact behavior cannot be revealed even using the advanced techniques of Calculus and Analysis. An analytic approach provides an exact and complete solution when it is applicable, but it can be much more difficult than approximate incomplete methods, and in many cases it cannot be fulfilled completely. Rephrasing, more accurate information about a function we want to obtain, more restricted class of functions we are forced to consider. In practice, when completely analytic approach does not work, we can combine it with other methods, like numerical evaluation, approximation, graphing, etc.

The second part of the bibliography is related to some Calculus and Analysis books, which have very good (albeit terse and succinct) introductions, representing in depth such prerequisites as initial concepts of the set theory, the set of real numbers, coordinate systems, and general characterization of functions. We would like to highlight the following books:

1. Apostol T.M., *Calculus, Vol.1*. Wiley, Hoboken NJ, 1991.
2. Gaughan E.D., *Introduction to Analysis*. American Mathematical Society, Providence RI, 2009.
3. Spivak M., *Calculus*. Publish or Perish, Houston TX, 2008.
4. Zorich V.A., *Mathematical Analysis I*. Springer, Berlin, 2015.

All these books are classical and highly regarded expositions of the material, but, of course, their choice among many other acclaimed sources reflects our personal subjective preferences and tastes.

Although different in details and the modes of exploration, the core parts of the introductory chapters of these four books deal with the same preliminary topics important for development of the theory of Calculus/Analysis. Notice that when we refer to the introductions we mean Parts 1–4 together with sections 1.1–1.5 of the book by Apostol, Chapter 0 of the book by Gaughan, Chapters 1–4 of the book by Spivak, and Chapters 1 and 2 of the book by Zorich.

First, the authors provide some basic notions of the set theory, including an intuitive description of sets and their elements, operations with sets and their properties.

Next covered topic is the set of real numbers. All the four books use axiomatic approach to description of real numbers, which is quite traditional in Calculus/Analysis. In short words, it means that the real numbers are considered as a set of elements satisfying a number of properties, called axioms. The axioms are naturally divided into three groups. The first group, related to the properties of the operations of addition and multiplication, and the second group, related to the ordering properties, are familiar from elementary algebra of high school and just represent the well-known properties in a more formal way. The property of the last group, called the completeness axiom, as well as its different consequences, are not usually come into consideration in the high school courses, but they play an important role in Calculus/Analysis and are treated with some details in the four books. The consideration of real numbers is completed by geometric interpretation through equivalence relation between real numbers and the points of the coordinate line. All these topics of sets and real numbers are considered in Chap. 1 of our text.

The review of functions in the introductory chapters also follows the traditional style. It concerns with a definition of a function and its general properties such as image, graph, boundedness, symmetry, monotonicity, composition and invertibility. Additionally, the authors consider a formal definition of a function as a set of ordered pairs, which essentially identifies a function and its graph.

Except for the last formalization of functions, which does not play any essential role, but is more difficult for the younger readers, all the mentioned topics are included in Chap. 2 of our text, where they are explored in detail according to the purpose of this work. However, we encourage the reader to give a look at one of the four books (especially we recommend the book by Spivak) in order to compare the presentations of the same topics in different sources. We also recommend to solve some problems proposed in these books. Finally, each of these four books is an acclaimed source for further study of functions at the level of Calculus and Analysis.

Index

A. Bourchtein, L. Bourchtein, *Elementary Functions*,
https://doi.org/10.1007/978-3-031-29075-6